Mechanics of
Engineering Materials

WILEY SERIES IN
NUMERICAL METHODS IN ENGINEERING

Consulting Editors
R. H. Gallagher, *College of Engineering,*
University of Arizona
and
O. C. Zienkiewicz, *Department of Civil Engineering,*
University of College of Swansea

Rock Mechanics in Engineering Practice
Edited by K. G. Stagg and O. C. Zienkiewicz

Optimum Structural Design: Theory and Applications
Edited by R. H. Gallagher and O. C. Zienkiewicz

Finite Elements in Fluids
Vol. 1 Viscous Flow and Hydrodynamics
Vol. 2 Mathematical Foundations, Aerodynamics and Lubrication
Edited by R. H. Gallagher, J. T. Oden, C. Taylor, and O. C. Zienkiewicz

Finite Elements for Thin Shells and Curved Members
Edited by D. G. Ashwell and R. H. Gallagher

Finite Elements in Geomechanics
Edited by G. Gudehus

Numerical Methods in Offshore Engineering
Edited by O. C. Zienkiewicz, R. W. Lewis, and K. G. Stagg

Finite Elements in Fluids
Vol. 3
Edited by R. H. Gallagher, O. C. Zienkiewicz, J. T. Oden, M. Morandi Cecchi, and C. Taylor

Energy Methods in Finite Element Analysis
Edited by Glowinski, E. Rodin, and O. C. Zienkiewicz

Finite Elements in Electrical and Magnetic Field Problems
Edited by M. V. K. Chari and P. Silvester

Numerical Methods in Heat Transfer
Edited by R. W. Lewis, K. Morgan, and O. C. Zienkiewicz

Finite Elements in Biomechanics
Edited by R. H. Gallagher, B. R. Simon, P. C. Johnson, and J. F. Gross

Soil Mechanics—Transient and Cyclic Loads
Edited by G. N. Pande and O. C. Zienkiewicz

Finite Elements in Fluids
Vol. 4
Edited by R. H. Gallagher, D. Norrie, J. T. Oden, and O. C. Zienkiewicz

Foundations of Structural Optimization: A Unified Approach
Edited by A. J. Morris

Creep and Shrinkage in Concrete Structures
Edited by Z. Bažant and F. Wittmann

Hybrid and Mixed Finite Element Methods
Edited by S. N. Atluri, R. H. Gallagher and O. C. Zienkiewicz

Numerical Methods in Heat Transfer
Vol. II
Edited by R. W. Lewis, K. Morgan, B. A. Shrefler

Numerical Methods in Coupled Systems
Edited by R. W. Lewis, E. Hinton, P. Bettess

Optimum Structural Design
Edited by E. Atrek, R. H. Gallagher, K. M. Ragsdell, O. C. Zienkiewicz

Mechanics of Engineering Materials
Edited by C. S. Desai,R. H. Gallagher

Mechanics of Engineering Materials

Edited by

C. S. Desai

and

R. H. Gallagher

University of Arizona, Tucson, Arizona

A Wiley–Interscience Publication

JOHN WILEY & SONS

Chichester · New York · Brisbane · Toronto · Singapore

Library of Congress Cataloging in Publication Data:

Main entry under title:

Mechanics of engineering materials.

(Wiley series in numerical methods in engineering)
'A Wiley–Interscience publication.'
Includes index.
1. Strength of material—Addresses, essays, lectures.
I. Desai, C. S. (Chandrakant S.), 1936–
II. Gallangher, Richard H. III. Title.
TA405.M49 1984 620.1'12 83–12556

ISBN 0 471 90276 4

British Library Cataloguing in Publication Data:

Mechanics of engineering materials.—
(Wiley series in numerical methods in engineering)
1. Materials—Congresses
2. Mechanics, Applied—Congresses
I. Desai, Chandrakant S. 2. Gallagher, R.H.
620.1'12 TA405

ISBN 0 471 90276 4

Photosetting by Thomson Press (India) Limited, New Delhi and printed in Great Britain by Page Brothers, Norwich.

Contributing Authors

E. C. AIFANTIS	*Department of Mechanical Engineering, and Engineering Mechanics, Michigan Technological University, Houghton, Michigan 49931, USA*
G. Y. BALADI	*Geomechanics Division, US Army Engineer Waterways Experiment, Station, Vicksburg, Mississippi, USA*
M. L. BARON	*Weidlinger Associates, New York, New York, USA*
Z. P. BAŽANT	*Department of Civil Engineering, Northwestern University, Evanston, Illinois, USA*
J. CASEY	*Department of Mechanical Engineering, University of Houston, Houston, Texas, USA*
W. F. CHEN	*Department of Civil Engineering, Purdue University, West Lafayette, Indiana, USA*
S. C. COWIN	*Department of Biomedical Engineering, Tulane University, New Orleans, Louisiana, USA*
J. I. CURISKIS	*School of Textile Technology, University of New South Wales, Sydney, Australia*
D. R. CURRAN	*Shock Physics and Geophysics, Poulter Laboratory, SRI International, 333 Ravenswood Avenue, Menlo Park, California, USA*
Y. F. DAFALIAS	*Department of Civil Engineering, University of California, Davis, California, USA*
F. DARVE	*Institut de Mecanique de Grenoble, B. P. 53X-38041 Grenoble Cedex, France*
R. O. DAVIS	*Department of Civil Engineering, University of Canterbury, Christchurch, New Zealand*

J. Dein — *Shock Physics and Geophysics, Poulter Laboratory, SRI International, 333 Ravenswood Avenue, Menlo Park, California, USA*

C. S. Desai — *Department of Civil Engineering and Engineering Mechanics, university of Arizona, Tucson, Arizona, USA*

F. L. DiMaggio — *Department of Civil Engineering and Engineering Mechanics, Columbia University, New York, New York, USA*

D. C. Drucker — *College of Engineering, University of Illinois, Urbana, Illinois, USA*

M. O. Faruque — *Department of Civil Engineering University of Rhode Island, Kingston, RI, USA.*

W. D. L. Finn — *Department of Civil Engineering, University of British Columbia, Vancouver, British Columbia, Canada*

J. Ghaboussi — *Department of Civil Engineering, University of Illinois, Urbana, Illinois, USA*

W. Herrmann — *Solid Mechanics Research Department, Sandia National Laboratories, Albuquerque, New Mexico, USA*

M. Ikeda — *Japan Information Processing Service Co., Ltd, Tokyo, Japan*

K. Ishihara — *Department of Civil Engineering, University of Tokyo, Bunkyo-ku, Tokyo, Japan*

M. G. Katona — *Department of Civil Engineering, University of Notre Dame, Notre Dame, Indiana, USA*

T. Kawai — *Institute of Industrial Science, University of Tokyo, Tokyo, Japan*

K. J. Kim — *Department of Civil Engineering, University of Illinois, Urbana, Illinois, USA*

E. Krempl — *Department of Mechanical Engineering, Aeronautical Engineering and Mechanics, Rensselaer Polytechnic Institute, Troy, New York, USA*

P. V. Lade — *Mechanics and Structures Department, School of Engineering and Applied Science, University of California, Los Angeles, California, USA*

F. A. Leckie — *Department of Mechanical and Industrial Engineering, University of Illinois, Urbana, Illinois, USA*

C. F. Lee — *College of Engineering, University of Cincinnati, Cincinnati, Ohio, USA*

M. M. Mehrabadi — *Department of Mechanical Engineering, Tulane University, New Orleans, Louisiana, USA*

S. T. Montgomery — *Sandia National Laboratories, Albuquerque, New Mexico, USA*

Z. Mróz — *Institute of Fundamental Technological Research, Swletokrzyska 21, 00–049 Warsaw, Poland*

M. A. Mulert — *Department of Civil Engineering, University of Notre Dame, Notre Dame, Indiana, USA*

G. Mullenger — *Department of Civil Engineering, University of Canterbury, Christchurch, New Zealand*

P. M. Naghdi — *Department of Mechanical Engineering, University of California, Berkeley, California, USA*

S. Nemat-Nasser — *Department of Civil Engineering, Northwestern University, Evanston, Illinois, USA*

K. Niwa — *Japan Information Processing Service Co., Ltd, Tokyo, Japan*

A. Phillips — *Department of Mechanical Engineering, Yale University, New Haven, Connecticut, USA*

P. M. Pinsky — *Division of Engineering, Brown University, Providence, Rhode Island, USA*

K. S. Pister — *College of Civil Engineering, University of California, Berkeley, California, USA*

J. H. Prevost — *Department of Civil Engineering, Princeton University, Princeton, New Jersey, USA*

R. S. Rivlin — *Center for the Application of Mathematics, Lehigh University, Bethlehem, Pennsylvania, USA*

B. Rohani — *Geomechanics Division, US Army Engineer Waterways Experiment Station, Vicksburg, Mississippi, USA*

I. S. Sandler — *Weidlinger Associates, New York, New York, USA*

L. Seaman — *Shock Physics and Geophysics, Poulter Laboratory, SRI International, 333 Ravenswood Avenue, Menlo Park, California, USA*

S. P. SHAH — *Department of Civil Engineering, Northwestern University, Evanston, Illinois, USA*

R. L. TAYLOR — *Department of Civil Engineering, University of California, Berkeley, Berkeley, California, USA*

I. TOWHATA — *Department of Civil Engineering, University of Tokyo, Bunkyo-ku, Tokyo, Japan*

K. C. VALANIS — *College of Engineering, University of Cincinnati, Cincinnati, Ohio, USA*

S. VALLIAPPAN — *Department of Civil Engineering, University of New South Wales, Sydney, Australia*

W. R. WAWERSIK — *Sandia National Laboratories, Albuquerque, New Mexico, USA*

D. M. WOOD — *Engineering Department, Cambridge University, Cambridge, UK*

O. C. ZIENKIEWICZ — *Department of Civil Engineering, University of Wales, Swansea, UK*

Contents

Preface

The importance of constitutive laws for engineering materials, for use in obtaining reliable and realistic solutions from analytical and computational procedures, has been recognized by both researchers and practitioners. This recognition has resulted in significant recent activity towards theoretical and experimental research and the implementation of various laws in the solution procedures. The *International Conference on Constitutive Laws for Engineering Materials: Theory and Application* was held at Tucson, Arizona, USA, during 10–14 January 1983 in order to provide a forum for discussion, review, and identification of future needs for this important subject. The present volume contains full texts of the invited and theme papers presented during the conference; summaries of these papers were included in the volume of proceedings published at the time of the conference.

The thirty-three chapters contained herein cover constitutive models for a wide range of materials such as soils, rocks, concrete, aggregates, metals, composites, and discontinuous media. The models are based on various theories, such as elasticity, plasticity, viscoelasticity, viscoplasticity, endochronic, rate type, and micromechanical. Mathematical modelling, laboratory determination of parameters and implementation in numerical solution schemes are included among the phases discussed in these papers. Despite the apparent physical diversity of the models for different engineering materials, there appears a significant common thread among them. Moreover, many models used for a given material are sufficiently general so that they can be applied to other materials as well. In view of these considerations it was decided not to categorize papers under different topics, but to present the chapters alphabetically on the basis of the name of the first author.

We believe that the objectives of the conference, represented by useful discussions during the meeting, significant contributions in this volume and the preconference volume, have been achieved. The outcome can be considered a step towards providing a bridge between theoretical developments and applications and an indication of the unity of the physical basis of the various models.

We wish to express thanks to the authors for their cooperation and the timely completion of the manuscripts. The financial support to the conference by the National Science Foundation, Washington, DC, is gratefully acknowledged.

March 1983

C. S. DESAI
R. H. GALLAGHER

Mechanics of Engineering Materials
Edited by C. S. Desai and R. H. Gallagher

Chapter 1

Microscopic Processes and Macroscopic Response

E. C. Aifantis

1.1 INTRODUCTION

A general programme is outlined for considering in a unified manner the analysis of various types of microstructures and their effect on macroscopic response. It is emphasized that while phenomenological constitutive models—that have been proposed in 'myriads' (see, for example, the proceedings of this conference [1]) in the last ten or twenty years within the metal, soil, rock, polymer, concrete, wood, and biomechanics literature—can be descriptive for a relatively wide range of behaviour and therefore useful to certain engineering applications, predictive models can only be developed through physical analysis and consideration of the existing microstructures. In this connection, it is pointed out that the 'old dream' of establishing a systematic and efficient characterization of material behaviour in various constitutive classes can possibly be realized through a detailed classification of the prevailing microstructures and occurring microscopic processes. The importance and proper choice of suitable measures of stress and deformation tensors and their rates, the definition of appropriate internal variables and characterization of their temporal and spatial distribution, the significance of anisotropic and non-linear effects, and the relevance of statistical and probabilistic aspects can only be rigorously evaluated through microstructural analysis rather than phenomenological guessing.

A unifying framework for considering general microscopic processes and their influence on the macroscopic behaviour of materials should take into account the interaction and coupling of mechanical, thermal, and chemical factors. An effort towards a general characterization of bodies undergoing mechanical, thermal, chemical, and microstructural changes has been made earlier by the author in [2, 3]. A representative example of such general bodies are the so-called 'degrading bodies'. Roughly speaking, degrading bodies are those whose character changes in the presence of an appropriate chemomechanical and/or thermal environment. Thus, materials undergoing stress corrosion and cavitation damage are special classes of degrading bodies. The various questions were organized in four categories: axiomatics, constitution, thresholds, and applications. The mathematical apparatus employed was that of usual continuum thermomechanics but properly extended to include information from classical chemistry and modern materials science. Among the new concepts introduced, central was the possibility of viewing a general body as a mixture of chemo-mechanically interacting continua with appropriate chemical axioms and chemical constitutive equations governing these interactions. In discussing the constitution, the new notion of pseudo-quasi-elasticity was proposed and a distinction was also made between internal variables associated with the pre-existing inelasticity of the parent material and internal variables associated with the evolving microstructures induced by the combined action of chemomechanical and thermal environments. Another new idea was the proposition that the usual evolution laws for the internal variables should be replaced, in general, with complete balance laws containing both a rate and a flux term. Threshold conditions were also considered as part of the constitution and treated as the yield condition in plasticity theory. It was found that the programme is quite general to describe processes ranging from simple elasto-diffusive and flow through porous media phenomena to more complicated phenomena of embrittlement, dislocation-induced plasticity, fatigue, and chemical creep. Moreover, it was shown that the programme can encompass microstructural phenomena that occur at various scales ranging from lattice degradation (such as spinodal decomposition and dislocation-induced plasticity) to microscopic damage (such as embrittlement and microcrack-induced creep).

By considering special classes of microscopic processes and their effects on macroscopic response, it appeared recently [4, 5] that new interdisciplinary areas of research can be identified and elaborated upon usefully within the above programme. In particular, alternative theoretical means can be developed for the qualitative and quantitative description of observed behaviour (such as localization of damage and stability of dislocation patterns) which, strictly speaking, has not been interpreted before by predictive methods. These new theoretical means are based on recent advances in modern non-linear analysis and concepts of non-convexity, multiplicity, and bifurcation. A possibility is thus now being developed for applying modern mathematical tools in the analysis of microstructures, their patterns, and their effect on macroscopic behaviour.

These possibilities become apparent by reformulating the general mathematical structure of the earlier programme. This reformulation is based on the new physical assumption of distinguishing among two types of atoms†: 'excited' participating in the formation and evolution of the various microstructures and 'normal' accommodating these excitations to come about. Thus, instead of viewing a general body as a mixture of physically different constituents (such as parent material, corrodents, and products of reactions in a degradation process),

† Generally speaking, these can be either actual atoms or aggregates of them or even macroscopic particles (such as granules) and distinct material phases.

we view it as a superposition of states: those asociated with the particular microstructural mechanisms (or imperfect states) and the parent (or perfect) state. The various states can interact chemically, mechanically, and thermally by the transfer of atoms carrying mass, momentum, and energy from one state to another. Our results show that the new formulation is more general than the previous one and leads into a natural definition of the appropriate internal variables. Important topics—such as the use of proper measures for stress and strain and their rates in the case of large deformations, the implications of dynamic effects and wave propagation aspects, the non-local character of microstructural effects and the localization of microstructures to form slip or shear bands and plastic or failure zones—can be more conveniently discussed and analysed within this reformulation.

Moreover, the programme of specializations initiated during our earlier studies can be cast into a more systematic and unifying framework where a distinction to various classes of material behaviour is made according to prevailing microstructures and their interactions. Microscopic processes at the lattice level such as spinodal decomposition (up-hill diffusion) and phase separation is an instructive example because of its basic non-linear character and inherent tendency to pattern formation. This behaviour (which is central in understanding the nature of general microstructural processes) and the influence of stress on it, can be conveniently analysed within the proposed reformulation. With the introduction of an appropriate set of internal orientations the motion and generation of dislocations can be considered and the effect of these microscopic processes on the macroscopic phenomena of cyclic plasticity, creep, and fatigue can be evaluated. While lattice-induced microstructural processes is a basic mechanism occurring at the atomic scale, other microstructural processes associated with the nucleation and growth of chemically inert voids and microcracks occurring at a larger but still submacroscopic scale are of considerable interest. These can also be considered more conveniently within our modified framework and their role on the macroscopic phenomena of spallation and high-temperature creep can be investigated. Similarly, chemically-induced microscopic processes can be interpreted and their relation to embrittlement and environmental cracking can easily be established. Our earlier studies in this area emphasized subcritical steady-state environmental crack growth and the proposed theoretical models were in good agreement with experiments. Transient growth, however, is often occurring in structural components of modern technology but very little is known about it. Our preliminary investigations have produced encouraging results in this direction by successfully interpreting certain experimental findings. In addition to the practical importance of these studies, the theoretical analysis is quite challenging because of the coupling between chemical and structural microprocesses in the presence of a mechanical stress acting as driving force. Chemical creep (where the interaction effects between vacancy generation, dislocation motion, and precipitate growth are governing the overall macroscopic deformation) is another phenomenon of chemomechanical nature which can be conveniently and rigorously considered within our programme.

Finally, it is noted that within the proposed reformulation new methods and theoretical tools (not necessarily of a deterministic nature) can be employed. These include the introduction of statistical and probabilistic concepts at various stages of development of the deterministic programme discussed earlier. Distribution functions for the prevailing microstructures can be employed and concepts from percolation theory and effective medium theory can be utilized to model features of the topology and geometry of particular microstructures and allow calculations of various phenomenological coefficients. The problem of interrelation of scales can be addressed with emphasis on establishing contact between microstructural localization and macroscopic localization of deformation. Computer codes for various models advanced within the programme can be developed, solution of particular boundary value problems can be performed, and comparisons with experiment can be made. Moreover, a new approach to irreversible thermodynamics for general microscopic processes can be initiated aiming to address successfully questions and behaviour where traditional continuum thermodynamics have produced undesirable answers.

1.2 REVIEW OF PREVIOUS WORK

A review of the physics involved in three main forms of microstructure dependent processes—that is, stress corrosion, embrittlement, and material damage (e.g. creep or spall damage)—reveals that, in general, such processes depend on a combination of metallurgical, chemical and thermomechanical factors, and that the prevailing mechanism is likely to vary according to the particular situation. However, certain broad principles can be applied and important features can be singled out as characterizing any microstructure dependent process. The central issues that formed the basis of our earlier studies on media with general microstructures are:

(i) In general, the body is viewed as a mixture of mechanically, thermally, and chemically interacting materials. The various constituents of the body are the parent material, the corrodents, and the products of reaction and they are all free to support their own stress, to conduct their own heat, and obey their own diffusional and chemical path, so long as the basic laws of mechanics (conservation of mass and momentum), thermodynamics (conservation of internal energy and second law), and chemistry (permanence of atomic substances) are satisfied for the individual constituents separately, as well as for the body as a whole.

(ii) The microscopic configuration is important and should enter explicitly in the phenomenological description. The following aspects are central in any model development: (a) The particular chemistry involved. This is usually investigated by chemical analysis and incorporated within the formalism of chemical kinetics. (b) The microstructure of the general body as related to the constitutive model that will eventually be assumed for the parent material. That is, those aspects of the microstructure that in the absence of environmental conditions would classify the parent material as elastic, viscoelastic, plastic, viscoplastic, etc. This is investigated by microscopic and macroscopic observation with the aid of microscope and experiment of the parent material in the absence of environment. (c) The microstructure of the body as related to the particular chemically-induced microprocesses. For example, while grain boundaries need not be identified in modelling the parent material in the absence of environment, they should explicitly be considered in the presence of environment where diffusion and/or chemical reactions are localized. Such characteristics can be detected with the electron microscope.

(iii) In most forms of microstructure dependent processes, various threshold conditions (e.g. stress corrosion and hydrogen embrittlement thresholds) have been established experimentally. Often, however, thresholds are thought to be a result of the particular experimental setting rather than an actual physical occurrence. For practical purposes, thresholds are useful to understand and a general theory for microstructures should be featured with a 'threshold criterion'. This should not only be a formal mathematical generalization of various experimentally established threshold conditions but also enjoy a deeper physical basis associated with the activation mechanism of the particular microscopic process.

(iv) Any approach to understanding media with microstructures should be unifying in character. This is not only because the physical problem is of complex nature and requires the collaboration of specialists from different existing disciplines but also because of its potential to generate new fields of research such as the discipline of chemomechanics. Moreover, all previous models on the subject such as embrittlement and stress corrosion models, creep and spall damage models, etc. should find a rational position within a general theory for media with microstructures.

Guided by the above discussion a general phenomenological programme for elastic or inelastic bodies with general microstructures has been constructed in the last few years, by employing the techniques of modern continuum thermomechanics together with information provided by classical chemistry, modern materials science, and existing experimental data. The programme was mainly motivated by microstructure dependent processes relevant to metals (e.g. stress corrosion, embrittlement, creep damage, spall damage) but it can easily be extended to non-metals and other forms of microstructures. The whole effort was characterized by four main components within which new concepts and ideas have been incorporated as follows:

(i) *An axiomatic framework*. Introduction of (a) mixture formalism and (b) chemical axioms.

(ii) *A constitutive representation.* Introduction of (a) chemical kinetics, (b) notion of pseudo-quasi-elasticity (which includes usual elastic, viscoelastic, plastic, and viscoplastic models), (c) internal variables related to the particular inelasticity of parent material, (d) internal variables related to constitution of the particular process, (e) complete balance laws for the internal variables replacing the usual evolution laws.

(iii) *A threshold criterion.* Introduction of (a) universal activation energy for microscopic processes as a part of the constitution, and (b) concepts and analogies from internal barriers, absolute rate reaction, and yielding theories.

(iv) *A programme of specializations*, leading to engineering applications. Previous models of stress corrosion, embrittlement, creep damage, and spall damage have been recovered and generalized within our theory. Models of plasticity, cyclic plasticity, and fatigue have also found a position within our proposal. Nevertheless, the issue that has been emphasized most within the programme of specializations was the topic of 'chemical damage' and 'chemomechanics'.

The issue of 'chemomechanics' has been approached by considering in detail the examples of (a) diffusion, (b) embrittlement, (c) dislocation-induced plasticity, and (d) chemical creep.

(a) *Diffusion.* Here we are dealing with the most elementary microscopic process of chemical nature. The strains due to the evolving microstructure are the extra strains of the lattice due to diffusion. This strain is anelastic in nature, that it is not recovered immediately upon unloading but it returns slowly to its reference value in contrast to the permanent strain of plastic deformation. Our results here are conclusive and they show the interaction of elasticity, viscoelasticity, or plasticity of the solid matrix with the microscopic process which is simply diffusional in nature.

(b) *Embrittlement.* In this case the chemical effects are more explicitly shown. In addition to diffusion, chemical reaction of the diffusing agent and precipitation in the form of growing gas bubbles occurs. The microstructurally-induced strains are now due not only to diffusion but mainly to the coarsening of the bubbles. The rate of the chemical reaction, precipitation, and coarsening is stress controlled and we have already derived partial differential equations for a linearized theory under the assumptions of a fixed number of bubbles per unit volume which do not migrate and do not grow. These equations resemble the usual elasto-diffusive theories and possess simple analytic solutions. This suggestive theory is currently being generalized to allow removal of the above assumptions and yield a realistic continuum theory of embrittlement.

(c) *Dislocation-induced plasticity.* In the third case the chemical reactions are 'massless' or 'generalized' in nature and occur between mobile dislocations, immobile dislocations, and perfect lattice. A dislocation model of plasticity has thus resulted as an outgrowth of continuum theories of dislocations advanced by mechanicians and discrete models of dislocation dynamics proposed by materials scientists. While most difficulties which existed in previous dislocation theories of macroscopic behaviour have been removed, there are still important issues that are currently being elaborated. Among the advantages of this new approach to dislocations is the possibility of developing realistic models predicting the formation and evolution of slip bands and dislocation fabrics. Moreover, yielding is naturally interpreted in terms of the momentum balance for the immobile dislocation state and classical phenomenological models of plasticity, cyclic plasticity, and fatigue have been recovered and microstructurally substantiated. Large deformations, statistical and probabilistic aspects, averaging and polycrystals, localization of dislocation structures and its relation to macroscopic localization of deformation in shear and plastic zones are among the topics of current investigation.

(d) *Chemical creep.* The chemomechanical effects are even more explicitly shown in the fourth case where creep is induced because of the interaction of dislocations and $Ni_{\alpha}Al_{\beta}$ precipitates whose formation and coarsening is stress controlled. While the physics of such systems is well understood, we only have preliminary results towards their quantitative modelling. Among our future goals is a detailed elaboration towards the obtaining of unifying predictive models for these types of very interesting chemomechanical systems which govern the macroscopic behaviour of most alloys.

More details substantiating the discussion and claims of this section can be found in references [2 to 5, 6 to 8], and [9 to 14], where equations modelling the relevant microscopic processes and their effect on macroscopic response for small strains and inertialess situations are given.

1.3 REFORMULATION OF MATHEMATICAL FRAMEWORK

As was discussed in the previous section, in our earlier attempts towards modeling the response of media with general microstructures, it was proposed that concepts from mixture theory (see, for example, [15]) and the internal variable formalism (see, for example, [16]) could conveniently be utilized to characterize various classes of materials with microstructure. The mixture formulation was used to interpret the interaction between the parent material and the products of microscopic changes, while internal variables were introduced to represent the inelasticity of the parent material and the character of the particular structural mechanism. It was emphasized that internal variables should be determined in accordance with complete balance laws [2, 5] containing both a rate and a flux term as opposed to the traditional evolution laws [17, 18] which neglect the divergence term. Even though this generalization of the evolution law for the internal variables to include the flux of microstructures within the elementary material volume was formally introduced by the author in 1977 ([19a, b], see also [2]) it was only recently [20, 21] that the significance of this generalization has been noted and discussed by others.

The natural occurrence of the flux term (which does not necessarily lead to Laplacians and diffusive-type operators as postulated and investigated in [20, 21]) in the evolution equations for the internal variables becomes even more transparent within this presentation. We elaborate further on our earlier proposal and demonstrate briefly the physical significance of both the flux term and the nature of non-linearity modelling the generation of microstructures. In particular, a framework is provided to link the approaches of mixture and internal variable theories into a single formalism by distinguishing among 'normal' atoms and 'excited' atoms. The 'excited' atoms are those which participate in the nucleation and growth of the various microstructues, while the 'normal' atoms are accommodating the occurrence of such excitations. It is essentially this identification of the excited atoms and its direct relation to a proper definition of the internal variables that lead (through the balance of mass for the excited atoms) to a straightforward generalization of the usual evolution laws to produce complete balance laws containing both rate and flux terms.

Probabilistic effects and statistical aspects are often important and can be conveniently incorporated at various stages of this reformulation. Moreover, large deformations and dynamic effects can be important and these issues can also be rigorously considered. Comparison with existing experiments can be consistently made and used as a guide in the theoretical investigations. While it is believed that some type of thermodynamics is underlying general microstructural phenomena, we do not attempt to employ any existing formalism of thermodynamics as a starting point of the analysis. This is motivated by our recent findings [22] indicating that straightforward application of the usual thermodynamic relationships can lead to undesirable results when one attempts to trace the evolution of a system during instability and bifurcation. Moreover, explicit consideration of defects such as dislocations presents an additional complication [23]. We thus employ the principles of mechanics alone and anticipate to incorporate an appropriate thermodynamics at a later stage when more definite progress is made in this direction.

Media with general microstructures undergo continuous irreversible changes or transformations (living bodies) under the action of external stress. In this sense, an alloy undergoing spinodal decomposition and phase separation by up-hill diffusion of an initially uniformly distributed solute, a metal undergoing plastic deformation via generation, motion and interaction of dislocations, and a rock undergoing creep via nucleation and propagation of microcracks are particular types of media with microstructures. In the first case the 'excited' atoms are foreign species which by their up-hill diffusion lead to spinodal decomposition and phase separation under the action of external stress. In the second case the 'excited' atoms are dislocation-forming atoms which by their motion, annihilation, and generation can give rise to the formation of slip bands, plastic zones, and spatial patterns under the action of external stress. Finally, in the third case the 'excited' atoms are microcrack-forming atoms which under the action of external stress nucleate, propagate, dissociate, or recombine to form regions of various degrees of damage.

By adopting simple geometric arguments, the following relations can be readily shown for each one of the three cases:

Diffusion: $\rho = \text{const. } c$,
Dislocations: $\rho = \text{const. } n$,
Cracks (voids): $\rho = \text{const. } nl \; (\rho = \text{const. } nr)$,

In all these equations ρ denotes the density of the excited atoms, const. an explicit molecular constant, c concentration of diffusing species (e.g. in p.p.m.), n number of dislocations, cracks or voids per unit volume, l length of flat cracks, and r radius of spherical voids. The above derivation and explicit calculation of the molecular constants is based on certain simplifying hypotheses including the assumptions that in the case of dislocations the excited atoms are confined within the core, while in the case of voids and cracks they are confined within a small surface layer. An interesting feature here is the fact that the quantities c, n, l, and r appear naturally in this theory, while in other theories they are usually postulated at the outset as internal variables.

For illustrative purposes we assume below small deformations and linear elasticity for the parent material (or normal atoms). Then the macroelement of the body obeys the following familiar equations

$$\left.\begin{aligned} \operatorname{div} \mathbf{S} &= \tilde{\rho}\ddot{u} \\ \mathbf{E} &= \tfrac{1}{2}(\nabla \mathbf{u} + \nabla \mathbf{u}^{\mathrm{T}}) \\ \mathbf{E} &= \phi(\mathbf{E}^{\mathrm{e}}, \mathbf{E}^{\mathrm{m}}) \\ \mathbf{E}^{\mathrm{e}} &= \mathbf{CS} \end{aligned}\right\} \tag{1.1}$$

The first equation denotes dynamic equilibrium with $\mathbf{S}$ and $\tilde{\rho}$ being the total stress and density supported by both normal and excited atoms. The second equation is the usual relation between the total linear strain $\mathbf{E}$ and the displacement $\mathbf{u}$. The third equation is a functional relationship between the elastic strain $\mathbf{E}^{\mathrm{e}}$ (induced by reversible displacements of normal atoms) and the microstructure related strain $\mathbf{E}^{\mathrm{m}}$ (induced by irreversible displacements of excited atoms). In certain cases ϕ becomes simply a sum of $\mathbf{E}^{\mathrm{e}}$ and $\mathbf{E}^{\mathrm{m}}$, that is $\mathbf{E} = \mathbf{E}^{\mathrm{e}} + \mathbf{E}^{\mathrm{m}}$. Finally, the fourth relation is the assumption that the normal state is a linear elastic state with $\mathbf{C}$ being the compliance tensor.

In order that the system of equation (1.1) be solvable, an equation for $\mathbf{E}^{\mathrm{m}}$ is needed. It is in this place where probabilistic and statistical methods can be introduced to properly relate the macroscopic strain $\mathbf{E}^{\mathrm{m}}$ with the various irreversible displacements that can possibly occur at a lower or microscopic level.

If probabilistic aspects are to be burried in a functional relationship between the microstructurally induced strain $\mathbf{E}^m$ and the various parameters which define the relevant microscopic processes, we can write

$$\mathbf{E}^m \xrightarrow{f} [\rho, \mathbf{j}, (\hat{\mathbf{n}})] \tag{1.2}$$

where f is a rule of correspondence (function, differential, or integral operator), ρ is the density of excited atoms, $\mathbf{j}$ is their flux, and $(\hat{\mathbf{n}})$ is a set of unit vectors designating internal directions associated with a particular microscopic process.

It remains to introduce a set of internal balance laws and constitutive equations to determine the density and motion of excited atoms which are carriers of the irreversible displacement events. The following equations are relevant here

$$\left.\begin{aligned} &\dot{\rho} + \operatorname{div} \mathbf{j} = \hat{q} \\ &\operatorname{div} \mathbf{T} + \hat{\mathbf{f}} = 0 \\ &[\mathbf{T}, \hat{\mathbf{f}}, \hat{q}] \xrightarrow{f} [\rho, \mathbf{j}, (\hat{\mathbf{n}}), \mathbf{S}] \end{aligned}\right\} \tag{1.3}$$

The first equation denotes balance of mass of the excited atoms with $\hat{q}$ representing transfer of atoms between normal and excited states. The stress associated with the excited state (i.e. the stress exerted by the excited atoms on themselves) is denoted by $\mathbf{T}$ and for simplicity is assumed to be symmetric in this presentation. The exchange of momentum between normal and excited states gives rise to the resistance force vector $\hat{\mathbf{f}}$. The second equation is thus a statement of conservation of momentum where inertia effects associated with the excited atoms are included in the expression for $\hat{\mathbf{f}}$. Finally, the third set of equations consists of internal constitutive equations for the excited state; this is also a potential place for incorporating statistical concepts and introducing in a rigorous way the effect of an externally applied stress.

Our results show that this programme is capable of describing conveniently a variety of classes of microstructure dependent processes ranging from spinodal decomposition and dislocation-induced plasticity to microcrack-induced damage and chemical damage. The outlined formalism is particularly useful in considering the coupling between various types of microstructures that can nucleate and evolve simultaneously. All the quantities entering into the theory have a direct physical meaning and can easily be related to measurable quantities. For example, in the case of spinodal decomposition in an isotropic solid matrix we can have

$$(\hat{\mathbf{n}}) \equiv 0, \qquad \hat{q} \equiv 0, \qquad \mathbf{E}^m \equiv Ac\mathbf{1}$$

while in the case of lattice degradation by dislocation generation and motion we have

$$(\hat{\mathbf{n}}) = (\hat{\mathbf{n}}, \hat{\mathbf{v}}), \qquad \hat{q} = \hat{a}n - \hat{b}n^2 + \hat{c}n^3, \qquad \dot{\mathbf{E}}^m = (A\dot{n} + Bj)\mathbf{M}$$

with $j = \mathbf{j} \cdot \hat{\mathbf{v}}$ being the slip component of the flux $\mathbf{j}$; $\hat{\mathbf{n}}$ and $\hat{\mathbf{v}}$ denoting respectively unit vectors normal to the glide plane and in the slip direction, $(\hat{a}, \hat{b}, \hat{c})$ are stress dependent coefficients, the components of the orientation tensor $\mathbf{M}$ are $M_{ij} = \frac{1}{2}(n_i v_j + v_j n_i)$; and the coefficients A and B are viewed as constants even though in some cases are also stress dependent. Similar expressions hold for the case of cracks and voids without or with chemical reactions. When a single family of dislocations or microcracks is present, the internal directions are crystallographically specified. When the number of internal orientations is rapidly increasing, appropriate distribution functions should be introduced (e.g. [24 to 26, 27, 28, 29, 30]). In the case of infinitely many and randomly distributed orientations, certain averaging techniques can be employed which determine the internal directions in terms of the local stress tensor. These types of averaging procedures have led to a microstructural substantiation of classical yield criteria and plastic flow rules [3 to 5, 10] and illustrated the importance of Von-Mises stress in some theories of creep damage.

1.4 SPECIAL TOPICS

Particular topics that can be conveniently emphasized within the above reformulation are the consideration of large deformations, dynamic effects, non-local character of microstructures and their localization in submacroscopic and macroscopic deformation bands.

Large deformations can be considered by adopting a multiplicative decomposition of the deformation gradient as discussed for example, in [31 to 35] (see also [25, 36]). The appropriateness of various objective strain and stress rate measures, which has recently been a controversial subject, can be addressed on physical grounds by demonstrating that the selection of these measures is often an issue of convenience and that their precise form can possibly by determined only through microstructural, as opposed to phenomenological, reasoning (after possible elimination of the microstructural variables).

Dynamic effects can be considered and wave propagation aspects due to both normal and excited atoms can be discussed in connection with available one-dimensional experimental data. This can yield convenient procedures for evaluating certain phenomenological coefficients through non-destructive testing. Also, it is possible to correlate these studies with current studies on wave propagation in media with microstructure which, however, are based on mappings on linear elasticity with appropriate adjustments on the elastic constants (e.g. [39] and similar papers presented in the same volume; see also dynamic extensions of the methods discussed in [30, 44]). Along more applied lines it can be shown explicitly how dislocation acceleration is incorporated into our framework and how transient environmental cracking can be interpreted within our formalism. Preliminary results show that findings reported in [37] can be recovered and generalized and that our earlier steady-state environmental cracking studies [3, 12, 13] can be extended to the transient case in accordance with experiment [38].

The non-local character of microscopic processes is formally apparent from equation $(1.3)_1$. The flux term in this equation, which is commonly omitted from usual internal variable theories, establishes contact between the present programme and a recently developed non-local theory of fracture [40]. Indeed, it is possible to show that as in the case of memory materials the rate term in the evolution law for the internal variables can lead to time-integral relations between stress and strain, similarly in the case on non-local materials the flux term in the complete balance law for the internal variables can lead to space-integral relations, i.e. non-local constitutive equations.

Finally, the localization of microstructures can be elaborated upon in a unified and elegant manner. This approach can produce interesting results towards the obtaining of predictive models for the localization of general microstructures and its implications on macroscopic localization of deformation. For example, by confining attention to one dimension, it can be shown that for steady states under certain conditions the spatial distribution of a large class of microstructures is determined by the non-linear equation

$$u_{xx} = g(u) \tag{1.4}$$

where u denotes the concentration of a certain type of microstructure such as diffusing species ($u \equiv c$), dislocations ($u \equiv n$), or cracks and voids ($u \equiv n$); and g is a function with a single loop (see Figure 1.1 (a)). The loop occurs when the applied stress surpasses a critical value. The bounded solutions of equation (1.4) for the infinite line have one of the forms depicted in Figures 1.1(b)–(d) and they obviously resemble observed behaviour [41].

Some results prevailing to the analysis of equation (1.4) were reported in [4] and [22]. These results can be further elaborated and extended to more dimensions and more microscopic variables. Thus, u in equation (1.4) can be a vector (u_1, u_2) with u_1 representing, for example, number of dislocations per unit volume and u_2 number of voids per unit volume. A non-trivial extension of the mathematical analysis reported in [4, 22] can then provide information for the coupling of the two types of microstructures and their spatial distributions. These studies (with the aid of the computer) can eventually lead to predictions for the formation and stability of patterns, as related to shear localization, and their dependence on applied stress. Previous studies on this topic have been developed along different paths and are rather phenomenological in nature (e.g. [42]). While such studies have provided substantial insight to localization phenomena, they are also characterized by certain *ad hoc* hypotheses which obscure their predictive value. One immediate objective is to establish contact between the microstructural and phenomenological approaches to localization and substantiate microstructurally various phenomenological models. A connection to certain microscopic models [43] which, however, are different in philosophy and method of attack can also be made within our structure.

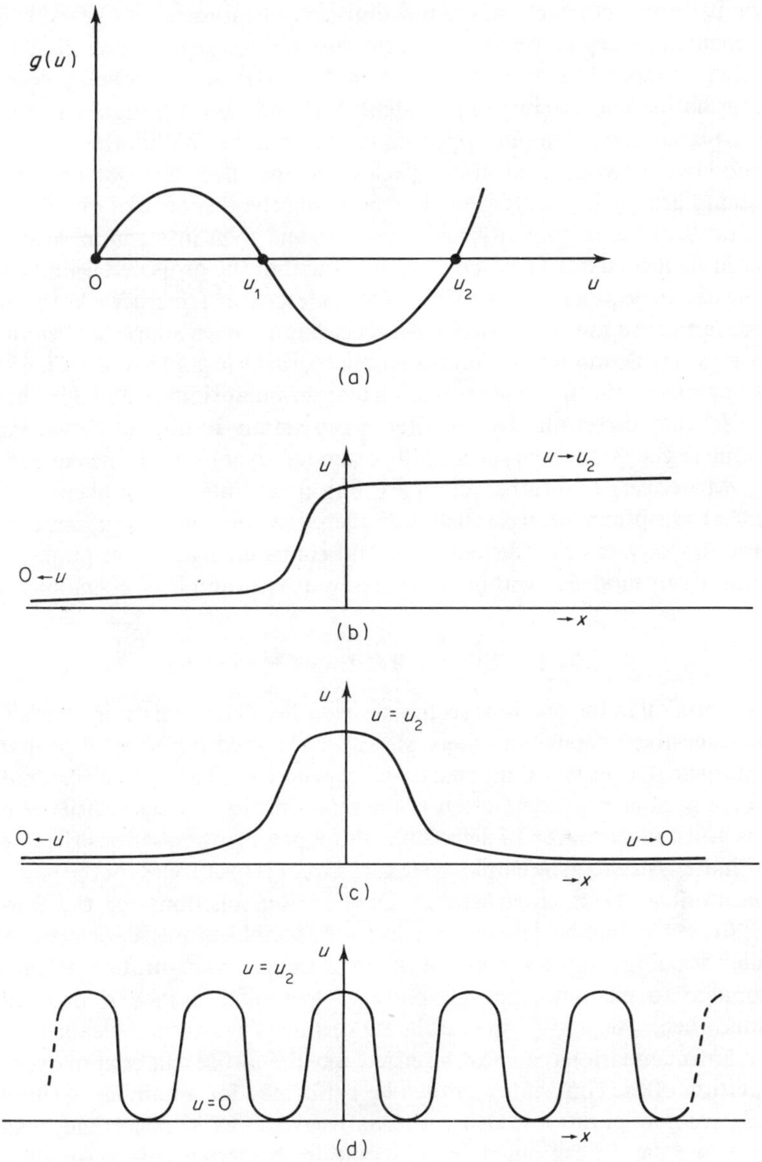

Figure 1.1

In summary, new directions for creative and interdisciplinary research are established within the proposed programme for media with microstructure. For example, non-linear and dynamic aspects of spinodal decomposition, as well as the effect of stress on the process, can be considered within purely mechanical grounds. In this connection, the metallurgical phenomena of Ostwald ripening and chemical creep can be offered new possibilities for interpretation within a strictly mechanical theory of stress and surface tension controlled precipitation. The formation and stability of persistent slip bands during fatigue can be given a new explanation within our approach to dislocations. Within this approach the kinetic laws between dislocation species and the effect of stress on the kinetic constants are the key factors in characterizing the response of materials under conditions of cyclic plasticity, fatigue, creep, and their interaction. Microcrack-induced inelasticity can also be considered within the proposed framework and emphasis can be placed on describing the coalescence of microcracks through the source term $\hat{q}$ in equation $(1.3)_1$. Coalescence has not been addressed before within internal variable models for voids and microcracks (e.g. [18 to 25, 27, 35]). It is thus hoped that this particular research may stimulate initial modelling attempts towards this direction. Our earlier steady-state results on environmental cracking (e.g. [13]) can be successfully extended to consider transient behaviour [38]. Moreover, the rather complex but quite interesting phenomenon of chemical creep can be rigorously considered within our programme and the interaction of vacancies, dislocations, and shape-changing precipitates, can be quantitatively modelled within this representative example of 'chemomechanics'.

1.5 FURTHER RESEARCH ISSUES

The discussion in the previous sections was rather deterministic in character and only suggestive propositions were made for the incorporation of probabilistic and statistical aspects. An immediate place where such arguments can naturally be introduced is in the derivation of the relationship between density of excited atoms and concentration of defects (i.e. those preceding equations (1.1)). A local coordinate system can be employed (e.g. [24 to 27]) to account for the preferential orientation of the microstructures. Distribution functions for the prevailing microstructures can be defined to reflect the special features associated with the detailed topology and geometry of the microscopic configuration at hand. It is reasonable to examine the suitability of 'percolation theory' and 'effective medium theory' (e.g. [30]) in calculating certain phenomenological coefficients. The relevant equations can be discretized, a bethe-lattice can be employed, and a calculation of the constants can then be performed for a number of simple and instructive configurations. Other alternatives such as Voronoi and hexagonal tessellations can be examined and percolation thresholds can be sought for the mobile microstructures. Moreover, the effect of deformation can be statistically examined (by properly introducing deformation dependent distribution functions) for simplified and instructive cases. Finally, the relation and applicability of the self-consistent [44] and the micromechanics [46] approaches and the homogenization theory [47, 48] to the above developments can be investigated.

Principles from molecular dynamics can be utilized in simulating the effect of microstructures on the macroscopic response. The atomic force–displacement relations of molecular dynamics will of course be replaced now by other macroscopic force–displacement relations between material elements with discretized microstructures. This technique has been applied successfully to soils and granular materials (e.g. [45]) and is now proposed to be extended to more complicated microstructures. This discrete modelling and numerical simulation approach is the only sensible method in the case where the microstructures (e.g. precipitates or voids) are sufficiently parted from one another; it is thus a necessary component for a complete study of the subject. Moreover, it is offered as a 'computer experiment' approach for validating the continuum assumptions. Finally, it could be used as a means of estimating the size of the 'neighbourhood' of the continuum approximation according to the particular microstructures present, and deduce numerical expressions for the average stress and deformation tensors.

Other additional problems can also be attacked; these are: the interrelation between scales with emphasis on the effect of localization of microstructures on macroscopic localization of deformation; the development of computer codes with emphasis on solutions to boundary value problems of practical importance; and the consideration of a new thermodynamics capable of describing complex physical processes characterized by small-scale instabilities, defects, and microstructures.

1.6 APPENDIX 1 MECHANICAL EXAMPLES OF MEDIA WITH MICROSTRUCTURE

To illustrate the general mathematical framework of this programme, we list below classes of differential equations modelling certain types of elementary microstructures. The details of the derivation are postponed for a future occasion. In the absence of dynamic effects, a procedure for deriving the appropriate parts counter of some of these equations was discussed in [2, 3].

(a) Diffuso-elasticity

On confining attention to small deformation of a linear elastic solid in which the only microstructure present is a solute of small concentrations whose motion is of a diffusive nature, we can derive

$$\left.\begin{aligned} G\Delta\mathbf{u} + (\lambda + G)\nabla(\nabla\cdot\mathbf{u}) - \gamma\nabla\rho &= \tilde{\rho}_{\mathrm{R}}\ddot{\mathbf{u}} \\ \dot{\rho} + \rho_{\mathrm{R}}\nabla\cdot\dot{\mathbf{u}} &= D\Delta\rho - L\Delta(\nabla\cdot\mathbf{u}) \end{aligned}\right\} \tag{1.5}$$

with $\mathbf{u}$ denoting the displacement of the medium and $\tilde{\rho}_{\mathrm{R}}$ its reference density; ρ the density of the solute and ρ_{R} its reference value $(\nabla, \nabla\cdot, \Delta)$ denote respectively gradient, divergence, and Laplacian; and the various coefficients are constants.

(b) Diffuso-viscoelasticity

If we are concerned with the same microstructure as above but the character of the parent material (normal atoms) is of linear viscoelastic nature (instead of linear elastic), the relevant equations take the form

$$\left.\begin{aligned} \hat{G} \circledast \Delta \mathbf{u} + (\hat{\lambda} + \hat{G}) \circledast \nabla(\nabla \cdot \mathbf{u}) - \hat{\gamma} \circledast \nabla \rho &= \rho_R \ddot{\mathbf{u}} \\ \dot{\rho} + \rho_R \nabla \cdot \dot{\mathbf{u}} &= \hat{D} \circledast \Delta \rho - \hat{L} \circledast \Delta(\nabla \cdot \mathbf{u}) \end{aligned}\right\} \tag{1.6}$$

where the various symbols have the same meaning as in equation (1.5) with the exception of the symbols with the superimposed ^ which now denote functions of the time variable s and the Boltzmann operator $\circledast$ defined for two functions $a(t)$ and $b(x, t)$ as follows

$$a \circledast b \equiv \mathring{a} b(\mathbf{x}, t) + \int_0^\infty \dot{a}(s) b(\mathbf{x}, t - s)\, ds$$

with $\mathring{a} \equiv a(0)$ denoting initial response and $\mathbf{x}$ the spatial coordinate.

(c) Gradient diffuso-elasticity (spinodal decomposition)

If we are still concerned with the same microstructure as in (a) but surface tension effects are now important (under certain circumstances these effects cause up-hill diffusion and spinodal decomposition), the appropriate equations become

$$\left.\begin{aligned} G\Delta \mathbf{u} + (\lambda + G)\nabla(\nabla \cdot \mathbf{u}) - \gamma \nabla \rho + \delta \nabla(\Delta \rho) &= \tilde{\rho}_R \ddot{\mathbf{u}} \\ \dot{\rho} + \rho_R \nabla \cdot \dot{\mathbf{u}} &= D\Delta \rho - E\Delta^2 \rho - L\Delta(\nabla \cdot \mathbf{u}) \end{aligned}\right\} \tag{1.7}$$

where the various symbols have their usual meaning, the newly introduced coefficients are constants, and the symbol Δ^2 is the biharmonic operator.

(d) Double porosity consolidation.

If the existing microstructure is in the form of a liquid filling completely two types of coexisting pores (fissures and pores), then the relevant equations read

$$\left.\begin{aligned} G\Delta \mathbf{u} + (\lambda + G)\nabla(\nabla \cdot \mathbf{u}) - \beta_1 \nabla p_1 - \beta_2 \nabla p_2 &= \tilde{\rho}_R \ddot{\mathbf{u}} \\ \alpha_1 \dot{p}_1 + \beta_1 \nabla \cdot \dot{\mathbf{u}} &= \frac{k_1}{\mu} \Delta p_1 + \kappa(p_2 - p_1) + f_1 \\ \alpha_2 p_2 + \beta_2 \nabla \cdot \dot{\mathbf{u}} &= \frac{k_2}{\mu} \Delta p_2 - \kappa(p_2 - p_1) + f_2 \end{aligned}\right\} \tag{1.8}$$

where now p_1 and p_2 denote respectively the pressure in the fissures and the pressure in the pores; the various coefficients are associated with either the fissure (assigned the index 1) or the pore (assigned the index 2) spaces; and the quantities

f_1 and f_2 are, in general, linear functions of the inertia and relative velocities of the solid and fluid phases.

(e) **Dislocations, cracks, and chemical damage**

Other more complicated microstructures such as dislocations and chemically-induced cracks can be elegantly modelled within the proposed framework. As a result of the momentum balance for the excited atoms given by equation $(1.3)_2$ and appropriate constitutive assumptions for their stress $\mathbf{T}$ and their momentum exchange with the normal atoms $\hat{\mathbf{f}}$ (which also includes inertia terms), the following equation can be derived for the motion of a dislocation

$$A\ddot{u} + B\dot{u} - Cu_{xx} = \sigma b \tag{1.9}$$

with u denoting displacement, σ the applied stress, and the rest of the symbols denote material coefficients. The derivation of equation (1.9) was postulated before (e.g. [37]) on purely phenomenological grounds.

Similarly, under certain conditions the following equation can be derived for the spatial distribution of dislocations in a monocrystal with a single slip system $(\hat{\mathbf{n}}, \hat{\mathbf{v}})$ loaded with a constant tensile stress

$$D_1 \Delta n + D_2 \operatorname{tr}(\mathbf{M}\nabla^2 n) = \hat{q} \tag{1.10}$$

where $\mathbf{M}$ is an appropriate orientation tensor, n is the density of dislocations (lines/unit volume), $\hat{q}$ denotes the production of dislocations within a unit volume, (D_1, D_2) are constants, tr denotes trace and $\nabla^2 \equiv \partial^2/\partial x_i \partial x_j$.

For a crack growing under steady-state environmental conditions it can be shown that for certain conditions its velocity V relates to the stress-intensity factor K_{I} by the formula

$$V = V_0 K_{\mathrm{I}}^{\alpha} \tag{1.11}$$

with V_0 and α denoting material constants. For transient growth (where dynamic effects are important) the dynamic counterpart of equation (1.11) reads

$$V + \lambda^{-1}\dot{V} = V_0 K_{\mathrm{I}}^{\alpha} \tag{1.12}$$

which can be readily solved for constant K_{I} to produce V versus t profiles in agreement with experiments [38]. The consistency of equation (1.11) with experimental data is discussed in [13]. Equations (1.11) and (1.12) can be used as a starting point for the development of theories with a continuum distribution of small environmental cracks. For example, equation (1.11) can lead, in consistency with some experimental data, to the following relationship

$$t_{\mathrm{f}} = \frac{\alpha^*}{c^*} l_0^{-\alpha^*} \tag{1.13}$$

between the time to failure t_{f} and the initial length of microcracks l_0 with α^* and c^* denoting material parameters.

1.7 APPENDIX 2 A THERMODYNAMIC EXAMPLE OF A MEDIUM WITH MICROSTRUCTURE

The examples discussed in the Appendix 1 were based on the mechanical principles of mass and momentum balances alone. Thermodynamics was not utilized because of lack of a definite thermomechanical structure for general microstructures. The construction of such generalized thermodynamics is currently under consideration and is based on a detailed elaboration of an earlier proposal of the author [19b]. To provide insight to the structure of this theory, we briefly consider here certain possibilities on thermodynamic models for fluid interfaces. Fluid interfaces can be considered as materials of a relatively simple microstructure which as discussed in [22] can be sufficiently reflected in the constitutive equations by introducing density gradients up to second order as independent variables.

As will be discussed in detail in a future publication the balance laws for fluid interfaces are assumed to be of the form of usual thermomechanics with the exception that the energy equation contains an extra term $\hat{\varepsilon}$ representing surface effects and long-range interactions. It then turns out that the appropriate Clausius–Duhem inequality reads

$$\rho(\dot{\psi} + \eta\dot{\theta}) - \mathbf{T}\cdot\mathbf{Dv} + \theta^{-1}\mathbf{q}\cdot\mathbf{D}\theta - \hat{\varepsilon} \leq 0 \tag{1.14}$$

where ρ is the density, ψ the free energy density, η the entropy density, θ the absolute temperature, $\mathbf{T}$ the stress tensor $\mathbf{v}$ the velocity, $\mathbf{q}$ the heat flux, and $D(\cdot)$ denotes the gradient operator with respect to the Eulerian frame. If we assume that $\hat{\varepsilon}$ is exclusively due to surface effects then we have

$$\hat{\varepsilon} = \operatorname{div}\mathbf{h} \tag{1.15}$$

and thus the appropriate constitutive structure is of the form

$$\{\psi, \eta, \mathbf{T}, \mathbf{q}, \mathbf{h}\} \xrightarrow{\text{fct.'s}} \{\rho, \theta, \mathbf{D}\rho, \mathbf{D}^2\rho, \mathbf{D}\theta, \mathbf{Dv}\} \tag{1.16}$$

it then turns that equations (1.14) and (1.16) are compatible only when

$$\begin{aligned}
&\psi = \psi(\rho, \theta, \mathbf{D}\rho), \qquad \eta = -\psi_\theta \\
&\mathbf{h}_{\mathbf{D}^2\rho}\cdot\mathbf{D}^3\rho = 0 \\
&\mathbf{h}_{\mathbf{D}\theta}\cdot\mathbf{D}^2\theta = \mathbf{0} \\
&\mathbf{h} = \rho\dot{\rho}\psi_{\mathbf{D}\rho} + \mathbf{h}^{\mathrm{R}}(\rho, \theta, \mathbf{D}\rho, \mathbf{D}^2\rho, \mathbf{D}\theta) \\
&\{\mathbf{T} + [\rho^2\psi_\rho - \rho\operatorname{div}(\rho\psi_{\mathbf{D}\rho})]\mathbf{1} + \rho\psi_{\mathbf{D}\rho}\otimes\mathbf{D}\rho\}\cdot\mathbf{Dv} + \operatorname{div}\mathbf{h}^{\mathrm{R}} - \theta^{-1}\mathbf{q}\cdot\mathbf{D}\theta \geq 0
\end{aligned} \tag{1.17}$$

While it is possible to seek explicit solutions of equation $(1.17)_{3,4}$ by utilizing results from the theory of null Lagrangeans and invariants, we consider for simplicity special solutions such that

$$\mathbf{h}_{\mathbf{D}^2\rho} = \mathbf{h}_{\mathbf{D}\theta} = 0 \Rightarrow \mathbf{h}^{\mathrm{R}}_{\mathbf{D}^2\rho} = \mathbf{h}^{\mathrm{R}}_{\mathbf{D}\theta} = 0 \tag{1.18}$$

Then evaluation of equation $(1.17)_6$ at equilibrium ($\mathbf{D}\mathbf{v} = \mathbf{D}\theta = 0$) yields

$$\mathbf{h}^{\mathrm{R}}_{\rho} = \mathbf{h}^{\mathrm{R}}_{\mathbf{D}\rho} = \mathbf{0} \Rightarrow \mathbf{h}^{\mathrm{R}} = \mathbf{h}^{\mathrm{R}}(\theta) \tag{1.19}$$

which in conjunction with invariance considerations reduces equation $(1.17)_5$ to the simple form

$$\mathbf{h} = \rho\dot{\rho}\psi_{\mathbf{D}\rho} \tag{1.20}$$

and equations $(1.17)_6$ to a residual dissipation inequality of the form

$$\{\mathbf{T} + [\rho^2\psi_{\rho} - \rho\,\mathrm{div}(\rho\psi_{\mathbf{D}\rho})]\mathbf{1} + \rho\psi_{\mathbf{D}\rho} \otimes \mathbf{D}\rho\}\cdot\mathbf{D}\mathbf{v} - \theta^{-1}\mathbf{q}\cdot\mathbf{D}\theta \geq 0 \tag{1.21}$$

Instead of listing the implications of (1.21) in the general case, we confine attention for simplicity to non-viscous isothermal interfaces. Then equation (1.21) implies

$$\mathbf{T} = [-\rho^2\psi_{\rho} + \rho\,\mathrm{div}(\rho\psi_{\mathbf{D}\rho})]\mathbf{1} - \rho\psi_{\mathbf{D}\rho} \otimes \mathbf{D}\rho \tag{1.22}$$

which under the classical assumption that the free energy density is a quadratic form

$$\psi \equiv \psi + \frac{c}{2\rho}|\mathbf{D}\rho|^2 \tag{1.23}$$

with $c = c(\rho)$ denoting the surface tension, becomes

$$\mathbf{T} = [-p(\rho) + \rho c\Delta\rho + \tfrac{1}{2}(\rho c)'|\mathbf{D}\rho|^2]\mathbf{1} - c\mathbf{D}\rho \otimes \mathbf{D}\rho \tag{1.24}$$

which is the result of Theorem 4.3 of [22]. It thus follows that this particular thermodynamic model restricts the mechanical theory of [22] by imposing

$$\alpha = \rho c, \qquad \beta = \tfrac{1}{2}(\rho c)', \qquad \gamma = 0, \qquad \delta = -c \tag{1.25}$$

which, among other things, imply necessarily Maxwell's rule.

Next, we note that the restriction $\gamma = 0$ (but not Maxwell's rule) can be removed by adopting a more general form of the extra energy term $\hat{\varepsilon}$,

$$\hat{\varepsilon} = \hat{\mathbf{T}}\cdot\mathbf{D}\mathbf{v} + \mathrm{div}\,\mathbf{h} \tag{1.26}$$

with $\mathbf{h}$ being as before and $\hat{\mathbf{T}}$ being a null stress, i.e.,

$$\mathrm{div}\,\hat{\mathbf{T}} = 0 \tag{1.27}$$

such that the momentum balance remains unchanged. On restricting attention as before to a quadratic free energy density function ψ and therefore to a quadratic null stress $\hat{\mathbf{T}}$, it follows that

$$\hat{\mathbf{T}} = -(\gamma\Delta\rho + \gamma'|\mathbf{D}\rho|^2)\mathbf{1} + \gamma\mathbf{D}^2\rho + \gamma'\mathbf{D}\rho \otimes \mathbf{D}\rho \tag{1.28}$$

Then, the expression for the total stress becomes

$$\mathbf{T} = [-p(\rho) + \alpha\Delta\rho + \beta|\mathbf{D}\rho|^2]\mathbf{1} + \gamma\mathbf{D}^2\rho + \delta\mathbf{D}\rho \otimes \mathbf{D}\rho \tag{1.29}$$

where

$$\alpha + \gamma = \rho c, \qquad \beta + \delta = \tfrac{1}{2}(\rho c' - c), c = \gamma' - \delta \tag{1.30}$$

which is the substance of Theorem 4.2 of [22]. Thus, this second thermodynamic model gives the same equilibrium results as the variational modified van der Waals theory and implies again Maxwell's rule.

Finally, we note that a thermodynamic model compatible with the mechanical theory (which implies neither $\gamma = 0$ nor Maxwell's rule) is possible by assuming the extra energy term $\hat{\varepsilon}$ to be of the form

$$\hat{\varepsilon} = a(\rho, \mathbf{D}\rho, \mathbf{D}^2\rho)\dot{\rho} + \mathbf{b}(\rho, \mathbf{D}\rho, \mathbf{D}^2\rho)\cdot\mathbf{D}\dot{\rho} + \hat{\mathbf{T}}(\rho, \mathbf{D}\rho, \mathbf{D}^2\rho)\cdot\mathbf{D}\mathbf{v} \tag{1.31}$$

A detailed elaboration on the above thermodynamic models will be attempted in future publications.

ACKNOWLEDGEMENT

The support of the Solid Mechanics and Geotechnical Engineering Programs of the National Science Foundation and the Program of Mechanics of Microstructures of the Michigan Technological University is gratefully acknowledged.

REFERENCES

1. C. S. Desai and R. H. Gallagher (Eds.), *Proc. Int. Conf. Const. Laws for Engrg. Matl's.*, Tucson, Arizona, 1983.
2. E. C. Aifantis, 'Preliminaries on degradation and chemomechanics', *Workshop on a Continuum Mechanics Approach to Damage and Life Prediction* (Eds. D. C. Stouffer, E. Krempl and J. E. Fitzgerald), pp. 159–173, Carrolton, 1980.
3. E. C. Aifantis, 'Elementary physicochemical degradation processes', *Mechanics of Structured Media* (Ed. A. P. S. Selvadurai), pp. 301–317, Elsevier, Amsterdam–Oxford–New York, 1981.
4. E. C. Aifantis, 'Dislocation kinetics and the formation of deformation bands', *Defects, Fracture and Fatigue* (2nd Int. Symp. and 7th Canadian Fracture Conference, Mont Gabriel, Quebec, 1982, Eds. G. C. Sih and J. W. Provan), pp. 75–84, Martinus-Nijhoff Publ., 1983.
5. E. C., Aifantis, 'Some thoughts on degrading bodies', *Proc. Workshop on Mechanics of Damage and Fracture* (Eds. S. N. Atluri and E. J. Fitzgerald), pp. 1–12, Atlanta, 1982.
6. E. C., Aifantis, 'The mechanics of diffusion in solids', *T. & A.M. Report No.* 440, University of Illinois, Urbana, 1980.
7. E. C., Aifantis, 'On the problem of diffusion in solids', *Acta Mechanica*, **37**, 265–296 (1980).
8. E. C. Aifantis, 'Continuum theories of diffusion and infiltration', *Engineering Science Perspective* Part I: Vol. 5, pp. 10–20, 1980, Part II: Vol. 6.
9. P. A. Taylor and E. C. Aifantis, 'On the theory of diffusion in linear viscoelastic media', *Acta Mechanica*, **44**, 259–298 (1982).
10. D. J. Bamman and E. C. Aifantis, 'On a proposal for a continuum with microstructure', *Acta Mechanica*, **45**, 91–121 (1982).
11. R. K. Wilson and E. C. Aifantis, 'On the theory of stress-assisted diffusion—I', *Acta Mechanica*, **45**, 273–296 (1982).
12. D. J. Unger and E. C. Aifantis, 'On the theory of stress-assisted diffusion—II', *Acta Mechanica*, **47**, 117–151 (1983).

13. D. J. Unger, W. W. Gerberich, and E. C. Aifantis, 'Further remarks on the implications of steady-state stress-assisted diffusion on environmental cracking', *Scripta Metallugica*, **16**, 1059–1064 (1982).
14. J. A. Colios and E. C. Aifantis, 'On the problem of continuum theory of embrittlement', *Res. Mechanica*, **5**, 67–85 (1982).
15. R. M. Bowen, Theory of mixtures', *Continuum Physics III* (Ed. A. C. Eringen), pp. 1–127, Academic Press, New York, 1976.
16. B. D. Coleman and M. E. Gurtin, 'Thermodynamics with internal state variables', *J. Chem. Phys.*, **47**, 597–613 (1967).
17. P. Perzyna, 'Memory effects and internal changes of a material', *Int. J. Non-Linear Mech.*, **6**, 707–716 (1971). Also, 'Thermodynamics of rheological materials with internal changes', *J. Mechanique*, **10**, 391–408 (1971).
18. J. R. Rice, 'Inelastic constitutive relations for solids: An internal variable theory and its application to metal plasticity', *J. Mech. Phys. Solids*, **19**, 433–455 (1971). Also, 'Continuum mechanics and thermodynamics of plasticity in relation to microscale deformation mechanisms', *Constitutive Equations in Plasticity* (Ed. A. S. Argon, pp. 23–79, MIT Press, 1975.
19. E. C. Aifantis, (a) 'Continuum theory of degradation', *Proposal to National Science Foundation*, No. ENG-7824178, Solid Mechanics Program, 1977. (b) A Proposal for Continuum with Microstructure, *Mech. Res. Comm.*, **5**, 139–145, (1978).
20. P. Perzyna, 'On constitutive modelling of dissipative solids for plastic flow, instability and fracture', *Proc. Int. Conf. Const. Laws for Engrg. Matl's* (Eds. C. S. Desai and R. H. Gallagher), pp. 13–19, 1983.
21. J. R. Rice, 'Stability of shear in plastic materials with diffusive transport of state parameters', Reference [15] quoted in [20].
22. E. C. Aifantis and J. B. Serrin, Towards a Mechanical Theory of Phase Transformations, *Technical Report*, Corrosion Research Center, University of Minnesota, 1980. Also 'The mechanical theory of fluid interfaces and Maxwell's rule' *J. Colloid Interface Sci.* **96**, 517–529 (1983). 'Equilibrium solutions in the mechanical theory of fluid microstructures', *J. Colloid Interface Sci.* **96**, 530–547 (1983).
23. K. H. Anthony, A New Approach Describing Irreversible Processes, *Continuum Models of Discrete Systems* 4 (Eds. O. Brulin and R. K. T. Hsieh), pp. 481–494, North-Holland 1981.
24. S. J. Sacket, J. M. Kelly, and P. P. Gillis, 'A probabilistic approach to polycrystalline plasticity', *J. Franklin Inst.*, **304**, 33–63 (1977).
25. L. Davison, A. L. Stevens, and M. E. Kipp, 'Theory of spall damage accumulation in ductile metals', *J. Mech. Phys. Solids*, **25**, 11–28 (1977).
26. M. Ortiz and E. Popov, 'A statistical theory of polycrystalline plasticity', *Proc. Roy. Soc. Lond.*, **A379**, 439–458 (1982).
27. E. T. Onat, 'Elastic and yield properties of internally damaged materials—a statistical approach', *Proc. Symp. Media with Microstructure and Wave Propagation* (Eds. E. C. Aifantis and L. Davison), To appear in: *Lett Appl. Sci. Engng.*, 1984.
28. Sh. Kh. Khannanov, 'Kinetics of dislocations and disclinations', *Phys. Met. Metall.*, **49**, 47–53 (1981).
29. R. B. Stout, 'Dislocation kinetics and the acoustic wave propagation in liquids', *Proc. Symp. Media with Microstructure and Wave Propagation* (Eds. E. C. Aifantis and L. Davison), To appear in: *Lett. Appl. Sci. Engng.*, 1984.
30. B. Hughes, M. Sahimi, L. E. Scriven, and H. T. Davis, 'Transport and conduction in random systems', *Proc. Symp. Media with Microstructure and Wave Propagation* (Eds. E. C. Aifantis and L. Davison), To appear in: *Lett. Appl. Sci. Engng.*, 1984.
31. N. Fox, 'On plastic strain', *Proc. IUTAM Symp. Mechanics Generalized Continua* (Ed.

E. Kroner), Springer-Verlag, 1968, pp. 163–165. Also, 'Some problems of finite plastic deformation', *Arch. Mech.*, **24**, 373–381 (1972).

32. E. H. Lee, 'Elastic-plastic deformations at finite strains', *J. Appl. Mech.*, **36**, 1–6 (1969). Also, 'Constitutive relations for dynamic loading and plastic waves', *Inelastic Behaviour of Solids* (Eds. M. F. Kanninen *et al.*), pp. 423–446, McGraw-Hill, New York, 1970.
33. C. Teodosiu, 'A dynamic theory of dislocations and its applications to the theory of the elastic–plastic continuum', *Fundam. Aspects of Disloc. Theory* (Eds. J. A. Simmons *et al.*), NBS Spec. Publ. 317, 1970, pp. 837–876. Also, 'A physical theory of the finite elastic–viscoplastic behavior of single crystals', *Engrg. Trans.*, **23**, 151–184 (1975).
34. J. Kratochvil, 'Finite-strain theory of crystalline elastic–inelastic materials', *J. Appl. Phys.*, **42**, 1104–1108 (1971). Also, 'Finite-strain theory of inelastic behavior of crystalline solids', *Foundations of Plasticity*, (Ed. A. Sawczuk), pp. 401–415, Noordhoff, 1972.
35. E. T. Onat, 'Representation of inelastic behavior in the presence of anisotropy and of finite deformations', *Recent Advances in Creep and Fracture of Engineering Materials and Structures* (Eds. B. Wilshire and D. R. Owen), pp. 231–264, Pineridge Press, 1982.
36. G. C. Johnson and D. J. Bamman, 'A discussion of stress rates in finite deformation problems', *Sandia Report SAND* 82–8821, 1982.
37. A. D. Brailsford, 'Some aspects of dislocation-point defect interactions and their significance to internal friction', *Int. Conf. on Dislocation Model. of Physical Systems* (Eds. M. F. Ashby *et al.*), Pergamon Press, 1981.
38. H. Chung, D. J. Unger, D. D. McDonald, and E. C. Aifantis, 'Steady and non-steady crack growth under environmental conditions', Presented in: *STP 16th. Nat. Symp. Fract. Mech.* (Ed. M. F. Kanninen), 1983.
39. V. Varadan, 'Multiple scattering formalism for wave propagation in discrete media', *Proc. Symp. Media with Microstructure and Wave Propagation* (Eds. E. C. Aifantis and L. Davison), To appear in: *Lett. Appl. Sci. Engng.*, 1984.
40. A. C. Eringen, 'Continuum mechanics at the atomic scale', *Crystal Lattice Defects*, **7**, 109–130 (1977). Also, 'Non-local nature of fracture', *Proc. Workshop on Mechanics of Damage and Fracture* (Eds. S. N. Atluri and E. J. Fitzgerald), pp. 39–46, 1984.
41. H. Mughrabi, 'Cyclic plasticity of matrix and persistent slip bands in fatigued metals', *Continuum Models of Discrete Systems* 4 (Eds. O. Brulin and R. K. T. Hsieh), pp. 241–257, North-Holland, 1981.
42. J. R. Rice, 'The localization of plastic deformation', *Proc. 14th IUTAM Congress* (Ed. W. T. Koiter), pp. 207–220, North-Holland, 1976.
43. T. H. Lin, 'Micromechanics of deformation of slip bands under monotonic and cyclic loading', *Rev. Def. Beh. Matl's*, **2**, 263–316 (1977).
44. J. J. McCoy, 'Macroscopic response of continua with random microstructures', *Mechanics Today* (Ed. S. Nemat-Nasser), **6**, 1–40 (1981).
45. O. Walton, 'Application of the principles of molecular dynamics to macroscopic particles', *Proc. Symp. Media with Microstructure and Wave Propagation* (Eds. E. C. Aifantis and L. Davison), To appear in: *Lett. Appl. Sci. Engng.*, 1984.
46. D. R. Axelrad, *Micromechanics of Solids*, Elsevier, New York, 1978. Also: J. W. Provan and D. R. Axelrad, 'Probabilistic micromechanics of solids', *Rev. Deform. Beh. Matl's*, **2**, 174–209 (1977).
47. A. Bensoussan, J. L. Lions, and G. Papanicolaou, 'Asymptotic analysis for periodic structures', *Studies in Mathematics and Its Applications* 5, North-Holland, Amsterdam–New York–Oxford, 1978.
48. E. Sanchez-Palencia, 'Non-homogeneous media and vibration theory', *Lecture Notes in Physics* 127, Springer-Verlag, Berlin–Heidelberg–New York, 1980.

Mechanics of Engineering Materials
Edited by C. S. Desai and R. H. Gallagher

Chapter 2

Development of an Elastic–Viscoplastic Constitutive Relationship for Earth Materials

G. Y. Baladi and B. Rohani

2.1 INTRODUCTION

The mechanical response of most earth materials subjected to high-intensity dynamic loadings, such as those produced by large explosive detonations and impact of high-velocity projectiles, differs considerably from what is usually observed under relatively low-intensity static loadings (very slow rates of deformation). In order to obtain realistic solutions for dynamic problems, it is essential to use a constitutive model that can account for the dependency of the stress–strain–strength properties of earth materials on the various rates of loading or deformation being applied. In addition, the constitutive model must account for other pertinent features of the stress–strain properties of earth materials observed under both dynamic and static loading conditions, such as (a) the dependency of the shearing strength of the material on hydrostatic stress, (b) shear-induced volume change, and (c) permanent strain during a load–unload cycle of deformation (for both deviatoric and hydrostatic loading conditions).

Incremental elastic–plastic constitutive models have been used successfully to simulate such pertinent features of the stress–strain properties of soil [1, 2, and 7]. It is therefore logical to adopt a physically realistic incremental elastic–plastic constitutive model for earth materials and introduce rate dependency in such a model. Generally, two different types of rate-dependent models can be constructed in this manner: viscoelastic–plastic models in which both the elastic and the plastic responses of the material are rate-sensitive; and elastic–viscoplastic models in which the plastic portion of the model is rate-dependent and the elastic portion is rate-independent. As pointed out by Perzyna [6], the viscoelastic–plastic models are mathematically very complicated and are not suitable for solving practical engineering problems. The elastic–viscoplastic models, because of their mathematical simplicity (relative to viscoelastic–plastic models) and their similarities with the inviscid theory of plasticity, are more appropriate for practical engineering application [6 and 8]. Also, viscous effects appear to be more evident in the plastic range for most soils. Thus, it is reasonable to adopt an elastic–viscoplastic type constitutive relationship to model the rate-dependent response of earth materials.

This chapter describes the fundamental basis of elastic–viscoplastic constitutive relationships and the development of a specific model of this type for earth materials. To demonstrate the application of the model, its behaviour under cylindrical states of stress is examined and correlated with experimental data for a clayey sand.

2.2 FUNDAMENTAL BASIS OF ELASTIC–VISCOPLASTIC CONSTITUTIVE RELATIONSHIPS

The basic premise of elastic–plastic constitutive relationships is the assumption that certain materials are capable of undergoing small plastic (permanent) as well as elastic (recoverable) strains at each loading increment.† In the case of elastic–viscoplastic materials, it is further assumed that the behaviour of the material in the plastic region is rate-dependent. Mathematically, the total strain rate is assumed to be the sum of the elastic components and the viscoplastic components, i.e.,

$$\frac{\mathrm{d}\varepsilon_{ij}}{\mathrm{d}t} = \frac{\mathrm{d}\varepsilon_{ij}^{\mathrm{E}}}{\mathrm{d}t} + \frac{\mathrm{d}\varepsilon_{ij}^{\mathrm{vp}}}{\mathrm{d}t} \tag{2.1}$$

where

$\frac{\mathrm{d}\varepsilon_{ij}}{\mathrm{d}t}$ = total components of the strain rate tensor.

$\frac{\mathrm{d}\varepsilon_{ij}^{\mathrm{E}}}{\mathrm{d}t}$ = components of the elastic strain rate tensor.

$\frac{\mathrm{d}\varepsilon_{ij}^{\mathrm{vp}}}{\mathrm{d}t}$ = components of the viscoplastic strain rate tensor.

2.2.1 Elastic strain rate tensor

Within the elastic range, the behaviour of the material can be described by an elastic constitutive relation of the type

$$\frac{\mathrm{d}\varepsilon_{ij}^{\mathrm{E}}}{\mathrm{d}t} = C_{ijkl}(\sigma_{mn})\frac{\mathrm{d}\sigma_{kl}}{\mathrm{d}t} \tag{2.2}$$

† In this chapter, compression is considered positive.

where

C_{ijkl} = material response function.

$\dfrac{d\sigma_{kl}}{dt}$ = components of stress rate tensor.

σ_{mn} = components of stress tensor.

For isotropic compressible elastic materials, the simplest form of equation (2.2) is

$$\frac{d\varepsilon_{ij}^{E}}{dt} = \frac{1}{9K}\frac{dJ_1}{dt}\delta_{ij} + \frac{1}{2G}\frac{dS_{ij}}{dt} \tag{2.3}$$

where

$J_1 = \sigma_{mm}$ = first invariant of stress tensor.

$S_{ij} = \sigma_{ij} - (J_1/3)\delta_{ij}$ = stress deviation tensor.

K = elastic bulk modulus.

G = elastic shear modulus.

δ_{ij} = Kronecker delta $= \begin{cases} 1 & i=j \\ 0 & i \neq j \end{cases}$

The bulk and shear moduli can be expressed as functions of the invariants of the stress tensor. However, in order not to generate energy or hysteresis within the elastic range, the elastic response must be path-independent. This condition can be met if and only if the bulk modulus is a function of the first invariant of the stress tensor and the shear modulus is a function of the second invariant of the stress deviation tensor [7]. Thus,

$$K = K(J_1); \quad G = G(\sqrt{J_2'}) \tag{2.4}$$

where

$J_2' = \frac{1}{2}S_{ij}S_{ij}$ = the second invariant of the stress deviation tensor.

2.2.2 Viscoplastic strain rate tensor

The behaviour of the material in the plastic range is assumed to be rate-sensitive. As indicated by Perzyna [6], the viscoplastic component of the strain rate tensor can be expressed as an arbitrary function of the 'excess stress' above the initial yield condition which is called the static yield criterion. The static yield criterion should satisfy all the known conditions of the inviscid theory of plasticity [3]. In general, the static yield surface may be expressed as

$$f_s(\sigma_{ij}, \kappa) = 0 \tag{2.5}$$

For isotropic materials the static yield surface may be expressed, for example, as

$$f_s[J_1, \sqrt{J_2'}, \kappa] = 0 \tag{2.6}$$

where κ is a hardening parameter and generally is a function of the viscoplastic (or plastic) deformation. The static yield surface (equation (2.6)) may expand or contract as κ increases or decreases, respectively (Figure 2.1).

Since the viscoplastic strain rate is an arbitrary function of the excess stress above the static yield criterion, the 'dynamic yield surface' can be defined as

$$f_d[J_1, \sqrt{J_2'}, \kappa, \beta] = \frac{f_s[J_1, \sqrt{J_2'}, \kappa] - \beta}{B} = 0 \tag{2.7}$$

and the flow rule for work-hardening and rate-sensitive plastic materials may be

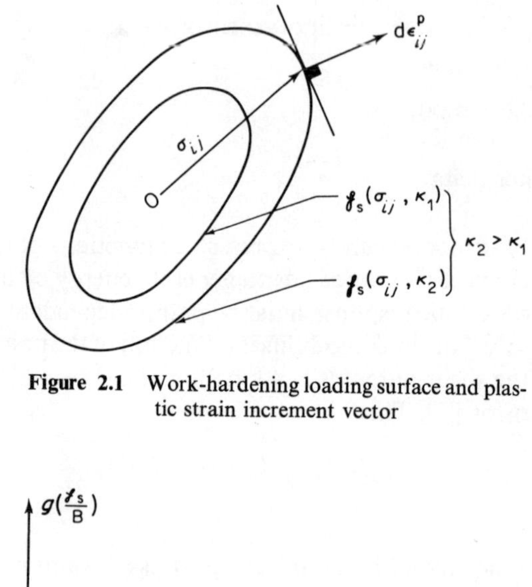

Figure 2.1 Work-hardening loading surface and plastic strain increment vector

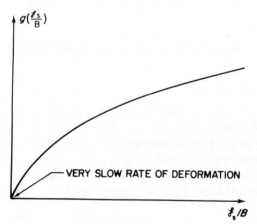

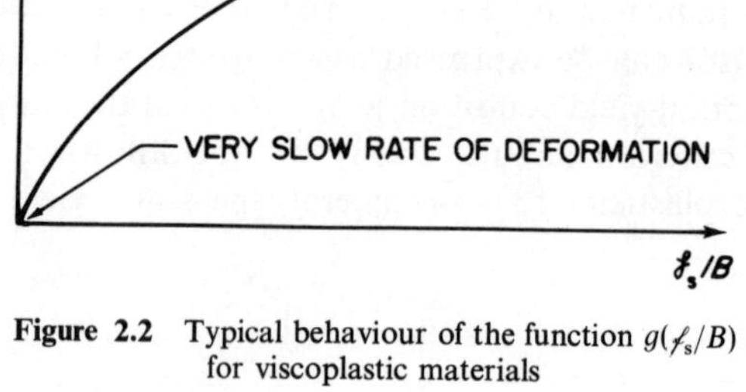

Figure 2.2 Typical behaviour of the function $g(f_s/B)$ for viscoplastic materials

written as

$$\frac{d\varepsilon_{ij}^{vp}}{dt} = \begin{cases} \gamma g\left(\frac{f_s}{B}\right)\frac{\partial f_s}{\partial \sigma_{ij}} & \text{if } f_s > 0 \\ 0 & \text{if } f_s \leq 0 \end{cases} \tag{2.8}$$

The parameter β/B in equation (2.7) is dimensionless and defines the rate of expansion of the yield surface. (B is a scaling factor which has the dimension of stress; it is introduced in equation (2.7) to make f_d dimensionless.) The function $g(f_s/B)$ in equation (2.8) is a dimensionless function which may be determined from the results of dynamic property tests for the material of interest (Figure 2.2). The parameter γ in equation (2.8) is a viscosity parameter associated with the viscoplastic response of the material and has the dimension of $(\text{time})^{-1}$. It should be pointed out that for very slow rates of deformation, β and $g(f_s/B)$ approach zero. Hence, f_d and f_s become identical and the viscoplastic flow rule (equation (2.8)) reduces to its inviscid counterpart.

The viscoplastic stress–strain relation can be expressed in terms of the hydrostatic and deviatoric components of strain. Applying the chain rule of differentiation to the right-hand side of equation (2.8) yields

$$\frac{d\varepsilon_{ij}^{vp}}{dt} = \gamma g\left(\frac{f_s}{B}\right)\left[\frac{\partial f_s}{\partial J_1}\delta_{ij} + \frac{1}{2\sqrt{J_2'}}\frac{\partial f_s}{\partial\sqrt{J_2'}}S_{ij}\right] \tag{2.9}$$

Multiplying both sides of equation (2.9) by δ_{ij} gives

$$\frac{d\varepsilon_{kk}^{vp}}{dt} = 3\gamma g\left(\frac{f_s}{B}\right)\frac{\partial f_s}{\partial J_1} \tag{2.10}$$

The deviatoric component of the viscoplastic strain rate tensor de_{ij}^{vp}/dt can be written as

$$\frac{de_{ij}^{vp}}{dt} = \frac{d\varepsilon_{ij}^{vp}}{dt} - \frac{1}{3}\frac{d\varepsilon_{kk}^{vp}}{dt}\delta_{ij} \tag{2.11}$$

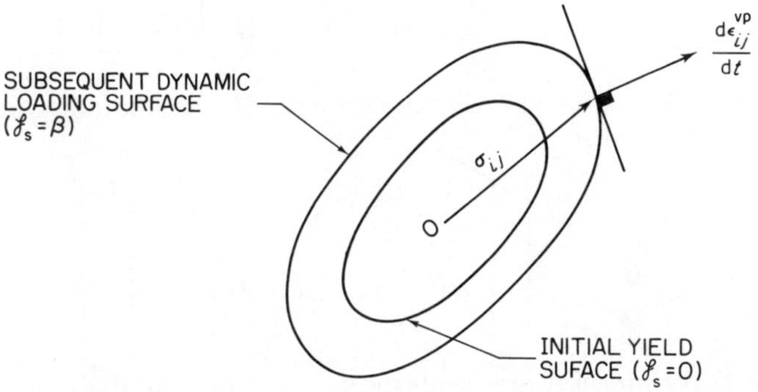

Figure 2.3 Dynamic loading surface and strain rate vector

Substitution of equations (2.9) and (2.10) into equation (2.11) yields

$$\frac{de_{ij}^{vp}}{dt}=\gamma g\left(\frac{f_s}{B}\right)\frac{S_{ij}}{2\sqrt{J_2'}}\frac{\partial f_s}{\partial\sqrt{J_2'}} \tag{2.12}$$

The associated flow rule is satisfied when equations (2.10) and (2.12) are used with equation (2.7); this shows that the vector $d\varepsilon_{ij}^{vp}/dt$ in the $(J_1,\sqrt{J_2'})$ space is always normal to the surface $f_s[J_1,\sqrt{J_2'},\kappa]=\beta$ (Figure 2.3). Squaring both sides of equation (2.9) gives

$$\gamma g\left(\frac{f_s}{B}\right)=\left[\frac{\dfrac{d\varepsilon_{ij}^{vp}}{dt}\dfrac{d\varepsilon_{ij}^{vp}}{dt}}{3\left(\dfrac{\partial f_s}{\partial J_1}\right)^2+\dfrac{1}{2}\left(\dfrac{\partial f_s}{\partial\sqrt{J_2'}}\right)^2}\right]^{1/2} \tag{2.13}$$

Substitution of equation (2.11) into equation (2.13) yields

$$\gamma g\left(\frac{f_s}{B}\right)=\left[\frac{\dfrac{1}{3}\left(\dfrac{d\varepsilon_{kk}^{vp}}{dt}\right)^2+2\left(\dfrac{d\bar{e}^{vp}}{dt}\right)^2}{3\left(\dfrac{\partial f_s}{\partial J_1}\right)^2+\dfrac{1}{2}\left(\dfrac{\partial f_s}{\partial\sqrt{J_2'}}\right)^2}\right]^{1/2} \tag{2.14}$$

where

$$\frac{d\bar{e}^{vp}}{dt}=\left[\frac{1}{2}\frac{de_{ij}^{vp}}{dt}\frac{de_{ij}^{vp}}{dt}\right]^{1/2}$$

= square root of the second invariant of the viscoplastic strain rate deviation tensor

Inverting equation (2.14) results in

$$\frac{f_s}{B}-g^{-1}\left\{\frac{1}{\gamma}\left[\frac{\dfrac{1}{3}\left(\dfrac{d\varepsilon_{kk}^{vp}}{dt}\right)^2+2\left(\dfrac{d\bar{e}^{vp}}{dt}\right)^2}{3\left(\dfrac{\partial f_s}{\partial J_1}\right)^2+\dfrac{1}{2}\left(\dfrac{\partial f_s}{\partial\sqrt{J_2'}}\right)^2}\right]^{1/2}\right\}=0 \tag{2.15}$$

Comparison of equation (2.7) with equation (2.15) yields the following expression for the parameter β/B

$$\frac{\beta}{B}=g^{-1}\left\{\frac{1}{\gamma}\left[\frac{\dfrac{1}{3}\left(\dfrac{d\varepsilon_{kk}^{vp}}{dt}\right)^2+2\left(\dfrac{d\bar{e}^{vp}}{dt}\right)^2}{3\left(\dfrac{\partial f_s}{\partial J_1}\right)^2+\dfrac{1}{2}\left(\dfrac{\partial f_s}{\partial\sqrt{J_2'}}\right)^2}\right]^{1/2}\right\} \tag{2.16}$$

Equation (2.16) implicitly represents the dependence of the dynamic yield surface on the viscoplastic strain rate tensor.

2.2.3 Total strain rate tensor

The total strain rate tensor can be obtained by combining equations (2.3) and (2.9); thus,

$$\frac{d\varepsilon_{ij}}{dt} = \frac{1}{9K}\frac{dJ_1}{dt}\delta_{ij} + \frac{1}{2G}\frac{dS_{ij}}{dt} + \gamma g\left(\frac{f_s}{B}\right)\left[\frac{\partial f_s}{\partial J_1}\delta_{ij} + \frac{1}{2\sqrt{J_2'}}\frac{\partial f_s}{\partial\sqrt{J_2'}}S_{ij}\right] \tag{2.17}$$

Similarly, the stress rate tensor can be written as

$$\frac{d\sigma_{ij}}{dt} = K\frac{d\varepsilon_{kk}}{dt}\delta_{ij} + 2G\frac{de_{ij}}{dt} - \gamma g\left(\frac{f_s}{B}\right)\left[3K\frac{\partial f_s}{\partial J_1}\delta_{ij} + G\frac{\partial f_s}{\partial\sqrt{J_2'}}\frac{S_{ij}}{\sqrt{J_2'}}\right] \tag{2.18}$$

Equation (2.17) or (2.18) is the general constitutive equation for an elastic–viscoplastic isotropic material. To use these equations, it is only necessary to specify the functional forms of K, G, f_s, and $g(f_s/B)$ and to determine experimentally the numerical values of γ and the coefficients in these functions. The development of a specific elastic–viscoplastic constitutive model for earth materials is presented in the following section.

2.3 ELASTIC–VISCOPLASTIC MODEL FOR EARTH MATERIALS

Before the development of the elastic–viscoplastic model is described, it may be useful to briefly discuss some of the salient features of the stress–strain properties of soil under dynamic loading. In the case of most soils, the strength of the

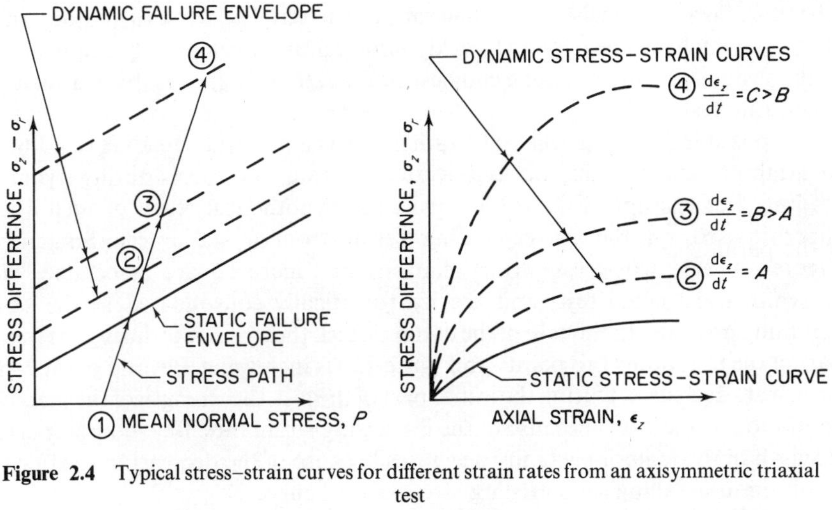

Figure 2.4 Typical stress–strain curves for different strain rates from an axisymmetric triaxial test

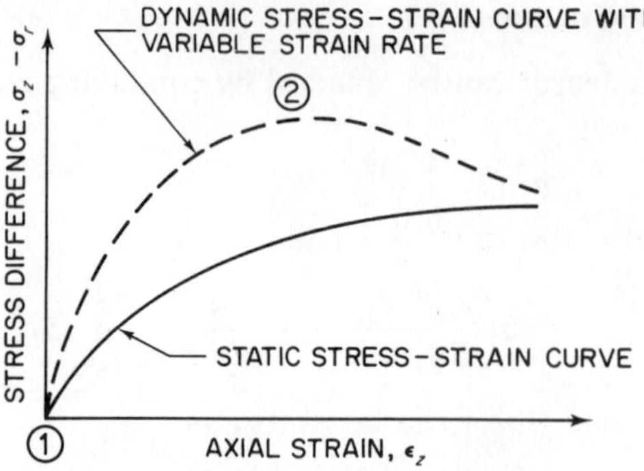

Figure 2.5 Dynamic stress–strain curve for a variable strain rate from an axisymmetric triaxial test

material increases with increasing rate of deformation. Because of this dependency on the rate of deformation, the overall character of the dynamic stress–strain curves often differs considerably from that of the corresponding curves obtained under static loading conditions [9].

Figure 2.4 depicts qualitatively typical stress–strain relations under axisymmetric triaxial test conditions for various rates of strain. As indicated in Figure 2.4, the soil specimens are hydrostatically consolidated to the same confining pressure (point 1, Figure 2.4). The samples are then sheared at different rates of strain by increasing the vertical stress σ_z while the radial stress σ_r is held constant. The important behaviour to be observed from Figure 2.4 is that, as the rate of strain $d\varepsilon_z/dt$ increases, the strength of the material also increases. Therefore, associated with each strain rate, the material possesses a unique failure envelope which may be referred to as 'dynamic failure envelope'. The implications of the dynamic failure envelope can best be realized from the results of a variable strain rate test.

It is possible for a material that strain hardens under static loading conditions to exhibit strain-softening behaviour due to strain rate effects during dynamic loading. For example, Figure 2.5 depicts the hypothetical result of such a test superimposed on the corresponding result from a static test. Similar to Figure 2.4, the two stress–strain relations in Figure 2.5 are associated with axisymmetric triaxial tests and are hydrostatically consolidated to the same confining pressure. In the case of the dynamic test, the strain rate during the initial part of the test (point 1 to point 2 in Figure 2.5) is increasing. Beyond point 2, the strain rate decreases. During the initial part of the test, the strength of the material continuously increases because of the increasing strain rate. Beyond point 2, the strength of the material actually decreases because of the decreasing strain rate, resulting in a 'falling' or softening stress–strain curve.

In the following section, the mathematical forms of the various response functions are developed for a proposed constitutive model for earth materials.

2.3.1 Elastic response functions

The behaviour of the model in the elastic (recoverable) range is described by the elastic bulk and shear moduli. The elastic bulk modulus is assumed to be a function of the first invariant of the stress tensor J_1 (Figure 2.6). The elastic shear modulus, on the other hand, is assumed to be a function of the second invariant of

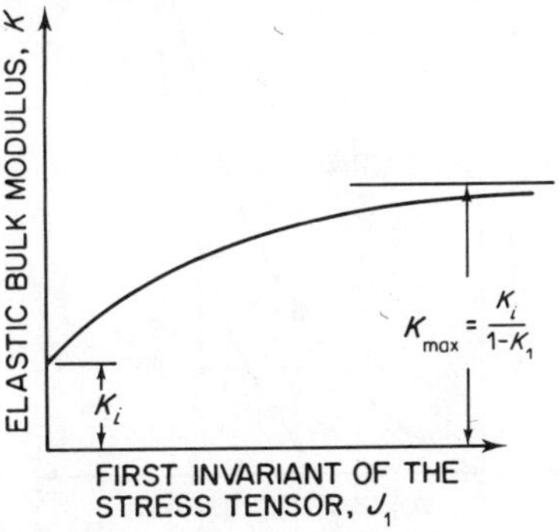

Figure 2.6 Elastic bulk modulus versus first invariant of the stress tensor

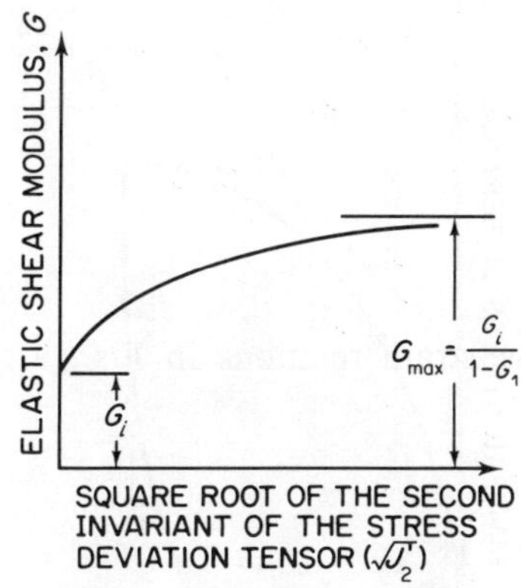

Figure 2.7 Elastic shear modulus versus second invariant of the stress deviation tensor

the stress deviation tensor J_2' (Figure 2.7), i.e.,

$$K = \frac{K_i}{1 - K_1}[1 - K_1 \exp(-K_2 J_1)] \tag{2.19}$$

$$G = \frac{G_i}{1 - G_1}[1 - G_1 \exp(-G_2 \sqrt{J_2'})] \tag{2.20}$$

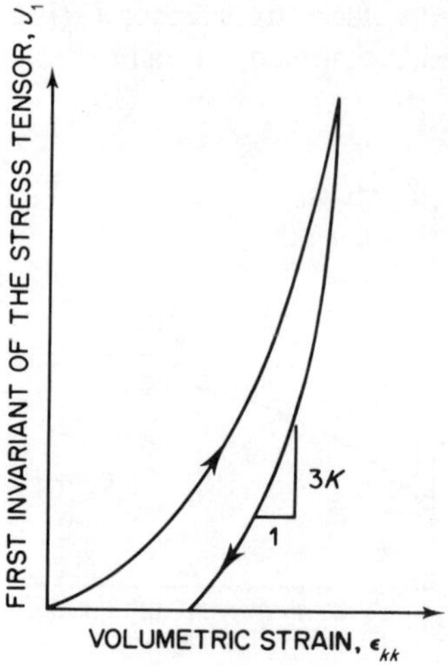

Figure 2.8 Proposed relationship for isotropic compression test

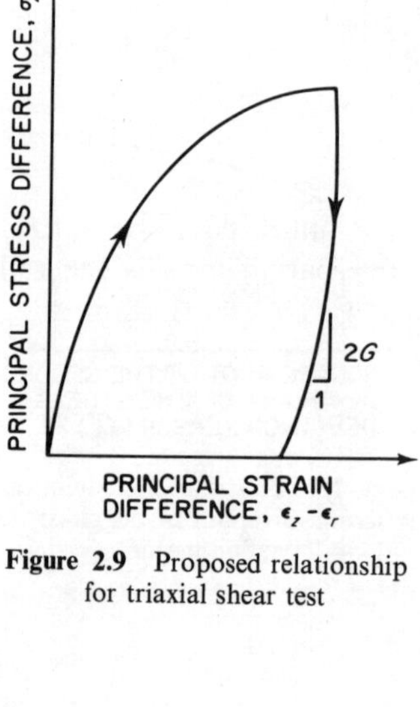

Figure 2.9 Proposed relationship for triaxial shear test

where

K_i = initial elastic bulk modulus.

K_1 and K_2 = material constants.

G_i = initial elastic shear modulus.

G_1 and G_2 = material constants.

The constants K_i, K_1, and K_2 can be determined from the characteristics of the unloading curve from an isotropic consolidation test (Figure 2.8). The constants G_i, G_1, and G_2 can be determined from the characteristics of the unloading stress difference–strain difference curves from triaxial shear tests (Figure 2.9).

2.3.2 Viscoplastic behaviour

For the plastic behaviour, the dynamic loading function f_d (equation (2.7)) is assumed to be isotropic and to consist of two parts (Figure 2.10): a rate-dependent ultimate failure envelope and a rate-dependent strain-hardening yield surface. The failure envelope portion of the loading function is assumed to be of the Prager–Drucker type and is denoted by

$$f_d[J_1, \sqrt{J_2'}, \beta] = \frac{f_s[J_1, \sqrt{J_2'}] - \beta}{B} = \frac{\sqrt{J_2'} - \alpha J_1 - k - \beta}{B} = 0 \qquad (2.21)$$

and the rate-dependent strain-hardening yield surface is assumed to be elliptical and of the following form

$$F_d[J_1, \sqrt{J_2'}, \kappa, \beta] = \frac{F_s[J_1, \sqrt{J_2'}, \kappa] - \beta}{B} = \frac{\sqrt{[(J_1 - L)^2/R^2 + J_2']} - (\kappa - L)/R - \beta}{B} = 0 \qquad (2.22)$$

The parameters k and α in equation (2.21) are material constants representing the static cohesive and frictional strength of the material, respectively. The parameter R (equation (2.23) below) in equation (2.22) is the ratio of the major to the minor axes of the elliptic yield surface (Figure 2.10). L and $\kappa + \beta$ in Figure 2.10 define the intersections of each elliptical yield surface (equation (2.22)) with the failure envelope $f_d[J_1, \sqrt{J_2'}, \beta]$ and the J_1 axis, respectively. The hardening parameter κ is generally a function of the history of the viscoplastic strain. For the present model, R and κ are chosen to be

$$R = \frac{R_0}{1 + R_1}[1 + R_1 \exp(-R_2 L)] \qquad (2.23)$$

$$\kappa = -\frac{1}{D}\ln\left(1 - \frac{\varepsilon_{kk}^{vp}}{W}\right) \qquad (2.24)$$

where R_0, R_1, R_2, D, and W are material constants.

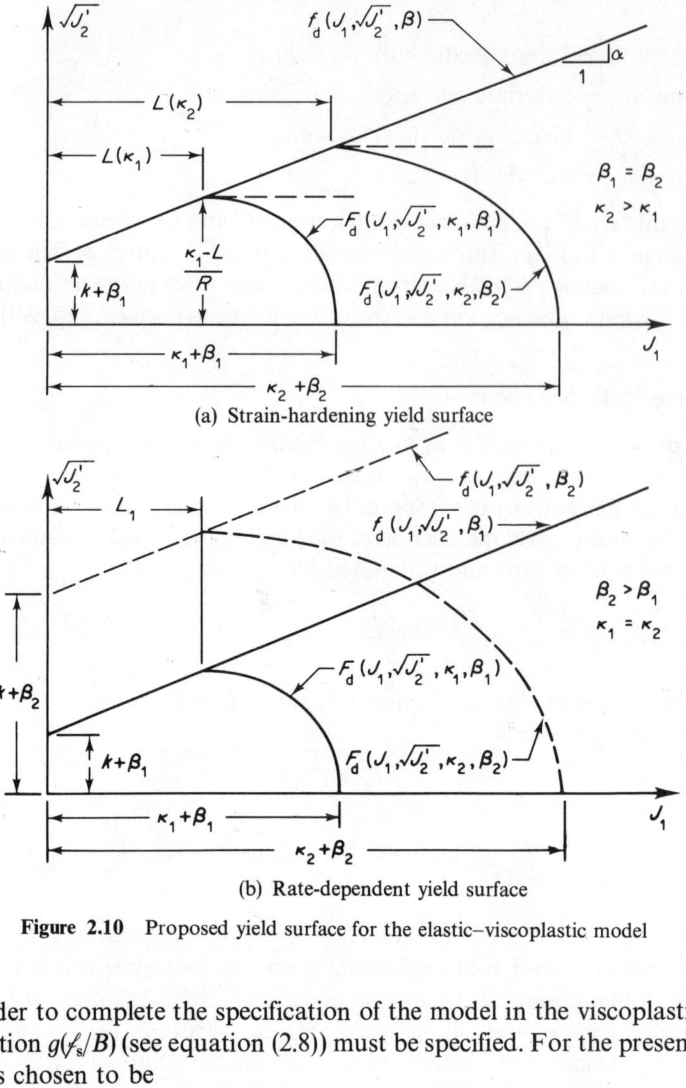

Figure 2.10 Proposed yield surface for the elastic–viscoplastic model

In order to complete the specification of the model in the viscoplastic range, the function $g(f_s/B)$ (see equation (2.8)) must be specified. For the present model, $g(f_s/B)$ is chosen to be

$$\left.\begin{aligned} g\left(\frac{f_s}{B}\right) &= \exp\left(\frac{B_1 f_s}{B}\right) - 1 \qquad \text{on the failure envelope} \\ g\left(\frac{F_s}{B}\right) &= \exp\left(\frac{B_1 F_s}{B}\right) - 1 \qquad \text{on the hardening surface} \end{aligned}\right\} \tag{2.25}$$

in which B_1 is a material constant controlling the rate of change in the value of the function g. Note in figure 2.10 that the elliptic yield surface has a horizontal tangent at its intersection with the failure envelope. This can be assured by the following relation:

$$L = \frac{\kappa - Rk}{1 + \alpha R} \tag{2.26}$$

In summary, the proposed model contains 16 material constants. Six of the constants (K_i, K_1, K_2, G_i, G_1, and G_2) are related to the elastic response of the material, and the remaining 10 constants (B, B_1, α, k, R_0, R_1, R_2, D, W, and γ) describe the material's viscoplastic response. The behaviour of the proposed model under conventional laboratory test conditions (cylindrical specimens) is examined in the following section.

2.4 BEHAVIOUR OF THE MODEL UNDER A CYLINDRICAL STATE OF STRESS

The rate-dependent nature of the proposed model can be demonstrated by examining the behaviour of the model under particular laboratory test boundary conditions. Since most mechanical testing of soils for engineering purposes is performed with the standard triaxial shear test apparatus, it is useful to investigate the behaviour of the model in cylindrical coordinates. Adopting the z-axis of a cylindrical coordinate system (z, r, and θ) as the axis of symmetry of the sample, the stress tensor and the strain tensor associated with this configuration become

$$\sigma_{ij} = \begin{bmatrix} \sigma_z & 0 & 0 \\ 0 & \sigma_r & 0 \\ 0 & 0 & \sigma_r \end{bmatrix} \tag{2.27}$$

and

$$\varepsilon_{ij} = \begin{bmatrix} \varepsilon_z & 0 & 0 \\ 0 & \varepsilon_r & 0 \\ 0 & 0 & \varepsilon_r \end{bmatrix} \tag{2.28}$$

The variables $P = J_1/3$ (mean normal stress), J_2' (the second invariant of the stress deviation tensor), and $\varepsilon_{kk}/3$ (mean volumetric strain) associated with the above stress and strain tensors take the following forms

$$P = \frac{J_1}{3} = \frac{\sigma_z + 2\sigma_r}{3} \tag{2.29}$$

$$J_2' = \frac{(\sigma_z - \sigma_r)^2}{3} \tag{2.30}$$

$$\frac{\varepsilon_{kk}}{3} = \frac{\varepsilon_z + 2\varepsilon_r}{3} \tag{2.31}$$

Generally, every triaxial test has two phases: the hydrostatic phase and the shear phase. These phases are discussed below.

2.4.1 Hydrostatic phase

During the hydrostatic phase of the triaxial test, all stresses and strains are equal. Thus,

$$\sigma_z = \sigma_r = J_1/3 = P \tag{2.32}$$

$$\varepsilon_z = \varepsilon_r = \frac{\varepsilon_{kk}}{3} \tag{2.33}$$

The relationship between the elastic volumetric strain rate and the rate of the first invariant of stress is given as (equation (2.3))

$$\frac{\mathrm{d}J_1}{\mathrm{d}t} = 3K\frac{\mathrm{d}\varepsilon_{kk}^{\mathrm{E}}}{\mathrm{d}t} \tag{2.34}$$

in which the elastic bulk modulus K is given in equation (2.19). Substituting equation (2.19) in equation (2.34) and integrating the resulting expression provides the following relation between the elastic volumetric strain and the first invariant of the stress tensor:

$$\varepsilon_{kk}^{\mathrm{E}} = \frac{1-K_1}{3K_2K_i}\ln\left[\frac{\exp(K_2J_1)-K_1}{1-K_1}\right] \tag{2.35}$$

The function β/B (equation (2.16)) for the hydrostatic phase takes the following form:

$$\frac{\beta}{B} = g^{-1}\left\{\frac{\dfrac{1}{\gamma\sqrt{3}}\left|\dfrac{\mathrm{d}\varepsilon_{kk}^{\mathrm{vp}}}{\mathrm{d}t}\right|}{\left[3\left(\dfrac{\partial F_s}{\partial J_1}\right)^2+\dfrac{1}{2}\left(\dfrac{\partial F_s}{\partial\sqrt{J_2'}}\right)^2\right]^{1/2}}\right\} \tag{2.36}$$

In view of equations (2.22) and (2.25), equation (2.36) results in

$$\frac{\mathrm{d}\varepsilon_{kk}^{\mathrm{vp}}}{\mathrm{d}t} = \frac{3\gamma}{R}\left\{\exp\left[\frac{B_1(J_1-\kappa)}{RB}\right]-1\right\} \tag{2.37}$$

Substitution of equation (2.24) into equation (2.37) yields

$$\frac{\mathrm{d}\varepsilon_{kk}^{\mathrm{vp}}}{\mathrm{d}t} = \frac{3\gamma}{R}\left\{\exp\left[\frac{B_1J_1}{RB}+\frac{B_1}{BDR}\ln\left(1-\frac{\varepsilon_{kk}^{\mathrm{vp}}}{W}\right)\right]-1\right\} \tag{2.38}$$

For a specified rate of loading, equation (2.38) can be integrated to yield a relationship between viscoplastic volumetric strain and time. For the same specified rate of loading, the elastic volumetric strain–time response can be calculated from equation (2.35). The total volumetric strain–time response can then be determined by adding the viscoplastic strain calculated from equation (2.38) and the elastic strain obtained from equation (2.35).

Note in equation (2.38) that when $\gamma = \infty$ (i.e., for an inviscid plastic material) equation (2.38) results in

$$\exp\left[\frac{B_1 J_1}{RB} + \frac{B_1}{BDR}\ln\left(1 - \frac{\varepsilon_{kk}^{\mathrm{vp}}}{W}\right)\right] - 1 = 0 \tag{2.39}$$

or

$$\varepsilon_{kk}^{\mathrm{vp}} = W[1 - \exp(-DJ_1)] \tag{2.40}$$

As a result of equations (2.35) and (2.40), the total volumetric strain for inviscid elastic–plastic materials takes the following form:

$$\varepsilon_{kk} = \frac{1-K_1}{3K_2K_1}\ln\left[\frac{\exp(K_2J_1) - K_1}{1-K_1}\right] + W[1 - \exp(-DJ_1)] \tag{2.41}$$

As expected, the volumetric strain in equation (2.41) is independent of the time rate of applied loading.

2.4.2 Shear phase

For the shear phase, a constant mean normal stress test will be considered. For this stress path,

$$\frac{\mathrm{d}J_1}{\mathrm{d}t} = 0 \tag{2.42}$$

$$\frac{\mathrm{d}\varepsilon_{kk}}{\mathrm{d}t} = \begin{cases} 0 \text{ in the elastic range} \\ \dfrac{\mathrm{d}\varepsilon_{kk}^{\mathrm{vp}}}{\mathrm{d}t} \text{ in the viscoplastic range} \end{cases} \tag{2.43}$$

$$\frac{\mathrm{d}\varepsilon_{kk}^{\mathrm{E}}}{\mathrm{d}t} = 0 \tag{2.44}$$

The relationship between the elastic strain deviation rate tensor and the rate of stress deviation tensor is given as (equation (2.3))

$$\frac{\mathrm{d}e_{ij}^{\mathrm{E}}}{\mathrm{d}t} = \frac{1}{2G}\frac{\mathrm{d}S_{ij}}{\mathrm{d}t} \tag{2.45}$$

where the elastic shear modulus G is given in equation (2.20). The function β/B (equation (2.16)) takes the following form:

$$\frac{\beta}{B} = g^{-1}\left\{\frac{1}{\gamma}\left[\frac{\frac{1}{3}\left(\frac{\mathrm{d}\varepsilon_{kk}^{\mathrm{vp}}}{\mathrm{d}t}\right)^2 + 2\left(\frac{\mathrm{d}\bar{e}^{\mathrm{vp}}}{\mathrm{d}t}\right)^2}{3\left(\frac{\partial F_{\mathrm{s}}}{\partial J_1}\right)^2 + \frac{1}{2}\left(\frac{\partial F_{\mathrm{s}}}{\partial\sqrt{J_2'}}\right)^2}\right]^{1/2}\right\} \tag{2.46}$$

In view of equations (2.12) and (2.22), equation (2.46) can be rewritten in the following form:

$$g\left(\frac{F_s}{B}\right)=\frac{1}{\gamma}\left[\frac{\dfrac{1}{3}\left(\dfrac{d\varepsilon_{kk}^{vp}}{dt}\right)^2+\frac{1}{2}\gamma^2\left[g\left(\dfrac{F_s}{B}\right)\right]^2\dfrac{J_2'}{(J_1-L)^2/R^2+J_2'}}{\dfrac{3(J_1-L)^2/R^4+\frac{1}{2}J_2'}{\dfrac{(J_1-L)^2}{R^2}+J_2'}}\right]^{1/2} \tag{2.47}$$

Equation (2.47), in conjunction with equation (2.25), leads to

$$\frac{d\varepsilon_{kk}^{vp}}{dt}=\frac{3(J_1-L)/R^2}{\sqrt{[(J_1-L)^2/R^2+J_2']}} \times\gamma\left\{\exp\left[\frac{B_1\sqrt{[(J_1-L)^2/R^2+J_2']}-B_1(\kappa-L)/R}{B}\right]-1\right\} \tag{2.48}$$

Equation (2.48) describes the coupling between the volumetric strain and the shear stress (J_2' in this case) during a constant mean normal stress test. For a specified time history of J_2', equation (2.48) can be integrated to determine the resulting time history of the viscoplastic volumetric strain. From equations (2.12), (2.22), and (2.25), the following expression is obtained for the viscoplastic strain–rate deviation tensor:

$$\frac{de_{ij}^{vp}}{dt}=\frac{1}{2}\frac{\gamma S_{ij}}{\sqrt{[(J_1-L)^2/R^2+J_2']}} \times\left\{\exp\left[\frac{B_1\sqrt{[(J_1-L)^2/R^2+J_2']}-B_1(\kappa-L)/R}{B}\right]-1\right\} \tag{2.49}$$

Equations (2.45) through (2.49) provide a complete specification for the behaviour of the material for constant mean normal stress triaxial tests. These equations, however, must be integrated numerically in order to relate stresses to strains during dynamic loading. A computer program has been developed for numerical integration of the governing equations of the proposed model for general three-dimensional states of stress. For the sake of brevity, the numerical implementation of the model is not described herein. In the following section, the behaviour of the model is correlated with test data for a clayey sand using this computer program.

2.5 CORRELATION OF TEST DATA WITH MODEL BEHAVIOUR

In this section, the behaviour of the model under states of uniaxial strain and triaxial shear are correlated with available test data [4] for a clayey sand classified as SC according to the Unified Soil Classification System. Static data consisted of (1) load–unload axial stress–axial strain relations (σ_z versus ε_z) for uniaxial strain, and the corresponding stress paths expressed in terms of principal stress difference versus mean normal stress ($\sigma_z - \sigma_r$ versus P), and (2) two triaxial shear test load–unload stress–strain relations (for two different confining stresses) presented in terms of principal stress difference versus principal strain difference ($\sigma_z - \sigma_r$ versus $\varepsilon_z - \varepsilon_r$), and a static failure envelope based on these two tests. The available dynamic data for this material consisted of several stress–strain curves from dynamic uniaxial strain tests [5].

The first step in correlating the behaviour of the model with test data is to simulate the static properties of the material and to determine the numerical

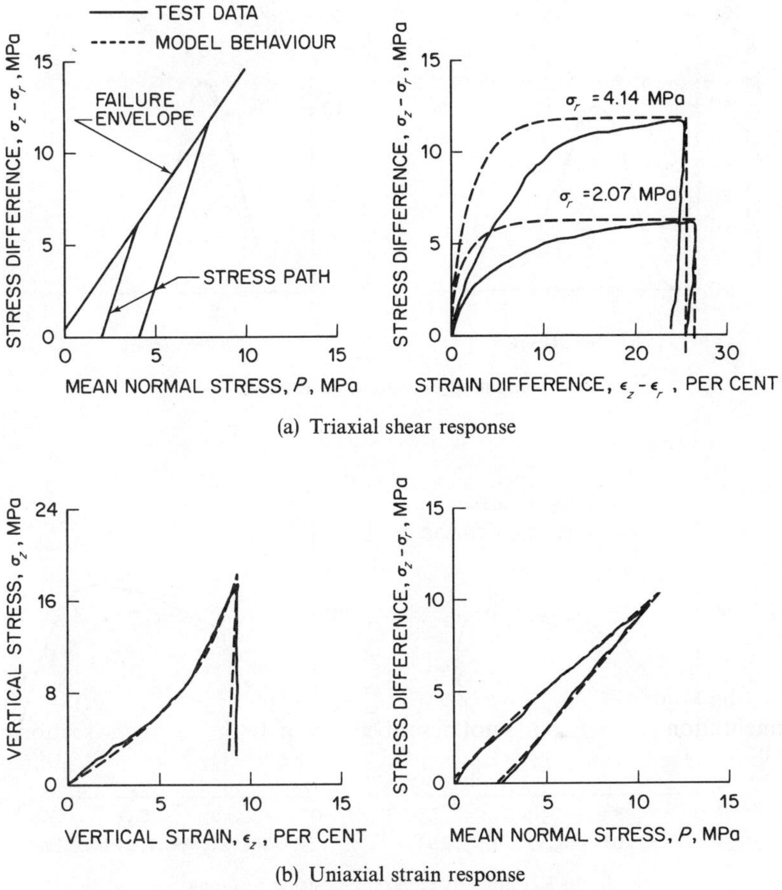

Figure 2.11 Static response of the material in triaxial shear and uniaxial strain tests; model behaviour versus test results

values of the material constant K_i, K_1, K_2, G_i, G_1, G_2, α, k, R_0, R_1, R_2, D, and W. The response of the material under static loading (or in the case of inviscid plasticity when rate dependency is neglected) is governed by these constants. Figure 2.11 portrays the static test data and the corresponding model behaviour for both the triaxial shear and the uniaxial strain test conditions. It can be noted from Figure 2.11 that the model has simulated the static response of the soil both qualitatively (triaxial shear response) and quantitatively (uniaxial strain response). The next step is to simulate the available dynamic stress–strain properties of the material, using the numerical values of the 13 constants above, and to determine the numerical values of the remaining three constants B, B_1, and γ. As indicated before, the available dynamic data for this material are limited to

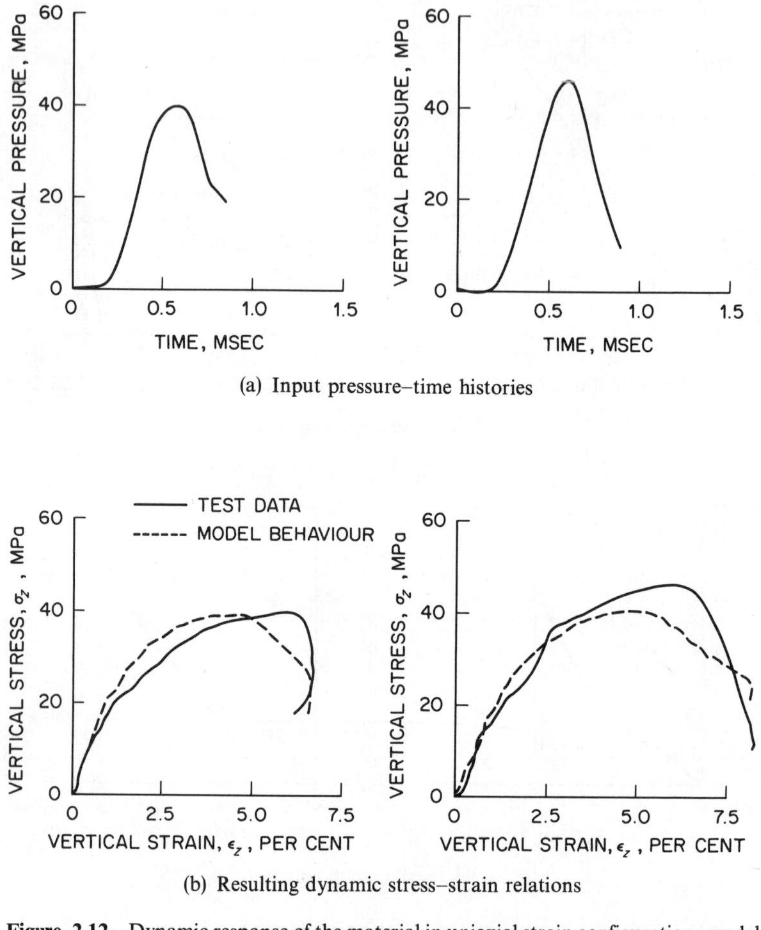

Figure 2.12 Dynamic response of the material in uniaxial strain configuration; model behaviour versus test results

several uniaxial strain stress–strain relations. The dynamic data were obtained for loading rise times on the order of a few tenths of a millisecond [5]. Two of the dynamic stress–strain curves were used for the purpose of model fitting. Figure 2.12(a) depicts the stress–time histories for the two tests. It should be noted that the entire load–unload cycles for these tests were completed in slightly less than 1 msec. These stress–time histories were used as input for driving the model. The experimental stress–strain curves and the corresponding model behaviour for the two dynamic tests are shown in Figure 2.12(b). The agreement between the dynamic test data and the model behaviour is very good, both quantitatively and qualitatively, especially during the early part of the test. It is of interest to compare these dynamic stress–strain relations with the corresponding static result in Figure 2.11(b). In the case of the static test, the stress–strain curve is concave to the stress axis (a 'stiffening' behaviour), whereas the dynamic curves are concave to the strain axis (a 'yielding' behaviour). The proposed constitutive model predicts this dramatic change in the overall character of the stress–strain curves remarkably well.

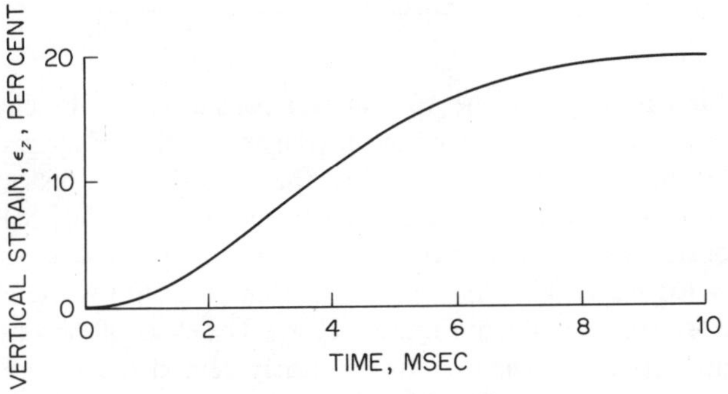

(a) Input strain–time history

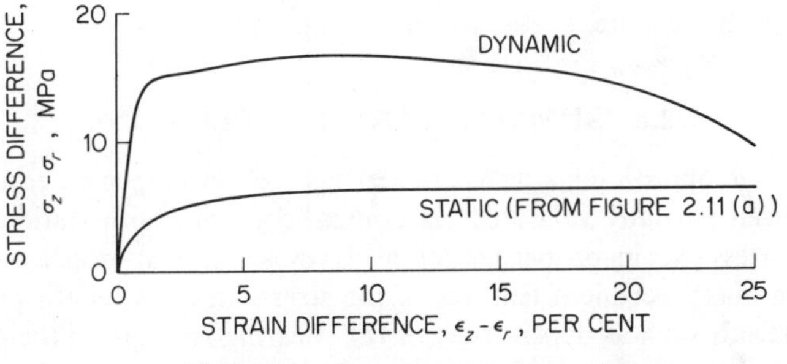

(b) Resulting dynamic stress–strain relation

Figure 2.13 Predicted dynamic stress–strain relation for triaxial shear test condition ($\sigma_r = 2.07$ MPa)

Table 2.1 Summary of the material constants for the clayey sand

Material constant	Unit of measure	Numerical value
K_i	MPa	800
K_1	—	0.5
K_2	MPa^{-1}	0.1
G_i	MPa	480.0
G_1	—	0.5
G_2	MPa^{-1}	0.65
α	—	0.2722
k	MPa	0.231
R_0	—	1.2
R_1	—	−0.5
R_2	MPa^{-1}	0.1
D	MPa^{-1}	0.05
W	—	0.0985
B	MPa	1.0
B_1	—	150 000.0
γ	$msec^{-1}$	0.1

The numerical values of the 16 material constants for the clayey sand are summarized in Table 2.1. To demonstrate the application of the proposed model further, the model was driven with the strain–time history shown in Figure 2.13(a) under triaxial shear test conditions ($\sigma_r = 2.07$ MPa) using the material constants given in Table 2.1. The resulting dynamic stress–strain curve is shown in Figure 2.13(b). Also shown in Figure 2.13(b) is the corresponding static stress–strain curve (from Figure 2.11(a)). The effect of the strain rate on the stress–strain response of the material is clearly demonstrated in Figure 2.13(b). As the strain rate increases during the early part of the simulated dynamic test, the stiffness and strength of the material also increases (relative to the static stiffness and strength). During the latter part of the test where strain rate decreases with time, the dynamic curve actually falls (apparent strain softening) and eventually coincides with the static stress–strain curve (at late times).

2.6 SUMMARY AND CONCLUSIONS

An elastic–viscoplastic constitutive model has been developed for earth materials and has been partially validated via comparison with both static and limited dynamic stress–strain properties for a clayey sand. The model is capable of simulating many pertinent features of the stress–strain–strength properties of earth materials such as dependency of the shearing strength of the material on hydrostatic stress and rate of deformation, shear-induced volume change, and permanent deformation under hydrostatic and deviatoric cyclic loadings. In its present form, the model contains 16 material constants that can be readily determined from the results of static and dynamic triaxial shear and uniaxial strain tests.

The model has been translated into a numerical algorithm for implementation into finite-difference or finite-element computer codes. The numerical algorithm is very versatile in that it embodies all classes of elastic–plastic constitutive models. Test data for several rates of deformation and test boundary conditions (other than those used to fit the model) are needed to further validate the accuracy and determine the range of application of the model.

REFERENCES

1. G. Y. Baladi, and B. Rohani, 'Elastic-plastic model for saturated sand', *Journal of the Geotechnical Engineering Division, ASCE*, **105**, No. GT4, Proc. Paper 14510, pp. 465–480, April 1979.
2. G. Y. Baladi, 'Numerical implementation of a transverse-isotropic inelastic, work-hardening constitutive model', *Transactions of the 4th International Conference on Structural Mechanics in Reactor Technology*, Vol. M, Methods for Structural Analysis, San Francisco, California, August 1977.
3. D. C. Drucker, 'On uniqueness in the theory of plasticity', *Quarterly of Applied Mathematics*, **14** (1956).
4. J. Q. Ehrgott, 'Loading response of a backfill along four different stress paths', Internal Data Report, U.S. Army Engineer Waterways Experiment Station, CE Vicksburg, Mississippi.
5. J. G. Jackson, J. Q. Jr., Ehrgott, and Rohani, B., 'Loading rate effects on compressibility of sand', *Journal of the Geotechnical Engineering Division, ASCE*, **106**, No. GT8, Proc. Paper 15640, pp. 839–852, August 1980.
6. P. Perzyna, 'Fundamental problems in viscoplasticity', *Advances in Applied Mechanics*, Vol. 9, pp. 243–377, Academic Press, New York, 1966.
7. Sandler, I. S., DiMaggio, F. L., and Baladi, G. Y., 'Generalized cap model for geological materials', *Journal of the Geotechnical Engineering Division, ASCE*, **102**, No. GT7, Proc. Paper 12243, pp. 683–699, July 1976.
8. R. S. Swift, 'Examination of the mechanical properties for a Kayenta sandstone from the MIXED COMPANY site', *Technical Report DNA 3683F*, Defense Nuclear Agency, Washington, DC, July 1975.
9. R. V. Whitman, 'The response of soils to dynamic loadings; final report', *Contract Report No.* 3–26, *Report* 26, U.S. Army Engineer Waterways Experiment Station, CE, Vicksburg, Mississippi, May 1970.

Mechanics of Engineering Materials
Edited by C. S. Desai and R. H. Gallagher

Chapter 3

Microplane Model for Strain-controlled Inelastic Behaviour

Z. P. Bažant

3.1 INTRODUCTION

Various heterogeneous brittle aggregate materials such as concretes, rocks, or sea ice, are inelastic but cannot be described as plastic, except at extremely high hydrostatic pressures. A characteristic property of such materials is that they exhibit strain-softening, i.e., a decline of stress at increasing strain, which results from progressive development of fracture. Since these materials can undergo strain-softening within a relatively large zone, a non-linear triaxial constitutive relation is needed for its description. There are, however, some important differences from the classical modelling of inelastic behaviour, i.e., from the theory of plasticity.

First, rather than determining the inelastic phenomena in terms of stresses, as in plasticity, one must determine them in terms of strains. This is because in terms of stresses the description is not unique, as two strains correspond to the same stress, in the case of strain-softening, while still only one stress corresponds to a given strain. Second, the normal inelastic strains are, in contrast to plasticity, important, in fact dominant. They describe the cumulative effect of microcracking. Third, the inelastic phenomena are highly oriented and happen almost independently on planes of various orientation within the material as a function of normal strains across the planes.

In the present work, it is proposed to describe this behaviour independently on planes of various orientations in the material, called microplanes, and then in a certain way superimpose the inelastic effects from all the planes. This type of approach has a long history. First proposed in 1938 by Taylor [1], the idea was exploited by Batdorf and Budianski in their slip theory of plasticity [2]. A number of subsequent investigators adopted this approach for plasticity of polycrystalline metals [2 to 6]. Zienkiewicz and Pande [7] and Pande *et al.* [8–9] developed an approach of this type in their multilaminate models for rocks and soils.

In the aforementioned models, the stress on each plane within the material is assumed to correspond to the same macroscopic stress and the inelastic stresses are superimposed. As mentioned, however, for certain materials the inelastic behaviour is predominantly strain-controlled, and it is then more appropriate to assume that strains, not stresses, correspond on the planes of all orientations to the same macroscopic strain. In this case it is necessary to superimpose in some way the inelastic stresses (relaxations) from the planes of all orientations. This approach was adopted for concrete and geomaterials in ref. [11] and was summarized in ref. [12]. In these works, the inelastic shear stresses on planes of all orientations within the material were neglected. However, although their role is no doubt secondary, in case of concrete and geomaterials at high hydrostatic pressures, they certainly have some effect. The purpose of this work is to generalize ref. [11] to include the effect of inelastic shear stresses.

3.2 BASIC HYPOTHESES

The macroscopic stress tensor will be denoted as σ_{ij}, and the macroscopic strain tensor as ε_{ij}. With regard to the interaction between the macro- and micro-levels, the following three hypotheses may be introduced.

Hypothesis I. The tensor of macroscopic stress, σ_{ij}, is a sum of a purely elastic macrostress σ^a_{ij} that is unaffected by inelastic processes on planes of various orientation, and an inelastic macrostress τ_{ij} which reflects the stress relaxations from microplanes of various orientations, i.e.,

$$\sigma_{ij} = \sigma^a_{ij} + \tau_{ij} \tag{3.1}$$

(latin lower case subscripts refer to cartesian coordinates $x_i, i = 1, 2, 3$).

Hypothesis II. The normal microstrain ε_N and the shear microstrain ε_T on each microplane of any orientation is the resolved component of the macroscopic strain tensor ε_{ij}.

Hypothesis III. There exist an independent stress–strain relation for each microplane of any orientation.

Hypothesis II is opposite to that made in the slip theory of plasticity, in which the stresses rather than strains on the planes of all orientations are assumed to be the resolved components of the macroscopic stress. One can offer three reasons for this. First, if the material state were characterized by stress rather than strain, the description would not be unique since, in the case of strain-softening, there are two strains corresponding to a given stress. Second, the relationship between the micro- and macro-levels would not be stable in the case of strain softening, which has been confirmed numerically. Third, the use of resolved strains, rather than stresses, appears to reflect the microstructure of a brittle aggregate material more realistically. In contrast to polycrystalline metals, brittle aggregate materials consist of hard inclusions embedded in a relatively soft matrix. The microstresses are far from uniform, having sharp extremes at the locations where the aggregate pieces are nearest. The deformation of the thin layer of matrix between two aggregate pieces, which is the chief source of inelastic behaviour, seems to be determined mainly by the relative displacements of the centroids of the aggregate pieces, which roughly correspond to the macroscopic strain. The microplanes may be imagined to represent the thin layers of matrix and the bond interfaces between the adjacent aggregate pieces (Figure 3.1(a)), since microcracking is chiefly concentrated there.

According to Hypothesis I (equation (3.1)), the virtual work of stresses per unit volume may be written as $\delta W = \varepsilon_{ij}\delta\sigma_{ij} = \varepsilon_{ij}^{a}\delta\sigma_{ij}^{a} + \varepsilon_{ij}^{m}\delta\sigma_{ij}^{m}$, in which ε_{ij}^{a} and ε_{ij}^{m} represent the strains associated with the additional elastic stress and the stress resulting from the microplanes. At the same time, $\delta W = \varepsilon_{ij}\delta\sigma_{ij}^{a} + \varepsilon_{ij}\delta\sigma_{ij}^{m}$. Since both expressions must hold for any $\delta\sigma_{ij}^{a}$ and any $\delta\sigma_{ij}^{m}$, we must have $\varepsilon_{ij}^{a} = \varepsilon_{ij}^{m} = \varepsilon_{ij}$.

According to Hypothesis II, the components of the strain vector $\boldsymbol{\varepsilon}^n$ on any microplane are

$$\varepsilon_j^n = n_i \varepsilon_{ji} \tag{3.2}$$

in which n_i are the cosines of the unit normal to the microplane. The normal microstrain, i.e., the normal component of strain vector $\boldsymbol{\varepsilon}^n$, may be denoted as ε_N, and the components of the vector of the shear component $\boldsymbol{\varepsilon}_T$ may be denoted as

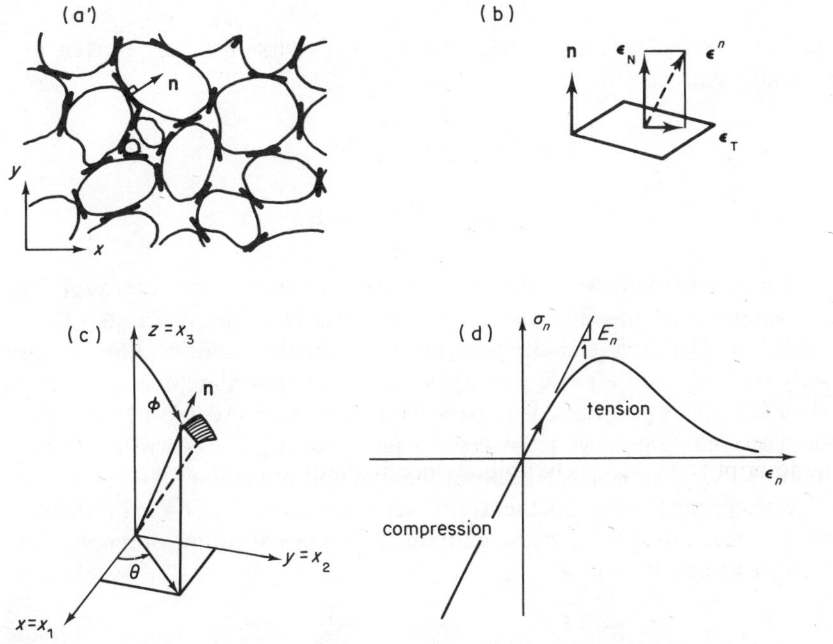

Figure 3.1 (a) Example of idealized microstructure, (b)–(c) explanation of notations, (d) stress–strain relation on a microplane ($\sigma_n = \tau_N, \varepsilon_n = \varepsilon_N$)

ε_{T_i}. With regard to the elastic parts of the shear components of stresses and strains on the microplane, we will now introduce an additional assumption, namely that the vectors of these shear components are parallel. This precludes anisotropic behaviour within each microplane, although overall anisotropy remains possible by considering different properties on various microplanes. According to Hypothesis III we may now write

$$\dot{\tau}_N = C\dot{\varepsilon}_N - \dot{\tau}''_N \qquad\qquad \dot{\tau}_{T_i} = B\dot{\varepsilon}_{T_i} - \dot{\tau}''_{T_i} \tag{3.3}$$

in which C and B are the elastic constants for the normal and shear response on the microplane, and τ''_N and τ''_{T_i} are the inelastic stress relaxations in the normal and tangential directions on the microplane. Superimposed dots denote time rates. For the magnitudes of the shear components, equation (3.3) implies $\dot{\tau} = B\dot{\varepsilon}_T - \dot{\tau}''_T$.

Further we need to specify the inelastic stress relaxations. For this purpose, we assume the existence of inelastic potentials f_β and loading surfaces $g_\beta(\beta = 1, \ldots, n)$ for each microplane. They must be defined in terms of strains rather than stresses, i.e.,

$$f_\beta(\varepsilon_N, \varepsilon_T) = 0 \qquad (\beta = 1, \ldots, n) \tag{3.4}$$

$$g_\beta(\varepsilon_N, \varepsilon_T) = 0 \tag{3.5}$$

The rates of inelastic stress relaxations may be assumed to be given by the normality rule

$$\dot{\tau}''_{ij} = \sum_{\beta=1}^{n} \frac{\partial f_\beta}{\partial \varepsilon_{ij}} \dot{\mu}_\beta \tag{3.6}$$

$$\dot{\mu}_\beta = h_\beta \frac{\partial g_\beta}{\partial \varepsilon_{km}} \dot{\varepsilon}_{km} H(\dot{\mu}_\beta) \tag{3.7}$$

in which h_β are material softening parameters depending on the current state of the material and possibly also its history, and H is Heaviside step function. Similarly to Drucker's stability postulate in plasticity, equations (3.6)–(3.7) can be easily derived from a more plausible hypothesis (postulate) for a strain cycle, called Il'yushin's postulate, as previously used for macroscopic inelastic theories based on loading surfaces in the strain space [13 to 20]. The Heaviside function in equation (3.7) distinguishes between loading and unloading.

By projecting the microplane strain vector (equation (3.2)) on to the direction $\mathbf{n}$ of the normal, we obtain the magnitude of the normal strain component on the microplane and its vector:

$$\varepsilon_N = n_j \varepsilon_j^n = n_j n_k \varepsilon_{jk}, \qquad (\boldsymbol{\varepsilon}_N)_i = n_i n_j n_k \varepsilon_{jk} \tag{3.8}$$

The magnitude of the strain vector on the microplane is $|\boldsymbol{\varepsilon}^n| = (\varepsilon_j^n \varepsilon_j^n)^{1/2} = (n_i \varepsilon_{ji} n_k \varepsilon_{jk})^{1/2}$. The vector of the tangential (shear) strain component is (see

Figure 3.1(c)) $\boldsymbol{\varepsilon}_T = \boldsymbol{\varepsilon} - \boldsymbol{\varepsilon}_N$, and its magnitude is $\varepsilon_T = [|\boldsymbol{\varepsilon}^n|^2 - (\varepsilon_N)^2]^{1/2}$ or $\varepsilon_T = |\boldsymbol{\varepsilon}_T| = (\varepsilon_{T_j}\varepsilon_{T_j})^{1/2}$. Thus, we obtain the following expressions for the vector and the magnitude of the shear strain component on the microplane:

$$\varepsilon_{T_i} = (n_k \delta_{ij} - n_i n_j n_k)\varepsilon_{jk} \tag{3.9}$$

$$\varepsilon_T = [n_i \varepsilon_{ji} n_k(\varepsilon_{jk} - n_j n_m \varepsilon_{km})]^{1/2} \tag{3.10}$$

According to equation (3.6), the normal and tangential components of the stress relaxation rate on the microplane can be expressed as

$$\dot{\tau}''_N = \sum_{\beta=1}^{n} \frac{\partial f_\beta}{\partial \varepsilon_N}\dot{\mu}_\beta, \qquad \dot{\tau}''_T = \sum_{\beta=1}^{n} \frac{\partial f_\beta}{\partial \varepsilon_T}\dot{\mu}_\beta, \qquad \dot{\tau}''_{T_i} = \sum_{\beta=1}^{n} \frac{\partial f_\beta}{\partial \varepsilon_{T_i}}\dot{\mu}_\beta \tag{3.11}$$

The derivatives of the inelastic potential in these equations may be calculated as

$$\frac{\partial f_\beta}{\partial \varepsilon_{T_i}} = \frac{\partial f_\beta}{\partial \varepsilon_T}\frac{\partial \varepsilon_T}{\partial \varepsilon_{T_i}} = \frac{\partial f_\beta}{\partial \varepsilon_T}\frac{\partial}{\partial \varepsilon_{T_i}}(\varepsilon_{T_j}\varepsilon_{T_j})^{1/2} = \frac{\partial f_\beta}{\partial \varepsilon_T}\frac{\varepsilon_{T_i}}{\varepsilon_T} \tag{3.12}$$

We see that the vector normal to the potential surface f_β is parallel to the vector of the tangential component of strain on the microplane, i.e.,

$$\dot{\boldsymbol{\varepsilon}}''_T \parallel \boldsymbol{\varepsilon}_T, \qquad \dot{\boldsymbol{\varepsilon}}_T \parallel \boldsymbol{\varepsilon}_T \tag{3.13}$$

The derivatives of the inelastic potentials and loading functions appearing in equations (3.6)–(3.7) may be calculated as

$$\frac{\partial f_\beta}{\partial \varepsilon_{ij}} = p_{ij}\frac{\partial f_\beta}{\partial \varepsilon_N} + q'_{ij}\frac{\partial f_\beta}{\partial \varepsilon_T}, \qquad \frac{\partial g_\beta}{\partial \varepsilon_{ij}} = p_{ij}\frac{\partial g_\beta}{\partial \varepsilon_N} + q'_{ij}\frac{\partial g_\beta}{\partial \varepsilon_T} \tag{3.14}$$

in which we introduce the notation

$$p_{ij} = \frac{\partial \varepsilon_N}{\partial \varepsilon_{ij}}, \qquad q'_{ij} = \frac{\partial \varepsilon_T}{\partial \varepsilon_{ij}} \tag{3.15}$$

From equation (3.8) we can further calculate

$$p_{ij} = \frac{\partial}{\partial \varepsilon_{ij}}(n_p n_q \varepsilon_{pq}) = \delta_{ip}\delta_{jq} n_p n_q = n_i n_j \tag{3.16}$$

while from equation (3.10) we obtain

$$q'_{ij} = \frac{1}{2\varepsilon_T}\frac{\partial}{\partial \varepsilon_{ij}}(n_p \varepsilon_{qp} n_r \varepsilon_{qr} - n_p n_q \varepsilon_{pq} n_r n_s \varepsilon_{rs})$$

which reduces to

$$q'_{ij} = \frac{1}{\varepsilon_T} n_j n_r(\varepsilon_{ir} - n_i n_s \varepsilon_{rs}) \tag{3.17}$$

The last tensor is non-symmetric. Later we will need its symmetric part, which reads

$$q_{ij} = \frac{1}{2\varepsilon_{\mathrm{T}}}[n_k(n_j\varepsilon_{ik} + n_i\varepsilon_{jk}) - 2a_{ijkm}\varepsilon_{km}] \tag{3.18}$$

in which we introduce the notation

$$a_{ijkm} = n_i n_j n_k n_m \tag{3.19}$$

We need to establish now the equilibrium relation between the microstresses on the microplanes of all orientations and the macroscopic stress tensor. We may use for this purpose the principle of virtual work, which requires that the virtual work of macroscopic stress rates on any macroscopic strain variations within a small unit sphere of unit radius be equal to the virtual work done on all the microplanes tangential to the unit sphere. This condition may be written as follows

$$\begin{aligned} \delta\dot{W} &= \frac{4\pi}{3}\dot{\tau}_{ij}\delta\varepsilon_{ij} = 2\int_S (\dot{\tau}_{\mathrm{N}}\delta\varepsilon_{\mathrm{N}} + \dot{\tau}_{\mathrm{T}}\delta\varepsilon_{\mathrm{T}})f(\mathbf{n})\,\mathrm{d}S \\ &= 2\int_S (C\dot{\varepsilon}_{\mathrm{N}}\delta\varepsilon_{\mathrm{N}} + B\dot{\varepsilon}_{\mathrm{T}_i}\delta\varepsilon_{\mathrm{T}_i} - \dot{\tau}''_{ij}\delta\varepsilon_{ij})f(\mathbf{n})\,\mathrm{d}S \end{aligned} \tag{3.20}$$

where S is the surface of a unit hemisphere, and $f(\mathbf{n})$ describes the frequency of microplanes as a function of orientation $\mathbf{n}$.

Substituting here from equations (3.8), (3.9), and (3.6), we may obtain the relation

$$\dot{\tau}_{ij}\delta\varepsilon_{ij} = \frac{3}{2\pi}\int_S \left(Cn_k n_m \dot{\varepsilon}_{km} n_i n_j + Bb'_{ijkm}\dot{\varepsilon}_{km} - \sum_{\beta=1}^{n} \frac{\partial f_\beta}{\partial \varepsilon_{ij}}\dot{\mu}_\beta \right)\delta\varepsilon_{ij} f(\mathbf{n})\,\mathrm{d}S \tag{3.21}$$

in which

$$\begin{aligned} b'_{ijkm} &= (n_m\delta_{rk} - n_r n_k n_m)(n_j\delta_{ri} - n_r n_i n_j) \\ &= n_j n_m \delta_{ik} - n_j n_k n_m n_i - n_m n_i n_j n_k + n_i n_j n_k n_m (n_r n_r) \\ &= \delta_{ik} n_j n_m - a_{ijkm} \end{aligned} \tag{3.22}$$

This fourth order tensor is symmetric when ij is interchanged with km but non-symmetric when i is interchanged with j or k is interchanged with small m. The tensor may be written as a sum of a symmetric part and an antisymmetric part,

$$b'_{ijkm} = b_{ijkm} + \tilde{b}_{ijkm} \tag{3.23}$$

in which the symmetric part is

$$b_{ijkm} = \tfrac{1}{4}(\delta_{ik}n_j n_m + \delta_{jk}n_i n_m + \delta_{im}n_j n_k + \delta_{jm}n_i n_k) - a_{ijkm} \tag{3.24}$$

For the antisymmetric part it is true that $\tilde{b}_{ijkm}\delta\varepsilon_{ij}\dot{\varepsilon}_{km} = 0$ for any $\delta\varepsilon_{ij}$. Therefore,

$$b'_{ijkm}\delta\varepsilon_{ij}\dot{\varepsilon}_{km} = b_{ijkm}\delta\varepsilon_{ij}\dot{\varepsilon}_{km} \tag{3.25}$$

Furthermore, using equations (3.14) and (3.7), we may express

$$\sum_{\beta=1}^{n} \frac{\partial f_\beta}{\partial \varepsilon_{ij}} \dot{\mu}_\beta \delta\varepsilon_{ij} = \sum_{\beta=1}^{n} h_\beta \left[p_{ij} \frac{\partial f_\beta}{\partial \varepsilon_N} + (q_{ij} + \tilde{q}_{ij}) \frac{\partial f_\beta}{\partial \varepsilon_T} \right] \times \delta\varepsilon_{ij} \left[p_{km} \frac{\partial g_\beta}{\partial \varepsilon_N} + (q_{km} + \tilde{q}_{km}) \frac{\partial g_\beta}{\partial \varepsilon_T} \right] \dot{\varepsilon}_{km} H(\dot{\mu}_\beta) \tag{3.26}$$

in which $\tilde{q}_{ij}$ is the antisymmetric part of q'_{ij}, *i.e.*, $\tilde{q}_{ij} = q'_{ij} - q_{ij}$. Noting that the antisymmetric parts give zero products with symmetric tensors, i.e., $\tilde{q}_{ij}\delta\varepsilon_{ij} = 0$, $\tilde{q}_{km}\delta\varepsilon_{km} = 0$, we find that

$$\sum_{\beta=1}^{n} \frac{\partial f_\beta}{\partial \varepsilon_{ij}} \dot{\mu}_\beta = R_{ijkm} \dot{\varepsilon}_{km} \tag{3.27}$$

in which

$$R_{ijkm} = \sum_{\beta=1}^{n} h_\beta \left(p_{ij} \frac{\partial f_\beta}{\partial \varepsilon_N} + q_{ij} \frac{\partial f_\beta}{\partial \varepsilon_T} \right) \left(p_{km} \frac{\partial g_\beta}{\partial \varepsilon_T} + q_{km} \frac{\partial g_\beta}{\partial \varepsilon_T} \right) H(\dot{\mu}_\beta) \tag{3.28}$$

Substituting equations (3.27) and (3.25) into the variational virtual work relation in equation (3.21), and noting that this relation must hold for any variation $\delta\varepsilon_{ij}$, we find that

$$\dot{\tau}_{ij} = D^{e}_{ijkm} \dot{\varepsilon}_{km} - \dot{\tau}''_{ij} \tag{3.29}$$

or

$$\dot{\tau}_{ij} = D^{m}_{ijkm} \dot{\varepsilon}_{km} \tag{3.30}$$

in which

$$D^{m}_{ijkm} = \frac{3}{2\pi} \int_S (a_{ijkm} C + b_{ijkm} B - R_{ijkm}) f(\mathbf{n}) \, dS,$$

$$D^{e}_{ijkm} = \frac{3}{2\pi} \int_S (a_{ijkm} C + b_{ijkm} B) f(\mathbf{n}) \, dS \tag{3.31}$$

$$\dot{\tau}''_{ij} = \frac{3}{2\pi} \int_S R_{ijkm} f(\mathbf{n}) \, dS \dot{\varepsilon}_{km} \tag{3.32}$$

Here D^{m}_{ijkm} is the tensor of tangential moduli corresponding to the microplanes, D^{e}_{ijkm} is the elastic part of this tensor, and $\dot{\sigma}''_{ij}$ is a tensor of the rate of inelastic stress relaxation. The integrals in equations (3.31) and (3.32) extend over the surface S of a unit hemisphere.

Consider now the special case of isotropic materials, for which $f(\mathbf{n}) = 1$. For this case the elastic stiffness matrix D^{e}_{ijkm} must be equivalent to an isotropic material stiffness matrix characterized by some shear modulus G^{m} and Poisson ratio ν^{m}. Their values may be easily calculated. To this end, consider a uniaxial strain rate $\dot{\varepsilon}_{33} = 1$ while all other components of $\dot{\varepsilon}_{ij} = 0$. We may now substitute $n_1 = \sin\phi\cos\theta$, $n_2 = \sin\phi\sin\theta$, $n_3 = \cos\phi$ and equations (3.29) and (3.31) for

$\dot{\tau}''_{ij} = 0$ then give

$$\dot{\tau}_{33} = \frac{3}{2\pi}\left[C\int_0^\pi\int_0^\pi \cos^4\phi\sin\phi\,\mathrm{d}\phi\,\mathrm{d}\theta + B\int_0^\pi\int_0^\pi \cos^2\phi\sin^3\phi\,\mathrm{d}\phi\,\mathrm{d}\theta\right]\dot{\varepsilon}_{33}$$

$$= \tfrac{1}{5}(3C + 2B) \tag{2.33}$$

$$\dot{\tau}_{11} = \frac{3}{2\pi}(C - B)\int_0^\pi\int_0^\pi \sin^3\phi\cos^2\phi\cos^2\theta\,\mathrm{d}\theta\,\mathrm{d}\phi\,\dot{\varepsilon}_{33} = \tfrac{1}{5}(C - B) \tag{2.34}$$

According to Hooke's law, $\dot{\tau}_{11}/\dot{\tau}_{33} = \nu^{\mathrm{m}}/(1 - \nu^{\mathrm{m}})$ for uniaxial strain, and so $\nu^{\mathrm{m}} = \dot{\tau}_{11}/(\dot{\tau}_{11} + \dot{\tau}_{33})$. Substituting from equations (3.33) and (3.34), we thus get

$$\nu^{\mathrm{m}} = \frac{C - B}{4C + B} \tag{2.35}$$

Furthermore, according to Hooke's law we have, for uniaxial strain, $\nu^{\mathrm{m}} = \dot{\tau}_{11}/(\dot{\tau}_{11} + \dot{\tau}_{33})$, from which we may solve

$$G^{\mathrm{m}} = \frac{1}{10}\frac{(1 - 2\nu^{\mathrm{m}})}{(1 - \nu^{\mathrm{m}})}(3C + 2B) \tag{3.36}$$

From equations (3.35) and (3.36) we may solve the constants C and B from desired values of G^{m} and ν^{m}. Equation (3.35) yields the following values of Poisson ratio:

$B/C = 0$	$\nu^{\mathrm{m}} = 0.25$
14/59	0.18
1	0
∞	-1

It is interesting to observe that Poisson ratios greater than 0.25 cannot be obtained. The range appears suitable for geomaterials.

In some situations, however, an adjustment of the Poisson ratio provided by the system of microplanes may be needed. For example, one might desire for some material an overall Poisson ratio $\nu > 0.25$, or one might simply need for the best fit of test data a different Poisson ratio for the microplane system than for the material as a whole. Such an adjustment of Poisson ratio is made possible by equation (3.1) (Hypotehsis I). Let the additional elastic stresses σ^{a}_{ij} be given by an isotropic stress–strain relation to ε_{ij}, characterized by shear modulus G^{a} and Poisson ratio ν^{a}. Then, for uniaxial strain $\dot{\varepsilon}_{33} = 1$ ($\dot{\varepsilon}_{11} = \dot{\varepsilon}_{22} = 0$) we have $\dot{\sigma}_{33} = 2G(1 - \nu)/(1 - 2\nu)$. Summing the stress from the microplane system and the additional elastic stress, we also have $\dot{\sigma}_{33} = 2G^{\mathrm{a}}(1 - \nu^{\mathrm{a}})/(1 - 2\nu^{\mathrm{a}}) + 2G^{\mathrm{m}}(1 - \nu^{\mathrm{m}})/(1 - 2\nu^{\mathrm{m}})$. From these equations we may solve

$$G^{\mathrm{a}} = \frac{1 - 2\nu^{\mathrm{a}}}{1 - \nu^{\mathrm{a}}}\left(\frac{G(1 - \nu)}{1 - 2\nu} - \frac{G^{\mathrm{m}}(1 - \nu^{\mathrm{m}})}{1 - 2\nu^{\mathrm{m}}}\right) \tag{3.37}$$

This relation permits us to choose the elastic constants of the material as a whole, as well as of the microplane system, and also choose the Poisson ratio for the additional elastic stress.

In general, the total tangential elastic moduli are

$$D_{ijkm} = D^{\mathrm{m}}_{ijkm} + D^{\mathrm{a}}_{ijkm} \tag{3.38}$$

For isotropic materials, D^{a}_{ijkm} represent the elastic moduli tensor of an isotropic material;

$$D^{\mathrm{a}}_{ijkm} = G^{\mathrm{a}}(\delta_{ik}\delta_{jm} + \delta_{jk}\delta_{im}) + \frac{2\nu^{\mathrm{a}}}{1-2\nu^{\mathrm{a}}} G^{\mathrm{a}}\delta_{ij}\delta_{km} \tag{3.39}$$

in which δ_{ij} = Kronecker delta = 1 if $i=j$, and 0 if $i \neq j$.

3.3 CASE OF ZERO SHEAR RELAXATIONS ON MICROPLANES

For tensile strain-softening of concrete, it seems that one may neglect the shear stress relaxations and consider only the normal stress relaxations on the microplanes, which correspond to the formation of microcracks in the direction of the microplanes. In this case $B = \tau_{\mathrm{T}_i} = \tau''_{\mathrm{T}_i} = 0$. One may consider here for each microplane only one loading surface ($\beta = 1$), $f_1 = g_1 = \varepsilon_{\mathrm{N}} =$ const. In this case we get

$$\dot{\tau}_{\mathrm{N}} = [C - h_1 H(\dot{\mu}_1)]\dot{\varepsilon}_{\mathrm{N}} \tag{3.40}$$

The relationship between the normal stress and the normal strain on the microplane may be conveniently described by the formula

$$\tau_{\mathrm{N}} = F(\varepsilon_{\mathrm{N}}) = C\varepsilon_{\mathrm{N}} \exp(-k\varepsilon^n_{\mathrm{N}}) \qquad (n \simeq 2) \tag{3.41}$$

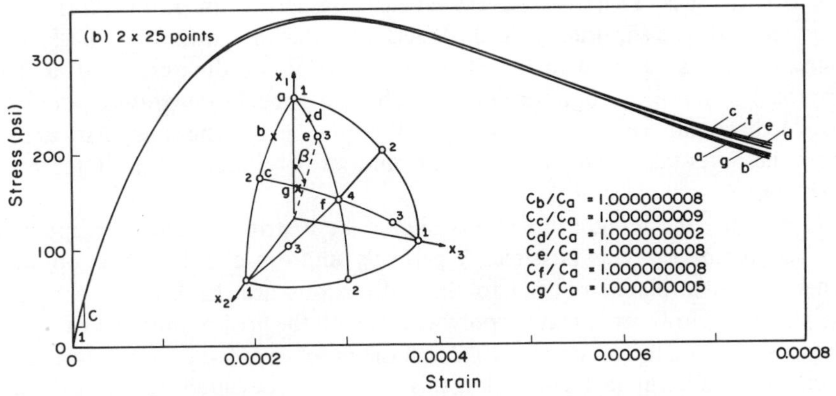

Figure 3.2 Distribution of integration points for 2 × 25-point formula defined in Table 3.1, and response curves for uniaxial stress applied at directions a, b,.. ,f.

in which C, k, and n are material constants. For large ε_N the value of τ_N becomes essentially zero ($n=2$ is a suitable exponent). Comparing this with equation (3.40), we have

$$h_1 = C - F'(\varepsilon_N) = C - C(1 - kn\varepsilon_N^n)\exp(-k\varepsilon_N^n) \tag{3.42}$$

The present special case has been considered in refs [11 and 12], and good fits of various tensile strain-softening test data have been demonstrated. Since, in absence of shear relaxations, the microplane model yields Poisson ratio 0.25, the additional elastic deformation was used to correct this value to 0.18, typical of concrete. There was a slight difference from the present formulation in that the superimposed additional elastic value was strain rather than stress. The present formulation is, however, more efficient. One of the comparisons with the test data of Evans and Marathe [21], made in ref. [11], is reproduced in Figure 3.2, and the associated stress–strain curve considered for the microplanes is shown in Figure 3.1(d).

3.4 NUMERICAL INTEGRATION ON THE SURFACE OF A SPHERE

In general situations, the integral in equation (3.31) over the surface of a unit hemisphere has to be evaluated numerically, approximating it by a finite sum:

$$D_{ijkm}^m = \sum_{\alpha=1}^{n} 6w_\alpha[(a_{ijkm}C + b_{ijkm}B - R_{ijkm})f(\mathbf{n})]_\alpha, \ \sum_\alpha w_\alpha = \tfrac{1}{2} \tag{3.43}$$

in which α refers to the values evaluated at certain numerical integration points on the spherical surface (i.e., certain characteristic directions), and w_α are the weights associated with the integration points. In finite element programmes for incremental loading, the numerical integration needs to be carried out a great number of times. Therefore, a very efficient numerical integration formula is required. For the slip theory of plasticity, a similar integration was performed using a rectangular grid in the plane of spherical coordinates θ and ϕ. This approach is, however, computationally inefficient since the integration points are crowded around the poles, and since, in the $\theta - \phi$ plane, the singularity arising from the poles takes away the benefit from a use of a higher-order integration formula.

Optimally, the integration points should be distributed over the spherical surface as uniformly as possible. A perfectly uniform distribution is obtained when the microplanes normal to the α-directions are the faces of a regular polyhedron. However, a regular polyhedron with the greatest number of sides is the icosahedron, for which $N = 10$ ($2N$ is the number of faces), and this number appears insufficient (a formula for this case was presented by Albrecht and Collatz [22]). The need for greater accuracy is indicated when the response curves in a uniaxial tensile test with strain-softening are calculated for various

Table 3.1 Direction cosines and weights for 2 × 25 points with error of 10th order (after Bažant and Oh [11],

α	x_1^α	x_2^α	x_3^α	w^α
1	1	0	0	0.012 698 410 58
2	0	1	0	0.012 698 410 58
3	0	0	1	0.012 698 410 58
4	0.707 106 781 2	0.707 106 781 2	0	0.022 574 956 12
5	0.707 106 781 2	−0.707 106 781 2	0	0.022 574 956 12
6	0.707 106 781 2	0	0.707 106 781 2	0.022 574 956 12
7	0.707 106 781 2	0	−0.707 106 781 2	0.022 574 956 12
8	0	0.707 106 781 2	0.707 106 781 2	0.022 574 956 12
9	0	0.707 106 781 2	−0.707 106 781 2	0.022 574 956 12
10	0.301 511 335 4	0.301 511 335 4	0.904 534 039 8	0.020 173 335 57
11	0.301 511 335 4	0.301 511 335 4	−0.904 534 039 8	0.020 173 335 57
12	0.301 511 335 3	−0.301 511 335 4	0.904 534 039 8	0.020 173 335 57
13	0.301 511 335 4	−0.301 511 335 4	−0.904 534 039 8	0.020 173 335 57
14	0.301 511 335 4	0.904 534 039 8	0.301 511 335 4	0.020 173 335 57
15	0.301 511 335 4	0.904 534 039 8	−0.301 511 335 4	0.020 173 335 57
16	0.301 511 335 4	−0.904 534 039 8	0.301 511 335 4	0.020 173 335 57
17	0.301 511 335 4	−0.904 534 039 8	−0.301 511 335 4	0.020 173 335 57
18	0.904 534 039 8	0.301 511 335 4	0.301 511 335 4	0.020 173 335 57
19	0.904 534 039 8	0.301 511 335 4	−0.301 511 335 4	0.020 173 335 57
20	0.904 534 039 8	−0.301 511 335 4	0.301 511 335 4	0.020 173 335 57
21	0.904 534 039 8	−0.301 511 335 4	−0.301 511 335 4	0.020 173 335 57
22	0.577 350 269 2	0.577 350 269 2	0.577 350 269 2	0.021 093 751 17
23	0.577 350 269 2	0.577 350 269 2	−0.577 350 269 2	0.021 093 751 17
24	0.577 350 269 2	−0.577 350 269 2	0.577 350 269 2	0.021 093 751 17
25	0.577 350 269 2	−0.577 350 269 2	−0.577 350 269 2	0.021 093 751 17

$\beta = 25.239401°$.

orientations of the uniaxial stress with regard to the α-directions. Ideally, the response curves for any orientation should be identical. However, large discrepancies are found for a ten-point formula.

Bažant and Oh [23] derived numerical integration formulas with more than 10 points, which give consistent results even in the strain-softening range. The most efficient formulas, with an almost uniform spacing of α-directions, are obtained by certain subdivisions of the faces of an icosahedron or a dodecahedron [23]. Such formulas do not exhibit orthogonal symmetries. Other formulas which do were also derived [23]. Taylor series expansions on a sphere were used and weights w_α were solved from the condition that the greatest possible number of terms of the expansion of the error would cancel out. The angular directions of certain integration points were further determined from the condition that the error term of the expansion be minimized. In this manner, formulas involving 16, 21, 25, 33, 37, and 61 points were established, with errors of 8th, 10th, and 12th order. Table 3.1 defines one of these numerical integration formulas, having 25 points for a hemisphere; this formula exhibits orthogonal symmetry [23]. The directions of the integration points are illustrated in Figure 3.2, and also shown are the stress–strain diagrams calculated for various directions of uniaxial tensile

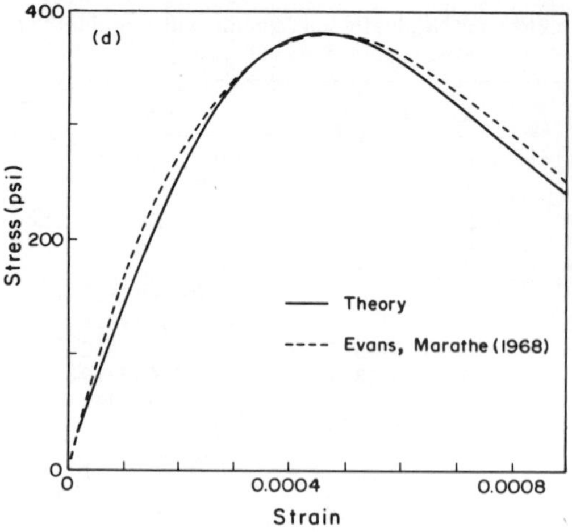

Figure 3.3 Comparison with test data of Evans and Marathe [21], after Božant and Oh [11]

stress with regard to the integration points (directions a, b, c, d, ...); the spread of the response curves characterizes the range of numerical errors. For crude calculations, the lowest required number of integration points is 16 [23].

3.5 APPLICATION TO ANISOTROPIC CREEP OF CLAY

As another application, we may demonstrate an adaptation of the microplane model to describe creep of an anisotropically consolidated clay and to correlate the stress–strain relation to known information about the distribution of the frequency of platelets of various orientations within the clay. This problem has been studied, for example, in ref. [24], using a micromechanics model in which triangular cells of mutually sliding clay platelets are constrained to the same macroscopic strains ε_{ij}, same as here. It appeared, however, that this approach becomes quite complicated in the three-dimensional case, although it is not very difficult for two-dimensional analysis. In three dimensions, the present type of microplane model seems appropriate.

In treating clay, the stress tensor σ_{ij} must be interpreted as the effective stress tensor, i.e., $\sigma_{ij} = t_{ij} - \delta_{ij}p$, in which p = pore-water pressure and t_{ij} = total stress in the solid-water system. Let us consider only the case of deviatoric creep, for which the normal stiffness on the microplanes may be neglected, i.e. $C = \tau''_{\mathrm{N}} = 0$. The microplanes of the present model may be interpreted, in the case of clay, as the planes of sliding in contact of adjacent clay platelets. As is well known, the sliding is governed by the rate-process theory, which yields the relation

$$\dot{\varepsilon}_{\mathrm{T}} = k_1 \sinh(k_2 \tau_{\mathrm{T}}) \tag{3.44}$$

in which $k_1 = 2A(kT/h)t^{-m}\exp(-Q/RT)$ and $k_2 = V_a/RT$. Here T is the absolute temperature, Q is the activation energy of creep, R is the universal gas constant, k is the Boltzmann constant, h is the Planck constant, V_a is the activation volume, and A, m are empirical constants. For the vectors of the tangential stress and strain components on the microplanes, equation (3.44) may be generalized, in the inverted form, as follows

$$\tau_{T_i} = \frac{1}{k_2}\frac{\dot{\varepsilon}_{T_i}}{\dot{\varepsilon}_T}\sinh^{-1}\left(\frac{\dot{\varepsilon}_T}{k_1}\right) \tag{3.45}$$

This equation now replaces equation (3.3) of the present model, and equations (3.6)–(3.11) become unnecessary. By following the same analysis as before, one obtains the macroscopic stress–strain relation, replacing equation (3.29) as follows

$$\sigma_{jk} = \eta_{jkrs}\dot{\varepsilon}_{rs}, \qquad \eta_{jkrs} = \frac{3}{2\pi}\int_S b_{jkrs}\frac{1}{k_2\dot{\varepsilon}_T}\sinh^{-1}\left(\frac{\dot{\varepsilon}_T}{k_1}\right)f(\mathbf{n})\,\mathrm{d}S \tag{3.46}$$

Here η_{jkrs} represents the fourth order tensor of current viscosities, and $f(\mathbf{n})$ represents the distribution function for the frequency of clay platelets of various orientations. For some clays, this distribution function has been measured experimentally, using X-ray scattering techique. Applicability of equation (3.43) to test results is presently being studied at Northwestern University by J. K. Kim.

Complete description of clays further requires superposing equations for the volume change. This may be best accomplished on the basis of the critical state theory, for example, in a manner recently described by Pande *et al.* [8–9].

3.6 CONCLUDING REMARKS

The microplane model allows great versatility in constitutive modelling. The present form of the model, in which the strains on microplanes of all orientations correspond to the same macroscopic strain, appears suitable for materials which exhibit progressive microcracking and tensile strain-softening. Constraining the microstructure to the same macroscopic strain is also important for numerical reasons, not merely for the purpose of stability and uniqueness of representation. The computational work required by a model of this kind is not as large as one might think. The work required is greatly reduced by the recent development of efficient numerical integration formulas for a spherical surface.

One advantage of the model is that the stress–strain relations are primarily defined on the microplane level, on which one does not need to heed the tensorial invariance requirements which are a source of great difficulty in constitutive modelling. Tensorial invariance is ensured subsequently, by combining the responses from microplanes of all orientations within the material.

ACKNOWLEDGEMENTS

Partial support during the writing of this work was obtained under Air Force Office of Scientific Research Grant No. 83–0009 to Northwestern University. The preceding development of the mathematical model was partially supported under US National Science Foundation Grant No. CEE800–9050 to Northwestern University. Mary Hill is thanked for her outstanding secretarial assistance.

REFERENCES

1. G. I. Taylor, 'Plastic strain in metals', *J. Inst. Metals*, **63**, 307–324 (1938).
2. S. B. Batdorf and B. Budiansky, 'A mathematical theory of plasticity based on the concept of slip', *NACA TN*1871, April, 1949.
3. T. H. Lin, *Theory of Inelastic Structures*, Chapter 4, John Wiley, New York, 1968.
4. T. H. Lin, and M. Ito, 'Theoretical plastic stress–strain relationships of a polycrystal', *Int. J. of Engng. Sci.*, **4**, 543–561 (1966).
5. T. H. Lin and M. Ito, 'Theoretical plastic distortion of a polytcrystalline aggregate under combined and reversed stress', *J. Mech. Phys. Solids*, **13**, 103–115 (1965).
6. J. R. Rice, 'Continuum plasticity in relation to microscale deformation mechanisms', in *Metallurgical effects at High Strain Rates* (Eds. R. W. Rohde *et al.*), Plenum Publishing Corp., New York, 1982.
7. O. C. Zienkiewicz and G. N. Pande, 'Time-dependent multi-laminate model of rocks—a numerical study of deformation and failure of rock masses', *Int. Journal of Numerical and Analytical Method in Geomechanics*, **1**, 219–247 (1977).
8. G. N. Pande and K. G. Sharma, 'Multi-laminate model of clays—a numerical evaluation of the influence of rotation of the principal stress axes', Report, Department of Civil Engineering, University College of Swansea, UK, 1982; see also *Proceedings, Symposium on 'Implementation of Computer Procedures and Stress–Strain Laws in Geotechnical Engineering*' (Eds. C. S. Desai and S. K. Saxena), pp. 575–590, held in Chicago, Aug. 1981, Acorn Press, Durham, NC, 1981.
9. G. N. Pande and W. Xiong, 'An improved multi-laminate model of jointed rock masses', *Proceedings, International Symposium on Numerical Model on Geomechanics* (Ed. by R. Dungar, G. N. Pande, and G. A. Studer), pp. 218–226, held in Zurich, Sept. 1982, Balkema, Rotterdam, 1982.
10. W. Wittke, 'New design concept for underground openings in rocks', Chapter 13 in *Finite Elements in Geomechanics* (Ed. G. Gudehus), John Wiley, 1977 (see also *Erzmetall*, **26**, 2, 66–74 (1973)).
11. Z. P. Bažant and B. H. Oh, 'Model of Weak Planes for progressive Fracture of Concrete and Rock', *Report No.* 83–2/428*m*, Center for Concrete and Geomaterials, Northwestern University, Evanston, Illinois, Feb. 1983.
12. Z. P. Bažant and B. H. Oh, 'Microplane model for fracture analysis of concrete structures', *Proc. Symposium on the Interaction of Non-nuclear Munitions with Structures*', held at US Air Force Academy, Colorado Springs, May 1983 (publ. by McGregor & Werner, Washington DC).
13. J. W. Dougill, 'On stable progressively fracturing solids', *ZAMP* (*Zeitschrift für Angewandte Mathematic und Physik*), Vol. 27, Fasc. 4, 423–437 (1976).
14. J. W. Dougill, 'Some remarks on path independence in the small in plasticity', *Quarterly of Appl. Math.*, **32**, 233–243 (1975).

15. Z. P. Bazant and S. S. Kim, 'Plastic-fracturing theory for concrete', *J. of the Engng. Mech Div., Proc. Am. Soc. of Civil Engrs.*, **105**, June 1979, pp. 407–428, with Errata in Vol. 106 (also as Preprint 3431, ASCE Annual Convention, Chicago, Oct. 1978).
16. W. D. Iwan and P. J. Yoder, 'Computational aspects of strain-space plasticity', *J. of Engineering Mechanics, Proc. ASCE*, **109**, 1, 31–243 (1983).
17. P. J. Yoder and W. D. Iwan, 'On the formulation of strain-space plasticity with multiple loading surfaces', *J. of Applied Mechanics, Trans. ASME*, **48**, 773–778 (1981).
18. P. M. Naghdi, 'Some constitutive restrictions in plasticity', in 'constitutive Equations in Viscoplasticity' in *Computational and Engineering Aspects, AMD, Am. Soc. of Mech. Engineers*, **20**, 79–93 (1976).
19. P. M. Naghdi and J. A. Trapp, 'The significance of formulating plasticity theory with reference to loading surfaces in strain space', *Intern. J. of Engng. Science*, **13**, 785–797 (1975).
20. J. Casey and P. M. Naghdi, 'On the characterization of strain-hardening plasticity', *J. of Applied Mechanics, Trans. ASME*, **48**, 285–296 (1981).
21. R. H. Evans and M. S. Marathe, Microcracking and stress–strain curves for concrete in tension', *Materials and Structures (Paris)*, No. 1, 61–64 (1968).
22. J. Albrecht and L. Collatz, 'Zur numerischen Auswertung mehrdimensionaler Integrale', *Zeitschrift Für Angewandte Mathematik und Mechanik*, Band 38, Heft 1/2, Jan./Feb., pp. 1–15.
23. Z. P. Bažant and B. H. Oh, 'Efficient numerical integration on the surface of a sphere', *Report*, Center for Concrete and Geomaterials, Northwestern University, Evanston, I11., 1982.
24. Z. P. Bažant, K. Ozaydin and R. J. Krizek, 'Micromechanics model for creep of anisotropic clays', *Journal of the Engineering mechanics Division, ASCE*, **101**, 57–78 (1975).
25. C. R. Calladine, 'A microstructural view of the mechanical properties of saturated clay', *Geotechnique*, **21**, 391–415 (1971).
26. D. N. Wood, 'Exploration of principal stress space with kaolin in a triaxial apparatus', *Geotechnique*, **25**, 783–797 (1975).
27. J. L. Sanders, 'Plastic stress–strain relations based on linear loading function', *Proceedings, 2nd US National Congress on Applied Mechanics, ASME*, 455–460 (1955).
28. J. F. W. Bishop and R. Hill, 'A theory of plastic distortion of polycrystalline aggregate under combined stress', *Philosophical Magazine*, **42**, 327, 414–427 (1951) (see also continuation on pp. 1298–1307).
29. A. J. M. Spencer, 'A theory of the kinematics of ideal soils under plane strain conditions', *J. Mech. Phys. Solids*, **12**, 337–351.

Mechanics of Engineering Materials
Edited by C. S. Desai and R. H. Gallagher

Chapter 4

Strain-Hardening Response of Elastic–Plastic Materials

J. Casey and P. M. Naghdi

4.1 INTRODUCTION

It is almost two decades since Green and Naghdi [1, 2] presented their rate-type theory of finitely deforming elastic–plastic materials. A number of advances in the foundations of this theory have occurred during the intervening period. For example, restrictions on constitutive functions have been obtained which have clarified the issues of normality of plastic strain rate and convexity of yield surfaces [3, 4]. More importantly, the necessity of a strain-space formulation of plasticity has been recognized [5, 6]. The present paper is an expository account of the current status of the purely mechanical form of the theory originally proposed in [1, 2], with particular emphasis on developments pertaining to strain-hardening behaviour.

By way of background, we recall that the theory in [1, 2] is developed within a general thermodynamical framework. The development in [3] incorporates in a purely mechanical setting those parts of [1, 2], specialized to the isothermal case, that are independent of the Clausius–Duhem entropy production inequality. It contains, in addition, a physically plausible assumption regarding the non-negativity of work in a closed cycle of homogeneous deformation. This assumption implies certain restrictions on the constitutive equations of elastic–plastic materials [3, 4].

We recall that the traditional formulation of plasticity theory employs yield surfaces in stress space, together with loading criteria which involve the time rate of stress. The theory of [1, 2] was therefore formulated in a stress (as well as temperature) space setting. However, as pointed out by Naghdi and Trapp [5], the stress-space formulation of plasticity necessarily leads to unreliable results in any region such as that corresponding to the maximum point of the engineering stress versus engineeering strain curve for uniaxial tension of a typical ductile metal. After also observing that the stress-space formulation does not reduce directly to the theory of elastic–perfectly plastic materials, and that a separate formulation for the latter is required, Naghdi and Trapp [5] proposed an alternative strain-space formulation of plasticity which (a) is valid for the full range of elastic–plastic deformation; and (b) includes, as a special case, the theory of elastic–perfectly plastic materials.

The strain-space formulation was further elaborated upon in [6], which also contains a summary of [3, 4]. A thermodynamical theory, developed in a strain-temperature space setting, is presented in [7].

Once a strain-space formulation is adopted, stress appears as a dependent variable, and it is conceivable that certain conditions in stress space might be induced by the conditions that are assumed in strain space. It was shown by Casey and Naghdi [8] that this is indeed the case. Specifically, the loading conditions in stress space are determined by those in strain space through the constitutive equations of the theory. However, the conditions induced in stress space are not identical to those of the strain-space formulation, nor do they imply the loading conditions of the strain-space formulation. The stress-space and strain-space formulations of plasticity are therefore not equivalent [8, 9].

The different types of loading conditions that can exist in stress space in conjunction with a postulated condition of loading in strain space suggest a natural classification of strain-hardening behaviour into three distinct types: hardening, softening, and perfectly plastic [8]. Geometrically, the yield surface in strain space is always moving outwards during loading, whereas the yield surface in stress space may concurrently be moving outwards, inwards, or may be stationary depending on whether the material is hardening, softening, or exhibiting perfectly plastic behaviour.

It is instructive to discuss the foregoing conditions with reference to the engineering stress versus engineering strain diagram for uniaxial tension of a typical ductile metal (see, for example, Fig. 1 of [8]). Corresponding to a given value of plastic strain, the yield surface in strain space is represented by a point on the (horizontal) strain axis, while the yield surface in stress space is represented by the associated value of stress on the vertical axis. Outward motion of the yield surface in stress space concurrent with outward motion of the yield surface in strain space corresponds to a rising stress–strain curve: this is hardening behaviour. Similarly, softening behaviour corresponds to a falling stress–strain curve, and perfectly plastic behaviour to a flat stress–strain curve.

Algebraically, the three types of strain-hardening behaviour of an elastic–plastic material may be defined [8] in terms of a rate-independent quotient $\hat{f}/\hat{g}$, where $\hat{f}$ derives from the loading function f in stress space, and $\hat{g}$ from the loading function g in strain space. Roughly speaking, $\hat{f}/\hat{g}$ measures the ratio of the outward velocities with which the yield surfaces in stress space and strain space are moving during loading.

The definition given in [8] for hardening, softening and perfectly plastic behaviour presupposes a condition of loading (i.e., $g=0$, $\hat{g}>0$) from an elastic–plastic state. It is possible to phrase an alternative definition in which this restriction is removed [10]. The new definition is actually simpler than that of [8], but not as intuitively obvious. It reduces to the previous definition during loading. Its main advantage is that in regions of hardening behaviour, it enables us to prove that the loading conditions of strain space and stress space imply one another. Of course, for regions of softening or perfectly plastic behaviour, no such equivalence can be established.

The work assumption of Naghdi and Trapp [3,4] which has already been mentioned, implies that a certain second-order tensor involving response functions is parallel to the normal to the yield surface in strain space. This result in turn leads to a geometrically revealing expression for the function Φ (see equation (4.39)).

We now summarize the contents of the paper. A strain-space formulation of plasticity is summarized in Section 4.2 and the induced conditions in stress space are outlined in Section 4.3. Section 4.4 is concerned with the consequences of the work assumption of Naghdi and Trapp [3,4]. Finally, in order to demonstrate the predictive capabilities of the theory, special constitutive equations are utilized in Section 4.5 to discuss a number of experimentally observed phenomena.

4.2 THE ELASTIC–PLASTIC CONTINUUM. STRAIN–SPACE FORMULATION

Let X be any particle in a deformable body $\mathscr{B}$. Choose a fixed origin in a three-dimensional Euclidean space, and denote by $\mathbf{X}$ the position vector of X in a fixed reference configuration of $\mathscr{B}$. The position vector $\mathbf{x}$ of X in the present configuration of $\mathscr{B}$ at time t is then given by the motion $\boldsymbol{\chi}$ so that

$$\mathbf{x} = \boldsymbol{\chi}(\mathbf{X}, t). \tag{4.1}$$

The deformation gradient relative to the reference position and its determinant are

$$\mathbf{F} = \frac{\partial \boldsymbol{\chi}}{\partial \mathbf{X}}(\mathbf{X}, t), \qquad J = \det \mathbf{F} > 0. \tag{4.2}$$

Also, the (relative) Lagrangian strain tensor $\mathbf{E}$ is defined by

$$\mathbf{E} = \tfrac{1}{2}(\mathbf{F}^{\mathrm{T}}\mathbf{F} - \mathbf{I}), \tag{4.3}$$

where a superscript T denotes transpose and $\mathbf{I}$ is the identity tensor. Geometrically, we regard $\mathbf{E}$ as a point in six-dimensional Euclidean *strain space*. Similarly, we regard the symmetric Piola–Kirchhoff stress tensor $\mathbf{S}$ as a point in six-dimensional *stress space*.

We now summarize the main elements of a purely mechanical rate-type theory of a finitely deforming elastic–plastic solid. Our treatment is based on the papers of Green and Naghdi [1,2], Naghdi and Trapp [5], Naghdi [6], and Casey and Naghdi [8]. In addition to the strain tensor $\mathbf{E}$, we assume the existence of a symmetric second-order tensor $\mathbf{E}^p$, called the plastic strain at $(\mathbf{X}, t)$ and a scalar κ, called a measure of work-hardening at $(\mathbf{X}, t)$. Adopting the abbreviated notation

$$\mathscr{U} = (\mathbf{E}, \mathbf{E}^p, \kappa), \tag{4.4}$$

we assume that the stress $\mathbf{S}$ is given by a constitutive equation of the form

$$\mathbf{S} = \hat{\mathbf{S}}(\mathscr{U}), \tag{4.5}$$

and that for fixed values of $(\mathbf{E}^p, \kappa)$, $\hat{\mathbf{S}}$ may be inverted to give

$$\mathbf{E} = \hat{\mathbf{E}}(\mathscr{V}), \tag{4.6}$$

where

$$\mathscr{V} = (\mathbf{S}, \mathbf{E}^p, \kappa). \tag{4.7}$$

The response functions $\hat{\mathbf{S}}$ and $\hat{\mathbf{E}}$ are taken to be smooth.

We admit the existence of a smooth scalar-valued yield (or loading) function g such that for fixed values of $(\mathbf{E}^p, \kappa)$ the equation

$$g(\mathscr{U}) = 0 \tag{4.8}$$

represents a closed orientable hypersurface $\partial\mathscr{E}$ of dimension five, enclosing an open region $\mathscr{E}$ in strain space. The function g is chosen such that $g(\mathscr{U}) < 0$ for all points in $\mathscr{E}$. The hypersurface $\partial\mathscr{E}$ is called the yield (or loading) surface in strain space. Corresponding to a motion (4.1), we may associate with each particle of the body $\mathscr{B}$ a smooth oriented curve C_e which lies in strain space and is parametrized by time; the curve C_e will be referred to as a strain trajectory. The strain trajectories are restricted to lie initially in $\mathscr{E}$ or on $\partial\mathscr{E}$ (i.e., $g(\mathscr{U}) \leq 0$ initially for all C_e).

The (six-dimensional) tangent vector at a point of C_e is $\dot{\mathbf{E}}$, where a superposed dot signifies material time differentiation. Writing the inner product of any two tensors $\mathbf{A}$ and $\mathbf{B}$ as $\mathbf{A}\cdot\mathbf{B} = \text{trace } \mathbf{A}^T\mathbf{B}$ we define $\hat{g}$ as

$$\hat{g} = \frac{\partial g}{\partial \mathbf{E}} \cdot \dot{\mathbf{E}}, \tag{4.9}$$

where $\partial g/\partial\mathbf{E}$ stands for the symmetric form $\frac{1}{2}\{\partial g/\partial\mathbf{E} + (\partial g/\partial\mathbf{E})^T\}$. When $g = 0$ and $\partial g/\partial\mathbf{E} \neq \mathbf{0}$, $\hat{g}$ is the inner product of the tangent vector to C_e and the outward normal to the yield surface $\partial\mathscr{E}$.

Adopting a strain-space formulation as primary, we may express the constitutive equations for the rate of plastic strain and the rate of work-hardening as [5]:

$$\dot{\mathbf{E}}^p = \begin{cases} \mathbf{0} & \text{if } g < 0, \quad \text{(a)} \\ \mathbf{0} & \text{if } g = 0 \text{ and } \hat{g} < 0, \quad \text{(b)} \\ \mathbf{0} & \text{if } g = 0 \text{ and } \hat{g} = 0, \quad \text{(c)} \\ \lambda\boldsymbol{\rho}\hat{g} & \text{if } g = 0 \text{ and } \hat{g} > 0, \quad \text{(d)} \end{cases} \tag{4.10}$$

and

$$\dot{\kappa} = \mathscr{C} \cdot \dot{\mathbf{E}}^{\mathrm{p}}, \tag{4.11}$$

where λ is a scalar-valued function of $\mathscr{U}$ (or equivalently of $\mathscr{V}$), and $\boldsymbol{\rho}$ and $\mathscr{C}$ are symmetric tensor-valued functions of $\mathscr{U}$. The conditions involving g and $\hat{g}$ in (4.10) are the loading criteria of the strain-space formulation. These conditions correspond, respectively, to (a) an elastic state; (b) unloading from an elastic–plastic state; (c) neutral loading from an elastic–plastic state; and (d) loading from an elastic–plastic state. If we write the material derivative of g as

$$\dot{g} = \hat{g} + \frac{\partial g}{\partial \mathbf{E}^{\mathrm{p}}} \cdot \dot{\mathbf{E}}^{\mathrm{p}} + \frac{\partial g}{\partial \kappa} \dot{\kappa}, \tag{4.12}$$

then it can be readily deduced that the following geometrical interpretation holds [8]: In an elastic state the strain trajectory lies in $\mathscr{E}$—which is therefore referred to as the elastic region in strain space—and the yield surface $\partial\mathscr{E}$ remains stationary. During unloading, C_{e} intersects $\partial\mathscr{E}$ and points into the elastic region, while $\partial\mathscr{E}$ remains stationary. During neutral loading, C_{e} continues to lie in $\partial\mathscr{E}$ which is again stationary. Finally, during loading C_{e} intersects $\partial\mathscr{E}$ and is pointing outwards. The constitutive equations (4.10) and (4.11) are of the *rate-type* since they contain time derivatives of $\mathbf{E}, \mathbf{E}^{\mathrm{p}}$, and κ. However, the values of $\dot{\mathbf{E}}^{\mathrm{p}}$ and $\dot{\kappa}$ are independent of the time scale used to calculate the rates, because the equations are linear in the rates. Furthermore, neither the yield function g, nor the stress response $\hat{\mathbf{S}}$ in (4.5) depend on rates. For these reasons, the theory under discussion is also said to be *rate-independent*. It is intended to apply to

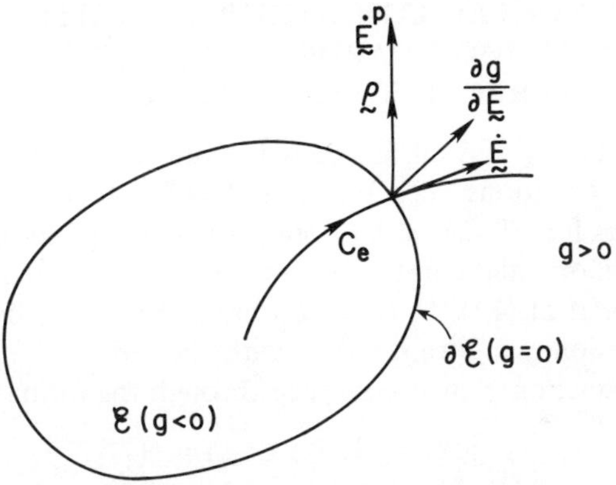

Figure 4.1 A sketch of the yield surface $\partial\mathscr{E}$ and the elastic region $\mathscr{E}$ in six-dimensional strain space. Also shown are a strain trajectory C_{e}, its tangent vector $\dot{\mathbf{E}}$, the normal vector $\partial g/\partial\mathbf{E}$ to $\partial\mathscr{E}$, as well as vectors representing $\boldsymbol{\rho}$ and the rate of plastic strain $\dot{\mathbf{E}}^{\mathrm{p}}$. During loading, the yield surface $\partial\mathscr{E}$ is pushed outwards by the strain trajectory C_{e}

inviscid media in which time effects such as strain rate sensitivity, creep, and relaxation may be ignored.

It is stipulated that during loading $\partial\mathscr{E}$ is pushed outwards by C_e, so that g remains equal to zero, and hence $\dot{g}=0$ during loading.[†] A further implication of this condition is that $\dot{\mathbf{E}}^p \neq \mathbf{0}$ during loading, for otherwise (4.12) would be violated. The coefficient of $\hat{g}$ in (4.10d) is consequently non-zero and without loss in generality we may take $\lambda > 0$.

It is clear from the foregoing that positive values of the function g can never be attained and that the strain trajectories always remain in the elastic region $\mathscr{E}$ or on the yield surface $\partial\mathscr{E}$. Figure 4.1 illustrates various quantities associated with the strain-space formulation given above.

From (4.12), (4.10d) and the fact that $\dot{g}=0$ during loading, it follows that at all points[§] on $\partial\mathscr{E}$

$$1+\lambda\boldsymbol{\rho}\cdot\mathscr{G}=0\,, \tag{4.13}$$

where for brevity we have set

$$\mathscr{G}=\frac{\partial g}{\partial \mathbf{E}^p}+\frac{\partial g}{\partial \kappa}\mathscr{C}\,. \tag{4.14}$$

Thus, it is unnecessary to provide a constitutive equation for λ, once constitutive equations are given for $g, \boldsymbol{\rho}$, and $\mathscr{C}$. It should be noted that the constitutive equations (4.5) for the stress response, or its inverse (4.6), hold during all four types of loading conditions in (4.10).

4.3 INDUCED CONDITIONS IN STRESS SPACE. DEFINITION OF HARDENING, SOFTENING, AND PERFECTLY PLASTIC BEHAVIOUR

For a given motion $\boldsymbol{\chi}$ and an associated strain trajectory C_e, we may utilize the constitutive equations (4.5), (4.10), and (4.11), together with appropriate initial conditions for $\mathbf{E}^p$ and κ, to obtain the corresponding stress trajectory C_s, a continuous oriented curve in stress space. In a similar fashion, with the use of (4.6) instead of (4.5), C_e may be obtained from C_s. Furthermore, for a given yield function g in strain space, with the aid of (4.6), we can obtain a corresponding function f in stress space through the formula

$$g(\mathscr{U})=g(\hat{\mathbf{E}}(\mathscr{V}),\mathbf{E}^p,\kappa)=f(\mathscr{V})\,. \tag{4.15}$$

Conversely, (4.5) may be used to obtain g from f. For fixed values of $(\mathbf{E}^p,\kappa)$,

[†] This is the so-called 'consistency' condition: loading from an elastic–plastic state leads to another elastic–plastic state.

[§] It is assumed here that loading can be effected at any point of $\partial\mathscr{E}$. The condition (4.13) must then hold at all points which the yield surface can traverse.

the equation

$$f(\mathscr{V}) = 0 \tag{4.16}$$

represents a hypersurface $\partial\mathscr{S}$ in stress space having the same geometrical properties as the yield surface $\partial\mathscr{E}$ in strain space. The open region enclosed by $\partial\mathscr{S}$ is denoted by $\mathscr{S}$. It is clear from the identity (4.15) that a point in strain space belongs to the elastic region $\mathscr{E}$ if and only if the corresponding point in stress space belongs to $\mathscr{S}$. Likewise, a point in strain space lies on the yield surface $\partial\mathscr{E}$ if and only if the corresponding point in stress space lies on $\partial\mathscr{S}$. In view of these properties, we refer to f as the yield (or loading) function in stress space, to $\partial\mathscr{S}$ as the yield (or loading) surface in stress space, and to $\partial\mathscr{S}$ as the elastic region in stress space. Also, as shown in Figure 4.2, we may draw an illustration representing the various quantities in stress space, which is similar to that of Figure 4.1.

Taking the material time derivative of (4.15), we find that

$$\dot{g} = \dot{f} = \hat{f} + \frac{\partial f}{\partial \mathbf{E}^{\mathrm{p}}} \cdot \dot{\mathbf{E}}^{\mathrm{p}} + \frac{\partial f}{\partial \kappa}\dot{\kappa}, \tag{4.17}$$

where

$$\hat{f} = \frac{\partial f}{\partial \mathbf{S}} \cdot \dot{\mathbf{S}} \tag{4.18}$$

is the inner product of $\partial f/\partial\mathbf{S}$ and the tangent vector to the stress trajectory C_s.

It follows at once from (4.10a, b, c), (4.11), (4.15), and (4.17) that for any particle

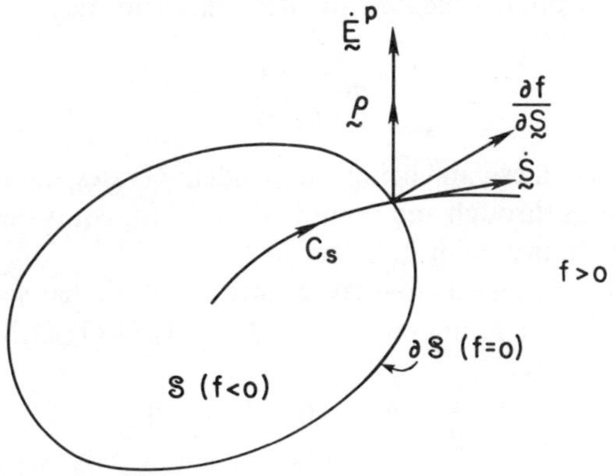

Figure 4.2 A sketch of the yield surface $\partial\mathscr{S}$ and the elastic region $\mathscr{S}$ in six-dimensional stress space. Also shown are a stress trajectory C_s, its tangent vector $\dot{\mathbf{S}}$, the normal vector $\partial f/\partial\mathbf{S}$ to $\partial\mathscr{S}$, as well as vectors representing $\boldsymbol{\rho}$ and the rate of plastic strain $\dot{\mathbf{E}}^{\mathrm{p}}$

X of the elastic–plastic body $\mathscr{B}$:

(a) if X is in an elastic state, then

$$f < 0, \tag{4.19a}$$

the stress trajectory C_s lies in the elastic region $\mathscr{S}$ in stress space and the yield surface $\partial\mathscr{S}$ is stationary;

(b) if X is experiencing unloading from an elastic–plastic state, then

$$f = 0, \qquad \hat{f} < 0, \tag{4.19b}$$

the stress trajectory C_s is intersecting the stationary yield surface $\partial\mathscr{S}$ and is pointing into the elastic region $\mathscr{S}$; and

(c) if X is experiencing neutral loading from an elastic–plastic state, then

$$f = 0, \qquad \hat{f} = 0, \tag{4.19c}$$

while C_s continues to lie in the stationary yield surface $\partial\mathscr{S}$. Thus, the loading criteria in (4.10a, b, c) of the strain-space formulation imply (4.19a, b, c). We refer to the latter type of conditions as loading conditions in stress space.

Preparatory to discussing the case of loading from an elastic–plastic state, we observe that at each value of $\mathscr{U}$ or $\mathscr{V}$ we may define a function Φ by

$$\Phi = -\lambda\boldsymbol{\rho}\cdot\boldsymbol{\mathscr{F}} \tag{4.20}$$

where

$$\boldsymbol{\mathscr{F}} = \frac{\partial f}{\partial \mathbf{E}^{\mathrm{p}}} + \frac{\partial f}{\partial \kappa}\boldsymbol{\mathscr{C}} \tag{4.21}$$

is the analogue in stress space to the quantity $\mathscr{G}$ in (4.14). From (4.20) and (4.13) it follows that at any point which $\partial\mathscr{E}$ (or $\partial\mathscr{S}$) may traverse,

$$\Phi = \frac{\boldsymbol{\rho}\cdot\boldsymbol{\mathscr{F}}}{\boldsymbol{\rho}\cdot\boldsymbol{\mathscr{G}}}\,. \tag{4.22}$$

Clearly, Φ is dimensionless, and being independent of rates, has the same value for all strain trajectories through any point in $\mathscr{E}$ or $\partial\mathscr{E}$ (or equivalently, for all stress trajectories through any point in $\mathscr{S}$ or $\partial\mathscr{S}$).

During loading from an elastic–plastic state, we recall that $\dot{g} = 0$ and hence, by (4.15), $\dot{f} = 0$ also. It then follows from (4.10d), (4.11), (4.17), (4.20), and (4.21) that

$$\frac{\hat{f}}{\hat{g}} = \Phi, \qquad (g = 0, \hat{g} > 0). \tag{4.23}$$

Thus, during loading from an elastic–plastic state, the quotient $\hat{f}/\hat{g}$ has the same value as the function Φ.

Now, Φ may be positive, negative, or zero; and, in order to distinguish between these cases we adopt the definition: a particle of an elastic–plastic material exhibits

$$\begin{cases} \text{hardening behaviour if and only if } \Phi > 0, & \text{(a)} \\ \text{softening behaviour if and only if } \Phi < 0, & \text{(b)} \\ \text{perfectly plastic behaviour if and only if } \Phi = 0. & \text{(c)} \end{cases} \quad (4.24)$$

In view of (4.23), during loading from an elastic–plastic state, a particle of an elastic–plastic material exhibits

$$\begin{cases} \text{hardening behaviour if and only if } \hat{f}/\hat{g} > 0, & \text{(a)} \\ \text{softening behaviour if and only if } \hat{f}/\hat{g} < 0, & \text{(b)} \\ \text{perfectly plastic behaviour if and only if } \hat{f}/\hat{g} = 0. & \text{(c)} \end{cases} \quad (4.25)$$

In [8], the definition (4.25) was used and the definition (4.24) was proposed in [10]. Definition (4.24) has the advantage of not presupposing a condition of loading.

It is clear from (4.24) and (4.23) that if a particle of an elastic–plastic material is exhibiting

$$\begin{cases} \text{hardening behaviour,} & \text{(a)} \\ \text{softening behaviour,} & \text{(b)} \\ \text{perfectly plastic behaviour,} & \text{(c)} \end{cases} \quad (4.26)$$

then $g = 0$ and $\hat{g} > 0$ imply that

$$\begin{cases} f = 0 \quad \text{and} \quad \hat{f} > 0, & \text{(a)} \\ f = 0 \quad \text{and} \quad \hat{f} < 0, & \text{(b)} \\ f = 0 \quad \text{and} \quad \hat{f} = 0, & \text{(c)} \end{cases} \quad (4.27)$$

respectively, and the following geometrical interpretation can be immediately established: During loading—when the yield surface $\partial\mathscr{E}$ in strain space is moving outwards locally—concurrently, the yield surface $\partial\mathscr{S}$ in stress space moves outwards in the case of hardening behaviour, inwards in the case of softening behaviour and is stationary in the case of perfectly plastic behaviour.

Consider next the outward normal velocities v_e and v_s of the yield surfaces $\partial\mathscr{E}$ and $\partial\mathscr{S}$, respectively, during loading. These are given by

$$v_e = \hat{g}/\|\partial g/\partial \mathbf{E}\|, \qquad v_s = \hat{f}/\|\partial f/\partial \mathbf{S}\|, \quad (4.28)$$

where for any second order tensor $\mathbf{A}, \|\mathbf{A}\| = (\mathbf{A}\cdot\mathbf{A})^{1/2}$ is the magnitude of $\mathbf{A}$. Therefore, as noted in [9],

$$\frac{\hat{f}}{\hat{g}} = \frac{v_s}{v_e} R, \qquad R = \|\partial f/\partial \mathbf{S}\|/\|\partial g/\partial \mathbf{E}\|. \quad (4.29)$$

In addition, let $t_1 \leq \tau \leq t_2$ be any time interval during which only loading from an elastic–plastic state occurs. The displacements of the yield surfaces along their normals may be calculated from the expressions

$$u_e = \int_{t_1}^{t_2} v_e \, d\tau, \qquad u_s = \int_{t_1}^{t_2} v_s \, d\tau . \tag{4.30}$$

In view of (4.29) and (4.30)

$$\frac{\hat{f}}{\hat{g}} = \frac{\dot{u}_s}{\dot{u}_e} R = \frac{du_s}{du_e} R. \tag{4.31}$$

It is clear from (4.10), (4.19), (4.26), and (4.27) that the stress space and strain space formulations of plasticity are not equivalent.† Indeed, the conditions $f = 0$ and $\hat{f} \leq 0$ correspond both to unloading and softening during loading, while $f = 0$ and $\hat{f} = 0$ correspond both to neutral loading and perfectly plastic behaviour during loading. However, for hardening behaviour, we have the following result [10]: If a particle of an elastic–plastic body is exhibiting hardening behaviour, then

$$\begin{cases} f = 0 \quad \text{and} \quad \hat{f} < 0, & \text{(a)} \\ f = 0 \quad \text{and} \quad \hat{f} = 0, & \text{(b)} \\ f = 0 \quad \text{and} \quad \hat{f} > 0, & \text{(c)} \end{cases} \tag{4.32}$$

imply that

$$\begin{cases} g = 0 \quad \text{and} \quad \hat{g} < 0, & \text{(a)} \\ g = 0 \quad \text{and} \quad \hat{g} = 0, & \text{(b)} \\ g = 0 \quad \text{and} \quad \hat{g} > 0, & \text{(c)} \end{cases} \tag{4.33}$$

respectively. The proof is immediate: conditions (4.33a, b) always imply (4.32a, b), respectively. In addition, for hardening behaviour, (4.33c) implies (4.32c). Suppose now that (4.32a) holds. Then $g = 0$ by (4.15). The quantity $\hat{g}$ is either positive, zero, or negative. If $\hat{g} > 0$, then (4.33c) implies (4.32c), which leads to a contradiction. Likewise, $\hat{g}$ cannot be zero and we conclude that $\hat{g} < 0$. Thus (4.32a) implies (4.33a). Similarly, (4.32b, c) imply (4.33b, c), respectively.

Thus, for hardening behaviour ($\Phi > 0$), the loading conditions of the stress-space and strain-space formulations imply one another. Of course, for softening and perfectly plastic behaviour, the loading conditions of the two formulations do not imply one another. A summary of the relationships between the two sets of conditions is provided in Table 4.1.

With the use of (4.25a, b), we can obtain an expression for the rate of plastic strain which involves $\hat{f}$ rather than $\hat{g}$, but which is valid only for hardening and softening behaviour. Thus, in view of (4.10d), (4.23), (4.25a, b), and (4.20),

$$\dot{\mathbf{E}}^p = \lambda \frac{\hat{f}}{\hat{f}/\hat{g}} \boldsymbol{\rho} = -\frac{\boldsymbol{\rho}\hat{f}}{\boldsymbol{\rho}\cdot\boldsymbol{\mathscr{F}}} \neq \mathbf{0} \tag{4.34}$$

† For a further discussion of this point, see [9].

Table 4. Summary of the relationship between the loading criteria in strain space and the associated loading conditions in stress space for the cases of hardening, softening, and perfectly plastic behaviour.

(a) Hardening behaviour ($\Phi > 0$)	
Elastic	$g < 0 \Leftrightarrow f < 0$
Unloading	$g = 0,\ \hat{g} < 0 \Leftrightarrow f = 0,\ \hat{f} < 0$
Neutral loading	$g = 0,\ \hat{g} = 0 \Leftrightarrow f = 0,\ \hat{f} = 0$
Loading	$g = 0,\ \hat{g} > 0 \Leftrightarrow f = 0,\ \hat{f} > 0$
(b) Softening behaviour ($\Phi < 0$)	
Elastic	$g < 0 \Leftrightarrow f < 0$
Unloading	$g = 0,\ \hat{g} < 0 \Rightarrow f = 0,\ \hat{f} < 0$
Neutral loading	$g = 0,\ \hat{g} = 0 \Leftrightarrow f = 0,\ \hat{f} = 0$
Loading	$g = 0,\ \hat{g} > 0 \Rightarrow f = 0,\ \hat{f} < 0$
(c) Perfectly plastic behaviour ($\Phi = 0$)	
Elastic	$g < 0 \Leftrightarrow f < 0$
Unloading	$g = 0,\ \hat{g} < 0 \Leftrightarrow f = 0,\ \hat{f} < 0$
Neutral loading	$g = 0,\ \hat{g} = 0 \Rightarrow f = 0,\ \hat{f} = 0$
Loading	$g = 0,\ \hat{g} > 0 \Rightarrow f = 0,\ \hat{f} = 0$

for both hardening and softening behaviour. For perfectly plastic behaviour $\dot{\mathbf{E}}^{\mathrm{p}}$ cannot be expressed in the form (4.34) because of (4.25c).

We close this section by recording some further expressions for the quantity Φ. Thus, from (4.20) and (4.13), we find that

$$\begin{aligned}\Phi &= 1 + \lambda\boldsymbol{\rho}\cdot\{\mathscr{G} - \mathscr{F}\}\\ &= 1 + \lambda\boldsymbol{\rho}\cdot\left\{\left(\frac{\partial g}{\partial \mathbf{E}^{\mathrm{p}}} - \frac{\partial f}{\partial \mathbf{E}^{\mathrm{p}}}\right) + \left(\frac{\partial g}{\partial \kappa} - \frac{\partial f}{\partial \kappa}\right)\mathscr{C}\right\}.\end{aligned} \tag{4.35}$$

With the aid of (4.15) and the chain rule of differentiation, $(4.35)_2$ may be written as

$$\Phi = 1 + \lambda\boldsymbol{\sigma}\cdot\frac{\partial f}{\partial \mathbf{S}}, \tag{4.36}$$

where the symmetric second order tensor $\boldsymbol{\sigma}$ is defined by

$$\boldsymbol{\sigma} = \left\{\frac{\partial \hat{\mathbf{S}}}{\partial \mathbf{E}^{\mathrm{p}}} + \frac{\partial \hat{\mathbf{S}}}{\partial \kappa} \otimes \mathscr{C}\right\}[\boldsymbol{\rho}] \tag{4.37a}$$

and the symbol $\otimes$ denotes tensor product. Also, the product between the fourth order tensors and $\boldsymbol{\rho}$ is such that relative to a basis $\mathbf{e}_M \otimes \mathbf{e}_N$ of tensors constructed from a fixed orthonormal basis $\mathbf{e}_M$ of vectors, the tensor $\boldsymbol{\sigma}$ has a component representation

$$\sigma_{MN} = \left\{ \frac{\partial \hat{S}_{MN}}{\partial E^{p}_{KL}} + \frac{\partial \hat{S}_{MN}}{\partial \kappa} \mathscr{C}_{KL} \right\} \rho_{KL}, \tag{4.37b}$$

the usual summation convention being employed.

4.4 CONSEQUENCES OF A PHYSICALLY PLAUSIBLE WORK ASSUMPTION

With particular reference to elastic–plastic materials, the work assumption of Naghdi and Trapp [3] may be stated as: *the external work done on any elastic–plastic body in any smooth homogeneous cycle of deformation is non-negative.* From this postulate it follows that the tensor $\boldsymbol{\sigma}$ in (4.37) is parallel to the normal to the yield surface $\partial\mathscr{E}$ in strain space and points into the elastic region $\mathscr{E}$ [3, 8]:

$$\boldsymbol{\sigma} = -\gamma^* \frac{\partial g}{\partial \mathbf{E}}, \qquad \gamma^* \geq 0, \qquad (g = 0), \tag{4.38}$$

where the scalar function γ^* depends only on $\mathscr{U}$ (or $\mathscr{V}$). We emphasize that (4.38) holds even for a motion that is not homogeneous [3].

In view of (4.38), (4.36) becomes

$$\Phi = 1 - \lambda\gamma^*\Lambda, \qquad (g = 0), \tag{4.39}$$

where

$$\Lambda = \frac{\partial g}{\partial \mathbf{E}} \cdot \frac{\partial f}{\partial \mathbf{S}} \tag{4.40}$$

is the inner product of the normal to the yield surface in strain space and the normal to the yield surface in stress space.

For certain purposes, it is convenient to express the constitutive equation (4.5) in terms of an equivalent set of kinematical variables in the form

$$\mathbf{S} = \bar{\mathbf{S}}(\mathbf{E} - \mathbf{E}^p, \mathbf{E}^p, \kappa). \tag{4.41}$$

Suppose that the response function $\hat{\mathbf{S}}$ satisfies the symmetry condition

$$\left(\frac{\partial \hat{S}_{MN}}{\partial E_{KL}} - \frac{\partial \hat{S}_{KL}}{\partial E_{MN}} \right) \mathbf{e}_M \otimes \mathbf{e}_N \otimes \mathbf{e}_K \otimes \mathbf{e}_L = \mathbf{0}. \tag{4.42}$$

It then follows from (4.38) that [3, 8]:

$$-\boldsymbol{\rho} + \frac{\partial \hat{\mathbf{E}}}{\partial \mathbf{S}} \left[\left(\frac{\partial \bar{\mathbf{S}}}{\partial \mathbf{E}^p} + \frac{\partial \bar{\mathbf{S}}}{\partial \kappa} \otimes \boldsymbol{\mathscr{C}} \right) [\boldsymbol{\rho}] \right] = -\gamma^* \frac{\partial f}{\partial \mathbf{S}}, \tag{4.43}$$

with $g = f = 0$. If the response function $\bar{\mathbf{S}}$ is independent of its second and third arguments, i.e., if

$$\frac{\partial \bar{\mathbf{S}}}{\partial \mathbf{E}^p} = \mathbf{0}, \qquad \frac{\partial \bar{\mathbf{S}}}{\partial \kappa} = \mathbf{0}, \tag{4.44}$$

then by (4.43)

$$\boldsymbol{\rho} = \gamma^* \frac{\partial f}{\partial \mathbf{S}} \neq 0, \qquad \gamma^* > 0, \qquad (g = f = 0) \tag{4.45}$$

and hence during loading

$$\dot{\mathbf{E}}^p = \lambda \gamma^* \hat{g} \frac{\partial f}{\partial \mathbf{S}} \neq \mathbf{0}, \tag{4.46}$$

where (4.10d) has been used. Thus, the rate of plastic strain is in this case directed along the normal to the yield surface $\partial \mathscr{S}$ in stress space.

When $\boldsymbol{\rho}$ satisfies (4.45), (4.13) can be written as

$$1 + \lambda \gamma^* \frac{\partial f}{\partial \mathbf{S}} \cdot \mathscr{G} = 0 \tag{4.47}$$

which can be used to solve for the product $\lambda\gamma^*$ that occurs in the flow rule (4.46). We note that in the case being discussed, i.e., when the conditions (4.44) hold, it is unnecessary to provide a constitutive equation for $\boldsymbol{\rho}$.

From (4.45) and (4.20), it follows that

$$\Phi = \lambda \gamma^* \Gamma, \qquad (g = f = 0), \tag{4.48}$$

where

$$\Gamma = -\frac{\partial f}{\partial \mathbf{S}} \cdot \mathscr{F} \tag{4.49}$$

Also, in view of (4.39), (4.48), and the inequality in (4.45),

$$0 < \frac{1}{\lambda \gamma^*} = \Gamma + \Lambda, \qquad \Phi = \frac{\Gamma}{\Gamma + \Lambda}, \qquad (g = f = 0). \tag{4.50}$$

We further note that

$$\Gamma \begin{cases} > 0 \text{ for hardening behaviour,} & \text{(a)} \\ < 0 \text{ for softening behaviour,} & \text{(b)} \\ = 0 \text{ for perfectly plastic behaviour,} & \text{(c)} \end{cases} \tag{4.51}$$

where (4.24), (4.48), and the inequality in (4.50) have been used.

Finally, from (4.34), (4.45), and (4.49), it is clear that for hardening behaviour and for softening behaviour

$$\dot{\mathbf{E}}^p = \frac{\hat{f}}{\Gamma} \frac{\partial f}{\partial \mathbf{S}} \neq \mathbf{0}, \qquad (g = f = 0, \hat{g} > 0). \tag{4.52}$$

Correspondingly, for perfectly plastic behaviour it follows from (4.10d), (4.45), $(4.50)_1$, and (4.51c) that

$$\dot{\mathbf{E}}^{\mathrm{p}} = \frac{\hat{g}}{\Lambda}\frac{\partial f}{\partial \mathbf{S}} \neq \mathbf{0}. \tag{4.53}$$

4.5 BEHAVIOUR OF SPECIAL CLASSES OF MATERIALS

In this section, we discuss the strain-hardening behaviour of some special classes of elastic–plastic materials. Our purpose is to illustrate how the basic theory of the previous sections may be utilized in a more concrete setting and to demonstrate its usefulness in relation to some experimentally observed phenomena.

In Subsections 4.5.1 and 4.5.2, we examine two well-known 'hardening rules', namely isotropic and kinematic hardening. For background information and references concerning these rules, the reader is referred to a paper by Naghdi [11] and to a report by Pugh *et al.* [12].

4.5.1 Isotropic hardening

During loading, the yield surface $\partial\mathscr{S}$ in stress space will, in general, continuously change its size, its shape, and its position relative to the origin in stress space. For isotropic hardening, the yield function $f(\mathscr{V})$ in (4.15) has the special form

$$f(\mathscr{V}) = f_1(\mathbf{S}) - \kappa, \qquad (\kappa > 0). \tag{4.54}$$

Let $\mathbf{S}_0$ and κ_0 denote the values of $\mathbf{S}$ and κ at initial yield. Then the initial yield surface $\partial\mathscr{S}_0$ is described by

$$f_1(\mathbf{S}_0) = \kappa_0. \tag{4.55}$$

Next, suppose that a 'line' drawn from the origin O in stress space pierces the initial yield surface $\partial\mathscr{S}_0$ at a point A (with OA not tangent to $\partial\mathscr{S}_0$). Let ζ_0 (>0) be the distance from O to A, measured as the norm (or the magnitude) of the stress at A. Then, for isotropic hardening, OA will pierce the subsequent yield surface at a point B which is a distance ζ from O. The family of yield surfaces may be parametrized by κ or by ζ, either of which determines the current location of the yield surface relative to O. (However, ζ varies with choice of line OA, whereas κ does not.) Writing the relationship between the two parameters as

$$\kappa = \hat{\kappa}\left(\frac{\zeta}{\zeta_0}\right)\kappa_0, \tag{4.56}$$

is isotropic hardening we have the condition

$$f_1\left(\frac{\zeta}{\zeta_0}\mathbf{S}_0\right) = \hat{\kappa}\left(\frac{\zeta}{\zeta_0}\right)\kappa_0, \tag{4.57}$$

i.e., for any $\mathbf{S}_0$ on the initial yield surface the state of stress $\zeta/(\zeta_0)\mathbf{S}_0$ is on the yield surface[†] $\partial\mathscr{S}$. Thus, subsequent yield surfaces are identical in shape to, but may have different sizes than the initial yield surface. The family of yield surfaces has a well-defined centre, which coincides with the origin in stress space. The surfaces need not be symmetric about the origin, although in practice they are usually taken to be so.

From the consistency condition, (4.15), (4.54), and (4.18), it is clear that during loading

$$\hat{f} = \dot{\kappa}. \tag{4.58}$$

Therefore, during any interval of loading, κ is monotonically increasing in the case of hardening behaviour, monotonically decreasing in the case of softening behaviour, and constant in the case of perfectly plastic behaviour. If, as usual, the family of yield surfaces is parametrized in such a way that larger values of κ correspond to yield surfaces that are more distant from the origin in stress space, then $\partial\mathscr{S}$ expands uniformly for hardening, contracts uniformly for softening and is stationary for perfectly plastic behaviour.

It follows from (4.21) and (4.54) that the expression (4.20) reduces to

$$\Phi = \lambda \boldsymbol{\rho} \cdot \boldsymbol{\mathscr{C}} \tag{4.59}$$

so that a material point in isotropic strain-hardening exhibits

$$\begin{cases} \text{hardening behaviour if and only if } \boldsymbol{\rho}\cdot\boldsymbol{\mathscr{C}} > 0, & \text{(a)} \\ \text{softening behaviour if and only if } \boldsymbol{\rho}\cdot\boldsymbol{\mathscr{C}} < 0, & \text{(b)} \\ \text{perfectly plastic behaviour if and only if } \boldsymbol{\rho}\cdot\boldsymbol{\mathscr{C}} = 0, & \text{(c)} \end{cases} \tag{4.60}$$

where (4.24) has been used (and also the fact that $\lambda > 0$).

If conditions (4.44) are satisfied, then the function Γ in (4.49) reduces to

$$\Gamma = \frac{\partial f_1}{\partial \mathbf{S}} \cdot \boldsymbol{\mathscr{C}} \tag{4.61}$$

which in accordance with (4.51) determines the type of strain-hardening behaviour that will occur.

It is possible to define a modified form of isotropic hardening in which the initial yield surface is centred not at O but at some other fixed point 0*, where the state of stress is $\mathbf{S}^*$, say. The family of yield surfaces may then be described by an equation of the form

$$f_1(\mathbf{S} - \mathbf{S}^*) = \kappa, \tag{4.62}$$

and (4.57) is replaced by

$$f_1\left(\frac{\zeta}{\zeta_0}(\mathbf{S}_0 - \mathbf{S}^*)\right) = \hat{\kappa}\left(\frac{\zeta}{\zeta_0}\right)\kappa_0, \tag{4.63}$$

[†] The condition (4.57) is usually not written explicitly, but seems to be understood.

where now ζ measures the distance from O* to $\partial\mathscr{S}$. The results (4.58) through (4.61) still hold.

4.5.2 Kinematic hardening

In kinematic hardening, the yield surface $\partial\mathscr{S}$ retains its original size and shape but moves about in stress space. In the context of the present development,† we assume that there exists a symmetric second order tensor-valued function $\boldsymbol{\alpha}(\mathbf{S}, \mathbf{E}^p)$ such that

$$f(\mathscr{V}) = f_2(\mathbf{S} - \boldsymbol{\alpha}(\mathbf{S}, \mathbf{E}^p)) - \kappa_0, \qquad (\kappa_0 > 0),$$
$$\boldsymbol{\alpha}(\mathbf{S}, \mathbf{0}) = \mathbf{0}. \tag{4.64}$$

At initial yield (assuming a zero value of $\mathbf{E}^p$),

$$f_2(\mathbf{S}_0) = \kappa_0. \tag{4.65}$$

Geometrically, $\boldsymbol{\alpha}$ may be interpreted as the translation of the centre of the initial yield surface $\partial\mathscr{S}_0$. It is not necessary that $\partial\mathscr{S}_0$ be symmetric about the origin, although this is the case most commonly discussed.

Adopting the notation

$$\mathbf{Z} = \mathbf{S} - \boldsymbol{\alpha}(\mathbf{S}, \mathbf{E}^p), \tag{4.66}$$

it is clear that

$$\frac{\partial f}{\partial \mathbf{S}} = \frac{\partial f_2}{\partial \mathbf{Z}} - \left(\frac{\partial \boldsymbol{\alpha}}{\partial \mathbf{S}}\right)^T \left[\frac{\partial f_2}{\partial \mathbf{Z}}\right] = \left\{\frac{\partial f_2}{\partial Z_{KL}} - \frac{\partial f_2}{\partial Z_{MN}} \frac{\partial \alpha_{MN}}{\partial S_{KL}}\right\} \mathbf{e}_K \otimes \mathbf{e}_L,$$
$$\frac{\partial f}{\partial \mathbf{E}^p} = -\left(\frac{\partial \boldsymbol{\alpha}}{\partial \mathbf{E}^p}\right)^T \left[\frac{\partial f_2}{\partial \mathbf{Z}}\right] = -\frac{\partial f_2}{\partial Z_{MN}} \frac{\partial \alpha_{MN}}{\partial E^p_{KL}} \mathbf{e}_K \otimes \mathbf{e}_L, \tag{4.67}$$
$$\hat{f} = \frac{\partial f_2}{\partial \mathbf{Z}} \cdot \dot{\mathbf{S}} - \frac{\partial f_2}{\partial \mathbf{Z}} \cdot \frac{\partial \boldsymbol{\alpha}}{\partial \mathbf{S}} [\dot{\mathbf{S}}] = \frac{\partial f_2}{\partial Z_{MN}} \left\{\delta_{MK}\delta_{NL} - \frac{\partial \alpha_{MN}}{\partial S_{KL}}\right\} \dot{S}_{KL},$$

where the chain rule of calculus, $(4.64)_1$, and (4.18) have been used. From the consistency condition, (4.15), $(4.64)_1$, and $(4.67)_2$, it follows that during loading

$$\frac{\partial f_2}{\partial \mathbf{Z}} \cdot \frac{\partial \boldsymbol{\alpha}}{\partial \mathbf{E}^p} [\dot{\mathbf{E}}^p] = \hat{f} \tag{4.68}$$

and, therefore, with the help of (4.10d) and (4.23)

$$\lambda \frac{\partial f_2}{\partial \mathbf{Z}} \cdot \frac{\partial \boldsymbol{\alpha}}{\partial \mathbf{E}^p} [\boldsymbol{\rho}] = \Phi. \tag{4.69}$$

† A more general approach is to introduce $\boldsymbol{\alpha}$ as an independent variable with its own rate-type constitutive equation.

Consequently,

$$\frac{\partial f_2}{\partial \mathbf{Z}} \cdot \frac{\partial \boldsymbol{\alpha}}{\partial \mathbf{E}^{\mathrm{p}}}[\boldsymbol{\rho}] \begin{cases} > 0 \text{ for hardening behaviour,} & \text{(a)} \\ < 0 \text{ for softening behaviour,} & \text{(b)} \\ = 0 \text{ for perfectly plastic behaviour.} & \text{(c)} \end{cases} \tag{4.70}$$

In the remaining three subsections, we consider in detail the strain-hardening response of classes of constitutive equations which are appropriate for ductile metals undergoing small strains and rotations. We suppose that the elastic–plastic body initially occupies a reference configuration in which the plastic strain vanishes, and also that the material is homogeneous and isotropic relative to this configuration.

Let $\boldsymbol{\gamma}, \boldsymbol{\gamma}^{\mathrm{p}}$, and $\boldsymbol{\tau}$ be the deviatoric parts of $\mathbf{E}, \mathbf{E}^{\mathrm{p}}$, and $\mathbf{S}$, respectively, and $\bar{e}\mathbf{I}$, $\bar{e}^{\mathrm{p}}\mathbf{I}$ and $\bar{s}\mathbf{I}$ their spherical parts. Thus, for example,

$$\mathbf{S} = \boldsymbol{\tau} + \bar{s}\mathbf{I}, \qquad \bar{s} = \tfrac{1}{3}\operatorname{tr}\mathbf{S} = \tfrac{1}{3}S_{KK}, \tag{4.71}$$

tr being the trace operator. We assume now that the stress response function in (4.41) is specified by generalized Hooke's law in the form

$$\boldsymbol{\tau} = 2\mu(\boldsymbol{\gamma} - \boldsymbol{\gamma}^{\mathrm{p}}), \qquad \bar{s} = 3k(\bar{e} - \bar{e}^{\mathrm{p}}), \tag{4.72}$$

where $\mu(>0)$ is the shear modulus and $k\ (>0)$ is the bulk modulus of elasticity.

4.5.3 Special type of isotropic hardening

Consider a special yield function of the type (4.54), for which

$$f_1(\mathbf{S}) = \bar{\tau}^2 + 3\psi\bar{s}^2, \tag{4.73}$$

where $\bar{\tau} = \pm\sqrt{(\boldsymbol{\tau}\cdot\boldsymbol{\tau})}$ and its absolute value represents the magnitude of the deviatoric stress tensor, and ψ is a positive material constant. The yield function associated with (4.73) has a strain-space representation

$$g(\mathscr{U}) = 4\mu^2\|\boldsymbol{\gamma} - \boldsymbol{\gamma}^{\mathrm{p}}\|^2 + 27\psi k^2(\bar{e} - \bar{e}^{\mathrm{p}})^2 - \kappa, \tag{4.74}$$

where (4.15), (4.54), (4.72), and (4.73) have been used. Suppose additionally that the response function $\mathscr{C}$ in (4.11) has the special form

$$\mathscr{C} = \beta\boldsymbol{\tau} + \phi\bar{s}\mathbf{I}, \tag{4.75}$$

where β and ϕ are material constants. Constitutive equations of the type (4.73) and (4.75) were utilized previously in [13, 8]. The equation $f(\mathscr{U}) = 0$ or

$$\bar{\tau}^2 + 3\psi\bar{s}^2 = \kappa \tag{4.76}$$

describes an ellipse in the $\bar{s}$–$\bar{\tau}$ plane. It is convenient for present purposes to draw this ellipse in the $\sqrt{(3)}\bar{s}$–$\bar{\tau}$ plane as shown in Figure 4.3. For ductile metals, ψ is typically much less than unity, resulting in an ellipse which is very much elongated in the $\bar{s}$-direction. Also drawn in Figure 4.3 is the ellipse which

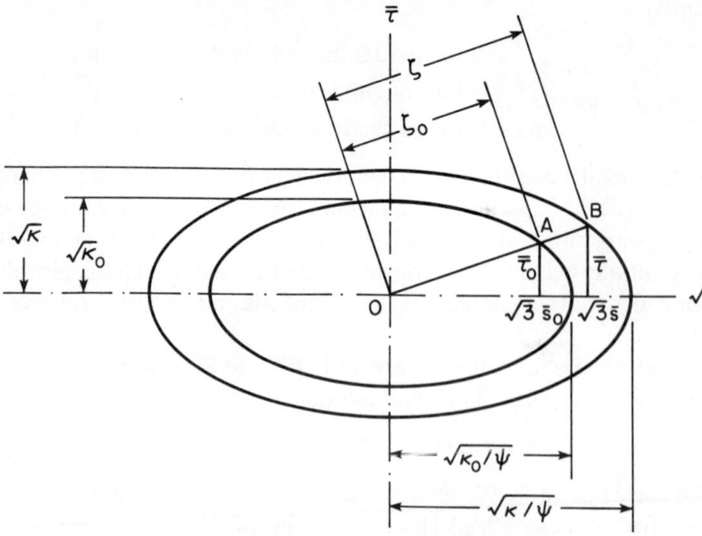

Figure 4.3 Sketch of the initial and a subsequent yield surface, as represented by ellipses in the plane of $\sqrt{(3)}\bar{s}$ and $\bar{\tau}$. For hardening behaviour, the ellipses appear as shown. For softening behaviour OB is less than OA, and for perfectly plastic behaviour OB = OA

represents the initial yield surface, and is described by

$$\bar{\tau}_0^2 + 3\psi\bar{s}_0^2 = \kappa_0, \tag{4.77}$$

where $\bar{\boldsymbol{\tau}}_0 + \bar{s}_0\mathbf{I} = \mathbf{S}_0$. The distances from O to $\mathbf{S}_0$ and $\mathbf{S}$, namely

$$\begin{aligned} \zeta_0 &= \|\mathbf{S}_0\| = \sqrt{(\bar{\tau}_0^2 + 3\bar{s}_0^2)} \\ \zeta &= \|\mathbf{S}\| = \sqrt{(\bar{\tau}^2 + 3\bar{s}^2)} \end{aligned} \tag{4.78}$$

are also indicated in Figure 4.3. In view of (4.76), (4.77), and (4.78), the function $\hat{\kappa}$ in (4.57) is

$$\hat{\kappa}\left(\frac{\zeta}{\zeta_0}\right) = \left(\frac{\zeta}{\zeta_0}\right)^2. \tag{4.79}$$

Recalling (4.58), (4.26), and (4.27), for hardening behaviour subsequent yield surfaces are represented by ellipses that lie outside the ellipse which corresponds to the initial yield surface; for softening behaviour subsequent ellipses lie inside the initial ellipse; and for perfectly plastic behaviour subsequent ellipses coincide with the initial ellipse.

It is worth mentioning that instead of the usual inner product

$$\mathbf{S}\cdot\mathbf{S}^* = \operatorname{tr}\mathbf{S}\mathbf{S}^* = \operatorname{tr}\boldsymbol{\tau}\boldsymbol{\tau}^* + 3\bar{s}\bar{s}^* \tag{4.80}$$

between the two symmetric stress tensors $\mathbf{S}$ and $\mathbf{S}^* = \boldsymbol{\tau}^* + \bar{s}^*\mathbf{I}$, we could choose

$$\mathbf{S}\cdot\mathbf{S}^* = \operatorname{tr}\boldsymbol{\tau}\boldsymbol{\tau}^* + \bar{s}\bar{s}^*. \tag{4.81}$$

The distance ζ would then be given by

$$\zeta = \sqrt{(\bar{\tau}^2 + \bar{s}^2)} \tag{4.82}$$

instead of $(4.78)_2$, and a figure similar to Figure 4.3 could be drawn in the $\bar{s}$–$\bar{\tau}$ plane.

For the yield function being considered (see (4.54) and (4.73)),

$$\begin{aligned} \frac{\partial f}{\partial \mathbf{S}} &= 2(\boldsymbol{\tau} + \psi \bar{s}\mathbf{I}), \\ \Gamma &= 2(\beta \bar{\tau}^2 + 3\psi\phi\bar{s}^2), \\ \Lambda &= 4(2\mu\bar{\tau}^2 + 9\psi^2 k\bar{s}^2) > 0. \end{aligned} \tag{4.83}$$

The measure of strain-hardening Φ can then be found from $(4.50)_2$.

In [8], we discussed the strain-hardening behaviour in uniaxial tension of a material satisfying the constitutive equations of the present subsection. Relative to a fixed orthonormal basis $\{\mathbf{e}_1, \mathbf{e}_2, \mathbf{e}_3\}$, uniaxial tension along the $\mathbf{e}_1$ direction is specified by

$$\boldsymbol{\tau} = \frac{s}{3}\mathbf{b}, \qquad \bar{\mathbf{s}} = \frac{s}{3}, \tag{4.84}$$

where

$$\mathbf{b} = 2\mathbf{e}_1 \otimes \mathbf{e}_1 - \mathbf{e}_2 \otimes \mathbf{e}_2 - \mathbf{e}_3 \otimes \mathbf{e}_3, \tag{4.85}$$

and $s = \hat{s}(t) \geq 0$ is a given function of time. For uniaxial tension, with the help of $(4.50)_2$, $(4.83)_{2,3}$, (4.84), and (4.85), the measure Φ can be written as

$$\Phi = \frac{2\beta + \psi\phi}{2(\beta + 4\mu) + \psi(\phi + 6\psi k)}. \tag{4.86}$$

Also, in view of (4.86) and the definition (4.24),

$$\begin{cases} \text{hardening,} & \text{(a)} \\ \text{softening,} & \text{(b)} \\ \text{perfectly plastic,} & \text{(c)} \end{cases} \tag{4.87}$$

behaviour occurs if and only if

$$2\beta + \psi\phi \begin{cases} > 0, & \text{(a)} \\ < 0, & \text{(b)} \\ = 0. & \text{(c)} \end{cases} \tag{4.88}$$

From (4.76), (4.84), and (4.78), it follows that

$$\begin{aligned} \kappa &= \tfrac{1}{3}(2 + \psi)s^2, \qquad \zeta = s, \\ \kappa_0 &= \tfrac{1}{3}(2 + \psi)s_0^2, \qquad \zeta_0 = s_0, \end{aligned} \tag{4.89}$$

where s_0 is the value of the function $\hat{s}$ at initial yield. During loading, by (4.26), (4.27), (4.58), and $(4.89)_{1,2}$, all three derivatives $\dot{\kappa}, \dot{s}, \dot{\zeta}$ are positive for hardening behaviour, negative for softening behaviour, and zero for perfectly plastic behaviour.

From (4.72) and (4.84) it follows that

$$\bar{e} - \bar{e}^{\mathrm{p}} = \frac{s}{9k}, \qquad \gamma - \gamma^{\mathrm{p}} = \frac{s}{6\mu}\mathbf{b}, \tag{4.90a}$$

or

$$\mathbf{E} - \mathbf{E}^{\mathrm{p}} = \frac{s}{E}\{\mathbf{e}_1 \otimes \mathbf{e}_1 - \nu(\mathbf{e}_2 \otimes \mathbf{e}_2 + \mathbf{e}_3 \otimes \mathbf{e}_3)\}, \tag{4.90b}$$

where E is Young's modulus of elasticity and ν is Poisson's ratio.

For hardening or softening behaviour, the plastic strain tensor may be obtained with the help of (4.52), (4.58), $(4.83)_{1,2}$, and $(4.89)_1$, and may be expressed in the form [8]:

$$\bar{e}^{\mathrm{p}} = \frac{s - s_0}{9k^*}, \qquad \gamma^{\mathrm{p}} = \frac{s - s_0}{6\mu^*}\mathbf{b}, \tag{4.91a}$$

or

$$\mathbf{E}^{\mathrm{p}} = \frac{s - s_0}{E^*}\{\mathbf{e}_1 \otimes \mathbf{e}_1 - \nu^*(\mathbf{e}_2 \otimes \mathbf{e}_2 + \mathbf{e}_3 \otimes \mathbf{e}_3)\} \tag{4.91b}$$

where the constants ν^*, E^*, μ^*, k^*, which are analogous to the corresponding constants of linear elasticity, are defined by

$$\begin{aligned} \nu^* &= \frac{1 - \psi}{2 + \psi}, \qquad E^* = \frac{3(2\beta + \psi\phi)}{2(2 + \psi)^2}, \\ \mu^* &= \frac{E^*}{2(1 + \nu^*)} = \frac{2 + \psi}{6}E^*, \\ k^* &= \frac{E^*}{3(1 - 2\nu^*)} = \frac{2 + \psi}{9\psi}E^*. \end{aligned} \tag{4.92}$$

From (4.87), (4.88), and $(4.92)_{2,3,4}$ it follows that

$$\begin{cases} \text{hardening,} & \text{(a)} \\ \text{softening,} & \text{(b)} \\ \text{perfectly plastic} & \text{(c)} \end{cases} \tag{4.93}$$

behaviour occurs if and only if

$$E^*, \mu^*, k^* \begin{cases} > 0, & \text{(a)} \\ < 0. & \text{(b)} \\ = 0. & \text{(c)} \end{cases} \tag{4.94}$$

In addition, observing that

$$\frac{\mathrm{d}E_{11}}{\mathrm{d}s} = \frac{1}{E} + \frac{1}{E^*}, \tag{4.95}$$

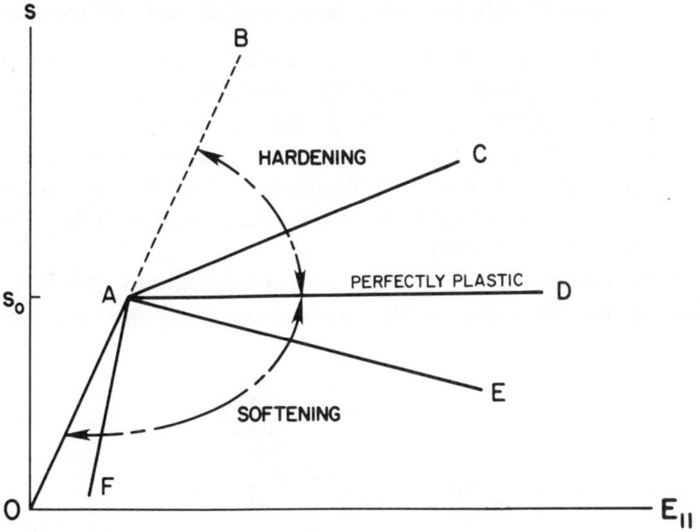

Figure 4.4 Strain-hardening behaviour in uniaxial tension for the constitutive equations of Subsection 4.5.3. Following initial yield at A, perfectly plastic behaviour corresponds to the line AD; hardening behaviour is represented by a line such as AC; and softening behaviour by lines such as AE and AF

which follows from (4.90b) and (4.91b), it may be concluded that [8]

$$\begin{cases} \infty > \dfrac{dE_{11}}{ds} > \max\left\{\dfrac{1}{E}, \dfrac{1}{E^*}\right\} \text{ if the material is hardening,} \\ \\ \dfrac{1}{E} > \dfrac{dE_{11}}{ds} > \dfrac{1}{E^*} > -\infty \text{ if the material is softening.} \end{cases} \tag{4.96}$$

Moreover,

$$\begin{cases} \infty > \dfrac{dE_{11}}{ds} > \dfrac{1}{E} \text{ implies that the material is hardening,} \\ \\ -\infty < \dfrac{dE_{11}}{ds} < \dfrac{1}{E} \text{ implies that the material is softening.} \end{cases} \tag{4.96}$$

Also, in view of (4.87c), (4.88c), and $(4.92)_{2,3,4}$, $E^* = \mu^* = k^* = 0$ if and only if the material is exhibiting perfectly plastic behaviour. In this case, it is necessary that $\dot{s} = 0$, as mentioned before. Thus, the slope of the stress–strain diagram during loading† in uniaxial tension may be used to characterize the strain-hardening behaviour of materials that satisfy the constitutive equations of this subsection. The three different types of response are illustrated in Figure 4.4.

† Experimentally, loading may be guaranteed by controlling the strain rather than the load.

4.5.4 Special type of modified isotropic hardening. Strength-differential effect

The yield strength of certain elastic–plastic materials, most notably high-strength steels, is appreciably greater in uniaxial compression than in tension [14, 15]. This phenomenon, referred to as the strength-differential or S–D effect, is accompanied by plastic volume expansion in both tension and compression and persists throughout the entire range of plastic strain. As will be seen presently, the S–D effect can be accounted for by allowing a suitable dependency of the yield function on mean normal stress.

In order to describe the S–D effect, Casey and Jahedmotlagh [16] proposed the following modifications to the constitutive equations that appear in Subsection 4.5.3:

$$f(\mathscr{V}) = \bar{\tau}^2 + 3\psi\left(\bar{s} - \frac{\eta}{3}\right)^2 - \kappa,$$

$$\mathscr{C} = \beta\boldsymbol{\tau} + \phi\left(\bar{s} - \frac{\eta}{3}\right)\mathbf{I}, \qquad (4.97)$$

where η is a material constant. The yield function in $(4.97)_1$ is of the type discussed at the end of Subsection 4.5.1, with f_1 given by (4.73) and

$$\mathbf{S}^* = \frac{\eta}{3}\mathbf{I},$$

$$\zeta^2 = \|\mathbf{S} - \mathbf{S}^*\|^2 = \bar{\tau}^2 + \left(\sqrt{(3)}\bar{s} - \frac{\eta}{\sqrt{3}}\right)^2, \qquad (4.98)$$

$$\zeta_0^2 = \bar{\tau}_0^2 + \left(\sqrt{(3)}\bar{s}_0 - \frac{\eta}{\sqrt{3}}\right)^2.$$

The function $\hat{\kappa}$ in (4.63) is given by (4.79). The yield surface may be represented by an ellipse similar to that in Figure 4.3, except that the origin of the ellipse is translated to a point O* on the $\sqrt{(3)}\bar{s}$ axis whose coordinate is $\eta/\sqrt{3}$ (see Figure 4.5). In this case, instead of (4.83), we have

$$\frac{\partial f}{\partial \mathbf{S}} = 2\left\{\boldsymbol{\tau} + \psi\left(\bar{s} - \frac{\eta}{3}\right)\mathbf{I}\right\},$$

$$\Gamma = 2\left\{\beta\bar{\tau}^2 + 3\psi\phi\left(\bar{s} - \frac{\eta}{3}\right)^2\right\}, \qquad (4.99)$$

$$\Lambda = 4\left\{2\mu\bar{\tau}^2 + 9\psi^2 k\left(\bar{s} - \frac{\eta}{3}\right)^2\right\} > 0.$$

The function Φ can be found by substituting $(4.99)_{2,3}$ in $(4.50)_2$.

In the rest of this subsection, we briefly indicate the nature of the predictions of the development in [16] as compared with experimental results reported in [15]

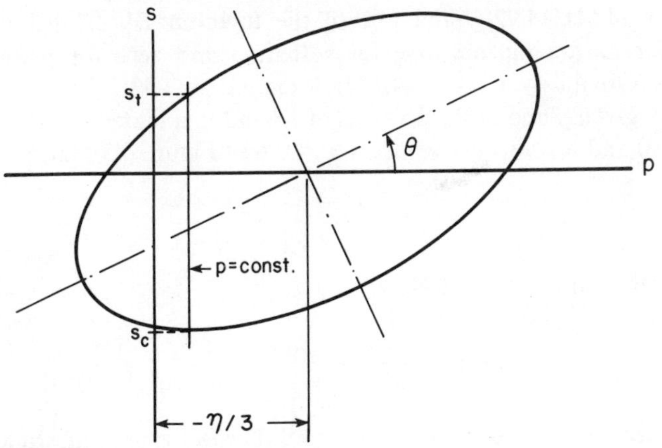

Figure 4.5 Sketch of the yield surface described by (4.101). At a constant value of p, the material yields at a tensile stress s_t and a compressive stress s_c

for a combination of uniaxial loading s and pressure p. Let

$$\tau = \frac{s}{3}\mathbf{b}, \qquad \bar{s} = \frac{s}{3} - p, \tag{4.100}$$

where p is a positive constant and $\mathbf{b}$ was defined in (4.85). During loading, the yield function $(4.97)_1$ becomes

$$2s^2 + \psi(s - 3p - \eta)^2 - 3\kappa = 0. \tag{4.101}$$

Equation (4.101) describes an ellipse in the p–s plane which is centred at $(p, s) = (-(\eta/3), 0)$, and is titled at an angle θ to the p-axis, with

$$\tan 2\theta = \frac{3\psi}{1 - 4\psi}, \tag{4.102}$$

as shown in Figure 4.5.

During hardening and softening behaviour it can be shown that

$$\frac{\mathrm{d}\mathbf{E}}{\mathrm{d}s} = \frac{1}{E}\{\mathbf{e}_1 \otimes \mathbf{e}_1 - \nu(\mathbf{e}_2 \otimes \mathbf{e}_2 + \mathbf{e}_3 \otimes \mathbf{e}_3)\}$$
$$+ \frac{1}{E^*}\{\mathbf{e}_1 \otimes \mathbf{e}_1 - \nu^*(\mathbf{e}_2 \otimes \mathbf{e}_2 + \mathbf{e}_3 \otimes \mathbf{e}_3)\}, \tag{4.103}$$

where as in [16]

$$E^* = E^*(s, p) = \frac{3}{2}\frac{2\beta s^2 + \psi\phi(s - 3p - \eta)^2}{\{2s + \psi(s - 3p - \eta)\}^2},$$
$$\nu^* = \nu^*(s, p) = \frac{s - \psi(s - 3p - \eta)}{2s + \psi(s - 3p - \eta)}. \tag{4.104}$$

In view of (4.51), $(4.99)_2$, and (4.100) the function E^* defined in (4.104) is positive for hardening, negative for softening and zero for perfectly plastic behaviour. Obviously, if $\eta = 0$, $(4.104)_{1,2}$ reduce to $(4.92)_{2,1}$.

If at any given p and at fixed values of $\mathbf{E}^{\mathrm{p}}$ and κ, a material yields at a tensile stress $s_{\mathrm{t}} > 0$ and a compressive stress $s_{\mathrm{c}} < 0$ (see Figure 4.5), the S–D effect is

$$\text{S–D} = 2\frac{|s_{\mathrm{c}}| - |s_{\mathrm{t}}|}{|s_{\mathrm{c}}| + |s_{\mathrm{t}}|} = 2\frac{s_{\mathrm{c}} + s_{\mathrm{t}}}{s_{\mathrm{c}} - s_{\mathrm{t}}}. \tag{4.105}$$

From (4.101) and (4.105), it follows that

$$\text{S–D} = -\frac{2\psi(3p + \eta)}{\{3(2 + \psi)\kappa - 2\psi(3p + \eta)^2\}^{1/2}}. \tag{4.106}$$

As can be seen from (4.106), with $\psi = 0$ no S–D effect is present, while with $\eta = 0$ there is an absence of S–D effect at zero pressure. Thus, in order to fully account for an S–D effect, it is necessary that both ψ and η be non-zero.

The rate of plastic volume change per unit initial volume is equal to $\operatorname{tr}\dot{\mathbf{E}}^{\mathrm{p}}$. Denoting the ratio of this quantity to $\dot{E}^{\mathrm{p}}_{11}$ by γ^{p}, it follows from (4.103) that

$$\gamma^{\mathrm{p}} = \frac{\mathrm{d}E^{\mathrm{p}}_{KK}}{\mathrm{d}E^{\mathrm{p}}_{11}} = 1 - 2\nu^*. \tag{4.107}$$

Unless ψ is non-zero, $\gamma^{\mathrm{p}} = 0$.

The material coefficients η, ψ, β, and ϕ were determined in [16], using the experimental data of [15], and the functions E^*, ν^*, etc., were calculated. The predicted results are in good agreement with experiment, except for the values of γ^{p} which are too large in magnitude.

4.5.5 Long-term responses. Saturation with erasure of memory. Special type of limiting kinematic hardening

After an initial period of variable strain-hardening, a material may settle down to a response that is very nearly constant. Of particular interest in this respect is the phenomenon of saturation hardening which has been found to occur for example in stainless steels, aluminium alloys, and annealed copper [12, 17, 18, 19]. These materials, after several cycles of uniaxial tension–compression loading, tend to a limiting stress–strain response.

In this subsection we discuss a special set of constitutive equations utilized by Caulk and Naghdi [17] for the description of saturation hardening, and subsequently discussed in [8].

First, we observe that the function Φ of Section 4.3 is an index of instantaneous strain-hardening behaviour at any point on a given strain trajectory. By examining the manner in which Φ varies with time along the trajectory, it is possible to analyse long-term and limiting strain-hardening responses. Thus, we may say that the strain-hardening response of a material saturates to a value K_h along a strain (or stress) trajectory, if

$$\lim_{t\to\infty} \Phi = K_h > 0. \tag{4.108}$$

Corresponding definitions may be given for saturation to softening, and to perfectly plastic responses.

The phenomenon of saturation hardening appears to involve one additional feature, namely an 'erasure of memory', to employ the terminology of Lamba and Sidebottom [19]. This aspect of saturation hardening is related to the long-term behaviour of the work-hardening parameter κ. Thus, we suppose that there is a 'saturation' value κ_s of κ such that $\kappa \leq \kappa_s$ and along a strain (or stress) trajectory

$$\lim_{t\to\infty} \kappa = \kappa_s. \tag{4.109}$$

The mean-value theorem of differential calculus then implies that

$$\lim_{t\to\infty} \dot{\kappa} = 0 \tag{4.110}$$

and it should be noted that this does not imply (4.108), nor vice versa.

The condition (4.110) was utilized by Caulk and Naghdi [17], while Casey and Naghdi [8] adopted (4.108). For the class of materials for which detailed comparisons with experiments were made in [17], both (4.108) and (4.109) are satisfied. In the remainder of this discussion we assume that both (4.108) and (4.109) are met.†

Consider now the following constitutive equations:

$$f(\mathscr{V}) = \tfrac{1}{4}\|2\boldsymbol{\tau} - \alpha\boldsymbol{\gamma}^{\mathrm{p}}\|^2 - \kappa,$$
$$\mathscr{C} = \tfrac{1}{2}\hat{\beta}(\kappa)\{2\boldsymbol{\tau} - \alpha\boldsymbol{\gamma}^{\mathrm{p}}\}, \tag{4.111a}$$

subject to the condition

$$\hat{\beta}(\kappa_s) = 0 \tag{4.111b}$$

where α is a constant and $\hat{\beta}$ is a constitutive function which depends on κ only. It is easily seen from (4.111a) and (4.11) that the condition (4.110) is satisfied. We note that at saturation, $(4.111\text{a})_1$ is of the form (4.64), and therefore corresponds to limiting kinematic hardening.

The equations (4.111) are a special case of those utilized in [17, 8] but include those that were used in [17] for comparison of theory and experiment.

† It is worth mentioning that for isotropic strain-hardening if (4.110) is satisfied, then by (4.58) and (4.23) during loading $\Phi \to 0$; and this, in turn, implies that perfectly plastic behaviour is the only type of saturation that can occur.

Corresponding to (4.111), we find that [8]

$$\frac{\partial f}{\partial \mathbf{S}} = 2\boldsymbol{\tau} - \alpha\boldsymbol{\gamma}^{\mathrm{p}},$$

$$\Lambda = 2\mu \|2\boldsymbol{\tau} - \alpha\boldsymbol{\gamma}^{\mathrm{p}}\|^2 > 0,$$

$$\Gamma = \frac{\alpha + \hat{\beta}(\kappa)}{2} \|2\boldsymbol{\tau} - \alpha\boldsymbol{\gamma}^{\mathrm{p}}\|^2. \tag{4.112}$$

Recalling that $f=0$ on the yield surface $\partial\mathscr{S}$, it is clear from $(4.111\mathrm{a})_1$ and $(4.112)_{2,3}$ that Λ and Γ may be written as

$$\Lambda = 8\mu\kappa > 0,$$

$$\Gamma = 2(\alpha + \hat{\beta}(\kappa))\kappa. \tag{4.113}$$

The measure Φ of strain-hardening may be obtained from $(4.50)_2$ and $(4.112)_{2,3}$ in the form

$$\Phi = \frac{\alpha + \hat{\beta}(\kappa)}{4\mu + \alpha + \hat{\beta}(\kappa)} < 1. \tag{4.114}$$

Thus, for a given material the strain-hardening behaviour depends on κ only. Also, since Φ cannot exceed the value 1, only a limited degree of hardening behaviour is possible. The type of strain-hardening is determined by $\alpha + \hat{\beta}(\kappa)$, so that

$$\left\{\begin{array}{ll} \text{hardening,} & \text{(a)} \\ \text{softening,} & \text{(b)} \\ \text{perfectly plastic} & \text{(c)} \end{array}\right. \tag{4.115}$$

behaviour occurs if and only if

$$\alpha + \hat{\beta}(\kappa) \left\{\begin{array}{ll} > 0, & \text{(a)} \\ < 0, & \text{(b)} \\ = 0. & \text{(c)} \end{array}\right. \tag{4.116}$$

In what follows we confine attention to hardening behaviour only. Then, by (4.108), (4.109), (4.111b), (4.114), and (4.116a), saturation hardening with erasure of memory will occur and the saturation constant K_{h} is

$$K_{\mathrm{h}} = \frac{\alpha}{4\mu + \alpha} < 1, \qquad \alpha > 0 \tag{4.117}$$

Also, Λ and Γ are then given by

$$\Lambda = 8\mu\kappa_{\mathrm{s}} > 0, \qquad \Gamma = 2\alpha\kappa_{\mathrm{s}} > 0 \tag{4.118}$$

where (4.109), (4.113), (4.111b), and (4.51a) have been invoked.

We now consider the case of uniaxial tensile loading, given by (4.84). For the constitutive equations of the present subsection, $\bar{e}^{\mathrm{p}} = 0$ and hence $\bar{e}$ is equal to the right-hand side of $(4.90\mathrm{a})_1$. For convenience, let $E_{11} = e$, $E_{11}^{\mathrm{p}} = e_{\mathrm{p}}$. Then, the deviatoric plastic strain tensor may be expressed as

$$\gamma^{\mathrm{p}} = \tfrac{1}{2} e_{\mathrm{p}} \mathbf{b}. \tag{4.119}$$

Next, with the help of (4.27a), $(4.84)_1$, $(4.111\mathrm{a})_1$, (4.112), (4.113), (4.116a), (4.119), (4.18), and (4.68), during loading we have [17, 8]:

$$\begin{aligned} &f = \tfrac{3}{8}(\tfrac{4}{3}s - \alpha e_{\mathrm{p}})^2 - \kappa = 0, \qquad \frac{\partial f}{\partial \mathbf{S}} = \tfrac{1}{2}(\tfrac{4}{3}s - \alpha e_{\mathrm{p}})\mathbf{b} \neq \mathbf{0}, \\ &\hat{f} = (\tfrac{4}{3}s - \alpha e_{\mathrm{p}})\dot{s} > 0, \qquad \dot{s} > 0, \\ &\Lambda = 3\mu(\tfrac{4}{3}s - \alpha e_{\mathrm{p}})^2 = 8\mu\kappa > 0, \\ &\Gamma = \tfrac{3}{4}(\alpha + \hat{\beta}(\kappa))(\tfrac{4}{3}s - \alpha e_{\mathrm{p}})^2 = 2(\alpha + \hat{\beta}(\kappa))\,\kappa > 0, \end{aligned} \tag{4.120}$$

and

$$\dot{e}_{\mathrm{p}} = \frac{4}{3}\frac{\dot{s}}{\alpha + \hat{\beta}(\kappa)}. \tag{4.121}$$

Using (4.71), (4.72), $(4.120)_4$, and (4.121) we find that

$$\begin{aligned} &\frac{\mathrm{d}e_{\mathrm{p}}}{\mathrm{d}s} = \frac{4}{3(\alpha + \hat{\beta}(\kappa))} > 0, \\ &\frac{\mathrm{d}e}{\mathrm{d}s} = \frac{1}{E} + \frac{\mathrm{d}e_{\mathrm{p}}}{\mathrm{d}s} > 0. \end{aligned} \tag{4.122}$$

It then follows from (4.114) and $(4.122)_1$ that

$$1 > \Phi = \left(1 + 3\mu\frac{\mathrm{d}e_{\mathrm{p}}}{\mathrm{d}s}\right)^{-1} > 0. \tag{4.123}$$

An examination of (4.122) and (4.123) indicates that a knowledge of μ, E and the slope $\mathrm{d}s/\mathrm{d}e$ of the uniaxial stress–strain curve suffices to determine the value of Φ for all stress states that correspond to the same value of κ.

Differentiating $(4.120)_1$ with respect to t and recalling $(4.120)_4$, we obtain a differential equation for κ in terms of s:

$$\frac{\mathrm{d}\kappa}{\mathrm{d}s} = (\tfrac{4}{3}s - \alpha e_{\mathrm{p}})\frac{\hat{\beta}(\kappa)}{\alpha + \hat{\beta}(\kappa)} = \frac{2\hat{\beta}(\kappa)}{\alpha + \hat{\beta}(\kappa)}\sqrt{\frac{2\kappa}{3}}. \tag{4.124}$$

At saturation, the expressions $(4.120)_1$, $(4.122)_2$, and (4.124) may be reduced with the help of (4.109) and (4.111b) to

$$\tfrac{4}{3}s - \alpha e_{\mathrm{p}} = 2\sqrt{\frac{2\kappa_{\mathrm{s}}}{3}},$$

$$\frac{\mathrm{d}e}{\mathrm{d}s} = \frac{1}{E} + \frac{4}{3\alpha} > 0,$$

$$\frac{\mathrm{d}\kappa}{\mathrm{d}s} = 0, \tag{4.125}$$

while K_{h}, Λ, and Γ are given by (4.117) and (4.118).

In [17], the function $\hat{\beta}(\kappa)$ was specified by

$$\hat{\beta}(\kappa) = \frac{\kappa - \kappa_{\mathrm{s}}}{\kappa_0 - \kappa_{\mathrm{s}}}\beta, \tag{4.126}$$

where β is a material constant. The constants α and β may then be obtained by using experimental data in the equations given above, and the various differential equations can be integrated numerically. For the stainless steel and aluminium alloy studied in [17], good agreement was found between theory and experiment.

In addition to the type of saturation response elaborated upon in this subsection, it is also of interest to consider certain other phenomena observed in cyclic loading. A discussion of such cyclic responses is contained in the paper of Drucker and Palgen [18]. In the context of the developments of the present paper, these phenomena can be easily accommodated by a simple modification of constitutive equations (4.111). In this connection, it may be noted that Naghdi and Nikkel [20] have recently shown that by appropriately modifying the functions f and $\mathscr{C}$ in (4.111a) the following types of behaviour can then be described: (1) ratcheting of strain caused by stress cycling between fixed values of stress in tension and compression (not necessarily equal in magnitude) and (2) progressive relaxation to zero of the mean stress caused by strain cycling between any two fixed values of strain.

ACKNOWLEDGEMENT

The results reported here were obtained in the course of research supported by the US Office of Naval Research under Contract N00014-75-C-0148, Project NR 064-436 with the University of California. Also, during the preparation of this paper, the work of J.C. was supported by the Engineering Foundation under Grant RC-A-79-1C with the University of Houston.

REFERENCES

1. A. E. Green and P. M. Naghdi, 'A general theory of an elastic–plastic continuum', *Archive for Rational Mechanics and Analysis*, **18**, 251–281 (1965).

2. A. E. Green and P. M. Naghdi, 'A thermodynamic development of elastic–plastic continua', *Proceedings of the IUTAM Symposium on Irreversible Aspects of Continuum Mechanics and Transfer of Physical Characteristics in Moving Fluids* (eds. H. Parkus and L. I. Sedov), pp. 117–131, Springer-Verlag, 1966.
3. P. M. Naghdi and J. A. Trapp, 'Restrictions on constitutive equations of finitely deformed elastic–plastic materials', *Quarterly Journal of Mechanics and Applied Mathematics*, **28**, 25–46 (1975).
4. P. M. Naghdi and J. A. Trapp, 'On the nature of normality of plastic strain rate and convexity of yield surfaces in plasticity', *Journal of Applied Mechanics*, **42**, 61–66 (1975).
5. P. M. Naghdi and J. A. Trapp, 'The significance of formulating plasticity with reference to loading surfaces in strain space', *International Journal of Engineering Science*, **13**, 785–797 (1975).
6. P. M. Naghdi, 'Some Constitutive restrictions in plasticity', *Proceedings of the Symposium on Constitutive Equations in Viscoplasticity: Computational and Engineering Aspects*, ASME-AMD Vol. 20, 79–93 (1976).
7. A. E. Green and P. M. Naghdi, On thermodynamical restrictions in the theory of elastic–plastic materials', *Acta Mechanica*, **30**, 157–162 (1978).
8. J. Casey and P. M. Naghdi, 'On the characterization of strain-hardening in plasticity', *Journal of Applied Mechanics*, **48**, 285–296 (1981).
9. J. Casey and P. M. Naghdi, 'On the nonequivalence of the stress space and strain space formulations of plasticity theory', *Journal of Applied Mechanics*, **50**, 350–354 (1983).
10. J. Casey and P. M. Naghdi, 'A remark on the definition of hardening, softening and perfectly plastic behavior', *Acta Mechanica*, **48**, 91–94 (1983).
11. P. M. Naghdi, Stress–strain relations in plasticity and thermoplasticity', *Proceedings of the Second Symposium on Naval Structural Mechanics* (Brown University, 1960), pp. 121–167, Pergamon Press, 1960.
12. C. E. Pugh, K. C. Liu, J. M. Corum, and W. L. Greenstreet, 'Currently recommended constitutive equations for inelastic design analysis of FFTF components', Oak Ridge National Laboratory, *ORNL-TM*-3602, 1972.
13. A. E. Green and P. M. Naghdi, 'A comment on Drucker's postulate in the theory of plasticity', *Acta Mechanica*, **1**, 334–338 (1965).
14. D. C. Drucker, 'Plasticity theory, strength-differential (S-D) phenomenon, and volume expansion in metals and plastics', *Metallurgical Transactions*, **4**, 667–673 (1973).
15. W. A. Spitzig, R. J. Sober, and O. Richmond, 'Pressure dependence of yielding and associated volume expansion in tempered martensite', *Acta Metallurgica*, **23**, 885–893 (1975).
16. J. Casey and H. Jahedmotlagh, 'The strength-differential effect in plasticity', International Journal of Solids and Structures, in press.
17. D. A. Caulk and P. M. Naghdi, 'On the hardening response in small deformation of metals', *Journal of Applied Mechanics*, **45**, 755–764 (1978).
18. D. C. Drucker and L. Palgen, 'On stress–strain relations suitable for cyclic and other loading', *Journal of Applied Mechanics*, **48**, 479–485 (1981).
19. H. S. Lamba and O. M. Sidebottom, 'Cyclic plasticity for nonproportional paths: Part I—Cyclic-hardening, erasure of memory, and subsequent strain hardening experiments' and 'Part II—Comparison with predictions of three incremental plasticity models', *Journal of Engineering Materials and Technology*, **100**, 96–103 and 104–111 (1978).
20. P. M. Naghdi and D. J. Nikkel, Jr., Calculations for uniaxial stress and strain cycling in plasticity', Journal of Applied Mechanics, to appear.

Mechanics of Engineering Materials
Edited by C. S. Desai and R. H. Gallagher

Chapter 5

Constitutive Modelling in Soil Mechanics

W. F. Chen

5.1. INTRODUCTION

5.1.1 Constitutive equations

Soil mechanics, as the name implies, is a branch of mechanics of solids. A valid solution to any problem in soil mechanics must therefore satisfy the following three basic equations:

(1) The equations of equilibrium of stress σ_{ij} and body and surface forces F_i, T_i, or of motion;
(2) The equations of compatibility of strains ε_{ij} and displacements U_i, or of geometry and;
(3) The relations between stresses and strains of material or the constitutive equations of soils.

A solution which does not satisfy these three equations or their equivalent is merely a guess.

There are three *equations of equilibrium* relating the six components of stress tensor, σ_{ij}, for an infinitesimal element of the body. In linear problems these equations do not contain strains, ε_{ij}, or displacements, U_i; in non-linear problems they often do. In problems of dynamics, the equilibrium equations are replaced by the equations of motion, which contain second-order time derivatives of the displacements, U_i. These are the first set of equations.

There are six *equations of kinematics* expressing the six components of strain tensor, ε_{ij}, in terms of the three components of displacements, U_i. These are known as the *strain–displacement relations* and are the second set of equations.

Clearly, both the equations of equilibrium and the equations of kinematics are independent of the particular material of which the body is made. The influence of this material is expressed by a third set of equations, the *constitutive equations.* They describe the relations between stresses and strains. In the simplest case, there are six equations expressing the strain components in terms of stress components, or vice versa. If they are linear, they are known as Hooke's law.

The six stress components, σ_{ij}, six strain components, ε_{ij}, and three displacement components, U_i, are connected by the three equilibrium equations, six kinematic equations, and six constitutive equations. These 15 unknown quantities of stresses, strains, and displacements inside a body are determined from the system of 15 equations expressing laws of nature.

Once the material constitutive equations are established, the general formulation for the solution of a solid mechanics problem can be completed. With the present state of development of finite element computer programs, we can confidently say that an almost unlimited range of solutions in solid mechanics can now be obtained. They are not limited to linear elastic problems but can be extended to include problems of various kinds involving material and geometric non-linearities [6].

5.1.2 Idealizations

For a long time, soil mechanics has been based on Hooke's law of linear elasticity for stress and deformation analysis for a soil mass under a footing, or behind a retaining wall, when no failure of the soil is involved. These are known as the *elasticity problems* in soil mechanics. On the other extreme, the theory of perfect plasticity is used to deal with the conditions of ultimate failure of a soil mass. Problems of earth pressure, retaining walls, bearing capacity of foundations, and stability of slopes are all considered in the realm of perfect plasticity. These are called the *stability problems*. Long-term settlement problems and consolidation problems, however, are treated in soil mechanics as essentially viscoelastic problems.

Partly for simplicity in practice and partly because of the historical development of mechanics of solids, the elasticity problems and the stability problems in soil mechanics are treated separately and in unrelated ways. The essential connection between the elasticity problems and the stability problems are the problems known as the progressive failure problems. The progressive failure problems deal with the elastic–plastic transition from the initial linear elastic state to the ultimate state of the soil by plastic flow. The essential set of equations for the solutions of progressive failure problems is the constitutive equations of soils.

It is well known that the stress–strain behaviour of soils is not linearly elastic for the entire range of loading of practical interest. In fact actual behaviour of soils is very complicated and they show a great variety of behaviour when subjected to different conditions. Drastic idealizations are therefore essential in order to develop simple mathematical models for practice applications. No one mathematical model can completely describe the complex behaviour of real soils under all conditions. Each soil model is aimed at a certain class of phenomena, captures their essential features, and disregards what is considered to be of minor importance in that class of applications. Thus, this constitutive model meets its limits of applicability where a disregarded influence becomes important. As mentioned previously, Hooke's law has been used successfully in soil mechanics to describe the general behaviour of soil media under short term working load conditions, but it fails to predict the behaviour and strength of a soil–structure interaction problem near ultimate strength condition, because plastic deformation at this load level attains a dominating influence, while elastic deformation becomes of minor importance.

Under short-term loading, soil behaviour may be idealized as *time-independent* where the effects of time can be neglected. This time-independent idealization of soils can be further idealized as *elastic* behaviour and *plastic* behaviour. For an elastic material there exists a one-to-one coordination between stress and strain. Thus, a body that consists of this idealized material returns to its original shape whenever all stresses are reduced to zero. This reversibility is not the case for a plastic material. As a first step in constitutive modelling of soils, it is therefore logical to utilize and refine the classical theories of elasticity and plasticity as developed for an idealized material. However, there are in many cases considerable differences between the properties of soils and those of the idealized bodies. These differences may have a significant influence on the solution of some boundary value problems in soil mechanics. In such cases, the classical theories must be modified and extended so that the special properties of soils in certain practical applications are taken into consideration. This is described in the present chapter.

The first part of the chapter (Section 5.3 to 5.5) is concerned with the general techniques used in the discussion of the elasticity-based stress–strain laws for soils. The second part (Section 5.6 to 5.10) is concerned with the extension of these techniques to the modern development of the theories of work-hardening plasticity for soils. Stress–strain models of *endochronic* type are not considered in the present discussion.

5.1.3. Elasticity and modelling

In a more restricted sense, an elastic material must also satisfy the energy equation of thermodynamics. The elastic material characterized by this additional requirement is known as hyperelastic (Section 5.4). On the other hand, the minimal requirement for a material to qualify as elastic in any sense is that there exists a one-to-one coordination between stress increment and strain increment. Thus, a body that consists of this material returns to its original state of deformation whenever all stress increments are reduced to zero. This reversibility in the infinitesimal sense justifies the use of the term hypoelastic for elastic materials satisfying only this minimal requirement (Section 5.5). The incremental constitutive formulations based on hypoelastic models have been increasingly used in recent years by geotechnical engineers for soils in which the state of stress is generally a function of the current state of stresses and strains as well as of the stress path followed to reach that state.

In spite of its obvious shortcomings, the linear theory of elasticity is the most commonly used constitutive model for soils. This linear elastic model can be significantly improved by assuming a bilinear or higher polynomial type of non-linear fit for the stress–strain relationship of soil in the form of *scant formulations*. This is known as the Cauchy type of elastic formulation. This type of elastic model must be combined with criteria defining 'failure' of the soil. The sections that follow present the failure criteria (Section 5.2) and the constitutive equations of hyperelastic and Cauchy elastic- and hypoelastic-based models which possess the hallmarks of elasticity in decreasing measure (Sections 5.3 to 5.5).

5.1.4 Plasticity and modelling

The elastic modelling in the form of scant formulations can be quite accurate for soils sustaining proportional loading. However, these formulations fail to identify plastic deformations when unloading occurs. This can, to some extent, be rectified by introducing *loading criteria* as in the *deformation theory of plasticity*. Although the deformation theory and the existence of loading function are incompatible even in the most limited sense, it is still a very attractive alternative for solution of large classes of soil and soil–structure interaction problems, because of its simplicity. This type of modelling is described in Section 5.6.

The *flow theory of plasticity* represents a necessary and correct extension of elastic stress–strain relations into the plastic range at which *permanent plastic* strain is possible in addition to elastic strain. This plastic strain remains when the stresses are removed. Thus, the strain in a plastic material may be considered as the sum of the reversible *elastic strain* and the permanent irreversible *plastic strain*. Since an elastic stress–strain law as described in the first part of the chapter is assumed to provide the relation between the incremental changes of stress and elastic strain, the stress–strain law for a plastic material reduces, essentially, to a relation involving the current state of stresses and strains and the incremental changes of stress and plastic strain. This relation is generally assumed to be homogeneous and linear in the incremental changes of the components of stress and plastic strain. This assumption precludes viscosity effects and thus constitutes the time-independent idealization.

The first step towards such a mathematical model is to establish the *yield limit* of an elastic material. This is known as the *yield function* which is a certain function of the stress components. A plastic material is called *perfectly plastic* or *work-hardening* or *softening* according to whether the yield function as represented by a certain hypersurface in six-dimensional stress space is a fixed surface or it admits changes (expansion or contraction) as plastic strain develops. For moderate strains, mild steel behaves approximately like a perfectly plastic material. It is therefore not surprising that in the early years (1950–1965) this perfect plasticity model was used almost exclusively and extensively in the analysis and design of steel structures because of its simplicity. The general theorems of limit analysis, developed on the basis of perfect plasticity, furnish simple, direct and realistic estimates of the load-carrying capacity of these structures. However, perfect plasticity is not nearly as appropriate for soils as for metals. Some of the troubles and their justifications for adoption of this idealization for practical use in soil mechanics are given in Section 5.7.

The use of *strain-* or *work-hardening plasticity* theories to describe the stress–strain behaviour of soils under cyclic loading conditions has been highly developed in recent years in soil mechanics. As a result of these advances, the theory of soil plasticity is now in a position to lead that of metal plasticity. Some of these developments will be described in Sections 5.8 to 5.10 for soils idealized as elastic–plastic materials with work-hardening.

5.2 FAILURE CRITERIA

5.2.1 General

The best known material property for soil appears to be that of the failure condition. There exist a number of failure criteria that reflect some important features of soil strength. Various of these criteria have been discussed in the book by Chen and Saleeb [9] where they were classified as one-paramter models including Tresca, von Mises, and Lade–Duncan criteria; and two-parameter models including the well-known Mohr–Coulomb criterion, extended Tresca and Drucker–Prager models, and the two-parameter failure model of Lade. The general shape of a failure surface in a three-dimensional stress space can best be described by its cross-sectional shapes in the deviatoric planes (or octahedral shear stress τ_0 planes) and its meridians in the hydrostatic planes (or octahedral normal stress σ_0 planes) as shown schematically in Figure 5.1 for one-parameter models and in Figure 5.2 for two-parameter models.

In all these strength models, two basic postulates are adopted; isotropy, and convexity in the principal stress space. The first assumption is mainly introduced because of the inherent simplification of the failure model. It is certainly true some clays exhibit significant anisotrophy with respect to their undrained strength which requires the formulation of the failure surface in the six-dimensional stress space instead of the three-dimensional space of principal stress. For many soils, however, the assumption of isotropy is reasonable. On the other hand, convexity is an assumption which is supported by global stability arguments in plasticity [9]. Clearly, there are some questions on the validity of this postulate, and in fact, there is experimental eveidence that the failure envelope for sands over a wide range of hydrostatic (confining) pressures is non-convex with respect to the hydrostatic axis (Bishop [3]).

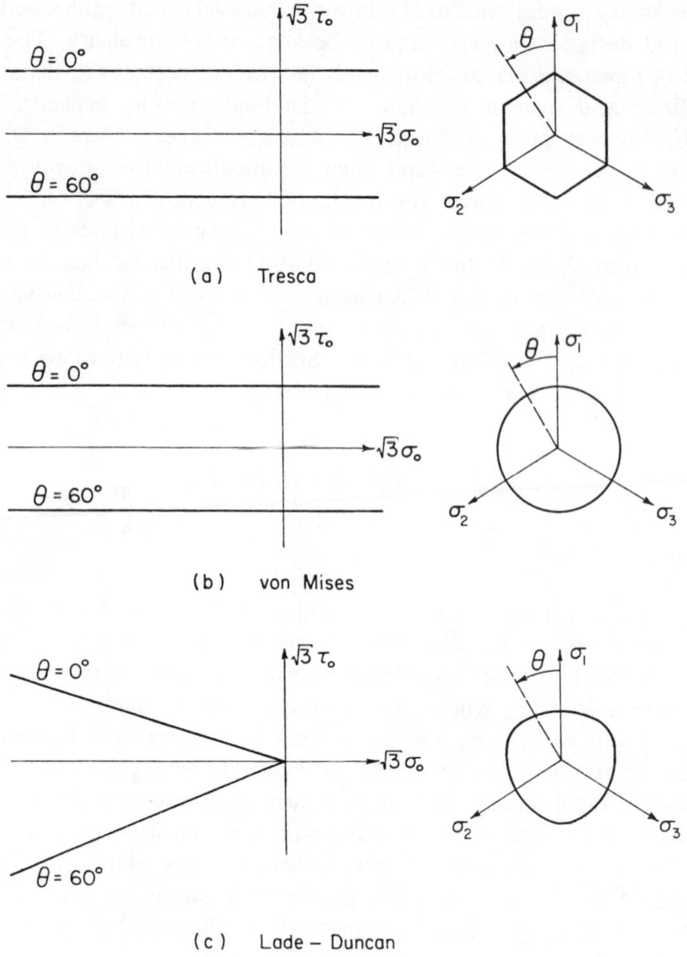

Figure 5.1 One-parameter models

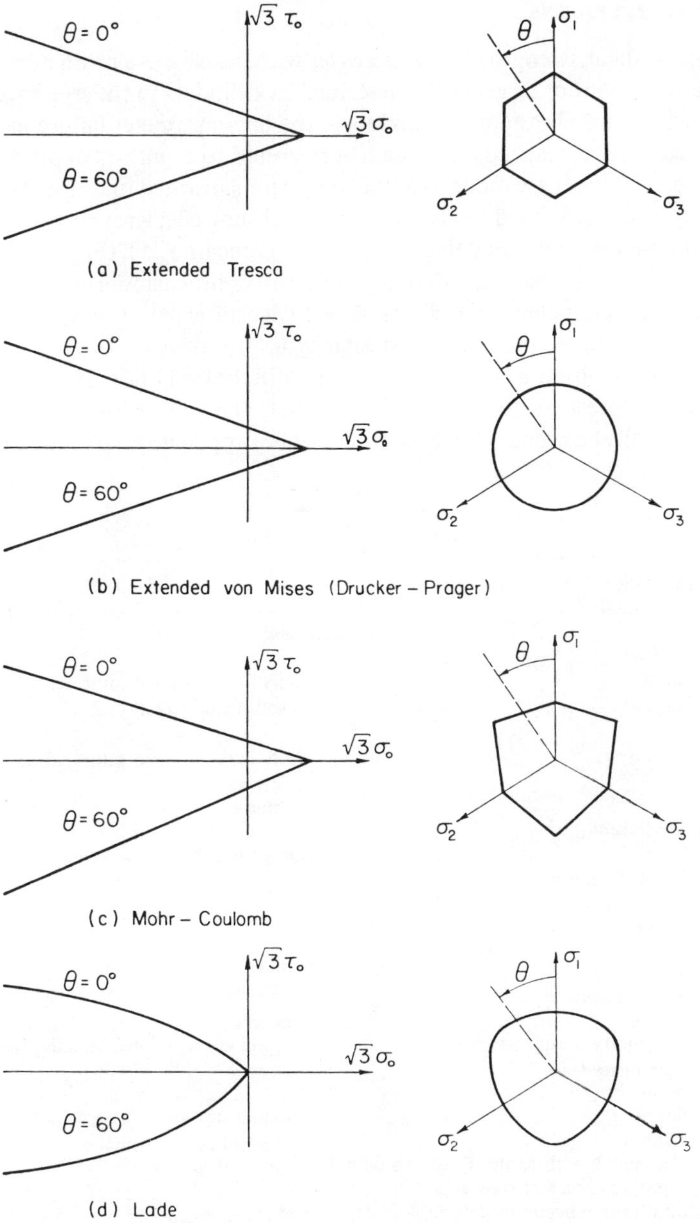

Figure 5.2 Two-parameter models

5.2.2 Strength models

Clearly, the shear strength of soils increases with the effective mean normal stress, and therefore cannot be generally described by cylinders in the stress space, such as the circular and hexagonal cylinders of the one-parameter failure models von Mises and Tresca, respectively, which correspond to a pure τ_0 dependence. The only case in which these models can be used is for saturated undrained soils when the analysis is performed in terms of the total (not effective) stresses.

The one-parameter model of Lade and Duncan [25] (Fig. 5.1(c)) is very efficient for cohesionless soils under general three-dimensional stress conditions. It is simple, and it includes the effects of the hydrostatic pressure, σ_0 on τ_0, and the dependence of τ_0 on the angle of similarity θ.

The Mohr–Coulomb criterion is certainly still the best known failure model for the isotropic pressure-sensitive soils (Fig. 2(c)). This criterion is, however, not mathematically convenient in three-dimensional applications due to the presence

Table 5.1 Strength models

	Advantages	Limitations
	One-parameter models	
1.	*von Mises*	
	Simple	Only for undrained saturated soils (total stress)
	Smooth	
2.	*Tresca*	
	Simple	Only for undrained saturated soils (total stress)
		Corners
3.	*Lade-Duncan*	
	Simple	Only for cohesionless soils
	Effect of intermediate principal stress	
	Smooth	
	Two-parameter models	
1.	*Mohr–Coulomb*	
	Simple	Corners
	Its validity is well established for many soils	Neglect effect of intermediate principal stress
2.	*Drucker–Prager*	
	Simple	Circular deviatoric trace which contradicts experiments
	Smooth	
	Can match with Mohr–Coulomb with proper choice of constants	
3.	*Lade's two-parameter model*	
	Simple	Only for cohesionless soils
	Smooth	
	Curve meridian	
	Wider range of pressures than the other criteria	

of corners (singularities). A reasonable smooth generalization of the Mohr–Coulomb into three-dimensional situations may be obtained by the Drucker–Prager cone (1952) (Figure 5.2(b)). There are basically two shortcomings of the Drucker–Prager surface in connection with soil-strength modelling; the linear dependence of τ_0 on σ_0 and the independence of τ_0 on the angle of similarity θ. It is known that τ_0–σ_0 relations is not linear in general and the trace of the failure surface on deviatoric planes is not circular. The generalized Lade–Duncan surface proposed by Lade [24] overcomes these two shortcomings (Figure 5.2(d)).

The two-parameter failure model of Lade [24] has been found adequate for a wide range of pressures for both sands and normally consolidated clays. It takes into account the curvature of the trace of the failure surface in the meridian plane indicating decreasing friction angle with increasing confining pressures, and a θ-dependence of τ_0 on the deviatoric planes. It should, however, be emphasized that this refinement requires additional computer effort. The advantages and limitations of these failure models are summarized in Table 5.1.

5.3 CAUCHY ELASTICITY AND MODELLING

5.3.1 General

For a Cauchy elastic material, the current state of stress, σ_{ij}, depends *only* on the current state of deformation, ε_{ij}; that is, stress is a function of strain (or vice versa). The constitutive relation of this material has the general form

$$\sigma_{ij} = F_{ij}(\varepsilon_{kl}) \tag{5.1}$$

Where F_{ij} is the elastic response function of the material. The behaviour of such material is both reversible and path independent in the sense that stresses are uniquely determined by the current state of strain (or vice versa). In general, although stresses are uniquely determined from strains (or vice versa), the converse is *not* necessarily true. Furthermore, reversibility and path dependency of the strain energy and complementary energy density functions, $W(\varepsilon_{ij})$ and $\Omega(\sigma_{ij})$ respectively, are *not* generally guaranteed. In fact, the Cauchy type of elastic models may generate energy for certain loading–unloading cycles (see, for example, Chen and Saleeb, [9]). That is, the models may violate the laws of thermodynamics, which is not acceptable on physical grounds. This has led to the consideration of the secant type of formulation called the Green hyperelastic type (Section 5.4).

5.3.2 Cauchy elastic models

In spite of these theoretical reservations, some simplified versions of non-linear Cauchy elastic constitutive models have been proposed for practical use in soil

Table 5.2 Modifications of linear elastic models

Advantages	Limitations
Conceptually and mathematically simple	Path-independent, reversible
Easy to determine the constants and wide data base is established for many parameters	No coupling between volumetric and deviatoric responses
	For arbitrary functions of the moduli, energy generation may occur for certain stress cycles

mechanics. In the following, the popular approach based on a simple modification of linear elasticity model is discussed.

Linear elasticity for isotropic materials constitutes the oldest and simplest approach to modelling the stress–strain behaviour of soils. It seems logical, therefore, that the linear isotropic model has formed the basis of developing various non-linear soil models. The simplest approach to formulate such non-linear models is to simply replace the elastic constants in the linear stress–strain relations with secant moduli dependent on the stress and/or strain invariants. Non-linear models of this type have been discussed in the papers by Boyce, 1980 [3a]; and Hardin and Drnevich [19]; among others.

These models are mathematically and conceptually very simple. The models account for two of the main characteristics of soil behaviour; non-linearity and the dependence on the hydrostatic stress.

The main disadvantage of the models is that they describe path-independent behaviour. Therefore, their application is primarily directed toward monotonic or proportional loading regimes. For arbitrary assumed functions for the secant moduli, there is no guarantee that the energy functions W and Ω will be path-independent and energy generation may be indicated in certain stress cycles, which is physically not acceptable.

The advantages and limitations of Cauchy elastic models based on modifications of the linear elasticity are summarized in Table 5.2.

5.4 HYPERELASTICITY AND MODELLING

5.4.1 General

To the contrary of the engineering or emperical approach described above, the classical hyperelasticity theory provides a more rational approach in formulating secant stress –strain models for soils. Here, the constitutive relations are based on the assumption of the existence of a strain energy function, W, or a complementary energy function, Ω, such that

$$\sigma_{ij} = \frac{\partial W}{\partial \varepsilon_{ij}}; \qquad \text{or} \qquad \varepsilon_{ij} = \frac{\partial \Omega}{\partial \sigma_{ij}} \tag{5.2}$$

in which W and Ω functions of the current components of the strain or stress tensors, respectively. Equation (5.2) yields a one-to-one relation between actual states of stress and strain. In addition to the reversibility and path independence of stresses and strains in hyperelastic type of elastic models, thermodynamic laws are always satisfied, and no energy can be generated through load cycles.

For an initially isotropic elastic material, W and Ω are expressed in terms of any three independent invariants of strain tensor, ε_{ij}, or stress tensor, σ_{ij}, respectively. Based on an assumed functional relation of W in terms of the strain invariants, or Ω in terms of the stress invariants, equation (5.2) can be used to obtain various non-linear elastic stress–strain relations in the form of secant formulation.

$$\dot{\sigma}_{ij} = \frac{\partial^2 W}{\partial \varepsilon_{ij} \partial \varepsilon_{kl}} \dot{\varepsilon}_{kl} = H_{ijkl} \dot{\varepsilon}_{kl} \tag{5.3}$$

$$\dot{\varepsilon}_{ij} = \frac{\partial^2 \Omega}{\partial \sigma_{ij} \partial \sigma_{kl}} \dot{\sigma}_{kl} = H'_{ijkl} \dot{\sigma}_{kl} \tag{5.4}$$

where the symmetrical matrices of the components of the fourth-order tensors H_{ijkl} and H'_{ijkl} are known mathematically as the Hessian matrices of function W and Ω, respectively.

From equations (5.3) and (5.4), it is observed that tangent moduli are identical for loading and unloading. Thus, the hyperelastic model yields a constitutive relation which is incapable of describing load history-dependence and rate-dependence. Hyperelasticity exhibits strain-induced anisotrophy in the material. Material instability occurs when

$$\det|H_{ijkl}| = 0 \quad \text{or} \quad \det|H'_{ijkl}| = 0 \tag{5.5}$$

Despite its shortcomings, hyperelasticity type of models has been utilized as non-linear constitutive relations for soil.

5.4.2 Hyperelastic models

In the early applications of the finite element method to soil mechanics problems, simplified forms of hyperelasticity were generated and used through a simple extension of the linear theory of elasticity. In this approach, strain or stress dependent and coupled or uncoupled bulk and shear moduli are used to construct a secant type of constitutive relation for coupled or uncoupled volumetric and deviatoric stresses and strains. A third-order model, based on the classical theory of hyperelasticity, has been formulated by Evans and Pister [18] and subsequently used by Ko and Masson [21], and Saleeb and Chen [41] among others in soil mechanics.

The hyperelastic formulation can be quite accurate for soils straining in proportional loading. Moreover, use of these models in such cases satisfies the rigorous theoretical requirements of continuity, stability uniqueness, and energy

Table 5.3 Hyperelastic models

Advantages	Limitations
Satisfy stability and uniqueness	Path-independent, reversible
Shear-dilatancy, and effect of all stress invariants may be included	Difficult to fit and requires large number of tests
Attractive from programming and computer economy points of view	Most models confined to small regions of applications

consideration of continuum mechanics. However, as noted previously, models of the hyperelastic type fail to identify the inelastic character of soil deformations, a shortcoming that becomes apparent when the material experiences unloading.

The main objection to the hyperelastic formulation is the complications involved with the material constants. Even when initial isotropy is assumed, a non-linear hyperelastic model often contains to many material parameters. For instance, a third-order model requires nine constants; while 14 constants are needed for fifth-order hyperelastic model. A large number of tests are generally required to determine these constants, which limits the practical usefulness of the models.

The advantages and limitations of hyperelastic models are summarized in Table 5.3

5.5 HYPOELASTICITY AND MODELLING

5.5.1 General

An obvious shortcoming in both of the previous types of non-linear elasticity models is the path-independent behaviour implied in the secant stress–strain formulation, which is certainly not true for soils in general. A more improved description of the soil behaviour is provided by the hypoelastic formulation in which the incremental stress and strain tensors are linearly related through variable material response moduli that are functions of the stress or strain state [43]; e.g.

$$\dot{\sigma}_{ij} = C_{ijkl}(\sigma_{\mathrm{mn}})\,\dot{\varepsilon}_{kl} \tag{5.6}$$

in which the material tangential response function C_{ijkl} describes the instantaneous behaviour directly in terms of the time rates of stress $\dot{\sigma}_{ij}$. These incremental stress–strain relations provide a natural mathematical model for materials with limited memory. This can be seen by an integration of equation (5.6).

$$\sigma_{ij} = \int_0^t C_{ijkl}(\sigma_{\mathrm{mn}}) \frac{\partial \varepsilon_{kl}}{\partial \tau}\,\mathrm{d}\tau + \sigma_{ij}^0 \tag{5.7}$$

The integral expression clearly indicates the path-dependency and irreversibility of the process. The hypoelastic response is therefore stress history (path) dependent. In the linear case for which $C_{ijkl}(\sigma_{mn})$ is a constant, the hypoelasticity degenerates to hyperelasticity, which corresponds to the history independent secant modulus formulation. The integration in equation (5.7) can be carried out explicitly and leads to the hyperelastic formulation, equation (5.3).

As observed from equation (5.6), the tangential stiffnesses are identical in loading and unloading. This reversibility requirement only in the infinitesimal (or incremental) sense justifies the use of the term hypoelastic or *minimum elastic*. Material instability or failure occurs when

$$\det|C_{ijkl}(\sigma_{mn})| = 0 \tag{5.8}$$

Equation (5.8) leads to an eigenvalue problem of which the eigenvectors span a surface, the failure surface, in the stress space.

There are two problems associated with the hypoelasticity modelling. The first problem is that, in the non-linear range, the hypoelasticity-based models exhibit stress induced anisotropy. This anisotropy implies that the principal axes of stress and strain are different, introducing a coupling effect between normal stresses and shear strains. As a result, a total of 21 material moduli for general triaxial conditions have to be defined for every point of the material loading path. This is a difficult task for practical application.

The second problem is that, under the uniaxial stress condition, the definition of loading and unloading is clear. However, under multiaxial stress conditions, the hypoelastic formulation provides no clear criterion for loading or unloading. Thus, a loading in shear may be accompanied by an unloading in some of the normal stress components. Therefore, additional assumptions are needed for defining the loading–unloading criterion.

5.5.2 Hypoelastic models

In the simplest class of hypoelastic models, the incremental stress–strain relations are formulated directly as a simple extension of the isotropic linear elastic model with the elastic constants replaced by variable tangential moduli which are taken to be functions of the stress and/or strain invariants. This approach has been commonly used in many geotechnical applications. Various simple incremental models have been discussed in the papers by Duncan and Chang [17], and Kondner [22], among others.

Models of this type are attractive from both computational and practical viewpoint. They are well suited for implementation of finite element computer codes. The material parameters involved in the models can be easily determined from standard laboratory tests using well defined procedures; and many of these parameters have a broad data base.

The early incremental finite element analyses were conducted with these simplified forms of hypoelasticity. In the simplest approach, the incremental constitutive model is based on an isotropic formulation using test data from a single parameter load set-up, resulting in, for example, a stress or strain dependent modulus of elasticity. To this end, three classes of formulations have emerged: Hyperbolic, parabolic, and exponential relations. In spite of the theoretical reservations against isotropic modelling with identical moduli in the principal directions and no coupling with the shear response, the hyperbolic type of models and their generalizations have been applied extensively in the past and used successfully in the finite element solution of non-linear soil mechanics problems (see Chen and Saleeb [9]).

A more sophisticated model is based on the decoupling of volumetric and deviatoric stress and strain rates with two parameters. In this case, the non-linear deformation model is developed on the basis of an isotropic formulation with variable bulk moduli and shear moduli.

The application of this type of hypoelastic model should be confined to monotonic loading situations which do not basically differ from the experimental tests from which the material constants were determined or curve fitted. Thus, the isotropic models should not be used in cases such as non-homogeneous stress states, non-proportional loading paths or cyclic loadings.

Examples of the classical formulations and applications of the first-order hypoelastic models can be found in the papers by Davis and Mullenger [12]; and Tokuoka [42], among others. Again, as for the hyperelastic models, the practical usefulness of the hypoelastic models is limited by the nature and number of tests

Table 5.4 Hypoelastic models

Advantages	Limitations
Modification of the linear elastic models	
Conceptually and mathematically simple	Incrementally reversible
Ideal for finite element implementation	No coupling between volumetric and deviatoric responses
Easy to fit	When E_t and ν_t are used, the behaviour near failure cannot be described adequately
Many of the parameters have wide data base	Possible energy generation for certain stress cycles if arbitrary functions for the moduli are used
Have been used successfully in many practical applications	
First-order hypoelastic models	
Stress-path dependency	Incrementally reversible
Stress-induced anisotropy	Tangent stiffness matrix is generally unsymmetric; thus requires increased storage and computation
	Difficult to fit and requires large numbers of tests
	Possible energy generation for certain stress cycles
	No uniqueness proof in general

required to determine the material constants. There is no unique way to determine these constants. Also, the material tangent stiffness matrix for a hypoelastic model is generally unsymmetric which results in considerable increase in both storage and computational time. Further, in such cases, uniqueness of the solution of boundary value problems cannot generally be assured.

The advantages and limitations for two of the hypoelastic models are summarized in Table 5.4.

5.6 DEFORMATION PLASTICITY AND MODELLING

5.6.1 General

The fundamental difference between elasticity and plasticity models lies in the distinction in the treatment of loading and unloading in plasticity theories. This is achieved by introducing the concept of a loading function. In addition, the total deformations ε_{ij} are decomposed into elastic and plastic components ε_{ij}^{e} and ε_{ij}^{p} by simple superposition:

$$\varepsilon_{ij} = \varepsilon_{ij}^{e} + \varepsilon_{ij}^{p} \tag{5.9}$$

The plastic strain is obtained from

$$\varepsilon_{ij}^{p} = \phi \frac{\partial F}{\partial \sigma_{ij}} \tag{5.10}$$

where ϕ is a scalar function relating to a one-dimensional test curve, positive during loading and zero during unloading, and F is a scalar function of the stress state and possible also of some hardening parameters.

In the deformation theories of plasticity for work-hardening materials, it postulates that the state of stress determines the state of strain uniquely as long as plastic deformation continues. Thus, they are identical with non-linear elastic stress–strain relations of the secant type as long as unloading does not occur.

If any of the elasticity models described above are to be used to describe soil behaviour under general loading conditions involving loading and unloading, it must be accompanied by special unloading treatment based on a criterion defining loading–unloading. Such a formulation is closely related to the deformation theory of plasticity [6].

In general, it has been clearly demonstrated that, except for certain special cases of loading (e.g. increasing proportional loading), the deformation type of theories cannot lead to meaningful results, and sometimes they lead to contradictions. For example, these types of models do not satisfy the *continuity* requirement for loading conditions near or at neutral loading. Basically, the difficulty lies in the fact that the deformation theory and the existence of the loading function, f, even in the most limited sense, are incompatible. This has led naturally to the consideration of the second type of formulation based on incremental theory of plasticity. This theory is based on three fundamental assumptions, the shape of an *initial yield surface*, the evolution of subsequent *loading surfaces* (*hardening rule*), and the formulation of an appropriate *flow rule*. In addition, the total strain increments, $\dot{\varepsilon}_{ij}$, are assumed to be the sum of the elastic and plastic strain components $\dot{\varepsilon}^{e}_{ij}$ and $\dot{\varepsilon}^{p}_{ij}$, respectively. Different constitutive models based on incremental theory of plasticity are described in the following sections.

5.6.2 Variable moduli models

A generalization of the deformation theory of plasticity for the case of incremental stress–strain models is that now known as the *variable moduli models* [32]. In these later models, different forms for the material response functions apply in initial loading, and in subsequent unloading and reloading, i.e. the models are generally irreversible, even for incremental loading.

The *variable-moduli models* have been extensively used to describe the behaviour of soils in ground shock studies [31]. The mathematical description of the variable moduli model is given in terms of the incremental stress–strain relations

$$\dot{p} = K\dot{\varepsilon}_{kk}; \quad \text{and } \dot{s}_{ij} = 2G\dot{e}_{ij} \tag{5.11}$$

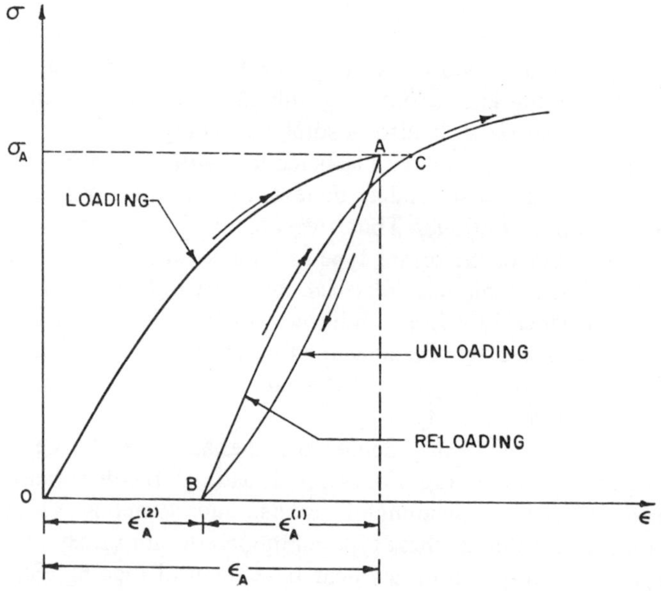

Figure 5.3 Typical uniaxial stress–strain relation for a variable moduli model

Table 5.5 Deformational plastic models

Advantages	Limitations
Deformation theory of plasticity	
Simple formulation Allow hysteretic behaviour	Continuity problem near or at neutral loading With the exception of unloading behaviour is still path-independent
Variable moduli models	
Simple Good fit of data Allow hysteretic behaviour Easy to fit Suitable for finite element implementation	Continuity problem near or at neutral loading

where $\dot{p}$ and $\dot{\varepsilon}_{kk}$ are the mean hydrostatic stress and the volumetric strain increments, respectively, and $\dot{s}_{ij}$ and $\dot{e}_{ij}$ are the deviatoric stress and strain increments. Generally, different functions for shear modulus G and bulk modulus K apply in initial loading, and in subsequent loading and reloading. This is illustrated in Figure 5.3. These relations are incrementally isotropic.

Variable moduli models have many advantages. They give good overall fit to the full set of tests available, and they are capable of fitting repeated hysteretic data in cyclic loading. In addition, they are computationally simple and relatively easy to fit to data. However, phenomena such as dilatation, cross effects, and non-coincidence of principal axes of stress and strain increments cannot be described by such relations as equations (5.11). The other problem associated with this type of formulation is that the model may not satisfy all rigorous theoretical requirements for all stress histories. For instance, at or near neutral loading conditions in shear, the model fails to satisfy the continuity conditions.

The advantages and limitations of constitutive models based on deformational type of plasticity theory are summarized in Table 5.5.

5.7 PERFECT PLASTICITY AND MODELLING

5.7.1 General

The development of the incremental stress–strain relations in the flow theory of plasticity is based on three fundamental assumptions [4, 6]:

(1) The existence of initial and subsequent yield (loading) surfaces, $f=0$.
(2) The formulation of a hardening rule that describes the evolution of subsequent loading surfaces.
(3) A flow rule which specifies the general form of the plastic stress–strain relation.

For an elastic–perfectly plastic model, no hardening occurs and all the subsequent yield surfaces coincide with the initial surface and remain fixed in the stress space.

According to the flow rule, the plastic strain increment is given by

$$\dot{\varepsilon}^{p}_{ij} = d\lambda \frac{\partial g}{\partial \sigma_{ij}} \tag{5.12}$$

where g is the plastic potential function, and $d\lambda$ is a loading parameter. When $g = f$, equation (5.12) is known as the associated flow (normality) rule, otherwise it is a non-associated flow rule.

Many proposals of geometrical forms of the yield surface $f(\sigma_{ij}) = 0$ can be found in the literature [9]. Some of the classical models are shown in Figures 5.1 and 5.2.

Drucker–Prager type of elastic–perfectly plastic models were discussed and evaluated in the book by Chen [4], among others. These models are computationally simple. With the proper selection of the material constants, the Drucker–Prager criterion can be matched with Mohr–Coulomb condition (e.g. to give the same collapse load under the plane strain condition [27].The models reflect some of the important characteristics of soil behaviour such as: elastic response at lower loads, small material stiffness near failure, failure condition, and elastic unloading after yielding. However, the main disadvantages of the models are the excessive amount of plastic dilatation at yielding as a result of the normality rule used, and the inability to describe hysteretic behaviour within the failure surface. A model of this type can be considered a fair first approximation in the progressive failure analysis of soil media and soil-structure interaction problems.

From the consistency condition $\dot{f} = 0$ and associated or non-associated flow rule, the general form of an incremental plastic stress–strain relation can be written as

$$\dot{\sigma}_{ij} = C^{ep}_{ijkl} \dot{\varepsilon}_{kl} \tag{5.13}$$

where the elastic–plastic constituent tensor C^{ep}_{ijkl} is given by [6].

$$C^{ep}_{ijkl} = C^{e}_{ijkl} - \frac{H^{*}_{ij} H_{kl}}{H} \tag{5.14}$$

where C^{e}_{ijkl} is elastic stiffness tensor. Note the lack of symmetry if non-associated flow rules are used $H^{*}_{ij} \neq H_{ij}$, but it becomes a symmetric one for the associated flow rule case $H^{*}_{ij} = H_{ij}$.

Comparing equation (5.13) with equation (5.6), a marked similarity can be seen between hypoelastic theory and incremental plastic theory, but the difference is extremely important. For the case of unloading, $f = 0, \dot{f} < 0$, there is no change in any component of plastic strain $\dot{\varepsilon}^{p}_{ij}$, all the deformations are elastic with the elastic material stiffness tensor C^{e}_{ijkl}.

Table 5.6 Drucker–Prager perfectly plastic models

Advantages	Limitations
Simple to use Easy to use Can be matched with Mohr–Coulomb by a proper selection of constants Computer codes available Limit analysis techniques can be used Satisfy uniqueness requirements (associated flow rule)	Excessive plastic dilatancy at yielding Cannot reproduce the hysteretic behaviour within the failure surface Cannot predict the pore pressure build-up during undrained subfailure cyclic shear loading

The flow theory of plasticity exhibits both stress and strain history dependence and stress-induced anisotropy, but age- and rate-independence. This can be seen clearly if components of the foregoing stress–strain relation are written out with the stress increments appearing explicitly in the form

$$d\varepsilon_x = D_1\, d\sigma_x + D_2\, d\sigma_y + D_3\, d\sigma_z + D_4\, d\tau_{xy} + D_5\, \tau_{yz} + D_6\, d\tau_{zx} \tag{5.15}$$

where the D's are functions of the state of stress and include both the elastic and plastic behaviour. They look like a highly anisotropic incremental generalized Hooke's law. An increment of shearing stress may produce an elongation or contraction, and similarly an increment of normal stress may cause a shearing strain. However, the anisotropy is produced by the existing state of stress and is not intrinsic. Removal of the stress leaves a material isotropic in the usual sense. Similarly, rotation of the state of stress with respect to the material rotates the anisotropy.

The advantages and limitations of Drucker–Prager type of perfectly plastic models are summarized in Table 5.6.

5.7.2 Limit analysis in soil mechanics

For the most part, the concept of perfect plasticity has been used extensively in different fields of engineering applications in order to determine the collapse load. This includes a wide variety of applications of limit analysis for metals, concrete, and soils. Examples of such applications for metallic and concrete structures are summarized in ASCE–WRC [1], and Chen [6], respectively. For stability problems in soil mechanics, different widely known techniques have been used to obtain the collapse load, such as the slip-line method, the limit equilibrium method, and the modern limit analysis method [4].

A perfectly plastic idealization for soil is of real value for many stability problems in soil mechanics. Nevertheless, the idealization is severe and it is necessary to guard against improper interpretation. Since the perfect plastic idealization ignores the real work-hardening or softening of the soil beyond the arbitrarily chosen yield stress level (Figure 5.4), it must therefore be interpreted as

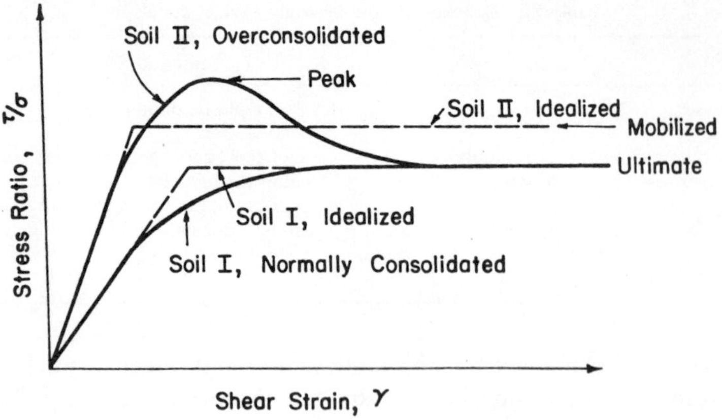

Figure 5.4 Typical stress–strain curves and perfectly plastic idealizations

an average value with the meaning that no more than small plastic deformation takes place in the so-called elastic range but large plastic deformation occurs in the collapse state.

Although the application of limit analysis to soil mechanics are relatively recent, there have been an enormous number of practical solutions available [4]. Many of the solutions obtained by the method are remarkably good, when comparing with the existing results for which satisfactory solutions already exist. The limit analysis approach is not only rigorous from the theoretical viewpoint, but the techniques are competitive with those of the limit equilibrium approach. It can be used as a working tool for design engineers to solve everyday problems in soil mechanics. The discussion on the validity of limit analysis in application to soils can be found elsewhere [4, 6, 8].

5.8 ISOTROPIC HARDENING PLASTICITY AND MODELLING

5.8.1 Basic concept

The general technique used in the previous section in the discussion of strees–strain relations for an elastic–perfectly plastic material can readily be extended to the material with work-hardening. The concept of a yield surface fixed in stress space for a perfectly plastic material is generalized for a hardening material to the idea of a *loading surface*. The *hardening* rule is used to define the motion of the yield surface during plastic loading.

There are a number of hardening rules, such as isotropic hardening, kinematic hardening, and mixed hardening [6]. The isotropic hardening rule allows for a uniform expansion of the loading surface, while the kinematic hardening rule permits the loading surface to move as a rigid body in stress space. Mixed hardening, on the other hand, combines both of these types of hardening into a single rule where the user selects a continuous scalar variable to be zero for kinematic hardening or one for isotropic hardening or some non-integer for a combination of the two. It is well known that isotropic models apply mainly to monotonic loading while for cyclic loadings and pronounced Bauschinger effects, kinematic hardening rules would be more appropriate.

Non-metallic materials such as soils exhibit hardening as well as softening effects in the post-failure regime. As a first step to construct a stress–strain model for such a complicated material as soil, it is logical to utilize the classical theories of work-hardening plasticity as well developed for materials such as metals. Using this approach, Roscoe and his students have found that 'wet' clays can be described remarkably well by a simple isotropic work-hardening idealization [39, 40]. It must be noted, however, that any such explicit description in phenomenological or mathematical terms is bound to be a drastic idealization of actual behaviour of soils and cannot be expected to be vaild over a wide range of conditions. Thus, when comparing the simple isotropic model with experiments, we should accordingly be satisfied with agreement in trends and not expect agreement in details in general.

In examining the advantages and shortcomings of the isotropic hardening formulation for soils, Drucker [14] concludes that yield surfaces are a matter of definition and the choice is not an absolute one, but is determined by the most significant features of the problem to be solved. Thus, if a small-strain–offset definition of yield is used, the comparison of actual yield surfaces for soils with isotropic hardening models is found rather poor in a detailed look at the yield surfaces and at the corresponding stress–strain relations. However, if a large offset definition of yielding is taken instead, a much less complicated picture emerges for a variety of loading paths which does not differ much from an isotropic hardening model. For practical design purposes, the crude measure of yielding may be more relevant than the refined small strain definition. This has been demonstrated clearly in the previous section for the case of a more drastic idealization of perfect plasticity, where a choice of a fixed yield value in the middle of the stress range of interest, as indicated on the simple compression diagram of Figure 5.4, is found of real value for many problems in soil mechanics.

The fact that the loading surface at large strains is almost independent of loading path leads to the concept of bounding surface models where the subsequent loading surfaces are allowed to move isotropically inside a *limit surface* or *bounding surface*, which grows and moves independently and encloses all the subsequent loading surfaces. Indeed, the assumption made concerning the hardening rule introduces the major distinction among various advanced plasticity models developed for soils in recent years. In the following sections, several plasticity models of soils with more complex hardening rules combining the concepts of kinematic and isotropic hardening are presented for the description of soils under monotonic and cyclic loading conditions. These work-hardening models are considered to be the major advances in recent years toward a more realistic representation of soil behaviour. A comprehensive discussion of the historical development of the modern theory of soil plasticity is given elsewhere [7].

5.8.2 General description of Cap Models

Apparently, Drucker *et al.* [15] were the first to suggest that soil might be modelled as an elastic–plastic work-hardening material. They proposed that successive yield functions might resemble Drucker–Prager cones with convex end caps. As the soil work-hardens, both the cone and cap expand. There are two important innovations in this reference. The first is the introduction of the idea of a spherical cap fitted to the cone. The second is the use of current soil density (or voids ratio, or plastic compaction) as the work-hardening parameter to determine the successive loading surfaces. There will be a succession of such surfaces, all geometrically similar, but of different sizes, for different specific volumes or densities.

Based on the same idea of using a cap as part of the yield surface, various types of cap models have been developed at Cambridge University (Cam-clay and modified Cam-clay models). In the modified Cam-clay model, only the volumetric strain is assumed to be partially recoverable; i.e., elastic shear strain component is assumed to be identically zero. Elastic volumetric strain is non-linearly dependent on hydrostatic stress and is independent of deviatoric stresses. For certain stress histories, this model strain-softens rather than strain-hardens. The additional feature which has been the integral part of all of these Cambridge models has been the concept of critical states. Discussion of various versions of Cambridge models are summarized in Palmer [33] and Parry [34].

The movement of the cap is controlled by the increase or decrease of plastic volumetric strain. Work-hardening is therefore reversed. It is this mechanism that leads to an effective control on dilatancy, which can be kept quite small (effectively zero) as required for many soils.

DiMaggio and Sandler [13] proposed a generalized cap model, where the functional forms for the cap model may be quite general and would allow for the fitting of a wide range of material properties. Their model has also been adapted for rocks by allowing only expansion of the cap (i.e. hardening.) In this variation of the model, the cap is not reversible. It, therefore, allows representation of a relatively large amount of dilatancy which is often observed during failure of rocks at low pressures. Anisotropy may be achieved by introducing pseudo-stress-invariants which are similar in form to stress-invariants but include weighting factors on the stress component in different directions. These modified cap models are now widely used in ground shock computations in soil mechanics.

The advantages and limitations of cap type of hardening models are summarized in Table 5.7.

Table 5.7 Cap type of isotropic hardening models

Advantages	Limitations
Satisfy all the theoretical requirements of stability, uniqueness, and continuity	Trial-and-error procedure needed to fit the data
Give a proper control on plasticity dilatancy	Relatively complicated
Allow hysteretic compaction during hydrostatic load–unload cycles	Cannot predict the pore pressure build-up during undrained sub-failure cyclic shear loading (if a kinematic cap is used this may be adequately described)

5.9 KINEMATIC HARDENING PLASTICITY AND MODELLING

5.9.1 Concept of nested yield surface

An alternative approach to the isotropic hardening type of models described above is provided by the kinematic type of strain-hardening rules. Recently, this approach has been employed by several researchers to provide for a more realistic representation of soil behaviour under reversed, and particularly cyclic, loading conditions.

Iwan [20], following the related work of Masing [26], proposed one-dimensional plasticity models which consist of a collection of perfectly elastic and rigid-plastic or slip elements arranged in either a series-parallel or a parallel-series combination. The model can contain a very large number of elements, and the properties of these elements can be distributed such that they can match the particular form of hysteretic behaviour of a certain type of soil. Such models are known as the overlay or mechanical sublayer models (e.g. Zienkiewicz *et al.* [44]). The concept of the sublayer model was originally introduced by Duwez [16] and further developed later by Besseling [2].

In order to extend the one-dimensional model to three-dimensional situations, and extended formulation of the classical incremental theory of plasticity has been proposed by Iwan [20]. Instead of using a single yield surface in stress space, he postulated a family (nest) of yield surface (Figure 5.5) with each surface translating independently in a pure kinematic manner, or individually obeying a linear work-hardening model. Their combined action, in general, gives rise to a piecewise linear work-hardening behaviour for the material as a whole. The approach leads to a realistic Bauschinger effect of a type that could not be obtained by using a single yield surface and a non-linear work-hardening rule even with kinematic hardening. The same concept of using a field of nested yield surfaces was also proposed independently by Mróz [28]. All the nested yield surface models have a feature common to the early sublayer models;

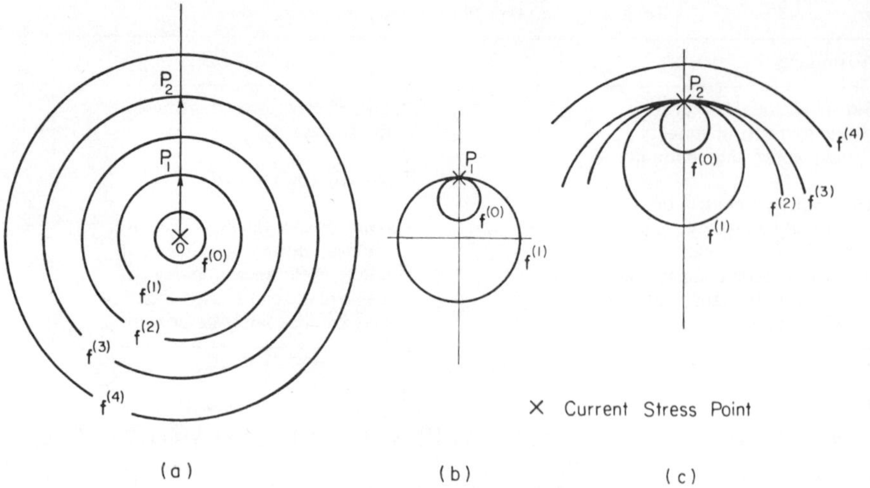

Figure 5.5 Series of nested yield surfaces in Mróz/Iwan type of strain-hardening models

namely, a piecewise linear stress–strain behaviour under proportional loading. These models have been used recently for soils and are usually known as nested yield surface or multi-surface plasticity models. In all the proposed models of this type, associated flow rule has been utilized.

5.9.2 An illustrated example

Figure 5.5 demonstrates the qualitative behaviour of a multisurface model with pure kinematic hardening. The initial positions of the yield surface $f^{(0)}$, $f^{(1)}$, $f^{(2)}$, and $f^{(3)}$ are shown in Figure 5.5 (a). When the stress point moves from 0 to P_1, elastic strains first occur until the surface $f^{(0)}$ is reached, where the plastic flow begins and the surface $f^{(0)}$ starts to move towards the surface $f^{(1)}$. Before their contact, the hardening modulus $H^{(0)}$ associated with $f^{(0)}$ governs the plastic flow according to the normality flow rule. However, when $f^{(0)}$ engages $f^{(1)}$ at P_1 (Figure 5.5(b)), the first nesting surface $f^{(1)}$ becomes the active surface and, upon further loading, the hardening modulus $H^{(1)}$ applies in the flow rule. Both $f^{(0)}$ and $f^{(1)}$ are then translated by the stress point, and they remain tangent to each other on the stress point and on the stress path until they touch the yield surface $f^{(2)}$ which then becomes the active surface. For subsequent contacts of consecutive surfaces, the process is repeated with new corresponding values of hardening moduli apply. The situation when $f^{(3)}$ is reached is illustrated in Figure 5.5(c).

5.9.3 Mixed-hardening nested yield surface model

Recently, Prevost [35, 36, 37] has extended the Iwan/Mróz model for the undrained behaviour of clays under monotonic and cyclic loading conditions. In

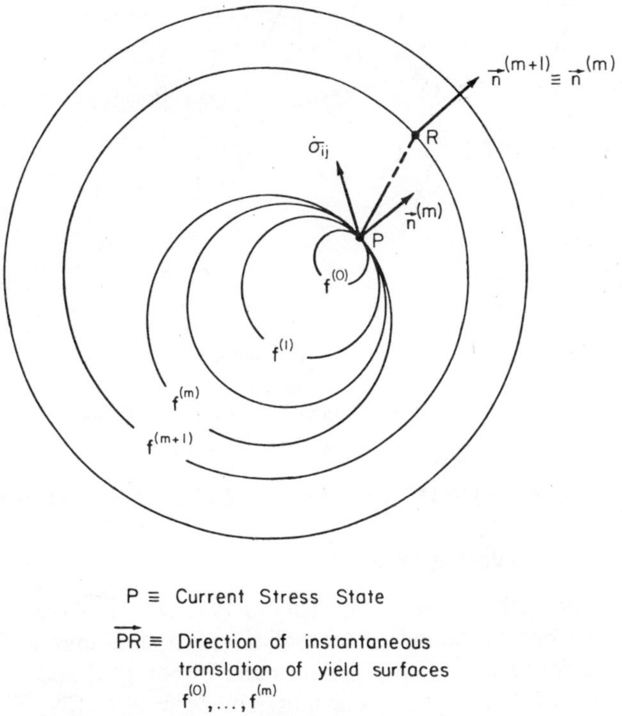

Figure 5.6 Translation rule of nested yield surfaces in stress space—Prevost type of strain-hardening models

this development, the basic work-hardening rule is still of the kinematic type but a simultaneous isotropic hardening (or softening) is allowed. The rule used to govern the translation of the nested yield surfaces during plastic loading was that suggested by Mróz [28]. According to this rule, when the stress point P (Figure 5.6) reaches the yield surface $f^{(m)}$, then, upon further loading (i.e. when the stress increment $\dot{\sigma}_{ij}$ is applied), the instantaneous translation of $f^{(m)}$ towards the next yield surface $f^{(m+1)}$ will be along **PR**, where R (known as the conjugate point) is the point on $f^{(m+1)}$ with outward normal in the same direction as the normal to $f^{(m)}$ at P(i.e., $\mathbf{n}^{(m+1)} = \mathbf{n}^{(m)}$) in Figure 5.6.

This model has been adopted very recently by Prevost *et al.* [38] in the finite element analyses of soil–structure interaction of centrifugal models under both monotonic and cyclic loadings simulating the situation encountered in the analysis of offshore gravity structure foundations under waves forces. The results obtained from the analysis agree quite well with those measured experimentally in the centrifuge. This study has demonstrated the ability of the multi-surface model to provide realistic representation of soil behaviour under complex loadings.

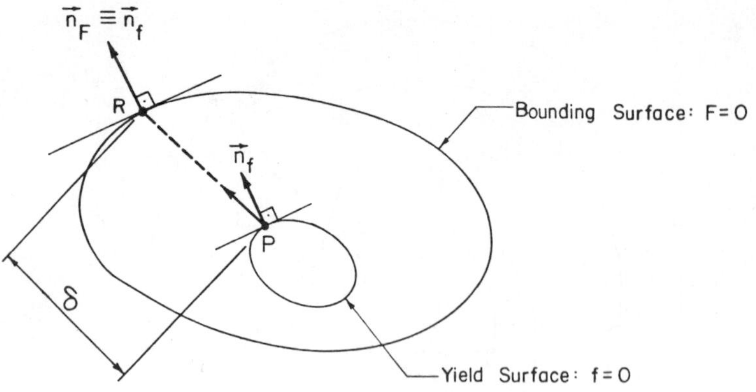

Figure 5.7 Yield and bounding surfaces in stress space

5.10 MIXED HARDENING PLASTICITY AND MODELLING

5.10.1 Concept of bounding surface

Various types of work-hardening plasticity models have been recently employed for soils based on the bounding (consolidation or limiting) surface concept introduced earlier for metals (e.g. Dafalias and Popov [10] and Krieg [23]). A two-surface model of this type was proposed by Mróz *et al.* [29, 30] for clays. A bounding surface, $F=0$, representing the consolidation history of the soil, and yield surface, $f=0$, defining the elastic domain within the bounding surface (Figure 5.7) were employed in the model.

The bounding surface was assumed to expand or contract isotropically, but the yield surface was allowed to translate, expand or contract within the domain enclosed by the bounding surface. The translation of the yield surface is governed by the same rule as the multi-surface models described earlier (i.e., f will translate towards the bounding surface along **PR** in Figure 5.7).

The hardening modulus on the yield surface is assumed to vary depending on the relative configuration of the yield and bounding surfaces. It was this last assumption that distinguished the present formulation from the previous multi-surface models. Here, instead of using a field of nested yield surfaces with associated hardening moduli, an interpolation rule is utilized to define the variation of the hardening moduli between the yield and the bounding surfaces. In the interpolation rule used by Mróz *et al.* [29, 30], the hardening modulus, H, was taken as a function of the distance δ (Figure 5.7) between the current stress point, P, on the yield surface and its conjugate point, R, on the bounding surface. Very detailed discussions of the present model and its application to represent the behaviour of clays under monotonic and cyclic triaxial test conditions have been given by Mróz *et al.* [30].

Dafalias and Herrmann [11] have also used the bounding surface formulation to describe the behaviour of clay under cyclic loadings. However, no explicit yield surface was postulated. Associated flow rule was utilized for the bounding surface. In this model, the variation of the plastic (hardening) modulus within the boundary surface is defined based on a very simple radial mapping rule. For each actual stress point within (or on) the bounding surface, a corresponding 'image' point on the surface is specified as the intersection of the surface with the straight (radial) line connecting the origin with the current stress point (the origin was assumed to be always within the bounding surface). The actual hardening modulus is then assumed to be a function of the hardening modulus on the bounding surface, at the 'image' point, and the distance between the actual stress point and its 'image'.

5.10.2 Concluding remarks

Unlike the nested yield surface models, the bounding surface models have the definite advantage of a smooth transition from elastic to full plastic state for general reversed loading which is generally observed experimentally on soils. In fact, the bounding surface model can be considered as a smooth extension of the nested yield surface model with an analytical function to replace the nest of several loading surfaces.

As mentioned previously, the large-strain–offset definition of yield leads to subsequent yield surfaces that resemble far more a uniformly expanded version of the initial yield surface than the translating and shrinking yield surfaces defined by the small-strain–offset definition. This insight leads one directly to the two-surface model concept with the usual concept of a loading surface enclosed by a 'limit' or 'bounding' surface. The loading surface is allowed to harden independent of the hardening of the bounding surface. Thus, the loading surface may be controlled by the mixed hardening rule for small strains, while the limit or bounding surface can harden independently in an isotropic manner as is observed for some soils at large strains.

ACKNOWLEDGEMENTS

This material is based in part upon work supported by the National Science Foundation under Grant No. PFR-7809326 to Purdue University.

REFERENCES

1. ASCE–WRC, *Plastic Design in Steel—A Guide and Commentary*, Manual No. 41, ASCE, New York, 1971.
2. J. F. Besseling, 'A theory of elastic, plastic and creep deformation of an initially

isotropic material showing strain hardening, creep recovery and secondary creep'. *Journal of Applied Mechanics*, **25**, *Transaction, ASME*, **80**, 529 (1958).

3. W. W. Bishop 'Shear strength parameters for undisturbed and remoulded soil specimens', *Proceedings of the Roscoe Memorial Symposium: Stress–Strain Behaviour of Soils* (Ed. R. H. G. Parry), pp. 3–58, Foulis, Henley-on-Thames, 1972.

3a. H. R. Boyce, 'A non-linear model for the elastic behaviour of granular materials under repeated loading', *International Symposium on Soils Under Cyclic and Transient Loading*, Swansea, Vol. 1, pp. 285–294 1980.

4. W. F. Chen, '*Limit Analysis and Soil Plasticity*', Scientific Publishing Co., Elsevier, Amsterdam, The Netherlands, 1975.

5. W. F. Chen, 'Plasticity in soil mechanics and landslides', *Journal of the Engineering Mechanics Division, ASCE*, **106**, No. EM3, 443–464 (1980).

6. W. F. Chen, '*Plasticity in Reinforced Concrete*', McGraw-Hill Book Company, Inc., New York, NY, 1982.

7. W. F. Chen, 'Soil mechanics, plasticity and landslides', *In Mechanics of Inelastic Materials* (Eds. G. J. Dvorak and R. T. Shield), Elsevier, Amsterdam, The Netherlands, 1984.

8. W. F. Chen and M. F. Chang, 'Limit analysis in soil mechanics and its applications to lateral earth pressure problems', *Solid Mechanics Archives*, **6**, Issue 3, pp. 331–399, Sijthoff & Noordhoff International Publishers, July, 1981.

9. W. F. Chen and A. F. Saleeb, '*Constitutive Equations for Engineering Materials*', Vol. 1—Elasticity and Modeling, February, 1982, Vol. 2—Plasticity and Modeling, 1984, to appear, John Wiley Interscience, New York.

10. Y. F. Dafalias and E. P. Popov, 'A model of nonlinearly hardening materials for complex loadings', *Acta Mechanics*, **21**, 173–192 (1975).

11. Y. F. Dafalias and R. Herrmann, 'A boundary surface soil plasticity model', *International Symposium on Soils Under Cyclic and Transient Loading*, Swansea, UK, Vol. 1, 1980.

12. R. O. Davis and G. Mullenger, 'A simple rate-type constitutive representation for granular media', *Proceedings of the 3rd International Conference on Numerical Methods in Geomechanics*, Aachen, Germany, Vol. 1, pp. 415–421, 1979.

13. F. L. Dimaggio and I. S. Sandler, 'Material model for granular soils', *Journal of the Engineering Mechanics Division*, ASCE, **97**, No. EM3, Proc. Paper 8212, 935–950 (1971).

14. D. C. Drucker, 'Concept of path independence and material stability for soils *Rheol. Mecan. Soils Proc. IUTAM Sym.* (Eds. J. Kravtchenko and P. M. Sirieys), pp. 23–43, Grenoble, Springer, Berlin, 1966.

15. D. C. Drucker, R. E. Gibson, and D. J. Henkel, 'Soil mechanics and work hardening theories of plasticity', *Transactions, ASCE*, **122**, Paper No. 2864, 338–346 (1957).

16. P. Duwez, 'On the plasticity of crystals', *Physical Review*, **47**, 494 (1953).

17. J. M. Duncan and C. Y. Chang, 'Nonlinear analysis of stress and strain in soils', *Journal of the Soil Mechanics and Foundation Division*, ASCE, **96**, No. SM5, Proc. Paper 7513, 1629–1653 (1970).

18. R. J. Evans and K. S. Pister, 'Constitutive equations for a class of nonlinear elastic solids', *International Journal of Solids and Structures*, **2**, 3, 427–445 (1966).

19. B. O. Hardin and V. P. Drnevich, 'Shear modulus and damping in soils: measurements and parameter effects', *Journal of the Soil Mechanics and Foundations Division, ASCE*, **98**, No. SM6, 603–624 (1972).

20. W. D. Iwan, 'On a class of models for the yielding behaviour of continuous and composite systems', *Journal of Applied Mechanics*, **34**, 612–617 (1967).

21. H. Y. Ko and R. M. Masson, 'Nonlinear characterization and analysis of sand', *Numerical Methods in Geomechanics, ASCE*, 294–304 (1976).
22. R. L. Kondner, 'Hyperbolic stress–strain response: cohesive soils', *Journal of the Soil Mechanics and Foundations Division, ASCE*, **89**, No. SM1, 115–143 (1963).
23. R. D. Krieg, 'A practical two-surface plasticity theory', *Journal of Applied Mechanics*, **42**, 641–646 (1975).
24. P. V. Lade, 'Elasto-plastic stress–strain theory for cohesionless soil with curved yield surfaces', *International Journal of Solids and Structures*, **13**, 1014–1035 (1977).
25. P. V. Lade and J. M. Duncan, 'Elastoplastic stress–strain theory for cohesionless soil', *Journal of the Geotechnical Engineering Division, ASCE*, **101**, No. GT10, 1037–1053 (1975).
26. G. Masing, 'Eigenspannungen and Verfestigung heim Messing', *Proceedings of the 2nd International Congress of Applied Mechanics*, Zurich, Switzerland, 1926.
27. E. Mizuno and W. F. Chen, 'Analysis of Soil response with different plasticity models', ASCE National Convention in Hollywood, FL, October 1980.
28. Z. Mróz, 'On the description of anisotropic hardening', *Journal of Mechanics and Physics of Solids*, **15**, 163–175 (1967).
29. Z. Mróz, V. A. Norris, and O. C. Zienkiewicz, 'An anisotropic hardening model for soils and its application to cyclic loading', *International Journal for Numerical and Analytical Methods in Geomechanics*, **2**, 203–221 (1978).
30. Z. Mróz, V. A. Norris, and O. C. Zienkiewicz, 'Application of an anisotropic hardening model in the analysis of elastoplastic deformation of soils', *Geotechnique*, **29**, 1, 1–34 (1979).
31. I. Nelson and M. L. Baron, 'Application of variable moduli models to soil behaviour', *International Journal of Solids and Structures*, **7**, 399–417 (1971).
32. I. Nelson, M. L. Baron and I. Sandler, 'Mathematical models for geological materials for wave propagation studies', *Shock Waves and the Mechanical Properties of Solids*, Syracuse University Press, Syracuse, NY, 1971.
33. A. C. Palmer (ed.), *Proceedings of the Symposium on the Role of Plasticity in Soil Mechanics*, Cambridge University, England, 1973.
34. R. H. G. Parry, (ed.), *Roscoe Memorial Symposium:* Stress–Strain Behaviour, of Soils, Henley-on-Thames, Cambridge University, England, 1972.
35. J. Prevost, 'Mathematical modeling of monotonic and cyclic undrained clay behaviour', *International Journal for Numerical and Analytical Methods in Geomechanics*, **1**, 195–216 (1977).
36. J. H. Prevost, 'Anisotropic undrained stress–strain behaviour of clays', *Journal of the Geotechnical Engineering Division, ASCE*, **104**, No. GT8, Proc. Paper 13942, 1075–1090 (1978).
37. J. H. Prevost, 'Plasticity theory for soil stress behaviour', *Journal of the Engineering Mechanics Division, ASCE*, **104**, EM5, Proc. Paper 14069, 1177–1194 (1978).
38. J. H. Prevost, B. Cuny, T. J. R. Hughes, and R. F. Scott, 'Offshore gravity structures: analysis', *Journal of the Geotechnical Engineering Division, ASCE*, **107**, No. GT2, February, 1981.
39. K. H. Roscoe, A. N. Schofield, and C. P. Wroth, 'On the yielding of soils', *Geotechnique*, **8**, No. 1, 22–52 (1958).
40. K. H. Roscoe, A. N. Schofield and A. Thurairagh, 'Yielding of clays in state wetter than critical', *Geotechnique*, **13**, No. 3, 211–240 (1963).
41. A. F. Saleeb and W. F. Chen, 'Nonlinear Hyperelastic (Green) Constitutive Models for Soils, Part I—Theory and Calibration, Part II—Predictions and Comparisons', *Proceedings of the North American Workshop on Limit Equilibrium, Plasticity, and*

Generalized Stress–Strain in Geotechnical Engineering, McGill University, Montreal, Canada, May 1980.
42. T. Tokuoka, 'Yield conditions and flow rules derived from hypoelasticity', *Archive of Rational Mechanics and Analysis*, **42**, 239–252 (1971).
43. C. Truesdell, 'Hypo-elasticity', *Journal of Rational Mechanics and Analysis*, **4**, No. 1, 83–133 (1955).
44. O. C. Zienkiewicz, V. A. Norris, and D. J. Naylor, 'Plasticity and viscoplasticity in soil mechanics with special reference to cyclic loading problems', *Finite Elements in Non-Linear Mechanics* (Eds. P. G. Bergan *et al.*), pp. 455–485 Tapir, Trondheim, Vol. 2, 1977.

Mechanics of Engineering Materials
Edited by C. S. Desai and R. H. Gallagher

Chapter 6

The Theory of Adaptive Elasticity

S. C. Cowin

6.1 INTRODUCTION

There are three major portions of this chapter. In the first portion the observations and experiments of stress adaptation of overall shape and of local microstructure in living bones, living trees, and certain saturated porous geological materials are reviewed. This material is presented in the first three sections following the introduction. The second portion of the chapter describes the development of a continuum model for the stress adaptation process. This material is contained in Sections 6.5, 6.6, and 6.7. The third and final portion of the chapter concerns applications of the theory to some elementary problems in bone mechanics. These applications are described in the latter part of Section 6.6 and in Section 6.8.

6.2 STRESS ADAPTATION IN BONE

Functional adaptation is the term used to describe the ability of organisms to increase their capacity to accomplish their function with an increased demand and to decrease their capacity with lesser demand. Living bone is continually undergoing processes of growth, reinforcement and resorption which are collectively termed 'remodelling'. The remodelling processes in living bone are the mechanisms by which the bone adapts its overall structure to changes in its load environment. The time scale of the remodelling processes is on the order of months or years. Changes in life style which change the loading environment, for example taking up jogging, have remodelling times on the order of many months. Bone remodelling associated with trauma has a shorter remodelling time, on the order of weeks in humans. The time scales of these remodelling processes should be distinguished from developments in bone due to growth, which have a time scale on the order of decades in humans, and the developments due to natural selection which have a time scale of many lifetimes.

It is necessary to describe the nature of bone as a material before describing the remodelling processes that occur in bone. Experiments have shown that bone can be modelled as an inhomogeneous transversely isotropic or orthotropic elastic material, with the degree of anisotropy varying inhomogeneously also.

There are two major classes of bone tissue which significantly contribute to the structural strength of the skeletal system. They are called cancellous and cortical bone. Cortical bone is the hard tissue on the outer surface or cortex of the femur (i.e. thigh bone). It is dense, it contains no marrow and its blood vessels are microscopically small. Cancellous bone occurs in the interior of the femur. It consists of a network of hard, interconnected filaments called 'trabeculae' interspersed with marrow and a large number of small blood vessels. Cancellous bone is also called trabecular bone or spongy bone. Generally cancellous bone is structurally predominant in the neighbourhood of the joints and cortical bone is structurally predominant in the central sections of a femur away from the joints. Bone tissue contains an abundant amount of extracellular material or matrix. The volume fraction of the matrix is orders of magnitude larger than the volume fraction of bone cells. The matrix accounts for virtually all the structural strength of bone.

The concept of stress or strain induced bone remodelling was first publicized by the German anatomist Julius Wolff [1], and is often called Wolff's law. Basset and Becker [2], Shamos and Lavine [3], Justus and Luft [4], and Somjen *et al*. [5] have proposed various mechanisms for bone remodelling in terms of certain electrical and chemical properties of bone. The distinction made by Frost [6] between surface and internal remodelling is followed here. *Surface* remodelling refers to the resorption or deposition of the bone material on the external surface of the bone. The details of the process of deposition of new lamina at the surface of a bone are described by Currey [7]. *Internal* remodelling refers to the resorption or reinforcement of the bone tissue internally by changing the bulk density of the tissue. The study of Kazarian and von Gierke [8] very graphically illustrated internal remodelling in cancellous bone. In this study 16 male Rhesus monkeys were immobilized in full body casts for a period of sixty days. Another set of 16 male Rhesus monkeys were used as control and allowed the freedom of movement possible in a cage. A subsequent comparison of the bone tissue of the immobilized monkeys with the tissue of the control monkeys showed considerable resorption of the bone tissue of the immobilized monkeys. Qualitative radiographic techniques demonstrated increased bone resorption in the metaphysis of the axially-loaded long bones, as well as the loss of cortical bone. Mechanical testing of the bone tissue also reflected the remodelling loss in the immobilized monkeys.

The effect of an increased loading environment on the remodelling of bone tissue is illustrated in a Latvian study reported by Shumskii, Merten, and Dzenis [9]. In this study the acoustic velocity in tibia of nine groups of individuals was determined ultrasonically. The nine groups were highly trained athletes (masters of sports, candidates for master of sports, and first degree athletes), swimmers, biathlon athletes, middle distance runners, high jumpers, hurdlers, track and field athletes, second and third degree middle distance runners, and some non-athletic individuals. The data from this study is presented in Table 6.1. It is easy to see that

Table 6.1 The acoustic velocity in the human tibia. This data is from Shumskii, Merten, and Dzenis [9]

Group no.	Group characteristics	Acoustic velocity in m/s	
		Right leg	Left leg
1	Individuals not participating in sports	1257	1270
2	Third degree middle distance runners	1315	—
3	Second degree middle distance runners	1430	—
4	First degree middle distance runners	1656	1710
5	Swimmers, 2 masters of sports, 2 candidates for master of sports, and 6 first-degree athletes	1346	1365
6	Biathlon athletes, candidates for master of sports	1502	1490
7	High jumpers, 4 candidates for master of sports, 6 first-degree athletes	1775	1860
8	First-degree hurdlers	1702	1576
9	First-degree track-and-field athletes	1876	1820

the acoustic velocity in the tibia increases with the group's athletic expertise. Acoustic velocity is proportional to the square root of the Young's modulus and inversely proportional to the square root of the bulk density; thus there is an implication of greater modulus which implies bone deposition and increased density of the bone tissue with increasing athletic expertise.

Surface remodelling can be induced in the leg bones of animals by superposing axial and/or bending loads. Woo *et al.* [10] has shown that increased physical activity (jogging) in pigs can cause the periosteal surface of the leg bone to move out and the endosteal surface to move in. Meade *et al.* [11] superposed a constant compressive force along the axis of the canine femur by an implanted spring system. This study showed a quantifiable increase in cross-section area with increasing magnitude of the superposed compressive force. Liskova and Hert [12] have shown that intermittent bending applied to the rabbit tibia can cause the periosteal surface to move out. Surface remodelling can also be induced in the leg bones of animals by reducing the loads on the limb. In two studies Uhthoff and Jaworski [13] and Jaworski *et al.* [14] immobilized one of the forelimbs of beagles. In the study of Uhthoff and Jaworski [13], young beagles were used and it was found that the endosteal surface showed little movement while there was much resorption on the periosteal surface. However, in the study with older beagles

(Jaworski *et al.* [14]), it was observed that the periosteal surface showed little movement while on the endosteal surface there was much resorption.

The feedback mechanism by which the bone tissue senses the change in load environment and initiates the deposition or resorption of bone tissue is not well understood. The two candidates are a piezoelectric effect that occurs in bone and the calcium ion concentration. These mechanisms are described in further detail by Cowin and Hegedus [15]. A fairly comprehensive survey of the electromechanical properties of bone was accomplished recently by Guzelsu and Demiray [16].

6.3 STRESS ADAPTATION IN TREES

Functional adaptation also occurs in trees and plants. Trees adapt their shape and structure to their environmental loading in a manner qualitatively similar to that observed in bones. The exact mechanism of stress adaptation in trees is unknown, but a combination of mechanical stress and the hormone auxin have been suggested by a number of studies [17, 18].

It has been observed that trees growing in dense forest strands have smaller trunk diameter than trees growing at the edge of the strand and are more inclined to be blown over than those at the edge. It has also been observed that nursery trees grown close together in containers are tall and spindly while those placed further apart are greater in trunk diameter. An interesting experiment which quantified this phenomenon was reported by Neel and Harris [19, 20]. These environmental horticulturists obtained eight matching pairs of young sweet-gum trees (Liquidambar). The trees were placed in four gallon cans in a greenhouse. Each morning at 8:30 for 27 mornings, one tree in each pair was shaken for 30 seconds. At the end of the 27-day period the shaken trees had reached a height which was only 20 per cent of the height of the unshaken trees. However, the trunks of the shaken trees were larger than those of the unshaken trees. At a distance of 5 cm from the ground the diameter of the shaken trees had increased by 8.3 mm while those of the unshaken had increased by only 6.8 mm. The wood fibre length and the vessel member length were significantly shorter in the shaken trees. Although the authors did not measure the elastic moduli, the changes in the wood tissue microstructure suggests that the elastic moduli are different in the shaken and unshaken trees. One would reasonably expect the moduli to be higher in the unshaken trees.

The wood tissue that is deposited on an external surface of a tree trunk in response to a superposed bending moment in a particular direction is called reaction wood. Reaction wood is visible when a transverse section of the tree trunk is viewed because it distorts the growth rings. In a tree that has been bent in a particular direction the growth rings will not be circular, but they will be distorted ovals in the particular direction associated with the bending. The additional wood tissue deposited will increase the area moment of inertia of the trunk cross-section and decrease the stress experienced by the wood tissue.

The elastic properties of wood are best modelled by orthotropic linear elasticity theory. The Young's modulus in the direction of the wood grain is the largest Young's modulus. The Young's moduli are the largest at the base of the tree and they decrease in magnitude for wood located at a distance up the tree away from the ground, generally increasing with limb diameter.

6.4 STRESS ADAPTATION IN CERTAIN POROUS SOLID GEOLOGICAL MATERIALS

The effect known as 'pressure solution' in the geophysical literature suggests that functional adaptation to stress is not restricted to organic materials. Pressure solution is described as the increasing solubility of a saturated porous solid matrix with increasing solid matrix strain. Three recent reports by Sprunt and Nur [21, 22, 23] attempt to quantify this effect. The materials studied were all porous and saturated and included sandstone, limestone, dolomitic limestone, marble, quartzite, and novaculite. These materials all adapted their microstructure by surface resorption or by changing their porosity. These microstructual changes were found to be proportional to the strain in the solid matrix and not the effect of a transport mechanism.

Although Sprunt and Nur report on experimental situations in which there was resorption of the solid matrix material, they indicate that they believe that this result is due to the fact that they employed an open system. They state 'If our system were closed, we would expect solution only in the regions of large compressive stress and deposition in the regions of tension or small compressive stress. Examples of pressure solution in both open and closed systems are well known in nature. Large net reductions in rock mass, where material is dissolved along stylolites and removed from the rock fromation (e.g. [24, 25]). Pressure shadows which form by solution of material at points of high stress around a stiff inclusion accompanied by transfer and recrystallization of the same material at points of lower stress around the same inclusion, [26, 27], are examples of natural closed systems'. The 1935 experiments of Russell [28] support the deposition aspect of the pressure solution effect. Russell used a smooth-surfaced crystal of ammonium alum in a saturated alum solution at constant temperature to demonstrate that material dissolved from one part of a crystal because of local stress may be redeposited on another part of the same crystal where the strain is relatively less.

The sandstone experiments reported by Sprunt and Nur [21, 22] employed a different geometry from their experiments on the other geological materials reported in [23]. The sandstone experiments were performed on a hollow circular cylinder where the matrix stress was induced by mechanical pressure applied to the external cylindrical surface. The experiments on the other materials were performed on specimens in the shape of rectangular parallelopipeds with a crylindrical hole. The matrix stress was induced by a compressive load in one direction. The experimental observations reported in the sandstone experiments were porosity changes while the observations reported in the experiments with other materials were surface resorption. It would then appear that there is a difference between surface and internal pressure solution effects. Most likely the difference in these effects is not in the basic mechanics of the reaction, but in the rate at which the reaction occurs. One would suspect that the reaction can proceed at a more rapid rate on the free surface because of the greater mobility of the solvent.

6.5 MODELLING OF THE STRESS ADAPTATION PROCESS

At this early stage of development, the stress adaptation processes in bone, trees, and saturated porous solid geological materials can be described by the same model. A description of this model is presented in this section. To make the presentation of the model easier, we develop it in two parts and then combine the parts. Thus, in the following section a theory of surface remodelling is presented and, in the section following, a theory of internal remodelling is presented.

The theories of surface and internal remodelling use a simple two constituent model. The solid matrix is modelled as a porous anisotropic linear elastic solid. The basic model is then a porous, anisotropic linear elastic solid perfused with a fluid. In the model of the stress adaptation process chemical reactions convert the fluid into the porous solid matrix and viceversa. As a result of the chemical reactions mass, momentum, energy, and entropy are transferred to or from the porous solid matrix. The rates of these chemical reactions depend on matrix strain and are very slow compared to the characteristic time of inertia effects. Thus, inertia effects are neglected and the stress in the matrix considered here is the actual stress averaged over a time period greater then any inertia effects.

At this point the discussion naturally bifurcates into the consideration of the two parts which constitute the complete model, namely the surface remodelling theory and the internal remodelling theory. The distinction between the two theories is made upon the locations at which the chemical reactions occur and the way in which mass is added or removed from a material body. In the theory of surface remodelling the chemical reactions occur only on the external surfaces of the body and mass is added or removed from the body by changing the external shape of the body. During surface remodelling the interior of the body remains at constant bulk density. In the theory of internal remodelling the chemical reactions occur everywhere within the porous solid matrix of the body and mass is added by changing the bulk density of the matrix and without changing the exterior dimensions of the body. In both cases the rate and direction of the chemical reaction at a point are determined by the strain at the point. It is important to note that these two theories are neither contradictory nor incompatible, but combine easily for a single body in which there are both overall shape changes and density changes.

The theory of surface remodelling acknowledges the observed fact that external changes in body shape are induced by changes in the loading

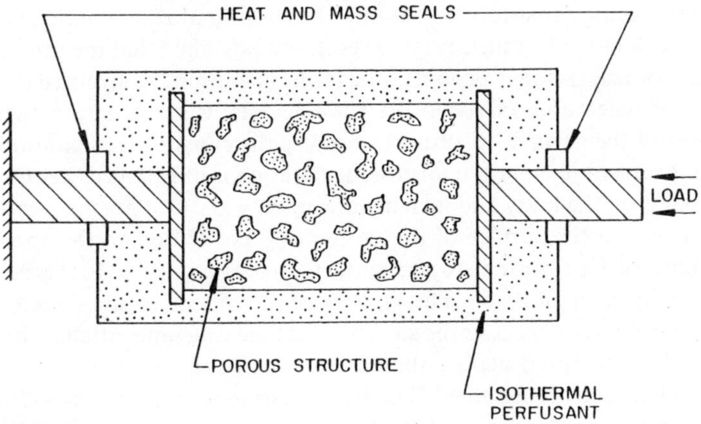

Figure 6.1 A schematic model of the remodelling mechanism postulated

environment of the body. This theory postulates a causal relationship between the rate of surface deposition or resorption and the strain in the surface of the body. The body is considered to be an open system with regard to mass transport and the mass of the body will vary as the external shape of the body varies. This theory is described by Cowin and Van Buskirk [29] and Cowin and Firoozbakhsh [30]. The theory of internal remodelling postulates a causal relationship between the rate of deposition or resorption of the solid matrix at any point and the strain at that point in the solid matrix. A schematic diagram of this model is shown in Figure 6.1. The fact that living bone and living wood tissue are encased in a living organism and that the geological materials are saturated is reflected in the model by setting the elastic porous solid in a bath of the perfusant. The mechanical load is applied directly to the porous structure across the walls of the perfusant bath as illustrated. The system consisting of the elastic porous solid and its perfusant bath is considered to be closed with respect to mass, heat energy, and entropy transfer, but open with respect to momentum transfer from loading. The system consisting of only the elastic porous solid without its entrained perfusant is open with respect to momentum transfer as well as mass, energy, and entropy transfer. The solid matrix is taken as the control system since the changes in the mechanical properties of the solid matrix alone determine the changes in the mechanical properties of the whole body.

The theory of internal remodelling is developed in a series of papers: Cowin and Hegedus [15], Hegedus and Cowin [31], Cowin and Nachlinger [32], and Cowin and Van Buskirk [33].

6.6 THE THEORY OF SURFACE REMODELLING

The model for surface remodelling employed assumes that solid matrix can be modelled as a linear elastic body whose free surfaces move according to an additional specific constitutive relation. The additional constitutive relation for the movement of the free surface is the result of a postulate that the rate of surface deposition or resorption is proportional to the change in the strain in the surface from a reference value of strain. At the reference value of strain there is no movement of the surface. In order to express the constitutive equation for the surface movement in equation form some notation is introduced. Let Q denote a surface point on the body and let $\mathbf{n}$ denote an outward unit normal vector of the tangent plane to the surface of the body at Q. Let U denote the speed of the remodelling surface normal to the surface, that is to say U is the velocity of the surface in the $\mathbf{n}$ direction. The velocity of the surface in any direction in the tangent plane is zero because the surface is not moving tangentially with respect to the body. Let $E_{ij}(Q)$ denote the cartesian components of the strain tensor at Q. Small strains are assumed. The hypothesis for surface remodelling is that the speed of the remodelling surface is linearly proportional to the strain tensor,

$$U = C_{ij}(\mathbf{n}, Q)[E_{ij}(Q) - E_{ij}^0(Q)] \tag{6.1}$$

where $E_{ij}^0(Q)$ is a reference value of strain where no remodelling occurs and $C_{ij}(\mathbf{n}, Q)$ are surface remodelling rate coefficients which are, in general, dependent upon the point Q and the normal $\mathbf{n}$ to the surface at Q. The surface remodelling rate coefficients and the reference values of strain are phenomenological coefficients of the body surface and must be determined by experiment. It is assumed here that the surface remodelling rate coefficients C_{ij} are not site specific, that is to say they are independent of the position of the surface point Q. It is also postulated above that surface remodelling rate coefficients are independent of strain. Equation (6.1) gives the normal velocity of the surface at the point Q as a function of the existing strain state at Q. If the strain state at Q, $E_{ij}(Q)$, is equal to the reference strain state $E_{ij}^0(Q)$, then the velocity of the surface is zero and no remodelling occurs. If the right-hand side of the equation (6.1) is positive, the surface will be growing by deposition of material. If, on the other hand, the right-hand side of (6.1) is negative, the surface will be resorbing. Equation (6.1) by itself does not constitute the complete theory. The theory is completed by assuming that the body is composed of a linearly elastic material. Thus, the complete theory is a modification of linear elasticity in which the external surfaces of the body move according to the rule prescribed by equation (6.1). A boundary value problem will be formulated in the same manner as a boundary value problem in linear elastostatics, but it will be necessary to specify the boundary conditions for a specific time period. As the body evolves to a new shape, the stress and strain states will be varying quasistatically. At any instant the body will behave exactly as an elastic body, but moving boundaries will cause local stress and strain to redistribute themselves slowly with time.

This theory has been applied to the problem of a hollow circular cylinder subjected to an axial load by Cowin and Firoozbakhsh [30]. The results suggest that the stable response of the cylinder to an increased compressive axial load is to increase its cross-sectional area by movement of the external surface of the cylinder outward and the internal surface inward. On the other hand, the stable response of the cylinder to a decreased compressive axial load is to decrease its cross-sectional area by movement of the external surface of the cylinder inward and the internal surface outward. When the theory is applied to the bending of a rod, it predicts deposition of material on the concave side and resorption on the convex side.

6.7 THE THEORY OF INTERNAL REMODELLING

The rationale underlying the theory of internal remodelling was outlined in the section entitled Modelling the Stress Adaptation Process. The adapting body is modelled as a chemically reacting elastic porous medium in which the rate of reaction is strain controlled (Cowin and Hegedus [15]). The porous medium has two components: a porous elastic solid representing the matrix structure and a perfusant. Mass is transferred from the porous elastic solid to the fluid perfusant and viceversa by the chemical reaction whose rate is strain controlled. The mass of the porous elastic solid is changed by increasing or decreasing its porosity, but not by changing the overall dimensions of the body.

The small strain theory of internal bone remodelling is an adaptation of the theory of equilibrium of elastic bodies. The theory models the bone matrix as a chemically reacting porous elastic solid. The bulk density ρ of the porous solid is written as the product of γ and ν,

$$\rho = \gamma \nu \tag{6.2}$$

where γ is the density of the material that composes the matrix structure and ν is the volume fraction of that material present. Both γ and ν are considered to be field variables. We let ξ denote the value of the volume fraction ν of the matrix material in an unstrained reference state. The density γ of the material composing the matrix is assumed to be constant, hence the conservation of mass will give the equation governing ξ. It is also assumed that there exists a unique zero-strain reference state for all values of ξ. Thus ξ may change without changing the reference state for strain. One might imagine a block of porous elastic material with the four points, the vertices of a tetrahedron, marked on the block for the purpose of measuring the strain. When the porosity changes, material is added or taken away from the pores, but if the material is unstrained it remains so and the distances between the four vertices marked on the block do not change. Thus ξ can change while the zero-strain reference state remains the same. The change in volume fraction from a reference volume fraction ξ_0 is denoted by e,

$$e = \xi - \xi_0. \tag{6.3}$$

The basic kinematic variables, and also the independent variables, for the theory of internal bone remodelling are the six components of the strain matrix E_{ij} and the change in volume fraction e of the matrix material from a reference value ξ_0.

The governing system of equations for this theory are (Hegedus and Cowin [31])

$$2E_{ij} = (u_{i,j} + u_{j,i}) \tag{6.4}$$

$$T_{ij,j} + \gamma(\xi_0 + e)b_i = 0 \tag{6.5}$$

$$T_{ij} = (\xi_0 + e)C_{ijkm}(e)E_{km} \tag{6.6}$$

$$\dot{e} = a(e) + A_{ij}(e)E_{km} \tag{6.7}$$

where $a(e)$, $A_{ij}(e)$, and $C_{ijkm}(e)$ are material coefficients dependent upon the change in volume fraction e of the adaptive elastic material from the reference volume fraction ξ_0, and where the superimposed dot indicates the material time derivative. Equation (6.4) represents the strain–displacement relations for small strain, valid in the present theory as well as in the theory of elasticity. Equation (6.5) represents the condition of equilibrium in terms of stress. Equation (6.6) is a generalization of Hooke's law, $T_{ij} = C_{ijkm}E_{km}$, in which the elastic coefficients C_{ijkm} now have a dependence upon the change in the reference volume fraction e and are denoted by $(\xi_0 + e)C_{ijkm}(e)$. In the case when the change in volume fraction e vanishes and the reference volume fraction ξ_0 is one, (6.6) coincides with the generalized Hooke's law. Equation (6.7) is the remodelling rate equation and it specifies the rate of change of the volume fraction as a function of the volume fraction and strain. A positive value of $\dot{e}$ means the volume fraction of elastic material is increasing while a negative value means the volume fraction is decreasing. Equation (6.7) is obtained from the conservation of mass and the constitutive assumption that the rate of mass deposition or absorption is dependent upon only the volume fraction e and the strain state E_{ij}. The linear dependence upon strain shown in (6.7) occurs as a result of the small strain assumption. A uniqueness theorem for this theory was given by Cowin and Nachlinger [32].

The system of equations (6.6) and (6.7) is an elementary mathematical model of the volumetric stress adaptation process. Equation (6.6) is a statement that the moduli occurring in Hooke's law actually depend upon the volume fraction of solid matrix material present. Equation (6.7) is an evolutionary law for the volume fraction of matrix material. We will now describe how the model works in terms of equations (6.6) and (6.7) using, to fix ideas, a hollow circular cylinder subjected to an axial compressive load. Suppose for $t < 0$ the hollow circular cylinder has been under a constant stress for a long time. The body is then in remodelling equilibrium and the volume fraction field e is steady ($\dot{e} = 0$). From equation (6.7) we can see that this means there is a particular steady strain associated with the steady stress and from equation (6.6) we can see that the elastic coefficients C_{ijkm} are steady. Now, at $t = 0$ the axial applied compressive stress is increased to a new value and held at the new value for all $t > 0$. This stress history is illustrated in Figure 6.2. From equation (6.6) it follows that the strain will jump to a new value at $t = 0$ and from equation (6.7) it follows that $\dot{e}$ will jump to a non-zero value. These jumps are also illustrated in Figure 6.2. Since $\dot{e}$ is non-

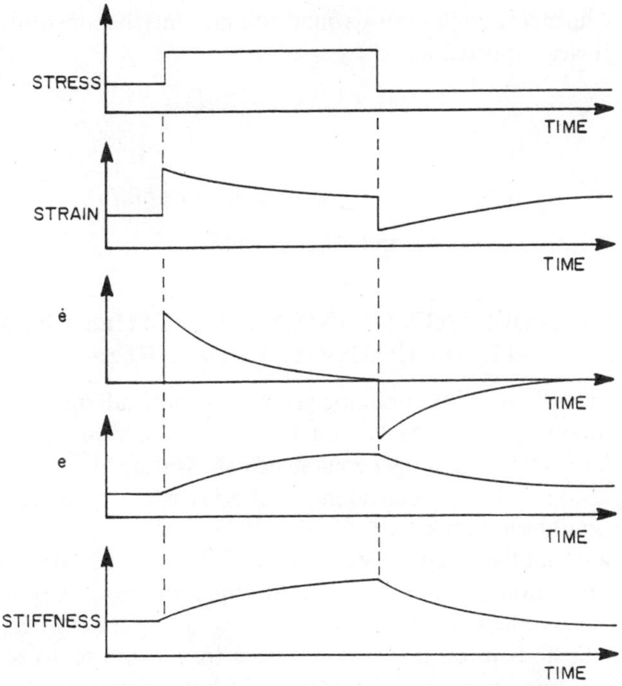

Figure 6.2 An illustration of the internal remodelling response to jumps in the constant applied stress. The response of the strain, $\dot{e}$, e, and bone stiffness to the illustrated stress history is shown

zero for $t > 0$, equation (6.7) shows that e will change with time and the normal physics of remodelling suggests that e will increase as a result of the increased strain. Since e is increasing for $t > 0$, the elastic coefficients $C_{ijkm}(e)$ will increase (see Figure 6.2). It follows from (6.6) that if the stress is constant and the elastic coefficients are increasing, then the strain must decrease in time. Since the strain is decreasing in time, it follows from (6.7) that one remodelling rate $\dot{e}$ will decrease and that e will continue to change, but more and more slowly. Thus for very large times $\dot{e}$ will tend to zero and the body will evolve to a new value of e and a new value of strain that is compatible with the increased stress state applied at $t = 0$. Remodelling of the cylinder is then complete. The process just described as well as process associated with a jump reduction in stress are illustrated in Figure 6.2.

The theory described above involves the functions $a(e)$, $A_{ij}(e)$ and $C_{ijkm}(e)$ characterizing the material properties. There is no data in the literature on the values of the functions $a(e)$ and $A_{ij}(e)$ and the data on $C_{ijkm}(e)$ suggests that it can be approximated as a linear function of e. Hegedus and Cowin [31] introduced an approximation scheme that gave $C_{ijkm}(e)$ as a linear function of e. This scheme involved a series expansion in which terms of the order e^3, $|E|e^2$, and $|E|^2e$ were neglected and terms of the order e, $|E|$, $|E|e$ and e^2 retained. The scheme showed that

$A_{ij}(e)$ was also linear in e while $a(e)$ was quadratic in e, thus the constitutive relations (6.6) and (6.7) were approximated by

$$T_{ij} = (\xi_0 C^0_{ijkm} + eC^1_{ijkm})E_{km} \tag{6.8}$$

and

$$\dot{e} = c_0 + c_1 e + c_2 e^2 + A^0_{ij}E_{ij} + eA^1_{ij}E_{ij} \tag{6.9}$$

where C^0_{ijkm}, C^1_{ijkm}, c_0, c_1, c_2, A^0_{ij}, and A^1_{ij} are constants.

6.8 DEVOLUTION OF INHOMOGENEITIES UNDER STEADY HOMOGENEOUS STRESS

To study the devolution of an inhomogeneous body into a homogeneous one we subject an inhomogeneous body which is in mechanical and remodelling equilibrium for $t < 0$ to a homogeneous steady stress state for $t > 0$. We consider an inhomogeneous, adaptive elastic material body which is in mechanical and remodelling equilibrium for time $t < 0$. By mechanical and remodelling equilibrium it is meant that there exists a steady stress $T^0_{ij}(\mathbf{x})$, a steady strain $E^0_{ij}(\mathbf{x})$ and a steady volume fraction field $e_0(\mathbf{x})$ which satisfy equations (6.4), (6.5), (6.8), and (6.9) identically. At $t = 0$ the stress field $T^0_{ij}(\mathbf{x})$ is changed to the steady homogeneous field T^*_{ij} and the body force, if it was not zero for $t < 0$, is set to zero for $t > 0$. Since the body force is zero and the stress is homogeneous, the equation of equilibrium (6.5) is satisfied identically. The strain tensor for $t > 0$ can be determined by inversion of (6.8), thus

$$E_{ij}(\mathbf{x}, t) = (K^0_{ijkm} - eK_{ijrs}C^1_{rspq}K^0_{pqkm})T^*_{km} \tag{6.10}$$

where K^0_{ijkm} is defined by

$$\xi_0 K^0_{ijkm} C^0_{kmpq} = \delta_{ip}\delta_{jq}. \tag{6.11}$$

The representation (6.10) for the strain tensor is then substituted into the remodelling rate equation (6.9) to obtain the differential equation governing the evolution of $e(\mathbf{x}, t)$, thus

$$\dot{e}(\mathbf{x}, t) = A\{e^2(\mathbf{x}, t) - 2Be(\mathbf{x}, t) + C\}, \qquad e(\mathbf{x}, 0) = e_0(\mathbf{x}) \tag{6.12}$$

where A, B, and C are constants, $A = c_2$,

$$B = -\frac{1}{2c_2}(c_1 + A^1_{ij}K^0_{ijmk}T^*_{mk} - A^0_{ij}K^0_{ijrs}C^1_{rspq}K^0_{pqmk}T^*_{mk}) \tag{6.13}$$

and

$$C = \frac{1}{c_2}(c_0 + A^0_{ij}K^0_{ijmk}T^*_{mk}) \tag{6.14}$$

The initial condition indicated as the second part of (6.12) requires that the inhomogeneity at time $t = 0$ be that associated with the mechanical and remodelling equilibrium state that existed for $t < 0$.

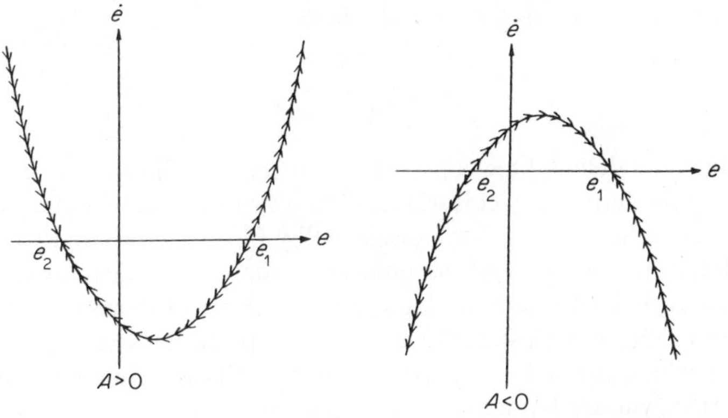

Figure 6.3 Phase plane representations of the remodelling rate equation. The arrowheads indicate the evolution of the solution in positive time

The solution to the differential equation (6.12) is presented by Firoozbakhsh and Cowin [34]. In order to discuss this solution we introduce the notation e_1, e_2 for the solutions to the quadratic equation

$$\begin{gathered} e^2 - 2Be + C = 0, \\ e_1, e_2 = B \pm (B^2 - C)^{1/2} \end{gathered} \tag{6.15}$$

and we employ the convention $e_1 \geq e_2$ when e_1 and e_2 are real. Insight into the behaviour predicted by the differential equation (6.12) at a fixed value of **x** can be obtained by considering its representation in the phase plane. In Figure 6.3 the solution to the remodelling rate equation (6.12) is plotted in the case when e_1 and e_2 are real and distinct, for both $A > 0$ and $A < 0$. In this case the remodelling rate equation is a parabola in $\dot{e}$ and e which crosses the e axis at two points and opens up or down depending on the sign of A. These parabolas are sketched in Figure 6.3. The arrowheads on the parabolas indicate the direction a solution will evolve in a positive time for a given value of e. Thus, for example, the fact that $e \to e_2$ in infinite time for e_1 and e_2 real, $A > 0$ and $e_1 \geq e_0$ is indicated by the arrowheads on the parabola to the left and the right of e_2 being directed towards e_2. The arrowheads to the right of e_1 are oppositely directed for $A>0$ indicating that $e \to \infty$ in finite time for e_1 and e_2 real and distinct. A completely analogous description holds for the case $A < 0$ shown in Figure 6.3. Hegedus and Cowin [31] have shown that the solution to (6.12) is stable only if e_1 and e_2 are real and distinct and, under those conditions on e_1 and e_2, $e(\mathbf{x}, t)$ is given by

$$\dot{e}(\mathbf{x}, t) = \frac{e_1[e_0(\mathbf{x}) - e_2] + e_2[e_1 - e_0(\mathbf{x})] \exp\left[(\operatorname{sgn} A)\dfrac{t}{\tau}\right]}{[e_0(\mathbf{x}) - e_2] + [e_1 - e_0(\mathbf{x})] \exp\left[(\operatorname{sgn} A)\dfrac{t}{\tau}\right]} \tag{6.16}$$

where sgn A means the sign of A and where

$$\tau = \frac{1}{|A|(e_1 - e_2)} \tag{6.17}$$

is called the remodelling time constant. The precise conditions under which the differential equation (6.12) yields stable solutions which tend to finite, physiologically possible values for $e(\mathbf{x}, t)$ are discussed by Firoozbakhsh and Cowin [34]. The result (6.16) shows that an inhomogeneous adaptive elastic body will become homogeneous under the action of a steady homogeneous stress field.

In order to illustrate this result we consider a particular case. Specifically we consider a cylindrical body of length $2l$ which is initially inhomogeneous along the axis of the cylinder, but which is homogeneous in each transverse plane of the cylinder. For the purposes of this illustration the initial inhomogeneity is assumed to be sinusoidal,

$$e(\mathbf{x}, 0) = 0.1 \sin\frac{\pi x_3}{l} \tag{6.18}$$

where we have taken the axis of the cylinder to be in the x_3 direction. The steady homogeneous stress this body is subjected to for all $t > 0$ is a constant compressive stress of magnitude P along its axis, thus

$$T^*_{33} = -P, \qquad \text{all other } T^*_{ij} = 0 \tag{6.19}$$

The orthotropic elastic constants for human cortical bone are taken from Knets and Malmeister [35] and the dependence of these constants upon the change in volume fraction e is estimated from the microstructure dependence data given by Wright and Hayes [36], thus

$$\frac{1}{E_3} = 0.055(1-11e), \qquad \frac{\nu_{31}}{E_3} = 0.0165(1-11e), \qquad \frac{\nu_{32}}{E_3} = 0.033(1-11e) \tag{6.20}$$

where the units are $(\text{GPa})^{-1}$. The reference volume fraction ξ_0 is 0.892 for (6.20). We assume that $A > 0$ and that the rate coefficients have the following values:

$$c_0 = 1.5 \times 10^{-8}\,\text{sec}^{-1}, \qquad c_1 = -15 \times 10^{-8}\,\text{sec}^{-1}, \qquad c_2 = 2.5 \times 10^{-7}\,\text{sec}^{-1}$$
$$A^0_{11} = A^0_{22} = A^0_{33} = A^1_{11} = A^1_{22} = A^1_{33} = 10^{-4}\,\text{sec}^{-1} \tag{6.21}$$

Since $A > 0$ the solution for $e(x_3, t)$ is

$$e(x_3, t) = \frac{0.129\left(4.4 - \sin\dfrac{\pi x_3}{l}\right) + 0.44\left(\sin\dfrac{\pi x_3}{l} - 1.29\right)\mathrm{e}^{-t/\tau}}{\left(4.4 - \sin\dfrac{\pi x_3}{l}\right) + \left(\sin\dfrac{\pi x_3}{l} - 1.29\right)\mathrm{e}^{-t/\tau}} \tag{6.22}$$

where τ is 12.73×10^6 sec or about 147 days. This result is plotted in Figure 6.4 for various values of time. From this illustration of the temporal evolution of the sine

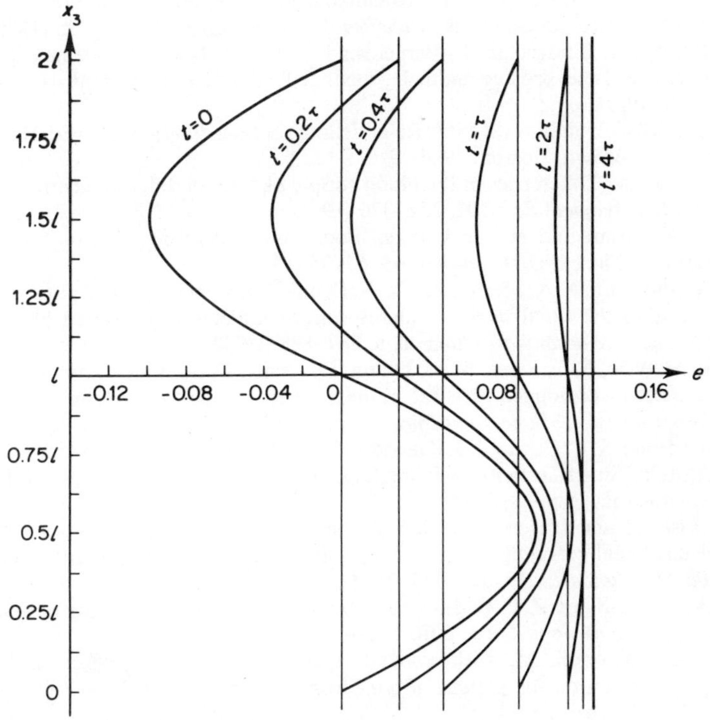

Figure 6.4 Devolution of an initial inhomogeneity. A graphical representation of the solution $e(x_3, t)$ given by equation (6.22)

wave inhomogeneity one can see that, as time progresses, the amplitude of the sine wave decreases, rapidly at first and then more slowly. At large times the sine wave becomes a straight line signifying that the cylinder has become homogeneous.

ACKNOWLEDGEMENT

This work was partially supported by the Solid Mechanics Program of the National Science Foundation and by the Musculoskeletal Diseases Program of the National Institute of Arthritis, Diabetes, Digestive, and Kidney Diseases.

REFERENCES

1. J. Wolff, *Das Gesetz der Transformation der Knochen*, A. Hirschwald Berlin, 1892.
2. C. A. L. Basset, and R. O. Becker, 'Generation of electric potentials in bone in response to stress', *Science*, **137**, 1063–1064 (1962).
3. M. H. Shamos and L. S. Lavine, 'Piezoelectric effect in bone', *Nature*, **197**, 81 (1963).

4. R. Justus and J. H. Luft, 'A Mechanochemical hypothesis for bone remodeling induced by mechanical stress', *Calcified Tissue Research*, **5**, 222–235 (1970).
5. T. Somjen, I. Binderman, E. Berger, and A. Harell 'Bone remodelling induced by physical stress in prostaglandin E_2 mediated', *Biochimica et Biophysica Acta*, **627**, 91–100 (1981).
6. H. M. Frost, 'Dynamics of Bone Remodelling', in *Bone Biodynamics* (Ed. H. M. Frost), Little and Brown, Boston, 1964.
7. J. D. Currey, 'Differences in the blood supply of bone of different histological types', *Q. J. Microscopical Sci.*, **101**, 351–370 (1960).
8. L. E. Kazarian and H. von Gierke, 'Bone loss as a result of immobilization and chelation', *Clinical Orthopaedics*, **65**, 67–75 (1969).
9. V. V. Shumskii, A. A. Merten, and V. V. Dzenis, 'Effect of the type of physical stress on the state of the tibial bones of highly trained athletes as measured by ultrasound techniques', *Mekhanika Polimerov*, **5**, 884–888 (1978).
10. S. L. Y. Woo, S. C. Kuei, W. A. Dillon, D. Amiet, F. C. White, and W. H. Akeson, 'The effect of prolonged physical training on the properties of long bone—a study of Wolff's law', *J. Bone Joint Surg.*, in press, 1981.
11. J. B. Meade, S. C. Cowin, J. J. Klawitter, W. C. Van Buskirk, H. B. Skinner and A. M. Weinstein, 'Short Term Remodeling Due to Hyperphysiological Stress', *J. Bone Joint Surg.*, abstract in press, 1981.
12. M. Liskova and J. Hert, 'Reaction of bone to mechanical stimuli: part 2. Periosteal and endosteal reaction of tibial diaphysis in rabbit to intermittent loading (sic)', *Folia Morphologica*, **19**, 301–317 (1971).
13. H. K. Uhthoff and Z. F. G. Jaworski, 'Bone loss in response to long term immobilization', *J. Bone Joint Surg.*, **60B**, 420–429 (1978).
14. Z. F. G. Jaworski, M. Liskova-Kiar, and H. K. Uhthoff, 'Effect of long term immobilization on the pattern of bone loss in older dogs', *J. Bone Joint Surg.*, **62B**, 104–110 (1980).
15. S. C. Cowin and D. H. Hegedus, 'Bone remodeling I: theory of adaptive elasticity', *J. Biomechanics*, **6**, 313–326 (1976).
16. N. Guzelsu and H. Demiray, 'Electromechanical properties and related models of bone tissues', *Int. J. Engrg. Sci.*, **17**, 813–851 (1979).
17. M. H. Zimmermann and C. L. Brown, *Trees, Structure and Function*, Springer-Verlag, 1970.
18. M. H. Zimmermann, (Ed), *The Formation of Wood in Forest Trees*, Academic Press, 1964.
19. P. L. Neel and R. W. Harris, 'Motion induced inhibition of elongation and induction of dormancy in Liquidambar', *Science*, **173**, 58–59 (1971).
20. Anonymous, The Shaken Trees, *Time*, **42**, September 6, 1971.
21. E. S. Sprunt and A. Nur, 'Reduction of porosity by pressure solution: experimental verification', *Geology*, **4**, 463–466 (1976).
22. E. S. Sprunt and A. Nur, 'Destruction of porosity through pressure solution', *Geophysics*, **42**, 726–741 (1977).
23. E. S. Sprunt and A. Nur, 'Experimental study of the effects of stress on solution rate', *J. Geophysical Research*, **82**, 3013–3022 (1977).
24. B. W. Logan and V. Semeniuk, 'Dynamic metamorphism: processes and products in Devonian carbonate rocks, Canning Basin, Western Australia', *Spec. Publ.*, **6**, 138 pp., Geol. Soc. of Aust., Sydney, 1976.
25. R. H. Groshong, Jr. 'Strain fractures and pressure solution in natural single layer folds', *Geol. Soc. Amer. Bull.*, **86**, 1363–1376 (1975).

26. H. R. Burger, 'Pressure-solution: How important a role?' *Geol. Soc. Amer. Abstr. Programs*, **6**, 1026–1027 (1974).
27. D. W. Durney, 'Pressure-solution and crystallization deformation', *Phil. Trans. Roy. Soc. London, Ser. A*, **283**, 229–240, 1976.
28. G. A. Russell, 'Crystal growth under local stress', *Amer. Mineral*, **20**, 733–737 (1935).
29. S. C. Cowin and W. C. Van Buskirk, 'Surface bone remodeling induced by a medullary pin', *J. Biomechanics*, **12**, 269–276 (1979).
30. S. C. Cowin and K. Firoozbakhsh, 'Bone remodeling of diaphyseal surfaces under constant load: theoretical predictions', *J. Biomechanics*, **14**, 471–484 (1981).
31. D. H. Hegedus and S. C. Cowin, 'Bone remodeling II: small strain adaptive elasticity', *J. Elasticity*, **6**, 337–352 (1976).
32. S. C. Cowin and R. R. Nachlinger, 'Bone remodeling III: uniqueness and stability in adaptive elasticity theory', *J. Elasticity*, **8**, 285–295 (1978).
33. S. C. Cowin and W. C. Van Buskirk, 'Internal bone remodeling induced by a medullary pin', *J. Biomechanics*, **11**, 269–275 (1978).
34. K. Firoozbakhsh and S. C. Cowin, 'Devolution of inhomogeneities in bone structure—predictions of adaptive elasticity theory', *J. Biomechanical Engr.*, **102**, 287–293 (1980).
35. I. V. Knets and A. Malmeisters, 'The deformability and strength of human compact bone tissue'. In *Mechanics of Biological Solids*, Proceedings of the Euromech Colloquium 68, Varna, Bulgaria, 1977.
36. T. M. Wright and W. C. Hayes, 'Tensile testing of bone over a wide range of strain rates; effects of strain rate, microstructure and density', *Medical and Biological Engineering*, **14**, 671–679 (1976).

Mechanics of Engineering Materials
Edited by C. S. Desai and R. H. Gallagher

Chapter 7

Constitutive Laws for Failing Material

D. R. Curran, L. Seaman, and J. Dein

7.1 INTRODUCTION

Fracture is a rate process in which voids nucleate, grow, and coalesce in previously intact material. This process can be discussed at many descriptive levels. At the most primitive level, atomic or molecular bonds are broken to form voids with sizes on the order of 10^{-7} cm. In a polycrystalline material these submicroscopic voids may concentrate on grain boundaries to form microscopic voids with sizes on the order of the grain size, perhaps 10^{-4} to 10^{-5} cm. In composite materials individual components may debond to form voids with sizes equal to the composite unit cell size, perhaps 0.1 to 1 cm or more. Finally, such 'microscopic' voids may coalesce to form voids or cracks on the continuum scale with sizes ranging from centimetres to metres, and in geologic materials, from millimetres to kilometres.

How should we approach constructing constitutive relations for such a wide range of material damage? In this chapter we present an approach that we believe is both consistent and useful.

First, we wish to begin at the microscopic level, the size scale at which the intact material no longer appears as a continuum. Our approach therefore differs from that of continuum fracture mechanics, which deals with the conditions for instability of a single macroscopic crack [1]. The continuum limit is reached at the size scale at which the 'graininess' of the material first appears: the unit cell size for a composite, the grain size for a polycrystalline metal, and the atomic spacing for a perfect crystal.

We therefore select as our basic material element for analysis a mass of material that contains a statistical number of 'grains'. The above argument thus forces us at the onset to introduce a material-specific scale size into the constitutive relations, a factor that is absent in most theories of elasticity or plasticity. We shall refer to this unit of material as the 'material element'.

Second, we shall define material damage as microcracks or voids with sizes on the order of the 'grain' size. Submicroscopic-sized damage will not be considered explicitly, but will be lumped into the nucleation process in a manner to be described later.

A third constraint that we now introduce is that each material element must potentially contain a statistical number of microcracks or voids. This will normally be the case because the material element has been chosen large enough to contain many 'grains' or heterogeneities that form nucleation sites. For example, in a typical heat of A533B pressure vessel steel the continuum limit is determined by arrays of MnS and Al_2O_3 inclusions several μm in size and spaced about 0.1 mm apart. Thus, the continuum limit is determined by the inclusion spacing, and by our above rules the material element must be several tenths of millimetre on a side so as to contain many inclusions. Under sufficient loading the inclusions begin to debond to 'nucleate' voids.

In summary, our approach to constitutive modelling of damaged material rests on three assumptions:

(1) There is a material-specific material element of irreducible size that is determined by the material continuum limit or 'graininess'.
(2) Damage is defined as microscopic cracks or voids with characteristic sizes on the order of the size of the material element.
(3) Each material element potentially contains a statistical number of microscopic cracks or voids.

7.2 APPROACH

7.2.1 Definition of damage

A microscopic damage measure must be defined that will serve as an internal state variable in the constitutive relations. Kachanov simply defined a scalar damage function D that is zero for intact material and unity for completely fractured material [2]. The damage function D is made a function of stress and strain history. In a similar approach Norris and colleagues chose the scalar D to be an integral function of mean tensile stress and plastic strain corresponding to theoretical void growth laws in plastic material [3].

A disadvantage of a scalar damage function is that it cannot describe the anisotropy produced by planar microcracks or ellipsoidal voids. Davison and Stevens handled such anisotropy by introducing a vector damage function $\mathbf{D}$ to represent both the average area and orientation of planar microcracks at each material point [4].

We have chosen a detailed description of the microscopic damage similar to that of Davison and Stevens [4]. The description is based on actual counts and measurements of microcrack numbers, sizes, and orientation in specimens that have been exposed to known stress and strain histories.

We first characterize each microcrack or void with a size R and orientation $\mathbf{n}$. For basically planar microcracks, R is the average radius and $\mathbf{n}$ is unit vector normal to the crack plane. For ellipsoidal voids, R can be chosen to be the average of the major and minor axes, and **n** can be chosen to be a unit vector normal to the plane of maximum cross-sectional area.

The description of damage is then chosen to be the function $\rho_f(\mathbf{X}, t, R, \mathbf{n})$, where ρ_f is the concentration of active flaws, t is the time, and $\mathbf{X}$ is the Lagrangian coordinate that specifies the material point, the centre of the material element of interest. (The size of the material element will be labelled $\delta\mathbf{X}$.)

For given values of $\mathbf{X}$ and t, the total number of active flaws per unit volume (in the reference system defined by $\mathbf{X}$) is the integral of ρ_f over all values of R and $\mathbf{n}$:

$$N_t = \int_R \int_{\mathbf{n}} \rho_f \, \mathrm{d}\mathbf{n} \, \mathrm{d}R \tag{7.1}$$

The distribution function ρ_f represents the transition from microscopic, discrete processes to continuum mechanics. A discussion of the relation of such microscopic deformation processes to continuum mechanics has been given by Rice [5].

Another useful function is the total number of microcracks or voids $N_g(R)$ with sizes greater than or equal to R, obtained by integrating ρ_f over all orientations $\mathbf{n}$ and from R to infinity:

$$N_g(\mathbf{X}, t, R) = \int_R^\infty \int_{\text{all } \mathbf{n}} \rho_f(\mathbf{X}, t, R, \mathbf{n}) \, \mathrm{d}\mathbf{n} \, \mathrm{d}R \tag{7.2}$$

Experimental observations, as well as theoretical predictions by McClintock [6] and Batdorf [7], show that N_g often has the form

$$N_g(\mathbf{X}, t, R) = N_t(\mathbf{X}, t) \exp[-R/R_0(\mathbf{X}, t)] \tag{7.3}$$

where N_t is the total number of microcracks or voids per unit volume (in the reference configuration) and R_0 is a characteristic size for this exponential size distribution. Equation (7.3) requires that

$$\rho_f(\mathbf{X}, t, R) = -[N_t(\mathbf{X}, t)/R_0(\mathbf{X}, t)] \exp[-R/R_0(\mathbf{X}, t)] \tag{7.4}$$

where

$$\rho_f(\mathbf{X}, t, R) \equiv \int_{\text{all } \mathbf{n}} \sigma_f(\mathbf{X}, t, R, \mathbf{n}) \, \mathrm{d}\mathbf{n} \tag{7.5}$$

7.2.2 Damage kinetics

The constitutive relations for damage must be based on the evolution of ρ_f in time in response to applied continuum (remote) variables such as stress, strain, strain rate, temperature, etc. That is,

$$\dot{\rho}_f = \dot{\rho}_f(\boldsymbol{\sigma}, \varepsilon, \dot{\varepsilon}, T, \ldots) \tag{7.6}$$

In a given material element $\delta\mathbf{X}$, $\dot{\rho}_f$ will have one contribution from the nucleation of new microcracks or voids and another contribution from the growth of previously nucleated ones.

To see how one may attempt to separate the contribution of nucleation from that of growth, consider the example given by equations (7.3) and (7.4). For size distributions of this exponential type, nucleation will increase the value of N_t. If the nucleation process maintains the size distribution shape, i.e., equal numbers of new cracks are added to each crack size, then, R_0 would be unchanged by the nucleation process. That is, from equation (7.4)

$$\dot{\rho}_f(\text{nuc.}) = -(\dot{N}_t/R_0)\exp(-R/R_0)$$

or

$$\dot{\rho}_f(\text{nuc.})/\rho_f = \dot{N}_t/N_t \tag{7.7}$$

On the other hand, microcrack growth does not change N_t. If the growth also preserves the form given by equation (7.3) or (7.4), then

$$\dot{\rho}_f(\text{growth})/\rho_f = (R/R_0 - 1)\dot{R}_0/R_0 \tag{7.8}$$

That is, the rate of change in ρ_f for a given $\mathbf{X}$ and R due to growth would depend on $\dot{R}_0$, and would be negative for $R>R_0$ and positive for $R>R_0$.

It is usually more convenient to deal with $N_g(\mathbf{X}, t, R)$ than with $\rho_f(\mathbf{X}, t, R)$ because the experimental data is more easily displayed in this form and, as we shall see, it is easier to deduce nucleation and growth information from the N_g function.

Similar to the approach that led to equations (7.7) and (7.8), we write

$$\dot{N}_g \equiv (\partial N_g/\partial t)_{\mathbf{X},\mathbf{R}} = \dot{N}_g^{\mathrm{N}} + \dot{N}_g^{\mathrm{G}} \tag{7.9}$$

where the superscripts N and G refer to nucleation and growth respectively. If we assume a distribution given by equation (7.3), we obtain

$$\dot{N}_g^{\mathrm{N}}/N_g = \dot{N}_t/N_t \tag{7.10}$$

and

$$\dot{N}_g^{\mathrm{G}}/N_g = (R/R_0)(\dot{R}_0/R_0) \tag{7.11}$$

Solving equations (7.9), (7.10), and (7.11) for $\dot{R}_0/R_0$ yields

$$(\dot{R}_0/R_0) = (R_0/R)[(\dot{N}_g/N_g) - (\dot{N}_t/N_t)] \tag{7.12}$$

Thus, to obtain the nucleation and growth rates, consider the experimental data shown schematically in Figure 7.1. The intercept of the $N_g(R)$ curve with the $R = 0$ axis is N_t. Thus the rate this intercept moves up in time gives us $\dot{N}_t$. Then, for any other value of R (say, $R = R_0$) we measure the rate of increase of $N_g(R)$ to get $\dot{N}_g(R)$. Substitution into equation (7.12) gives us $\dot{R}_0/R_0$, the growth rate.

Furthermore, it is easy to see that preserving the form of equation (7.3) implies that cracks of all sizes grow according to a viscous growth equation

$$\dot{R}/R = A$$

That is, if no nucleation occurs, a particular crack of size R is specified by the value of N_g (the 10 000th largest crack, for example). Thus, according to equation (7.3),

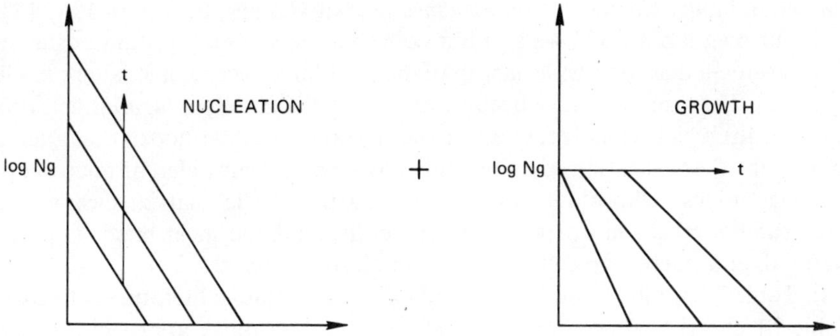

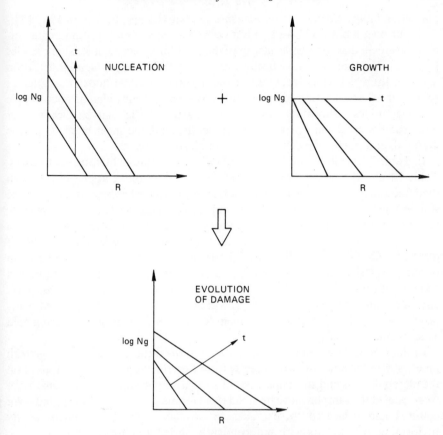

Figure 7.1 Evolution of microscopic damage

during growth of a crack of size R the corresponding value of N_g is constant, so

$$\begin{aligned} R/R_0 &= \text{constant} \\ \dot{R}/R &= \dot{R}_0/R_0 \end{aligned} \tag{7.13}$$

Thus, every crack grows with the relative rate given by $\dot{R}_0/R_0$.

7.3. EXAMPLES OF CONSTITUTIVE MODELS FOR FAILING MATERIAL

7.3.1 Summary of previous work

During the past ten years there has been a growing number of publications on the subject of constitutive models of failing material, many of them by participants in this conference. A recent overview can be obtained from the *Proceedings of the*

5th International Conference on Fracture held at Cannes, France in 1981 [8].

In our own laboratory, we and our colleagues have developed microstatistical fracture models for a wide variety of materials and microscopic failure modes [9 to 16]. These models have become known by the acronym NAG (nucleation and growth), which actually represents a family of constitutive models that attempt to describe the nucleation, growth, and coalescene of four different microscopic damage modes: ellipsoidal voids in an elastic–plastic matrix, cleavage or intergranular cracks in elastic–plastic material, adiabatic shear bands in plastic material, and microscopic shear cracks in elastic material.

In Table 7.1 we list these models and cite representative literature references. Table 7.1 shows that we have attempted to use the above approach on both dynamic and quasistatic problems. The degree of success has been greatest for the simplest geometries, for example for impact or radiation loads that generate uniaxial strain conditions in the structures of interest. For such cases the data from these models typically agree in the detailed description of damage distributions throughout the material specimens to within a factor of two in number and sizes of microscopic voids per unit volume. More complicated geometries pose greater problems, although we have achieved reasonably good agreement also for quasistatic void growth in a steel in the centre-cracked panel geometry [10, 11, 12] and fair agreement for large high explosive cratering field experiments in basalt [16].

In the work referenced in Table 7.1, we described the nucleation and growth kinetics with formulas derived directly from experimental data. We imposed on the material stress and strain histories of known amplitudes and durations. Then we sectioned the samples and examined them for microscopic cracks or voids. We counted and measured these cracks to produce $N_g(\mathbf{X}, R, \mathbf{n}, t)$ functions. By performing experiments with independently varied amplitudes and durations, we could construct evolving damage plots like that in Figure 7.1 for each amplitude level and infer from the data formulas for the nucleation and growth rates discussed earlier.

This procedure can be followed even if we have no theoretical understanding of the basis of the inferred formulas. However, it is clearly valuable to understand the underlying phenomenology, both to increase confidence in the formulas and to provide guidelines for extrapolating the results to load regions outside those exercised by the model calibration experiments.

In the following paragraphs we choose as an example one microscopic failure mode, microscopic ellipsoidal void nucleation and growth in an elastic–plastic material, and discuss the available theoretical guidance for formulating and understanding observed damage evolution.

7.3.2 Ellipsoidal voids in elastic–plastic material, nucleation

We have previously reviewed theories of void nucleation [17] and summarize them briefly here.

Table 7.1 Microstatistical NAG failure models

Microscopic failure mode	Name of computational subroutine	Application	References
Ellipsoidal void growth in elastic–plastic matrix	DFRACT	Dynamic response to impact or radiation loads, hot spots	Seaman *et al.* [9]
	DFRACTS	Quasistatic failure of engineering structures	Shockey *et al.* [10, 11, 12]
Brittle cleavage or intergranular cracking	BFRACT	Dynamic response to impact or radiation loads	Seaman *et al.* [9]
	Unnamed, under development	Quasistatic stress corrosion cracking	Caligiuri *et al.* [13]
Adiabatic shear bands	SHEAR 1 SHEAR 2 SHEAR 3 SHEAR 4	Fragmenting munitions, armour penetration and fragmentation, high pressure phase transitions, hot spots.	Erlich *et al.* [14]
Elastic shear cracks	'Z' model, under development	Quasistatic fracture and dilatancy of geologic materials.	Moss and Gupta [15]

Under load conditions of high triaxial tensile stress or high temperature, atomic vacancy diffusion can theoretically nucleate microscopic voids on grain boundaries or triple-point grain boundary junctions. The nucleation rate is driven by Arrhenius rate theory, and many investigators have developed models for creep rupture based on such theories [18].

On the other hand, under conditions of lower triaxial stress and temperature as well as significant plastic deformation, nucleation of microscopic voids can be expected to be driven by dislocation pile ups at grain boundaries or other obstacles, or by mechanical debonding at inclusions. That is, local plastic strain causes voids to appear at microscopic heterogeneities.

Conceptually the simplest nucleation mechanism is debonding of inclusions. As discussed in our previous review [17] there are a variety of theories for plastic deformation-driven debonding of inclusions in an elastic–plastic matrix. At present, these theories are limited by the lack of an analytical solution for the stress and strain fields around an elastic inclusion in a plastic matrix. Eshelby's solution for an elastic inclusion in an elastic matrix [19] can be generalized to the case of a plastic inclusion in an elastic matrix [20], but a solution has not yet been obtained for the opposite case, which is unfortunately the one of most practical interest.

Despite these difficulties, some guidelines for nucleation rate functions can be obtained for inclusion debonding. As discussed in our previous review [17], one can assume that inclusion debonding will occur when some combination of local shear stress and normal tensile stress is exceeded at the interface. We further assume that this combination will translate remote from the inclusion to an equivalent combination of critical mean stress σ_m and equivalent shear stress $\bar{\sigma}$. The critical equivalent shear stress could of course be replaced by a critical equivalent plastic strain through the appropriate hardening law. In our previous review [17], we presented a plausibility argument that resulted in several nucleation functions, all of which were similar to the form

$$F = \exp(-\sigma_{m0}/\sigma_m) + \exp(-\bar{\sigma}_0/\bar{\sigma}) \qquad (7.14)$$

where F is the fraction of the inclusions that have been nucleated, and σ_{m0} and $\bar{\sigma}_0$ are the critical mean stress and effective shear stress, respectively, to alone debond an inclusion of the characteristic size R_0 [see equation (7.3)]. The isonucleation curves of equation (7.14) form a yield surface for the onset of damage. The surface moves outward with increasing amounts of nucleation.

Since stress-free voids are formed by the nucleation (debonding) process, stress relaxation (softening) will also occur, and will move the curves for high values of N_i/N_t closer to the origin. However, usually the relative inclusion volume is so small in typical metals that the softening effect during nucleation may be neglected.

7.3.3 Growth

Growth of microscopic voids in an elastic–plastic matrix can also occur by Arrhenius-type atomic diffusion processes, for example in high temperature creep experiments, or by local plastic flow around the voids in room temperature and high plastic deformation experiments. We concentrate on the latter case.

The growth due to local plastic flow is, analogous to the debonding of inclusions, conceptually simple but analytically difficult. In response to remotely applied deformation, plastic zones will first form at the void surface, and then expand until they intersect the plastic zones from neighbouring voids. At this point, the material element becomes 'fully plastic', defining a yield surface for unbounded void growth.

Gurson [21] performed the most complete analysis of this process, deducing for voids in a rigid-plastic material an approximate form of the yield surface given by the equation:

$$\phi = (3/2)\sigma'_{ij}\sigma'_{ij}/\bar{\sigma}_m^2 + 2f\cosh(\sigma_{kk}/2\bar{\sigma}_m) - (1 + f^2) = 0 \tag{7.15}$$

where ϕ is the yield function, the σ'_{ij}s are the continuum (macroscopic) deviator stresses, f is the void volume fraction, the σ_{ij}'s are the continuum (macroscopic) true stress components, and $\bar{\sigma}_m$ is the equivalent tensile flow strength of the unvoided (microscopic) material. (The summation convention is used.)

In the absence of nucleation, an expression for void growth can be obtained by applying the normality condition, as discussed by Yamamoto [22]. The resultant expression is

$$\frac{d\varepsilon_{kk}^{p}}{d\bar{\varepsilon}^{p}} = \frac{3\alpha}{\sqrt{w + 2\alpha^2}} = \frac{\dot{f}}{1 - f} \tag{7.16}$$

where $\alpha = \frac{1}{2} f \sinh\left(\frac{\sigma_{kk}}{2\bar{\sigma}_m}\right)$

$$w = \left[(1 + f^2) - 2f\cosh\left(\frac{\sigma_{kk}}{2\bar{\sigma}_m}\right)\right]$$

$$d\bar{\varepsilon}^{p} \equiv \sqrt{\left(\tfrac{2}{3} d\varepsilon_{ij}^{p} d\varepsilon_{ij}^{p}\right)}$$

Since equation (7.16) defines $d\bar{\varepsilon}^{p}$ in terms of total plastic strain components rather than deviator components, it will reduce to the identity

$$d\varepsilon_{kk}^{p} = d\varepsilon_{kk}^{p}$$

under pure triaxial tension, as desired. (A common procedure is to define $d\bar{\varepsilon}^{p}$ in terms of deviators which then, however, cannot provide an equation like equation (7.16) that applies to all stress states.)

Equation (7.16) can thus be used to obtain the increment in relative void volume as a function of the increment in plastic strain and the current state of stress, hardening modulus, and relative void volume.

If nucleation occurs simultaneously with growth, the normality condition cannot be used. However, the nucleation itself usually causes only very small increments in macroscopic strain although there is an increment in f. Thus, to a certain approximation the nucleation and growth processes can be decoupled, and normality can be used for the growth process.

On the other hand, at dynamic loading rates where viscous forces become important the void growth rate will be limited to a maximum value given by

$$\dot{f} \propto f/\eta$$

where η is a viscosity [23–26]. In this case, the value of $\dot{f}$ deduced from equation (7.16) would only be the limit toward which the viscous growth relaxes. Furthermore, growth by diffusive processes would impose another, slower growth rate.

Thus, in general, rate processes will prevent the growth rate from obeying the normality condition, but for a wide range of strain rates where viscous and diffusive effects are unimportant, equation (7.16) may be a good approximation.

Damage-induced anisotropy is another major complication in constructing growth rates for ellipsoidal voids in elastic–plastic material. Under combined hydrostatic and shear stresses the voids will flatten or elongate [26], thereby directionally softening the material. The Gurson yield surface [equation (7.15)] is a function of stress invariants and is therefore isotropic, as is the growth rate given by equation (7.16).

How can one account for the damage-induced anisotropy? The growth under combined stresses of an ellipsoidal void in an elastic–plastic, hardening material is a special case of the inclusion problem referred to earlier, and is currently under attack, both analytically and numerically, by many researchers. Probably, however, for some time we shall have to make do with approximate solutions in the spirit of Gurson's results for the isotropic case.

7.3.4 Coalescence

The final stage of failure for ductile voids is that of void coalescence, a form of 'strong void interaction'. Initially the voids grow autonomously as described by Rice and Tracey [27]. 'Weak interaction' occurs when the plastic strain fields around individual voids coalesce to form Gurson's yield surface. 'Strong interaction' may occur by near impingement with ligament stretching for very ductile materials, or by formation of shear bands between voids for materials less resistant to plastic instabilities.

There have been many attempts to predict the onset of strong void interaction and resultant coalescence, as discussed, for example, by Yamamoto [22] and Tvergaard [28]. The current status appears to be that the conditions for coalescence depend strongly on the geometrical arrangement of the voids as well as on possible clusters of voids with a resultant high local relative void volume.

7.3.5 Computer simulation

The ultimate practical reason for constructing constitutive relations for microscopic failure processes is to allow prediction of failure in engineering test specimens or structures. In the following paragraphs we show some preliminary computer simulations of test specimens failing by the nucleation, growth, and coalescence of ductile microvoids.

Smooth round bar tensile specimens of ductile metals begin the failure process by nucleation and growth of microscopic voids in the centre of the neck. On the other hand, a notched round bar will initiate failure by the same process at the notch root if the notch is sharp enough. It is of practical interest to understand how sharp a notch must be before fracture initiation begins at the notch root rather than elsewhere in the specimen. That is, when is a notch a dangerous 'stress raiser' and when can it be ignored? Upon what material properties does the answer depend?

In previous publications [10, 11, 12] we have reported on modelling of microvoid kinetics in A533B pressure vessel steel at a temperature of 355° K. We now briefly summarize the applications of that work to the above notch problem.

We first performed interrupted tests on smooth round bar specimens, sectioned them, and measured and counted the nucleation of voids at inclusions to obtain evolving $N_g(R)$ curves.

A nucleation function similar to function (7.14) was constructed from the data. However, since the value of the mean tensile stress $P(=\frac{1}{3}\sigma_{kk})$ was fairly constant, only a strain criterion was used, namely

$$\begin{aligned} N &= N_1(\bar{\varepsilon}^{\mathrm{P}} - \varepsilon_1), \qquad \varepsilon_1 < \bar{\varepsilon}^{\mathrm{P}} < \varepsilon_2 \\ N &= N_1(\varepsilon_2 - \varepsilon_1) + N_2(\bar{\varepsilon}^{\mathrm{P}} - \varepsilon_2), \qquad \varepsilon_2 < \bar{\varepsilon}^{\mathrm{p}} \end{aligned} \tag{7.17}$$

where N_1 and N_2 are material specific void densities, and ε_1 and ε_2 are strain thresholds determined from the data. The results of the tensile tests suggested values of $5 \times 10^6\,\mathrm{cm}^{-3}$ and $1 \times 10^5\,\mathrm{cm}^{-3}$ for N_1 and N_2, respectively, and 0.11 and 0.13 for ε_1 and ε_2.

Once nucleated, voids grow gradually by plastic flow, elastic strain, and thermal expansion. The incremental plastic growth law,

$$\mathrm{V_v} = \mathrm{V_0} \exp\left[-T_1 \frac{P_{\mathrm{s}}}{\bar{\sigma}_{\mathrm{m}}} (\bar{\varepsilon}^{\mathrm{p}} - \bar{\varepsilon}_0^{\mathrm{p}}) \right] \tag{7.18}$$

where P_s is the average stress in the solid material, V_0 is initially the inclusion volume, and T_1 is a dimensionless coefficient determined from a plot of void volume as a function of strain, was obtained from the combined measurements and computer simulations of the smooth round bar tests. This equation is very similar to the theoretically derived McClintock [25] equations or the Rice–Tracey [27] growth law for stress triaxialities less than one.

The elastic expansion of the void was also calculated, and added to equation (7.18) to get the total void growth.

As voids form and grow in the material, the load-bearing capacity of the specimen decreases. This reduction in strength was calculated from a procedure consistent with Gurson's yield surface.

We did not attempt a detailed model of the void coalescence. Instead, we assumed that void growth continues until a relative void volume of about 0.01, indicated by the round-bar tension experiments, is reached, at which point coalescence and specimen failure occur.

The above computational model, although derived from experimental data, is thus consistent with the theoretical notions of the nucleation and growth processes discussed earlier.

The next step was to use the model in computational simulations of smooth and notched round bar tensile tests. TROTT, a finite difference, Lagrangian explicit two-space dimensional code, was used [29].

Since the smooth bar tests were used to calibrate the model, the smooth bar simulations not surprisingly gave the correct history of void nucleation and growth: begining at the centre of the neck and expanding outward to nearly fill the neck region.

On the other hand, an experiment with a notch root radius of 0.025 cm had been seen experimentally to initiate damage at the notch root, Encouragingly, the calculated load–displacement curve agreed with experiment, and the calculated void nucleation and growth began at the notch root and spread out at about 45°, as observed.

What are the key material and structural properties that govern this plastic void notch response? The above calculation suggests that the answer lies in the plastic void growth law, equation (7.18). In that equation the key parameters are the inclusion volume V_0, the stress triaxiality $P_s/\bar{\sigma}_m$, and the equivalent plastic strain $\bar{\varepsilon}^p$. V_0 is a microstructural material property whereas $P_s/\bar{\sigma}_m$, and $\bar{\sigma}^m$ are largely structural properties, i.e., they depend on the notch radius and other specimen dimensions. If a specimen is microstructurally homogeneous and isotropic so that V_0 is constant throughout the specimen, then for a given V_0 the notch behaviour is governed by the trade-off between stress triaxiality and plastic strain. For a blunt notch both $\bar{\varepsilon}^p$ and $P_s/\bar{\sigma}_m$ are relatively small at the notch root, whereas $P_s/\bar{\sigma}_m$ is larger in the specimen interior. Therefore, the nucleation begins there. On the other hand when the notch becomes sharper, both $\bar{\varepsilon}^p$ and $P_s/\bar{\sigma}_m$ become greater close to the notch root and the first nucleation shifts to the notch root region.

The above discussion also enlightens us as to the nature of the fracture toughness, i.e., the resistance of a notch to macroscopic fracture initiation and growth. We see that fracture toughness of ductile materials also depends on the microstructural material property, V_0, and two structural properties, $P_s/\bar{\sigma}_m$ and $\bar{\varepsilon}^P$. Further consequences of this deduction for fracture scaling have been recently discussed by Giovanola [30].

7.4 CONCLUSIONS

We conclude that a combined experimental and theoretical approach to constructing constitutive relations for failing materials is productive. The basic experiments are interrupted laboratory tests in simple geometries over a wide range of load amplitudes and durations. The post-test specimens are examined microstructurally to obtain damage distribution functions $\rho_f(R, \mathbf{n})$ and their evolution as a function of the histories of continuum stresses and strains. These experimental data are supplemented by theories of microscopic crack and void nucleation, growth, and coalescence that add confidence to the models and allow extrapolation of the models to a wide range of engineering applications.

Future work should significantly improve each link in this chain of experimental and theoretical procedures.

ACKNOWLEDGEMENTS

This work was partially supported by the National Science Foundation under grant number MEA-8108186.

REFERENCES

1. H. Liebowitz (ed.), *Fracture*, vols. I–VII, Academic, New York (1968–1972).
2. L. M. Kachanov, 'Time of the rupture process under creep conditions', *IZV. Akad. Nauk. SSR, Otd. Tekh, Nauk, No.* **8**, 2631 (1958).
3. D. M. Norris, Jr., J. C. Reaugh, B. Moran, and D. F. Quinones, 'Computer model for ductile fracture: application to the Charpy V-notch test', *Lawrence Livermore Laboratory Report NP*–961, Research Project 603, prepared for Electric Power Research Institute (Jan. 1979).
4. L. Davison and A. L. Stevens, *J. Appl. Phys.*, **44**, No. 2 668 (Feb. 1973).
5. J. R. Rice, in *Constitutive Equations in Plasticity* (Ed. A. S. Argon), Chapter 2, MIT Press, Cambridge, Mass., (1974).
6. F. A. McClintock, in *Fracture Mechanics of Ceramics* (Eds. R. C. Bradt, D. P. H. Hasselman and F. F. Lange), Vol. 1, Plenum Press, 1973.
7. S. B. Batdorf, *Nucl. Eng. and Design*, **35**, 349 (1975).
8. D. Francois (ed.), 'Advances in Fracture Research' (in 5 volumes), *Proceedings of 5th International Conference on Fracture*, Cannes, France, 29 March–3 April 1981.
9. L. Seaman, D. R. Curran, and D. A. Shockey, *J. Appl. Phys.*, **47**, 4814–4826 (1976).
10. D. A. Shockey, K. C. Dao, L. Seaman, R. Burback, and D. R. Curran, 'Computational modeling of microstructural failure processes in A533B pressure vessel steel', *Report NP*-1398, Research Project 1023–1, prepared by SRI International for Electric Power Research Institute (May 1980) (1980a).
11. D. A. Shockey, L. Seaman, K. C. Dao, and D. R. Curran, 'Kinetics of void development in fracturing tensile bars', *Trans ASME J. Pressure Vessel Tech.*, **102**, 14–21 (Feb. 1980) (1980b).

12. D. A. Shockey, L. Seaman, R. L. Burback, and D. R. Curran, 'J_{IC} calculations for pressure vessel steel from microvoid kinetics', to be published in special ASM volume, *Fracture Mechanics of Ductile and Tough Materials and its Applications to Energy Related Structure* (1980) (1980c).
13. R. D. Caligiuri, L. E. Eiselstein, and D. R. Curran, 'Microkinetics of Stress Corrosion Cracking in Steam Turbine Disc Alloys', SRI Interim Report submitted to Electric Power Research Institute, EPRI Project No. 1929–8 (March 1982).
14. D. C. Erlich, L. Seaman, R. D. Caligiuri, and D. R. Curran, SRI Annual Report to Ballistic Research Laboratory, Contract No. DAAK11–78–C–0115 (Nov. 1980).
15. W. C. Moss and Y. M. Gupta, 'A constitutive model describing dilatancy and cracking in brittle rocks', *J. Geophysics Res.*, **87**, 2985–2998 (1982).
16. D. R. Curran, D. A. Shockey, L. Seaman, and M. Austin, 'Mechanisms and models of cratering in earth media', in *Proceedings of the Symposium on Planetary Cratering Mechanics—Impact and Explosion Cratering*, (Eds. D. J. Roddy, R. O. Pepin and R. B. Merrill), Pergamon Press, New York, (1977).
17. D. R. Curran, L. Seaman, and D. A. Shockey, 'Linking dynamic fracture to microstructural processes', in *Shock Waves and High-Stain-Rate Phenomena in Metals* (Eds. Marc. A. Meyers and Lawrence E. Murr), Plenum, New York, 1981.
18. R. Raj and M. F. Ashby, *Acta Met.*, **23**, 653–666 (1979).
19. J. D. Eshelby, *Proc. Royal Soc. London, A*, **241**, 376 (1957).
20. Y. M. Gupta, 'Analysis and modeling of piezoresistance response', SRI Final Report to Defence Nuclear Agency, Contract No. DWA-001–79–C–0180 (1980).
21. A. L. Gurson, 'Continuum theory of ductile rupture by void nucleation and growth: Part I—Yield criteria and flow rule for porous ductile media', *Journal of Engineering Materials and Technology, Transactions of the AIME*, 2–15 (1977).
22. H. Yamamoto, 'Conditions for shear localization in the ductile fracture of void-containing materials', *Int. Journ. of Fracture*, **14**, 4, 347–365 (Aug. 1978).
23. H. Poritsky, 'The Collapse or Growth of a Spherical Bubble or cavity in a Viscous Field', *Proceedings of the First U.S. National Congress of Applied Mechanics*, ASME, New York, 813 (1952).
24. C. A. Berg, *Proc. 4th U.S. National Congress of Applied Mechanics*, **2**, 885 (1962).
25. F. A. McClintock, 'A criterion for ductile fracture by the growth of holes', *Journ. Appl. Mechanics*, **35**, 363–371 (1968).
26. B. Budiansky, J. W. Hutchinson, and S. Slutsky, 'Void Growth and Collapse in Viscous Solids', *Mechanics of Solids, The Rodney Hill 60th Anniversary Volume* (Ed. H. G. Hopkins and M. J. *Sewell*), *Pergamon Press*, 1982.
27. *J. R. Rice and D. M. Tracey*, '*On the ductile enlargement of voids in triaxial stress fields*', *J. Mech. Phys. Solids*, **17**, 201–217 (1969).
28a. V. Tvergaard, 'Influence of voids on shear band instabilities under plane strain conditions', *Technical University of Denmark, Report No.* 159 (June 1979).
28b. V. Tvergaard, 'Material failure by void coalescence in localized shear bands', *Int. Journal Solids Structures*, **18**, 659–672 (1982).
28c. V. Tvergaard, 'On localization in ductile materials containing spherical voids', *Int. Journal of Fracture*, **18**, No. 4, 257–259 (April 1982).
28d. V. Tvergaard, 'Influence of Void Nucleation on Ductile Shear Fracture at a Free Surface', *Technical University of Denmark, Report No.* 236 (April 1982).
29. L. Seaman, 'TROTT computer program for two-dimensional stress wave propagation', *SRI International Final Report*, Vol. III, submitted to Ballistic Research Laboratory, Report No. ARBRL–CR–00428 (April 1980).
30. J. H. Giovanola, 'The scaling of fracture phenomena', *Poulter Laboratory Technical Report* 001–82, SRI International (Feb. 1982).

Mechanics of Engineering Materials
Edited by C. S. Desai and R. H. Gallagher

Chapter 8

Modelling Cyclic Plasticity: Simplicity Versus Sophistication

Y. F. Dafalias

8.1 INTRODUCTION

The realistic analytical description of the response of engineering materials under symmetric, unsymmetric, or random cyclic stress reversals in the inelastic range is one of the very difficult and important subjects of constitutive modelling. In this chapter an effort is undertaken to critically examine and analyse the methods employed in a macroscopic approach towards the construction of constitutive models within the framework of classical rate independent elastoplasticity. Depending on the level of sophistication, these models can describe a spectrum of different material response characteristics under monotonic and cyclic loading from the simplest to the very complex. Three points must be noted in this context. First, the number of loading cycles, symmetric or not, is restricted to a few hundred at most, thus precluding fatigue consideration for metals although they may induce internal damage and degradation effects on geological media. Second, attention is focused not on the foundation of elastoplasticity theory, which is taken for granted, but rather on the more applied and practical character of the actual process of constructing a specific model, always within a firmly established theoretical basis. Third, in order to focus on the material nonlinearities only, small deformations will be considered. It is also necessary to emphasize that in view of the limited space available no extensive literature review of the subject is possible, and only selected references are given, some of them on the basis of their illustrative features for the content under discussion.

Although the general framework applies to any material which exhibits rate independent incremental irreversibility, specific observations on detailed material response characteristics can be clearly presented and easily understood in relation to specific materials. It is beyond the scope of this work to examine all these different aspects for numerous engineering materials. Therefore, emphasis is placed primarily on the response of structural metals with a much briefer exposition of the relevant features for soils, concrete, and rocks. Nevertheless, many of the phenomenological observations made on metals can apply to other materials on the basis of their common elastoplastic features even if their substructures differ. Metals were preferred as the main sample class of engineering materials because they are simpler to describe, better understood and there is a plethora of detailed experimental data under cyclic loading.

A stress space elastoplasticity formulation will be adopted in this chapter properly modified in order to consider simultaneously stable and unstable material response (falling uniaxial stress–strain curve). All corresponding models of classical rate independent elastoplasticity can be put into a common framework by defining the state in terms of external variables, here the stress (temperature is omitted for simplicity), and a set of internal variables which will be called plastic internal variables or piv for abbreviation. The name piv was chosen in order to emphasize their association with plasticity in particular, while the name internal variables is associated with inelasticity in general. The piv are usually scalar or second order tensor quantities entering the analytical expressions of the loading surface, the elastic potential and other constitutive entities. They embody the past loading history as it is manifested by means of phenomenological material properties and their change. The law of evolution and the values of piv provide the hardening (softening) and degradation (or internal damage) features of the stress rate–strain rate relations which are the final objective of any constitutive model. Examples of piv are the plastic strain tensor, the plastic work, a scalar measure of cumulative plastic strain, the back-stress tensor, etc. In a stress–space formulation, the common feature of piv is that their evolution is rate independent and their rate is different than zero for a stress point on a loading surface only when a scalar quantity, called the loading index and being a function of the state and the stress rate, is positive.

There are many authors who prefer to identify the internal variables as entities associated with the microstructure of the material. Examples are the density and distribution of dislocations for metals or the porosity and orientation of particles and their contact planes for particulate media. Along these lines, quantities such as plastic strain and plastic work are not considered as proper piv. The difficulty associated with the above microstructural identification of piv is twofold: first, the piv can only be quantified by special experimental procedures (electron microscopy, thin section studies, etc.). Second, it is a formidable and often impossible task to relate the evolution of these piv with the stress–strain rate relations which are of final interest. On the other hand, the phenomenological approach adopted in this article introduces as piv quantities which can be easily quantified on the basis of macroscopic stress–strain experimental data on homogeneous samples, and whose values and laws of evolution express in a global sense the effect of the microstructure and its change on the constitutive relations for the macroelement. For example, the change of the 'size' of the yield surface can be considered an indirect measure of the change in dislocation density. Similarly, the deviatoric plastic strain of a soil sample in triaxial compression can be considered an indirect measure of the change in soil particles' orientation towards the horizontal, thus it can be used to describe induced plastic transverse isotropy effecting the shape of the yield surface. In both examples note that the word 'change' has been used, implying that an initial value for every piv must be specified. This value can be either determined experimentally, as for example the 'size' of the initial yield surface, or assumed if the piv expresses a relative quantity as in the case of plastic strain which is put equal to zero initially. It is the relative character of the plastic strain (compares geometrically an initial and a current relaxed configuration) which has raised some questions, especially in the case of large deformations, about the appropriateness of its use as a piv which is assumed to define the current state. In view of the previous discussion on initial values and the fact that plastic strain represents a geometrical and measureable manifestation of substructural changes from one configuration to the other, there is no reason why it cannot be used as a piv in a general theory. On the other hand, it may not be the best piv to use for direct correlation with material properties which change monotonically even under cyclic loading where the plastic strain oscillates. In fact, more often it is not the plastic strain but its rate which is used by being properly related to the rates of other piv, e.g. the plastic work or the back-stress tensor.

8.2 GENERAL FORMULATION OF RATE INDEPENDENT ELASTOPLASTICITY

Tensorial quantities will be presented in direct notation and symbolized by bold-faced characters. A single subscript in such notation for the piv (usually the letter n) is not a tensorial index but merely identifies any one of the many piv. Juxtaposition of tensors in direct notation implies proper contraction of tensorial indices on the neighbouring sides of the juxtaposed tensors referred to a common cartesian coordinate system, and compatible with the tensorial order of the terms in which these tensors are juxtaposed. The presence of the same subscript, e.g. n, in two juxtaposed tensors signifies repetition of the implied contraction (if any) for each n and subsequent summation over all values of n. For small deformations the usual additive decomposition of the total strain tensor $\boldsymbol{\varepsilon}$ into an elastic $\boldsymbol{\varepsilon}^{e}$ and a plastic part $\boldsymbol{\varepsilon}^{p}$ is assumed, i.e.

$$\boldsymbol{\varepsilon} = \boldsymbol{\varepsilon}^{e} + \boldsymbol{\varepsilon}^{p} \tag{8.1}$$

Let $\boldsymbol{\sigma}$ represent the stress tensor, $\mathbf{q}_n$ the piv which include the $\boldsymbol{\varepsilon}^{p}$ in general and g, ψ a pair of dual elastic potentials interrelated through the Legendre transformation with respect to their active variables $\boldsymbol{\sigma}, \boldsymbol{\varepsilon}^{e}$, respectively, as follows

$$g(\boldsymbol{\sigma}, \mathbf{q}_n) + \psi(\boldsymbol{\varepsilon}^{e}, \mathbf{q}_n) = \boldsymbol{\sigma}\boldsymbol{\varepsilon}^{e} \tag{8.2}$$

with $\mathbf{q}_n$ the common passive variables. The elastic relations are given by

$$\boldsymbol{\varepsilon}^{e} = \frac{\partial g}{\partial \boldsymbol{\sigma}}, \qquad \boldsymbol{\sigma} = \frac{\partial \psi}{\partial \boldsymbol{\varepsilon}^{e}} \tag{8.3}$$

The dependence of g and ψ on $\mathbf{q}_n$ indicates in general the presence of elastoplastic coupling. Assuming linear dependence on the rates and a smooth loading surface in stress space not necessarily identical with the yield surface [1], the plastic constitutive relations for a state on a loading surface are given by:

Loading surface: $$f(\boldsymbol{\sigma}, \mathbf{q}_n) = 0 \tag{8.4}$$

Scalar loading index: $$L = \frac{1}{K_p} \mathbf{n}\dot{\boldsymbol{\sigma}} \tag{8.5}$$

Rate equations: $$\dot{\boldsymbol{\varepsilon}}^p = \langle L \rangle \boldsymbol{\rho} \tag{8.6a}$$

$$\mathbf{q}_n = \langle L \rangle \mathbf{r}_n \tag{8.6b}$$

Consistency condition: $$\dot{f} = 0 \Rightarrow K_p = -\frac{1}{g^*}\frac{\partial f}{\partial \mathbf{q}_n}\mathbf{r}_n \tag{8.7}$$

where a superposed dot indicates the rate, $\mathbf{n} = (1/g^*)(\partial f/\partial \boldsymbol{\sigma})$ is the outward unit normal along the gradient of $f = 0$ whose norm is symbolized by g^*, $\mathbf{r}_n$ and $\boldsymbol{\rho}$ depend on the state ($\boldsymbol{\rho}$ can be considered a unit 'vector' in the superposed strain–stress space, with $\boldsymbol{\rho} = \mathbf{n}$ implying the classical associated flow rule), K_p symbolizes the plastic modulus and the brackets $\langle\ \rangle$ imply the operation $\langle A \rangle = AH(A)$ with H the Heavyside step function. Although the $\mathbf{q}_n$ include in general the $\boldsymbol{\varepsilon}^p$, the separate rate equation (8.6a) for the latter was presented in order to emphasize the importance of $\dot{\boldsymbol{\varepsilon}}^p$ in relation to deformation. Not all the $\mathbf{q}_n$ (including $\boldsymbol{\varepsilon}^p$) whose rate are given by equations (8.6) enter necessarily the expressions for g, ψ, and f. The above set of equations completes the elastoplastic constitutive relation in a direct form. It is instructive to present in terms of the rates the direct and inverse form. From equation (8.1), $(8.3)_1$, and (8.6) one has

$$\dot{\boldsymbol{\varepsilon}} = \dot{\boldsymbol{\varepsilon}}^e + \dot{\boldsymbol{\varepsilon}}^p = \dot{\boldsymbol{\varepsilon}}^r + \dot{\boldsymbol{\varepsilon}}^c + \dot{\boldsymbol{\varepsilon}}^p = \frac{\partial^2 g}{\partial \boldsymbol{\sigma}\,\partial \boldsymbol{\sigma}}\dot{\boldsymbol{\sigma}} + \langle L \rangle \left(\frac{\partial^2 g}{\partial \boldsymbol{\sigma}\,\partial \mathbf{q}_n}\mathbf{r}_n + \boldsymbol{\rho} \right) \tag{8.8}$$

where $\dot{\boldsymbol{\varepsilon}}^e = \dot{\boldsymbol{\varepsilon}}^r + \dot{\boldsymbol{\varepsilon}}^c$ represents the decomposition of the elastic strain rate into an incrementally reversible part $\dot{\boldsymbol{\varepsilon}}^r = (\partial^2 g/\partial \boldsymbol{\sigma}\,\partial \boldsymbol{\sigma})\dot{\boldsymbol{\sigma}}$ and a coupling part $\dot{\boldsymbol{\varepsilon}}^c = (\partial^2 g/\partial \boldsymbol{\sigma}\,\partial \mathbf{q}_n)\dot{\mathbf{q}}_n$. By a straightforward procedure equation (8.8) can be inverted to yield:

$$L = \frac{\mathbf{n}\dfrac{\partial^2 \psi}{\partial \boldsymbol{\varepsilon}^e \partial \boldsymbol{\varepsilon}^e}\dot{\boldsymbol{\varepsilon}}}{K_p + \mathbf{n}\dfrac{\partial^2 \psi}{\partial \boldsymbol{\varepsilon}^e \partial \boldsymbol{\varepsilon}^e}\boldsymbol{\rho} - \mathbf{n}\dfrac{\partial^2 \psi}{\partial \boldsymbol{\varepsilon}^e \partial \mathbf{q}_n}\mathbf{r}_n} \tag{8.9}$$

$$\dot{\sigma} = \frac{\partial^2 \psi}{\partial \boldsymbol{\varepsilon}^e \partial \boldsymbol{\varepsilon}^e}\dot{\boldsymbol{\varepsilon}} - \langle L \rangle \left(\frac{\partial^2 \psi}{\partial \boldsymbol{\varepsilon}^e\,\partial \boldsymbol{\varepsilon}^e}\rho - \frac{\partial^2 \psi}{\partial \boldsymbol{\varepsilon}^e\,\partial \mathbf{q}_n}\mathbf{r}_n \right) \tag{8.10}$$

By chain rule differentiation and using equations (8.1) and $(8.3)_2$, it can be shown that the numerator of equation (8.9) equals $(1/g^*)(\partial F/\partial \boldsymbol{\varepsilon})\dot{\boldsymbol{\varepsilon}}$, where $F(\boldsymbol{\varepsilon}, \mathbf{q}_n) = 0$ is the expression of the loading surface in strain space. The equivalent expressions (8.5) and (8.9) for L are well defined when their denominators are different than zero. This includes unstable material response manifested by a falling stress–strain curve in a uniaxial case (uniaxial refers henceforth to both stress and strain) and an inward local 'shrinking' of the loading surface in multiaxial stress space. In this case $\mathbf{n}\dot{\boldsymbol{\sigma}} < 0$ but also $K_p < 0$ so that $L > 0$. If K_p was not included in the definition of L by equation (8.5), such a case could not be described because it would imply necessarily elastic unloading due to $\mathbf{n}\dot{\boldsymbol{\sigma}} > 0$. Observe, however, that even with the inclusion of K_p in equation (8.5), the $L > 0$ at an unstable point does not necessarily imply plastic loading since the $\mathbf{n}\dot{\boldsymbol{\sigma}} < 0$ at such a point may also signify elactic unloading. This ambiguity, inherent in a stress space formulation, is easily resolved by referring to the sign of the equivalent expression (8.9) for L in terms of the total strain rate $\dot{\boldsymbol{\varepsilon}}$. Two limiting cases can now be distinguished. The perfectly plastic response is obtained in the limit as $\mathbf{n}\dot{\boldsymbol{\sigma}} \rightarrow 0$ and simultaneously $K_p \rightarrow 0$, with L specified by equation (8.9) (horizontal uniaxial stress–strain curve and locally stationary $f = 0$). The so-called 'critical softening' is obtained in the limit as $(\partial F/\partial \boldsymbol{\varepsilon})\dot{\boldsymbol{\varepsilon}} \rightarrow 0$ (recall this is the numerator of equation (8.9) multiplied by g^*) and simultaneously the denominator of equation (8.9) tends to zero, with L specified by equation (8.5) (vertically falling uniaxial stress–strain curve and locally stationary $F = 0$). The critical and subcritical softening (negative denominator of equation (8.9)) do not represent common material response and in general one can impose the restriction of positive value for the denominator of equation (8.9). In this case a stress or a strain space formation are equivalent with L defined by equations (8.5) or (8.9)). For further discussion on these points the reader is referred to [2, 3].

The construction of a specific constitutive model within the framework of the above general formulation requires the specification of g or $\psi, f, \mathbf{r}_n$ and $\boldsymbol{\rho}$ on the basis of experimental data. This implies of course that a proper identification of the corresponding $\mathbf{q}_n$ is necessary as the first step. The problem, however, is that a proper choice of $\mathbf{q}_n$ is not as straightforward as it may appear, since the typical experimental data in terms of stress and strain certainly do not reveal any particular set of $\mathbf{q}_n$ in a straightforward way. It is there where engineering intuition, creativity, and imagination must blend with the rational requirements of the analytical formulation (restrictions imposed by material symmetries, invariance requirements, etc.) to produce what one may call the 'art of modelling' with the final purpose to construct an analytical constitutive model. In this effort two basic guidelines must always be kept in mind. First, the model must be parsimonious in the sense of combining maximum simplicity, especially in reference to easy numerical implementation, with maximum predictive capabilities. Second, the set of model constants must be easily obtainable from typical experimental data. The following section will illustrate further these points.

8.3 OBSERVATIONS ON CYCLIC UNIAXIAL LOADING DATA OF METALS

In order to be specific, in this section a detailed examination of certain characteristic cyclic uniaxial loading data on metals will be presented. Figures 8.1

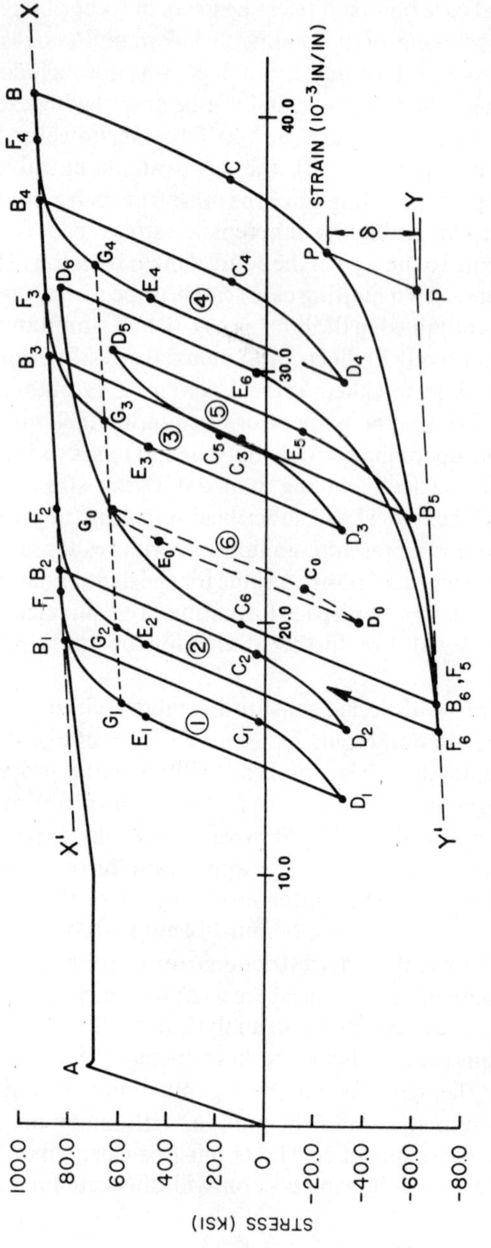

Figure 8.1 Experimental results for uniaxial random cyclic loading on grade 60 steel specimen, after [4]

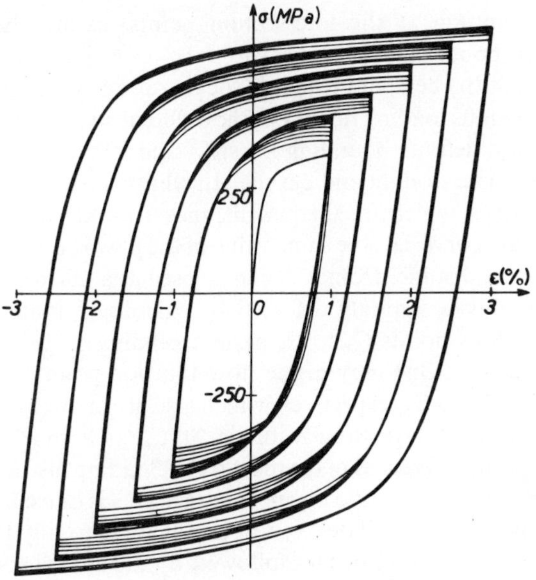

Figure 8.2 Experimental results for uniaxial symmetric cyclic loading at increasing strain level on 316 L steel specimen. Reproduced with permission from Chaboche *et al.* [5]

and 8.2 show such experimental data for grade 60 and 316 L steel specimens taken from references [4] and [5], respectively.

Beginning with Figure 8.1 a striking first impression is the complexity of the material response under random cyclic loading with intersecting loops of different sizes and curvatures which at first appears chaotic defying any obvious rule of behaviour. The response is definitely elastoplastic with the elastic modulus E remaining constant (no elastoplastic coupling) as observed by the unloading paths (a small change of E occurs between the virgin elastic loading path and the subsequent ones, but is neglected). With the exception of the initial yield point (followed by the 'plateau'), all other curves show a smooth elastoplastic transition rendering the determination of the corresponding yield points largely a matter of subjective interpretation. Such points are indicated by the Cs and Es in a way that the elastic range between the corresponding Bs, Cs, and Ds, Es is kept constant. Another characteristic is the strong presence of Bauschinger effect with reverse yielding initiating before even the tensile stress changes to compressive. Comparing the loops 1, 2, 3, and 4 one can observe a strong similarity of their shape despite the fact that they correspond to drastically different levels of total or cumulative plastic strain, which therefore cannot be used as an appropriate piv. Equally interesting is the repeatedly observed intersection of different loops along tensile or compressive loading, indicating that stress and plastic strain in combination (common at the intersection points) cannot be used alone to describe such response characteristics.

A closer look at the behaviour reveals the first orderly and persisting feature: whatever the previous loading history is, the different stress–strain curves tend to converge with two definite bounding linesX′X and Y′Y which in this particular example appear to be straight and parallel. In addition a comparsion of a typical set of points G_1 to G_4 which are at equal distance from X′X shows the same slope of the stress–strain curve, i.e., the same value of the plastic modulus. A rush to the conclusion that the distance from X′X can be used as a macroscopic piv defining the plastic modulus is strongly defeated by examining point G_0 at the same distance from X′X as points G_1 to G_4 at an ascending curve, but with a totally different slope at G_0. One may argue now that for points G_1 to G_4 yielding initiated at points E_1 to E_4 respectively which also are at about the same distance from X′X, while for G_0 yielding initiated at C_6 at a much larger distance, essentially reflecting differences of the most recent loading history associated with the previous excursion in reverse plastic loading. This is indeed a very important conclusion, as it will be seen later, but does not solve totally the problem. For example, if at G_0 unloading occurs followed by partial reverse loading along C_0D_0 and reloading initiating at E_0 (all these are shown by a dashed loop and were not actually performed in the above expaeriment, but are typical of what is expected), the following is observed: a smooth elastoplastic transition is observed at E_0, but despite the fact that E_0 is approximately at the same distance from X′X as the E_1 to E_4 are, when the subsequent curve meets the C_6D_6 curve at about the point G_0 it suddenly bends and changes slope following the C_6D_6 instead of behaving similarly to the curves E_iF_i, $i = 1$ to 4. This shows that the material exhibits a more sophisticated and extended memory of discrete events of unloading–reloading which is not associated only with the immediately preceding reverse loading.

Examining now Figure 8.2 for symmetric cyclic loading under piecewise increasing strain amplitudes, the response appears more orderly with the following basic features. The elastoplastic transition is again smooth and the Bauschinger effect is again evident. For each strain amplitude the peak stress increases with the number of cycles stabilizing at a level which increases with the subsequent strain amplitude for the next set of cycling. This indicates an increase in the elastic range not observed in Figure 8.1. The stress–strain curves tend again to converge with bounds similar to X′X, Y′Y of figure 8.1, but with the additional observations that the bounds appear to increase their distance although they remain parallel. One additional feature which is not shown in Figure 8.2 but can be found in [5] is that the level of peak stress stabilization for each strain amplitude appears to be independent of previous history as far as this history included stabilization under strain amplitudes smaller than the current one.

Finally let us simply refer to three other important material features under cyclic loading, not shown in Figure 8.1 and 8.2. For cyclic loading under a fixed stress amplitude but non-zero constant means stress, one observes the phenomenon of cyclic creep. This can be seen to a certain extent in the loops 1 to 4 of Figure 8.1 where the above conditions are approximately satisfied. The reason for cyclic creep is the totally different variation of the values of the plastic modulus during plastic loading in opposite directions. A second important feature not shown, is that under fixed strain amplitude cyclic loading with non-zero mean strain, the mean stress relaxes to zero for most cases. Finally a third feature is cyclic softening (for work hardened as opposed to annealed specimen which harden), with a response quite the opposite to the one shown in Figure 8.2, i.e. a decreasing peak stress stabilization level under cycling.

8.4 THE IMPORTANCE OF THE PLASTIC MODULUS

Although the conclusions reached in this section are based on the observations made for metal response, a great part of the methodology and way of thinking can be applied to other materials as well. The analysis of the experimental data of the previous section has shown a highly complex response, if one desires to account for all the observed characteristics. A small reflection on what we have been exposed to, reveals that essentially we are faced with the problem of describing by means of an analytical model a set of experimental curves, a typical case in any phenomenological modelling. The important point to emphasize, is that our analytical modelling is far from being simply a curve fitting approach for a very simple reason: the analytical model which will be calibrated on the basis of a given set of experimental data for a given material, must be capable of predicting reasonably well any other set of data in the form of stress–strain curves under loading histories drastically different than those which were used for calibration. Thus, the constitutive model must have an internal structure which can describe the *modus operandi* of the material under any or many loading conditions. This internal material code of operation is represented phenomenologically by the proper choice and evolution laws of the piv.

The fact remains, however, that from a practical point of view we are obliged to think in terms of how to obtain the curve fitting of experimental data without making particular concessions to particular data which will hinder the generality and predictive capability of the model. Recalling the observations of the previous section, successful curve fitting with given initial conditions implies essentially successful description of the slope of a curve and vice versa. This is very simply a practical statement of the obvious fact that the differential equation of a curve describes the curve itself. But with the elastic modulus constant, the slope of a stress–strain curve is a function of the plastic modulus. Thus, the plastic modulus can be recognized as the main phenomenological quantity whose evolution is of cardinal importance for the successful modelling. Indeed, that was essentially our line of thinking when the loops of figure 8.1 and 8.2 were analysed. It is pertinent to mention here that much effort in the literature has been directed towards the exact definition of the onset of yielding and, thus of the yield surface, reaching a point of diminishing return for a very simple reason: shifting the emphasis from the exact determination of the yield point to the proper determination of the plastic modulus so that among other things a smooth elastoplastic transition can be described, it is not important if yielding initiates a little bit above or below on the stress–strain curve. The same good fit would be obtained anyway due to the smooth transition. Of course this is not true for cases where an abrupt yield occurs, as in Figure 8.1 for initial yield before the 'plateau'.

One may object to the importance attributed here to the details of the plastic modulus variation on the basis that essentially it is responsible only for a good fit during the transition from a purely elastic response to a fully developed elastoplastic response, e.g. the convergence with the bounds X′X, Y′Y of Figure 8.1. Along this line of reasoning, its influence is restricted to this transient behaviour only and does not deserve full attention since peak stresses and peak strains are not effected by these intermediate variations of the plastic modulus. We strongly disagree with this way of thinking because the close prediction of the stress–strain curves at the transient stage from elastic to fully developed plastic is not just of academic interest. For example, two important phenomena depend heavily on it. To illustrate the first, consider the curve BCB_5 in Figure 8.1. The eventual convergence with the bound Y′Y can be obtained with or without an accurate description of the transident stage CB_5 which is mostly in compression. But in an actual situation of a uniaxially loaded truss member, for example, the exact slope along this transient stage is directly proportional to the buckling load and if it is over-estimated it will have catastrophic effects on the structure. A second phenomenon which depends on the realistic description of the plastic modulus in the transient stages, is the response to cyclic loading with small excursions in the plastic range for both directions and non-zero mean stress. Referring to Figure 8.1 again, while the exact description of the curves C_1D_1, E_1F_1 may not be of great importance for a single loop, the cumulative effect of the corresponding deviation under a repetition of such loops will indeed greatly over- or under-estimate the resulting cyclic creep.

Reference now to equation (8.7), clearly shows that the evolution of the plastic modulus K_p is directly related to the dependence of f on $\mathbf{q}_n$ and the evolution of the latter. Thus, focusing attention on the plastic modulus does not eliminate the importance of choosing the proper $\mathbf{q}_n$ and defining their law of evolution. On the other hand, this attention provides the guidelines for a successful identification of the $\mathbf{q}_n$ from a more practical perspective. Whether aiming directly at the heart of the problem of a successful modelling which is the variation of K_p' or obtaining K_p indirectly through equation (8.7) by first specifying the evolution of all the $\mathbf{q}_n$ entering $f=0$, is a matter of difference in tactics rather than strategy, but an important one.

8.5 A GENERAL CLASS OF KINEMATIC/ISOTROPIC HARDENING MODELS FOR METALS

Assuming elastic isotropy, absence of elastoplastic coupling and linear elasticity, the problem of specifying g (or ψ) for a typical polycrystalline metal is resolved by specifying two elastic constants. The associated flow rule $\boldsymbol{\rho} = \mathbf{n}$, the independence of the yield stress on hydrostatic pressure and the assumption of initial isotropy dictates the dependence of f, equation (8.4), on the direct and mixed isotropic invariants of the deviatoric stress $\mathbf{s}$ and $\mathbf{q}_n$. The observations of Section 8.3 suggest that two predominant hardening characteristics are those of kinematic and isotropic hardening. Thus, omitting possible distortions of the shape of the loading surface (which could not anyway be depicted from uniaxial data) a typical Mises type kinematic/isotropic hardening class of models is adopted. Let $\boldsymbol{\alpha}$, k be the corresponding $\mathbf{q}_n$ representing the shifted centre (back-stress) and size of the loading surface in deviatoric stress space, respectively, $\mathbf{v}$ a unit 'vector' in stress space indicating the direction of $\dot{\boldsymbol{\alpha}}$, and K_α a kinematic hardening modulus giving the magnitude of $\dot{\boldsymbol{\alpha}}$ for a given $\dot{\boldsymbol{\sigma}}$ in such a way that $\dot{\boldsymbol{\alpha}}\mathbf{n} = K_\alpha\langle L\rangle$. Assuming that k depends on the effective cumulative plastic strain $\bar{\varepsilon}^{\mathrm{p}}$ obtained by integration of its rate $\dot{\bar{\varepsilon}}^{\mathrm{p}} = [(2/3)\dot{\boldsymbol{\varepsilon}}^{\mathrm{p}}\dot{\boldsymbol{\varepsilon}}^{\mathrm{p}}]^{1/2}$ and denoting by k' the derivative $\mathrm{d}k/\mathrm{d}\bar{\varepsilon}^{\mathrm{p}}$, the complete set of constitutive relations corresponding to the general set of equations (8.4)–(8.7) becomes:

Loading surface: $$f = \tfrac{3}{2}(\mathbf{s} - \boldsymbol{\alpha})(\mathbf{s} - \boldsymbol{\alpha}) - k^2 = 0 \tag{8.11}$$

Loading index: $$L = \frac{1}{K_{\mathrm{p}}}\mathbf{n}\dot{\boldsymbol{\sigma}} = \frac{1}{K_{\mathrm{p}}}\mathbf{n}\dot{\mathbf{s}} \tag{8.12}$$

Rate equations: $$\dot{\boldsymbol{\varepsilon}}^{\mathrm{p}} = \langle L\rangle\mathbf{n} \tag{8.13a}$$

$$\dot{\boldsymbol{\alpha}} = \langle L\rangle\frac{1}{\mathbf{v}\mathbf{n}}K_\alpha\mathbf{v} \tag{8.13b}$$

$$\dot{k} = \langle L\rangle\sqrt{(\tfrac{2}{3})}k' \tag{8.13c}$$

Consistency condition: $$K_{\mathrm{p}} = K_\alpha + \tfrac{2}{3}k' \tag{8.14}$$

with $\mathbf{n} = (3/2)^{1/2}(\mathbf{s} - \boldsymbol{\alpha})/k$. With σ, α, and ε^{p} being the uniaxial (tension–compression) counterparts of $\boldsymbol{\sigma}, \boldsymbol{\alpha}$, and $\boldsymbol{\varepsilon}^{\mathrm{p}}$ respectively, $E_{\mathrm{p}} = (3/2)K_{\mathrm{p}}$, $E_\alpha = (3/2)K_\alpha$ for radial loading [4], and with loading occurring when $L = \dot{\sigma}(\sigma - \alpha)/E_{\mathrm{p}}k > 0$ on $f = 0$, the uniaxial counterpart of the above equations are

Loading surface: $$f = (\sigma - \alpha)^2 - k^2 = 0 \tag{8.15}$$

Rate equations: $$\dot{\varepsilon}^{\mathrm{p}} = \frac{\dot{\sigma}}{E_{\mathrm{p}}}, \qquad \dot{\alpha} = E_\alpha\dot{\varepsilon}^{\mathrm{p}}, \qquad \dot{k} = k'|\dot{\varepsilon}^{\mathrm{p}}| \tag{8.16}$$

Consistency condition: $$E_{\mathrm{p}} = E_\alpha + k' \tag{8.17}$$

Within the framework of kinematic/isotropic hardening the above equations are fairly general. As a matter of fact equation (8.13b) is the most general expression for a kinematic hardening rule, since any such rule can be brought to the above form for any f once $\boldsymbol{v}$ is specified. In this respect it is important to observe that the key equation (8.14) holds true for any $\boldsymbol{v}$. The isotropic hardening could have assumed different expressions, as for example by rendering k a function of the plastic work instead of $\bar{\varepsilon}^{p}$. For most cases this has little influence on the evolution of k and the most important point for isotropic hardening is the specification of k'.

8.6 SPECIFIC MODELS FOR METALS UNDER CYCLIC LOADING

Within the general class presented analytically in the previous section, a brief but critical review of specific models will be presented. In order to have a common basis for comparison among the different models and recalling the extensive observations of section 8.3, the following set of five requirements for a model will be stated, four of them presented in a recent paper by Drucker and Palgen [6]. Requirements related primarily to kinematic hardening:

(1) Unsymmetric stress cycles will cause cyclic creep in the direction of mean stress.
(2) Unsymmetric strain cycles will cause progressive relaxation to zero mean stress.
(3) The model must predict as accurately as possible the variation of the plastic modulus during random cyclic loading.

Requirement related primarily to isotropic hardening:

(4) Under symmetric stress or strain cycles, the material hardens or softens towards a properly defined stabilized state with only kinematic hardening.

General requirement:

(5) Extensive plastic loading of almost fixed direction overwhelms and wipes out many of the past history effects, if not all.

It is clear from the analytical formulation of Section 8.5 that attention should be focused on the three quantities K_p or E_p, K_α or E_α and k' interrelated by equations (8.14) or (8.17). Referring to the discussion of Section 8.4, the key phenomenological quantity is K_p or E_p. Assuming that a consensus has been reached concerning k', and recalling the general concluding remarks of Section 8.4, two basic approaches have been followed: either $K_\alpha(E_\alpha)$ is specified and then $K_p(E_p)$ is computed or vice versa. Although equivalent in essence, these two approaches have shaped accordingly the particular forms of different models by shifting the emphasis of the primary goal towards the specification of K_α or K_p. In the following the above distinction will be maintained in presenting the models, and let us also note that while the requirement (4) associated with isotropic hardening will be also discussed, the modelling of some very particular features of such hardening will be postponed for the next section.

8.6.1 Models with emphasis on the determination of K_α

The kinematic hardening was first introduced by Edelman and Drucker [7], Ishlinskii [8], Prager [9], Ziegler [10], and Eisenberg and Phillips [11] with $\mathbf{v} = \mathbf{n}$ for most of them. A serious deficiency of all these models is that K_α or E_α is assumed to be either constant (linear kinematic hardening) or a function of $\bar{\varepsilon}^p$ (non-linear kinematic hardening) and, therefore, for a given k' the plastic moduli K_p or E_p found from equations (8.14) or (8.17) assumed the same value in reverse loading or continued loading. This drastically violates the first three requirements stated above, while no attempt was made to satisfy the fourth requirement, although this is possible within their framework. The first three requirements are also violated by the specific form of a model by Caulk and Naghdi [12] for the same reason, although it must be credited that the fourth requirement is successfully satisfied and the derivation of the basic equations follows a very rigorous and systematic approach from the general to the particular, rarely found at other works where *a priori* assumptions are made.

To overcome this difficulty Mróz *et al.* [13], Chaboche [14], and Chaboche *et al.* [5] have introduced more sophisticated kinematic hardening rules which allow the K_α to vary along the loading surface. With small differences, in essence all these rules are variations of a basic theme initially proposed by Armstrong and Frederick [15] which introduces an evanescent memory by means of a kinematic rule of the form

$$\dot{\boldsymbol{\alpha}} = \tfrac{2}{3}c_1\dot{\boldsymbol{\varepsilon}}^p - c_2\dot{\bar{\varepsilon}}^p\boldsymbol{\alpha} \tag{8.18}$$

with c_1, c_2 functions of $\bar{\varepsilon}^p$ in general. If the above equation is brought to the general form (8.13b) for comparison, it yields

$$K_\alpha = \tfrac{2}{3}c_1 - \sqrt{(\tfrac{2}{3})}c_2\boldsymbol{\alpha}\mathbf{n} \tag{8.19}$$

which indicates that K_α depends on the inner product $\boldsymbol{\alpha}\mathbf{n}$, i.e. the position of the current stress point on the loading surface. For extensive monotonic loading $K_\alpha \to 0$ as the magnitude of $\boldsymbol{\alpha}$ increases. The dependence of K_α on the position of the stress on $f = 0$ transfers directly to the plastic modulus K_p by means of equation (8.14). Chaboche *et al.* [5] extended equation (8.18) by assuming $\boldsymbol{\alpha} = \boldsymbol{\alpha}_1 + \boldsymbol{\alpha}_2 + \cdots + \boldsymbol{\alpha}_n$, with each one of the $\boldsymbol{\alpha}_i$ obeying an equation of the form (8.18) with constant but different coefficients c_1, c_2, for each $\boldsymbol{\alpha}_i$ in order to model better the smooth elastoplastic transition. The essence of equation (8.18) is that it removes the back stress $\boldsymbol{\alpha}$ upon reverse loading faster than it builds it up due to the proper K_α variation on the loading surface. This is schematically illustrated in Figure 8.3 where the uniaxial response in the $\sigma - \varepsilon^p$ space along the pathe $ABCC'$ is considered. It is straightforward to obtain from equation (8.19)

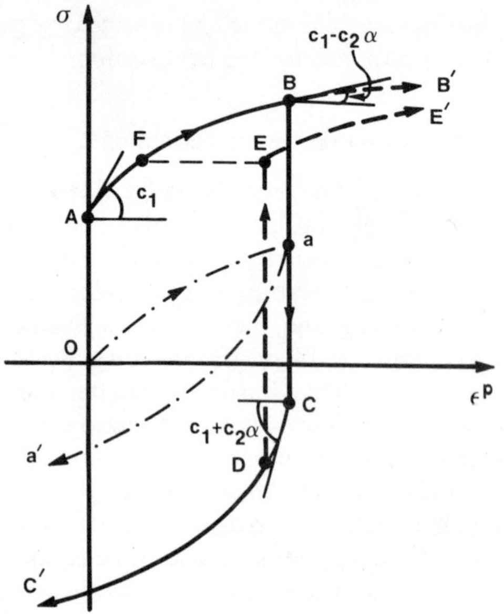

Figure 8.3 Schematic illustration of the response and deficiency of the evanescent memory kinematic hardening model

that the uniaxial value of the kinematic modulus is given by $E_\alpha = c_1 \mp c_2\alpha$ with minus for loading and plus for reverse loading. Assuming no isotropic hardening, i.e. $k' = 0$, equation (8.17) yields $E_p = E_\alpha$ for the slope of the σ–ε^p curve. The slopes $c_1, c_2 - c_2\alpha$ and $c_1 + c_2\alpha$ are eloquently shown for points A (initial yield), B and C, respectively, in Figure 8.3. If loading continued towards B', the $E_\alpha \to 0$ as $\alpha \to c_1/c_2$. The trace of the back stress α is shown by the dashed/dot line $0aa'$.

Clearly this model describes better the loading/unloading/reverse loading curves, although does not predict smooth elastoplastic transition since c_1, c_2 are finite and cannot even assume very large values if an overall good curve fitting is desired. But there is a hidden deficiency which has not been appreciated so far, and shown schematically by the path $CDEE'$ in Figure 8.3. The rapid decrease of α during the partial reverse loading path CD (recall $E_\alpha = c_1 + c_2\alpha$), cannot be compensated fast enough during the subsequent unloading–reloading path DEE' and the predicted elastoplastic stress–strain curve EE' undershoots the actual one which should merge fast with BB' as observed in corresponding experimental data like the ones shown in Figure 8.1. As a matter of fact the slope E_p at E is identical to the one at point F on the same stress level, and the curve EE' is parallel translation of the FB' to the right by the segment FE. The preceding shortcoming can be ameliorated in the case that many $\boldsymbol{\alpha}_i$ are used as in [5]. This deficiency can severely over-estimate the cyclic creep phenomenon for stress cycling with non-zero mean stress. Thus, the above model cannot satisfy the requirements (1) and (3), although it satisfies the remaining ones and notably requirement (2), where the zero mean stress relaxation response can easily be satisfied as seen by the above detailed discussion of the uniaxial response.

8.6.2 Models with emphasis on the determination of K_p

Extending Besseling's idea of the overlay model [16], Iwan [17], and mainly Mróz [18] introduced the concept of a field of work-hardening moduli by means of nested surfaces. Mróz [18] took the first step in recognizing the importance of the plastic modulus determination as the primary goal of a constitutive model, thus he first defined the variation of K_p in a discrete way associated with each one of the nested surfaces, and then let K_α to be specified by means of equation (8.14). The ν of equation (8.13b) was specified as the unit vector along the line connecting conjugate points of two consecutive surfaces characterized by the same **n**. Still, requirements (1) and (2) are not fully met [6], and if requirement (3) is to be satisfied a very large number of such surfaces must be employed which is prohibitive for large scale computations.

Recently, Drucker and Palgen [6] proposed a model which employs a plastic modulus inversely proportional to a power of the normalized second deviatoric stress invariant, with two options for a sharp or rounded corner of the stress–strain curve associated with a large or small elastic range, respectively. Essentially they follow the approach of first defining K_p and then obtaining K_α by an equation similar to equation (8.14) (although they do not follow exactly this sequence in formulation), with zero variation of k for the rounded option or a variable k for the sharp corner option as a function of plastic work rather than $\bar{\varepsilon}^p$. This model exhibits an incorrect variation of K_p if reverse plastic loading occurs before the stress becomes zero as seen to occur in the experimental data of Figure 8.1. Most of the requirements are satisfied except number (3) which is partially satisfied, as a result of the necessary compromise between the two options. The relative simplicity of the model justifies, to a certain extent, this compromise.

The structure of the model proposed by Eisenberg [19] appears to satisfy most of the requirements, although in a concrete application on symmetric cyclic loading of 304 stainless steel the rate of saturation of the isotropic hardening and the rounding of the stress–strain curves do not fit very accurately the experimental data. Again emphasis is placed primarily on the determination of the plastic modulus first. What distinguishes this model, however, is the use of updated discrete memory parameters of the most recent significant event of unloading–reloading. This is done in conjunction with a definition of reversed loading, partial reversed loading and continuation of original loading on the basis of the negative, positive, or unit value of the inner product of the unit normals on the loading surface at the points of unloading and new loading. The updating procedure, however, breaks down if loading along a reversed direction occurs in the multiaxial stress space not by a definite unloading–loading event which would classify it accordingly, but by a continuous rotation of the stress around the loading surface remaining in a state of almost neutral loading keeping the above mentioned inner product always positive.

Finally, in this category belong the models which employ the concept of the bounding surface. This concept was originally proposed by Dafalias and Popov [20] and subsequently elaborated by these and other authors [21 to 24] along similar if not identical perspectives. Its essence can be illustrated in Figure 8.1, where a typical σ–ε curve converges with bounds X′X, Y′Y under monotonic or cyclic random loading as discussed in Section 8.3. For a current stress P, an 'image' stress $\bar{P}$ on the corresponding bound is associated by a proper mapping rule, such that E_p depends on the distance δ between $\bar{P}$ and P, and the slope $\bar{E}_p$ of the bound at $\bar{P}$. In [4, 21] the form

$$E_p = \bar{E}_p + h(\delta_{in})\frac{\delta}{\delta_{in} - \delta} \tag{8.20}$$

was proposed as a possible choice, where δ_{in} is the value of δ at initiation of a yielding process playing the role of a discrete memory parameter of the most recent event of unloading–reloading, and h is a model parameter function of δ_{in} controlling the 'steepness' of the stress–strain curves. Certain provisions were made to maintain $\delta_{in} - \delta$ non-negative. If the bounds remain straight parallel lines then $\bar{E}_p = E_p^0 =$ constant, but in general $\bar{E}_p$ will vary with the hardening of the bounds. Observe that at $\delta = \delta_{in} \Rightarrow E_p = \infty$ (smooth elastoplastic transition) and at $\delta = 0 \Rightarrow E_p = \bar{E}_p$. A direct generalization in multiaxial stress space of the 'image' stress $\bar{P}$ yields the concept of the bounding surface enclosing and hardening in a coupled way with the yield surface, where now the mappping rule associates P and $\bar{P}$ with the same $\mathbf{n}$ on the two surfaces and δ is measured by the Euclidean norm. The generalized plastic moduli K_p and $\bar{K}_p$ are now related in the same way as E_p and $\bar{E}_p$. The steps for obtaining the constitutive relations are essentially executed in two levels. In the first level one obtains the isotropic–kinematic hardening of the bounding surface by specifying the change $\bar{k}'$ of its size and the kinematic modulus $K_\beta = (2/3)E_\beta$ associated with its centre $\boldsymbol{\beta}$. Subsequently $\bar{K}_p$ or $\bar{E}_p$ are obtained by the corresponding consistency condition for the bounding surface in direct analogy to equations (8.14) or (8.17). In the second level the inverse approach is followed where the $K_p = (2/3)E_p$ is obtained from equation (8.20), and $K_\alpha = (2/3)E_\alpha$ follows from equations (8.14) or (8.17). This two-level execution offers a very powerful decoupling between the phenomena of saturation, peak stresses, etc. associated with the evolution of the bounding surface (or the bounds in uniaxial space), and the phenomena associated with the proper description of the stress–strain curve in the transient stage before convergence with the bounds occurs.

The efficiency of the bounding surface model was demonstrated in [4, 21] by predicting almost perfectly the experimental data of Figure 8.1. The updating of h

with δ_{in} lies in the heart of this successful prediction, because it accounts for the accurate description of the intrinsic geometry of a stress–strain curve depending on the recent past history by means of δ_{in}. This benevolent feature, however, has its disadvantage for a case of partial reverse loading followed by reloading, as discussed in relation to the loop $G_0C_0D_0E_0G_0$ of Figure 8.1. In this case the δ_{in} at E_0 is smaller than the δ_{in} at C_6 and, therefore, according to equation (8.20) when the curve E_0G_0 meets the curve C_6D_6 (the G_0 does not have to be exactly the same with the point of unloading) upon reloading, the former does not bend sharply towards the latter but instead it overshoots it. The actual sharp bending shows that the material has not 'forgot' the past extensive loading C_6G_0 because the reverse loading along C_0D_0 is of small magnitude and, therefore, has little influence in erasing the material memory. The remedy is to store in the material memory the value of δ_{in} at C_6 and that of δ at the moment of unloading at G_0, and when the stress level reaches again the same δ at around G_0 the material memory code will choose the maximum δ_{in}/δ for the same δ and the two different δ_{in} for points C_6 and E_0. This proposition is not as simple to implement as it appears for the case of multiaxial loading and repeated events of partial reverse loading followed by reloading, but it is not discussed further here. It is sufficient to mention that there is no known constitutive model which can describe the above intricate material response, unless a rather complex discrete memory code of unloading–reloading events is implemented in the model. The reader is referred to a more recent version of the bounding surface model, called the 'radial' mapping model [25, 26], which essentially eliminates the explicit consideration of an enclosed yield surface and responds better to particular loading events as above, although again an updating of h is found to be necessary for improved predictions. In its application, so far, the bounding surface models have fully satisfied all requirements except the second, simply because a linear kinematic hardening was assumed for the bounding surface (but not for the loading surface). It is a simple matter to incorporate, instead, a kinematic hardening for the bounding surface identical or similar to the form of equation (8.18), where now the centre $\boldsymbol{\beta}$ must replace $\boldsymbol{\alpha}$. With this modification the mean stress relaxation to zero level is realized for the 'image' stress on the bounding surface and, therefore, transfers this property to the actual stress on the enclosed yield surface satisfying the second requirement. Closing this section let us observe that the fifth general requirement is satisfied by all the above models.

8.7 ISOTROPIC HARDENING AND CYCLIC STABILIZATION UNDER SYMMETRIC AND UNSYMMETRIC STRAIN CYCLES

What follows applies to any one of the previously discussed plasticity models, and is essentially a further elaboration on the fourth requirement of the previous section by studying the effect of symmetric and unsymmetric strain cycles on the cyclic saturation hardening or softening and the corresponding peak stresses. This feature of the material response is basically represented by the isotropic hardening part which is characterized by k. Within the framework of the class of analytical models presented in Section 8.5, the isotropic hardening is specified by k'. A possible expression for k' is

$$k' = c(k_s - k) \tag{8.21}$$

with k_s a saturation value for k and c a material constant controlling the pace of saturation. Equation (8.21) can be easily integrated which is very convenient for the specification of c and k_s by curve fitting. As $k \to k_s \Rightarrow k' \to 0$ and the material exhibits only kinematic hardening as can be seen from equations (8.14), (8.17), reflecting the fourth requirement. While the material in Figure 8.1 exhibited an almost constant k (excluding the 'plateau' region), the material in Figure 8.2 not only exhibited a variable k but in addition a variable k_s as well, depending on the corresponding strain amplitude.

The strain amplitude $\Delta\varepsilon$ is the major factor which determines the peak cyclic stress σ_c of the stabilized hysteresis loops under symmetric strain cycle loading. The k_s is directly related to σ_c upon subtraction of the effect of kinematic hardening. In an increasing strain level test with zero mean strain as shown in Figure 8.2, σ_c decreases or increases with respect to the σ associated with monotonic loading at $\Delta\varepsilon/2$, as the material softens or hardens respectively. More complex situations may also arise with softening following hardening, but they will not be considered here. A $\sigma_c - \Delta\varepsilon/2$ relation is provided by the experimentally determined cyclic curve (locus of tips of stabilized hystersis loops). For such loading, $\Delta\varepsilon$ is approximately equal to the maximum plastic strain range $\Delta\varepsilon^p_{max}$ that the material has experienced so far. This motivated Chaboche *et al.* [5,27] to introduce a memory hypersphere in plastic strain space which hardens isotropically and kinematically in such a way that its diameter is always equal to $\Delta\varepsilon^p_{max}$ and controls the stabilization of the hystersis loops by specifying σ_c and consequently k_s from the cyclic curve. This assumption implies of course that at non-zero mean strain the σ_c will be a function of $\Delta\varepsilon \simeq \Delta\varepsilon^p_{max}$ only. For example, if the material is uniaxially cycled from 0 to 0.04 strain, apart from stress relaxation which will produce a zero mean stress, the σ_c will be identical to the one obtained for cycling at ± 0.02 strain. This appears to be a reasonable conclusion, but poses a question. After saturation at the 0–0.04 range assume that subsequent cycling occurs at ± 0.01. Although now $\Delta\varepsilon$ is smaller, according to [5] $\Delta\varepsilon^p_{max} \simeq 0.05$ and further hardening must occur corresponding to this new $\Delta\varepsilon^p_{max}$. This does not seem entirely reasonable and, therefore, an extension of the idea in [5] was proposed by Dafalias and Seyed-Ranjbari [28] and independently by Ohno [29] to account for such cases. The plastic strain space hypersphere is described by

$$S = \tfrac{2}{3}(\boldsymbol{\varepsilon}^p - \boldsymbol{\gamma})(\boldsymbol{\varepsilon}^p - \boldsymbol{\gamma}) - p^2 = 0 \tag{8.22}$$

with $\boldsymbol{\gamma}$ being the deviatoric plastic strain coordinates of its centre and p its radius.

With η a material parameter assume that $\dot{p} \sim \eta\dot{\bar{\varepsilon}}^{\mathrm{p}}$ only when $S=0$ and $\dot{\boldsymbol{\varepsilon}}^{\mathrm{p}}$ points towards the domain outside the sphere. Then, employing the consistency condition $\dot{S}=0$ for $\boldsymbol{\varepsilon}^{\mathrm{p}}$ on $S=0$, one has:

$$\dot{p} = \sqrt{(\tfrac{2}{3})}\eta\langle L\rangle\langle \mathbf{nn}^*\rangle \tag{8.23}$$

$$\dot{\gamma} = (1-\eta)\langle L\rangle\langle \mathbf{nn}^*\rangle\mathbf{n}^* \tag{8.24}$$

with $\mathbf{n}^* = (2/3)^{1/2}(\boldsymbol{\varepsilon}^{\mathrm{p}} - \gamma)/p$ the unit normal to $S=0$. For $\eta = 1/2$ the relations of [5] are retrieved and $p = \Delta\varepsilon^{\mathrm{p}}_{\mathrm{max}}/2$. For $0<\eta<1/2$, however, the diameter $2p$ of the hypersphere is less than $\Delta\varepsilon^{\mathrm{p}}_{\mathrm{max}}$ and will become asymptotically equal only when the imposed $\Delta\varepsilon^{\mathrm{p}} = \Delta\varepsilon^{\mathrm{p}}_{\mathrm{max}}$. That is if the material has experienced once a large $\Delta\varepsilon^{\mathrm{p}}_{\mathrm{max}}$ by a monotonic excursion deep into the plastic range but then is cycled at smaller ranges, its excursion will be only partially imprinted in its memory by means of a $\Delta\varepsilon^{\mathrm{p}} < \Delta\varepsilon^{\mathrm{p}}_{\mathrm{max}}$. As another similar example for further illustration, after saturation at the 0–0.04 range a subsequent cycling at ± 0.03 will lead to a saturation level corresponding to a $\Delta\varepsilon^{\mathrm{p}}$ somewhere between 0.06 and 0.07, depending on the value of η, but not exactly to $\Delta\varepsilon^{\mathrm{p}}_{\mathrm{max}} = 0.07$ according to the model in [5] for $\eta = 1/2$.

The uniaxial counterpart of equations (8.22), (8.23), (8.24) for plastic strain rate pointing outside the hypersphere becomes

$$S = (\varepsilon^{\mathrm{p}} - \gamma)^2 - p^2 = 0 \tag{8.25}$$

$$\dot{p} = \eta|\dot{\varepsilon}^{\mathrm{p}}| \tag{8.26}$$

$$\dot{\gamma} = (1-\eta)\dot{\varepsilon}^{\mathrm{p}} \tag{8.27}$$

and provides a convenient way for calibration of η. Recall that the ultimate objective of the above scheme is to find a proper dependence of k_s on p, the latter expressing the material memory of all or part of the experienced $\Delta\varepsilon^{\mathrm{p}}_{\mathrm{max}}$.

Note that the bounding surface models are particularly adjustable to the above sophisticated approach accounting for the intricate phenomena of isotropic hardening, because its incorporation will affect only the evolution of the size of the bounding surface ('bounds' in σ–ε space), without disturbing the other parts of the constitutive model associated with the kinematic hardening (the enclosed yield surface can usually be assumed similar and proportional to the bounding surface).

8.8 CYCLIC LOADING FOR GEOLOGICAL MEDIA

The subject of this section can only be very briefly discussed within the space limitation of the present article. A great part of the basic methodology presented so far for metals does apply for geological media, but due to the higher complexity of the response of such media one should expect a less ambitious accomplishment than the almost perfect prediction which can sometimes be achieved for metals. This is particularly true for the response under complex cyclic loading conditions and the goal should be to model realistically the cummulative result of the cyclic response rather than expecting an exact modelization of all individual cyclic stress–strain curves. The lack of satisfactory reproducibility of experimental data, in fact the rather large scatter of them for identical samples and loading, renders meaningless to try an exact fit of any or all such available data from a macroscopic point of view.

Three basic phenomenological features distinguishes geological media from metals especially in relation to cyclic loading: the strong dependence of the material response on hydrostatic pressure, the lack of a well defined purely elastic domain and the presence of stiffness degradation and internal damage. All three are obviously a result of the corresponding particulate nature of the microstructure. The first renders the elastic bulk modulus and the yield or loading surface a function of the mean pressure. The second necessitates the use of more unorthodox concepts than those of a classical yield surface, and the third effects both the elastic and plastic constitutive relations by coupling their evolution.

The effect of the almost non-existing purely elastic domain is nowhere more seriously felt than in the case of cyclic deviatoric loading of saturated clay or sand samples under undrained conditions. As cycling proceeds within fixed stress limits, a continuous increase of pore water pressure is observed which can only be interpreted within the framework of elastoplasticity by the interchange between elastic and plastic volumetric strains for undrained conditions. And this is true for even very small, if not infinitesimal, stress amplitudes, which indicates that any attempt to use a classical yield surface plasticity formulation is doomed to fail in predicting such cyclic response. That is because immediately after the first semi-cycle and for sufficiently small stress amplitude the material will remain purely elastic according to such classical models without the possibility to predict the cyclic pore water pressure increase. This may have catastrophic effects by overestimating the material strength which actually decreases as the effective stress (total stress minus pore water pressure) decreases (compression is conventionally considered positive in soil mechanics). To overcome this and related phenomena associated with non-existing or very small purely elastic range, a series of models were developed using more than one surface in stress space. They include the multisurface [30, 31] or the two-surface models [32], in the latter case the outer surface playing essentially the role of the bounding surface with a stress distance dependent plastic modulus as introduced originally for metals [20]. In both cases kinematic hardening together with a small size of the inner surface can realistically predict the above described phenomena. Two simpler approaches using the bounding surface concept without the explicit introduction of one or more inner surfaces were proposed in [33, 23]. In the first approach [33], the elastic range 'shrinks' to zero along the lines of an earlier development for metals [34]. Its predictive capabilities have not yet been investigated in detail. The second approach which is based on the radial mapping version of the bounding surface model [25], introduces a quasi-elastic domain

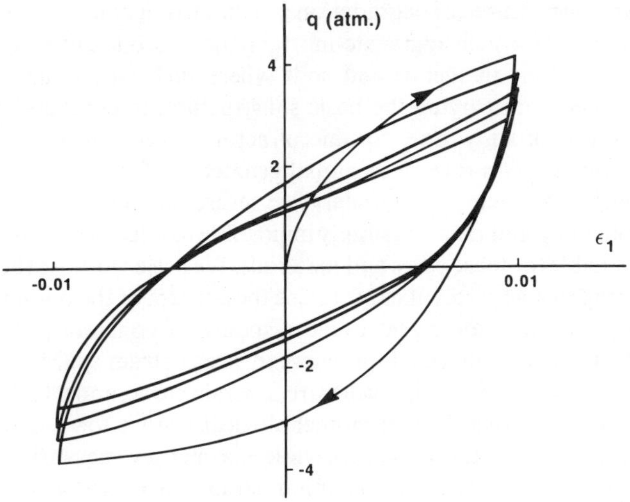

Figure 8.4 Experimental results for undrained triaxial cyclic loading on kaolinite, after [35]

within the bounding surface [23] where unloading leads to another plastic state unless the stress eventually enters the elastic nucleus, a domain where the plastic modulus is infinite. This model is indeed very simple and efficient with excellent predictions for monotonic loading under any conditions, and reasonably good predictions for cyclic. The lack of kinematic hardening, however, which adds much to the simplicity, is also responsible for a less accurate prediction of the cyclic stress–strain curves compared to the classical bounding yield surface combination as in [32].

Due to their particulate structure, geological media are subjected to the effects of degradation and internal damage for a much smaller number of cycles than metals which require indeed thousands of cycles. A typical example is shown in Figure 8.4 where the experimental data of triaxial undrained cyclic deviatoric loading of a sample normally consolidated at about 400 kPa under ± 0.01 strain cycles [35] are shown in the q (triaxial deviatoric stress measured in atm)–ε_1 space. Observe that not only the peak stress but also the stiffness in terms of the tangent modulus decreases with cycling, what is usually referred to as degradation. Degradation is different from the simple softening (decrease of peak stress only for fixed strain amplitude) which can also be observed in metals, and for clays is due to the rupture of interparticle electrochemical bonds under the predominantly shear strain cycling. A proper piv to account phenomenologically for the effects of degradation is the cumulative deviatoric plastic strain which can influence both the elastic moduli as well as the size of the bounding surface and the coefficients relating the actual plastic modulus to the one associated with the 'image' stress on the bounding surface [36].

An even stronger degree of degradation is exhibited in concrete and rock. Let us first observe that for such aggregate-mortar materials, one cannot interpret the plastic strain as done in metals and soils where such strain can be reversed without, in general, destroying the basic substructure. In concrete and rocks a 'plastic' strain is mainly due to microfractural mechanisms and even if macroscopically one can reverse its global geometric effect on the deformation, the underlying substructure has undergone severe damage. This is intensified under cyclic loading and here it is simply impossible to refer to the vast amount of literature available for these cases and materials. We prefer to close the section by simply referring to a very recent effort to use the concept of the bounding surface in describing the monotonic and cyclic response of concrete [37]. The very characteristic feature of an 'envelope' curve in stress–strain space which cannot be exceeded by any monotonic or cyclic stress–strain curve, and which falls after a peak stress, provides the ideal phenomenological feature for a description by means of a 'shrinking' bounding surface while a proper piv, measuring damage or degradation (e.g. accumulated or maximum strain), increases. This is shown in Figure 8.5 as predicted by the uniaxial model developed in [37]. The falling peak stresses for consecutive cycles, are below the envelope curve (dashed line) obtained by monotonic loading due to the increased cyclically induced damage. In addition, the non-linear elastic response is coupled with the induced plastic deformation as seen by the decreasing average slope of the loops, making it necessary to use the uniaxial counterpart of equations (8.9), (8.10) including the elastoplastic coupling terms. In the above model a kinematic/isotropic hardening

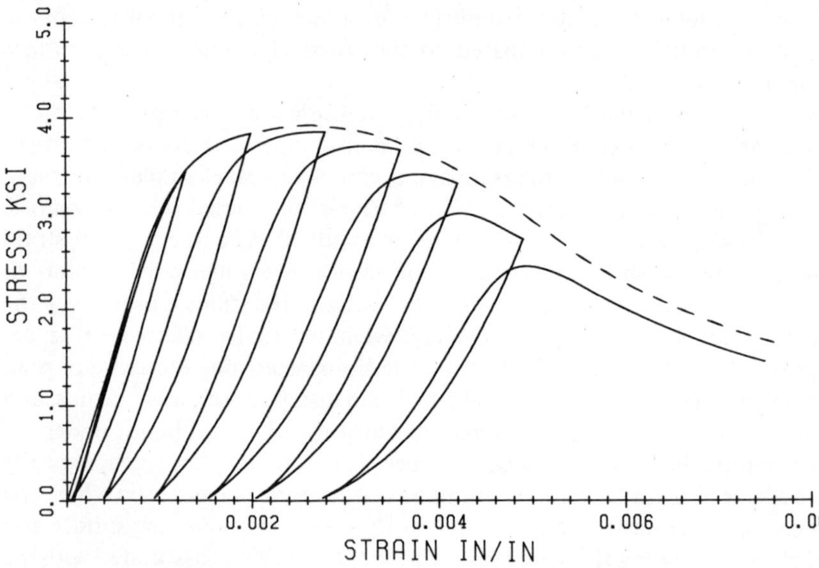

Figure 8.5 Schematic illustration of the uniaxial cyclic response of a bounding surface model for concrete, after [37]

yield surface is enclosed in the bounding surface which softens isotropically as damage measured by a weighted cumulative plastic strain increases. The degradation of the non-linear elastic response is controlled by another damage parameter, the maximum plastic strain. It should be mentioned that a similar effort to use the concept of the bounding surface for a concrete plasticity model is made in [38], along different lines using the concept of a vanishing elastic range [34].

8.9 CONCLUSION

A detailed qualitative analysis of experimental data on metals has shown a highly complex material response under symmetric, unsymmetric, and random cyclic loading conditions. Kinematic and isotropic hardening are the basic phenomenological features, but in their common form can describe realistically only part of the complexities involved in cyclic loading. This complexity is largely due to the fact that abrupt changes of the loading direction, not necessarily obtained by unloading followed by reverse loading, require a discrete material memory of the past states at the instance of such changes, expecially the most recent if not just the last one.

Subsequently, a general phenomenological stress space formulation of rate independent elastoplasticity by means of plastic internal variables (piv) is specialized to a typical class of kinematically/isotropically hardening constitutive models for metals. The plastic modulus is recognized as the key phenomenological parameter for macroscopic modelling, and its evolution is directly related to the dependence of the analytical expression for the loading surface on properly chosen piv. A series of particular constitutive models within the above kinematic/isotropic hardening class are discussed and presented at increasing levels of sophistication, as it becomes desirable to describe more material response characteristics under cyclic loading such as smooth elastoplastic transition, cyclic hardening/softening under symmetric and unsymmetric stress or strain cycles, cyclic creep, cyclic stress relaxation, accurate response under random cyclic stress reversals including partial reverse loading/unloading/reloading, etc. The description of each additional response characteristic requires additional piv and the strife is towards constructing a parsimonious model. It is there where the art of modelling, requiring a combination of engineering intuition and creativity with analytical rigour, reigns supreme. A successful model is one which can be modularized, i.e. which posesses a basic structure on which different characteristic structural moduli can be superimposed describing corresponding phenomena, but which can easily be omitted if desired to have a simpler model without any consequence for the basic structure.

Along these lines it was recognized that the constitutive relations can be executed at two interlocking levels: the first is primarily concerned with the evolution of peak stresses under cyclic loading, what could be called a bounding state. This was discussed in detail including an improved version of the concept of a hypersphere in the plastic strain space whose evolving size and position controls the saturation stress limits. The second level refers to the transition from an intermediate to the bounding state (i.e. reaching the peak stress level) and is basically influenced by random stress reversals. The concept of the bounding surface reflects exactly this two-level decomposition of the execution of the constitutive relations, and this is primarily the reason for its successful prediction of what can be considered the most difficult challenge for any constitutive model: the prediction of the material response under random cyclic loading, like the one shown in Figure 8.1.

The cyclic response for geological media was only briefly considered, emphasizing two distinguishing features: the lack of a well defined purely elastic range and the strong presence of degradation (or elastoplastic coupling) and internal damage. A preliminary attempt to account for both by means of a bounding surface formulation for concrete was demonstrated in the case of cyclic uniaxial loading.

Although not explicitly mentioned in this chapter, it goes without saying that the easy numerical implementation of any elastoplasticity model is of cardinal importance for its usefulness in practice. The sophistication required to model complex cyclic loading response and the modern numerical capabilities should not detract us from the necessity for simplicity, which from the numerical point of view addresses the following requirement: the model must include as small as possible a number of internal variables which must be kept in memory and updated at each step. This is more important, for obvious memory space requirement reasons, than the desirability to have a small number of material (or model) constants. Nevertheless, it must be possible to easily calibrate these constants, a topic which has not been discussed in this chapter.

The development in this chapter was clearly phenomenological and may not find many workers in the field of constitutive relations in agreement, from a microscopic point of view. In this respect it is important to recall that classical mechanics is essentially a geometrical science in conjunction with the fundamental concepts of mass, time, and force (and the derivative concept of stress), whose formulation is placed at different scales of magnitude but it is essentially the same. The proper scale is dictated by the corresponding problem, more specifically the nature of the boundary conditions. It is doubtful that it will ever be possible or efficient to analyse an earth dam or a steel plate by imposing boundary conditions on individual grains or dislocations. Therefore, whatever the point of departure in constructing a material constitutive model for the analysis of structures of the above scale and continuum nature, the end result must be that of relations between the stress, strain and their rates. This is not to say, however, that the detailed search of the microstructure should not be pursued, not only for the obvious reason of its own value for microstructural analysis, but also because it can rationally motivate qualitatively and in some cases provide quantitatively the structure of proper macroscopic constitutive relations.

Returning to the phenomenological description, one may say that the difficulties associated with the identification of the proper piv which can represent the memory of past loading history and model all the complex material response characteristics under cyclic loading, lead to the conclusion of this study: the ever existing dilemma of compromising between simplicity and sophistication in our effort to model the real physical phenomena from a macroscopic point of view within the framework of continuum mechanics, and particularly in association with rate independent elastoplasticity.

REFERENCES

1. M. A. Eisenberg and A. A. Phillips. 'Theory of plasticity with non-coincident yield and loading surfaces', *Acta Mech.*, **11**, 247–260 (1971).
2. Y. F. Dafalias, 'Il'iushin's postulate and resulting thermodynamics conditions on elasto-plastic coupling' *Int. J. Solids Structure*, **13**, 239–251 (1977).
3. G. Maier and T. Hueckel, 'Nonassociated and coupled flow rules of elastoplasticity for geotechnical media' *Int. J. Rock Mech. Min. Sci. and Geomech. Abstr.*, **16**, 77–92 (1979).
4. Y. F. Dafalias, *On Cyclic and Anisotropic Plasticity:* (*i*) *A General Model Including Material Behaviour Under Stress Reversals,* (*ii*) *Anisotropic Hardening for Initially Orthotropic Materials*, Ph.D. thesis, Dept. of Civil Engineering, University of California, Berkeley, 1975.
5. J. L. Chaboche, K. Dang-Van, and G. Cordier, 'Modelization of strain memory effect on the cyclic hardening of 316 stainless steel', *Trans. 5th SMiRT*, L 11/3, Berlin, 1979.
6. D. C. Drucker and L. Palgen, 'On stress–strain relations suitable for cyclic and other loadings', *J. App. Mech.*, **48**, 479–485 (1981).
7. F. Edelman and D. C. Drucker, 'Some Extensions of elementary plasticity theory', *J. Franklin Inst.*, **251**, 581–605 (1951).
8. A. Iu. Ishlinskii, 'General theory of plasticity with linear strain hardening', *Ukr. Mat. Zh.*, **6**, 314 (1954).
9. W. Prager, 'A new method of analyzing stresses and strains on work-hardening plastic solids', *J. App. Mech.*, **23**, 493–496 (1956).
10. H. Ziegler, 'A modification of Prager's hardening rule', *Quart. App. Math.*, **17**, 55–65 (1959).
11. M. A. Eisenberg and A. Phillips, 'On nonlinear kinematic hardening', *Acta Mech.*, **5**, 1–13 (1958).
12. D. A. Caulk and P. M. Naghdi, 'On the hardening response in small deformation of metals', *J. App. Mech.*, **45**, 755–764 (1978).
13. Z. Mróz, H. P. Shrivastava, and R. N. Dubey, 'A nonlinear hardening model and its application to cyclic loading', *Acta Mech.*, **25**, 51–61 (1976).
14. J. L. Chaboche, Sur l' Utilisation des Variables d' Etat Interne pour la Description du Comportement Viscoplastique et de la Rupture par Endommagement, *Symposium Franco-Polonais, Problemes Non-Lineaires de Mecanique*, Cracovie, 1977, Varsovie, 137–159, (1980).
15. P. J. Armstrong and C. O. Frederick, 'A mathematical representation of the multi-axial Bauschinger effect', *CEGB Report No. RD/B/N*731, 1966.
16. J. F. Besseling, 'A theory of elastic, plastic and creep deformations of an initially isotropic material', *J. App. Mech.*, **25**, 529–536 (1958).
17. W. D. Iwan, 'On a class of models for the yielding behaviour of continuous and composite systems', *J. App. Mech.*, **34**, 612–617 (1967).
18. Z. Mróz, 'On the description of anisotropic work hardening', *J. Mech. Phys. Solids*, **15**, 163–175 (1967).
19. M. A. Eisenberg, 'A generalization of plastic flow theory with application to cyclic

hardening and softening phenomena', *J. Eng. Mater. Techn.*, **98**, 221–228 (1976).

20. Y. F. Dafalias and E. P. Popov, 'A model of nonlinearly hardening materials for complex loading', *Proc. 7th U.S. National Congress of App. Mech.*, 149, 1974, and *Acta Mech.*, **21**, 173–192 (1975).
21. Y. F. Dafalias and E. P. Popov, 'Plastic internal variables formalism of cyclic plasticity', *J. App. Mech.*, **43**, 645–650 (1976).
22. R. D. Krieg, 'A practical two surface plasticity theory', *J. App. Mech.*, **47**, 641–646 (1975).
23. Y. F. Dafalias and L. R. Herrmann, 'Bounding surface formulation of soil plasticity' Chapter 10 in *Soil Mechanics—Transient and Cyclic Loads* (Eds. G. N. Pande and O. C. Zienkiewicz), pp. 253–282 John Wiley and Sons, 1982.
24. K. Hashiguchi, 'Constitutive equations of elastoplastic materials with elastic-plastic transition', *J. App. Mech.*, **47**, 266–272 (1980).
25. Y. F. Dafalias, 'The concept and application of the bounding surface in plasticity Theory', *Physical Non-Linearities in Structural Analysis* (Eds. J. Hult and J. Lemaitre), 56–63, IUTAM Symposium Senlis, France, 1980, Springer-Verlag, 1981.
26. Y. F. Dafalias, 'A novel bounding surface constitutive law for the monotonic and cyclic hardening response of metals', *Trans. 6th SMiRT*, L 3/4, Paris 1981.
27. J. L. Chaboche and G. Rousselier, 'On the plastic and viscoplastic constitutive equations. Part I: Rules developed with internal variable concept. Part. II: Application of internal variable concepts to the 316 stainless steel', *J. of Pressure Vessel Technology*, **105**, 153–164 (1983).
28. Y. F. Dafalias and M. S. Seyed-Ranjbari, 'Constitutive modeling of cyclic metal plasticity', Keynote lecture in the *Proc. of 2nd Cairo University Conference on Current Advances in Mechanical Design and Production*, pp. 429–438, December 1982, Cairo, Egypt.
29. N. Ohno, 'A constitutive equation of cyclic plasticity with a non-hardening strain region', *J. App. Mech.*, **49**, 721–727 (1982).
30. J. H. Prevost, 'Plasticity theory for soil stress–strain behaviour', *ASCE J. Engin. Mech.*, **104**, 1177–96 (1978).
31. Z. Mróz, V. A. Norris, and O. C. Zienkiewicz, 'An anisotropic hardening model for soils and its application to cyclic loading', *Int. J. Num. Meth. Geomech.*, **2**, 203–221 (1978).
32. Z. Mróz, V. A. Norris, and O. C. Zienkiewicz, 'Application of an anisotropic hardening model in the analysis of elastoplastic deformation of soils', *Geotechnique*, **29**, 1–34 (1979).
33. Y. F. Dafalias, 'A model for soil behaviour under monotonic and cyclic loading conditions', *Trans. 5th SMiRT*, K 1/8, Berlin, 1979.
34. Y. F. Dafalias and E. P. Popov, 'Cyclic loading for Materials with a Vanishing Elastic Region', *Nuclear Engin. and Design*, **41**, 293–302 (1977).
35. S. Jafroudi, Unpublished work, Department of Civil Engineering, University of California, Davis, 1982.
36. Z. Mróz, V. A. Norris, and O. C. Zienkiewicz, 'An Anisotropic, critical state model for soils subject to cyclic loading', *Geotechnique*, **31**, 451–469 (1981).
37. C. F. Wong, *Monotonic and Cyclic Uniaxial Loading of Concrete Using the Bounding Surface Concept*, Master's thesis, Department of Civil Engineering, University of California, Davis, 1982.
38. M. N. Fardis, B. Alibe, and J. L. Tassoulas, 'Monotonic and cyclic constitutive law for concrete', *J. of Engineering Mechanics, ASCE*, **109**, 516–536 (1983).

Mechanics of Engineering Materials
Edited by C. S. Desai and R. H. Gallagher

Chapter 9

An Incrementally Non-linear Constitutive Law of Second Order and its Application to Localization

F. Darve

9.1 INTRODUCTION

The analysis of the failure of soils for engineering works as well as for samples in the laboratory exhibits most often shear zones with strong kinematic discontinuities. In a triaxial apparatus the sample was approximately homogeneous at the beginning of the test. During the loading, the strain mode of the sample changed from the initial mode which is close to the homogeneous one (the sample remaining approximately cylindrical) to the final mode which is strongly heterogeneous by strain localization along certain surfaces.

In our study, we will assume that this localization is due to an instability of the constitutive law of the material: we follow here the usual way for studying the rupture of metallic structures (for instance, Hill [1], Rice [2]). This point of constitutive instability produces the change in the mode of diffuse strains into the strictly localized one although the limit conditions of sample remain identical. Therefore this point can be called the 'bifurcations point' of the strain mode (an application of the theory of bifurcations to mechanical science is given in Thompson and Hunt [3]). This concept of bifurcation is also a particular case of the general notion of 'catastrophe' (see, for example, Thom [4]).

For a sample of soils in a triaxial apparatus the localized mode of strains will be reached if the 'localization condition' is verified for a given loading path before the homogeneous plastic limit condition. Following Hill [1] and Rice [2], we will give the expression of the localization condition which characterizes the appearance of a shear surface in a medium. We will see how this condition also gives us the local direction of the shear surface.

The localization condition depends on the constitutive behaviour of the material by the means of the 'tangential' tensor of the constitutive law written in an incremental form. It is well known that the kind of constitutive law has a great influence on the quality of the results (for example, Rice and Rudnicki [5], Christoffersen and Hutchinson [6], Hutchinson and Neale [7]). We have developed an incrementally non-linear constitutive law of second order which is a generalization of our previous 'eightfold-linear' law (Darve *et al.* [8,9,10] and which allows us to determine the tangential constitutive tensor varying with the direction of the stress rate. We will present here the main features of this law.

Finally, we will give an application of the localization condition (related to our constitutive law) to structured clays (from east Canada) whose rupture had been studied experimentally in the Department of Civil Engineering of the University of Sherbrooke by Lefèbvre *et al.* [11].

9.2 THE LOCALIZATION CONDITION

Let us consider an homogeneous sample with uniform fields of Cauchy stress $[\sigma_{ij}]$ and strain in the coordinate system (x_1, x_2, x_3). A shear band appears when the field of the gradient of small displacements is non-uniform in a direction perpendicular to the band (Figure 9.1). Let $\Delta(\mathbf{T})$ be the difference between the value of the tensor $\mathbf{T}$ inside the band and its uniform value outside.

If $[d\sigma_{ij}]$ is the increment of the current stress in the coordinate system (x_1, x_2, x_3) the continuity of the incremental stress vector implies the following 'statical' condition

$$n_i \Delta(d\sigma_{ij}) = 0 \tag{9.1}$$

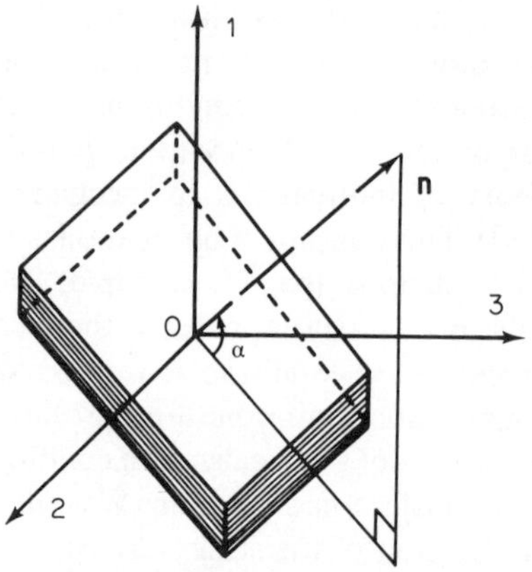

Figure 9.1 Shear band and its normal vector **n** in the coordinate system (x_1, x_2, x_3).

If **v** is the current velocity vector, **g** an arbitrary vector whose value is null outside of the band and dt is the time increment, the 'kinematical' condition takes the form

$$\Delta\left(\frac{\partial v_i}{\partial x_j}\,\mathrm{d}t\right) = g_i n_j\,\mathrm{d}t \tag{9.2}$$

With $\mathrm{d}\varepsilon_{kl} = \frac{1}{2}[(\partial v_k/\partial x_l) + (\partial v_l/\partial x_k)]\,\mathrm{d}t$, it follows

$$\Delta(d\varepsilon_{kl}) = \tfrac{1}{2}(g_k n_l + g_l n_k)\,\mathrm{d}t \tag{9.3}$$

We must add to the equations (9.1) and (9.3) the ones relating $d\sigma_{ij}$ to $d\varepsilon_{kl}$. This is the constitutive law of the material

$$D\sigma_{ij} = F_{ij}(d\varepsilon_{kl}) \tag{9.4}$$

F is an homogeneous function of degree one (in a restricted sense to positive values of the multiplicative parameter) because of the rate independency and at the same time a non-linear function because of plastic irreversibilities. $\mathbf{D}\boldsymbol{\sigma}$ is the Jaumann's differential of $\boldsymbol{\sigma}$ and we have

$$D\sigma_{ij} = d\sigma_{ij} - d\omega_{ik}\sigma_{kj} - d\omega_{jk}\sigma_{ik} \tag{9.5}$$

with

$$d\omega_{ij} = \tfrac{1}{2}\left(\frac{\partial v_i}{\partial x_j} - \frac{\partial v_j}{\partial x_i}\right)\mathrm{d}t \tag{9.6}$$

After the equations (9.1), (9.2), (9.5), and (9.6) it becomes

$$n_i\Delta(D\sigma_{ij}) = -\tfrac{1}{2}n_i(g_i n_k - g_k n_i)\sigma_{kj}\,\mathrm{d}t - \tfrac{1}{2}n_i(g_j n_k - g_k n_j)\sigma_{ik}\,\mathrm{d}t \tag{9.7}$$

Equation (9.4) implies

$$\Delta(D\sigma_{ij}) = L^{\mathrm{T}}_{ijkl}\Delta(d\varepsilon_{kl}) \tag{9.8}$$

in which $L^{\mathrm{T}}_{ijkl} = [\partial(D\sigma_{ij})/\partial(d\varepsilon_{kl})] = [\partial F_{ij}/\partial(d\varepsilon_{kl})]$ and indice 'T' signifies that we must consider the tangential constitutive tensor defined by $D\sigma_{ij} = L^{\mathrm{T}}_{ijkl}d\varepsilon_{kl}$ after the Euler's identity for homogeneous functions of degree one. Obviously since our law is incrementally non-linear $\mathbf{L}^{\mathrm{T}}$ will depend on $\mathbf{d}\boldsymbol{\varepsilon}$ (or its inverse on $\mathbf{D}\boldsymbol{\sigma}$).

Equations (9.8) and (9.3) imply

$$n_j\Delta(D\sigma_{ij}) = \tfrac{1}{2}n_i L^{\mathrm{T}}_{ijkl}(g_k n_l + g_l n_k)\,\mathrm{d}t \tag{9.9}$$

After equations (9.7) and (9.9)

$$[n_i L^{\mathrm{T}}_{ijkl} n_l + \tfrac{1}{2}(n_p\sigma_{pj}n_k + n_p\sigma_{pq}n_q\delta_{jk} - n_p\sigma_{pk}n_j - \sigma_{jk})]g_k = 0 \tag{9.10}$$

A necessary and sufficient condition for the existence of a non-null solution in g_1, g_2, g_3, is that the determinant of the homogeneous system (9.10) with the

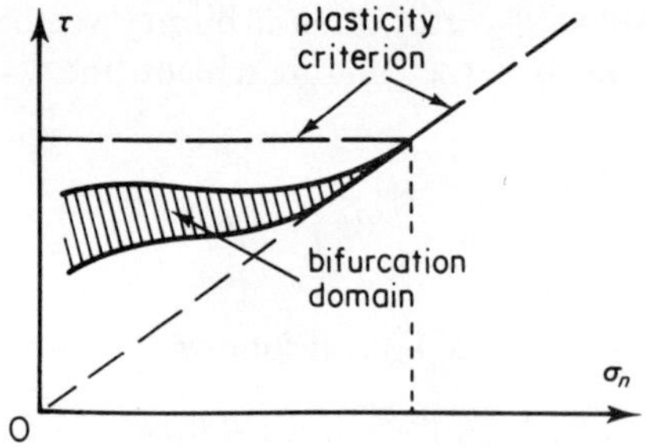

Figure 9.2 Plasticity criterion and bifurcation domain in the Mohr–Coulomb plan for a structured clay

three unknowns g_1, g_2, g_3, is null. Thus

$$\det[n_i L^{\mathrm{T}}_{ijkl} n_l + \tfrac{1}{2}(n_p \sigma_{pj} n_k + n_p \sigma_{pq} n_q \delta_{ik} - n_p \sigma_{pk} n_j - \sigma_{jk})] = 0 \tag{9.11}$$

There will be bifurcation when for a given loading path the equation (9.11) will admit one or several real solutions in n_1, n_2, n_3. The stress–strain state defined by this condition is a point of bifurcation. The set of these points constitutes the 'criterion of bifurcation' of the material for the given set of loading paths. We see here the essential difference between the criterion of plasticity or limit condition, which is independent of the path and intrinsic, and the criterion of bifurcation which is dependent on the previous strain history (see Figure 9.2).

9.3 THE NON-LINEAR INCREMENTAL CONSTITUTIVE LAW OF SECOND ORDER

By using the symmetry of the tensors $\mathbf{d\varepsilon}$ and $\mathbf{D\sigma}$ the study of the tensorial function $\mathbf{F}$ (equation (9.4)) or of its inverse is identical to the study of the vectorial function $\mathbf{G}$ defined in a six-dimensional space

$$d\varepsilon_\alpha = G_\alpha(D\sigma_\beta) \qquad (\alpha, \beta = 1, \ldots, 6) \tag{9.12}$$

(for the definition of $d\varepsilon_\alpha$ and $D\sigma_\beta$, see, for instance, Darve and Labanieh [10]).

Because of the rate-independency $\mathbf{G}$ is homogeneous of degree one in a restricted sense to the positive values of the multiplicative parameter.

After Euler's identity for the homogeneous functions

$$d\varepsilon_\alpha = \frac{\partial G_\alpha}{\partial(D\sigma_\beta)} D\sigma_\beta = M^{\mathrm{T}}_{\alpha\beta}(u_\gamma) D\sigma_\beta \tag{9.13}$$

in which $M^{\mathrm{T}}_{\alpha\beta}$ is an homogeneous function of degree zero in $D\sigma_\beta$; thus it is a function of $\mathbf{u}$ which is the unit vector in the direction of $\mathbf{D\sigma}$:

$$\mathbf{u} = \mathbf{D\sigma}/\|\mathbf{D\sigma}\|, \qquad \|\mathbf{D\sigma}\| = \sqrt{(D\sigma_\alpha D\sigma_\alpha)}$$

The non-linearity of **G**, due to the plastic irreversibilities, implies the directional dependency of **M** (equation (9.13)).

The first constitutive hypothesis is to assume the orthotropy of the law. We will see later that **M** depends on the current values of $\boldsymbol{\sigma}$ and $\boldsymbol{\varepsilon}$. We must also assume the frame in which the law is orthotropic. After numerical experiments of simulation of simple shear tests it seemed to us (see Darve [12]) that the principal axes of total irreversible strains could be possible orthotropic directions.

We note by $\overline{\mathbf{d\varepsilon}}$, $\overline{\mathbf{D\sigma}}$, $\bar{\mathbf{u}}$ the following vectors:

$$(d\varepsilon_{11}, d\varepsilon_{22}, d\varepsilon_{33}),\ (D\sigma_{11}, D\sigma_{22}, D\sigma_{33}),\ \text{and}\ (u_1, u_2, u_3),$$

expressed in the orthotropy frame.

Our second assumption concerns the directional dependency of **M(u)** which is restricted to $\bar{\mathbf{u}}$ for the matrix **M** expressed in the orthotropy frame. With the first and the second assumption the constitutive law takes the following form when it is expressed in the orthotropy frame

$$\begin{cases} \overline{\mathbf{d\varepsilon}} = \mathbf{N}(\bar{\mathbf{u}})\overline{\mathbf{D\sigma}} \\ d\varepsilon_{23} = \dfrac{1}{2G_1} d\sigma_{23}; d\varepsilon_{31} = \dfrac{1}{2G_2} d\sigma_{31}; d\varepsilon_{12} = \dfrac{1}{2G_3} d\sigma_{12} \end{cases} \tag{9.14}$$

For G_1, G_2, G_3 we have chosen

$$G_1 = \frac{\sigma_{22} - \sigma_{33}}{2(\varepsilon_{22} - \varepsilon_{33})} \quad \text{(same formulae for } G_2, G_3\text{)} \tag{9.15}$$

We must now study the 3 × 3 matrix, $\mathbf{N}(\bar{\mathbf{u}})$ defined in the orthotropy axes by

$$\begin{Bmatrix} d\varepsilon_{11} \\ d\varepsilon_{22} \\ d\varepsilon_{33} \end{Bmatrix} = \mathbf{N}(\bar{\mathbf{u}}) \begin{Bmatrix} d\sigma_{11} \\ d\sigma_{22} \\ d\sigma_{33} \end{Bmatrix}$$

We have assumed a linear variation of **N** versus $\bar{\mathbf{u}}$ of following kind:

$$\mathbf{N}(\bar{\mathbf{u}}) = \mathbf{A} + \mathbf{B}\begin{bmatrix} u_1 & 0 & 0 \\ 0 & u_2 & 0 \\ 0 & 0 & u_3 \end{bmatrix} \tag{9.16}$$

With $u_i = D\sigma_{ii}/\|\mathbf{D\sigma}\|$ (without summation on i) and

$$\|\mathbf{D\sigma}\| = \sqrt{(D\sigma_{ij}D\sigma_{ji})} = \sqrt{(D\sigma_\alpha D\sigma_\alpha)}.$$

The main idea of this constitutive model is to substitute the conventional framework of the elastoplastic theories for an interpolation between known behaviours. It is well known that the 'bilinear' laws (as the conventional elastoplastic ones) or the 'multilinear' ones (with several plastic potentials) must be related with microslip mechanisms in particular directions (as for instance

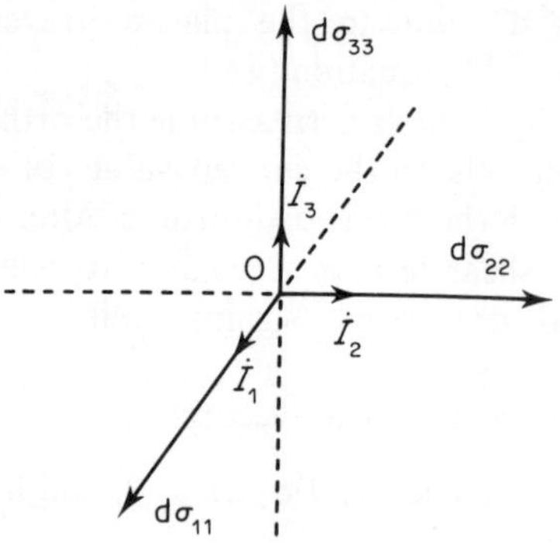

Figure 9.3 Definition of the fitting paths of the law: $\pm\dot{\mathbf{I}}_1$, $\pm\dot{\mathbf{I}}_2$, $\pm\dot{\mathbf{I}}_3$

in monocrystals whose structure is regular) which are difficult to exhibit in granular media. It is thus interesting to use a law which is incrementally 'non-linear', but at the same time we must be able to develop theories which remain realistic. Our constitutive law is based on a non-linear interpolation (equation (9.16)) between known behaviours for particular paths: the 'generalized' triaxial paths.

These generalized triaxial paths are constituted by compressions and extensions for which the principal axes of stress and strain are fixed and confounded, only one principal stress varying (increasing or decreasing), the other two remaining constant but not necessarily equal. The behaviour of the material is assumed to be known for the six incremental paths:

$$\overline{\mathbf{D}\boldsymbol{\sigma}} = \begin{Bmatrix} \pm 1 \\ 0 \\ 0 \end{Bmatrix}, \begin{Bmatrix} 0 \\ \pm 1 \\ 0 \end{Bmatrix}, \begin{Bmatrix} 0 \\ 0 \\ \pm 1 \end{Bmatrix} \text{ (see Figure 9.3)}$$

This behaviour is given analytically by the three families of functions f, g, h, in compression (f^+, g^+, h^+) and in extension (f^-, g^-, h^-)

$$\begin{cases} \sigma_k = f(\varepsilon_k, \sigma_j, \sigma_l) & (k, j, l = 1,2,3) \\ \varepsilon_j = g(\varepsilon_k, \sigma_j, \sigma_l) & (k \neq j \neq l) \\ \varepsilon_l = h(\varepsilon_k, \sigma_j, \sigma_l) & (\sigma_j, \sigma_l \text{ constant parameters}) \end{cases}$$

We introduce the following notations

$$U_k = \left(\frac{\partial f}{\partial \varepsilon_k}\right)_{\sigma_j, \sigma_l}; \qquad V_k^j = -\left(\frac{\partial g}{\partial \varepsilon_k}\right)_{\sigma_j, \sigma_l}; \qquad V_k^l = -\left(\frac{\partial h}{\partial \varepsilon_k}\right)_{\sigma_j, \sigma_l}$$

$U_k^+, V_k^{j+}, V_k^{l+}$ are related to f^+, g^+, h^+ and $U_k^-, V_k^{j-}, V_k^{l-}$ to f^-, g^-, h^-

Finally let $\mathbf{N}^+$ be

$$\begin{bmatrix} \dfrac{1}{U_1^+} & -\dfrac{V_2^{1+}}{U_2^+} & -\dfrac{V_3^{1+}}{U_3^+} \\ -\dfrac{V_1^{2+}}{U_1^+} & \dfrac{1}{U_2^+} & -\dfrac{V_3^{2+}}{U_3^+} \\ -\dfrac{V_1^{3+}}{U_1^+} & -\dfrac{V_2^{3+}}{U_2^+} & \dfrac{1}{U_3^+} \end{bmatrix}$$

and $\mathbf{N}^-$ in a same way.

Equations (9.14) and (9.16) imply

$$\begin{Bmatrix} d\varepsilon_{11} \\ d\varepsilon_{22} \\ d\varepsilon_{33} \end{Bmatrix} = \mathbf{A} \begin{Bmatrix} D\sigma_{11} \\ D\sigma_{22} \\ D\sigma_{33} \end{Bmatrix} + \frac{1}{\|\mathbf{D}\boldsymbol{\sigma}\|} \mathbf{B} \begin{Bmatrix} (D\sigma_{11})^2 \\ (D\sigma_{22})^2 \\ (D\sigma_{33})^2 \end{Bmatrix} \tag{9.17}$$

By identifying the behaviour given by equation (9.17) and the known behaviour for the generalized triaxial paths characterized by the known matrices $\mathbf{N}^+$ and $\mathbf{N}^-$ we obtain

$$\mathbf{A} + \mathbf{B} = \mathbf{N}^+ \quad \text{and} \quad \mathbf{A} - \mathbf{B} = \mathbf{N}^-$$

Finally it becomes

$$\begin{Bmatrix} d\varepsilon_{11} \\ d\varepsilon_{22} \\ d\varepsilon_{33} \end{Bmatrix} = \tfrac{1}{2}(\mathbf{N}^+ + \mathbf{N}^-) \begin{Bmatrix} D\sigma_{11} \\ D\sigma_{22} \\ D\sigma_{33} \end{Bmatrix} + \frac{1}{2\|\mathbf{D}\boldsymbol{\sigma}\|} (\mathbf{N}^+ - \mathbf{N}^-) \begin{Bmatrix} (D\sigma_{11})^2 \\ (D\sigma_{22})^2 \\ (D\sigma_{33})^2 \end{Bmatrix} \tag{9.18}$$

Equations (9.14), (9.15), and (9.18) define the constitutive law which can be characterized as incrementally non-linear of second order.

The formation of the law allows us to exhibit a 'secant' constitutive matrix $\mathbf{M}^s$: $\mathbf{d}\boldsymbol{\varepsilon} = \mathbf{M}^s(\mathbf{u})\mathbf{D}\boldsymbol{\sigma}$. But we need (see equation (9.11)) the tangential one. We will give the expression of $\mathbf{M}^T(\mathbf{u})$ in the case more simply (in relation with the application presented here) for which

$$D\sigma_{23} = D\sigma_{31} = D\sigma_{12} = 0 \tag{9.19}$$

Equation (9.18) can be written

$$d\varepsilon_i = A_{ij} D\sigma_j + \frac{1}{\|\mathbf{D}\boldsymbol{\sigma}\|} B_{ij} (D\sigma_j)^2 \tag{9.20}$$

By differentiation and with equation (9.19) we obtain

$$d\varepsilon_i = \left[A_{ij} + \frac{2}{\|\mathbf{D}\boldsymbol{\sigma}\|} B_{ik} D\sigma_k \delta_{kj} - \frac{1}{\|D\boldsymbol{\sigma}\|^3} B_{ik} (D\sigma_k)^2 D\sigma_j \right] D\sigma_j \tag{9.21}$$

Equation (9.21) gives the expression of $\mathbf{N}^{\mathrm{T}}(\bar{\mathbf{u}})$ whereas (9.20) could give only $\mathbf{N}^{\mathrm{s}}(\bar{\mathbf{u}})$. One can verify that

$$\mathbf{N}^{\mathrm{T}}(\bar{\mathbf{u}})\overline{\mathbf{D}\boldsymbol{\sigma}} \equiv \mathbf{N}^{\mathrm{S}}(\bar{\mathbf{u}})\overline{\mathbf{D}\boldsymbol{\sigma}}$$

The shear moduli G_1, G_2, G_3, are 'directionally' constant. Thus with (9.19) they are the same for $\mathbf{M}^{\mathrm{s}}(\mathbf{u})$ and $\mathbf{M}^{\mathrm{T}}(\mathbf{u})$.

9.4 APPLICATION TO TRIAXIAL TESTS

We now restrict our study to the case of axisymmetrical triaxial tests. The axial direction is denoted by 1 and the radial direction by 2. The sign conventions are those of soil mechanics (stress and strain positive in compression). α is the angle of the local direction of the shear band with the axial direction 1.

In the case of an axisymmetrical problem with fixed principal axes, equation (9.11) gives a pair equation of degree six in $\mathrm{tg}\,\alpha$ (written explicitly in Darve and Lefèbvre [13]). The elimination of an always negative solution in $\mathrm{tg}^2\,\alpha$ simplifies this equation into the following

$$\begin{aligned} & L_{1111}\left(L_{1212} - \frac{\sigma_1 - \sigma_2}{2}\right)\mathrm{tg}^4\alpha \\ & + \left(L_{1111}L_{2222} - L_{1122}L_{2211} - L_{1122}\left(L_{1212} + \frac{\sigma_1 - \sigma_2}{2}\right)\right. \\ & \left. - L_{2211}\left(L_{1212} - \frac{\sigma_1 - \sigma_2}{2}\right)\right)\mathrm{tg}^2\,\alpha + L_{2222}\left(L_{1212} + \frac{\sigma_1 - \sigma_2}{2}\right) = 0 \end{aligned} \tag{9.22}$$

This last equation is of second degree in $\mathrm{tg}^2\alpha$. We find here the conventional discussion (Rudnicki and Rice [14], Rice [2], Vardoulakis [15, 16]) with elliptic regime (no real solution) and an hyperbolic one (two distinct real solutions) separated by the parabolic boundary (one double solution). The appearance of a shear band for a given stress–strain path is to be related with the first real solution. The parabolic boundary will thus characterize the bifurcation points (Figure 9.4). For a given set of loading paths the bifurcation criterion will be given by the nullity of the discriminant of equation (9.22) and the directions of shear bands by the double solution of (9.22). In particular cases this theoretical maximum stress will be reduced by the imperfections of the initial sample and of the boundary conditions of the test (Desrues [17], Boulon and Cichy [18]).

The correction due to the objective derivation of Jaumann appears in equation (9.22) only in relation with the shear modulus L_{1212}.

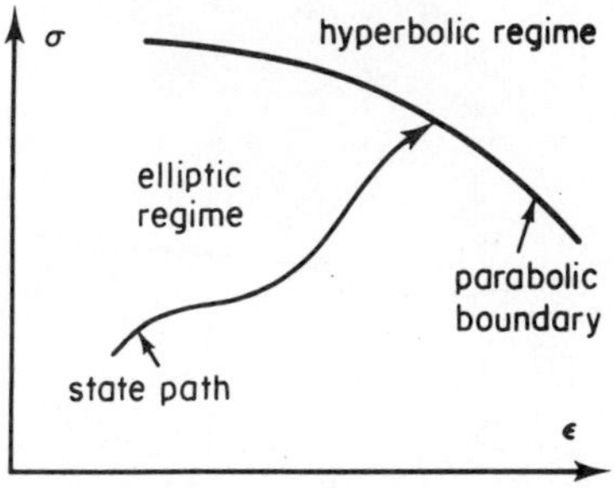

Figure 9.4 The bifurcation appears on the curve of parabolic regime between the elliptic domain and the hyperbolic one

For a given material and a given loading path, the integration of the constitutive law increment by increment will give us the tangential constitutive matrix $\mathbf{M}^{\mathrm{T}}(\mathbf{u})$ then easily the values of the elements L_{ijkl} simply related with the inverse of $\mathbf{M}^{\mathrm{T}}(\mathbf{u})$. The computation of the discriminant of equation (9.22) will let us find the stress–strain state at which the bifurcation occurs.

9.5 APPLICATION TO CLAY 'OLGA'

The studied clay comes from the territories of James Bay at 700 km in the north-west of Montreal (Quebec). It is a varved clay normally consolidated from a geological point of view. This natural clay presents an important 'effect of structure'.

In oedometric tests the behaviour of the clay is the same as for an over-consolidated clay and one can define a crushing pressure of the structure.

In triaxial tests and for a consolidation pressure smaller than the crushing pressure the strains remain small and 'moduli' have big values until the failure which appears abruptly and implies an important reduction of strength (Lefèbvre and La Rochelle [19], Lefèbvre [20]) in particular for small values of lateral stress. The volume variations in drained triaxial tests are always contractant. An eventual dilatancy is due to the development of small voids along the shear band.

Figures 9.5 to 9.8 (Lefèbvre *et al.* [11]) present typical results for drained triaxial tests. From these figures one can conclude:

(a) The failure appears for an axial strain of 0.01 and a 'deviatoric' stress $\sigma_1 - \sigma_3$ of 60 kPa.
(b) The moduli before the failure have big values: the homogeneous plastic failure with small moduli is 'far'.
(c) This rupture occurs while the clay is contractant.
(d) The failure occurs by one or several shear planes appearing in the sample. These planes have an inclination of 25° in respect to the axial direction approximately.

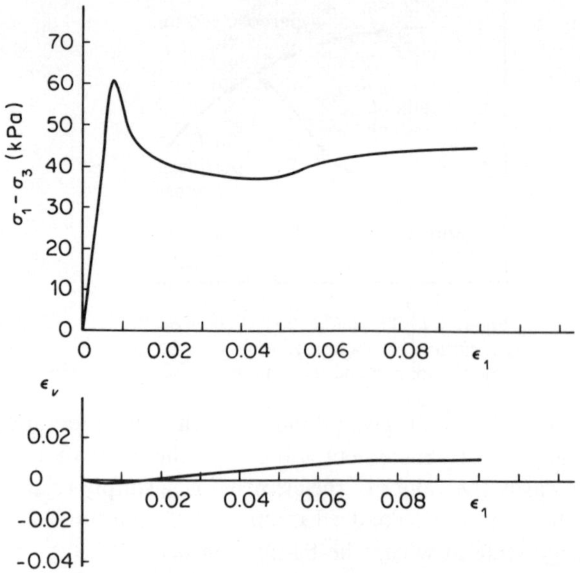

Figure 9.5 Failure of the clay in drained compression for a lateral stress of 5 kPa

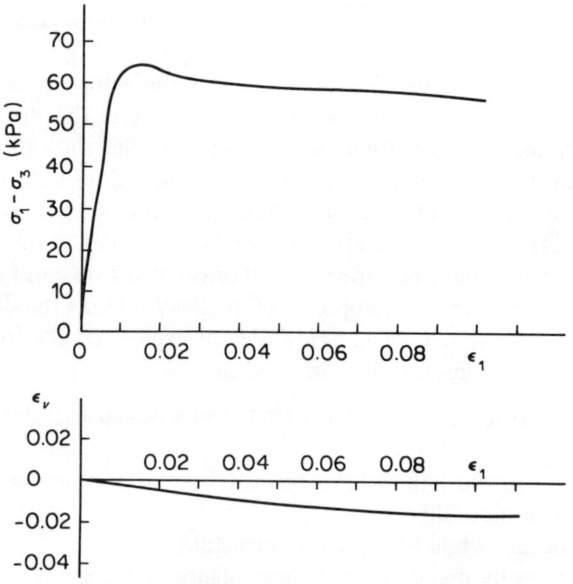

Figure 9.6 Failure of the clay in drained compression for a lateral stress of 15 kPa

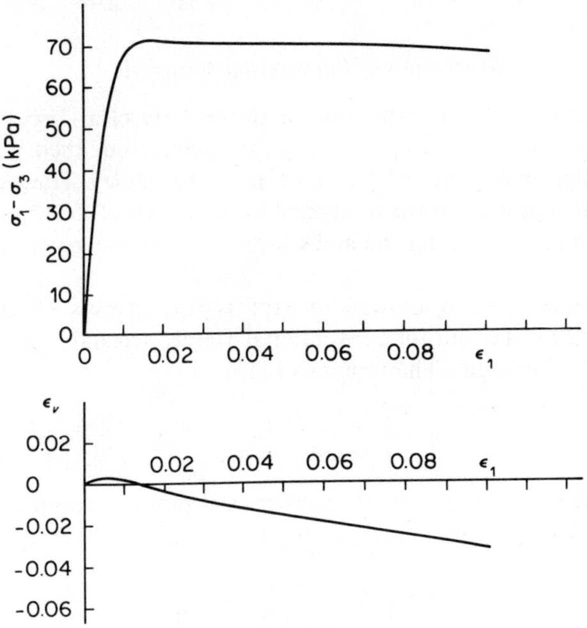

Figure 9.7 Failure of the clay in drained compression for a lateral stress of 30 kPa

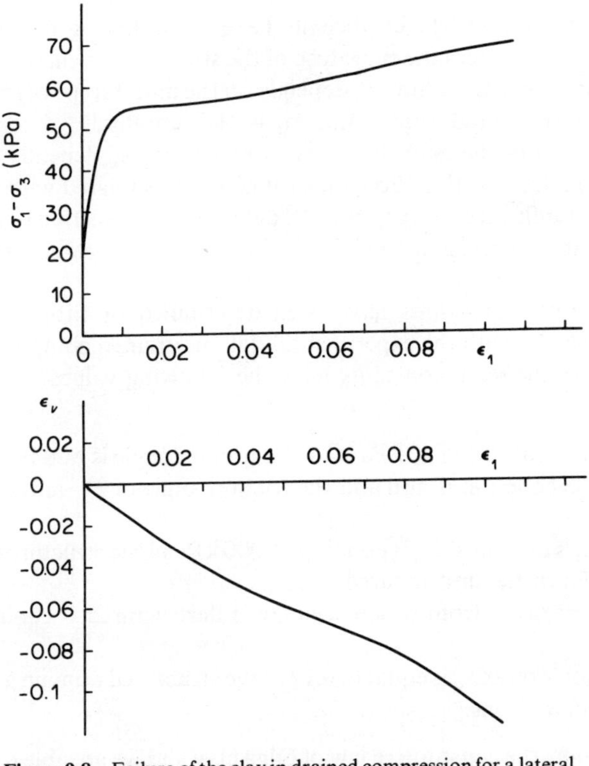

Figure 9.8 Failure of the clay in drained compression for a lateral stress of 50 kPa

This structured clay which presents a particularly clear process of strain localization was thus a good opportunity for applying our theory.

As a first step, we determined the constants of Olga clay. These constitutive parameters are appearing in the analytical formulations of the three families of functions f, g, h characterizing the soil's behaviour for the generalized triaxial paths.

Some assumptions have allowed to express f, g, h versus f^* and k which describe the soil's behaviour for conventional drained triaxial paths (Darve and Labanieh [10], Darve and Thanopoulos [21])

$$\begin{cases} \sigma_1 = f^*(\varepsilon_1, \sigma_2) \\ e = k(\varepsilon_1, \sigma_2) \end{cases}$$

in which e is the void ratio. For drained triaxial paths in compression and in extension suitable analytical formulations are

$$\begin{aligned} \sigma_1 = f^*(\varepsilon_1, \sigma_2) &= \sigma_2\left(1 + A_s \frac{\exp(A_1|\varepsilon_1|) - 1}{\exp(A_1|\varepsilon_1|) + B_s}\right) \\ e = k(\varepsilon_1, \sigma_2) &= e_0 - e_M[1 - \exp(-A_e\varepsilon_1)] + B_e\varepsilon_1 \\ &\quad - C_e(\varepsilon_1)^2 \exp(-D_e\varepsilon_1) \end{aligned} \tag{9.23}$$

In these formulations the coefficients have the following meanings: A_1 is a constant which influences the curvature of the stress–strain curve, A_s is the value of $(\sigma_1 - \sigma_2)/\sigma_2$ in plastic failure, B_s depends on the initial slope of $f^*(\ )$ (denoted by U_0), e_0 is the initial void ratio, e_M is the amplitude of contractancy in compression and of the light dilatancy in extension, A_e depends on the initial slope of $k(\ )$ related with a 'Poisson's ratio' ν_0, B_e is related with the dilatancy angle (without influence here), C_e and D_e determine the position of the minimum of void ratio in compression and of its light maximum in extension in the (e, ε_1) plane.

The constitutive constants have been determined by fitting the analytical formulations (9.23) with the experimental data in drained compression. For the structured clay the main constants have the following values:

$A_1 = 100$.

The cohesion value is of 40 kPa and the friction angle is null in the structured domain; the cohesion is null and the friction angle of 24° in the destructured domain.

The 'Young's modulus' U_0 is equal to 15 000 kPa in the structured domain and to 7500 kPa in the destructured one.

The mean pressure from which the clay is destructured is equal to 90 kPa.

$e_0 = 0.8$.

The 'Poisson's ratio' ν_0 is equal to 0.12 in the structured domain and to 0.024 in the destructured one.

We now know the constitutive law of Olga clay and we are able to simulate the tests presented on the Figures 9.5 to 9.8 by analysing for each increment of computations whether the discriminant of equation (9.22) is positive. The computation is stopped when this condition is verified and the double solution of equation (9.22) is determined at this point for exhibiting the theoretical inclination of the shear band.

The double solution of (9.22) is equal to:

$$\operatorname{tg}^2\alpha = \sqrt{\left(\frac{L_{2222}\left(L_{1212} + \dfrac{\sigma_1 - \sigma_2}{2}\right)}{L_{1111}\left(L_{1212} - \dfrac{\sigma_1 - \sigma_2}{2}\right)}\right)}$$

thus to an approximate value of: $\sqrt{(L_{2222}/L_{1111})}$. In the particular case of this contractant clay in drained triaxial compression, this last value itself can be

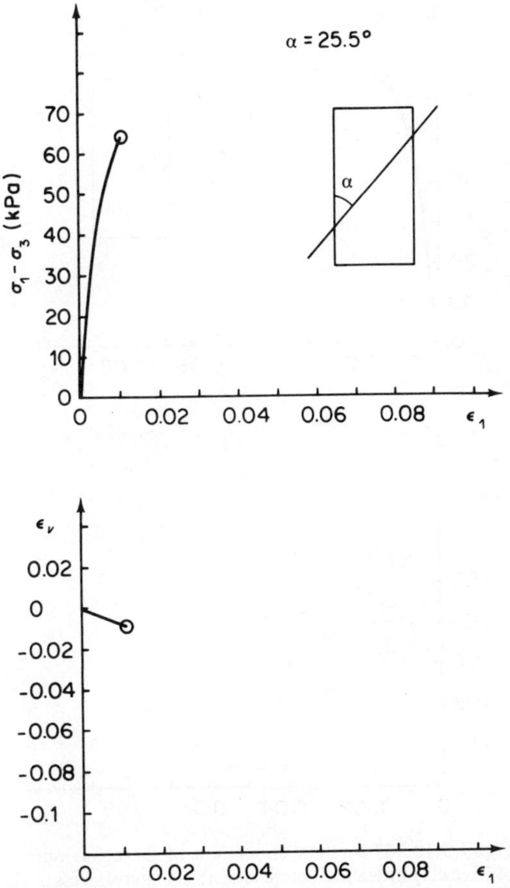

Figure 9.9 Theoretical simulation of the behaviour of Olga clay in drained compression for a lateral stress of 5 kPa until the bifurcation point

approximated by the following more simple expression:

$$\mathrm{tg}\alpha \simeq \sqrt[4]{\left(\frac{2U_2^-}{U_1^+}\right)} \tag{9.24}$$

with the notations appearing in the formulations of $\mathbf{N}^+$ and $\mathbf{N}^-$. The relationship (9.24) is interesting because only variables of type 'moduli' influence the bifurcation.

The moduli U_1^+ and U_2^- by definition depend mainly on the cohesion value which thus will influence the formation of shear bands. In previous studies (Darve [22] and Darve *et al.* [23]) we had showed that the angle of dilatancy had considerable influence on the bifurcation of sands. These opposite results for a

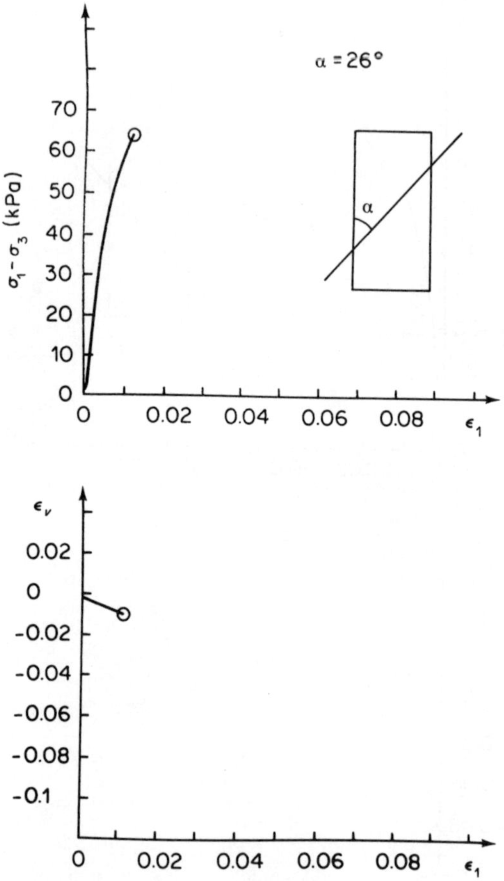

Figure 9.10 Theoretical simulation of the behaviour of Olga clay in drained compression for a lateral stress of 15 kPa until the bifurcation point

sand and a structured clay have been obtained by means of the generality and the flexibility of our constitutive law.

The five following figures (9.9 to 9.13) present the theoretical results obtained and can be compared with the experimental data given on Figures 9.5 to 9.8. The end points of curves are bifurcation points. The isotropic path preceding the drained triaxial compressions is plotted at the beginning of the curves. The fourth first loading paths (Figures 9.9 to 9.12) are placed in the structured domain of clay, the fifth on the contrary is out. We have not obtained any bifurcation in this last case which agrees with the experiments.

9.6 CONCLUSION

We have presented here first results concerning the study of the failure of clays as a bifurcation problem. This is an application of a non-linear incremental

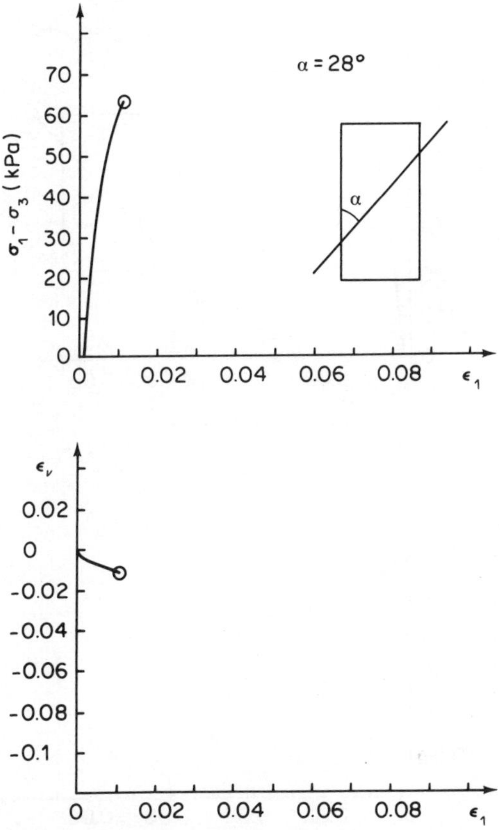

Figure 9.11 Theoretical simulation of the behaviour of Olga clay in drained compression for a lateral stress of 30 kPa until the bifurcation point

constitutive law to this specific problem. This application is interesting in that it allows us a better understanding of shear bands appearing in the soils and also to establish precisely the domain within which the Law operates. Other applications have been given recently to describe also the viscous behaviour of clays (Darve and Vuaillat [24]) and the complex cyclic behaviour of sands (Darve and Thanopoulos [21], Darve and Labanieh [25]).

ACKNOWLEDGEMENTS

The author would like to thank Professor Guy Lefèbvre, Director of the Civil Engineering Department in the University of Sherbrooke, for his welcome, the Centre National de la Recherche Scientifique in France and also the National Research Council of Canada for their financial support.

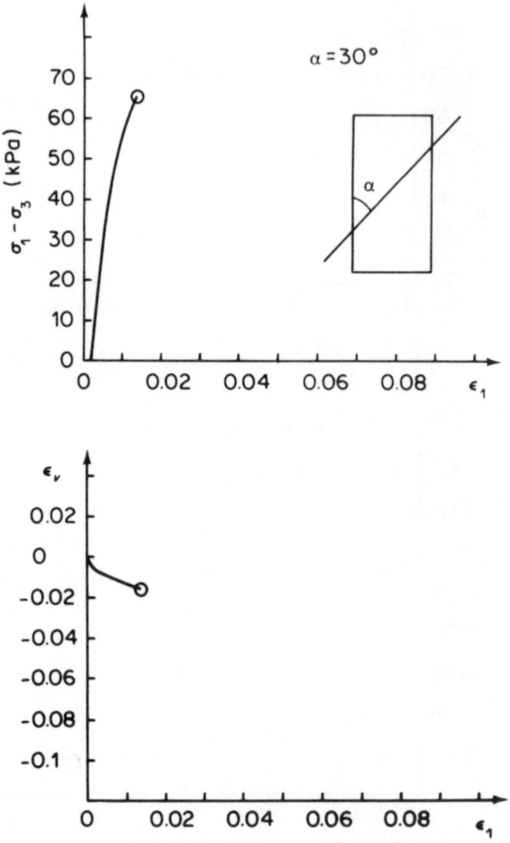

Figure 9.12 Theoretical simulation of the behaviour of Olga clay in drained compression for a lateral stress of 50 kPa until the bifurcation point

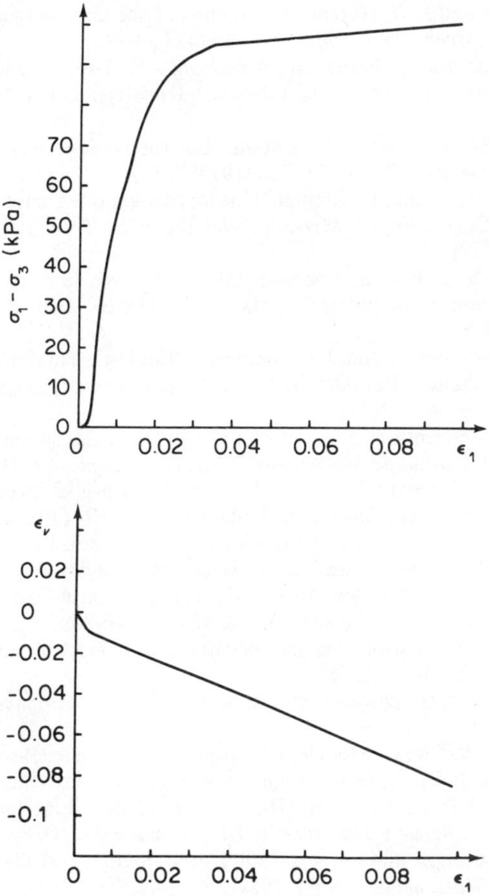

Figure 9.13 Theoretical simulation of the behaviour of Olga clay in drained compression for a lateral stress of 70 kPa until the bifurcation point

REFERENCES

1. R. Hill, 'Acceleration waves in solids', *J. Mech. Phys. Solids*, **10**, 1–16 (1962).
2. J. R. Rice, 'The localization of plastic deformation', *Theoretical and Applied Mechanics* (Ed. W. T. Koiter), North Holland Publishing Company, pp. 207–220, 1976.
3. J. M. T. Thompson and G. W. Hunt, *A General Theory of Elastic Stability*', John Wiley, London, 1973.
4. R. Thom, *Stabilité structurelle et morphogenese: essai d'une théorie génerale des modeles* W. A. Benjamin Inc., S. A. Ediscience, Paris, 1972.
5. J. R. Rice and J. W. Rudnicki, 'A note on some feature of the theory of localization of deformation', *Int. J. Solids Structures* **16**, 597–605 (1980).

6. J. Christoffersen and J. W. Hutchinson, 'A class of phenomenological corner theories of plasticity', *J. Mech. Phys. Solids*, **27**, 465–487 (1979).
7. J. W. Hutchinson and K. Neale, 'Sheet necking—II. Time independent behaviour', *Mechanics of Sheet Metal Forming* (Eds. D. P. Koistinen and N. M. Wang), pp. 127–153, 1978.
8. F. Darve, M. Boulon, and R. Chambon, 'Loi rhéologique incrementale des sols', *Journal de Mecanique*, **17**, (5) 679–716 (1978).
9. F. Darve, E. Flavigny, and P. Vuaillat, 'Une loi rhéologique complete pour matériaux argileux', *7éme Conf. Europ. de Mécanique des Sols et des Travaux de Fondation*, vol. 1, pp. 119–124 (1979).
10. F. Darve and S. Labanieh, 'Incremental constitutive law for sands and clays. Simulations of monotonic and cyclic tests', *In. J. Num. and Anal. Methods in Geomech.*, **6**, 243–275 (1982).
11. G. Lefèbvre, J. P. Bosse, and J. G. Beliveau, 'Etude de l'argile du site Olga sous sollicitations cycliques', Rapport GEO–79–06 présenté a la Société d'énergie de la Baie James, Sherbrooke, 1979.
12. F. Darve, 'Une description du comportement cyclique des solides non visqueux', A paraitre dans le *Journal de Mécanique Théorique et appliquée*, 1983.
13. F. Darve and G. Lefebvre, 'Etude de la rupture des argiles structurées comme un problème de bifurcation', *Journée de Rhéologie* 1981, ENTPE, Lyon, 1981.
14. J. W. Rudnicki and J. R. Rice, 'Conditions for the localization of deformation in pressure-sensitive dilatant materials', *J. Mech. Phys. Solids*, **23**, 371–394 (1975).
15. I. Vardoulakis, 'Shear band inclination and shear modulus of sand in biaxial tests', *Int. J. Num. Anal. Methods in Geomech.*, no. 4, 103–119 (1980).
16. I. Vardoulakis, 'Bifurcation analysis of the triaxial test on sand samples', *Acta Mechanica*, no. 32, 35–54 (1979).
17. J. Desrues, 'Rupture surfaces in soils mechanics', *Colloque Euromech* 134, Copenhagen, 1980.
18. M. Boulon and W. Cichy, 'Modele numérique de rupture localisée dans les solides', *Colloque Franco-Polonais de Mécanique Non Lineaire*, Marseille, 1980.
19. G. Lefebvre, and P. La Rochelle, 'The analysis of two slope failures in cemented Champlain clays', *Revue Canadienne de Géotechnique*, **11**, (1), 89–108 (1974).
20. G. Lefebvre, 'Strength and slope stability in Canadian soft clay deposits', *Revue Canadienne de Géotechnique*, **18** (3) (1981).
21. F. Darve and I. Thanopoulos, 'Description of cyclic behaviour of sand by a non-linear incremental constitutive law', *Proceedings of the IUTAM Symposium*, Delft, 1982.
22. F. Darve, 'Une formulation incrémentale des lois rhéologiques. Application aux sols', These de Doctorat d'Etat, Grenoble, 1978.
23. F. Darve, J. Desrues, and M. Jacquet, 'Les surfaces de rupture en mecanique des sols en tant qu'instabilite de deformation', *Cahiers du Groupe Français de Rheologie*, **5**, (3), 93–106 (1980).
24. F. Darve and P. Vuaillat, 'A visco elasto-plastic law for clays and its use', *Proceedings of the IVth Int. Conf. on Num. Meth. in Geomechanics*, vol. 1, pp. 131–138, Edmonton, 1982.
25. F. Darve and S. Labanieh, 'An incremental non-linear constitutive law and cyclic behaviour of sands', *Proceedings of the Int. Symp. on Num. Models in Geomechanics*, Zürich, 1982.

Mechanics of Engineering Materials
Edited by C. S. Desai and R. H. Gallagher

Chapter 10

Some Simple Boundary Value Problems for Dilatant Soil in Undrained Conditions

R. O. Davis and G. Mullenger

10.1 INTRODUCTION

In this chapter we consider some simple problems concerning deformation of a rate-type material model. The model used is designed to represent soil. It can respond in a number of ways which are typical of real soil behaviour. In particular, it can either dilate or compact when subjected to shearing, depending upon whether it is in a dense or loose state, exactly as do the critical state plasticity models. In a loose state, it exhibits strain hardening; in a dense state, strain softening. The model has been developed as a rate-type counterpart to a particular critical state model.

The problems treated here all concern *undrained* deformation, which, in regard to a saturated soil, means incompressible. Some similarity between the results obtained here and those found in the finite theory of incompressible elastic materials will be evident. The actual material response will generally be quite different, however, since the rate-type material exhibits yielding and flow. At first glance it may appear that compaction or dilatation effects will not be evident since the deformations are incompressible, but this is not so. Instead, the undrained nature of the deformation leads to the development of pore pressures which depend on the dilatational nature of the material model. Dense materials will generally exhibit decreasing pore pressure; loose materials, increasing pore pressure. Like the indeterminate pressure found in the theory of incompressible elasticity, the pore pressure may generally only be determined by matching a specified boundary traction, but the change in pore pressure may be found.

We will treat five different problems. These are (i) homogeneous biaxial deformation such as is usually invoked to represent the conventional undrained triaxial test, (ii) simple shearing, (iii) the closely related problem of pure torsion, (iv) expansion of a cylindrical cavity, and (v) expansion of a spherical cavity. We consider only monotonic loading processes. In all cases analytic solutions are found for the effective stress state. Total stress and pore pressure may be determined in closed form in some cases but not in others. In line with common soil mechanics terminology we take stress and deformation to be positive in compression.

10.2 CONSTITUTIVE EQUATIONS

We consider materials which obey the following rate-type constitutive relation

$$\mathring{\mathbf{s}} = (g_1 \operatorname{tr} \mathbf{D} + g_2 \operatorname{tr} \mathbf{sD})\mathbf{I} + (g_4 \operatorname{tr} \mathbf{D} + g_5 \operatorname{tr} \mathbf{sD})\mathbf{s} + g_{10}\mathbf{D} \tag{10.1}$$

Here $\mathbf{s}$ denotes the effective stress tensor, $\mathbf{D}$ denotes the rate of deformation tensor, $g_1, \ldots, g_{10}$ are functions of invariants of stress and the material specific volume, and $\mathring{\mathbf{s}}$ denotes the co-rotational stress rate

$$\mathring{\mathbf{s}} = \dot{\mathbf{s}} + \mathbf{sW} - \mathbf{Ws} \tag{10.2}$$

where $\mathbf{W}$ is the spin tensor. Equation (10.1) is a special case of the general hypoelastic constitutive relationship proposed by Truesdell [1], and was used as the starting point for the soil constitutive models of Romano [2] and Davis and Mullenger [3, 4]. The stress $\mathbf{s}$ is the usual effective stress of soil mechanics. It is related to the total stress $\boldsymbol{\sigma}$, and the pore pressure u, by

$$\mathbf{s} = \boldsymbol{\sigma} - u\mathbf{I} \tag{10.3}$$

The specific volume, denoted v, represents the volume of soil per unit mass of solid particles, equal to the reciprocal of the equivalent dry density.

The model is fully defined once the functional forms of the coefficients $g_1, \ldots, g_{10}$ are specified. This will be accomplished in such a way that the model response must reflect the primary features of real soil response. To begin, we decompose the stress $\mathbf{s}$ into mean and deviatoric parts

$$p = \tfrac{1}{3} \operatorname{tr} \mathbf{s}, \qquad \mathbf{s}^* = \mathbf{s} - p\mathbf{I} \tag{10.4}$$

and let q denote the square root of the second invariant of $\mathbf{s}^*$

$$q^2 = \operatorname{tr} \mathbf{s}^*\mathbf{s}^* \tag{10.5}$$

Now let $\mathbf{t}$ and $\boldsymbol{\phi}$ be the vectors

$$\mathbf{t} = \frac{1}{2}\begin{bmatrix} p^2 \\ q^2 \end{bmatrix}, \qquad \boldsymbol{\phi} = \begin{bmatrix} p \operatorname{tr} \mathbf{D} \\ \operatorname{tr} \mathbf{s}^*\mathbf{D} \end{bmatrix} \tag{10.6}$$

The components of $\boldsymbol{\phi}$ represent the rates of working of the mean and deviatoric stresses. Equation (10.1) can now be represented by the following linear relationship between $\dot{\mathbf{t}}$ and $\boldsymbol{\phi}$

$$\dot{\mathbf{t}} = \mathbf{G}(\mathbf{t}, v)\boldsymbol{\phi} \tag{10.7}$$

where **G** is the square matrix with components

$$\begin{aligned} G_{11} &= g_1 + (g_2 + g_4)p + g_5 p^2 + \tfrac{1}{3} g_{10} \\ G_{12} &= g_2 p + g_5 p^2 \\ G_{21} &= (g_4 + g_5 p) q^2/p \\ G_{22} &= g_{10} + g_5 q^2 \end{aligned} \tag{10.8}$$

We assume p is never zero.

Rate-type constitutive representations do not explicitly employ the concept of a yield surface, but an implicit representation of yield is embedded in the constitutive equations. If we look upon yield as the condition in which deformations may grow without change in stress, then this condition occurs when the matrix **G** in (10.7) becomes singular [5, 6]. Thus we define the rate-type yield surface as the set of vectors **t** and the values of specific volume v which make **G** singular. Setting det $\mathbf{G} = 0$ we find that

$$q^2 = \alpha_1 + \alpha_2 p + \alpha_3 p^2 \tag{10.9}$$

where

$$\begin{aligned} \alpha_1 &= -g_{10}(g_1 + \tfrac{1}{3} g_{10})/h \\ \alpha_2 &= -g_{10}(g_2 + g_4)/h \\ \alpha_3 &= -g_5 g_{10}/h \\ h &= g_5(g_1 + \tfrac{1}{3} g_{10}) - g_2 g_4 \end{aligned} \tag{10.10}$$

These expressions define the implicit yield surface in terms of the as yet unknown coefficients $g_1, \ldots, g_{10}$.

Next we wish to specify the functions α_1, α_2, and α_3 in such a way that the yield surface (10.9) is a reasonable representation for granular media. We will, in fact, employ an elliptical yield surface exactly like that suggested by Roscoe and Burland [7]. To do so we introduce a function of specific volume called the critical state pressure, $p_c = p_c(v)$, and a failure stress $M = M_0 p_c$, where M_0 is a dimensionless constant. If the actual material stress state is such that $p = p_c$ and $q^2 = M^2$, then the soil is said to be at the critical state. The functions p_c and M are common to all critical state models. The yield surface is specified by setting

$$\begin{aligned} \alpha_1 &= 0 \\ \alpha_2 &= 2M_0^2 p_c \\ \alpha_3 &= M_0^2 \end{aligned} \tag{10.11}$$

so that (10.9) becomes

$$q^2 = M_0^2(2p_c p - p^2) \tag{10.12}$$

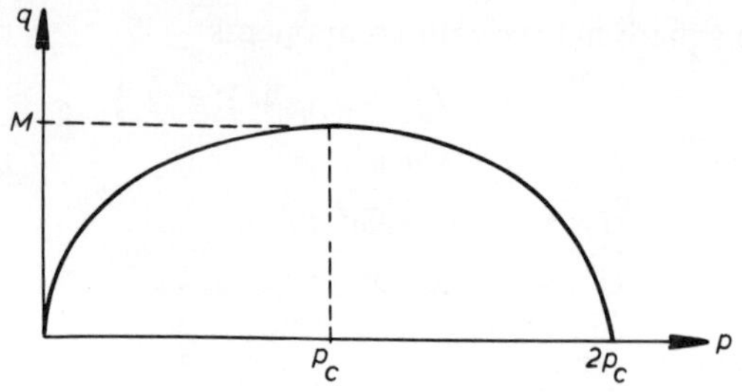

Figure 10.1 The yield surface

Geometrically, (10.12) represents an ellipse of revolution as illustrated in Figure 10.1.

Equations (10.11) now represent three relationships between the five unknown coefficients $g_1, \ldots, g_{10}$. Two more relationships are necessary before all the coefficients can be determined. One relationship is obtained by noting that the spherical and deviatoric response should uncouple if the material is at the critical state. That is, we require $G_{12} = G_{21} = 0$ in (10.8) whenever $p = p_c$. This is accomplished by setting

$$g_2 = g_4 = -g_5 p_c \tag{10.13}$$

Finally, we note that for small deformations from a stress free state, the coefficient g_{10} may be interpreted as twice the shear modulus of a linearly elastic material. Thus we set

$$g_{10} = 2\mu \tag{10.14}$$

where μ is a material constant with dimensions of stress called the elastic shear modulus.

Equations (10.13) and (10.14) supply two additional relationships between the coefficients. Using (10.10), (10.11), (10.13), and (10.14) we find

$$\begin{aligned} g_1 &= -2\mu/3 \\ g_2 &= g_4 = 2\mu p_c/M^2 \\ g_5 &= -2\mu/M^2 \end{aligned} \tag{10.15}$$

Then using (10.14) and (10.15) in (10.1) we obtain the following general constitutive equation

$$\mathring{\mathbf{s}} = \frac{2\mu}{M^2}[(-\tfrac{1}{3}M^2 \operatorname{tr} \mathbf{D} + p_c \operatorname{tr} \mathbf{sD})\mathbf{I} + (p_c \operatorname{tr} \mathbf{D} - \operatorname{tr} \mathbf{sD})\mathbf{s} + M^2\mathbf{D}] \tag{10.16}$$

This equation defines the effective stress response in loading processes.

In the remainder of this article we will consider the special case of undrained or incompressible deformation. Thus we now specialize (10.16) by setting

$$\operatorname{tr} \mathbf{D} = 0 \tag{10.17}$$

to have

$$\overset{\circ}{\mathbf{s}} = \frac{2\mu}{M^2}[(p_c\mathbf{1} - \mathbf{s})\operatorname{tr}\mathbf{sD} + M^2\mathbf{D}] \tag{10.18}$$

It will be convenient to recast (10.18) in dimensionless form. This is accomplished by defining a dimensionless effective stress $\mathbf{S}$ according to

$$\mathbf{S} = \frac{\sqrt{2}}{M}(\mathbf{s} - p_c\mathbf{1}) \tag{10.19}$$

Then we have, using (10.17)

$$\operatorname{tr}\mathbf{SD} = \frac{\sqrt{2}}{M}\operatorname{tr}\mathbf{sD}$$

and (10.18) becomes

$$\overset{\circ}{\mathbf{S}} = \beta[-\mathbf{S}(\operatorname{tr}\mathbf{SD}) + 2\mathbf{D}] \tag{10.20}$$

where $\dot{p}_c$ vanishes due to incompressibility ($v = \text{constant}$) and β is given by

$$\beta = \sqrt{2}\mu/M \tag{10.21}$$

In the following sections we will use (10.20) to determine the effective stress response of the rate-type material.

10.3 FORMULATION OF PROBLEMS

We consider only the case of static equilibrium, in the absence of body forces, so that the total stress $\boldsymbol{\sigma}$ must obey

$$\operatorname{div}\boldsymbol{\sigma} = 0 \tag{10.22}$$

Using (10.3) this implies that

$$\operatorname{div}\mathbf{s} + \operatorname{grad}u = 0 \tag{10.23}$$

In the first two problems treated below, the stress field is homogeneous and these equations are trivially satisfied.

In all the problems we assume the stress field is initially homogeneous and hydrostatic.

10.4 BIAXIAL DEFORMATION

We consider a homogeneous deformation with

$$\mathbf{D} = \begin{bmatrix} \dot{\zeta} & 0 & 0 \\ 0 & -\frac{1}{2}\dot{\zeta} & 0 \\ 0 & 0 & -\frac{1}{2}\dot{\zeta} \end{bmatrix}, \qquad \mathbf{W} = 0 \tag{10.24}$$

and the associated stress fields

$$\boldsymbol{\sigma} = \begin{bmatrix} \sigma_{ZZ} & 0 & 0 \\ 0 & \sigma_{rr} & 0 \\ 0 & 0 & \sigma_{rr} \end{bmatrix}, \qquad \mathbf{s} = \begin{bmatrix} s_{ZZ} & 0 & 0 \\ 0 & s_{rr} & 0 \\ 0 & 0 & s_{rr} \end{bmatrix} \tag{10.25}$$

Here ζ represents to the natural compressive axial strain

$$\zeta = -\ln(L/L_0) \tag{10.26}$$

where L and L_0 are axial lengths in the deformed and undeformed configurations. Using (10.24) and (10.25) the constitutive equation (10.20) can be reduced to two scalar equations

$$\begin{aligned} \dot{S}_{rr} &= \beta[-S_{rr}(S_{ZZ} - S_{rr}) - 1]\dot{\zeta} \\ \dot{S}_{ZZ} &= \beta[-S_{ZZ}(S_{ZZ} - S_{rr}) + 2]\dot{\zeta} \end{aligned} \tag{10.27}$$

and these equations may be integrated to yield

$$\begin{aligned} S_{rr} &= S_0 \operatorname{sech} \sqrt{(3)}\beta\zeta - \frac{1}{\sqrt{3}} \tanh \sqrt{(3)}\beta\zeta \\ S_{ZZ} &= S_0 \operatorname{sech} \sqrt{(3)}\beta\zeta + \frac{2}{\sqrt{3}} \tanh \sqrt{(3)}\beta\zeta \end{aligned} \tag{10.28}$$

We have assumed that at $\zeta = 0$ the initial condition

$$S_{rr} = S_{ZZ} = S_0 = \sqrt{2}(s_{rr_0} - p_c)/M \tag{10.29}$$

applies. Here s_{rr_0} is the initial confining stress.

To proceed further we must have a boundary condition for total stress. In a conventional triaxial test, for example, σ_{rr} is held constant. From the effective stress principle, we have, at the initiation of deformation

$$\sigma_{rr} = s_{rr_0} + u_0 = \text{constant} \tag{10.30}$$

where u_0 is the initial pore pressure. At any other stage of the deformation the pore pressure is

$$u = \sigma_{rr} - s_{rr} = M(S_0 - S_{rr})/\sqrt{(2)} + u_0 \tag{10.31}$$

Thus if we define the dimensionless pore pressure increase by

$$U = \sqrt{2}(u - u_0)/M \tag{10.32}$$

we have from (10.31) and(10.28)

$$U = S_0(1 - \operatorname{sech} \sqrt{(3)}\beta\zeta) + \frac{1}{\sqrt{3}} \tanh \sqrt{(3)}\beta\zeta \tag{10.33}$$

The total axial stress follows easily from (10.28).

$$\sigma_{ZZ} = \sigma_{rr} + \sqrt{(\tfrac{3}{2})}M \tanh \sqrt{(3)}\beta\zeta \tag{10.34}$$

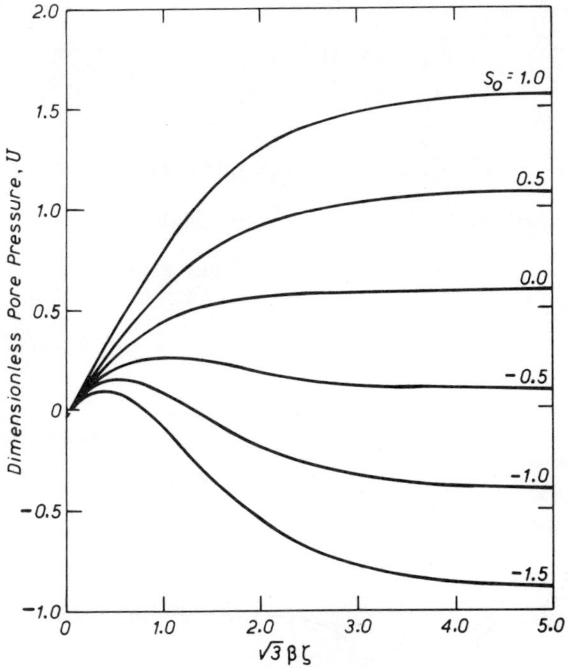

Figure 10.2 Pore pressure response in biaxial compresion

In Figure 10.2, graphs of U versus ζ for various values of S_0 are shown. Note that S_0 defines the initial stress state with reference to the critical state pressure. It may be positive or negative depending upon whether the soil is loose or dense. From (10.33) it is clear that if $S_0 < -1/\sqrt{3}$, then U will ultimately become negative.

10.5 SIMPLE SHEAR

Next consider the homogeneous deformation with

$$\mathbf{D} = \begin{bmatrix} 0 & \dot{\chi} & 0 \\ \dot{\chi} & 0 & 0 \\ 0 & 0 & 0 \end{bmatrix}, \qquad \mathbf{W} = \begin{bmatrix} 0 & \dot{\chi} & 0 \\ -\dot{\chi} & 0 & 0 \\ 0 & 0 & 0 \end{bmatrix} \tag{10.35}$$

where $\dot{\chi}$, the shearing rate, depends only upon time. The stress fields are given by

$$\boldsymbol{\sigma} = \begin{bmatrix} \sigma_{xx} & \sigma_{xy} & 0 \\ \sigma_{yx} & \sigma_{yy} & 0 \\ 0 & 0 & \sigma_{zz} \end{bmatrix}, \qquad \mathbf{s} = \begin{bmatrix} s_{xx} & s_{xy} & 0 \\ s_{yx} & s_{yy} & 0 \\ 0 & 0 & s_{zz} \end{bmatrix} \tag{10.36}$$

and $\sigma_{xy} = \sigma_{yx}$, $s_{xy} = s_{yx}$. Here, the constitutive equation (10.20) reduces to four scalar equations

$$\begin{aligned}\dot{S}_{xx} &= 2S_{xy}(1-\beta S_{xx})\dot{\chi}\\ \dot{S}_{yy} &= S_{xy}(-1-\beta S_{yy})\dot{\chi}\\ \dot{S}_{ZZ} &= -2\beta S_{xy}S_{ZZ}\dot{\chi}\\ \dot{S}_{xy} &= [S_{yy}-S_{xx}+2\beta(1-S_{xy}^2)]\dot{\chi}\end{aligned} \tag{10.37}$$

If we use the initial condition that S_{xy} is zero, while

$$S_{xx}=S_{yy}=S_{ZZ}=S_0 \quad \text{at} \quad \chi=0 \tag{10.38}$$

then (10.37) may be integrated to give

$$\begin{aligned}S_{xx} &= [(\beta S_0-1)\psi+1]/\beta\\ S_{yy} &= [(\beta S_0+1)\psi-1]/\beta\\ S_{ZZ} &= S_0\psi\\ S_{xy} &= [\beta^2-1+2\psi-(\beta^2+1)\psi^2]^{1/2}/\beta\end{aligned} \tag{10.39}$$

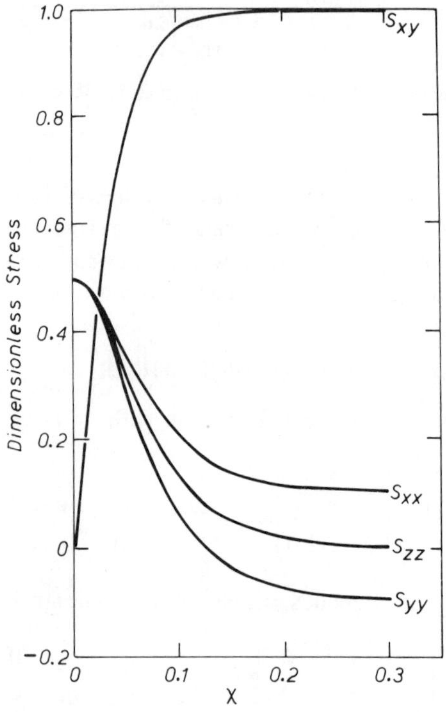

Figure 10.3 Simple shear, $\beta=10$, $S_0=0.5$

where

$$\psi = \frac{\beta^2 - 1}{\beta^2 \cosh 2\sqrt{(\beta^2 - 1)}\chi - 1} \tag{10.40}$$

Solutions with initial stress states which are non-hydrostatic are slightly more complex. Typical response with S_0 equal to 0.5 and β equal to 10 is illustrated in Figure 10.3. These values would correspond to a soft, lightly over-consolidated clay. The positive value for S_0 indicates that the material is loose rather than dense and the decreasing effective normal stresses confirm this. If one component of total normal stress were held constant, then clearly a positive pore pressure would be generated. Note, however, that the three effective normal stresses all take on different values, making clear that *only* one component of total normal stress may be constant in this deformation.

The ultimate or failure values of the dimensionless effective stresses occur when the left-hand sides of (10.37) approach zero. Thus we see that for large values of χ the stresses become

$$S_{xx} \to 1/\beta, \qquad S_{yy} \to -1/\beta, \qquad S_{ZZ} \to 0, \qquad S_{xy} \to \sqrt{(1 - 1/\beta^2)} \tag{10.41}$$

Note that the intermediate effective normal stress S_{ZZ} will equal the average of S_{xx} and S_{yy}.

10.6 PURE TORSION

Consider a cylindrical coordinate system r, θ, z in which the deformation of pure torsion is described by

$$\mathbf{D} = \begin{bmatrix} 0 & 0 & 0 \\ 0 & 0 & \dot{\chi}r \\ 0 & \dot{\chi}r & 0 \end{bmatrix}, \qquad \mathbf{W} = \begin{bmatrix} 0 & -2\dot{\chi}Z & 0 \\ 2\dot{\chi}Z & 0 & \dot{\chi}r \\ 0 & -\dot{\chi}r & 0 \end{bmatrix} \tag{10.42}$$

Without loss of generality we can consider the plane $Z = 0$, so that the rotation components $W_{r\theta}$ and $W_{\theta r}$ vanish. Then this deformation is locally equivalent to a simple shear of amount $\dot{\chi}r$. The analysis of Section 10.5 applies with χ replaced by χr and the following equivalences between components of dimensionless stress

$$S_{xx} \leftrightarrow S_{\theta\theta}, \qquad S_{yy} \leftrightarrow S_{ZZ}, \qquad S_{ZZ} \leftrightarrow S_{rr}, \qquad S_{xy} \leftrightarrow S_{\theta Z} \tag{10.43}$$

In each of these, the left-hand component refers to simple shear, the right-hand to pure torsion. The effective stresses may now be taken directly from (10.39).

The equivalence between simple shear and pure torsion is not complete, however, for the torsion stress field is not homogeneous and hence equilibrium must be satisfied. The only equilibrium equation not trivially satisfied is

$$\frac{\partial \sigma_{rr}}{\partial r} + \frac{1}{r}(\sigma_{rr} - \sigma_{\theta\theta}) = 0 \tag{10.44}$$

Note from (10.3) and (10.19) that

$$\sigma_{rr} - \sigma_{\theta\theta} = s_{rr} - s_{\theta\theta} = \frac{M}{\sqrt{2}}(S_{rr} - S_{\theta\theta}) \tag{10.45}$$

and from (10.39) and (10.42),

$$S_{rr} - S_{\theta\theta} = (\psi - 1)/\beta \tag{10.46}$$

Combining the last three equations and integrating we find

$$\sigma_{rr}(r,\chi) = \sigma_{rr}(r_0,\chi) + \frac{M}{\sqrt{2}}\int_{r_0}^{r} \frac{1-\psi}{\beta\rho}\,\mathrm{d}\rho \tag{10.47}$$

Here ψ is given by (10.40) with χ replaced by $\chi\rho$. In general, (10.47) must be evaluated numerically. Note that the total radial stress σ_{rr} may be made to vanish at any single radius r_0, but it will not in general vanish at two radii, such as the inner and outer surfaces of a cylinder.

If (10.47) is evaluated to give σ_{rr}, then the pore pressure follows from

$$u(r,\chi) = \sigma_{rr}(r,\chi) - s_{rr}(r,\chi) \tag{10.48}$$

The remaining components of total stress are then fully determined by

$$\sigma_{\theta\theta} = s_{\theta\theta} + u, \qquad \sigma_{ZZ} = s_{ZZ} + u \tag{10.49}$$

10.7 EXPANSION OF A CYLINDRICAL CAVITY

We retain the cylindrical geometry from the preceding section, but now consider the deformation field

$$\mathbf{D} = \begin{bmatrix} \dot{\xi} & 0 & 0 \\ 0 & -\dot{\xi} & 0 \\ 0 & 0 & 0 \end{bmatrix}, \qquad \mathbf{W} = \mathbf{0} \tag{10.50}$$

where ξ is given by

$$\xi = \xi(r,t) = -\tfrac{1}{2}\ln\left(1 - \frac{a^2 - a_0^2}{r^2}\right) \tag{10.51}$$

This deformation corresponds to an infinite medium containing a cylindrical cavity of initial radius a_0 which has been expanded to radius $a = a(t)$. The quantity ξ represents the natural compressive radial strain, the decrease in length per unit length in the deformed configuration. The corresponding total and effective stress fields are

$$\mathbf{s} = \begin{bmatrix} s_{rr} & 0 & 0 \\ 0 & s_{\theta\theta} & 0 \\ 0 & 0 & s_{ZZ} \end{bmatrix}, \qquad \boldsymbol{\sigma} = \begin{bmatrix} \sigma_{rr} & 0 & 0 \\ 0 & \sigma_{\theta\theta} & 0 \\ 0 & 0 & \sigma_{ZZ} \end{bmatrix} \tag{10.52}$$

The constitutive equations (10.20) now reduce to three scalar equations

$$\begin{aligned} \dot{S}_{rr} &= \beta[2 - S_{rr}(S_{rr} - S_{\theta\theta})]\dot{\xi} \\ \dot{S}_{\theta\theta} &= \beta[-2 - S_{\theta\theta}(S_{rr} - S_{\theta\theta})]\dot{\xi} \\ \dot{S}_{ZZ} &= -\beta S_{ZZ}(S_{rr} - S_{\theta\theta})\dot{\xi} \end{aligned} \tag{10.53}$$

Integrating these we obtain

$$\begin{aligned} S_{rr} &= S_0 \operatorname{sech} 2\beta\xi + \tanh 2\beta\xi \\ S_{\theta\theta} &= S_0 \operatorname{sech} 2\beta\xi - \tanh 2\beta\xi \\ S_{ZZ} &= S_0 \operatorname{sech} 2\beta\xi \end{aligned} \tag{10.54}$$

where we have assumed $S_{rr} = S_{\theta\theta} = S_{ZZ} = S_0$ when $\xi = 0$.

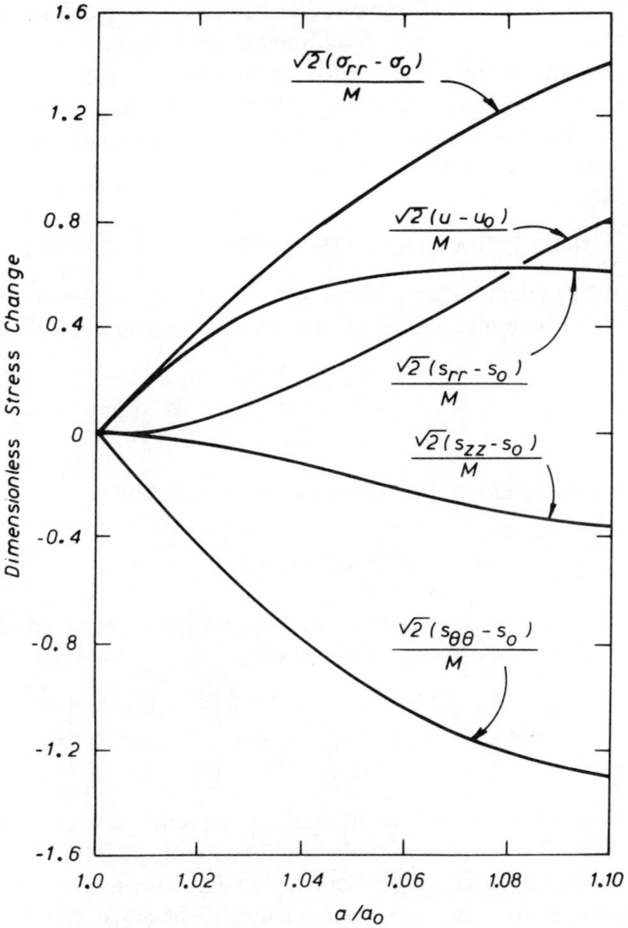

Figure 10.4 Cylindrical cavity expansion, $\beta = 10$, $S_0 = 0.5$

Once again the stress field is non-homogeneous and equilibrium must be considered. As in the case of pure torsion, the only equilibrium equation not trivially satisfied is (10.44). Then combining (10.3), (10.19), (10.44), and (10.54) we have

$$\frac{\partial \sigma_{rr}}{\partial r} + \frac{\sqrt{(2)}M}{r} \tanh 2\beta\xi = 0 \tag{10.55}$$

If we let σ_0 denote the value of σ_{rr} as $r \to \infty$, then

$$\sigma_{rr}(r) = \sigma_0 - \sqrt{(2)}M \int_{\infty}^{r} \tanh 2\beta\xi \frac{\mathrm{d}\rho}{\rho} \tag{10.56}$$

gives the total radial stress at any radius r. The pore pressure follows from the difference of the total and effective radial stresses. The remaining components of total stress are then obtained by adding the pore pressure to the appropriate effective stress. In general, (10.56) must be integrated numerically.

In Figure 10.4 we illustrate this problem by plotting stress increases or decreases evaluated at the cavity wall as the cavity expands. All stresses were calculated with $\beta = 10$ and $S_0 = 0.5$.

10.8 EXPANSION OF A SPHERICAL CAVITY

For our final problem we consider a spherical cavity expanding in a infinite body. The deformation is described in spherical coordinates by

$$\mathbf{D} = \begin{bmatrix} \dot{\varepsilon} & 0 & 0 \\ 0 & -\frac{1}{2}\dot{\varepsilon} & 0 \\ 0 & 0 & -\frac{1}{2}\dot{\varepsilon} \end{bmatrix}, \qquad \mathbf{W} = 0 \tag{10.57}$$

where ε represents the natural compressive radial strain

$$\varepsilon = \varepsilon(r, t) = -\tfrac{1}{3}\ln\left[1 - \frac{a^3 - a_0^3}{r^3}\right] \tag{10.58}$$

and a_0 and $a = a(t)$ denote the undeformed and deformed cavity radii respectively. The associated stress fields are

$$\boldsymbol{\sigma} = \begin{bmatrix} \sigma_{rr} & 0 & 0 \\ 0 & \sigma_{\theta\theta} & 0 \\ 0 & 0 & \sigma_{\theta\theta} \end{bmatrix}, \qquad \mathbf{s} = \begin{bmatrix} s_{rr} & 0 & 0 \\ 0 & s_{\theta\theta} & 0 \\ 0 & 0 & s_{\theta\theta} \end{bmatrix} \tag{10.59}$$

Note that the azimuthal and meridional components of stress must be equal for equilibrium. Comparison of these equations with (10.24) and (10.25) shows that this problem is locally equivalent to biaxial deformation with ζ replaced by ε. Effective stress response may be taken directly from (10.28) with S_{rr} replaced by $S_{\theta\theta}$ and S_{ZZ} replaced by S_{rr}.

Equilibrium requires that

$$\frac{\partial \sigma_{rr}}{\partial r}+\frac{2}{r}(\sigma_{rr}-\sigma_{\theta\theta})=0 \tag{10.60}$$

Thus from the effective stress principle and (10.28) we find that

$$\sigma_{rr}=\sigma_0-\sqrt{(6)}M\int_{\infty}^{r}\tanh\sqrt{(3)}\beta\varepsilon\frac{\mathrm{d}\rho}{\rho} \tag{10.61}$$

where σ_0 denotes the value of σ_{rr} as $r\to\infty$. Pore pressure and the circumferential component of total stress follow immediately from

$$u=\sigma_{rr}-s_{rr},\qquad \sigma_{\theta\theta}=s_{\theta\theta}+u \tag{10.62}$$

10.9 DISCUSSION

We first note that it is the undrained or incompressible character of the problems considered here which in general allows the determination of analytical solutions. In 1972 Palmer and Mitchell [8] noted that the possibility of using critical state soil mechanics to solve boundary value problems had received little attention, despite the well developed plasticity theories then extant. Since then computer solutions to a variety of problems have appeared, but there still exists a paucity of analytical solutions. This is largely due to the complexity of critical state models when volumetric deformations are allowed. Undrained conditions, on the other hand, offer great simplification without masking the essential complex shearing character of the material model.

The deformations considered above are practically the simplest one might imagine. Nevertheless, the variety of material response is very large. We have illustrated solutions in only a few cases and then only for very special initial conditions. Non-isotropic initial stress states lead to considerably more complex behaviour, even though analytic solutions may still be obtained.

Use of the dimensionless effective stress definition (10.19) also leads to simplification. The non-dimensionalizing factor $M/\sqrt{(2)}$ is approximately equal to the undrained shear strength c_u in simple shear. In fact, combining (10.19), (10.21) and the last of (10.41) we see that the actual shear strength in simple shear is

$$c_u=\lim_{\chi\to\infty}s_{xy}=\frac{M}{\sqrt{(2)}}\sqrt{\left(1-\frac{M^2}{2\mu^2}\right)} \tag{10.63}$$

The term $M^2/2\mu^2$ results from the finite nature of the deformation. It will generally be small and the conventional representation $c_u=M\sqrt{(2)}$ is very nearly correct.

Finally, we note that all of the solutions given above properly account for finite deformation. There are no restrictions to small strains or rotations. For the cases of cavity expansion, the initial cavity radius may be set equal to zero, resulting in infinite strains at the cavity wall.

REFERENCES

1. C. Truesdell, 'Hypoelasticity', *J. Rational Mech. Anal.*, **4**, 83–133 (1955).
2. M. Romano, 'A continuum theory for granular media with a critical state', *Arch. Mech.*, **26**, 1011–1028 (1974).
3. R. O. Davis and G. Mullenger, 'A rate-type constitutive model for soil with a critical state', *Int. J. Numer. Anal. Methods Geomech.*, **2**, 255–282 (1978).
4. R. O. Davis and G. Mullenger, 'A simple rate-type constitutive representation for granular media', in *Numerical Methods in Geomechanics, Aachen* (Ed. W. Wittke), Vol. 1, pp. 415–421, Balkema, Rotterdam, 1979.
5. T. Tokuoka, 'Yield conditions and flow rules derived from hypoelasticity', *Arch. Rat. Mech. Anal.*, **42**, 239–252 (1971).
6. R. O. Davis and G. Mullenger, 'Derived failure criteria for granular media', *Int. J. Numer. Anal. Methods Geomech.*, **3**, 279–283 (1979).
7. K. H. Roscoe and J. B. Burland, 'On generalized stress–strain behaviour of wet clay', in *Engineering Plasticity* (Eds. J. Heyman and F. A. Leckie), pp. 535–609, Cambridge Univ. Press, 1968.
8. A. C. Palmer and R. J. Mitchell, 'Plane-strain expansion of a cylindrical cavity in clay', in *Stress Strain Behaviour of Soils* (Ed. R. H. G. Parry), pp. 588–599, Foulis, Henley-on-Thames, 1972.

Mechanics of Engineering Materials
Edited by C. S. Desai and R. H. Gallagher

Chapter 11

A Generalized Basis for Modelling Plastic Behaviour of Materials

C. S. Desai and M. O. Faruque

11.1 INTRODUCTION

A procedure for development of yield and plastic potential functions in the context of plasticity was proposed previously by Desai [1]. It was based on expressing the yield function as a complete polynomial in $J_1, J_2^{1/2}$, and $J_3^{1/3}$ or $J_1, J_{2D}^{1/2}$, and $J_{3D}^{1/3}$, where $J_i (i = 1, 2, 3)$ and $J_{iD} (i = 2, 3)$ are invariants of the stress and deviatoric stress tensors, respectively. Truncated forms of the complete polynomial were shown to yield various conventional and recent models used in plasticity. This chapter presents a brief review of this previous work as well as details of further development of one of the truncated forms for inclusion of hardening, non-associative behaviour, and induced anisotropy.

The constitutive function in relation to plasticity is expressed as

$$F(\sigma_{ij}, \varepsilon_{ij}^{p}, W^{p}, \alpha_m) = 0 \tag{11.1a}$$

where σ_{ij} is the stress tensor, ε_{ij}^{p} is the plastic strain tensor, W^{p} is the plastic work, and α_m $(m = 1, 2, \ldots, n)$ are other internal state variables. For initially isotropic material, equation (11.1a) can be written as

$$F(J_i, I_i^{p}, K_j, \alpha_m) = 0 \tag{11.1b}$$

where I_i^{p} $(i = 1, 2, 3)$ are invariants of the plastic strain tensor and K_j $(j = 1, 2, 3, 4)$ are joint invariants of stress and plastic strain tensor [2 to 4].

$$\begin{aligned} K_1 &= \sigma_{ij}\varepsilon_{ij}^{p} \\ K_2 &= \sigma_{ij}\varepsilon_{jk}^{p}\varepsilon_{ki}^{p} \\ K_3 &= \sigma_{ij}\sigma_{jk}\varepsilon_{ki}^{p} \\ K_4 &= \sigma_{ij}\sigma_{jk}\varepsilon_{kl}^{p}\varepsilon_{li}^{p} \end{aligned} \tag{11.1c}$$

A special form of equation (11.1b) for initially isotropic as well as isotropic during plastic straining can be expressed as a general complete polynomial in

$J_1, J_2^{1/2}$, and $J_3^{1/3}$. Consider such a polynomial up to third order:

$$\begin{aligned}F(J_1, J_2, J_3) = {} & a_0 + a_1 J_1 + a_2 J_2^{1/2} + a_3 J_3^{1/3} + a_4 J_1^2 \\ & + a_5 J_1 J_2^{1/2} + a_6 J_2 + a_7 J_2^{1/2} J_3^{1/3} \\ & + a_8 J_3^{2/3} + a_9 J_1 J_3^{1/3} + a_{10} J_1^3 \\ & + a_{11} J_1^2 J_2^{1/2} + a_{12} J_1 J_2 + a_{13} (J_2^{1/2})^3 \\ & + a_{14} J_2 J_3^{1/3} + a_{15} J_2^{1/2} J_3^{2/3} + a_{16} J_3 \\ & + a_{17} J_1 J_3^{2/3} + a_{18} J_1^2 J_3^{1/3} + a_{19} J_1 J_2^{1/2} J_3^{1/3}\end{aligned} \tag{11.2}$$

where a_i $(i = 0, 1, 2 \ldots) = a_i(I_i^{\mathrm{p}}, W^{\mathrm{p}}, \alpha_m)$ are parameters to be defined from appropriate test data. Here $J_1 = \sigma_{ii}$, $J_2 = \frac{1}{2}\sigma_{ij}\sigma_{ji}$ and $J_3 = \frac{1}{3}\sigma_{ij}\sigma_{jk}\sigma_{ki}$. The foregoing function can also be expressed in terms of $J_1, J_{2\mathrm{D}}$, and $J_{3\mathrm{D}}$. In this paper compressive stresses are taken as positive.

It was shown previously [1] that F in equation (11.2) contains, as special cases, various plasticity models such as von Mises and Drucker–Prager, and those evolved recently by Lade [5] and Matusoka and Nakai [6] based on test results for specific materials. In a paper in the proceedings of this conference, Schreyer and Babcock [7] have modified Lade's approach to formalize a continuous function for behaviour of concrete.

11.1.1 Comment

The difference between the procedure proposed in this study and previous works is that the present procedure starts from a general polynomial expansion and permits development of one or more appropriate models including the previous ones for any material. It was also shown that many of the foregoing models contained in the function, equation (11.2), can be derived from the singularity condition of the stress–strain matrices of various orders of hypoelastic models [1, 8, 9].

It is possible to find one or more truncated forms of equation (11.2) for a given material based on observed behaviour. Often, it may be possible to discard some of the terms at the outset; for instance, in the case of pressure insensitive materials, the terms with J_1 can be dropped.

Desai [1] choose a number of truncated forms of equation (11.2) to describe the total stress–strain response. He then examined their values at ultimate yield for materials such as Ottawa sand and artificial soil. Here ultimate yield corresponds to the asymptotic value(s) of stresses to the final region of the stress–strain response. Hence, 'failure' and 'critical state' may occur before the ultimate yield condition. It was found that they assumed nearly invariant values at the ultimate condition; however, the degree of invariance varied. One of the functions

$$a_6 J_2 + a_9 J_1 J_3^{1/3} = 0 \tag{11.3a}$$

showed the best degree of invariance compared to the other chosen functions. Its value near ultimate condition was found to be around 0.55 for a series of laboratory tests for a variety of stress paths followed. This value was found for the above two materials as well as other materials such as a medium dense sand, Florida Zircon sand, glass beads, and a sub-ballast. This function was found to possess the properties of convexity as well as continuity in the $\sqrt{(J_{2D})} - J_1$ space (see subsequent description and Figure 11.1). The latter property was considered important because it can provide a useful alternative to the currently used two surface models such as critical state and cap [10, 11]. In view of these, the function in equation (11.3a) was chosen for further study performed previously [12, 13] and is presented here in. It is, however, believed that there may exist a number of other such functions that can also be suitable for given material(s) and can be identified through process of optimization in conjunction with (laboratory) observations.

11.2 DETAILS OF FUNCTION

The function in equation (11.3a) is expressed in terms of J_1, J_{2D}, and J_3 as

$$F = J_{2D} + \alpha J_1^2 - \beta J_1 J_3^{1/3} - \gamma J_1 - k^2 = 0 \qquad (11.3b)$$

where α, β, γ, k are material parameters; k has meaning similar to cohesive strength, and β is called the hardening or growth function, which is expressed as

$$\beta = \beta(\xi) \qquad (11.4a)$$

where $\xi = \int (d\varepsilon_{ij}^p d\varepsilon_{ij}^p)^{1/2}$ is the trajectory of plastic strain vector and is called the hardening parameter. A form of β chosen herein, which was guided by experimental hardening behaviour of the silty sand considered subsequently, is given by

$$\beta = \beta_u \left(1 - \frac{\beta_a}{\xi^\eta}\right) \qquad (11.4b)$$

Eauation (11.4b) indicates $\beta = -\infty$ for $\xi = 0$. From a computational viewpoint, it was found that the predictions are not affected significantly by the magnitude of the initial value of β. In other words, an initial value of β less than about -100 can be used; in this study (for silty sand) a value equal to -1000 was used.

Equation (11.4b) considers the trajectory of the total plastic strain vector. If it is required to examine volumetric and deviatoric components separately, an alternative to equation (11.4b) can be defined in terms of two hardening parameters, ξ_v and ξ_D, where ξ_v and ξ_D, are the volumetric and the deviatoric parts

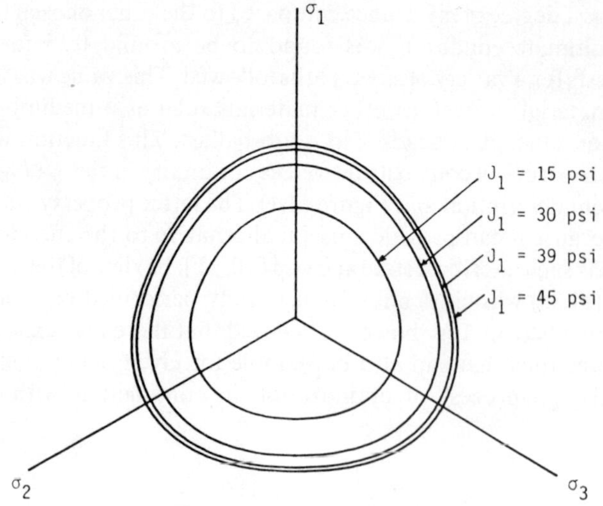

σ_1
J_1 = 15 psi
J_1 = 30 psi
J_1 = 39 psi
J_1 = 45 psi
σ_2
σ_3

(a)

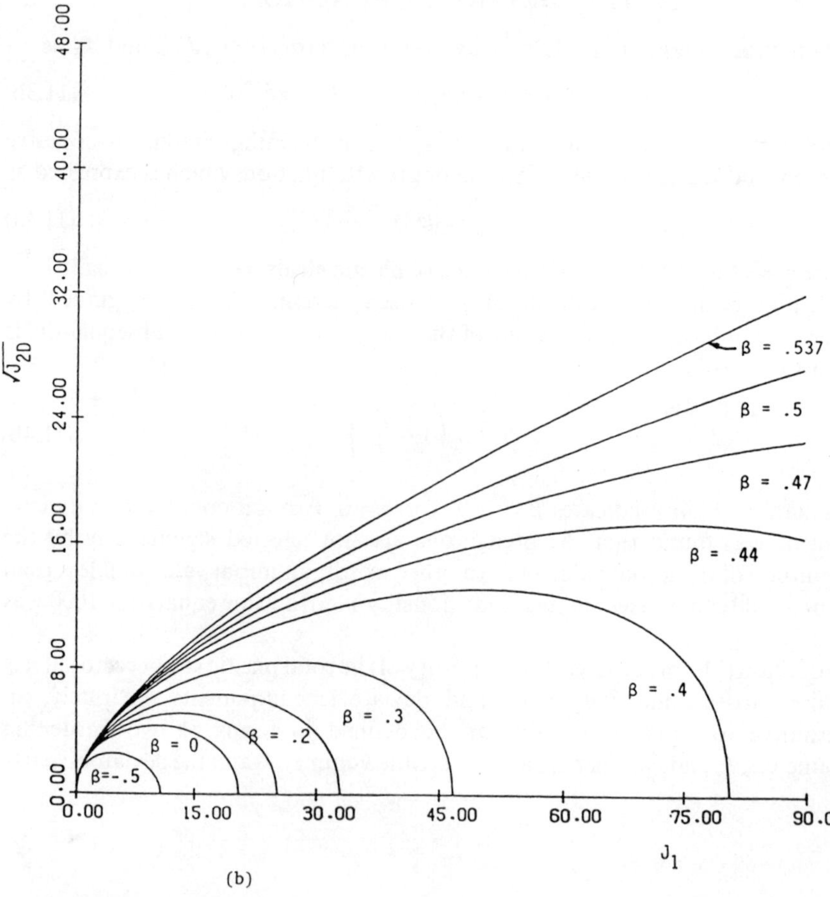

48.00
40.00
32.00
24.00
16.00
8.00
0.00
$\sqrt{J_{2D}}$
β = .537
β = .5
β = .47
β = .44
β = .4
β = .3
β = .2
β = 0
β=-.5
0.00
15.00
30.00
45.00
60.00
75.00
90.00
J_1

(b)

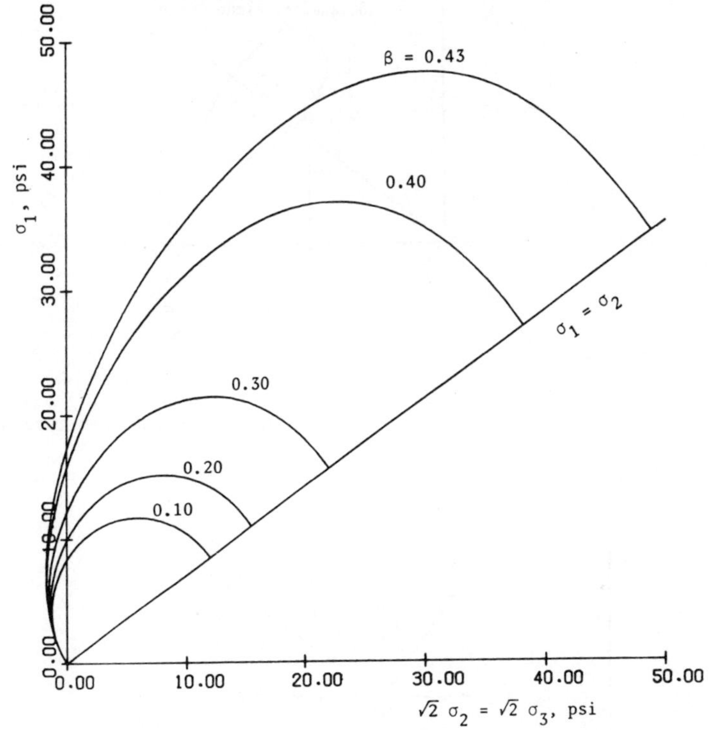

(c)

Figure 11.1 Plots of equation (11.3(b)) on octahedral, $\sqrt{(J_{2D})} - J_1$, and triaxial planes. (a) On octahedral planes: 1 p.s.i. = 6.89 kPa. (b) On $\sqrt{(J_{2D})} - J_1$ plane for given values of J_{3D}. (c) On triaxial stress space

of ξ, respectively. For example,

$$\beta = \beta_u \left[1 - \frac{1}{\bar{\beta}_a \cdot \xi_v^{\eta_1} (1 + \beta_b \cdot \xi_D^{\eta_2})} \right] \tag{11.4c}$$

where $\bar{\beta}_a, \eta_1, \beta_1$ and η_2 are the material constants associated with hardening.

Alternative forms of β expressed as functions of the ratio of trajectory of deviatoric plastic strain to that of total plastic strain can provide improved representation of the growth function. It is possible to incorporate softening behaviour by treating the growth function as the softening function beyond the peak state of stress.

The relation between ξ_v, ξ_D, and ξ is given by

$$\xi = [\mathrm{d}\xi_v^2 + \mathrm{d}\xi_D^2]^{1/2} \tag{11.4d}$$

where ξ_v and ξ_D are volumetric and deviatoric plastic strains, respectively.

As indicated before, the function plots convex and continuous in the stress space. Figures 11.1(a) and (b) show such plots in octahedral planes with different

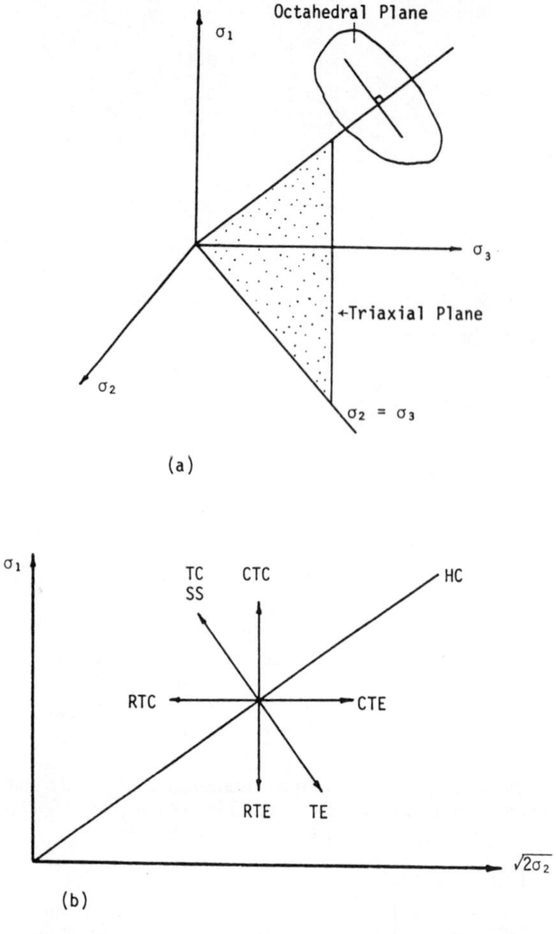

Figure 11.2 Commonly used stress paths. (a) Principal stress space. (b) Projections of stress paths on triaxial plane

values of J_1 and in the $\sqrt{(J_{2D})} - J_1$ space, respectively. The latter shows that the hardening behaviour is defined by a single continuous function. The curves in Figure 11.1(b) are orthogonal to the J_1-axis. For $\beta = 0$, the functions assume elliptical shape and for $J_{3D} = 0$, they are circular in the octahedral plane.

Figure 11.1(c) shows plots of the function on the triaxial stress space. The initial region in these plots indicates that there can occur tensile σ_2 and/or σ_3; however, the mean pressure remains compressive. This may be consistent with a number of materials following shear and extension stress paths. Figure 11.2 (at low confining pressures) that travel to this region. If a material does not experience tensile stresses under above conditions, additional constraints may be required in the model.

11.2.1 Ranges of β

The function β can grow with values ranging from $-\infty$ to 3α. For very large values of $-\beta$, the function will tend to be a singular surface at the origin. The value of β at ultimate yield can be defined from the intersection of the function with the J_1-axis:

$$J_1 = \frac{3\gamma}{2(3\alpha-\beta)} + \sqrt{\left(\left[\frac{3\gamma}{2(3\alpha-\beta)}\right]^2 + \frac{3k^2}{(3\alpha-\beta)}\right)} \tag{11.5}$$

This indicates that for $\beta = 3\alpha$, the function will intersect the J_1-axis at infinity; for the values of β greater than 3α, the function may not be convex. As will be seen subsequently, this is consistent with the value of β at ultimate (for a silty sand) to be about 0.55, which is equal to 3α. For the yield surface to intersect on the positive side of J_1-axis (compression positive), it is required that $\beta_u \leq 3\alpha$. For finite values of k, the surface will intersect the J_1-axis on the negative side showing tensile strength.

11.3 PARAMETERS

For the hardening behaviour, the foregoing function with β as in equation 11.4(b) can be defined with five parameters $\alpha, \gamma, k, \beta_a$, and η. With two elastic parameters, Young's modulus E and Poisson's ratio, ν, the total number of parameters is seven. The elastic parameters can be obtained from unloading behaviour in hydrostatic compression (HC) and conventional triaxial compression (CTC) tests, Figure 11.2. Determination of the plasticity parameters is illustrated by using laboratory test data for a silty sand. The silty sand is well graded, has a specific gravity of 2.59 and the optimum water content of 9 per cent. Average density of the samples used for testing was about 2 gm/cc. Further details of this sand and of testing are available in ref. [14]. Figures 11.3(a) to 11.3(g) show typical test results for the soil obtained by using a truly triaxial device [15, 16] under various stress paths, Figure 11.2.

11.3.1 Parameters α, γ, k

The test results, Figures 11.3(b)–11.3(g), are used to obtain an (average) ultimate yield envelope, Figure 11.4. For the soil considered, the value of k is found to be about zero. The values of α and γ are found to be 0.179 and 3.695, respectively. The value of β at ultimate is 3α; that is, $\beta_u \cong 0.54$.

11.3.2 Parameter β

The relation between β and ξ for the silty sand for the CTC test, Figure 11.3(b), is shown in Figure 11.5(a).

The expression in equation (11.4(b)) can be transformed as

$$\ln(\beta_a) - \eta \cdot \ln(\xi) = \ln\left(1 - \frac{\beta}{\beta_u}\right) \tag{11.6}$$

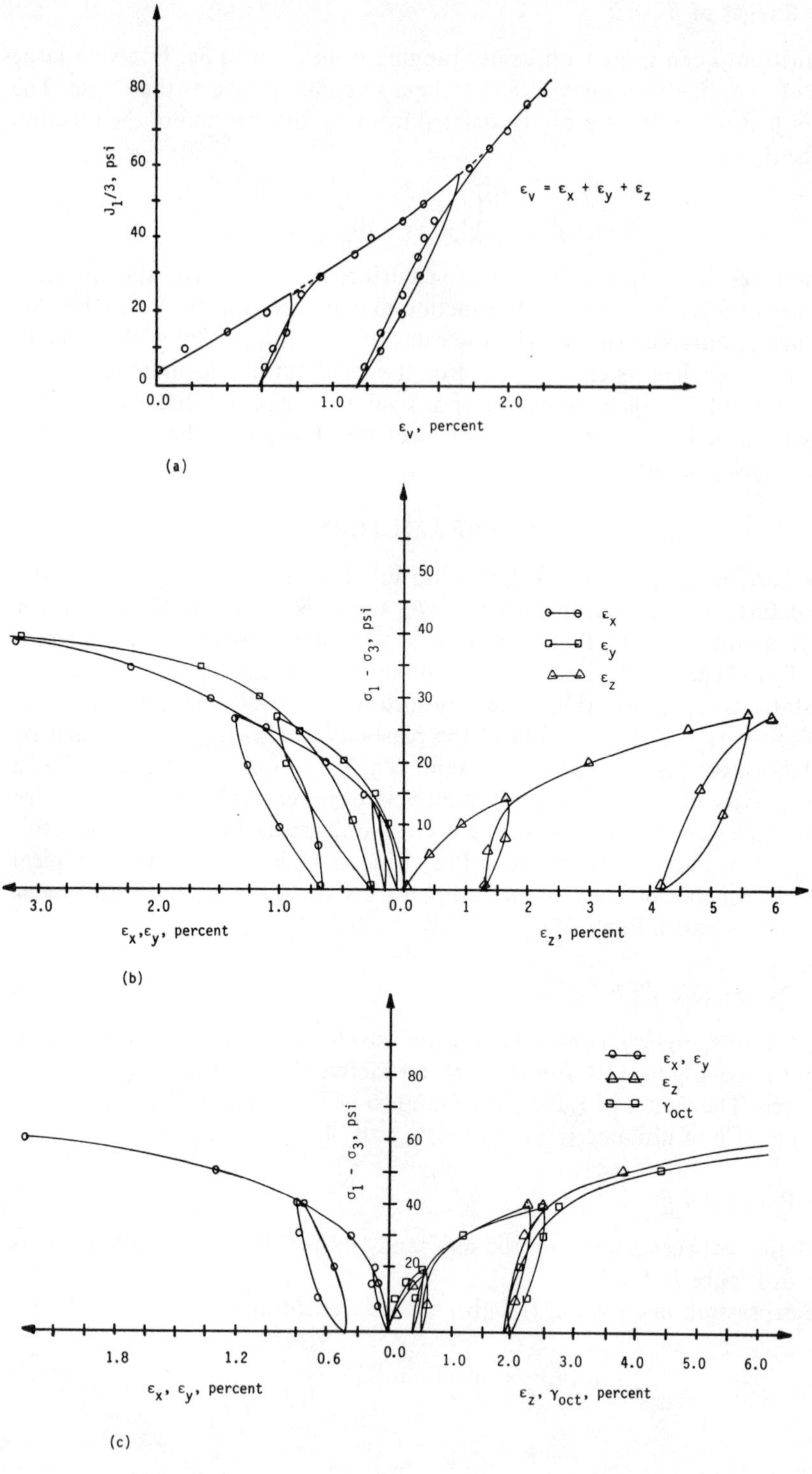

J1/3, psi
0
20
40
60
80
0.0
1.0
2.0
εv, percent
εv = εx + εy + εz
(a)
σ1 - σ3, psi
10
20
30
40
50
εx
εy
εz
3.0
2.0
1.0
0.0
1
2
3
4
5
6
εx,εy, percent
εz, percent
(b)
σ1 - σ3, psi
20
40
60
80
εx, εy
εz
γoct
1.8
1.2
0.6
0.0
1.0
2.0
3.0
4.0
5.0
6.0
εx, εy, percent
εz, γoct, percent
(c)

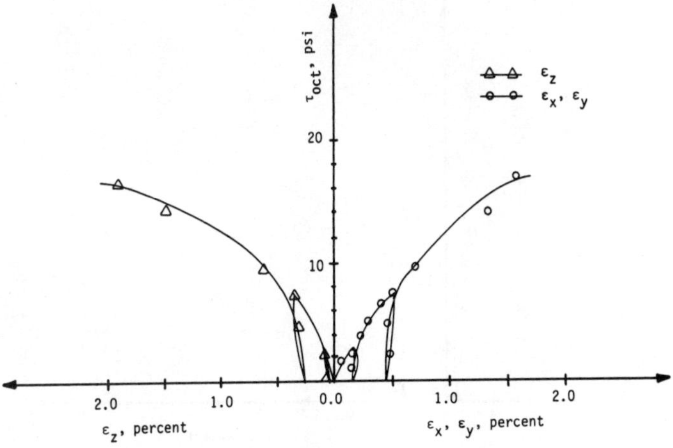
τoct, psi
εz
εx, εy
20
10
2.0
1.0
0.0
1.0
2.0
εz, percent
εx, εy, percent

(d)

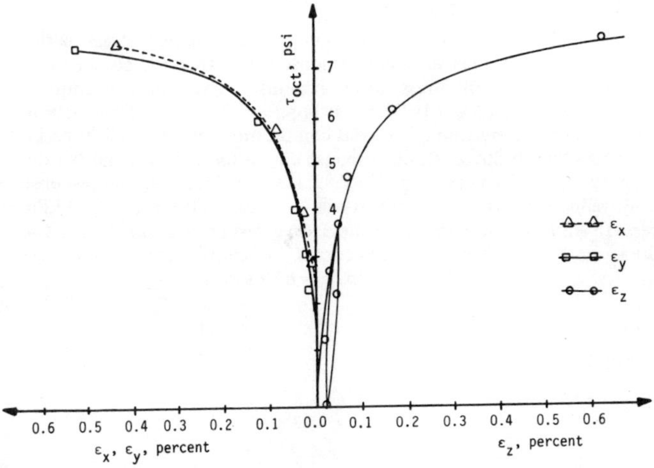
τoct, psi
7
6
5
4
εx
εy
εz
0.6
0.5
0.4
0.3
0.2
0.1
0.0
0.1
0.2
0.3
0.4
0.5
0.6
εx, εy, percent
εz, percent

(e)

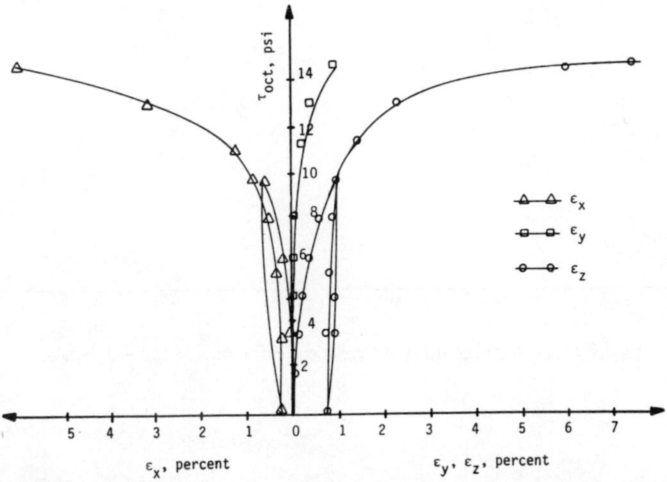
τoct, psi
14
12
10
8
6
4
2
εx
εy
εz
5
4
3
2
1
0
1
2
3
4
5
6
7
εx, percent
εy, εz, percent

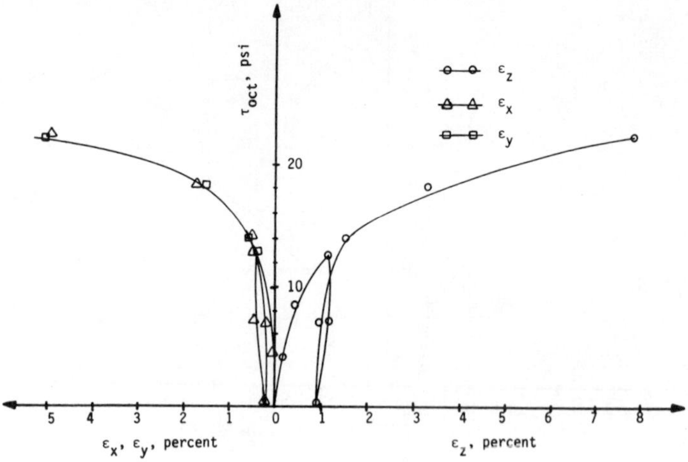

(g)

Figure 11.3 Test results for silty and sand under different stress paths. (a) Mean pressure–Volumetric strain response for hydrostatic compression test: 1 p.s.i. = 6.89 kPa. (b) Stress–strain response curves for conventional triaxial compression test ($\sigma_0 = 10.00$ p.s.i.): 1 p.s.i. = 6.89 kPa. (c) Stress–strain response curves for conventional triaxial compression test ($\sigma_0 = 20.00$ p.s.i.): 1 p.s.i. = 6.89 kPa. (d) Stress–strain response curves for conventional triaxial extension test ($\sigma_0 = 20$ p.s.i.): 1. p.s.i. = 6.89 kPa. (e) Stress–strain response curves for reduced triaxial compression test ($\sigma_0 = 20$ p.s.i.): 1 p.s.i. = 6.89 kPa. (f) Stress–strain response curves for simple shear test ($\sigma_0 = 20$ p.s.i.): 1 p.s.i. = 6.89 kPa. (g) Stress–strain response curves for triaxial compression test ($\sigma_0 = 25$ p.s.i.): 1 p.s.i. = 6.89 kPa

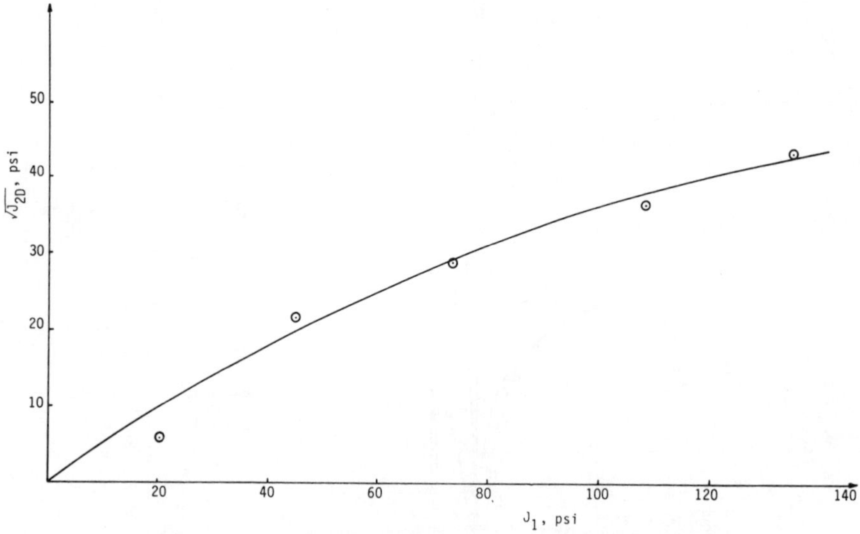

Figure 11.4 Average ultimate yield envelope on $\sqrt{(J_{2D})} - J_1$ space

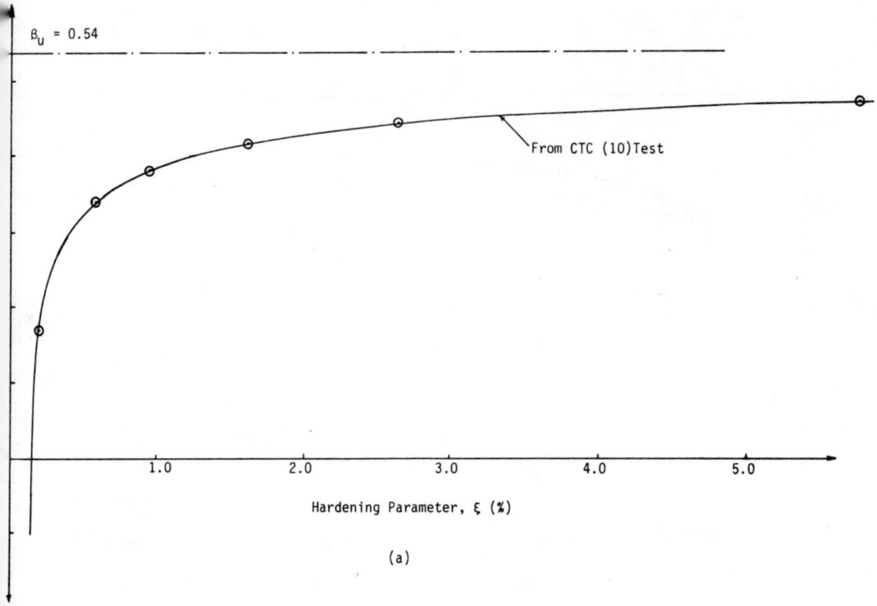

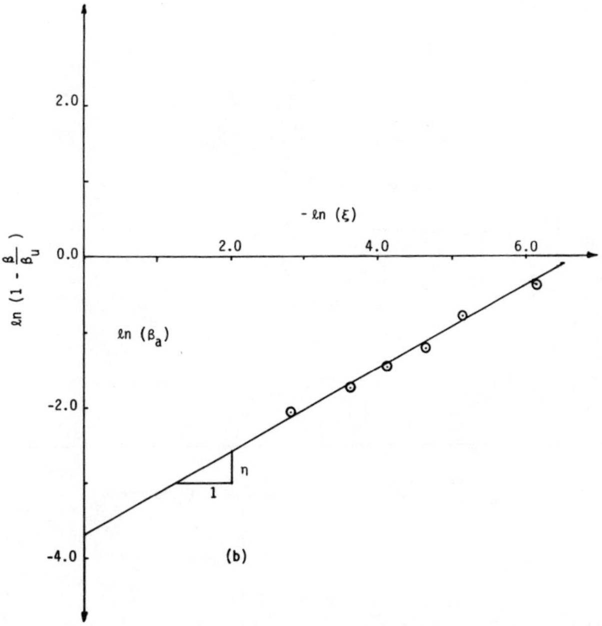

Figure 11.5 Variation of β with 'ξ' for CTC test, Figure 11.3(b). (a) β versus ξ. (b) Transformed plot, equation (11.6)

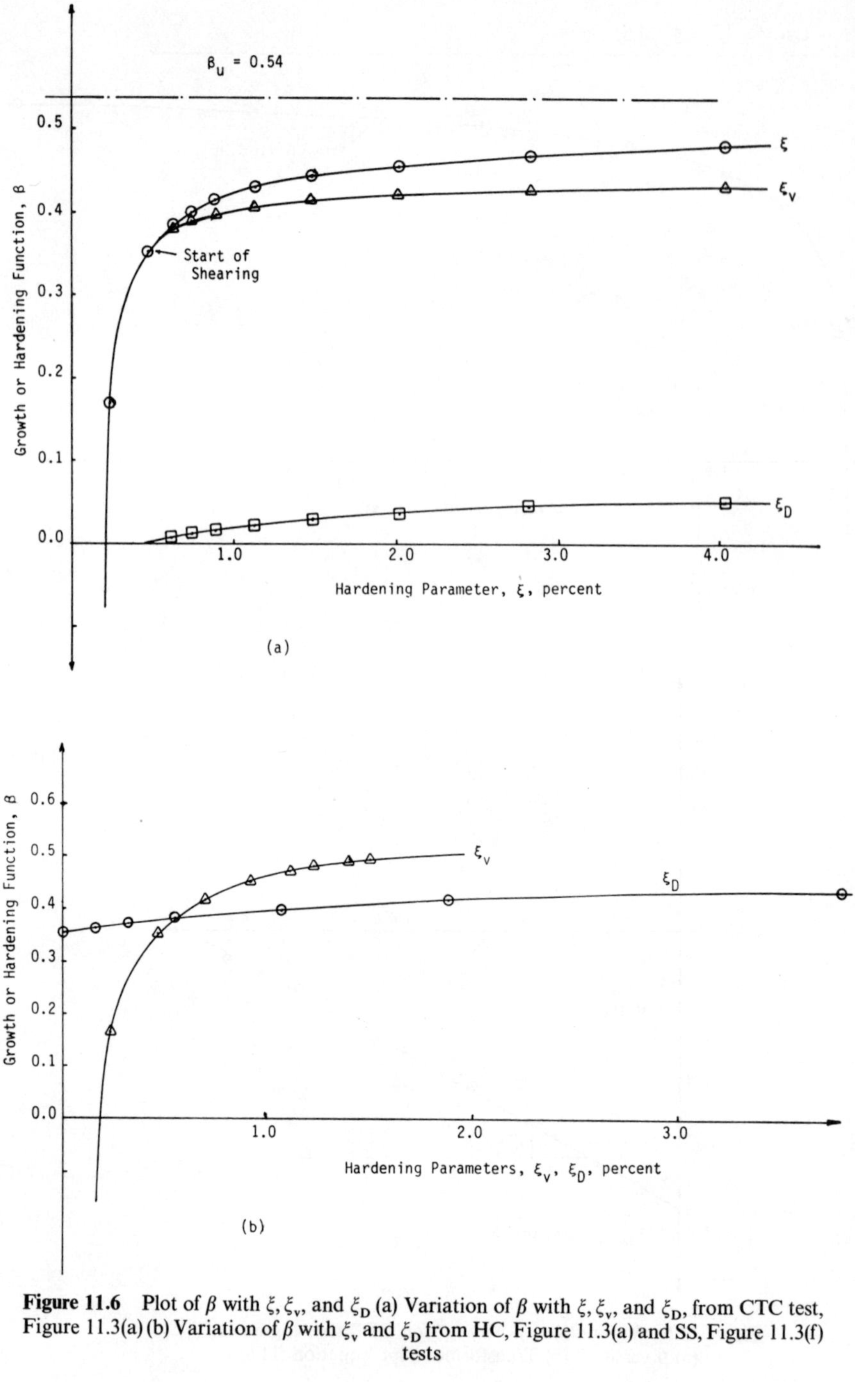

Figure 11.6 Plot of β with ξ, ξ_v, and ξ_D (a) Variation of β with ξ, ξ_v, and ξ_D, from CTC test, Figure 11.3(a) (b) Variation of β with ξ_v and ξ_D from HC, Figure 11.3(a) and SS, Figure 11.3(f) tests

which is plotted in Figure 11.5(b). The values of β_a and η from this figure are found to be 0.0255 and 0.545, respectively.

If β were expressed as in equation (11.4(c)), it will be appropriate to plot and analyse variations of β with respect to ξ_v and ξ_D separately. The variations of β with ξ, ξ_v, and ξ_D for the CTC test, Figure 11.3(c), are shown in Figure 11.6(a). It can be seen that although the magnitudes of the volumetric plastic strain predominates, there can be significant influence of the deviatoric plastic strain on the hardening behaviour.

Figure 11.6(b) shows variation of β with ξ_v and ξ_D for the HC tests, Figure 11.3(a) and the SS test, Figure 11.3(f), respectively. In the HC test, the values of ξ_D are small and for the SS test, ξ_v during the shearing remains nearly constant. It is interesting to note that for the SS test, the plastic behaviour is predominantly influenced by the deviatoric plastic strain.

From the foregoing, it appears that a model that considers only the volumetric plastic strain as the hardening parameter may not be able to predict behaviour under all significant stress paths.

Furthermore, many previous models such as critical state and cap assign a limit to the amount of volumetric plastic strain. It is believed that such a restriction may not be necessary. The proposed model can allow for such a condition as a special case.

11.3.3 Comments

Many previously proposed models [5, 10, 11] were derived on the basis of test results for specific material(s). As a result, their applicability may be limited to the specific material(s), and for other materials, they need modifications. On the other hand, the concept presented here can provide a general basis for developing suitable models for any material. It is possible to identify functions such as in equation (11.3b) that involve only a single function for both yielding and the ultimate yield behaviour. Such single-function models can be much simpler than the two-surface ones in terms of lesser number of parameters, ease of deriving the parameters and implementation in (numerical) solution procedures.

11.4 VERIFICATION

By using the flow rule of plasticity:

$$d\varepsilon_{ij}^{p} = \lambda \frac{\partial Q}{\partial \sigma_{ij}} \tag{11.7}$$

where λ is the scalar constant of proportionality and Q is the plastic potential function, the incremental form of the stress–strain relation can be obtained as

$$d\sigma_{ij} = C_{ijkl}^{ep} d\varepsilon_{kl} \tag{11.8}$$

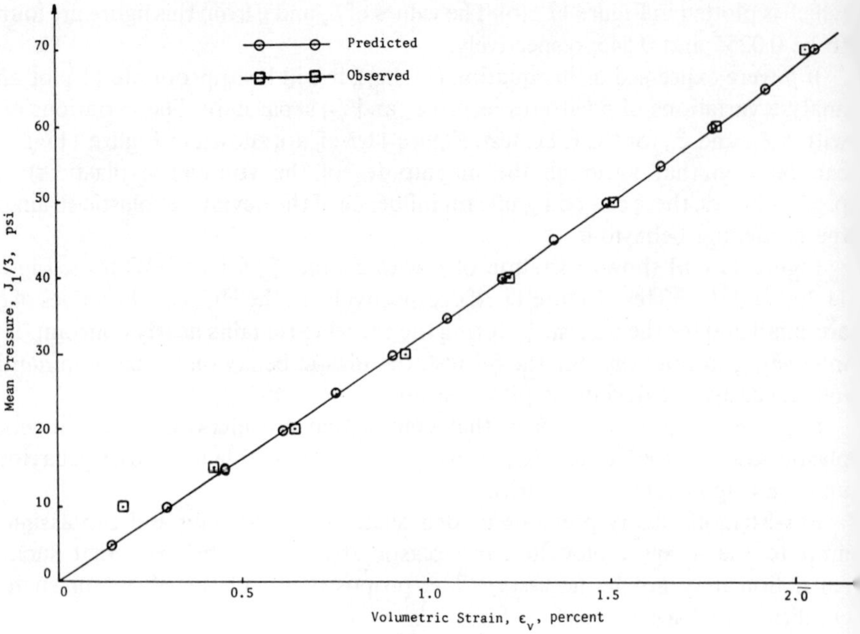
Predicted
Observed
Mean Pressure, $J_1/3$, psi
70
60
50
40
30
20
10
0
0
0.5
1.0
1.5
2.0
Volumetric Strain, ε_v, percent

(a)

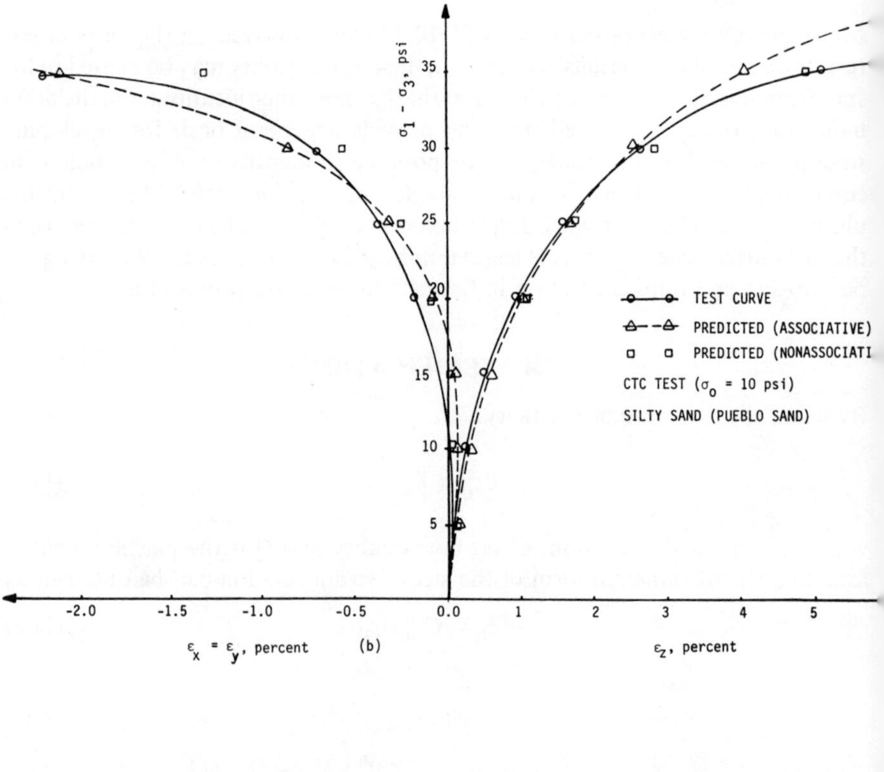
$\sigma_1 - \sigma_3$, psi
35
30
25
20
15
10
5
-2.0
-1.5
-1.0
-0.5
0.0
1
2
3
4
5
TEST CURVE
PREDICTED (ASSOCIATIVE)
PREDICTED (NONASSOCIATI
CTC TEST (σ_0 = 10 psi)
SILTY SAND (PUEBLO SAND)
$\varepsilon_x = \varepsilon_y$, percent
ε_z, percent

(b)

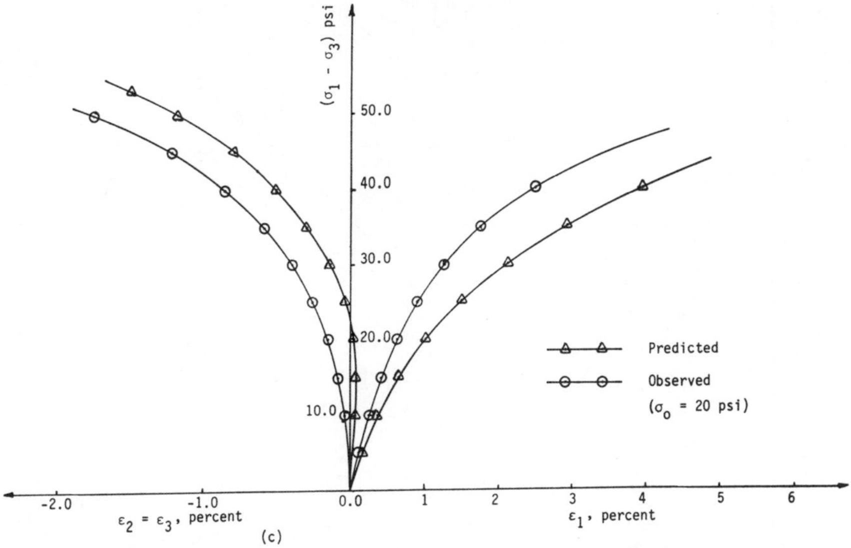

Figure 11.7 Comparison of predicted and observed behaviour. (a) HC test data. (b) CTC test, Figure 11.3(b). (c) CTC test, Figure 11.3(c)

where C^{ep}_{ijkl} is the constitutive tensor composed of the elastic and plastic parts, and is expressed in terms of the elastic and plastic parameters. The matrix differential equations can be integrated along various stress paths, Figure 11.2, starting from the initial conditions in the laboratory tests. It is first assumed that the material is associative; that is, $Q = F$. The parameters evaluated previously were used to predict typical stress paths for the silty sand, Figure 11.7(a)–11.7(c). It can be seen that the correlation between the predictions and observations is satisfactory.

It may be noted that β_a and η above were found only from a CTC test, Figure 11.3(b). In general, these parameters will depend upon the stress paths. Work toward this aspect and prediction of other stress paths is currently in progress.

11.5 NON-ASSOCIATIVE BEHAVIOUR

The foregoing concept can be modified to include non-associative behaviour by using the idea of correction functions [17]. Accordingly, the plastic potential, Q, can be expressed as

$$Q = F + h(J_i, \xi) \tag{11.9}$$

where $h(J_i, \xi)$ are correction functions. If a material exhibits deviatoric normality [18, 19, 20]; that is, if the non-associativeness is exhibited essentially in the volumetric behaviour, h can be expressed as a function of J_1 (and ξ) only [18].

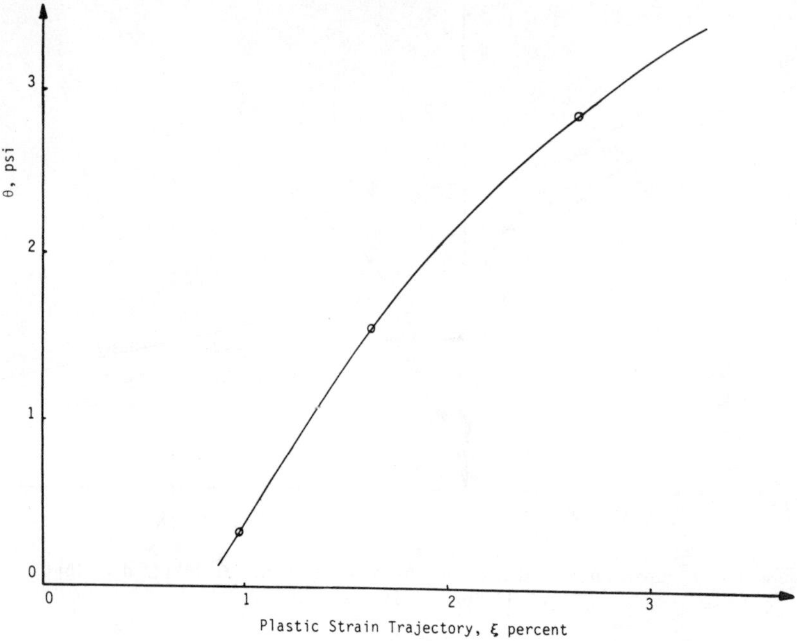

Figure 11.8 Variation of θ in correction term $h(J_1)$ with ξ

For the initial study herein, $h(J_1, \xi)$ is expressed as

$$h(J_1, \xi) = \theta J_1 \tag{11.10a}$$

where θ is assumed to be a function of ξ

$$\theta = \phi_0 + \phi_1 \xi + \phi_2 \xi^2 \tag{11.10b}$$

and ϕ_0, ϕ_1, ϕ_2 are material parameters. A plot of θ versus ξ for the silty sand is shown in Figure 11.8. It shows non-linear behaviour for which the parameters are found to be

$$\phi_0 = -2.033 \text{ psi}$$
$$\phi_1 = 280.82 \text{ psi}$$
$$\phi_2 = -3601.40 \text{ psi}$$

if the relation is approximated as linear, only two parameters will be needed. These parameters, together with the previous elastic and plastic parameters, were used to predict behaviour for various stress paths by integration of Equation (11.8) with $Q = F + h(J_1, \xi)$. Typical comparison between the predictions from the associative, non-associative models and the test results for a CTC test, Figure 11.3(b), are shown in Figure 11.7(b). It can be seen that the non-associative model shows closer correlation with the test data.

11.5.1 Comments

Further research will be needed to establish the ranges of applicability and limitations of this procedure. However, the major advantage of the correction function approach is that it is possible to use a given yield function, conventional or recently developed, and allow for non-associativeness in a simple manner. The resulting stiffness matrix with the correction function will be non-symmetric. An algorithm based on the residual or correction load method has been developed to implement the model using the traditional symmetric equation solver; brief details are presented in Appendix 11.1.

11.6 INDUCED ANISOTROPY

Induced anisotropy is an important factor that influences the behaviour during straining of many materials. For example, it can be significant in describing behaviour of geological materials under loading, unloading, and reverse loading during cyclic and static loads. A number of investigators have proposed various models to account for induced anisotropy [21, 22, 23]. An idea based on introduction of the joint invariants, K_j, (2, 3, and 4) equation 11.1(c), is proposed by Baker and Desai [24] to account for induced anisotropy. Here it was shown that in the context of plasticity, only certain classes of induced anisotropy can be accounted for, and that inclusion of one or more joint invariants can provide a method for including induced anisotropy, for instance,

$$\bar{F} = F + K_j \tag{11.11}$$

can be used. This concept is currently under investigation which includes laboratory testing, mathematical modelling, and verification with respect to boundary value problems. With most existing laboratory testing, it is difficult to measure induced anisotropy in the laboratory specimens. An approximate but simple method will be used to define a measure of anisotropy with the truly triaxial device. It involves a series of hydrostatic and shear loading and unloading cycles in which a measure of anisotropy at the end of each cycle is computed. Then the predictions from the models with induced anisotropy and without it are compared with the observations.

11.7 CONCLUSIONS

Representation of yield function as a polynomial in invariants of stress provides a generalized basis for deriving one or a set of appropriate plasticity models for constitutive behaviour of materials. One of the functions examined herein provides a potentially useful scheme for representing continuously yielding behaviour with a single function. The approach involves definition of ultimate yield and hardening behaviour and appears to be simpler than previous multi-surface models. It is applied for characterizing behaviour of a silty sand based on a comprehensive series of laboratory test data. The concept can be extended to include non-associative behaviour and induced anisotropy. Further research towards development of the proposed approach including implementation in finite element procedures is in progress. Although it is currently applied for geological materials, the procedure appears to be capable of representing behaviour of other engineering materials also.

ACKNOWLEDGEMENTS

A part of the investigations reported herein were performed under Grants Nos. CEE 81–16949 and CEE 82–15344 from the National Science Foundation. Useful comments from O. C. Zienkiewicz, M. G. Katona, A. D. Gupta, W. G. Pariseau, P.E. Sensey, and A. Fossum are gratefully acknowledged.

APPENDIX 11.1 ALGORITHM FOR NON-ASSOCIATIVE MODEL

During incremental iterative non-linear analysis, the (finite) element stiffness equations are expressed as

$$([K]_{si} + [K]_{nsi})\{\Delta q\}_{i+1} = \{\Delta Q\}_{i+1}. \tag{11.12}$$

where $[K]_s$ is the symmetric part of the matrix arising from yield function F, $[K]_{nsi}$ is the non-symmetric part arising from the correction function h, $\{\Delta q\}$ is the incremental displacement vector, $\{\Delta Q\}$ is the incremental external load vector and i denotes the increment.

Equation (11.12) can be written as

$$[K]^n_{si}\{\Delta q\}^{(n+1)}_{i+1} = \{\Delta Q\}_{i+1} - [K]^{(n)}_{nsi}\{\Delta q\}^n_{i+1} = \{\Delta Q\}_{i+1} - \{Q\}^n_c \tag{11.13}$$

where n denotes an iteration during the increment from i to $i+1$ and $\{Q\}_c$ = correction or residual load vector due to the non-symmetric part evaluated based on the previous iteration. The iterations can be continued until criteria based on displacements or correction loads is satisfied. For instance,

$$\frac{\| \{Q\}^{n+1}_c - \{Q\}^n_c \|}{\| \{Q\}^n_c \|} \leq \varepsilon \tag{11.14}$$

where ε is a small number.

REFERENCES

1. C. S. Desai, 'A general basis for yield, failure and potential functions in plasticity', *Int. J. Num. Analyt. Meth. in Geomech.*, **4**, 361–375 (1980).
2. R. S. Rivlin and J. I. Ericksen, 'Stress-deformation relations for isotropic materials', *J. Ratl. Mech. Analy.*, **4**, 323–425 (1955).
3. A. E. Green and P. M. Naghdi, 'A general theory of an elastic-plastic continuum', *Arch. of Rat. Mech. & Analysis*, **18**, (4), 252–281 (1965).
4. H. P. Shrivastava, Z. Mroz, and R. N. Dubey, 'Yield criterion and the hardening rule for a plastic solid,' Zeitschrift fur Angewandte Mathematik und Mechanik, Band 53, Heft, **10**, 625–633 (1973).
5. P. V. Lade, 'Elasto-plastic strain–theory for cohesionless soil with curved yield surfaces', *Int. J. Solids and Structures*, **13**, 1019–1035 (1977).
6. H. Matsuoka and T. Nakai, 'stress-deformation and strength characteristics of soil under three different principal stresses', *Proc. Japanese Soc. Civil Engrs., No.* 232, 59–70 (1974).
7. H. L. Schreyer and S. M. Babcock, 'A viscoplastic representation of concrete', *Proceedings International Conference on Constitutive Laws for Engineering Materials: Theory and Application*, pp. 451–456, Tucson Arizona, 1983.
8. T. Tukuoka, 'Yield conditions and flow rules derived form hypoelasticity', *Arch. Rat. Mech. Analysis*, **42**, 239–252 (1971).
9. G. Mullenger and R. O. Davis, 'A unified yield criterion for cohensionless granular materials', *Int. J. Num. Analyt. Methods in Geomech.*, pp. 285–294 (1981).
10. F. L. DiMaggio and I. S. Sandler, 'Material model for granular soils', *J. Eng. Mech. Div., ASCE*, **97**, EM3, 935–950 (1971).
11. K. H. Roscoe and J. B. Burland, 'On generalized stress–strain behaviour of wet clay', in *Engineering Plasticity* (Eds. J. Heyman and F. A. Leckie), Cambridge Univ. Press, Cambridge, 1968.
12. C. S. Desai and M. O. Faruque, 'Further developments of generalized basis for modelling of geological materials', *Report*, Dept. of Civil Eng. and Eng. Mech., Univ. of Arizona, Tucson, 1982.
13. C. S. Desai and H. J. Siriwardane, *Constitutive Laws for Engineering Materials: With Emphasis on Geological Materials*, Prentice-Hall, Inc., Englewood Chiffs, N. J., 1983.
14. C. S. Desai H. J. Siriwardane, and R. Janardhanam, 'Interaction and load transfer in track-guideway systems', Vol. 1 and Vol. 2, *Report to DOT, Office of Univ. Research*, Washington, DC, 1982.
15. S. Sture and C. S. Desai, 'Fluid cushion truly triaxial or multiaxial testing device', *J. Geotech. Testing, ASTM*, **2**, 1 (1979).
16. C. S. Desai, R. Janardhanam, and S. Sture, 'High capacity multiaxial testing device', *Geotech. Testing Journal, ASTM*, **5**, 1/2, 26–33 (1982).
17. C. S. Desai and H.J. Siriwardane, 'A concept of correction functions to account for nonassociative characteristics of geologic media', *Int. J. Num Analyt. Meth. in Geomech.*, **4**, (1980).
18. R. Baker and C. S. Desai 'Consequences of deviatoric normality in plasticity with isotropic strain hardening', *Short Comm., Int. J. Num. Analyt. Meth. in Geomech.*, **6**, 3 (1982).
19. R. D. Krieg, 'A simple constitutive description for soils and crushable foams', *Report SC-DR*-72-0883, Sandia Nat. Lab., Albuquerque, NM, 1972.
20. D. V. Swenson and L. M. Taylor, 'A finite element model for the analysis of tailored pulse stimulation of Boreholes', Under publication, *Int. J. Num. Analyt. Methods in Geomech.*

21. Z. Mróz, 'On the description of anisotropic work hardening', *J. of Mechanics and Physics of Solids*, **5**, 163–175, 1967.
22. J. H. Prevost, 'Mathematical modelling of soil stress–strain strength behaviour', *Proc. 3rd Int. Conf. Num. Meth. in Geomechanics*, Aachen, (Ed. W. Wittke) Vol. 1, pp. 347–361, Balkema Press, 1979.
23. Y. F. Dafalias and E. P. Popov, 'A model for nonlinearly hardening materials for complex loading', *Acta Mechanica*, **21**, 3, pp. 173–192 (1975).
24. R. Baker and C. S. Desai, 'Induced anisotropy during plastic straining', *Report*, Dept. of Civil Eng., Virginia Tech, Blacksburg, 1981.

Mechanics of Engineering Materials
Edited by C. S. Desai and R. H. Gallagher

Chapter 12

From Limited Experimental Information to Appropriately Idealized Stress–Strain Relations

D. C. Drucker

12.1 INTRODUCTION

Remarkably little experimental information is available on the multiaxial stress–strain behaviour of most materials in the inelastic range. The little that exists is far from unambiguous in its full implications. Nevertheless, a reasonably good understanding has been developed of the broad aspects of the small incremental plastic (time-independent) deformation of common structural metals and alloys from their as-received or moderately deformed state. The degree of stability of the test specimen and of the material provide a unifying approach.

Possible destabilizing effects of infinitesimal as well as large rotations and the obvious value of a fully correct mathematical representation have led to many efforts over many years to include finite as well as infinitesimal rotation and plastic strain properly and conveniently in constitutive relations [1, 2, 3]. Mathematical approaches and definitions entirely appropriate for finite elastic deformation have been carried over to finite elastoplasticity, but not always with the attention to physical behaviour required to do more than eliminate spurious effects of rigid body rotation.

A main point of this chapter is that misleading or incorrect inferences have been drawn for plastic stress–strain relations in general and for material stability under shear deformation in particular. The major difficulty arises from the usual but improper identification of the physical rotation associated with shear as the continuum rotation rather than the material rotation. This point has been emphasized by Mandel [4, 5]. Stability or instability in a shearing mode is a matter of experimental fact which should be incorporated properly in a correct mathematical idealization, not arise as an artefact of an over-simple or incorrect assumption.

The question of the meaning and significance of material rotation on the microscale and macroscale in the basic experiments and in applications is explored through a series of special cases. The limiting case of negligible elastic deformation in a homogeneous material provides an especially instructive reference. A unidirectional fibre reinforced composite exhibits the need to obtain direct experimental data in shear. It also serves to clarify the meaning of no average out-of-plane rotation in torsion, illustrates the possibility of large lattice plane rotation only when cross-sectional planes warp in a cooperative manner and so helps bring out the small lattice rotation compared with continuum rotation for a macroscopically homogeneous ductile material. The small but not entirely negligible effect of elastic deformation on plastic deformation in high strength structural metals and alloys also is reviewed along with the accompanying small but measurable SD effect of pressure on yield strength in shear [6, 7].

It appears likely that for all practical engineering predictions a proper finite deformation theory is a nicety at best rather than a necessity for metals and alloys in normal use. Even when the current orientation of the material in each region is used as reference, the usual infinitesimal forms with conventional stress components will match physical reality more conveniently.

Finally an historic look at the development of inelastic stress–strain relations emphasizes the need to keep clearly in mind the essential features of the behaviour to be modelled as successive idealizations are developed. Periodically, mathematical consistency of simple forms becomes confused with physical reality and conversely the absence of some aspect of physical reality leads to rejection of very useful simple forms. Unrealistic hopes rise and inevitably fall for mathematical forms suitable for a scientifically acceptable synthesis of the entire range of complex time-dependent and independent inelastic behaviour that is observed and yet require only modest computational effort for engineering calculations. Simplicity and complexity are relative terms whose meaning does change very much with the advance of knowledge and the ever-increasing ability to compute. Yet it remains true at all times that the appropriately idealized stress–strain relation for the solution of a problem of continuum mechanics is the simplest one that contains the essence of the physical behaviour of importance in the problem to be solved.

12.2 FUNDAMENTAL EXPERIMENTAL INFORMATION AND STABILITY

Experimental information always is limited, even when very modest objectives are pursued and when repeatable reversible processes are examined. Furthermore, plasticity is irreversible, and experiments are far from repeatable in all detail. There is no natural reproducible state of each material of interest. Worst of all, the tests that are possible are very limited, and remarkably few of those that can be done have been done. Yet despite all these impediments we do know a great deal about plastic stress–strain relations [8, 9, 10].

The conceptual jump in the last century to a useful mathematical theory of plasticity from the very limited experimental information on metals then available showed remarkable insight. Now constitutive relations are expected to be both more realistic and more fundamental in character with more incisive descriptions given for large displacements involving both large rotations and large strains. Instabilities as well as stable deformations of bodies and their surfaces are examined with these stress–strain relations that reduce to forms familiar in small displacement theory when strains are small and rotation is ignored [11, 12]. Care is taken to define stresses and strains or strain rates with proper work complementarity but most often there is no direct connection made to fundamental experimental information.

In part, this lack of connection reflects an excessive faith in the carryover from finite elasticity and infinitesimal plasticity of the predictive ability of quite simple analytic forms. Mainly, however, it reflects the fact that, despite the many valuable reports of experimental results, little or no data are available for complex paths of stress or strain with or without rotation in most materials. The effort needed to examine the inelastic response of a given alloy to just a few abrupt changes in a continuing path of loading in a two- or three-dimensional stress or strain space is enormous unless the alloy of interest is homogeneous on the macroscale and a thin-walled tube can be made that is axisymmetric. Such a statically determinate test specimen is required for direct determination of stress–strain relations, a specimen in which knowledge of the geometry and the loads give the stresses unambiguously, however they are defined, and the strains can be determined by measuring the geometry change. Unfortunately, fabrication of a tube with axisymmetric properties is not possible for most alloys after thermomechanical treatment of practical interest.

When there is axial symmetry and homogeneity of material along the length, the standard thin-walled cylindrical tube specimen, Figure 12.1, under combinations of twisting moment M_T, axial force F, and interior pressure p (or combined interior and exterior pressure) is uniquely suitable. Yet it has severe limitations. Reality is imperfect. Tube walls that are too thin will buckle too early in torsion. The most careful machining of specimens affects the properties of the material somewhat and annealing generally is not feasible when the structural alloys are to be examined in their mechanically deformed and heat-treated condition. Furthermore, the paths of loading that can be followed are limited to plane stress with or without superposed hydrostatic pressure.

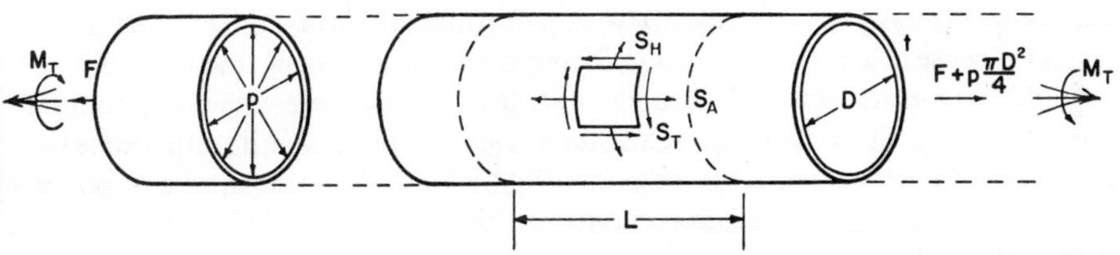

Figure 12.1 An axisymmetric homogeneous thin-walled tube is statically determinate

Why then is so much trouble taken for a few, rather special, materials under restricted states of stress? It is more than the inability to do better, it is the very reasonable hope that the essential features of response of wide classes of materials, whether time-independent or time-dependent, can be discovered. Knowledge and fundamental understanding of the stress–strain relation for any one material should carry over in principle although not in detail to materials of comparable qualities. Mathematical generalizations of the basic characteristics found experimentally should be transferable to initially anisotropic as well as isotropic materials and to general states of stress, not to plane stress alone. However, it is not to be expected that the fine details of behaviour within the established broad framework can be determined analytically. Experimental information alone can provide quantitative values (sometimes including sign) for the coefficients in the incremental relation between stress and strain or for a detailed description of the continually changing yield surfaces, etc.

Suppose M_T, F, and p are applied to explore stress–strain relations. Working displacements for the test section initially of length L_0, diameter D_0 (volume $V_0 = \pi D_0^2 L_0/4$) are the relative angle of twist $\delta\phi$ corresponding to M_T, the change in length δL corresponding to F, and the change in interior volume δV of the gauge section corresponding to p, all of which can be viewed as measured quantities. The change in thickness δt needed to specify the geometry completely does not enter yet and may be very troublesome to measure accurately, but suppose that difficulty to be overcome as needed.

Many equivalent methods of analysing the data can be devised but the most direct is simply to present results in terms of the applied generalized forces (M_T, F, and p) and corresponding displacements ($\delta\phi$, δL, δV) and so avoid any detailed interpretation to start. Suppose the response to be reasonably time-independent over a considerable range, as indeed it will be for most useful structural metals at low to moderate temperatures. Suppose further that the system is in the usual work-hardening range with its response stable for small changes in displacement in the strict sense of my definition, i.e. the work done by the increments of M_T, F, p on the displacement increments produced is positive for all equilibrium paths (including cycles of loading and unloading) involving small plastic strains [13, 14, 15].

Then as illustrated in Figure 12.2, the infinitesimal plastic displacement vector $\delta\phi^P, \delta L^P, \delta V^P$ or the rate vector $\dot{\phi}^p, \dot{L}^p, \dot{V}^p$ must be normal to the yield or loading surface at the current load point M_T, F, p (normality is to be interpreted in the extended sense at a corner or vertex). This plastic displacement rate is simply the total rate minus the elastically recoverable displacement rate when infinitesimal $\delta M_T, \delta F, \delta p$ are added to M_T, F, p, and then removed. In the small strain range the elastic response at each stage is indistinguishable from the initial elastic response. As plastic strains become appreciable in this stable range of homogeneous deformation the radius, cross-sectional area, and length of the material in the gauge section will change sufficiently that the current elastic response rates must

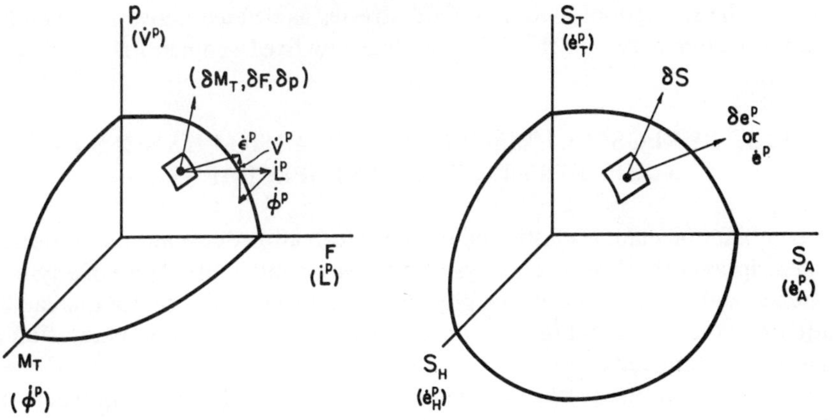

Figure 12.2 Normality and convexity

be subtracted from the total rates to get the plastic rates with the desired accuracy.

Another consequence of stability is that the initial yield and each subsequent loading surface in M_T, F, p space must be convex. The small degree of concavity possible when elastic strains are large or when small plastic deformation produces a large change in elastic response is unlikely to exist or to be observable if it does exist in structural metals [16].

When the initial geometry or any later geometry of the tube is taken as reference, and subsequent changes in configuration are treated as in small displacement elasticity and plasticity theory, convexity then follows in the resulting conventional stress space and the conventional plastic strain rate obeys normality Figure 12.2. The word 'conventional' is used here for the shear stress S_T, the axial stress S_A on the cross-section, and the hoop stress S_H along with the corresponding increments of shear, axial, and hoop strain δe_T, δe_A, δe_H or rates $\dot{e}_T$, $\dot{e}_A$, $\dot{e}_H$ and stress rates $\dot{S}_T$, $\dot{S}_A$, $\dot{S}_H$. Normality and convexity carry over from M_T, F, p space with $\dot{\phi}^p$, $\dot{L}^p$, $\dot{V}^p$ superposed to S_T, S_A, S_H space with $\dot{e}_T^p$, $\dot{e}_A^p$, $\dot{e}_H^p$ superposed, if the same geometry is used for the computation of stress and strain and their rates. Nominal stress, stress rate, and strain rate is the strength-of-materials terminology when the original dimensions D_0, L_0, t_0 are employed [e.g. $S_T = M_T/(\pi D_0^2 t_0/2)$, $\dot{S}_T = \dot{M}_T/(\pi D_0^2 t_0/2)$, $\dot{e}_T = (D_0/2)\dot{\phi}/L_0$, $(\dot{S}_T\dot{e}_T)\pi D_0 t_0 L_0 = \dot{M}_T\dot{\phi}$], true stress, stress rate, and strain rate when a set of current dimensions D, L, t are used; $(\dot{S}_T\dot{e}_T)\pi DtL = \dot{M}_T\dot{\phi}$. Ordinarily it will be the true rather than the nominal quantities that are directly translatable from the test specimen results to the response of another body of the same material.

The terms nominal and true as described are useful and meaningful. However, when the current elastic domain is traversed, greater precision of definition is needed if the purely elastic changes in geometry that occur are to be taken into account. The strength-of-materials 'true' stresses as defined above are actually nominal stresses based on the deformed but now fixed geometry D, L, t [1].

12.3 PURPOSEFUL NEGLECT OF ELASTIC CHANGES IN GEOMETRY OF THE TEST SPECIMENS

The combination of finite elastic and plastic deformation does cause confusion in the description of each aspect of physical behaviour and in the choice of a system of stresses and stress rates in which physical or mathematical assumptions can be made with confidence that they are appropriate. A metal bar in simple tension F illustrates the difficulty.

Suppose the strains are large but purely elastic. Nominal stress F/A_0, where A_0 is the initial area of cross-section at zero load, is a physically appropriate quantity because crudely the tensile load is carried on each cross-section by a constant number of atomic chains parallel to the axis of the bar as the bar elongates and contracts laterally. True stress F/A, where A is the current area of cross-section, is not physically appealing because the number of atoms per unit current area is higher in the stretched configuration.

When, instead, the elastic strains are very small and plastic deformation dominates, true stress is physically appealing because the current area does represent the number of atomic chains carrying the load. Atomic distances remain almost unchanged on average as the bar lengthens and contracts laterally in the process of slip or shear deformation on inclined planes.

If both elastic and plastic deformations are large, neither nominal nor true stress is a good physical measure for either the elastic or plastic response. Of course there is no problem in this simple example of following the geometry change and employing a stress–strain relation for any measure of stress to give the observed response correctly. The difficulty can be severe for more complex states of stress and geometric changes, even for the elastic response when the material is initially anisotropic or becomes so after plastic deformation. Following material orientation and rotation as distinguished from continuum rotation does require care in the plastic regime.

Returning to the thin-walled tube of Figure 12.1, suppose plastic deformation is occurring under increasing axial force F and twisting moment M_T and that the current state of stress is S_A^Y, S_T^Y with $S_H = 0$ based on the current dimensions. Suppose further that the current yield or loading surface in F, M_T space has been found experimentally. The yield surface then is determined directly as a yield curve in S_A,S_T space if the conventional stresses are computed using the geometry at $S_A^Y S_T^Y$, a nominal stress type of calculation with a fixed although a current geometry. If the elastic changes in geometry that accompany changes of load within the yield domain are taken into account, the yield values in this true conventional stress space S_A, S_T differ from those based on the S_A^Y, S_T^Y fixed geometry. Their ratios are of order $1\pm$ changes in elastic strain or $1\pm(S_A$ or $S_T)/E$ where E is an average elastic modulus [2, 16].

Greater complications are introduced if the S_A, S_T values are converted to true or Cauchy stresses in the set of axes that rotates with the continuum rotation $\delta e_T/2$ from the reference state S_A^Y, S_T^Y, Figure 12.3. Then the data in S_A, S_T, $S_H=0$ space translate to data in a σ_a, τ, σ_h space with $\sigma_a^Y=S_A^Y$, $\tau^Y=S_T^Y$, $\sigma_h^Y=0$ at the reference state but elsewhere generally $\sigma_h\neq 0$. The elastic rotation is of the order of S_T/G where G is an average elastic shear modulus. Terms of order S_A and S_T multiplied by the elastic rotation then add to or subtract from each of the primary terms $\sigma_a=S_A$, $\tau=S_T$, $\sigma_h=0$ in the rotated axes.

If the elastic strains and rotations are not negligible in comparison with unity, a familiar Mises form or kinematic hardening form in nominal S_A, S_T space will be converted to a complicated form in true S_A, S_T space and to a completely unrecognizable form in σ_a, τ, σ_h space with $\sigma_h\neq 0$. Conversely, a simple form in the rotated Cauchy stress system will be complex and unrecognizable in either conventional stress system.

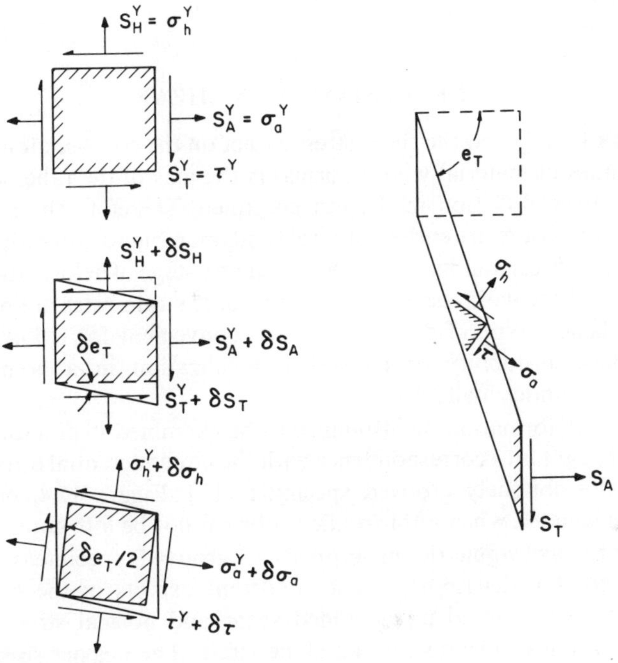

Figure 12.3 Conventional stress for no rotation of material and Cauchy stress referred to axes that follow continuum rotation

Historically, our thinking about the Mises and other yield criteria, the associated Prandtl–Reuss and other stress–strain relations, and most of the intuitive feeling about the inelastic response of material to stress has been restricted to materials with a very small elastic strain range. It therefore has not included such matters as the possible significance of the rotation $\delta e_T/2$ associated with the simple shear increment δe_T or for that matter any elastic dimensional changes in the tube test.

The combination of finite elastic strains and inelastic response is a fascinating and little explored subject with little guidance yet available on how to codify inelastic physical behaviour in appropriate mathematical form for use in the solution of problems. It is likely to be of great importance for polymeric materials, of some importance for very high strength ceramics and metals, and possibly significant for all materials under extremely high pressures and strain rates. However, for the common structural metals under ordinary conditions, elastic strains are very small compared to unity and elastic response is very close to linear at each stage of plastic deformation. The clearest understanding is achieved of finite plastic deformation itself and of appropriate incremental plastic stress–strain relations when all elastic changes in geometry of the test specimen are purposefully neglected and the elastic–plastic idealization approaches the rigid-plastic limit.

12.4 RESPONSE IN SHEAR

The normal stress S_A and the shear stress S_T act on the cross-sectional planes of the tube, planes of material whose normal is the axis of the tube, which is the reference or fixed direction in the test specimen. Therefore the conventional stresses and stress rates are entirely suitable and unambiguous descriptions of the existing state of stress and its rate of change at any stage of deformation, large or small. They and the work-associated conventional strain rates are precisely the directly applicable type of descriptors most convenient for a look at simple torsional shear instability or equivalent localization in a homogeneously deformed body under load.

If the shear deformation or instability to be examined is in a direction that cannot be brought into correspondence with the circumferential direction in the tube, then the obliquely grooved specimen [17], Figure 12.4, provides the analogue. In general, when a thin-walled tube cannot be fabricated, or if made would not be axisymmetric in properties, grooved specimens are useful alternatives for the determination of important aspects of the behaviour of already deformed material under added shear and normal stress in a fairly uniform condition of plane strain or plane stress. The heavier sections of the specimen remain close to rigid and do not rotate toward or away from each other, just as two nearby cross-sections of the thin-walled tube. End and edge effects of the transition from the much thicker constraining material to the thinned down

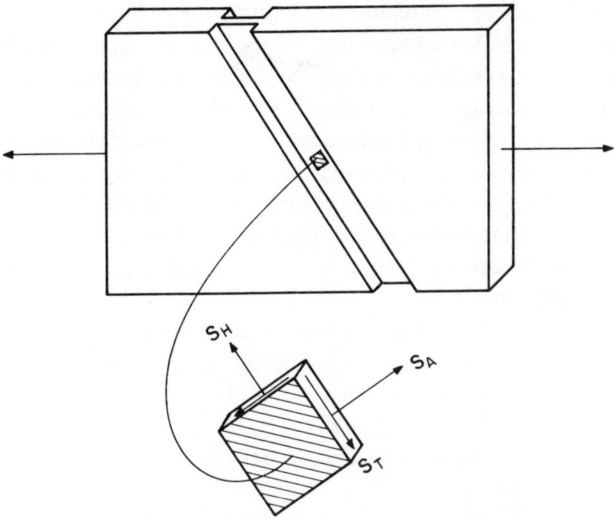

Figure 12.4 Obliquely grooved test specimen [17]

region do obscure the initial plastic response to added load. Also, the equivalent of S_H in the tube is not statically determined. These deficiencies probably account for the small use of this type of specimen in basic studies. However, as the additional plastic strains in the thinner region grow, the incremental plastic response of the deformed material to the known active components of the current stress state and their stress rates is observable with reasonable accuracy. Planes parallel to the midplane of the deforming region (parallel to the almost rigid constraint boundary planes of the thick regions) are analogous to the cross-sectional planes of the thin-walled tube. The stress and stress rates on those planes correspond to S_A, S_T, $\dot{S}_A$, $\dot{S}_T$. As for the tube, if there is no instability in the test on a grooved specimen, there will be none in the body under the same conditions.

For an initially isotropic material in the small deformation range, the stress–strain curve in shear S_T, or combined shear and normal stress in constant ratio S_A/S_T (proportional or radial loading), is determined by radial loading data in biaxial stress S_A, S_H. A rather simple form intermediate between Mises and Tresca usually will not be all that bad for small plastic strains. However, considerable divergence is likely for large plastic strains in the amount of work hardening when the direction of the axes of principal stress and strain remain fixed and coincident as contrasted with when the principal axes of strain rotate under torsion. It is not really the continuum rotation or its absence as described that governs but rather the slip on material planes that accommodates the deformation. A wide variety of intersecting slip planes are activated continually in the large deformation of ductile metals under increasing principal stresses S_A,

S_H while the most active slip planes in torsion tend to be those aligned with the cross-sectional planes. Work hardening in torsion is reported as much lower than in extension. Only the experiment itself can determine the stress–strain curve for torsion as the plastic strain becomes larger and larger. Similarly, as is true also for the small strain range, only experiment can give the effect of adding a combination of axial and shear stress S_A, S_T (as defined for the thin-walled tube or grooved specimen) after prior plastic deformation. Whenever the loading is not proportional or radial so that all stresses referred to key material (lattice) directions translated to the macroscale do not stay in ratio, initial isotropy provides essentially no guide at all.

Real behaviour is likely to vary enormously from one material to the next, especially for finite strains. An elastic–plastic matrix material reinforced with strong elastic fibres provides a well-known illustrative example. As shown in Figure 12.5 if the fibre direction is inclined to the generator of the tube in such a manner that it tends to become parallel when the tube is twisted without axial constraint, the matrix will resist the extension that would result from the straightening out of the fibres. The fibres will be in compression and the matrix in tension. If the axial dimension is held fixed, the tube as a whole will go into compression. On the other hand if the fibres are tilted the other way so that twisting tends to shorten the tube, a constrained tube will go into axial tension. If the fibres run circumferentially there will be no axial strain for small torsional displacement. A real homogeneous material may behave for small or for large strain in any one of these three ways. No thermodynamic preference and certainly no requirements

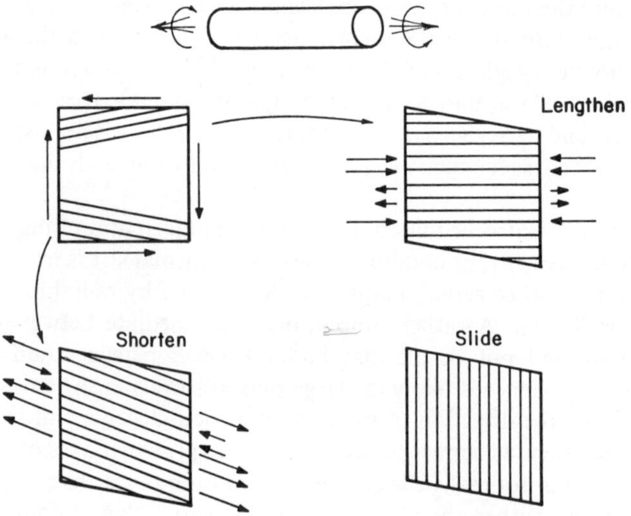

Figure 12.5 A twisted tube composed of strong fibres in a ductile matrix

exist for one or another of these very different behaviours. Only experiment can give the answer which then must be translated properly into the chosen set of stresses and corresponding strains. Obviously the choice of the system of stresses cannot influence the physical outcome. In one form or another this is the message Biot conveys in his book [18] and in his papers over the years.

12.5 ROTATION OF MATERIAL MICROSTRUCTURE VERSUS CONTINUUM ROTATION

It is worth repeating that the conventional stresses, strains, and their rates (or their equivalent) are the directly relevant quantities in a look at shear instability in test specimens or any body of material. When their fundamental applicable nature is not recognized and the aim is to take changes in geometry of the test specimen or body into account more fully in the calculation of stresses, strains, and their rates, many problems can arise. There is no difficulty with overall rigid body rotation because the axis of the specimen serves as reference. A variety of consistent stress and strain or displacement gradient definitions are permissible. Although the irreversible nature of plastic deformation is universally recognized, somehow the false feeling carries over from elasticity that the current state of deformation says a great deal about the current state of stress. Alternative forms of nominal and true stress or something in between have been proposed to take rotation into account when computing stress rates or increments. Each form has some mathematical or physical appeal which often relates to the response of initially isotropic and fully elastic material to large deformation rather than to inelastic response.

Experimental results from ideal thin-walled tube tests are translatable to each choice, whether physically appealing or not, without ambiguity because the values and changes of the loads and the geometry are completely known. Conversely, each consistent choice will reproduce the experimental results when employed properly in reverse fashion. Properly means with due regard to the orientation of key material directions, rather than with respect to some arbitrary assumption about stress–strain relations. This is a complicated algebraic requirement whenever the reference coordinate directions are forced to follow the continuum rotation as usually defined.

Suppose, for example, that the choice is made to describe the state of stress S_T and the changes in stress δS_T in pure torsion by the Cauchy (true stress) components in the coordinate system that continually rotates in the usual continuum sense, Figure 12.3. As the total shear angle e_T grows, the continuum rotation grows at half the rate in the early stages, and the ratio of the normal stress to the shear stress on the rotated coordinate planes then grows as 2 tan $(e_T/2)$. The proportional or radial loading in the material system S_T, S_A, S_H is a non-radial loading in the rotating continuum coordinate system for S_T alone or similarly for any $S_T \neq 0, S_A, S_H$ in fixed ratio. One set of stresses and stress rates

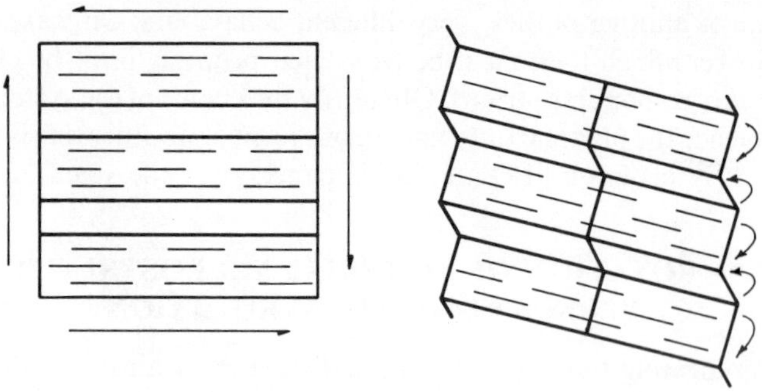

Figure 12.6 Zero average rotation of material in shear

can be transformed to the other. However, an assumption such as kinematic hardening in the rotating continuum system is physically and mathematically quite different from that same assumption for the fixed direction material system S_T, S_A, S_H in which the data are obtained directly.

The absence of material rotation in torsion or in torsion combined with axial force and interior pressure is the familiar strength of materials statement of plane cross-sections remaining plane [15]. It is strictly true for the cross-sectional planes if the material is perfectly homogeneous, and is true on a macroscale average for statistically homogeneous material with inhomogeneous microstructure. A ductile matrix, perfectly bonded to a high volume fraction of strong fibres initially parallel to the axis of the tube, Figure 12.6, serves as an extreme example of the meaning of no material rotation on average. Each cross-section warps locally in a cooperative manner with large clockwise rotation of fibres balanced by large counterclockwise rotation of the matrix. The simple shear planes of the matrix are constrained to be parallel to the fibre direction, longitudinal to start. When the fibres are circumferential, Figure 12.5, the simple shear directions are circumferential.

Unless cross-sections must respond in the cooperative local warping manner forced by the fibres in Figure 12.6, it is not possible for appreciable simple plastic shear to continue for long on planes parallel to the axis of the tube or at any angle far removed from 90°. A statistically homogeneous rather random microstructure can permit only modest local warping of each cross-section within the constraint of zero warping on the macroscale. Most particles cannot deviate appreciably from the path they would follow in a truly homogeneous continuum. Simple shearing motion or large slip is dominantly circumferential, parallel to the cross-section, in a tube whose length and diameter may change.

The individual grains of a ductile polycrystalline aggregate will rotate only a little one way and another compared with a large total shear angle and corresponding continuum rotation. When elastic strains are neglected, the individual grain (lattice) rotations must average to zero. Rigid inclusions, voids, and other such microstructural features can distort this simple picture in their vicinity. The matrix reinforced with longitudinal fibres shows that other pictures are possible in special circumstances. Nevertheless it must be true that the torsional strain picture of a ductile structural metal on most of the microscale as well as on the macroscale is dominated by slip on planes parallel to or at a moderate angle to the cross-sectional planes. Independently of the microscopic details, the behaviour in simple shear, combined with changes in dimension along and perpendicular to the shear direction, is given clearly and directly in the conventional system and rather obscurely in the rotating system.

12.6 INSTABILITY IN SHEAR, REAL OR ARTEFACT

Rice has emphasized [19] that no matter how small the plastic strain increment δe^p may be, it is not permissible in considerations of stability to ignore terms of magnitude $S\delta e^p$ in comparison with the stress increment δS. They are comparable when, as is usual, plastic moduli are comparable to stresses in magnitude. Such a term, the stress multiplied by the decrease in cross-sectional area, is included in the elementary analysis of necking of bars in tension. It accounts for this loss of stability despite continued work-hardening of the material. Similar destabilizing terms appear in the analysis of plastic torsional buckling of a cruciform section [20]. They are implicit if not explicit in all stability analyses which have the stress following an imagined or real fibre in structural elements, as in Figure 12.5. Little if any controversy arises because the geometry change including rotation of the material is obvious or at least agreed upon.

Disagreement can occur in continuum analyses when a simple stress–strain relation is employed for one particular choice of stresses and stress increments rather than another. Suppose, as before, that the stress–strain relations are expressed in terms of the Cauchy stresses in the coordinate system that follows the continuum rotation, but this time neglect the elastic changes in geometry. Now when the current deformed configuration of the thin-walled tube is taken as the reference configuration, Figure 12.3, the Cauchy stress components $\sigma_a^Y, \tau^Y, \sigma_h^Y$ coincide with the conventional true stress components S_A^Y, S_T^Y, S_H^Y at all points of the current yield surface. Additional plástic deformation produced by small stress increments $\delta S_A, \delta S_T, \delta S_H$ includes a small additional shear deformation δe_T, continuum rotation $\delta e_T/2$, and small changes in linear dimensions. To bring out the key point about rotation, suppose the small changes in area are included in the change of conventional true stress from S_A^Y to $S_A^Y + \delta S_A$, etc. Then for $\delta S \ll S$, the Cauchy stresses in the coordinate system rotated by the small angle $\delta e_T/2$ are:

$$\sigma_a^Y + \delta\sigma_a = S_A^Y + \delta S_A + S_T^Y \delta e_T$$
$$\sigma_h^Y + \delta\sigma_h = S_H^Y + \delta S_H - S_T^Y \delta e_T$$
$$\tau^Y + \delta\tau = S_T^Y + \delta S_T + (S_H^Y - S_A^Y)\delta e_T/2$$

A compressive (negative) S_A^Y with $S_H^Y = 0$ then increases the magnitude of $\delta\tau$ beyond the increase δS_T by a destabilizing term $(-S_A^Y)\delta e_T/2$. If $(-S_A^Y/2)$ exceeds the tangent modulus computed for additional plastic deformation in shear, a shear instability will be predicted, plastic deformation continuing at decreasing load. Similar calculations of stability or instability can be made for large additional shear deformations e_T and accompanying continuum rotations, Figure 12.3.

However, the question of stability or instability is answered directly by the experiment on the thin-walled tube. If the tube is work-hardening (stable) in F, M_T, p space and in S_A, S_T, S_H space, any calculated instability indicates no more than an inappropriate idealization.

Suppose the set of yield or loading functions that gives correct results in the fixed material orientation space represented by S_A, S_T, S_H is arbitrarily transformed to the rotated configuration by substitution of σ_a for S_A, τ for S_T, and σ_h for S_H. A false instability in shear then could be predicted from the $(-S_A)\delta e_T/2 = (-\sigma_a)\delta e_T/2$ destabilizing term and even more likely from the finite rotation associated with the finite shear strain e_T. Such an arbitrary and physically inappropriate substitution of Cauchy or other stresses referred to coordinates that follow continuum rotation is all too likely to be built into computer programs for finite plastic deformation. Instabilities may be predicted where none exist or not predicted when they will occur.

If all or almost all materials of interest continually work-harden in shear in the absence of material rotation, then other explanations are needed for many of the interesting observations of shear localization in the bulk of material. Adiabatic heating in metal processing with high local rates of deformation, time-dependent effects that are the equivalent of water migration in soil masses, effects of hydrostatic pressure, and other reasons for localized shearing in the absence of rotation of material or loading are needed.

Shear instabilities for a stable material are an artefact of such finite strain calculations when Jaumann spin terms are combined with kinematic hardening [21]. So also, despite its stability, is the curve obtained for torsion when the actual limited continuum rotation is employed instead of the unlimited cumulative spin [22].

12.7 A PRESSURE OR SD EFFECT IN STRUCTURAL METALS

With the recognition that plastic deformation is basically a shear deformation response to shear stresses, it is not unreasonable to expect that added hydrostatic pressure should raise the required shear stress level because it packs the material more closely together and inhibits dislocation motion. However, the effect should not be appreciable for a compact material without holes or cracks until the pressure is high enough to produce a significant change in atomic spacing, a finite elastic strain that cannot be neglected. Schmidt's law for single crystals, giving full credit to the shear stress on the slip plane, or the usual tension versus compression tests on common carbon structural steels or structural aluminium alloys, involve changes in hydrostatic pressure or of normal stress on the slip plane that range from miniscule to small in comparison with the elastic bulk or Young's modulus. Bridgman in his pioneering tests went to far higher pressures. He did not report that the effect of hydrostatic pressure on the yield stress in shear was zero. The 10 per cent or so increase registered in his apparatus, when no phase change took place, was quite appropriately viewed by him as not large enough to invalidate the idealization. Neglect of any effect of pressure on yielding was and is an entirely reasonable idealization for most structural metals and alloys under almost all practical situations.

There was great excitement in the materials community when reports began to circulate in the early 1970s of a giant strength differential or SD effect in martensitic steels, the yield strength in compression 30 to 40 per cent above the yield strength in tension. There was considerable confusion initially, and some still may remain, about whether or not such an effect was associated with hydrostatic pressure or instead was of the more familiar Bauschinger type, or was due to such known phenomena as the opening of cracks at interfaces or of voids around inclusions. Disappointingly, by the time genuine SD effects were separated out from the spurious ones based on inhomogeneity of stress and material on the microscale, it also turned out that the difference between tensile and compressive yield and flow strengths in the high strength martensitic steels was down to a less startling and more easily explained 10 per cent.

A side remark seems in order here. When some interesting and unexpected primary or secondary effect is discovered in a scientific study, considerable money and manpower is devoted to its elucidation and properly so. Scientific curiosity provides ample justification, as indeed it should. Strangely, when an effect of possible practical as well as scientific interest is uncovered in the course of engineering research, attention fades rapidly unless the practical effect is found to be large in current applications. Would it not be reasonable to expect that every highly competent materials engineering group would have a facility in which constitutive relations could be determined under high hydrostatic pressure and other extreme environments? It is obvious that with time there will be more and more important engineering applications of materials under what we now consider to be extreme conditions. Such a facility will be expensive, but not in comparison with a variety of routine tools available and used in many fields of basic science.

Fundamental questions remain unanswered. Is the effect of high pressure, one uniquely determined by a change in elastic modulus with atomic spacing, or the more complex elastic picture involving interstitials [6], or a continuum plasticity effect with an upper limit determined by the theoretical strength of the material [7], or some combination, or none of the above? Crucial tests are called for, but no one instead of everyone is in a position to do them.

If the effect of hydrostatic pressure in the absence of voids, even though it may be moderate, is to be placed within the domain of elastoplastic behaviour, the compartmentalization of earlier thinking that separated the behaviour of metals and alloys from that of soils, rocks, and polymeric materials becomes less comfortable. None of the physical and mathematical uncertainties of the application of plasticity theory to soil and rock mechanics appeared relevant to structural metals at the time. Flow strengths of structural steels of about 1400 MPa (200 000 p.s.i) must be reached before hydrostatic or normal pressure effects on tensile yield strength versus compressive yield strength become noticeable, say more than a 5 per cent difference. Now that such steels are in common use they raise some of the same issues of instability that are so troublesome for soils.

As for over-consolidated soils it is reasonable to expect that any permanent increase in volume will approach an asymptotic limit rather rapidly as shear deformation continues. Unlike over-consolidated soils, however, work-hardening is found to accompany the entire volume expansion. The response does appear to be stable in the small. Unfortunately, no data exist to confirm or disprove any proposed sketches of successive yield surfaces [7] on which idealizations then could be based. Suppose, in the absence of real information, that the response is stable or at least neutrally stable for small plastic deformations at all times. For this postulated well behaved material, just as for the corresponding idealization proposed for soils [23], the familiar stable pictures of successive yield surfaces piling up tangentially against a limit surface do hold in shear, but not for the combination of varying hydrostatic pressure and shear.

The real system suffers from a degree of instability in the large; the greater the pressure dependence the greater the degree of this instability. Work can be extracted from the system of a body composed of a pressure dependent inelastic material and the forces acting on it. Selective release of hydrostatic pressure may lead to a genuine and unwanted instability, not merely a computational instability, the same problem that must be faced in soil and rock mechanics. Fortunately, for metals and alloys in almost all practical situations, the pressure dependence is sufficiently small and the range of possible hydrostatic pressure sufficiently limited that it is possible to play safe by choosing the yield strength in shear at the low end of the permissible range and not pay too high a penalty for ignoring the increase of shear strength with hydrostatic pressure.

If the effect is large or if playing safe is not considered efficient, then, in reality and in idealization, initial stresses do matter; the path of loading is very important; and all the comforting implications of limiting theorems which hold in the deviator space disappear. Unfortunately, it is rare in practice that the initial state of stress is known or can be determined accurately; it is rare that the path of loading, as distinguished from the extremes or spectrum of loading, can be specified. Consequently, pressure dependence of inelastic behaviour is an extremely troublesome phenomenon that requires much additional study for both infinitesimal and finite elastoplasticity. When changes in hydrostatic pressure occur that are of the order of the theoretical strength of the material, changes in elastic strains are large. The caveats about possible lack of normality and convexity then do require attention. However in normal structural use, the elastic strain range of these high strength steels is but a few per cent and does not require a finite elastoplastic approach.

12.8 AN HISTORIC LOOK

Thirty years ago, when the theory of plasticity was undergoing a period of very rapid development, there was a clear need for a combined experimental and theoretical approach to the fundamental questions that were so troublesome then and seem so transparent now [24, 25]. For many years following a clarification of rather simple mathematical formalisms and a number of critical experiments, the mechanics community felt comfortable with the basic framework of the theory, its drastic simplifications, and its applications. Use was made only of the simplest of work-hardening forms. Most often, ideal or perfect plasticity was considered amply satisfactory for the solution of problems. Structures were designed and built on the basis of plastic limit theorems and shake-down theorems.

With the passage of time, as so often is the case, too much came to be expected of the elementary theory. Computing capability had advanced by many orders of magnitude. More and more experiments had been run. To the surprise of newcomers to the field, some of the idealizations of reality codified in the time-independent theory of plasticity were found to be in appreciable disagreement with the results of very sensitive tests, especially those performed on soft materials or materials at temperatures elevated for them. In part the unrest was stimulated by those of us who pointed out the need to take time effects into account in many problems, who made connection with dislocation motion and structure, who made distinctions between frictional and plastic behaviour and exhibited the existence of non-associated flow rules [26]. However, the greatest distress was displayed by those who took the 1940s and 1950s as their starting point in time and rediscovered what the pioneers from Tresca on through Huber, Hencky, and von Mises understood very well: time-independent plasticity is not reality itself but is a moderate to strong idealization of reality. The greatest contribution of these earliest workers was in fact to disregard the all-too-obvious time effects they saw and create a useful time-independent theory. Disappointment has been so extreme in some instances that those very aspects of plasticity are completely abandoned that make it of such great use in practical analysis and design. Yield surfaces often are thrown out and rate-dependent expressions occasionally are taken as the starting point even for quasi-static loading of materials exhibiting negligibly small time effects. There can be no quarrel with the replacement of one idealization by another either for convenience of computation or for an effort at fundamental understanding. However, when great confusion arises about the role and purpose of idealization for time-independent as well as time-dependent behaviour, practitioners are misled and the education of new entrants to the field is placed in jeopardy.

Each set of idealizations is valid that contributes positively to the current state of the art or to the future advancement of science or engineering. Unifying principles based on idealization are of great value when the idealization covers a useful range of physical behaviour. There is a need for more and more elaborate theory to comprehend and advance the continually growing body of knowledge about material response. There is an equal need for drastic idealization to permit approximate but satisfactory answers to be obtained to practical problems of tremendous complexity.

As pointed out earlier, the class of clear and incisive experiments that can be performed on materials is very limited and possible abstractions or generalizations are rarely unique. Conclusions drawn will depend upon the level and complexity of the experiments and the accuracy of measurement. As an example, when strain increments in a carefully performed thin-walled tube test are measured to within a small multiple of 10^{-6}, experiments at room temperature on stable structural alloys do demonstrate clearly in a two- or three-dimensional stress space that an incremental plasticity theory with at most sharply rounded yield or loading surfaces rather than corners is an excellent idealization for the test conditions. When strain increments are measured to 10^{-4}, corners are likely to be reported. Yet greater confusion can result from less fundamental experiments. For example the inference from the results of complicated and (almost) proportional loading applied to a complex structure might well be that a total or deformation theory gives satisfactory answers. Elastic plastic fracture mechanics takes advantage of this (almost) fact in its use of the J-integral.

Many of us remember the turmoil in plastic buckling theory resulting from apparently contradictory 'facts' and can only hope the lesson is not lost that experiments can only decide fundamental questions *at the level of the experiments.* Despite the great controversy then and the continuing difficulty today in matching buckling data easily with an accepted incremental plastic stress–strain relation, there really was no fundamental issue to be resolved. No matter how many tests are run that fit the predictions of a particular deformation theory well, and no matter how badly a particular incremental theory may do, the basic thermodynamic requirement for an incremental theory remains.

There is also no basic unresolved issue in the use of plastic stress–strain relations with conventional yield or loading surfaces as opposed to those with no yield surface (zero elastic response domain). If the least dislocation motion is thought of as significant plastic deformation, then indeed there is a zero elastic domain for real metals and alloys. If inelastic strains of order 10^{-5} are thought of as ignorable, the elastic domain of structural alloys is appreciable at room temperature and below. If 10^{-3} can be ignored, the diameters of initial and subsequent yield surfaces are very large. Whether or not a yield surface should be employed, and if it is what size it should be, is a matter of idealization, a matter of taste or convenience [27]. However, the essence of the physical behaviour to be modelled, as represented by the yield surfaces of conventional theory, must be incorporated in whichever choice is made. The incremental stress versus incremental strain picture must be acceptably close to reality for each allowable path for each idealization. If reverse as well as forward loading is to be included, then the (almost) elastic response domain and the very different tangent modulus for continued loading and reversed loading do have to be reproduced similarly in each acceptable idealization. It is only wishful thinking to imagine that, for complex paths of loading, the elimination of the yield surface will permit calculations of moderate accuracy without the equivalent of the 'bookkeeping' that is involved in keeping track of the yield surfaces in conventional plasticity. A record must be kept of the extent and direction of previous loading and taken into account properly if a reasonable estimate is to be made of the response to a radically new direction and extent of loading, whether or not the yield surface of the idealization is large or instead considered as shrunk to zero.

Of course, if the loading is close to proportional and monotonic, simpler forms without an elastic domain or its equivalent may well prove to be suitable despite their improper behaviour on unloading. No more should be expected of any such representation, however, than of deformation theory, which really is non-linear elastic theory. No effort, therefore, should be spent on a thermodynamic or on any other than pragmatic justification. A purist might wish to avoid forms that would violate the laws of thermodynamics for cycles of loading and unloading.

Some of the lessons of the past can serve as a guide, or at least a warning, as finite elastoplasticity develops for those classes of problems and materials where it may be needed. A recurring lesson to be heeded is that the following of concepts and mathematical procedures entirely suitable for elastic response is all too likely to lead to error in the physical interpretation or mathematical description of plastic response. The finite elastic response of initially isotropic fully elastic materials is especially misleading because the continuum rotation which plays so central a role in isotropic elasticity must be replaced by the very different rotation of material for plastic response.

12.9 CONCLUSIONS

The experimental information on which our understanding of plastic stress–strain behaviour of metals has been built is very limited. Elastic strains have been very small and there has been almost no rotation of crystal lattice orientations in test specimens. Both reflect the world outside the laboratory, however, so that results are directly applicable to most of the practical problems that must or should be solved. Conventional incremental stress–strain relations familiar in infinitesimal elasticity and plasticity apply in an orthogonal coordinate system that follows any changes in orientation of the material, not the continuum rotation so commonly used that plays so central a role in isotropic elasticity. Consequently much of the concern about the instability of bulk material in shear is misplaced and many of the calculations involving finite deformation in shear are in error in principle.

The value of a finite plastic deformation theory is less than obvious for the needed incremental stress–strain relations when elastic strains are very small and the rotation of physically appropriate reference axes is not related to the local continuum rotation computed from the displacement field.

REFERENCES

1. R. Hill, 'Some basic principles in the mechanics of solids without a natural time', *J. Mech. Phys. Solids*, **7**, 209–225 (1959).
2. R. Hill and J. R. Rice, 'Elastic potentials and the structure of inelastic constitutive laws', *SIAM J, Appl. Math.*, **25**, 448–461 (1973).
3. R. Hill 'Aspects of invariance in solid mechanics', in *Advances in Applied Mechanics*, vol. 18, pp. 1–75, Academic Press, 1978.
4. J. Mandel, 'Equations constitutives et directeurs dans les milieux plastiques et viscoplastiques', *Int. J. Solids Structures*, **9**, 725–740 (1973).
5. J. Mandel 'Définition d'un repère privilégié pour l'etude des transformations anelastiques du polycristal', *Jl. de Mécanique théorique et appliquée*, **1**, (1), 7–23 (1982).
6. J. P. Hirth and M. Cohen, 'On the strength-differential phenomenon in hardened steel', *Met. Trans.*, **1**, 3–8 (1970).
7. D. C. Drucker, 'Plasticity theory, strength-differential (SD) phenomenon, and volume expansion in metals and plastics', *Met. Trans.*, **4**, 667–673 (1973).
8. W. Olszak, Z. Mróz and P. Perzyna, *Recent Trends in the Development of the Theory of Plasticity*, Macmillan–Pergamon, New York, 1963.
9. J. B. Martin, *Plasticity: Fundamentals and General Results*, MIT Press, Cambridge, Massachusetts, 1975.
10. W. F. Chen, 'Limit analysis and soil plasticity', *Developments in Geotechnical Engineering*, vol. 7, Elsevier, Amsterdam, Oxford, New York, 1975.
11. J. W. Hutchinson and V. Tvergaard, 'Surface instabilities on statically strained plastic solids', *Int. J. Mech. Sci.*, **22**, 339–354 (1980).
12. J. W. Hutchinson and V. Tvergaard, 'Shear band formation in plane strain', *Int. J. Solids Structures*, **17**, 451–470 (1981).
13. D. C. Drucker, 'A more fundamental approach to stress–strain relations', *Proc. 1st U.S. National Congress for Applied Mechanics, ASME*, pp. 487–491, June 1951.
14. D. C. Drucker, 'Plasticity', in *Structural Mechanics*, Proc. First Symposium on Naval Structural Mechanics, August 1958, (Eds. J. N. Goodier and N. J. Hoff), pp. 407–455, Pergamon Press, 1960.
15. D. C. Drucker, *Introduction to Mechanics of Deformable Solids*, McGraw-Hill, New York, 1967.
16. L. Palgen, *The Structure of Stress–Strain Relations in Finite Elasto-Plasticity*, Ph. D. Thesis, University of Illinois at Urbana-Champaign, T & A.M. Report 452, Oct. 1981, UILU-ENG 81-6006.
17. R. Hill, 'A new method for determining the yield criterion and plastic potential of ductile metals', *J. Mech. Phys. Solids*, **1**, 271–276 (1953).
18. M. A. Biot, *Mechanics of Incremental Deformations*, Wiley, New York, 1965.
19. R. M. McMeeking and J. R. Rice, 'Finite-element formulations for problems of large elastic–plastic deformation', *Int. J. Solids Structures*, **11**, 601–616 (1975).

20. E. T. Onat and D. C. Drucker, 'Inelastic instability and incremental theories of plasticity', *J. Aero. Sci.*, **20**, 181–186 (1953).
21. J. C. Nagtegaal and J. E. de Jong, 'Some aspects of nonisotropic workhardening in finite deformation plasticity', *Proc. Workshop, Plasticity at Finite Deformation* (Ed. E. H. Lee), Stanford University, 1981.
22. E. H. Lee, R. L. Mallett and T. B. Wertheimer, 'Stress analysis for kinematic hardening in finite-deformation plasticity', SUDAM Rept. 81–11.
23. D. C. Drucker, R. E. Gibson and D. J. Hankel, 'Soil mechanics and work-hardening theories of plasticity', *Proc. ASCE*, **81**, Sep. No. 798, Sept. 1955, *Trans. ASCE*, **122**, 338–346 (1957).
24. D. C. Drucker, 'Relation of experiments to mathematical theories of plasticity', *J. Appl. Mech.*, **16**, *Trans. ASME*, **71**, 1949, pp. A349–A357.
25. R. Hill, *The Mathematical Theory of Plasticity*, Oxford Press. London, England, 1950.
26. D. C. Drucker, 'Coulomb friction, plasticity, and limit loads', *J. Appl. Mech.*, **21**, *Trans. ASME*, **76**, 71–74 (1954).
27. D. C. Drucker, 'Concepts of path independence and material stability for soils', in *Rheologie et Mechanique des Sols*, Proc. IUTAM Symposium, Grenoble, April 1964 (Eds. J. Kravtchenko and P. M. Sirieys), Springer, 1966, pp. 23–43.

Mechanics of Engineering Materials
Edited by C. S. Desai and R. H. Gallagher

Chapter 13

Dynamic Response Analysis of Soils in Engineering Practice

W. D. Liam Finn

13.1 INTRODUCTION

The development of constitutive relations for soils has been one of the most active research areas in geotechnics in recent years. In 1980, an international workshop [1] on generalized stress–strain relations in geotechnical engineering was held in Montreal to test the predictive capability of twelve selected constitutive relations on the same data base. The discussions of the predictions and the spirited probing of the foundations and applications of the constitutive relations by the participants resulted in a clearer understanding of the limitations of existing relations and the substantial inherent difficulties in developing truly general constitutive relations for soil. The many generalized stress–strain relations available may, for the most part, be classified into four categories: elastic–plastic with isotropic hardening, elastic–viscoplastic, elastic–plastic with combined isotropic and kinematic hardening, and endochronic theory. Many of these stress–strain relations have been incorporated in computer programs for the analysis of boundary value problems in geotechnical engineering.

A major stimulus for the development of general constitutive relations has been the need to analyse the response of soils and soil-structure systems to cyclic loading. Typical examples of important cyclic loading problems are the response of nuclear power plants and earth dams to earthquake loading and offshore oil and gas production platforms to storm wave and earthquake loading. Despite the importance of these problems and the need for more representative and comprehensive analyses of them, the vast amount of research on constitutive relations in recent years has had little impact on geotechnical engineering practice in these areas.

Against the background of active research in constitutive relations which has resulted in a plethora of new models of soil behaviour, it is instructive to review the slow evolutionary development of the methods of dynamic analysis used in geotechnical engineering practice.

13.2 ANALYSIS IN ENGINEERING PRACTICE

Dynamic response analysis as practised today had its origins in the pioneering attempts of Seed and his co-workers at the University of California at Berkeley to explain in a quantitative way the extensive liquefaction of saturated sands that occurred in 1964 during the earthquakes in Alaska and Niigata, Japan. The liquefaction failures occurred primarily on level ground and in formulating a procedure for analysis, Seed and Idriss [2] made two basic assumptions: (a) seismic excitation is primarily due to shear waves propagating vertically and (b) level ground conditions may be approximated by horizontal layers with uniform properties. Under these conditions the ground deforms in shear only and may be analysed by treating a vertical column of soil as a shear beam.

Seed and Idriss [2] included the non-linear hysteretic stress–strain properties of the sand by using an equivalent linear elastic method of analysis. The method was originally based on a lumped-mass mechanical model of a sand deposit resting on a rigid base to which the seismic motions were applied. Later the method was generalized to a wave-propagation model with an energy transmitting boundary. The seismic excitation could be applied at any level in the new model.

The generalized method was incorporated in the computer program, SHAKE [3]. This program is widely used to predict the distribution of ground motions due to seismic excitation for liquefaction studies and to provide input motions for soil–structure interaction analyses.

The success of the one-dimensional equivalent linear model in describing the response of level ground during earthquakes fixed the direction of development of dynamic response analysis in engineering practice for the next ten years. The basic model was generalized to two and three dimensions in the finite element computer programs QUAD-4 [4], LUSH [5], and FLUSH [6]. These finite element programs, all based on the concept of modelling soil by an equivalent linear solid, are the most frequently used programs for the dynamic response analysis of soil structures such as slopes and earth dams and for the solution of dynamic soil structure interaction problems such as the response of embedded nuclear reactor structures to earthquake loading.

13.3 THE EQUIVALENT LINEAR METHOD

The fundamental assumption of the equivalent linear method of analysis is that the dynamic response of soil, a non-linear hysteretic material, may be approximated satisfactorily by a damped elastic model if the properties of that model are chosen approximately. The appropriate properties are obtained by an iterative process.

In finite element analyses, the stress–strain properties of the soil are defined in each finite element by the Poisson ratio and shear strain dependent shear moduli

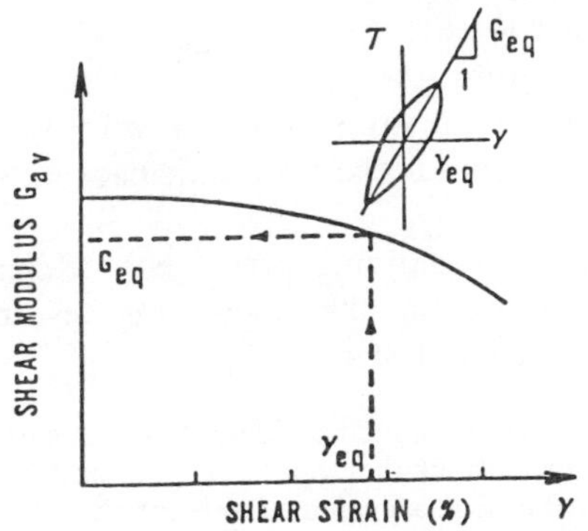

Figure 13.1 Secant modulus versus shear strains

and equivalent viscous damping ratios. An equivalent modulus and damping ratio at any strain level are determined from the slope of the major axis of the hysteresis loop corresponding to that strain (Figure 13.1) and the area of the loop respectively [2]. Initial values of moduli and damping ratios are selected corresponding to small shear strain values ($\gamma = 10^{-4}$ per cent) or to strain levels judged appropriate for the anticipated earthquake or cyclic loading and an elastic analysis is carried out for the entire duration of the earthquake. An average or effective strain (usually assumed to be about 65 per cent of the maximum value) is computed in each finite element and moduli and damping ratios are selected compatible with these average strains and used for the next iteration. The procedure is repeated until no significant changes in moduli or damping ratios are necessary. The response determined during this last iteration is considered to be a reasonable approximation to the non-linear response.

Since the final analysis with strain compatible soil properties is purely elastic, the permanent deformations caused by earthquake shaking or wave loading cannot be computed by this type of analysis. The computed strains may bear no relation to strains in the field and are used only for deriving the strain compatible properties. However, the stresses derived from these strains are assumed to be representative of stresses in the ground. The accelerations are also assumed to be reasonably representative of field values. Finn *et al.* [7, 8, 9] have shown by comparison of the results of iterative elastic and true non-linear dynamic response analyses of level ground that the above assumptions are reasonable for stable soils, that is, for soils which do not develop significant pore-water pressures during earthquake excitation and if pseudo-resonance does not occur during the equivalent linear analysis.

Since Newmark's [10] classic paper on seismically induced deformations in dams it has been accepted in engineering practice that the seismic performance of soil structures should be evaluated in terms of deformations rather than in terms of factors of safety. Since the iterative elastic method does not allow the direct computation of deformations, the performance of the soil structure must be deduced from the computed stress histories with the aid of appropriate laboratory test data. This is one of the more difficult aspects of dynamic response analysis.

The state of the art for analysing permanent deformations was recently assessed in a report on earthquake engineering research by the National Research Council of the United States [11]:

> Many problems in soil mechanics, such as safety studies of earth dams, require that the possible permanent deformations that would be produced by earthquake shaking of prescribed intensity and duration be evaluated. Where failure develops along well-defined failure planes, relatively simple elastoplastic models may suffice to calculate displacements. However, if the permanent deformations are distributed throughout the soil, the problem is much more complex, and practical, reliable methods of analysis are not available. Future progress will depend on development of suitable plasticity models for soil undergoing repetitive loading. This is currently an important area of research.

The equivalent linear methods of dynamic analyses described above are conducted in terms of total stresses. They do not contain a model for the generation and dissipation of pore water pressures during excitation. Soil properties such as strength and stiffness are dependent on effective (intergranular) stresses and may change significantly as pore water pressures increase during dynamic loading. Comparative studies by Finn *et al.* [9] on the response of saturated level sandy sites to seismic loading indicate that total stress analysis tends to overestimate the dynamic response when the pore water pressures exceed about 30 per cent of the effective overburden pressures. In such cases dynamic analysis should be conducted in terms of effective stresses. The dual needs to predict permanent deformations and to conduct analyses in terms of effective stresses has led to the development of non-linear effective stress methods which model directly the non-linear hysteretic stress–strain response of soils to cyclic loading. These methods have been fully developed for 1-D problems such as the response of level ground to earthquake excitation and are now being used in practice.

13.4 ONE-DIMENSIONAL NON-LINEAR ANALYSIS

Streeter *et al.* [12] produced the first true non-linear analysis of the seismic response of soils in 1973. They incorporated a Ramberg–Osgood representation of the stress–strain behaviour of soil into the equations of motion and solved the equations by the method of characteristics. The method of analysis is incorporated into the computer program, CHARSOIL.

In 1976, Finn *et al.* [7] developed a non-linear method of analysis in which the equations of motion were integrated directly. The stress–strain behaviour of sand was represented by the hyperbolic initial loading curve (skeleton curve) shown in

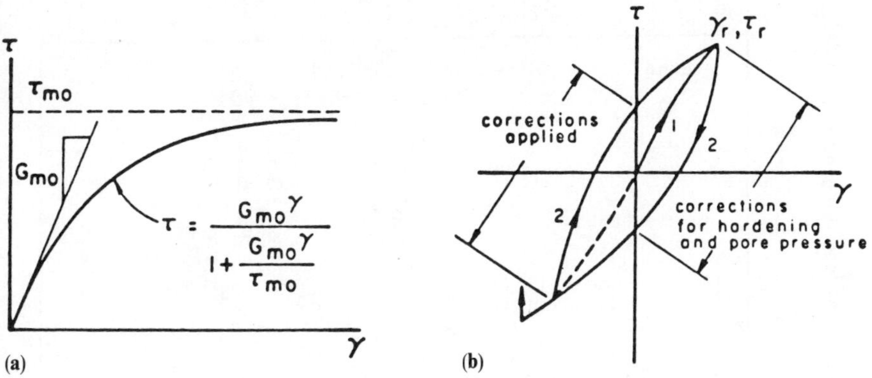

Figure 13.2 (a) Initial loading curve. (b) Masing stress strain curves for unloading and reloading [7]

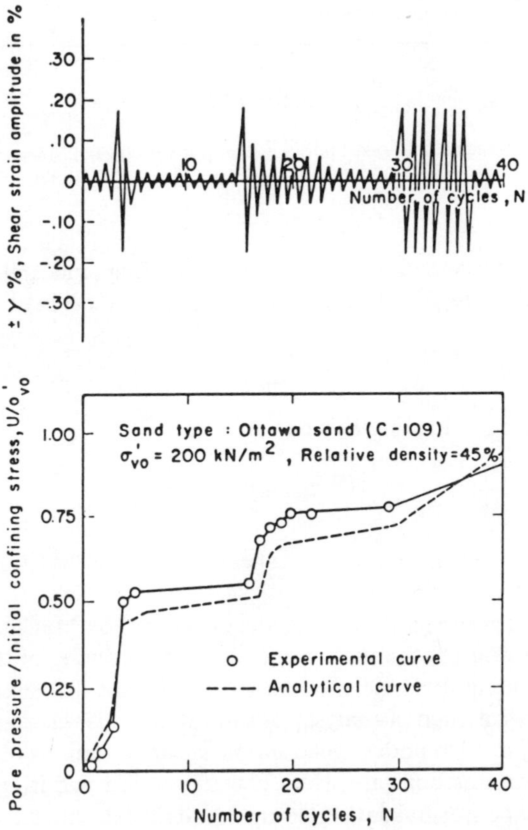

Figure 13.3 Predicted and measured pore water under irregular cyclic loading [15]

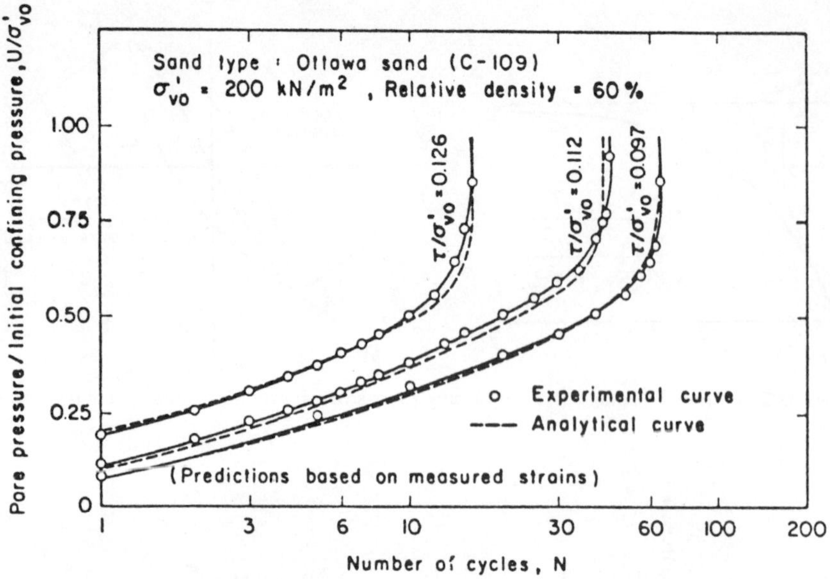

Figure 13.4 Predicted and measured pore water pressure in constant stress cyclic simple shear tests, $D = 60\%$ [15]

Figure 13.2(a). Stress–strain curves during loading and unloading (Figure 13.2(b)) were defined by the Masing criterion [13]; i.e., if the loading stress–strain curve is given by

$$\tau = f(\gamma) \tag{13.1}$$

then the unloading–reloading curve is given by

$$\frac{\tau - \tau_r}{2} = f\left(\frac{\gamma - \gamma_r}{2}\right) \tag{13.2}$$

in which τ, γ = current shear stress and shear strain respectively and (τ_r, γ_r) define the reversal point.

A porewater pressure generation model developed by Martin, *et al.* [14] was coupled with the non-linear equations of motion to provide, for the first time, the option of performing dynamic effective stress analyses if desired. An extensive verification of this method of analysis for saturated sands was conducted by Finn and Bhatia [15] in 1980 under cyclic simple shear conditions using a variety of cyclic loading patterns, constant stress, constant strain, and irregular loading. A comparison of experimental data and response data calculated using the effective stress method of analysis for an irregular strain-history pattern is given in Figure 13.3. A similar comparison for three constant stress cyclic simple shear tests is given in Figure 13.4. It appears that the effective stress method of analysis is capable of making quantitatively useful predictions of dynamic response. For further corroboration references [14] and [16] should be consulted.

The Finn *et al.* [7] method is incorporated in two computer programs, DESRA-1 [16] and DESRA-2 [17]; the latter includes an energy transmitting boundary. The programs may be operated in either the total or effective stress mode and contain various options regarding the diffusion and drainage of the pore water under the seismically induced pore water pressure gradients.

13.5 COMPARATIVE STUDIES

Finn *et al.* [7, 8] have investigated the validity of the equivalent linear method for determining the dynamic response of a non-linear hysteretic solid by analysing the response of level sites by SHAKE and the two non-linear programs CHARSOIL and DESRA. The situation most suited to equivalent linear analysis is the response to steady state sinusoidal motion. Maximum acceleration responses for a deep cohesionless site determined by SHAKE and DESRA are shown in Figure 13.5. The results are very similar except around a frequency of 1 Hz where SHAKE shows a tendency towards resonant response.

The acceleration response spectra of ground motions at a sandy site 15 m deep (Figure 13.6) which were computed by SHAKE, CHARSOIL, and DESRA are shown in Figure 13.7. The spectra all show strong response around a period of 0.5s but SHAKE shows much stronger response than the non-linear programs. This stronger response is also reflected in the magnitudes of computed dynamic shear

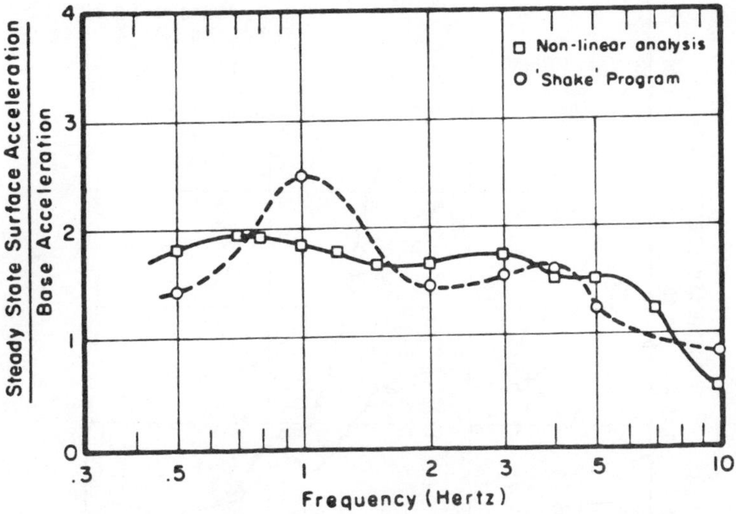

Figure 13.5 Acceleration responses for deep sandy site by non-linear and equivalent linear analysis [7]

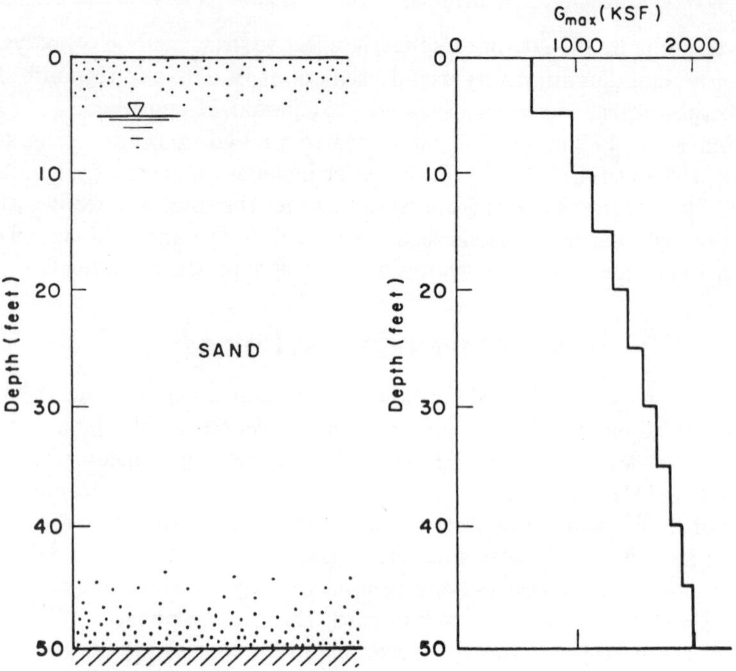

Figure 13.6 Soil profile used for response analysis [8]

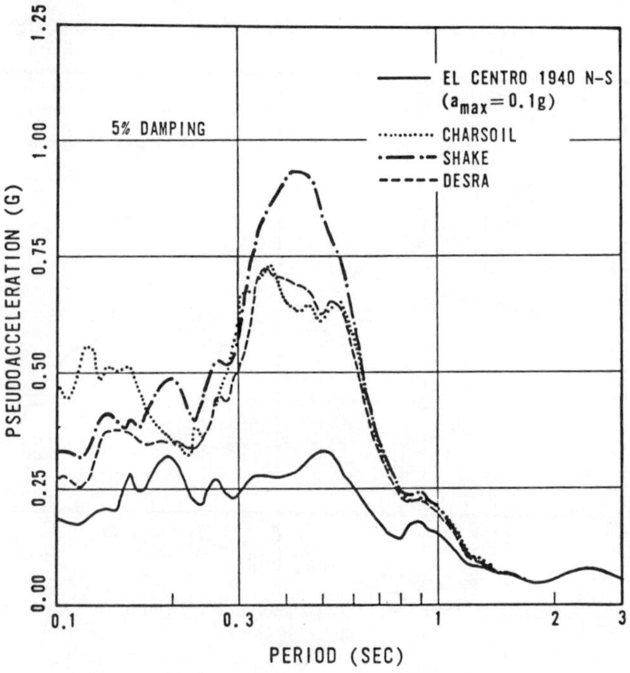

Figure 13.7 Acceleration response spectra—total stress analyses [8]

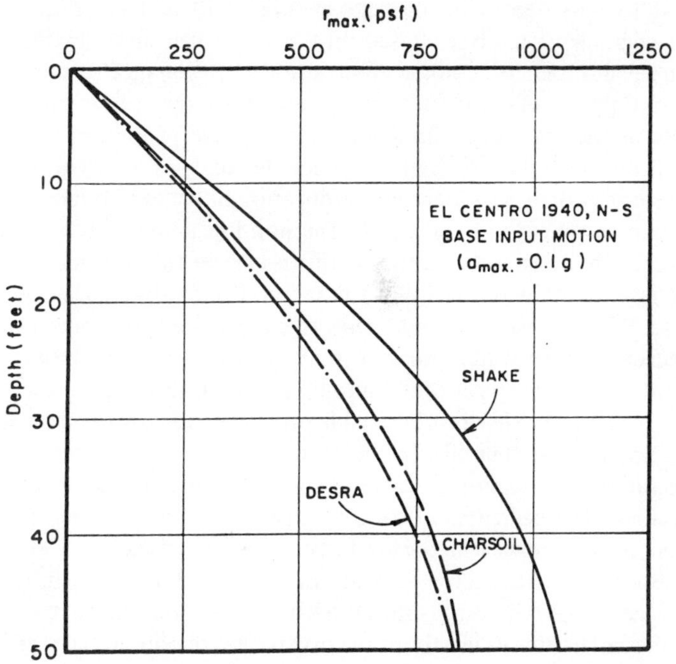

Figure 13.8 Maximum shear stress distributions—total stress analyses [8]

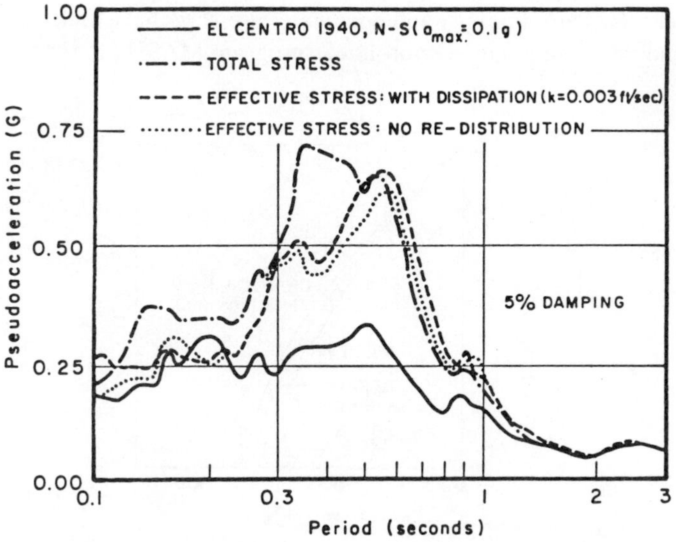

Figure 13.9 Acceleration response spectra—effective stress analyses versus total stress using DESRA [8]

stresses at various depths in the deposit (Figure 13.8). This tendency towards resonant response in analyses based on the equivalent linear method has been noted in a number of comparative studies. Resonance occurs when the fundamental period of the input motion corresponds to the fundamental period of the site as defined by the final set of compatible properties in the iterative equivalent linear method of analysis. Since the analysis is carried out with this constant set of properties for the entire duration of the earthquake, there is time for resonant response to build up. In the non-linear methods this tendency is controlled by the constantly changing stiffness properties. When strong resonant response is a function primarily of the method of analysis it is called pseudo-resonance. Pseudo-resonance may lead to exaggerated dynamic response.

The effect of pore water pressure may be seen from the plots of pseudo-acceleration spectra for 5 per cent damping in Figure 13.9. The spectrum for total stress analysis by DESRA reproduced from Figure 13.8 shows a maximum at a period of about 0.35 s. The effect of increasing pore water pressure on the moduli is not included in the analysis. Increases in pore water pressure during seismic excitation leads to decreases in effective stresses and a softening of the moduli which results in an increase in the fundamental period of the site to about 6 s. This shift in period is illustrated by the peaks in the spectra determined by non-linear effective stress analysis, also using DESRA. The various effective stress spectra are for different assumptions about the dissipation of pore water pressure during excitation.

Porewater pressure also affects the transmission of wave energy and alters the acceleration and shear stress responses in the ground [7].

These effects of non-linearity and pore water pressures have been confirmed by Martin and Seed using the 1-D non-linear program MASH [18]. A comparison

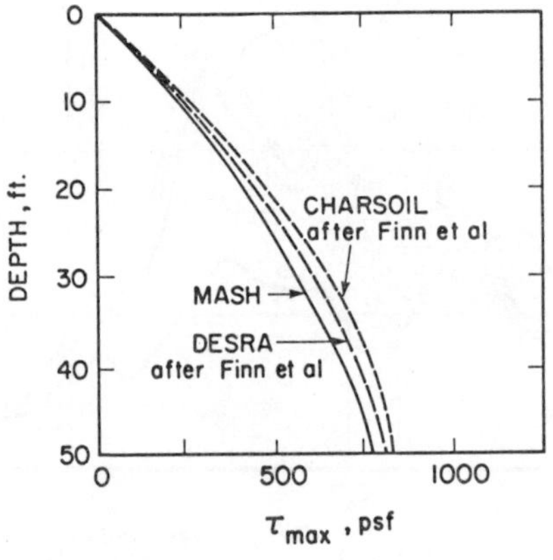

Figure 13.10 Shear stresses by three non-linear methods (Martin and Seed [18])

of shear stress distributions with depth for the site in Figure 13.6 computed by DESRA and MASH is shown in Figure 13.10.

A comprehensive verification of non-linear effective stress analysis using field data is described in the next section.

13.6 CASE HISTORY: SEISMIC RESPONSE OF SAND ISLAND [19, 20]

Owi Island No. 1 is an artificial island located on the west side of Tokyo Bay. The island is triangular in shape with side lengths of 4 km, 3.5 km, and 2.5 km and was constructed over a period of eight years from 1961–1969 with materials dredged from the nearby sea-bed. The depth of water was approximately 10 m. The

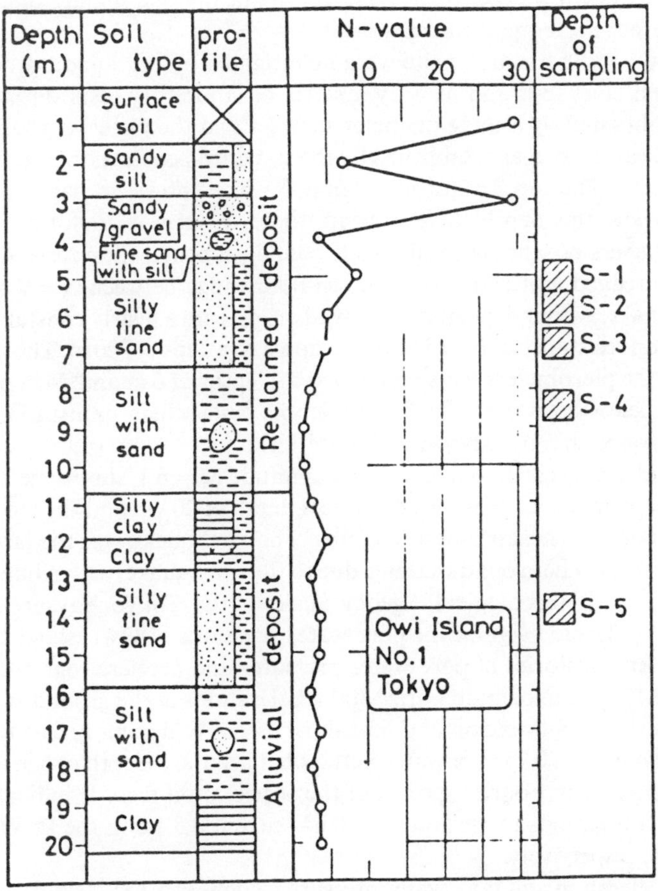

Figure 13.11 Soil profile at test site [19]

dredged materials were pumped through pipes on to the site until the water level was reached. At this point, waste materials from construction sites were dumped on top of the dredged fill.

A test site at the south end of the island is instrumented to record pore water pressures and ground accelerations during earthquakes. Pore water pressures are recorded by piezometers installed at depths of 6 m and 14 m. The transducer in each recorder is of the strain-gauge type with a full capacity of 200 kN/m^2. A two-component seismograph is installed on the ground surface to measure horizontal acceleration. The horizontal accelerometers are electromagnetic and self-starting with a full range of 1000 gals and a frequency range of 0.1 to 30 Hz. A vertical accelerometer provides the signal that triggers a strip chart recorder. After amplification the recorded acceleration and pore water signals are sent to an oscillograph to provide visual records that can be correlated readily with each other. The system was set to provide automatically one minute of continuous recording per earthquake shock.

The soil conditions at the site were investigated by the standard penetration test, Dutch cone test, and also by a series of laboratory tests on undisturbed samples obtained by a large diameter sampler and the Osterberg sampler. The soil profile at the site, as established by the standard penetration test, is shown in Figure 13.11. The top 3 m formed dumped waste shows no particular stratification. Below this depth there appears to be a series of uniformly deposited alternate layers of sandy and silty soils. Standard penetration resistances in the pumped dredged material above the sea-floor range between $N = 2$ and $N = 5$. The original sea-floor deposit is of alluvial origin with a nearly constant standard penetration resistance of $N = 3$ to $N = 5$ down to a depth of 20 m. The sand layers in which the piezometers were embedded at depths of 6 m and 14 m had almost identical below counts of $N = 5$. The depths from which undisturbed samples were recovered are also shown in Figure 13.11.

The Mid-Chiba earthquake, with a magnitude $M = 6.1$, shook the Tokyo Bay area on September 25, 1980. It had a focal depth of 20 km and the epicentre was located about 15 km south-east of Chiba. The earthquake was the largest in the area since 1929. The ground shaking, due to the earthquake, was of intensity V on the Japanese Meteorological Agency Scale in the Tokyo Bay area and was sufficient to develop significant pore water pressures in Owi Island No. 1. The recorded time histories of pore water pressures and accelerations are shown in Figure 13.12. The maximum horizontal accelerations at the ground surface were 95 gals in the N–S direction and 65 gals in the E–W direction. The rise in pore water pressure was 0.75 m of water in the sand layer at a depth of 6 m and 1.32 m at a depth of 14 m. Fourier spectra of the acceleration records indicate that the predominant periods of motion were 0.64 sec and 0.5 sec in the E–W and N–S directions, respectively.

The acceleration and pore water pressure responses of Owi Island No. 1 to the Mid-Chiba earthquake were investigated by dynamic effective stress analysis

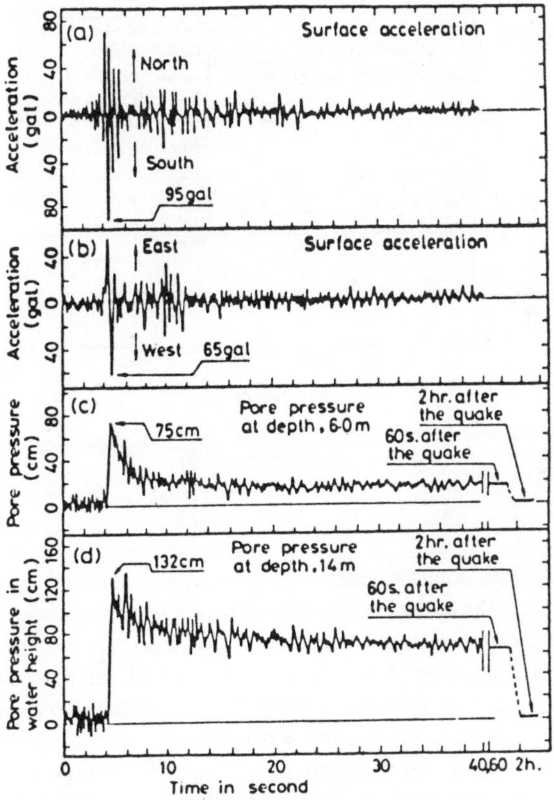

Figure 13.12 Accelerations and pore pressures recorded during Chiba earthquake [19]

using the computer program, DESRA-2. The required input for the program consists of stress–strain properties, shear strengths on horizontal planes, and parameter values for the pore water generation and dissipation model included in DESRA-2. Ideal procedures for determining these parameters have been described by Finn *et al.* [7]. Over the last few years of practical use, more approximate but still adequate procedures have evolved for developing the input to DESRA-2 from ordinary laboratory data. These procedures were followed here and will be described briefly.

Input motions were not available for the site. However, because the acceleration levels were modest and the pore water pressures were low enough not to affect soil properties to a significant extent, it was relatively easy to develop an input motion consistent with the ground acceleration. Shear wave velocities in the different layers were estimated from the standard penetration values using Imai's [21] correlations and small strain shear moduli were deduced from the

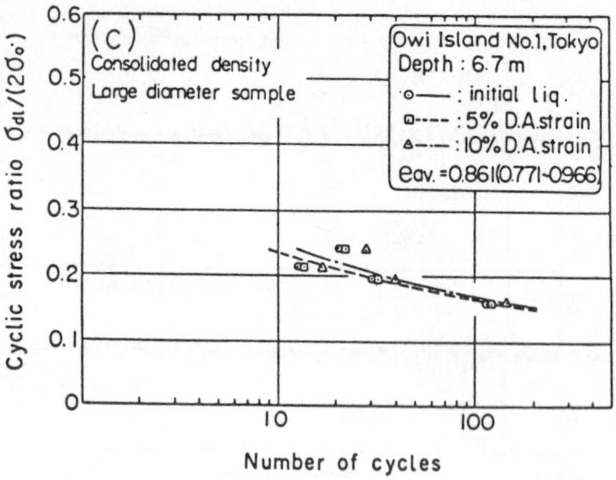

Figure 13.13 Liquefaction resistance of dredged sand at depth of 6.7 m [19]

shear wave velocities. Shear strengths of the sands were based on a friction angle of 30°. Shear strengths of the clay layers were deduced from correlations between moduli and shear strength given by Ishihara [22].

Resistance of the sands to liquefaction was investigated by means of cyclic triaxial tests on test specimens obtained from samples recovered by the large diameter sampler. Cyclic axial loads were applied under undrained conditions until the specimen deformed to a peak to peak axial strain of 10 per cent. Typical test results are shown in Figure 13.13 in terms of the cyclic stress ratio, $\sigma_{d1}/2\sigma'_0$, required to cause initial liquefaction, 5 per cent and 10 per cent double-amplitude axial strains. Initial liquefaction is defined as the state in which the pore water pressure is equal to the confining pressure. The cyclic stress ratio, $\sigma_{d1}/2\sigma'_0$, is the ratio of the cyclic axial load, σ_{d1}, to twice the confining pressure, σ'_0. The range in void ratio, e, of the test specimens and the average void ratio, e_{avg}, are also shown in Figure 13.13.

Pore water pressure parameters may be measured directly on samples of island material but current field practice with DESRA-2 was followed. This consists of selecting from the existing data files a sand with known pore water pressure parameters that had a liquefaction resistance curve of the same shape as that shown in Figure 13.13. These parameters were then scaled so that the liquefaction resistance curves matched and the rate of pore water development with cycles of loading were similar. The pore water pressures predicted by the model over the first cycle of cyclic loading were compared with those measured in triaxial tests. The results shown in Figure 13.14 indicate similar rates of pore water pressure development over a wide range of stress ratios.

Thus, all the properties required for the analysis of Owi Island No. 1 by

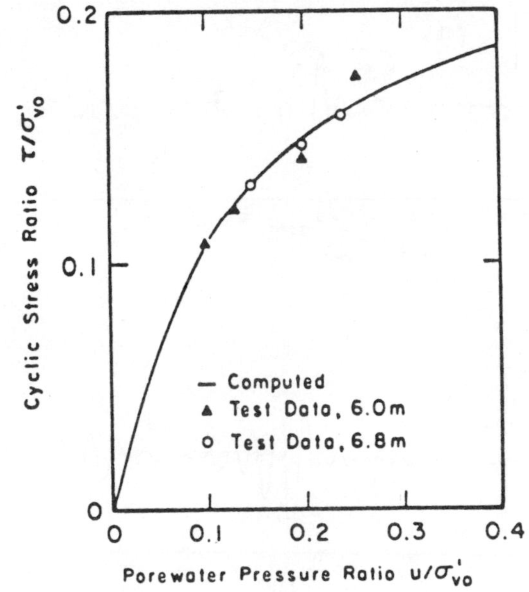

Figure 13.14 Rates of pore water pressure generation/first cycle [19]

DESRA-2 were obtained using data usually available from engineering site and laboratory investigations.

Full details of the instrumentation, recorded data, and the site investigations on Owi Island and the associated laboratory testing have been described by Ishihara *et al.* [23].

13.7 RESULTS FROM CASE HISTORY ANALYSIS

The first 10 s of the recorded ground accelerations in the N–S direction are shown to an expanded scale in Figure 13.15(a). During the first 4 s very low accelerations occurred. Significant accelerations developed between 4 and 6 s and, thereafter, only low level excitation was recorded. The ground motions computed using DESRA-2 are shown in Figure 13.15(b). Except for some minor differences in frequency and magnitude in the 8–10 s range, the computed record is very similar to the recorded motions.

The pore water pressures recorded at the 6 m depth on Owi Island No. 1 are shown to an expanded scale in Figure 13.16(a) and are typical of pore water pressures recorded in other locations during earthquakes. During the low level shaking of the first 4 s, the response was elastic and pore water pressures developed in instantaneous response to changes in the total applied stresses. Such pore water pressures result from the elastic coupling of soil and water. With the

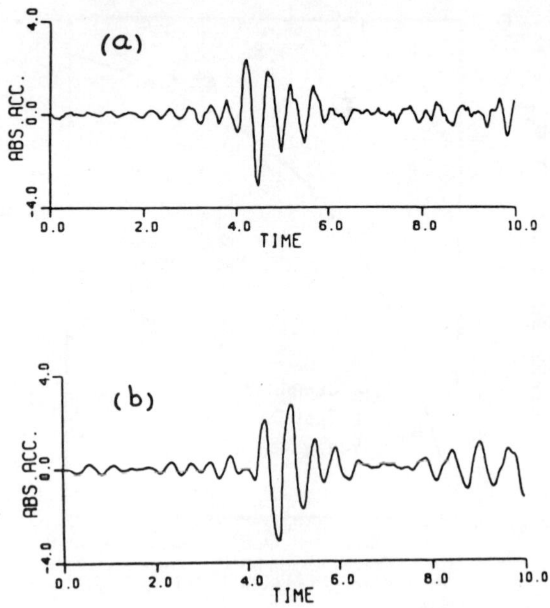

Figure 13.15 Measured (a) and computed (b) ground accelerations (acc. in ft/sec^2, time in sec) [19]

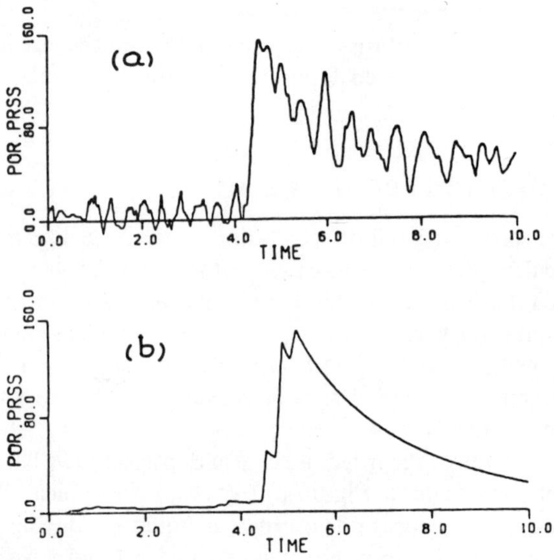

Figure 13.16 Measured (a) and computed (b) pore water pressures at a depth of 6 m (pore pressures in lb/ft^2, time in sec) [19]

onset of more severe shaking, plastic volumetric deformations are induced and these result in the development of residual pore water pressures which are independent of the instantaneous states of stress. These pressures accumulate with continued plastic volumetric deformation. Residual pore water pressure is indicated by the steep rise and sustained level in recorded pore water pressure in Figure 13.16(a). During shaking, the varying applied stresses continue to generate small instantaneous fluctuations in the pore water pressure which are superimposed on the larger residual pore water pressures. The gradual decay in the sustained level of pore water pressure is due to dissipation of pore water pressure by drainage. At this stage in the excitation, the dissipation of pore water pressure by drainage exceeds the generation by low level excitation.

The computed pore water pressures are shown in Figure 13.16(b). The residual pore water pressures are very similar to the recorded values. Very little residual pore water pressure is generated until the onset of severe shaking. The decay of the peak value during the period of subsequent low excitation is very similar to what was recorded. DESRA-2 computes only residual pore water pressures so there are no fluctuations due to changes in instantaneous stress levels in the DESRA output.

Recorded and computed pore water pressures for the sand layer at a depth of 14 m are shown in Figure 13.17(a) and 13.17(b), respectively. DESRA-2 results compare very favourably with the recorded values. It should be noted in this case that there is very little decay in the recorded pore water pressures over the 10 sec

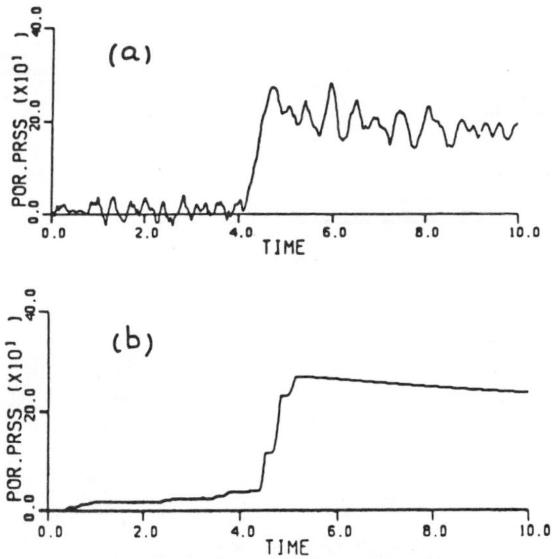

Figure 13.17 Measured (a) and computed (b) pore water pressures at a depth of 14 m (pore pressures in lb/ft^2, time in sec) [19]

period under analysis. This results from the fact that the sand layer is sealed above by a clay layer which impedes drainage. Therefore, dissipation of pore water pressure is negligible compared with that occurring in the upper sand layer which is capped by previous fill. The DESRA-2 program can take these different drainage conditions into account during the dynamic analysis and the computed results in Figure 13.17(b) show the same slow decay of residual pore water pressure.

It is probable that the drainage of the lower sand layer occurs after shaking due to horizontal drainage. This sand layer is part of the original seafloor. Larger pore water pressures are developed in this layer under the island than in the surrounding sea-floor. Eventual dissipation of these pore water pressures will occur under the seismically induced pore water pressure gradients.

13.8 MULTIDIMENSIONAL RESPONSE ANALYSIS

As discussed earlier a major difficulty in practice with the equivalent linear method is that it does not allow the direct computation of the permanent deformations. These are estimated from the static and dynamic stresses with the aid of strain data from cyclic triaxial tests. In principle, the static and dynamic

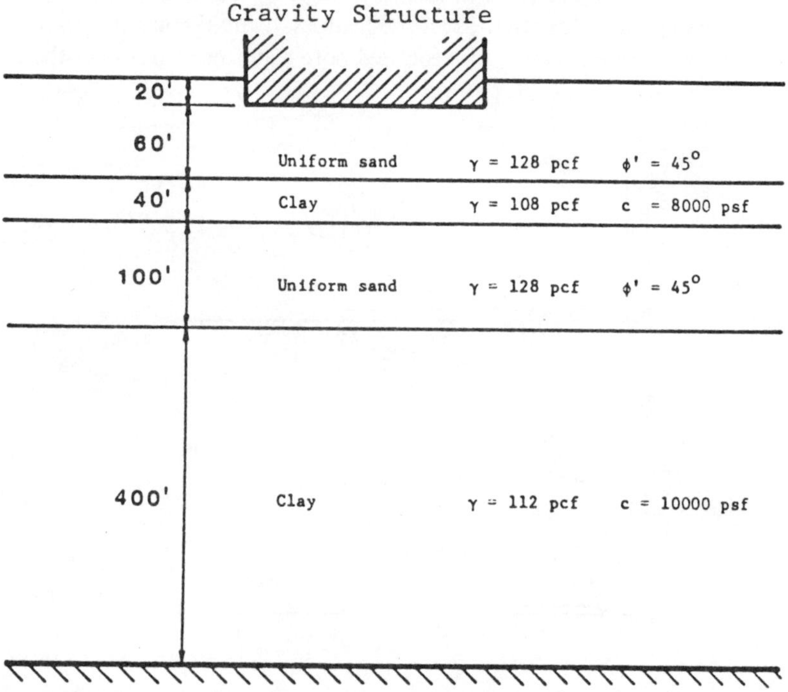

Figure 13.18 Foundation soils of offshore gravity structures [20]

stresses in a given finite element are simulated as closely as possible on a sample in a cyclic triaxial test, and the resulting axial strain is assumed to be the strain potential of the finite element. The strain potential is defined as the strain that develops in an unconstrained soil element under specified loading. The finite elements in a soil structure are all interconnected, so obviously strains in the elements obtained by the above procedure are not the strains that will develop in the structure but are an indication of its potential for straining under the given excitation.

The concept of strain potential, although quite a plausible basis for judgement, violates compatibility requirements and makes the resulting deformation field, to a large extent, arbitrary. This arbitrariness is not removed by the Serff [24] procedure for converting strain potential to compatible deformations. The compatibility must be satisfied when the strains are being computed, not afterwards. One approach to the direct computation of permanent deformations is the extension of the DESRA program to two dimensions as in the computer program TARA-2 developed by Siddharthan and Finn [25] which has the capability of performing both total and effective stress analyses. They assumed that the response of soil to hydrostatic pressure was non-linearly elastic and shearing deformation could be described by the non-linear hysteretic Masing model used in DESRA.

To illustrate the application of non-linear methods in displacement analysis, TARA was used to calculate directly the distribution of displacements of a gravity platform (Figure 13.18) to earthquake excitation. The resultant final permanent displacements are shown in Figure 13.19 as an illustration of the kind of results that may be obtained. Note that all these displacements are compatible with the computed stress field and are derived directly from the constitutive equations of the soils.

A major research programme has been initiated to evaluate the predictive capability of TARA using centrifuge tests.

There is increasing interest in the application of methods based on the theory of plasticity especially the anisotropic (kinematic) theory of plasticity to the problems of soil dynamics. Two particular formulations of the anisotropic theory of plasticity appear to have great potential for multidimensional analysis; the multi-yield surface model of Prevost [26, 27] and the two-surface model of Mróz *et al.* [28] and Dafalias and Herrmann [29]. The models are very new and still in the process of development, but progress to date has been impressive. An extensive review of these methods was recently conducted by Finn and Martin [30] as part of a study for an offshore firm to select a constitutive relation for inclusion in a general program for the analysis of the behaviour of offshore foundations under wave and earthquake loading.

The study concluded that both models represent the response of soils to static loading very well. Both have demonstrated the potential to model the phenomenological aspects of pseudo-static cyclic loading but verification has been limited.

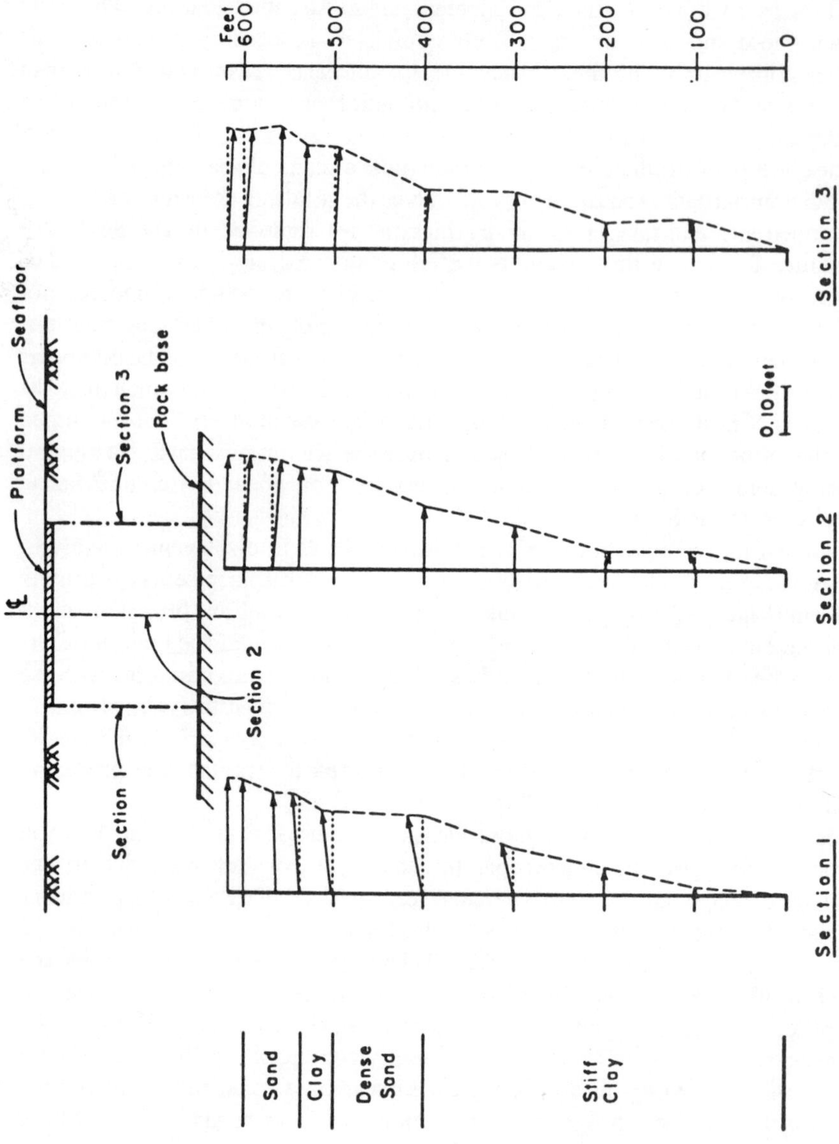

Figure 13.19 Displaced positions of three initially vertical sections [20]

None of the models has yet been verified for the case of strongly degrading clays or readily liquefiable sands under dynamic loading.

13.8 CONCLUSIONS

Soil is a very complex material and, in spite of the undoubtedly formidable potential of the theory of plasticity, one is still left with the impression that no one theory can model all aspects of its behaviour. There will always be room for special models for particular kinds of problems. The equivalent linear method of analysis appears to be particularly applicable to stable sites whose soil properties do not degrade significantly during cyclic loading and where significant permanent deformations are unlikely. For approximately level ground conditions the non-linear hysteretic methods of analysis appear best. DESRA has been used extensively for site response studies offshore [31, 32] and for predicting the performance of drilling and production islands in the Arctic during earthquake excitation [33]. For two- and three-dimensional dynamic analyses, equivalent linear methods are used almost exclusively.

As pointed out by the earthquake engineering research study [11] more sophisticated methods are required for deformation analyses in two and three dimensions. This is particularly so when soil properties degrade significantly during dynamic excitation. Extensions of successful 1-D methods such as TARA may prove useful but first they require substantial verification. Plasticity methods incorporating very generalized stress–strain relations have not yet found favour in geotechnical engineering practice.

A major reason for the very slow penetration of practice by the more generalized stress–strain relations is their complexity compared to what is currently in general use. Such complexity necessitates very effective communication to potential users of fundamental concepts and modes of application to geotechnical problems. For the most part, no special efforts have been made to educate potential users. Reports on the newer relations are often written from the analyst's point of view and published in journals read by other analysts. Very little has appeared in the traditional geotechnical journals on applications of these models to typical boundary value problems in geotechnical engineering.

Dynamic analysis in geotechnical practice is intimately related to the capability of measuring the necessary soil properties. In the presentation of new constitutive models, inadequate attention is usually given to the direct measurement or the indirect determination of the parameters in the model. But in practice, the crucial factors are: how are the parameters to be measured, can they be measured with conventional readily available laboratory equipment, and what procedures for data reduction must be followed?

Another requirement for acceptance is convincing verification of a constitutive relation or model. This implies something more than predicting response in a triaxial test or merely duplicating the phenomenological aspects of response without verifying their quantitative features. This is not very easy to do as most case histories of behaviour in the field frequently do not contain all the data required by a model. Nevertheless, though the environment for verification may not be perfect, a credible analysis of case histories is one of the most effective ways of displaying the potential of a new model and encouraging its use.

The search for a good constitutive model is not yet over. The development of such a model has been identified as one of the major research areas of soil dynamics for the next decade [11]. The prospects for success will be enhanced by close co-operation between researchers and practitioners. Such co-operation will facilitate verification of proposed models and ease the difficulties and resistance associated with the introduction of complex models in engineering practice.

REFERENCES

1. R. N. Yong and K. Hon-Yim, (Eds.), *Proc. of Workshop on Limit Equilibrium, Plasticity and Generalized Stress–Strain in Geotechnical Engineering*, McGill University, Montreal, Quebec, May 28–30, ASCE, New York, NY, 1980.
2. H. B. Seed and I. M. Idriss, 'Influence of soil conditions on ground motion during earthquakes', *Jour. of the Soil Mechanics and Foundations Divisions, ASCE*, **95**, No. SM1, Proc. Paper 6347, 99–137 (1969).
3. P. B. Schnabel, J. Lysmer, and H.B. Seed, 'SHAKE: a computer program for earthquake response analysis of horizontally layered sites', *Report No. EERC* 72–12, Earthquake Engineering Research Center, University of California, Berkeley, California, December, 1972.
4. I. M. Idriss, J. Lysmer, R. Hwang, and H. B. Seed, 'QUAD-4: a computer program for evaluating the seismic response of soil-structures by variable damping finite element procedures', *Report No. EERC* 73–16, Earthquake Engineering Research Center, University of California, Berkeley, California, July, 1973.
5. J. Lysmer, T. Udaka, H. B. Seed, and R. Hwang, 'LUSH: a computer program for complex response analysis of soil-structure system', *Report No. EERC* 74–4, Earthquake Engineering Research Center, University of California, Berkeley, California, 1974.
6. J. Lysmer, T. Udaka, C. F. Tsou, and H. B. Seed, 'FLUSH: a computer program for approximate 3-D analysis of soil-structure interaction problems', *Report No. EERC* 75–30, Earthquake Engineering Research Center, University of California, Berkeley California, 1975.
7. W. D. Liam, Finn, K. W. Lee, and G. R. Martin, 'An effective stress model for liquefaction'. *Proc. of ASCE Annual Convention and Exposition*, Philadelphia, Pa., Sept. 22–Oct. 1, 1976, Preprint 2752; also in *Jour. of the Geotechnical Engineering Division, ASCE*, **103**, No. GT6, Proc. Paper 13008, 517–533 (1977).
8. W. D. Liam, Finn, G. R. Martin, and M. K. W. Lee, 'Comparison of dynamic analyses for saturated sands', *Proc. of ASCE Geotechnical Engineering Division Specialty Conference*, pp. 472–491, June 19–21, Pasadena, California, 1978.
9. W. D. Liam, Finn, K. W. Lee, and P. M. Byrne, 'Response of saturated sands to wave and earthquake loading', *Numerical Methods in Offshore Engineering* (Ed. O. C. Zienkiewicz, *et al.*, pp. 515–553, John Wiley & Sons Ltd., London, 1977.
10. N. M. Newmark, 'Effects of earthquakes on dams and embankments', *Geotechnique*, **5** (2), (1965).

11. National Research Council, USA, 'Earthquake Engineering Research—1982', Report by Committee on Earthquake Engineering Research, National Academy Press, Washington, DC, 1982.
12. V. L. Streeter, E. B. Wylie, and F. E. Richart, 'Soil motion computations by characteristics method', *ASCE National Structural Engineering Meeting*, April 9–13, San Francisco, California, Preprint 1952, 1973.
13. G. Masing, Eigenspannungen und Verfestigung beim Messing. *Proc. of 2nd International Congress of Applied Mechanics*, Zurich, Switzerland, 1926.
14. G. R. Martin, W. D. Liam, Finn, and H. B. Seed, 'Fundamentals of liquefaction under cyclic loading', *Soil Mechanics Series Report, No.* 23, Department of Civil Engineering, University of British Columbia, 1974; also in *Jour. of the Geotechnical Engineering Division, ASCE*, **101**, No. GT5, Proc. Paper 11284, 423–438 (1975).
15. W. D. Liam, Finn, and S. K. Bhatia, 'Verification of non-linear effective stress model in simple shear', *Proc. of ASCE Fall Meeting*, Hollywood-by-the-Sea, Florida, October, 1980.
16. M. K. W. Lee and W. D. Liam, Finn, 'DESRA-1: program for the dynamic effective stress response analysis of soil deposits including liquefaction evaluation', *Soil Mechanics Series Report, No.* 36, Department of Civil Engineering, University of British Columbia, Vancouver, BC, 1975.
17. M. K. W. Lee and W. D. Liam, Finn, 'DESRA-2: dynamic-effective stress response analysis of soil deposits with energy transmitting boundary including assessment of liquefaction potential', *Soil Mechanics Series Report, No.* 38, Department of Civil Engineering, University of British Columbia, Vancouver, BC, 1978.
18. P. O. Martin and H. B. Seed, 'Simplified procedure for effective stress analysis of ground response', *Jour. of the Geotechnical Engineering Division, ASCE*, **105**, No. GT6, Proc. paper 14659, 739–758 (1979).
19. W. D. Liam, Finn, S. Iai, and K. Ishihara, 'Performance of artificial offshore islands under wave and earthquake loading: field data analyses', *Proc. of Offshore Technology Conference*, Houston, Texas, OTC Paper 4220, May, 1983.
20. W. D. Liam, Finn, 'Dynamic analysis and liquefaction—emerging trends', *Proc. of the 3rd Internal Earthquake Microzonation Conference*, Vol. II, pp. 909–927, 1982.
21. T. Imai, 'The relation of mechanical properties of soils to P and S wave velocities in Japan', *Proc. of the 4th Japan Earthquake Engineering Symposium*, Tokyo, 1970.
22. K. Ishihara, *Fundamentals of Soil Dynamics*. Kashima-shuppan-kai, Tokyo, 1976, in Japanese.
23. K. Ishihara, K. Shimizu, and Y. Yasuda, 'Porewater pressures measured in sand deposits during an earthquake', *Soils and Foundation Jour.*, **21** (4), December, 85–100 (1981).
24. N. Serff, H. B. Seed, F. I. Makdisi, and D. K. Chang, 'Earthquake induced deformation of earthdam', *Report No. EERC* 76–4, Earthquake Engineering Research Center, University of California, Berkeley, California, 1976.
25. R. Siddharthan and, W. D. Liam, Finn, 'TARA: two-dimensional non-linear static and dynamic response analysis', *Report to ERTEC Western, Inc.*, Long Beach, California, pp. 1–168, 1981.
26. J. H. Prevost, 'Anisotropic undrained stress–strain behaviour of clays', *Jour. of the Geotechnical Engineering Division, ASCE*, **104**, No. GT8, Proc. Paper 13942, 1075–1090 (1978).
27. J. H. Prevost, 'Mathematical modeling of soil stress–strain strength behaviour', *Proc. of the 3rd International Conference on Numerical Methods in Geomechanics*, pp. 347–361, Aachen, Germany, April 2–6, 1979.
28. Z. Mroz, V. A. Norris, and O. C. Zienkiewicz, 'Application of an anisotropic hardening

model in the analysis of elastoplastic deformation of soils', *Geotechnique*, **29** (1), 1–34 (1979).
29. Y. F. Dafalias and L. R. Herrmann, 'A boundary surface soil plasticity model', *Proc. of the International Symposium on Soils Under Cyclic and Transient Loading*, Swansea, Wales, Vol. 1, pp. 335–345, January 7–11, 1980.
30. W. D. Liam, Finn, and G. R. Martin, 'Soil as an anisotropic kinematic hardening solid', *Proc. of the ASCE Fall Meeting*, Hollywood-by-the-Sea, Floride, October, 1980.
31. W. D. Liam, Finn, 'Review of earthquake criteria for the Balder platform', *Report to Exxon Production and Research*, Houston, Texas, 1981.
32. W. D. Liam, Finn, 'Seismic response of the Pescado B-2 site', *Report to Exxon Production Research*, Houston, Texas, 1983.
33. W. D. Liam, Finn, 'Seismic response of Koakoak Tanker Island', *Report to Dome Petroleum Limited*, Calgary, Alberta, 1982.

Mechanics of Engineering Materials
Edited by C. S. Desai and R. H. Gallagher

Chapter 14

Quasistatic and Dynamic Analysis of Saturated and Partially Saturated Soils

J. Ghaboussi and K. J. Kim

14.1 INTRODUCTION

The linear elastic finite element formulation of saturated soils modelled as coupled two-phase media has been presented in [11] and [12] for quasistatic and dynamic problems. The quasistatic formulation has been extensively used in finite element analysis of consolidation problems and the dynamic formulation has been used in computation of seismically induced pore pressures and evaluation of liquefaction potential. In both categories of problems inelastic material behaviour plays an important role in pore pressure generation. Realistic finite element analysis of these problems requires accounting for the non-linear material behaviour. In fully saturated soils it is possible to ignore the compressibility of the pore fluid without loss of accuracy. However, in partially saturated soils the compressibility of pore air–water mixture must be taken into account. As will be described in a later section, the compressibility of pore air–water mixture is inherently non-linear.

A non-linear finite element formulation for coupled two-phase model of saturated and partially saturated soils is presented in this chapter, including the compressibility of pore fluid. The field equations are presented and various aspects of the volumetric coupling between granular solid and pore fluid are discussed. A constitutive model for compressibility of pore air–water mixture for partially saturated soils is presented. In the last sections an example problem of finite element analysis of construction pore pressures in earth dams is presented and discussed.

14.2 FIELD EQUATIONS

In coupled two-phase models for saturated and partially saturated soils, Terzaghi's effective stress equation plays a fundamental role. It defines the effective stresses in relating the total stresses and pore pressures. The total stresses σ_{ij} are defined as the stresses carried by the bulk of the fluid saturated porous media. The effective stresses σ'_{ij} are carried by granular solid skeleton. These stresses and the pore water pressure π are related in the following incremental form of the Terzaghi's effective stress equation,

$$d\sigma_{ij} = d\sigma'_{ij} + \delta_{ij}\, d\pi \tag{14.1}$$

in which δ_{ij} is the Kronecker's delta. The incremental form of this equation is required for non-linear material characterization.

The validity of the effective stress equation for partially saturated soils must be qualified. This equation is only valid for high degrees of saturation; at water contents equal to or above the standard optimum water content. It has been shown that at these high water contents the pore air is in occluded state, in the form of air bubbles in the pore water. The pore air is therefore not in contact with the solid grains and the Terzaghi's effective stress equation is valid. At lower water contents, the pore air may be in contact with the solid grain and the effective stress equation cannot be used in its present form.

The behaviour of the solid skeleton is modelled in drained condition by an elastoplastic material model [15, 21, 26]. Such models give a relation between the increments of effective stress and the increments of strain.

$$d\sigma'_{ij} = D^{\mathrm{ep}}_{ijkl}\, d\varepsilon_{kl} \tag{14.2}$$

The strains are assumed to be linear and infinitesimal.

$$\varepsilon_{ij} = \tfrac{1}{2}(u_{i,j} + u_{j.i}) \tag{14.3}$$

in which u_i are the components of the solid displacements. The components of pore water displacement are denoted by U_i and the components of the relative pore water displacement w_i are defined by the following equation,

$$w_i = n(U_i - u_i) \tag{14.4}$$

in which n is the porosity.

The constitutive model for the pore fluid (pore water or pore air–water mixture) is expressed through the storage equation. The following is the incremental form of the storage equation.

$$d\pi = (1 - n)\alpha\, d\varepsilon_v + n\alpha\, d\zeta \tag{14.5}$$

in which $\varepsilon_v = u_{i,i}$ is the bulk volumetric strain and $\zeta = U_{i,i}$ is the pore fluid volumetric strain.

The equivalent modulus α is defined by the following equation.

$$\frac{1}{\alpha} = (1 - n)C_{\mathrm{g}} + nC_{\mathrm{w}} \tag{14.6}$$

in which C_{g} is the compressibility of the solid grains and C_{w} is the compressibility of pore fluid. The detailed derivation of the storage equation (or coupled continuity equation) is given in next section.

For fully saturated soils, it is often assumed that the pore fluid is incompressible and the compressibility of solid grains are neglected. This results in the following storage equation

$$(1-n)\mathrm{d}\varepsilon_v + n\,\mathrm{d}\zeta = 0 \tag{14.7}$$

For partially saturated soils, the compressibility of the solid grains can still be ignored ($C_g = 0$). However, the compressibility of the pore fluid C_w must be taken into account, since it represents the compressibility of the pore air–water mixture. The determination of the compressibility of pore air–water mixture will be described in a later section.

The bulk equilibrium is expressed through the following equation

$$\sigma'_{ij,j} + \pi_{,i} + \rho b_i - \rho \ddot{u}_i - \rho_f \ddot{w}_i = 0 \tag{14.8}$$

in which ρ is the bulk mass density, ρ_f is the mass density of the fluid and b_i are the components of the body force, and $\ddot{u}_i$ and $\ddot{w}_i$ are the components of solid phase acceleration and relative acceleration of the fluid phase, respectively.

The flow of the pore fluid relative to the granular solid is assumed to be governed by the generalized Darcy flow law.

$$\dot{w}_i = k_{ij}\left(\pi_{,j} + \rho_f b_j - \rho_f \ddot{u}_j - \frac{1}{n}\rho_f \ddot{w}_j\right) \tag{14.9}$$

in which k_{ij} is the permeability tensor.

It should be pointed out that this equation is also valid for partially saturated soil only if the pore air is in occluded state, which is true for water contents equal to or greater than the standard optimum.

14.3 COUPLED VOLUMETRIC CONTINUITY EQUATION

The volumetric constitutive equations for the pore fluid and solid grains are given by the following two equations

$$\frac{\mathrm{d}\gamma_w}{\gamma_w} = -C_w\,\mathrm{d}\pi \tag{14.10}$$

$$\frac{\mathrm{d}\gamma_s}{\gamma_s} = -C_g\,\mathrm{d}\pi - C'_g\,\mathrm{d}p' \tag{14.11}$$

in which C_w is the compressibility of the pore fluid, C_g and C'_g are the solid grain compressibilities, π is the pore pressure, and $p' = \frac{1}{3}\sigma'_{ii}$. The mass densities of pore fluid and granular solid are denoted by γ_w and γ_s. Equivalent bulk mass densities for the pore fluid and granular solid, ρ_f and ρ_s, are defined through the following equations.

$$\rho_f = n\gamma_w, \qquad d\rho_f = n\,d\gamma_w + \gamma_w\,dn \tag{14.12}$$

$$\rho_s = (1-n)\gamma_s, \qquad d\rho_s = (1-n)d\gamma_s - \gamma_s\,dn \tag{14.13}$$

The conservation of mass for pore fluid and solid grains is expressed by the following equations

$$d\rho_f = -\rho_f\,d\zeta \tag{14.14}$$

$$d\rho_s = -\rho_s\,d\varepsilon_v \tag{14.15}$$

in which $d\zeta$ and $d\varepsilon_v$ are the volumetric strains of pore fluid and solid skeleton, respectively.

The expressions for volumetric strain can be obtained by combining equations (14.12) to (14.15)

$$d\zeta = -\frac{dn}{n} - \frac{d\gamma_w}{\gamma_w} \tag{14.16}$$

$$d\varepsilon_v = \frac{dn}{1-n} - \frac{d\gamma_s}{\gamma_s} \tag{14.17}$$

The coupled continuity equation is obtained by substituting equations (14.10) and (14.11) into equations (14.16) and (14.18) and eliminating dn between the resulting equations (14.16) and (14.18).

$$(1-n)d\varepsilon_v + n\,d\zeta = (1-n)C_g'\,dp' + [nC_w + (1-n)C_g]d\pi \tag{14.18}$$

This is the general form of the volumetric stress–volumetric strain relation for fluid saturated porous solids. The special limiting conditions for this equation are the fully drained and undrained cases.

In fully drained case $d\pi = 0$ and the total and effective pressures are equal, $dp' = dp$. A bulk compressibility C_d can be measured in a drained hydrostatic test.

$$C_d = \frac{d\varepsilon_v}{dp} \tag{14.19}$$

The coupled continuity relation of equation (14.18) will reduce to the following equation

$$(1-n)(C_d - C_g')dp + n\,d\zeta = 0 \tag{14.20}$$

It should be pointed out that, usually C_d is several orders of magnitude larger than C_g'. It is therefore reasonable to ignore C_g' in equation (14.20).

In the undrained case $d\varepsilon_v = d\zeta$ and the measured volumetric strains are solely due to the compression of pore fluid and granular solid. The measured compressibility in undrained condition will be denoted by $C_u = d\varepsilon_v/dp$. The coupled continuity relation of equation (14.18) for undrained case will reduce to

$$C_u \, dp - (1-n)C'_g \, dp' - [nC_w + (1-n)C_g] d\pi = 0 \tag{14.21}$$

In this equation also C_u is at least an order of magnitude greater than C'_g.

In general, the solid compressibility due to changes in effective pressure C'_g can be ignored. As a result equation (14.18) will reduce to the coupled continuity (or storage) equation, equation (14.5), given in the previous section.

14.4 COMPRESSIBILITY OF PORE AIR–WATER MIXTURE

In material models for partially saturated soils, the pore space is assumed to be filled with an equivalent fluid which has the same compressibility as the pore air–water mixture. A constitutive model for the compressibility of pore air–water mixture is presented in this section and the determination of the material constants from laboratory tests is discussed.

The physics of pore air–water mixture is complex. The compressibility depends mainly on the degree of saturation, S. The treatment in this section is restricted to ranges of degree of saturation corresponding to standard optimum or above optimum water contents. At these water contents the pore air is likely to exist in occluded state, in the form of air bubbles.

In earlier treatments of compressibility of air–water mixture by Hilf [25], Bishop and Eldin [1], and Skempton and Bishop [35], the basic principles of Boyle's law and Henry's solubility are used to arrive at an expression for compressibility, C_w.

$$C_w = (1 - S_0 + H_c S_0)\frac{\pi_{a0}}{\pi^2} \tag{14.22}$$

in which, C_w is the compressibility of air–water mixture, S_0 is the initial degree of saturation, H_c is the coefficient of solubility, π_{a0} is the initial pore air pressure, and π is the pore water pressure. For the sake of simplicity, the surface tension between the air and water phases has been neglected in equation (14.22). The surface tension, which is equal to one-half of the difference between pore air pressure and pore water pressure multiplied by the radius of the air bubble, has been explicitly introduced in an expression for compressibility of air–water mixture proposed by Schuurman [33]. Apart from complexity of this expression, the initial radius of the air bubbles has to be determined from the initial negative pore pressure under the assumption that the initial pore pressure is atmospheric. Thus, the accuracy of the whole expression depends on the degree of accuracy in the measurement of the initial negative pore pressure.

A simpler model has been proposed by Kim [27] which is more amenable to determination of parameters from laboratory experiments. The difference between the air and water pressures, $T = \pi_a - \pi$, is assumed to remain constant over the pressure ranges encountered in practice. The pressure difference T

slightly increases with increasing pore water pressure since the radius of the air bubbles decrease as the pressure in the surrounding pore water increases. Since the compressibility is approximately inversely proportional to the square of the pore air pressure, the slight variations in the pressure difference T will cause a very slight change in compressibility C_w. In short, the assumption of constant pressure difference is a reasonable assumption as long as the value of T can be determined from laboratory tests. With the assumption of constant T, Kim [27] has derived the following expression for the compressibility of air–water mixture

$$C_w = (1 - S_0 + H_c S_0)\frac{\pi_{a0}}{(\pi + T)^2} \tag{14.23}$$

The value of pressure difference T can be determined from trial simulation of undrained isotropic compression tests with pore pressure and volume change measurements.

The volume change in undrained isotropic test is solely due to contraction (or expansion) of the pore air space. Therefore, the increments of effective pressure dp' and pore pressure $d\pi$ can be related to the increment of volume change $d\varepsilon_v$ independently of each other.

$$dp' = (Bp')d\varepsilon_v \tag{14.24}$$

$$d\pi = \left(\frac{1}{nC_w}\right)d\varepsilon_v \tag{14.25}$$

in which n is the initial porosity and $B = B_c$ in virgin compression or $B = B_s$ in swelling and recompression. B_c and B_s are related to virgin compressibility index C_c and swelling index C_s, respectively.

$$B_c = 2.3(1 + e_0)\frac{1}{C_c} \tag{14.26}$$

$$B_s = 2.3(1 + e_0)\frac{1}{C_s} \tag{14.27}$$

in which e_0 is the initial void ratio.

Test results by Gilbert [24], Garlanger [10] and Campbell [3] show that, when the water content is on or above the optimum value, the pore air remains in occluded state and thus Terzaghi's effective stress principle is applicable.

$$dp = dp' + d\pi \tag{14.28}$$

The relation between applied pressure increment and the volumetric strain increment is obtained by substituting equations (14.24) and (14.25) into equation (14.28).

$$dp = \left(Bp' + \frac{1}{nC_w}\right)d\varepsilon_v \tag{14.29}$$

The following equations can be used to compute the increments of pore pressure and volumetric strain in terms of increments of applied pressure in undrained isotropic compression tests on partially saturated soils.

$$d\pi = \lambda \, dp \tag{14.30}$$

$$d\varepsilon_v = \lambda n C_w \, dp \tag{14.31}$$

$$\frac{1}{\lambda} = 1 + \frac{2.3np'}{1-n}\frac{C_w}{C_c}$$

The value of the pressure difference T is determined from undrained isotropic compression tests on samples of partially saturated soil. The results of such tests are presented in the form of plots of pore pressure π versus applied pressure p and volumetric strain ε_v versus applied pressure p. For a fully saturated sample

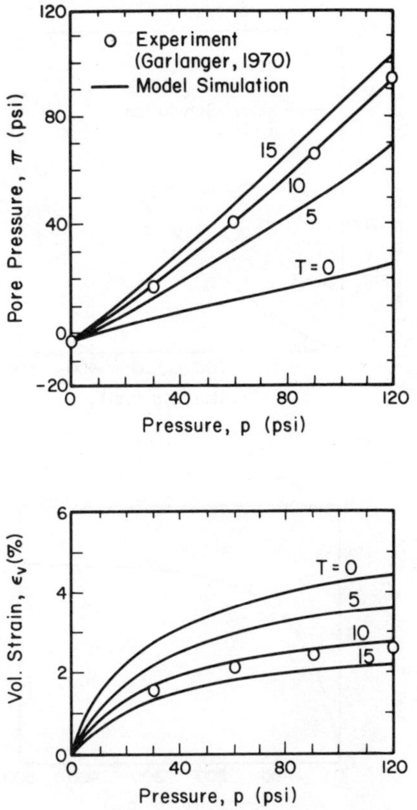

Figure 14.1 Simulation of undrained isotropic compression test on Champaign Till with water content, $w = 13\%$

the pore pressure will be equal to the applied pressure and volumetric strain will be zero. In partially saturated samples, the pore pressures are lower than the applied pressure and volumetric strains increase with applied pressure, approaching a constant value as the volume of pore air decreases. Eventually, at higher applied pressures, the air bubbles collapse and the sample behaves similar to a fully saturated one.

The results of an undrained isotropic compression test are shown in Figure 14.1 by open circles. This is one of a number of tests performed by Garlanger [10] on Champaign Till, which is a well graded sandy silty clay. The water content was $W = 13$ per cent (standard optimum $W = 12\%$) with a degree of saturation $S = 85$ per cent and porosity of $n = 29$ per cent. The soil was compacted by kneading under 90 p.s.i. confining pressure. Thus, the sample is recompressed up to 90 p.s.i. pressure. The weighted average of the

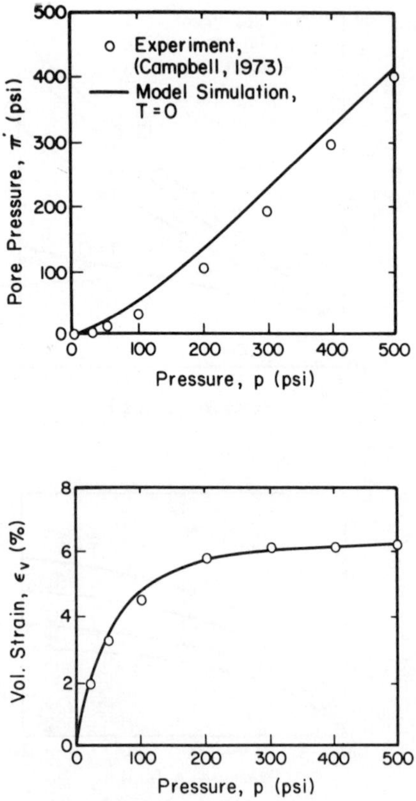

Figure 14.2 Simulation of undrained isotropic compression test on Peorian Loess with water content, $w = 21\%$

coefficient of recompression bulk modulus was determined to be $B_s = 82$. The sample exhibits an initial negative pore pressure of 3 p.s.i. under atmospheric confining pressure. Therefore, it can be expected that the pressure difference T would be greater than 3 p.s.i. The simulated results with four values of $T = 0, 5$, 10, 15 p.s.i. are shown in Figure 14.1. The curve for $T = 0$ corresponds to the case of neglecting the surface tension, which in this case is not a good assumption. It can be seen that the most likely value of pressure difference is $T = 10$ *p.s.i.*

Shown in Figure 14.2 are the results of one of the tests performed by Campbell [3] on Peorian Loess which is classified as inorganic silt. The sample water content is $W = 21$ per cent (standard optimum $W = 18$ per cent) with degree of saturation $S = 84$ per cent and a porosity of $n = 41$ per cent. The soil was statically compacted under 40 p.s.i. confining pressure and coefficients of bulk modulus were estimated to be $B_c = 41$ and $B_s = 105$. The sample does not exhibit any initial negative pore pressure. As can be seen in Figure 14.2 the simulation result with $T = 0$, corresponding to neglecting of surface tension is reasonably close to the measurements. This seems to indicate that, in absence of initial negative pore pressure, the surface tension can be neglected and the value of pressure difference $T = 0$ is a reasonable approximation. Additional experimental evidence is given by Kim [27] to support this conclusion.

14.5 FINITE ELEMENT FORMULATION: QUASISTATIC AND DYNAMIC

The quasi-static and dynamic finite element formulation for coupled fluid-saturated porous elastic media has been presented in [11, 12]. Extensions to non-linear solid behaviour have been presented in [15] for the quasistatic problems and in [17, 20] for the dynamic problems. A more general non-linear finite element formulation of the quasistatic and dynamic will be presented here.

For the quasistatic problem, the inertia effects are ignored ($\ddot{u}_i = \ddot{u}_i = 0$). The solid phase displacements u_i and the pore pressure π are treated as independent variables. For the solid phase, the equation of equilibrium is obtained from the virtual work principle.

$$\int_v \delta\varepsilon_{ij}\sigma_{ij}\,\mathrm{d}v = \int_v \delta u_i \rho b_i\,\mathrm{d}v + \int_s \delta u_i \bar{t}_i\,\mathrm{d}s \tag{14.32}$$

In this equation $\bar{t}_i$ are the components of surface tractions and $\delta(\ )$ denotes a virtual quantity.

The rate of complementary virtual work for the pore fluid is used to obtain the fluid balance equation

$$\int_v \delta\pi(\dot{\zeta} - \dot{\varepsilon}_v) n\,\mathrm{d}v + \int_v \delta\pi_{,i}(\dot{U}_i - \dot{u}_i) n\,\mathrm{d}v = \int_s \delta\pi \bar{Q}\,\mathrm{d}s \tag{14.33}$$

In this equation the boundary fluid flow is denoted by $\bar{Q}$. Using equations (14.4), (14.5), and (14.9), the rate of complementary work for the pore fluid can be written in the following form.

$$\int_v \delta\pi\left(\frac{1}{\alpha}\dot{\pi} - \dot{\varepsilon}_v\right)dv + \int_v \delta\pi_{,i}k_{ij}\delta\pi_{,j}\,dv = \int_s \delta\pi\bar{Q}\,ds - \int_v \delta\pi_{,i}\rho_f k_{ij}b_j\,dv \tag{14.33}$$

The displacements and pore pressures within each element are discretized by the same shape functions N_i.

$$\{u_i\} = [N]\{u\} \tag{14.35}$$

$$\pi = [N]\{\pi\} \tag{14.36}$$

By appropriate differentiation the following equation can be obtained

$$\{\varepsilon\} = [B]\{u\} \tag{14.37}$$

$$\{\pi_{,i}\} = [A]\{\pi\} \tag{14.38}$$

The discretization of equations (14.32) and (14.34) results in the following coupled equations.

$$\{I'\} + [C]\{\pi\} = \{P\} \tag{14.39}$$

$$[C]^{\mathrm{T}}\{\dot{u}\} - [E]\{\dot{\pi}\} - [H]\{\pi\} = -\{Q\} \tag{14.40}$$

The equivalent nodal load vector $\{P\}$ contains body force and surface traction contribution. Similarly, $\{Q\}$ is an equivalent nodal flow vector, containing body force and surface traction contributions. The other terms of the coupled equations are

$$\{I'\} = \sum \int_v [B]^{\mathrm{T}}\{\sigma'\}dv \tag{14.41}$$

$$[C] = \sum \int_v [B]^{\mathrm{T}}\{\delta\}[N]dv \tag{14.42}$$

$$[E] = \sum \int_v \frac{1}{\alpha}[N]^{\mathrm{T}}[N]dv \tag{14.43}$$

$$[H] = \sum \int_v [A]^{\mathrm{T}}[k][A]dv \tag{14.44}$$

in which $\{\delta\}$ is the vector equivalent of Kronecker's delta and $[k]$ is the permeability matrix. The summation sign in these equations denotes direct assembly of all element contributions. The incremental stress–strain relation is introduced into equation (14.39). The incremental form of equation (14.40) is obtained by assuming a linear variation of the variables within a time step Δt

and equation (14.40) is written at the middle of the time step. The incremental form of the coupled equations are

$$[K_t]\{\Delta u_n\} + [C]\{\pi_n\} = \{P_n\} - \{I'_{n-1}\} \tag{14.45}$$

$$[C]^{\mathrm{T}}\{\Delta u_n\} - \left([E] + \frac{\Delta t}{2}[H]\right)\{\pi_n\} = \{Q_n^*\} \tag{14.46}$$

and

$$[K_t] = \sum \int_v [B]^{\mathrm{T}}[D^{\mathrm{ep}}][B]\,\mathrm{d}v \tag{14.47}$$

$$\{Q_n^*\} = -\frac{\Delta t}{2}(\{Q_n\} + \{Q_{n+1}\}) - \left([E] - \frac{\Delta t}{2}[H]\right)\{\pi_{n-1}\} \tag{14.48}$$

The subscript n denotes the value of the variable at the nth time step.

In the finite element formulation of the dynamics of the coupled two-phase media, the solid phase displacements u_i and the relative displacements of pore fluid w_i are treated as the independent variables. The inertial forces are treated as body forces and the coupled equations of motion are derived from virtual work principle applied to the bulk of the saturated porous solid and the pore fluid.

$$\int_v \delta\varepsilon_{ij}\sigma_{ij}\,\mathrm{d}v + \int_v \delta u_i(\rho\ddot{u}_i + \rho_{\mathrm{f}}\ddot{w}_i)\,\mathrm{d}v = \int_v \delta u_i \rho b_i\,\mathrm{d}v + \int_s \delta u_i \bar{t}_i\,\mathrm{d}s \tag{14.49}$$

$$\int_v (\delta\zeta - \delta\varepsilon_{\mathrm{v}})\pi n\,\mathrm{d}v + \int_v (\delta U_i - \delta u_i)\pi_{,i} n\,\mathrm{d}v = \int_s (\delta U_i - \delta u_i)v_i\bar{\pi} n\,\mathrm{d}s \tag{14.50}$$

In equation (14.50) the components of outward normal to the boundary surface are denoted by v_i and $\bar{\pi}$ is the prescribed boundary surface fluid pressure. It is noted that $n(U_i - u_i) = w_i$ and the relative fluid volumetric strain $n(\zeta - \varepsilon_{\mathrm{v}}) = w_{i,i}$ is denoted by ξ. Therefore, equation (14.50) can be written in the following form after substituting for $\pi_{,i}$ from the generalized Darcy flow law of equation (14.9).

$$\int_v \delta\xi\pi\,\mathrm{d}v + \int_v \delta w_i \bar{k}_{ij}\dot{w}_j\,\mathrm{d}v + \int_v \delta w_i\left(\rho_{\mathrm{f}}\ddot{u}_i + \frac{1}{n}\rho_{\mathrm{f}}\ddot{w}_i\right)\mathrm{d}v$$

$$= \int_v \delta w_i \rho b_i\,\mathrm{d}v + \int_s \delta w_i v_i \bar{\pi}\,\mathrm{d}s \tag{14.51}$$

The components of the inverse of the permeability tensor are denoted by $\bar{k}_{ij}$.

Within each element, the independent variables u_i and w_i are interpolated with the same shape functions.

$$\{u_i\} = [N]\{u\} \tag{14.52}$$

$$\{w_i\} = [N]\{w\} \tag{14.53}$$

These shape functions are also assumed to interpolate the velocities and the accelerations. By proper differentiation of the shape functions, the following strain displacement relation can be obtained.

$$\{\varepsilon\} = [B]\{u\} \tag{14.54}$$

$$\{\xi\} = [G]\{w\} \tag{14.55}$$

Substitution of equations (14.52) to (14.55) into equations (14.49) and (14.51) leads to the following discretized coupled equations of motion.

$$[M_\mathrm{s}]\{\ddot{u}\} + [M_\mathrm{c}]\{\ddot{w}\} + \{I\} = \{P\} \tag{14.56}$$

$$[M_\mathrm{c}]\{\ddot{u}\} + [M_\mathrm{f}]\{\ddot{w}\} + [\bar{H}]\{\dot{w}\} + \{I_\mathrm{f}\} = \{R\} \tag{14.57}$$

The force vector $\{P\}$ and $\{R\}$ contain body force and surface traction contributions and

$$[M_\mathrm{s}] = \sum \int \rho [N]^\mathrm{T}[N]\,\mathrm{d}v \tag{14.58}$$

$$[M_\mathrm{c}] = \sum \int \rho_\mathrm{f} [N]^\mathrm{T}[N]\,\mathrm{d}v \tag{14.59}$$

$$[M_\mathrm{f}] = \sum \int \frac{1}{n} \rho_\mathrm{f} [N]^\mathrm{T}[N]\,\mathrm{d}v \tag{14.60}$$

$$[\bar{H}] = \sum \int [N]^\mathrm{T}[\bar{k}][N]\,\mathrm{d}v \tag{14.61}$$

$$\{I\} = \sum \int [B]^\mathrm{T}\{\sigma\}\,\mathrm{d}v \tag{14.62}$$

$$\{I_\mathrm{f}\} = \sum \int [G]^\mathrm{T}\pi\,\mathrm{d}v \tag{14.63}$$

The internal resisting force vectors I and I_f can be written in incremental form by introducing the following equations, where the subscript n denotes the value of the variable at the nth step.

$$\{\sigma_n\} = \{\sigma_{n-1}\} + \{\Delta\sigma'_n\} + \alpha\{\delta\}(\Delta\varepsilon_{vn} + \Delta\xi_n) \tag{14.64}$$

$$\pi_n = \pi_{n-1} + \alpha(\Delta\varepsilon_{vn} + \Delta\xi_n) \tag{14.65}$$

Using the elastoplastic effective stress–strain relations the incremental form of the equations of motion are

$$[M_\mathrm{s}]\{\ddot{u}_n\} + [M_\mathrm{c}]\{\ddot{w}_n\} + [\bar{K}_t]\{\Delta u_n\} + [\bar{C}]\{\Delta w_n\} = \{P_n\} - \{I_{n-1}\} \tag{14.66}$$

$$[M_\mathrm{c}]\{\ddot{u}_n\} + [M_\mathrm{s}]\{\ddot{w}_n\} + [\bar{H}]\{\dot{w}_n\} + [\bar{C}]^\mathrm{T}\{\Delta u_n\} + [\bar{E}]\{\Delta w_n\} = \{R_n\} - \{I_{\mathrm{f}n-1}\} \tag{14.67}$$

and

$$[\bar{K}_t] = \sum \int ([B]^T[D^{ep}][B] + \alpha[B]^T\{\delta\}\{\delta\}^T[B])dv \tag{14.68}$$

$$[\bar{C}] = \sum \int \alpha[B]^T\{\delta\}[G]dv \tag{14.69}$$

$$[\bar{E}] = \sum \int \alpha[G]^T[G]dv \tag{14.70}$$

Note that the tangent stiffness matrix in equation (14.68) is for undrained condition, whereas in quasistatic formulation the tangent stiffness matrix in equation (14.47) is for fully drained condition. The differences between the other matrices in quasistatic and dynamic formulation are due to different independent variables for the fluid phase used in the two formulations.

14.6 A QUASI-STATIC PROBLEM: PORE PRESSURES DURING CONSTRUCTION

The first finite element simulation of earth dam construction was reported in literature by Clough and Woodward [7]. Later, similar but independently developed methodology were proposed for finite element simulation of excavation. However, numerical problems have been reported in finite element simulation of excavation [6]. A simple method of analysis for finite element simulation of construction and excavation has been developed by the first author. This method treats the excavation and construction as two aspects of the more general process of construction and avoids the numerical problems encountered in the earlier method of simulation of excavation. This general method has been extensively applied by the first author to problems of simulation of underground excavation and construction [13, 14, 16, 18, 19]. This method differs from earlier method (for example, see Christian and Wong, [6]) in which excavation is simulated by applying, to the newly created boundary surface in post-excavation configuration, the negative of tractions which existed on such boundary in pre-excavation configuration.

In the more general approach proposed by the first author, construction and excavation process is treated as a form of non-linearity, the change in geometry being the source of non-linearity. Similar to other non-linear problems, an incremental approach is used. For the regions to be excavated or constructed elements are assigned initially. Construction is simulated by activating appropriate elements at specified steps in the process of incremental analysis. A reverse process is used for simulating excavation. Elements in the region to be excavated are initially present. At specified steps in the incremental analysis elements are deactivated to simulate excavation. The theoretical aspects of this method are described in Ghaboussi and Pecknold [22].

The same procedure can be used in simulation of excavation and construction when the saturated or partially saturated soils are modelled as two-phase media. However, some modifications are needed. Since the generation and dissipation of pore pressures are time-dependent processes, the state equations are integrated in real time and care must be taken in selection of construction increments and time intervals so that the analysis reflects the actual progress of construction as closely as possible. The actual construction is a continuous process corresponding to small increments. In finite element simulations the whole construction process is simulated in a few discrete stages, each corresponding to a finite increment. After the simulation of construction of each finite increment, the geometry of the system must be kept constant for a period of time approximately equal to the actual construction time for such an increment. As an example, the continuous process of pore pressure generation and dissipation as the height of a dam increases during the actual construction, is replaced in the finite element simulation by abrupt finite increases in the height of the dam, resulting in abrupt increases in the pore pressures. During the following interval of time the height of the dam is kept constant and pore pressures are allowed to dissipate. Special attention must also be paid to the boundary conditions. For example, in simulation of construction of an earth dam, the surface of a newly placed lift is a free drainage boundary surface (zero pore pressure). However, after adding the next lift, the nodes on such a surface become interior nodes, with their unknown pore pressures as independent variables.

The problem of pore pressures during the construction of earth dams has attracted the attention of many researchers. Various analytical methods for evaluation of such pore pressures have been proposed and comparisons have been made with field measurements. The stress redistribution resulting from dissipation of construction pore pressures may result in arching and zones of low stress in the core. Such low stress zones in the core may be vulnerable to hydraulic fracturing.

Probably the first analytical method for evaluation of construction pore pressures in earth dams was developed by Hilf [25]. Most of the early methods, including Hilf's formulation, are based on the assumption of one-dimensional vertical flow. Moreover, surface tension between pore air and pore water, as well as dissipation were neglected in Hilf's solution. Bishop [2] and Li [29] presented improved solution by allowing certain amounts of pore pressure dissipation. Li [29] divided a construction season into a number of incremental steps. At the beginning of each step pore pressures are generated with the assumption of no drainage and use of Hilf's formula. At the end of each step, the pore pressures are allowed to dissipate by a predetermined fraction of the current pore pressures, thus necessitating the subjective use of engineering judgement.

Gibson [23] proposed a more rigorous method by considering a moving boundary and allowing dissipation at such boundary. Terzaghi's consolidation theory was modified by including the moving boundary. Solutions for arbitrary rate of construction can be obtained from Gibson's equation by finite difference method. Sheppard and Aylen [34] noted that this one-dimensional solution gives goods agreement with field measurements, except towards the end of the construction when lateral flow effects are more pronounced.

The lateral flow effects which have been neglected in the earlier methods become specially important in case of dams with clayey cores. In such dams the high construction pore pressures are concentrated in the core and the geomerty of cores dictate significant lateral flows. A number of two-dimensional solutions have been proposed. Koppula [28] and Osaimi and Hoeg [31] proposed finite difference based methods, while Eisenstein *et al.* [8, 9] and Cavounidis and Hoeg [4, 5] have proposed finite element based methods. Various simplifying assumptions are implicit in these methods.

A method of analysis was recently proposed by Kim [27]. This method considers the coupling between solid and fluid phases and uses the previously described incremental construction method. The compressibility of pore air–water mixture, including the surface tension effects, as well as elastoplastic material behaviour for solid skelton are taken into account. This incremental solution uses the fully coupled equations.

Shown in Figure 14.3 is the finite element mesh for one-half of a hypothetical earth dam with clayey central core. The construction of this dam is simulated in five lifts, each layer of elements comprising a construction lift. The rate of

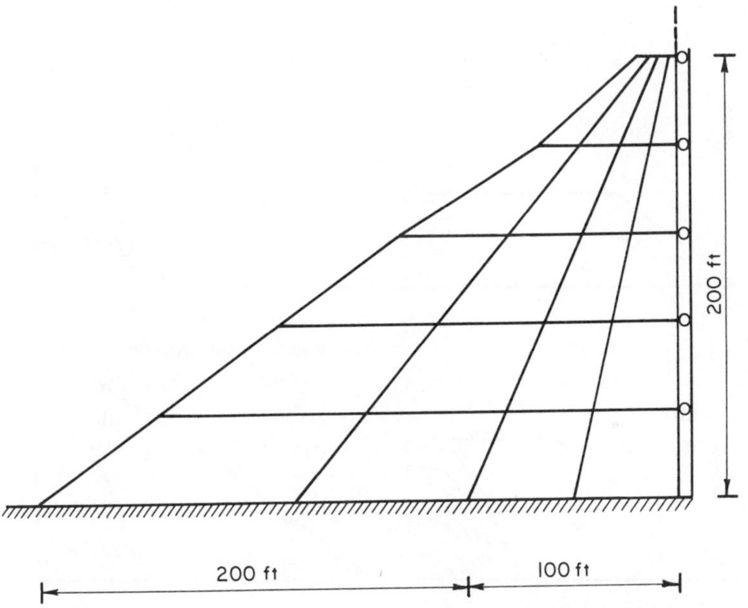

Figure 14.3 Finite element mesh for one-half of the dam with central core

construction is assumed to be $m = 0.5$ ft per day. After adding a layer of elements, representing a 40 ft lift, the height of the dam was kept constant for 80 days which is the actual construction time for a 40 ft lift at the assumed rate of construction. The material properties used in the analysis are summarized below:

Initial condition:
- Porosity, $n_0 = 0.30$
- Degree of saturation, $S_0 = 0.95$
- Pressure difference, $T = 8$ p.s.i.

Elastic properties:
- Modulus of elasticity, $E = E_0 p'$
- for core, $E_0 = 150$, $\nu = 0.4$
- for shell, $E_0 = 375$, $\nu = 0.3$

Coefficient of permeability:
- for core, $k = 0.2$ ft/year
- for shell, $k = 6.0$ ft/year

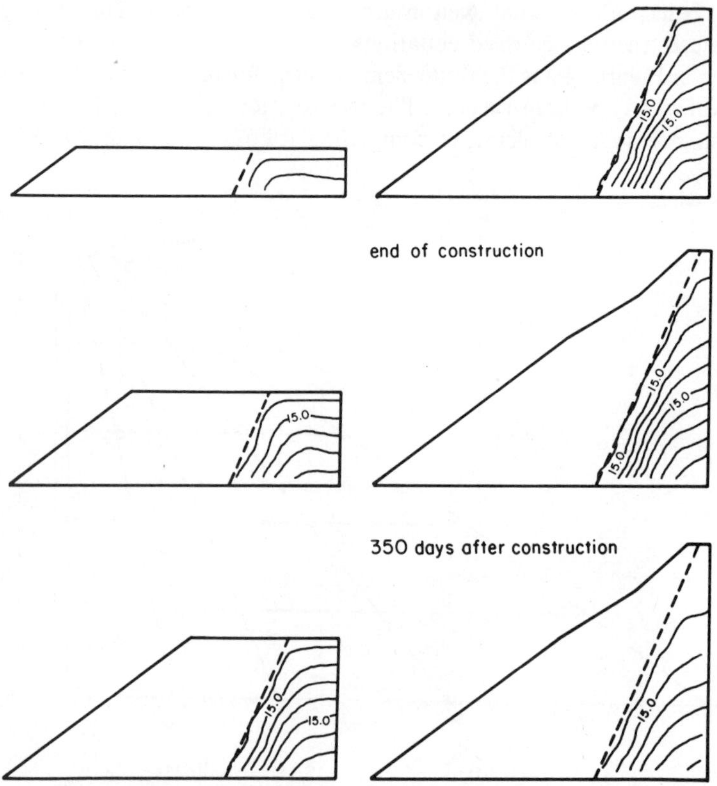

Figure 14.4 The distribution of pore pressures during the construction (five stages) and 350 days after the completion of construction

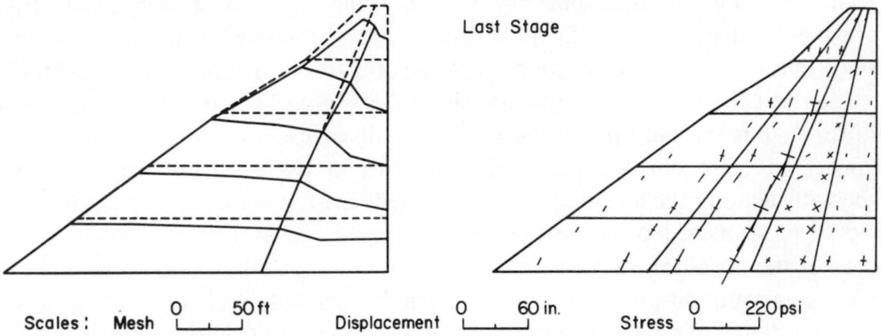

Figure 14.5 Displacements and principal stresses at the end of construction

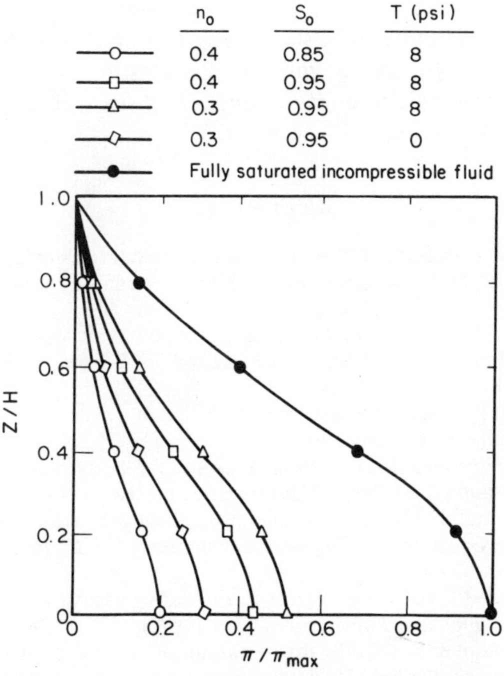

Figure 14.6 Effect of water content, porosity, and pressure difference on the pore pressures at the end of construction along the centerline of dam shown in Figure 16.3

The pore pressure distributions at 80 days after the addition of each lift are shown in Figure 14.4. Also shown in this figure is the pore pressure distribution at about one year after the completion of the construction. The effective principal stresses and the deformed shape are shown in Figure 14.5. It is clear from these figures that as the pore pressures in the core dissipate, more vertical stresses are transferred from core to the shells. Stresses in the core are generally low, especially around the mid-height of the dam, where the highest settlements occur. The stress transfer from core to shells also results in high shear stresses in the areas of the core adjacent to the shells.

The construction pore pressures in earth dams are strongly dependent on soil water content. This point is well illustrated in Figure 14.6, which shows the pore pressures at the end of construction along the centreline of the dam shown in Figure 14.3. For all the cases shown in Figure 14.6 the construction is simulated in five stages and all the material properties are the same except porosity, degree of saturation and pressure difference, T. The computed pore pressures are normalized with respect to those of fully saturated case ($S_0 = 100$ per cent), labelled 'incompressible fluid', which develops the highest pore pressures. It can be seen that a reduction of degree of saturation S_0 from 95 per cent to 85 per cent causes almost 50 per cent reduction in the pore pressures. The effect of variation in porosity is small. It can also be seen in Figure 14.6 that neglecting the surface tension effects ($T = 0$) causes significant underestimation of pore pressures. Therefore, it is reasonable to consider neglecting the surface tension effects an unconservative assumption.

REFERENCES

1. A. W. Bishop and A. K. G. Eldin, 'Undrained triaxial tests on saturated sands and their significance in the general theory of shear strength', *Geotechnique*, **2**, 13–32 (1950).
2. A. W. Bishop, 'Some factors controlling the pore pressure set-up during the construction of earth dams', *Proc. 4th Inter. Conf., Soil Mech. Found. Eng. (London)*, **2**, 294–300 (1957).
3. J. D. Campbell, *Pore pressures and volume changes in unsaturated soils*, Ph.D. Thesis, University of Illinois, Urbana Illinois, 1973.
4. S. Cavounidis, *Effective Stress–Strain Analysis of Earth Dams during Construction*, Ph.D. thesis, Stanford University, Stanford, California, 1975.
5. S. Cavounidis and K. Hoeg, 'Consolidation during construction of earth dams', *Journal of the Geotechnical Engineering Division, ASCE*, **103**, GT10, 1055–1067 (1977).
6. J. T. Christian and I. H. Wong, 'Errors in simulating excavations in elastic media by finite elements', *Soils and Foundations*, **13**, 1 (1972).
7. R. W. Clough and R. J. Woodward, 'Analysis of embankment stresses and deformations', *Journal of the Soil Mech. and Foundation Engineering Division, ASCE*, **93**, SM4, 529–549 (1967).
8. F. Eisenstein, A. V. G. Krishnayya, and T. C. Steven, 'Analysis of consolidation in cores of earth dams', *Proc. of the Second Inter. Conf. on Numerical Methods in*

Geomechanics, Engineering Foundation Conference, Blacksburg, VA., **1**, 1089–1107 (1976).
9. Z. Eisenstein and T. C. Steven, 'Analysis of consolidation, behavior of mica dam', *Journal of the Geotechnical Engineering Division, ASCE*, **103**, GT8, 879–895 (1977).
10. J. E. Garlanger, *Pore Pressures in Partially Saturated Soils*, Ph.D. thesis, University of Illinois, Urbana, Illinois, 1970.
11. J. Ghaboussi and E. L. Wilson, 'Variational formulation of dynamics of fluid saturated porous elastic solids', *Journal of Engineering Mechanics Division, ASCE*, **98**, EM4 (1972).
12. J. Ghaboussi and E. L. Wilson, 'Flow of compressible fluid in porous elastic media', *International Journal for Numerical Methods in Engineering*, **5**, 3 (1973).
13. J. Ghaboussi and R. E. Ranken, 'Interaction between two parallel tunnels', *International Journal for Numerical and Analytical Methods in Geomechanics*, **1**, 1, 75–103 (1977).
14. J. Ghaboussi and G. Gioda, 'On the time-dependent effects in advancing tunnels', *International Journal for Numerical and Analytical Methods in Geomechanics*, **1** (3) (1977).
15. J. Ghaboussi and M. Karshenas, 'On the finite element analysis of certain material nonlinearities in geomechanics', *Proceedings, International Conference on Finite Elements in Nonlinear Solids and Structures*, Geilo, Norway, 1978.
16. J. Ghaboussi, R. E. Ranken, and M. Karshenas, 'Analysis of subsidence over soft-ground tunnels', *Proceedings, International Conference on Evaluation and Prediction of Subsidence*, Pensacola, Florida, 1978.
17. J. Ghaboussi and U. S. Dikmen, 'Liquefaction analysis of horizontally layered sands', *Journal of Geotechnical Engineering Division, ASCE*, **104**, GT3 (1978).
18. J. Ghaboussi, R. E. Ranken, and A. J. Hendron, 'Time dependent behavior of solution caverns in salt', *Journal of Geotechnical Engineering Division, ASCE*, **107**, GT10, 1379–1401 (1981).
19. J. Ghaboussi and R. E. Ranken, 'Finite element simulation of underground construction', *Proceedings, Symposium on Implementation of Computer Procedures and Stress–Strain Laws in Geotechnical Engineering*, Chicago, Illinois, 1981.
20. J. Ghaboussi and S. U. Dikmen, 'Liquefaction analysis for multidirectional shaking', *Journal of Geotechnical Engineering Division, ASCE*, **107**, GT5 (1981).
21. J. Ghaboussi and H. Momen, 1983. 'Modelling and analysis of cyclic behaviour of sands', *Soils Mechanics—Cyclic and Transient Loading*, Chapter 12, (Eds. G. N. Pande and O. C. Zienkiewicz), John Wiley and Sons.
22. J. Ghaboussi and D. A. W. Pecknold. Incremental finite element analysis of geometrically altered structures. To appear in *Int. Jour. for Num. Methods in Eng.*
23. R. E. Gibson, 'The progress of consolidation in a clay layer increasing in thickness with time', *Geotechnique*, **8** (4), 171–182 (1958).
24. O. H. Gilbert, *The Influence of Negative Pore Water Pressures on the Strength of Compacted Clays*, Master of Science thesis, Massachusetts Inst. of Technology, Cambridge, Massachusetts.
25. J. W. Hilf, 'Estimating construction pore pressure in rolled earth dams, *Proc. 2nd International Conference on Soil Mechanics and Foundation Engineering*, **3**, 234–240 (1948).
26. M. Karshenas and J. Ghaboussi, 'Modeling and finite element analysis of soil behavior', *Report No. UILU-ENG*-79-2020, Dept. of Civil Engineering, University of Illinois, Urbana, Illinois, 1979.
27. K. J. Kim, *Finite Element Analysis of Nonlinear Consolidation*, Ph.D. thesis, University of Illinois, Urbana, Illinois, 1982.

28. S. D. Koppula, *The Consolidation of Soil in Two Dimensions and with Moving Boundary*, Ph.D. thesis, University of Alberta, Edmonton, Alberta, Canada, 1970.
29. C. Y. Li, 'Construction pore pressures in an earth dam', *Journal of the Soil Mechanics and Foundations Division, ASCE*, **85**, SM5, 43–59 (1959).
30. A. E. Osaimi, *Consolidation of earth dams during construction*, Ph.D. thesis, Standford University, Standford, California, 1973.
31. A. E. Osaimi and K. Hoeg, *Construction Pore Pressures in Earth Dams*, Ph.D. thesis, Standford University, Standford, California, 1977.
32. A. N. Schofield and C. P. Worth, *Critical State Soil Mechanics*, McGraw-Hill, London, 1968.
33. Ir. E. Schuurman, 'The compressibility of an air/water mixture and a theoretical relation between the air and water pressures', *Geotechnique*, **16**, 269–281 (1966).
34. G. A. R. Sheppard and L. B. Aylen, 'The Usk scheme for the water supply of Swansea', *Proc. Instn Civ. Engrs*, **7**, 246–265 (1957).
35. A. W. Skempton and A. W. Bishop, 'Building materials, their elasticity and inelasticity', *Soils*, Chap. 10, North-Holland Publ. Co., Amsterdam, 1954.

Mechanics of Engineering Materials
Edited by C. S. Desai and R. H. Gallagher

Chapter 15

Review of Creep Modelling for Rock Salt

W. Herrmann, W. R. Wawersik and S. T. Montgomery

15.1 INTRODUCTION

Repositories for nuclear waste and petroleum reserves have been proposed to be sited in salt deposits. The ductility of salt is expected to resist fracturing and to promote healing of fractures, thereby ensuring sealing. Since the requirements of repositories differ significantly from previous mining practice, there is an emphasis on engineering design by finite-element computer modelling. Adequate constitutive models of salt creep are central to this endeavour.

Previous laboratory investigations [1 to 4] of the creep of salt concentrated on relatively pure NaCl at high temperatures. Repository conditions involve natural rock salt, often with a very coarse grain structure and a variety of impurities. Moreover, low temperature creep is of paramount importance. Consequently, experimental work was begun at Sandia National Laboratories (SNL) and at RE/SPEC Inc. under contract to SNL, using relatively large specimens, and, for creep at low temperatures and stresses, relatively long testing times. Early expectations were that brittle response might be encountered. Therefore the test programme, included confining pressure as a test variable, and volumetric strain as an observed parameter. Specially designed creep test machines and test procedures were used to meet these requirements [5, 6].

While early quasi-static [7] and creep tests [8] provided important insights into the phenomena occurring in natural rock salt, material variability prevented clear quantitative deductions until a sufficiently large data base was built up to allow statistical analysis. Such analysis leads to 'creep laws' describing the behaviour of confined (triaxial) creep tests under nominally constant stress conditions. The creep laws which were used rest on an assumption that steady state creep exists and is approached asymptotically. This assumption is notoriously difficult to demonstrate conclusively by experiment, especially at low temperatures and stresses. Therefore observations of creep closure of existing mine openings over very long periods of time are appealed to in order to establish that long time creep at low temperatures is plausible.

Finally, creep laws are extended to constitutive equations for changing general stress states. This is done by invoking creeping viscous flow and elastic-viscous theories. These in turn are used to compute the closure of an existing, nearly isolated mine opening, for which creep closure measurements are available for a relatively long period, to determine the applicability of the constitutive equations derived from laboratory creep experiments to mine design.

This chapter is intended to provide an overview of the above programme at SNL. A comparable programme has been carried out concurrently in Germany [9].

15.2 CREEP CURVES

The experimental technique is described in detail elsewhere [5, 6]. Each experiment, carried out at nominally constant axial stress and confining pressure, resulted in a series of readings of axial and volumetric strains, as well as actual axial stress and confining pressure, at a large number of times as the test progressed. Digitized data were recorded, reduced to true shear stress and hydrostatic pressure, logarithmic axial and transverse strains, and plotted versus time [10]. A total of 68 separate tests on natural rock salt from the Salado formation in Southeast New Mexico were analysed, covering ranges of tempera-true from 22 to 200°C, shear stress from 1 to 6 k.s.i. (6.9 to 41 MPa) and pressures from 0 to 3 k.s.i. (0 to 20 MPa). In some cases, two or more tests at different constant stresses or temperatures were run in successive test stages on the same specimen.

Many specimens exhibited an initial small compaction upon loading. Some specimens tested at low temperatures exhibited some dilation, and showed other evidence of brittle deformation in terms of acoustic emission during creep, and post-test observations of cracking. However, a few specimens exhibited episodes of compaction during the test. Volume strains were always much smaller than shear strains, even at low temperatures, and were negligible at the higher end of the temperature range.

Some experiments, including several at room temperature, appeared to exhibit long nearly linear strain–time histories following an initial period of transient decelerating creep. However, many experiments, particularly at lower temperatures and shear stresses, seemed to approach linearity, but were not continued long enough to show long linear portions in their strain–time histories. Even though some tests seemed to show long linear histories, the question remains whether true steady-state creep exists, or whether creep would decelerate further, and maybe cease altogether, over tens or hundreds of years associated with repository lifetimes. On the basis of observations of continued creep closure over hundreds of years in existing mines, such as those in the Salzkammergut of Austria, for example, it would seem that creep continues for very long times even at low temperatures and stresses.

Since some tests appeared to approach steady-state creep, but did not reach linearity in their strain–time curves, a consistent method of estimating steady-state creep rates from the data was required. At higher temperatures, transient creep is often fitted by the empirical exponential creep law

$$e = e_0 + \dot{e}_s t + e_\infty\{1 - \exp(-\xi(t)\} \tag{15.1}$$

where e is the strain, t the time, and e_0, e_s, e_∞, and ξ are parameters. This expression was found to give good fits to the data for all tests, especially for data beyond the first hour or so [10]. Figures 15.1 and 15.2 show typical results at the highest and lowest temperatures which were tested.

In order to investigate the reliability of estimates of steady-state creep histories, a number of tests which exhibited long linear histories were reanalysed, successively truncating the data at earlier and earlier times. Even when the data were truncated well into the decelerating portion of the history, fitted values of e_s did not increase by more than a factor of two. This may be compared to a data scatter for tests at nominally identical stresses and temperatures of a factor of five or more.

At low temperatures, it is often argued that a logarithmic creep law should be valid

$$e = e_0 + \dot{e}_s t + \gamma \ln(1 + \eta t) \tag{15.2}$$

where e_0, $\dot{e}_s$, γ, and η are parameters. Fits were also made to the data using this

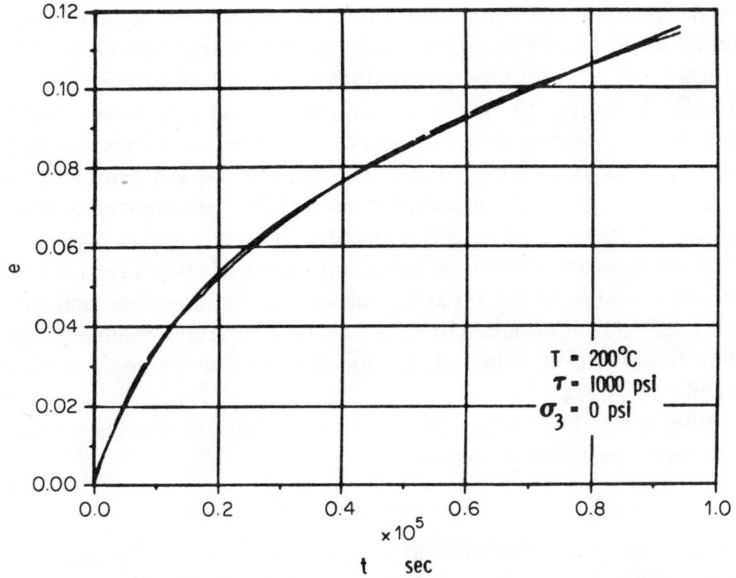

Figure 15.1 Creep data and exponential fit, high temperature

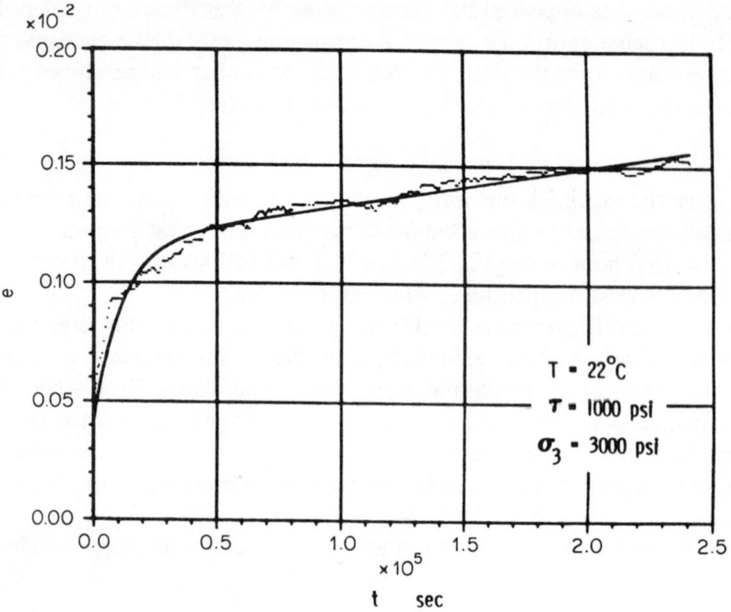

Figure 15.2 Creep data and exponential fit, low temperature

expression. Most of the data were fitted about equally well, but provided consistently lower steady-state creep rates (usually 20 to 50 per cent) [10]. Data for the first hour or so of low temperature tests were usually fitted somewhat better by (15.2) than by (15.1). It is often supposed that steady-state creep is not experienced at low temperatures. However, when the second term of (15.2) was omitted, very poor fits were obtained. Therefore, if the salt exhibits continued decelerating creep, it is not described at all well by the logarithmic creep law.

The strain on initial loading is not included in e_0, which was introduced as a free parameter in order to improve the fit. The very early time creep histories (for the first few hours) were not fitted as well as the later histories, and not much physical significance is attached to e_0 and initial creep rates deduced from either set of these fits. However, values of the time constant ξ or η respectively are more meaningful.

Two other creep laws were also used to fit many of the tests. One is the empirical time-hardening expression

$$e = e_0 + \beta t^{\alpha} \tag{15.3}$$

where steady-state creep is absent and α and β are parameters. This fitted the early time data rather better than did (15.1) and (15.2) [8, 11]. Another more complicated creep law proposed by Munson and Dawson [12] was also found to

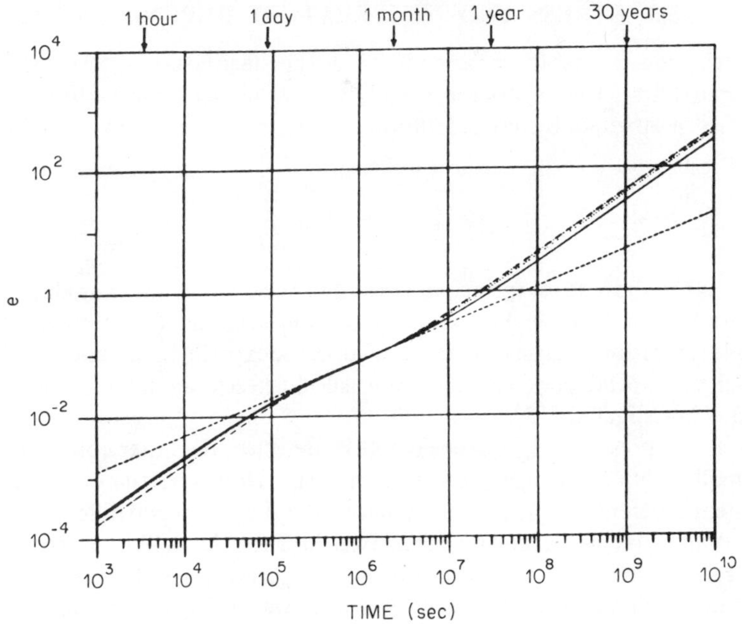

Figure 15.3 Extrapolation of creep fits, $T = 22°C$ and $r = 3$ k.s.i. Chain dotted curve: exponential, solid curve: logarithmic, dashed curve: power law, dotted curve: Munson and Dawson

fit the data as well as (15.1) and (15.2) [11]. In view of the fact that several expressions fit the data well, we hesitate to draw from them any conclusions about the micromechanical mechanisms of deformation which may be operative.

Since long time creep is of interest, extrapolations of the various creep laws were compared. A typical example for 22 °C and 3 k.s.i. (21 MPa) shear stress is shown in Figure 15.3.

All four curves overlay the data between about one day and two months. The three fits including steady-state terms approach unit slope after 2 to 6 months, thereafter showing only a small offset due to minor differences in the predicted long-time transient creep accumulation. However, the power law approaches the slope α which is less than unity, and predicts significantly smaller creep strains at long times. It is obviously difficult to resolve conclusively, from data of less than a year or two in duration, whether steady-state creep is attained, as described by (15.1) or (15.2), or whether creep continues to decelerate, as described by the power law (15.3).

Since all of the creep laws fitted the data about equally well after the first hour or so of the tests, further work was limited to the exponential creep law (15.1), which has a convenient mathematical form.

15.3 STRESS AND TEMPERATURE DEPENDENCE

The stress and temperature dependence of the steady-state creep rate $\dot{e}_s$ was investigated first. The expression usually used at higher temperatures [1 to 4], originally motivated by consideration of thermally activated climb of dislocations, is

$$\dot{e}_s = a\left(\frac{\tau}{\mu}\right)^n \exp\left(-\frac{Q}{RT}\right) \tag{15.4}$$

where τ is the shear stress (principal stress difference), μ the shear modulus, T the absolute temperature, R the gas constant, and a, n, and Q are parameters. A multiple regression analysis was made of all of the experimental data, using (15.4), in which the dependence on core location, specimen size and hydrostatic pressure was also investigated [13].

No size or pressure dependence could be detected, the latter correlating with the small volume changes observed in tests. However, there was a small consistent difference in the parameter a for salt taken from two different geologic levels. Values of the parameters are given in Table 15.1. Plots of steady-state creep rate versus shear stress for various temperatures, and creep rate versus temperature for various shear stresses are shown in Figures 15.4 and 15.5.

It may be seen that the agreement between lines representing the fit and points representing the data is satisfactory for purposes of an empirical fit, within the scatter of the data. The scatter of the data is such that more complicated fits with more than one term of the form (15.4), or with other types of terms, could easily be made to fit the data just as well. Therefore we again refrain from making any deductions from the fit regarding microstructural mechanisms of deformation.

Turning next to the stress and temperature dependence of transient creep, there is precedence in the high temperature metal creep literature [14] for the assumption that steady-state and transient creep should have similar stress and temperature dependencies, namely

$$\dot{e}_0 = k\dot{e}_s \tag{15.5}$$

where k is a constant, and where the initial creep rate $\dot{e}_0$ is found, by differentiating (15.1), to be $\dot{e}_0 = \dot{e}_s + \xi e_\infty$. Since this assumption leads to a

Table 15.1 Values of creep parameters

$\mu = 1.8 \times 10^6$ p.s.i.	$Q = 12.0 \pm 0.65$ kcal/mole
$= 12.4$ GPa	$= 50.2 \pm 2.72$ kJ/mole
$R = 1.987$ cal/mole	$k = 5.56 \pm 0.79$
$= 8.319$ J/mole	$B = 84 \pm 28$
$A = 6.7 \pm 1.42 \times 10^{14}$/s[†]	$c = 6.85 \pm 3.76 \times 10^{-6}$/s
$n = 4.9 \pm 0.27$	

[†] This value is for salt from the lower geologic level. For salt from the upper level $A = 2.5 \pm 0.53 \times 10^{14}$.

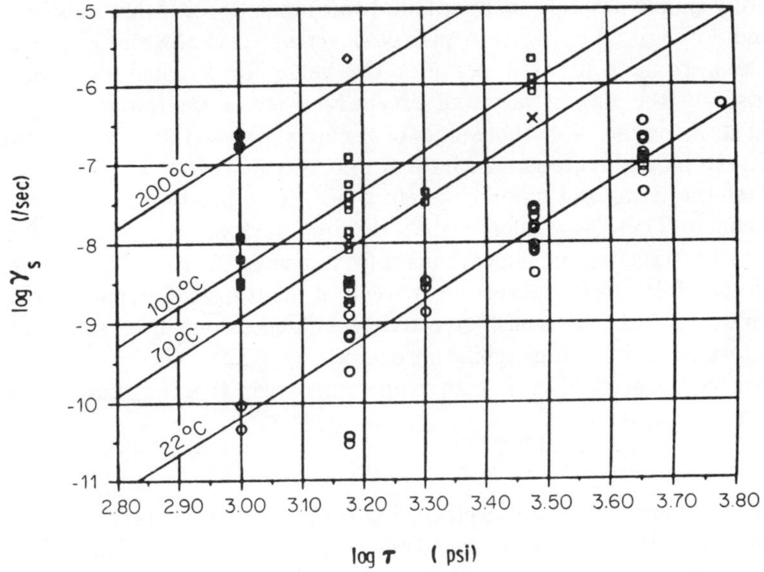

Figure 15.4 Steady-state creep rate versus shear stress. Circles: 22° C, crosses: 70° C, squares: 100°C, diamonds: 200° C

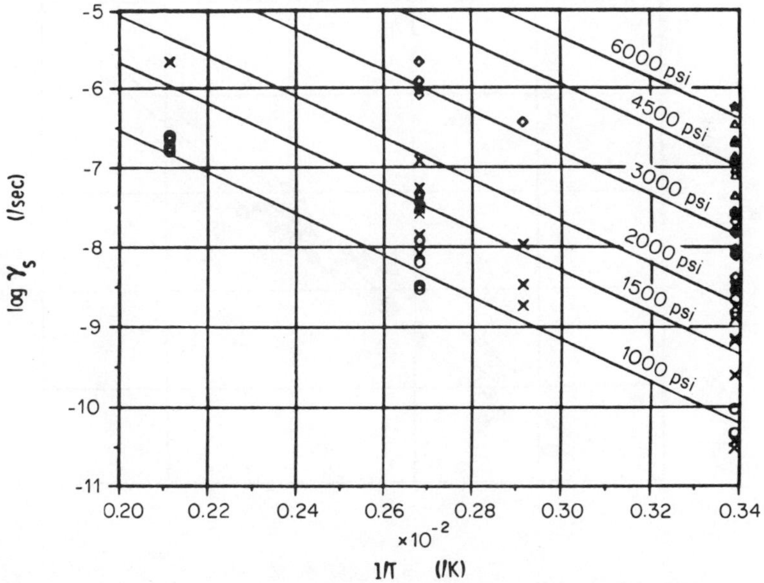

Figure 15.5 Steady-state creep rate versus temperature, Circles: 1 k.s.i., crosses: 1.5 k.s.i., inverted triangles: 2.0 k.s.i., diamonds: 3 k.s.i., triangles: 4.5 k.s.i., stars: 6 k.s.i.

relatively simple description, an attempt was made to see if the transient creep data could be fitted by (15.5). A plot of $\dot{e}_0$ versus $\dot{e}_s$ is shown in Figure 15.6.

A least squares fit [15] provides the value for k listed in Table 15.1. Surprisingly, the data fit very well, except for a few of the lowest temperature, lowest stress points. Note that values of $\dot{e}_0$ correspond to those derived from fits of (15.1) to individual tests, which do not necessarily reflect the very early time trend of the data, and should therefore not be imbued with great physical significance. There is evidence that the initial slope of the data for low temperature tests is in fact larger than the fits indicate (Figure 15.2). The empirical fit will therefore falsify the very early trend of the transient creep at the lowest temperatures, but this is not expected to influence long time predictions of repository behaviour to a significant degree.

It has been observed [14] in high temperature creep, that the creep frequency ξ is often proportional to $\dot{e}_s$

$$\xi = b\dot{e}_s \tag{15.6}$$

where b is a constant. Taken together, (15.5) and (15.6) imply that e_∞ is a constant $e_\infty = (k-1)/b$. Plots of ξ versus $\dot{e}_s$ and of e_∞ versus $\dot{e}_s$ (Figures 15.7 and 15.8) show much more scatter than Figure 15.6, especially at lower temperatures and stresses.

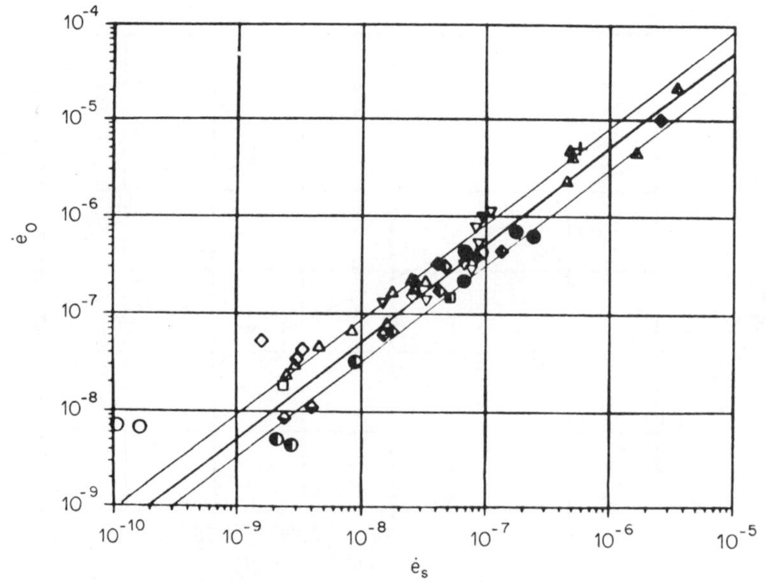

Figure 15.6 Initial versus steady-state creep rates. Circles: 1 k.s.i., diamonds: 1.5 k.s.i., squares: 2 k.s.i., triangles: 3 k.s.i., inverted triangles: 4.5 k.s.i., pluses: 5 k.s.i. Open symbols; 22 °C, horizontally split symbols: 70 °C, vertically split symbols: 100° C, solid symbols: 200° C. Solid lines represent the fit and one standard deviation

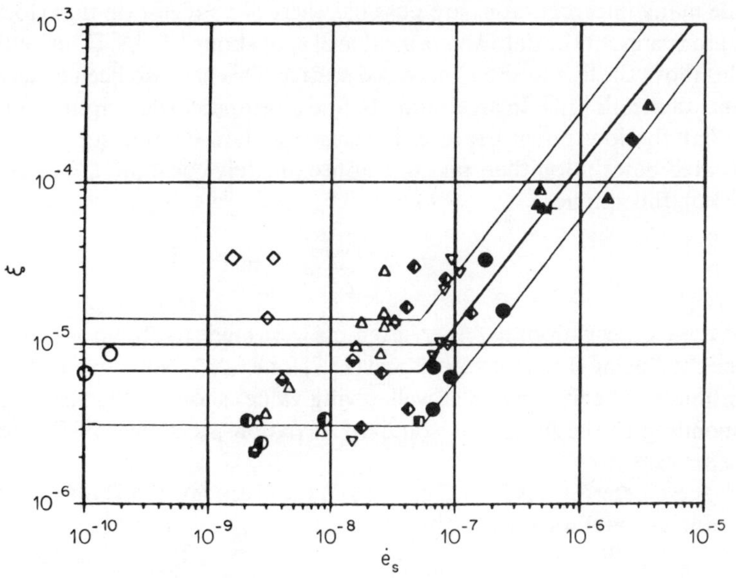

Figure 15.7 Creep frequency versus steady state creep rate. Same symbols as Figure 15.6 Solid lines represent the fit and one standard deviation

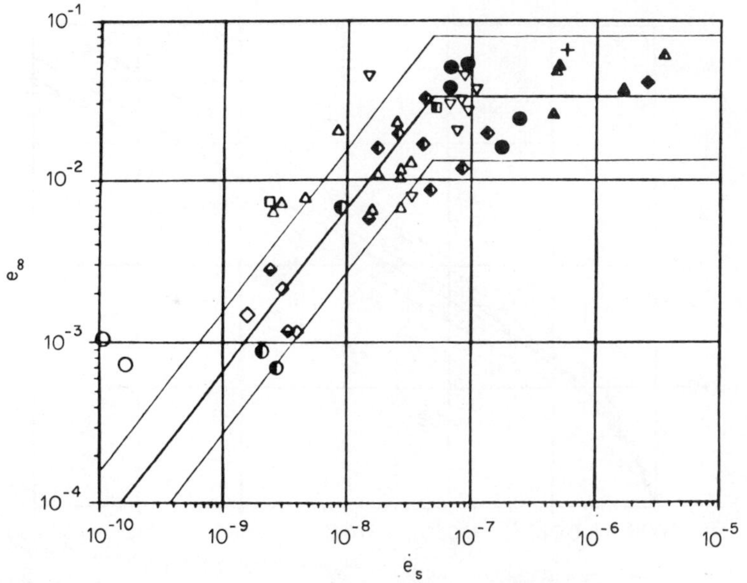

Figure 15.8 Total transient creep versus steady state creep rate. Same symbols as Figure 15.6 Solid lines represent the fit and one standard deviation

While many interpretations are possible, there is a suggestion that (15.6) may be used to represent the data above a value of $\dot{e}_s$ of about 10^{-7}/s. Data below this value lie above the fit and show increased scatter. This has also been observed for a number of metals [14]. In an attempt to find an empirical description, it may be noted that the lower temperature, lower stress data do not support a more complicated description than taking ξ approximately constant, $\xi = c$. Together with (15.5), this implies

$$e_\infty = \frac{k-1}{c}\dot{e}_s \tag{15.7}$$

Taking these observations together, attempts were made to obtain a least squares fit to the data, using (15.6) for $\dot{e}_s \geq \dot{e}^*$ and (15.7) for $\dot{e}_s \leq \dot{e}^*$. Note that at $\dot{e}^*, c = b\dot{e}^*$ for continuity. The fit converged well, giving values shown in Table 15.1. Lines corresponding to the fit and its standard deviation are shown in Figures 15.7 and 15.8.

With these descriptions for $\dot{e}_0$, ξ, and e_∞, the creep law (15.1) may be written in non-dimensional form as

$$\varepsilon = t_n + \frac{k-1}{b}\{1 - \exp(-bt_n)\} \tag{15.8}$$

where for $\dot{e}_s \geq \dot{e}^*$, $\varepsilon = (e - e_0)$ and $t_n = \dot{e}_s t$ while for $\dot{e}_s \leq \dot{e}^*$, $\varepsilon = (e - e_0)\dot{e}^*/\dot{e}_s$ and

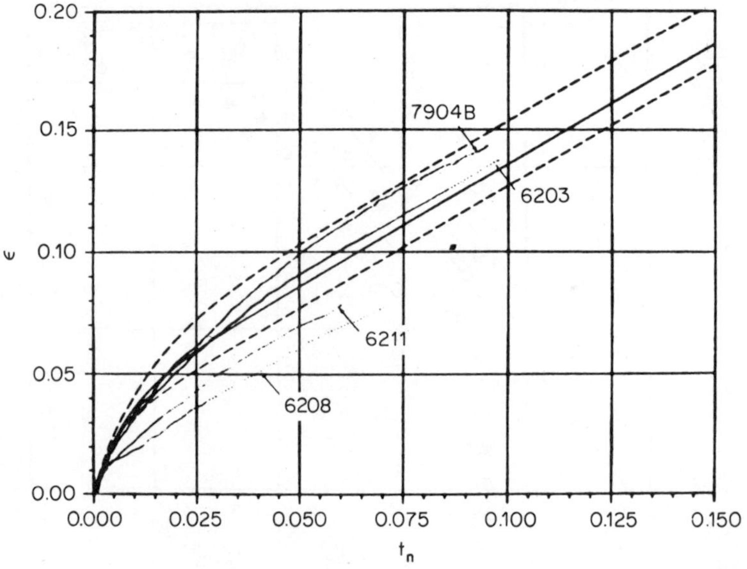

Figure 15.9 Normalized creep curves for $T = 100\,°C$ and $r = 1.5$ k.s.i.

$t_n = \dot{e}^* t$. This amounts to a universal creep curve, since the stress and temperature dependence, reflected only in $\dot{e}_s$, has been normalized out. Normalized data for tests at $T = 100\,°C$, $\tau = 1.5$ k.s.i. are shown in Figure 15.9. The solid curve represents (15.8) while the dashed curves on either side represent the envelope of curves, varying k, b, and c separately within their standard deviations. Data for other conditions are qualitatively similar, the scatter being somewhat larger at lower temperatures.

15.4 FINAL CREEP LAW

Each of the creep laws discussed in the previous sections, when differentiated, has the form

$$\dot{e} = \dot{e}_s(T, \tau) + \dot{e}_p(T, \tau, e_p) \tag{15.9}$$

where T is the temperature, τ the shear stress, and e_p the transient creep strain. For the exponential law (15.1) in particular, the second term has the form

$$\dot{e}_p = \xi(e_\infty - e_p) \tag{15.10}$$

While the creep law was constructed from data for creep tests at constant stress and temperature, the assumption is now made that in differential form (15.10) is valid for changing stresses and temperatures. This assumption should be checked by experiment, for example, constant strain rate or constant strain–stress relaxation experiments.

Partial confirmation of the use of (15.10) comes from tests in which successive stages at different constant stress and temperature were performed on the same specimen. It is possible to integrate (15.10) explicitly for this case. This leads to expressions relating creep parameters for successive test stages to those for a virgin specimen [15]. These expressions were, in fact, used in the original data analysis to interpret fits to successive stages involving increasing stresses or temperatures. No dependence on test stage could be found in the statistical analysis.

Explicit expressions also result from the integration of (15.10) for successive test stages when the stress or the temperature are decreased from the former stage [16]. Either no transient creep, or inverted transient creep with positive or negative initial slope is predicted, depending on the previous history and the present stress and temperature. Some tests involved subsequent stages at reduced stress. These show prolonged periods of very low creep rate, while (15.10) predicts an inverted transient with rapid recovery to the creep rate given by (15.4) for that stress and temperature. Figure 15.10 shows an example of such a test at 22 °C, in which the first stage involved a shear stress of 3 k.s.i., the second stage 4.4 k.s.i., and the third stage once again 3 k.s.i. Note that the stress exponent n in (15.4) can be evaluated directly from the steady-state creep rates of the first two stages, while the pre-exponential constant was evaluated by fitting the first-stage only. The prediction for the third stage is then determined. The data clearly do not exhibit the predicted exaggerated inverted transient, and, if recovery occurs at all, it must do so on a very much longer time scale.

Subsequent experiments with steps in stress or temperature have suggested that if the step is small, then recovery is more rapid. It is possible that under sufficiently slowly changing conditions (15.10) might be adequate, even if the stress or temperature are decreasing. In general, tests such as that in Figure 15.10 suggest that the salt exhibits a more complicated history dependence than that embodied in (15.10), and that further work is needed.

It is convenient to express the creep law in terms of stress and strain invariants. Choosing the stress invariant to be the usual effective stress $\tau = \sqrt{(3J_2')}$, then τ reduces to the difference in principal components under the triaxial conditions applying to the creep tests. We choose an identical definition for our strain invariant γ, contrary to usual practice, which uses an equivalent strain which is $\frac{2}{3}\gamma$. Under triaxial conditions, and assuming in addition that the material is incompressible, we find that $\gamma = \frac{3}{2}e$. The creep law, comprising (15.9) and (15.10),

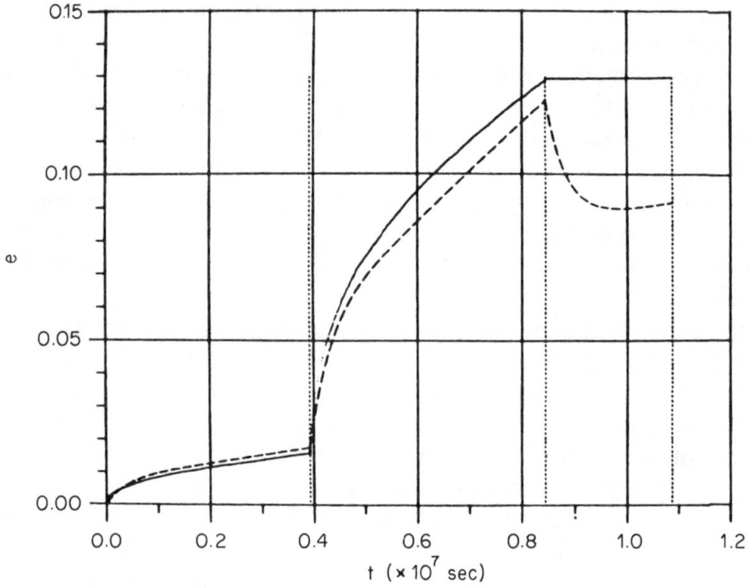

Figure 15.10 Strain-time history for multistage creep test, $T = 22\,°C$, first stage $r = 3$ k.s.i., second stage $r = 4.4$ k.s.i., third stage $r = 3$ k.s.i. Solid line: experiment, dashed line: prediction

with (15.4) to (15.7), then takes the form

$$\begin{aligned}
\dot{\gamma} &= \dot{\gamma}_s + \dot{\gamma}_p \\
\dot{\gamma}_s &= A\left(\frac{\tau}{\mu}\right)^n \exp\left(-\frac{Q}{RT}\right) \\
\dot{\gamma}_p &= (k - 1 - B\gamma_p)\dot{\gamma}_s \qquad \text{for } \dot{\gamma}_s \geq c/B \\
&= (k-1)\dot{\gamma}_s - c\gamma_p \qquad \text{for } \dot{\gamma}_s \leq c/B
\end{aligned} \tag{15.11}$$

where $A = \frac{3}{2}a$ and $B = \frac{2}{3}b$. The coefficients have values given in Table 15.1.

15.5 COMPARISON WITH OTHER DATA

Data for natural rock salt from a number of other sites, comprising both bedded and domal deposits, have been analysed in the same way [17]. It has been found that the data for bedded salt from Lyons, Kansas and for domal salt from Jefferson Island and Avery Island, Louisiana can be fitted to precisely the same equations as have been used for the data for salt from Southeast New Mexico. The data in each case scatter within the Southeast New Mexico data, although the data samples are smaller. Least squares fits provide values which are almost the same, within their standard deviations, as those for Southeast New Mexico salt. Steady-state creep data for salt from the Tatum dome, Mississippi, Grand Saline and Hockley mines, Texas, and West Hackberry dome, Louisiana are in fair to excellent agreement with the fit shown in Figures 15.4 and 15.5. However, data for salt from Bryan Mound and Bayou Chocktaw, Louisiana fit the same expressions, but with different fitting parameters.

Steady state creep rates have also been reported for salt from Asse, FRG [9]. They have been fitted to an expression identical with (15.4). Results are $A = 9.27 \times 10^{14}$/s, $n = 5.0$, and $Q = 12.9$ kcal/mole or 54.0 kJ/mole. These are in remarkably good agreement with the values given in Table 15.1.

Heard [2] has performed constant strain rate tests on relatively pure polycrystalline sodium chloride, from which steady-state creep rates were deduced. The data extended to 400 °C and to higher stresses than the Southeast New Mexico data. There is a well-known discrepancy [18] of approximately two orders of magnitude in creep rate between the data of Heard and those of Burke [1], and others. However, the data of Heard overlap the Southeast New Mexico data at 200 °C, with excellent agreement. For data from 200 to 400 °C, and shear stresses below about 4 k.s.i., Heard fitted the data to (15.4), finding a stress exponent $n = 5.5$ and an activation energy $Q = 23.9$ kcal/mole (105 kJ/mole). The stress exponent is in reasonably good agreement with the value given in Table 15.1. However, Heard's activation energy is much larger, and is in agreement with the activation energy for Na^+ diffusion. Munson and

Dawson [12] used two terms of the type (15.4), with different activation energies representing different creep mechanisms, to fit the sum of Heard and Southeast New Mexico data, placing the boundary between the two mechanisms roughly at 160 °C. For high shear stresses, Heard's data exhibit an increasing stress exponent, and terms representative of dislocation glide have been used in this region [2, 12, 18].

It is clear that the simple fit reported in the previous sections cannot be extrapolated to either higher temperatures or stresses. However, the simple fit extends to the highest temperatures and stresses normally encountered in repository design, and it is not warranted to attempt more complex fits with more terms without further information regarding the creep mechanisms which are active, especially in view of the scatter in the experimental creep data.

15.6 CONSTITUTIVE EQUATIONS

While the creep law (15.11) is couched in terms of invariants, it was derived from, and remains valid only for specific conditions, namely the confined compression or triaxial configuration. In order to extend this description to more general stress conditions, additional information is needed. This information is provided by making constitutive assumptions. The first is that the material is isotropic, and that the material behaviour enters only through invariants. In addition, we assume one of two constitutive theories, which, when specialized to triaxial conditions, result in (15.11).

The first constitutive theory is that of a non-Newtonian viscous incompressible fluid

$$\sigma' = \nu \dot{\mathbf{e}}' \tag{15.12}$$

where σ' is the deviatoric stress tensor whose components are $\sigma'_{ij} = \sigma_{ij} - p\delta_{ij}$, $\dot{\mathbf{e}}$ is the velocity strain rate tensor whose components are $\dot{e}_{ij} = (\partial u_i/\partial x_j + \partial u_j/\partial x_i)/2$ and ν is a viscosity coefficient which may be stress and temperature dependent. In order to force (15.12) to give the desired behaviour under triaxial conditions, we take $\nu = \tau/\dot{\gamma}$, where $\dot{\gamma}$ is given as a function of τ by (15.11). Inverting (15.12)

$$\dot{\mathbf{e}}' = \frac{\dot{\gamma}(T, \tau)}{\tau} \sigma' \tag{15.13}$$

Munson and Dawson [19] incorporated (15.12) into a finite-element solution method for quasi-static creeping flow of an incompressible medium, implemented in the COUPLEFLO computer program. While they used their own creep law [12], it has been shown above that predicted creep behaviour is almost indistinguishable from that of (15.11).

The second constitutive theory adds an elastic response to the viscous behaviour of (15.13). The total strain rate is divided into elastic and creep parts

$\dot{\mathbf{e}}' = \dot{\mathbf{e}}'_e + \dot{\mathbf{e}}'_c$, the latter being given by (15.13). The stress rate is related to the elastic strain rate by $\dot{\sigma}' = 2\mu\dot{\mathbf{e}}'_e$. Using these expressions

$$\dot{\sigma}' = 2\mu\left\{\dot{\mathbf{e}}' - \frac{\dot{\gamma}(T,\tau)}{\tau}\sigma'\right\} \tag{15.14}$$

Note that (15.14) is not only implicit in the stress, but the total primary strain γ_p is an internal state variable. An integration scheme has been developed [20] which converges fairly rapidly, so that (15.14) may be used in conventional finite-element computer programs, such as SANCHO.

15.7 MINE CLOSURE CALCULATIONS

In order to see if the creep law is applicable to field design calculations, it is desirable to compare predictions with actual mine closure data. Suitable data exist for the International Minerals and Chemical Corp. Esterhazy mine in Saskatchewan. One room cut in the shaft pillar is reasonably isolated in a thick bed of mixed halite and sylvinite. Measurements of room closure for a period of eight years have been reported by Mraz [21]. Measurements included convergence of the walls and floor of the nominally 6 m wide by 2.3 m high room, as well as displacements of bolt arrays in the walls and roof extending to different depths. The configuration is shown in Figure 15.11.

Finite-element calculations have been performed, using the viscous flow model (15.13) [19], as well as the elastic-viscous model (15.14) [22]. In all of these calculations the salt was taken, in first approximation, to be an effectively infinite homogeneous medium, neglecting the effects of geologic layering and the presence of clay seams. In view of the fact that a number of salts from different localities fitted the same creep law, it was assumed that the salt in this case also was well described by (15.11) with values of the parameters given in Table 15.1.

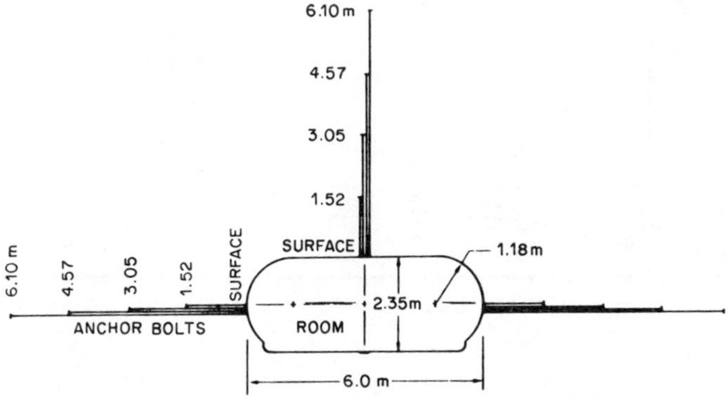

Figure 15.11 Cross-section of the room and instrumentation in the Esterhazy mine [21]

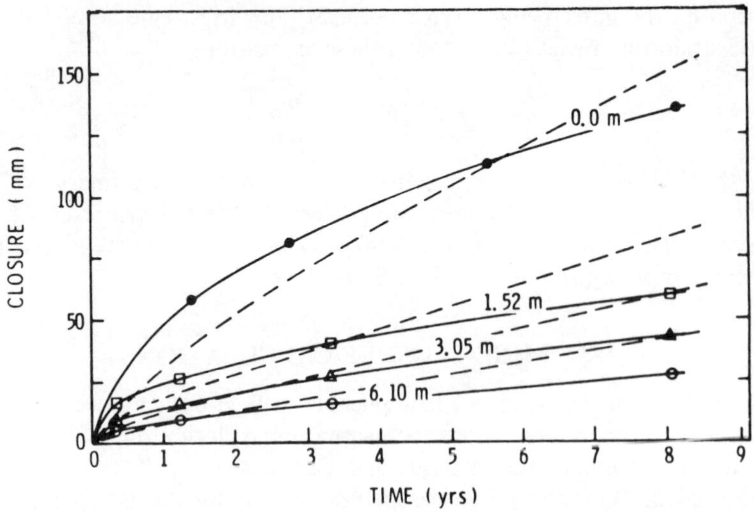

Figure 15.12 Horizontal closure, viscous model. Solid line: data, dashed line: prediction

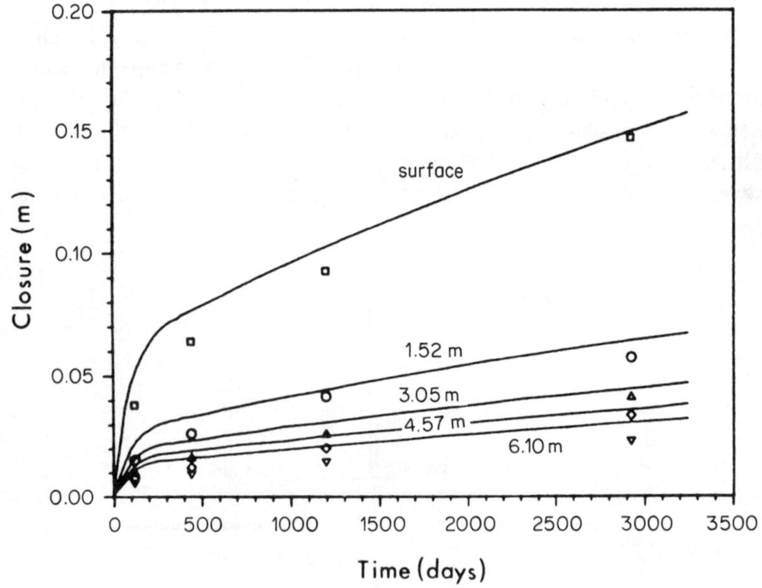

Figure 15.13 Horizontal closure, elastic-viscous model. Solid line: prediction, points: data. Squares: surface, circles: 1.52 m, triangles: 3.05 m, diamonds: 4.57 m, inverted triangles: 6.10 m

Predictions of horizontal closure are compared with measurements in Figures 15.12 and 15.13 for the viscous model and the elastic-viscous model respectively.

While the agreement in both cases might be considered satisfactory for mine design purposes, the elastic-viscous model provides a considerably better match, especially in magnitude at early times, and in the slope of the convergence curves at later times. This suggests greater confidence in predictions carried out with the elastic-viscous model when these are carried out to longer times. One might conclude that the inclusion of elastic response is important.

Calculations were also performed with the elastic-viscous model (15.14), in which the transient creep was suppressed by setting $\dot{\gamma}_p$ to zero in (15.11) [22]. Figure 15.14 shows predicted convergence with and without transient creep. It may be seen that the inclusion of the transient creep term affects only the early closure history, but that the two solutions converge after only one month. In the model, the total transient creep strain asymptotes to a relatively small value, which apparently has a minor effect on the long time solution. This suggests that reasonable predictions may be obtained using steady state creep alone. Since the inclusion of transient creep is expensive in terms of computer running times, the use of steady-state creep alone is very attractive, but further tests are indicated to ensure that the approximation is adequate for other cases.

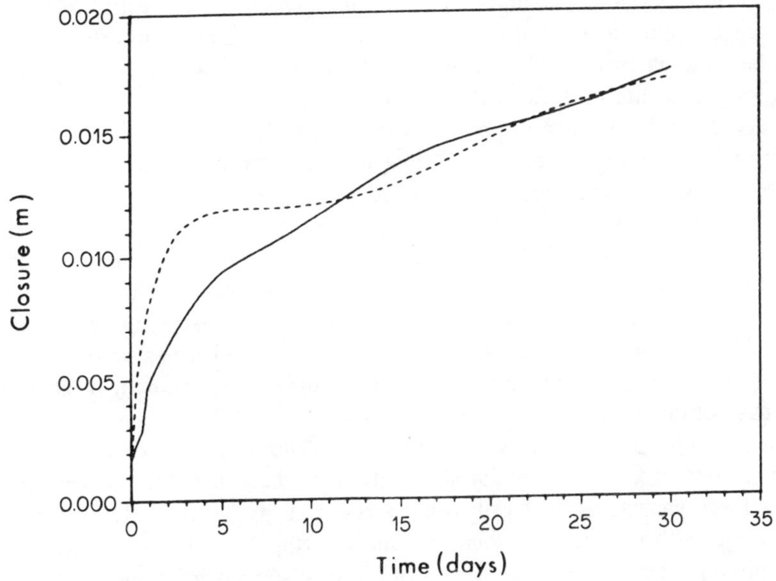

Figure 15.14 Comparison of predictions of room convergence at the wall with the elastic-viscous model. Solid line: steady state creep only, dotted line: transient and steady state creep

15.8 DISCUSSION

Laboratory creep data have been fitted to empirical expressions, resulting in a creep law. This in turn has been incorporated into constitutive relations, and used to predict creep closure of a mine opening, for comparison with experimental closure data.

Data from individual creep tests, comprising mostly transient decelerating creep, were fitted to various expressions. At least four expressions, including exponential and logarithmic dependence on time, were found to give comparable fits to the data. It is impossible to clearly distinguish between the fits, consequently it is not possible to deduce operative deformation mechanisms from the fits to the data alone. It is even impossible to tell whether steady-state creep exists, or is reached. Extrapolation of the exponential time dependence to very long times is therefore uncertain, based on the creep data.

It was assumed that steady-state creep exists, and values of steady-state creep rates were deduced from fits to the exponential creep expression. Steady-state creep rates for various tests were then fitted to a thermally activated power law in stress. The power law is motivated by evidence of dislocation climb in NaCl at high temperatures. However, the activation energy obtained by the present fit is smaller than the activation energies for various types of self-diffusion. The scatter in the data is such that other expressions could easily be made to fit the data. This makes it impossible to ascribe any physical significance to the fit. Other creep data suggest that the activation energy is larger above 200°C, and that the stress exponent increases at higher stresses. The empirical fit therefore cannot be extrapolated to higher temperatures and stresses.

It is possible that the low temperature creep tests did not approach steady state, and that the steady state creep rates deduced from the exponential time expression were overestimated, if indeed steady state creep exists. If so, the activation energy we have obtained is too low. More recent tests, in which the temperature is changed stepwise during creep testing on the same specimen, consistently give higher values of the activation energy. A careful activation analysis of such step tests is an important means of resolving this question. Analysis of creep tests, in which the stress is increased stepwise during creep testing on the same specimen, yield values of the stress exponent in agreement with that obtained by the statistical analysis.

Micrographic observations of dislocation structures are being made in an effort to determine the operative dislocation mechanisms. The as-received salt shows large well-developed polygonal structures, suggesting that the salt has been subjected to creep by dislocation climb during its *in situ* geological history. Preliminary observations of laboratory creep specimens suggest that these structures are merely refined during creep at 200 °C. At lower temperatures, slip bands are common, suggesting that dislocation glide plays a role. It appears that the lower temperature data can be made to fit an exponential stress dependence, appropriate to dislocation glide, about as well as the power law dependence used above. Extrapolations of the power law and the exponential law to lower stresses may not agree, and this may affect predictions of creep closure of mine openings. Certainly, the question cannot be resolved on a basis of the present creep data.

A careful and systematic micrographical study is needed to assist in mapping stress–temperature regions where different creep mechanisms dominate, and to suggest appropriate creep expressions in each region. This would lend confidence to the fits, and help establish their extents of validity. Some preliminary micrographic observations have been made in salt taken from the walls of actively closing mines. These suggest features very similar to those observed in salt from the creep tests. Systematic observations of this type would lend some confidence to the use of laboratory creep data for mine closure analysis, and perhaps reveal whether the same mechanisms occur at the typically lower stresses in the salt far removed from the mine walls, or whether still other mechanisms may have to be considered.

The empirical description of transient creep, which we have found, appears to capture the shape of transients in virgin specimens, or when stress or temperature are increased stepwise, about as well as the data allow. However, the very early creep behaviour at low temperatures is not fitted well, the initial creep rate, in particular, being grossly underestimated. The creep for the first few hours is relatively unimportant in mining. In fact, the time required by the mining operation itself masks this behaviour. The suggestion that transient creep may be unimportant for long time mine closure may relegate the modelling of transients to early time mine stability calculations, once suitable failure or creep rupture criteria are developed. However, much more work is required to delineate when transient creep can be omitted from calculations.

The disagreement between step tests, where the stress or temperature are decreased, and the empirical creep law suggests that the salt exhibits a more complex history effect than is allowed in the model. On the other hand, preliminary observations that recovery may keep up with slowly decreasing stress or temperature suggests that the empirical model may still be useful, but more work is required in this area as well.

Incorporation of the triaxial creep law into constitutive equations for arbitrary stress states involves sweeping assumptions which have not been checked in detail. In fact, the two constitutive theories we have used give qualitatively different solutions for mine convergence. The viscous fluid model provides a diffusion problem, while the elastic-viscous model begins with an elastic stress state which relaxes as viscous flow proceeds. Other more complex constitutive theories may also be constructed, such as unified creep-plasticity [23]. Investigation of the behaviour of these and other models is needed to delineate their behaviour.

It is possible that the restricted information obtainable from triaxial creep tests is not only insufficient, but may not dominate behaviour involved in mine closure.

Only comparison of closure predictions with actual mine closure data can illuminate the applicability of the model. The preliminary comparison reported here provides some indication that the model is applicable. However, the calculations are highly idealized, ignoring geologic layering and bands of anhydrite and clay seams, and only simple geostatic stresses were included. Other calculations have shown that these factors have an important effect on predicted closure rates. While encouraging, the present agreement may be fortuitous. Many more comparisons are needed. A search for closure data on existing mines, stretching over long periods of time, is needed to provide additional tests for the model. *In situ* experiments, again continued for very long periods, should be planned in connection with any repository construction. Large laboratory scale experiments, such as model pillar tests, or tests of blocks with openings, may also be helpful to establish phenomenology, because of the opportunity for extensive instrumentation and post-mortem examination.

Preliminary indications suggest that the creep law described above, the elastic-viscous constitutive theory, and its integration method, are useful for prediction of closure of mines or caverns used for repositories, and they have been proposed as an interim baseline model for design calculations. Many questions remain. Further work, along lines suggested above will undoubtedly lead to refinement of the model, and greater confidence in its use.

ACKNOWLEDGEMENT

This work performed by Sandia National Laboratories is supported by the US Department of Energy under contract DE-AC04-76-DP00789.

REFERENCES

1. P. M. Burke, Stanford University Ph.D. thesis, 1968.
2. H. C. Heard, in *Flow and Fracture of Rocks* (eds. H. C. Heard *et al.*), Am. Geophys. Union, 1972.
3. M. Guillope and J. P. Poirier, *J. Geophys. Res*, **84**, 5557 (1979).
4. W. Blum, *Phil. Mag.*, **28**, 245 (1973).
5. W. R. Wawersik, *Rock Mech.*, **7**, 231 (1975).
6. W. R. Wawersik, *Sandia National Labs.*, SAND 79-0114, 1979.
7. W. R. Wawersik and D. W. Hannum, *J. Geophys. Res.*, **85**, 891 (1980).
8. W. R. Wawersik and D. W. Hannum, *Sandia National Labs.*, SAND 79-0115, 1979.
9. H. Albrecht and U. Hunsche, *Fortschr. Miner.*, **58**, 212 (1980).
10. W. Herrmann, W. R. Wawersik, and H. S. Lauson, *Sandia National Labs.*, SAND 80-0087, 1980.
11. W. Herrmann and H. S. Lauson, *Sandia National Labs.*, SAND 81-0738, 1981.
12. D. E. Munson and P. R. Dawson, *Sandia National Labs.*, SAND 79-1853, 1979.
13. W. Herrmann, W. R. Wawersik, and H. S. Lauson, *Sandia National Labs.*, SAND 80-0558 1980.
14. G. A. Webster, A. P. D. Cox, and J. E. Dorn, *Metal Sci. J.*, **3**, 221 (1969).

15. W. Herrmann, W. R. Wawersik, and H. S. Lauson, *Sandia National Labs.*, SAND 80-2172, 1980.
16. W. Herrmann and H. S. Lauson, *Sandia National Labs.*, SAND 81-1612, 1981.
17. W. Herrmann and H. S. Lauson, *Sandia National Labs.*, SAND 81-2567, 1981.
18. R. A. Verral, R. J. Fields, and M. F. Ashby, *J. Am. Ceramics Soc.*, **60**, 211 (1977).
19. D. E. Munson and P. R. Dawson, *Sandia National Labs.*, SAND 80-0467, 1980.
20. S. T. Montgomery, *Sandia National Labs.*, SAND 81-1163, 1981.
21. D. Mraz, *Proc. 19th US Rock Mechanics Symposium*, U. Nevada, 1978.
22. S. T. Montgomery, *Sandia National Labs.*, SAND 82-2084, 1982.
23. R. D. Krieg, *Sandia National Labs.*, SAND 80-1195, 1980.

Mechanics of Engineering Materials
Edited by C. S. Desai and R. H. Gallagher

Chapter 16

Effects of Rotation of Principal Stress Directions on Cyclic Response of Sand

K. Ishihara and I. Towhata

16.1 INTRODUCTION

The deformation characteristics of sand under cyclic loading conditions have been a subject of concern for clarification of the mechanism of liquefaction and settlements of sandy grounds during earthquakes. However, most prior investigations have been focused on cyclic behaviour which is tested in the triaxial shear, simple shear, and torsional shear modes of deformation. In these modes of cyclic deformation, the axes of the principal stresses are suddenly changed by 90° when the direction of the shear application is reversed. In many of the modes of stress application such as those encountered in the seismic loading, the change in the principal stress directions as above is considered to represent the manner of actual stress changes occurring in soil deposits in the field. However, there are several other examples of cyclic loading environments where the mode of load application is associated with the continuous rotation of the principal stress direction. The change in shear stress induced in a sea-bed deposit by waves passing overhead is a typical example of such stress application. As will be proved theoretically in the following pages, the nature of cyclic stress changes occurring in the sea-bed deposit due to wave loading involves a continuous rotation of principal stress directions under the condition of constant deviator stress (difference between the major and minor principal stresses). Also, the shear stresses induced by traffic loads in the subgrade of road pavements or railways are associated with the rotation of the principal stress directions.

The effects of the principal stress axis rotation on the strength of clays under monotonic loading conditions were investigated by Broms and Casbarian [3] using a triaxial torsion shear apparatus. More comprehensive studies on this subject have been done by Arthur *et al.* [1,2] on clean sand by means of a specially designed flexible boundary shear test apparatus. Some of their test results showed that the deformation and strength characteristics of the sand under monotonic loading conditions are influenced significantly by the rotation of the principal stress directions.

In view of the importance of this aspect of the problem, an attempt has been made in this investigation to investigate the effects of the continuous rotation of the principal stress axes on the cyclic behaviour of saturated loose sand using a triaxial torsion shear test apparatus. The tests have been conducted by controlling the torsional shear stress and the stress difference between the axial and horizontal stresses in such a way that the deviator stress was kept constant while rotating the principal stress axes. The schemes, performances, and results of the tests will be described in detail in the following pages of this chapter.

16.2 LOADING INVOLVING ROTATION OF PRINCIPAL STRESS DIRECTIONS

Consider water waves propagating over a sea-bed. At the instant when the crest of a wave is positioned directly above the soil element being considered in the sea bottom, a positive vertical pressure will be exerted, but when the trough is located above the soil element, the resulting stress will be a negative vertical pressure. Consequently, there occurs a cyclic excursion of the vertical stress during wave propagation through a distance of one wavelength as seen in Figure 16.1(a). In the intermediate instant when the point of zero wave height comes right above the soil element, it will be subjected to a horizontal shear stress as illustrated in Figure 16.1(b). This horizontal shear stress also changes its direction back and forth in the course of the wave propagation, inducing another cyclic alteration of shear stress in the soil deposit. It should be noted here that the cyclic variation of the vertical stress is 90° out of phase with the cyclic change in the horizontal shear stress. Therefore, during the cyclic alteration of these two components of

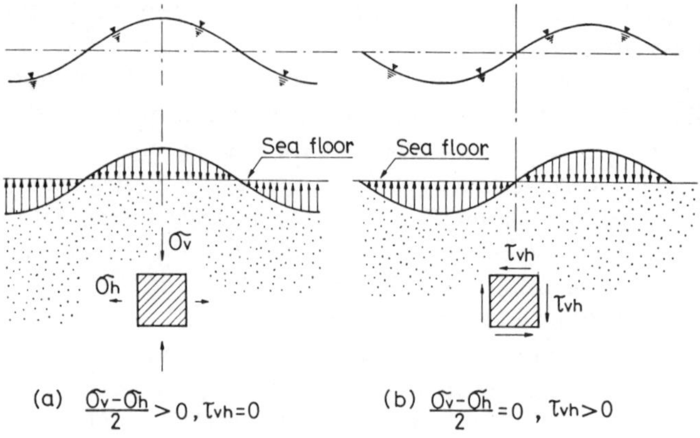

Figure 16.1 Changes in state of stress due to propagation of water waves over the seafloor

shear stress, the soil element is subjected to a continuous rotation of the principal stress directions. This aspect of stress change can be more clearly understood if a stress analysis is made for an elastic half-space subjected to a load sinusoidally distributed on the surface from minus to plus infinity. Consider a semi-infinite space with the coordinate axes, x and z, taken horizontally and vertically, as indicated in Figure 16.2. Assume that a harmonic load with a wavelength, L, and an amplitude, p_0, is distributed on the surface. The exact solution for this problem can be obtained by using the same analytical procedure as adopted by Madsen [4] and Yamamoto [7] as follows.

$$\sigma_v = p_0\left(1+\frac{2\pi z}{L}\right)e^{-(2\pi z/L)}\cos\left(\frac{2\pi x}{L}\right)$$

$$\sigma_h = p_0\left(1-\frac{2\pi z}{L}\right)e^{-(2\pi z/L)}\cos\left(\frac{2\pi x}{L}\right) \tag{16.1}$$

$$\tau_{vh} = p_0\frac{2\pi z}{L}e^{-(2\pi z/L)}\sin\left(\frac{2\pi x}{L}\right)$$

Hence, the stress difference is obtained as,

$$\frac{\sigma_v-\sigma_h}{2} = p_0\frac{2\pi z}{L}e^{-(2\pi z/L)}\cos\left(\frac{2\pi x}{L}\right) \tag{16.2}$$

Eliminating the variable, z, between the shear stress, τ_{vh}, and the stress difference, $(\sigma_v-\sigma_h)/2$, one obtains,

$$\tan\frac{2\pi x}{L} = \frac{2\tau_{vh}}{\sigma_v-\sigma_h} \tag{16.3}$$

The elementary stress analysis shows that the right-hand side of equation (16.3) is

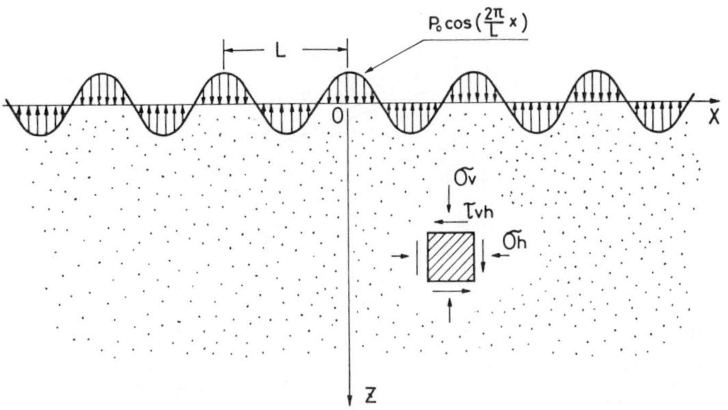

Figure 16.2 Semi-infinite elastic medium subjected to harmonic loading on the surface

equal to the tangent of twice the angle of the major principal stress direction to the vertical, β, as illustrated in Figure 16.3. Therefore, we have,

$$\beta = \frac{\pi}{L} x \tag{16.4}$$

This equation implies that when attention is drawn to the stress change along a horizontal line at a certain depth, the direction of the principal stress continuously rotates through 180° with an interval of distance, L. If the variable, x, is eliminated between the shear stress, τ_{vh}, and the stress difference, $(\sigma_v - \sigma_h)/2$, one obtains,

$$\left(\frac{\sigma_v - \sigma_h}{2}\right)^2 + \tau_{vh}^2 = p_0^2 \left(\frac{2\pi}{L}\right)^2 z^2 e^{-(4\pi z/L)} \tag{16.5}$$

The elementary stress analysis also shows that the left-hand side of equation (16.5) is equal to the square of the deviator stress defined as the difference between the major and minor principal stresses, σ_1 and σ_3, respectively, divided by two. Hence, one obtains,

$$\frac{\sigma_1 - \sigma_3}{2} = p_0 \frac{2\pi}{L} z e^{-(2\pi z/L)} \tag{16.6}$$

Equation (16.6) indicates that when attention is drawn to the stress change at a fixed depth, z, the deviator stress induced on the horizontal plane remains unchanged irrespective of the variable x.

In the analysis above, the time change in the stress system is not explicitly considered. However, if the point of question is viewed moving in the half-space continuously in a horizontal direction at a fixed depth, then the alteration in the

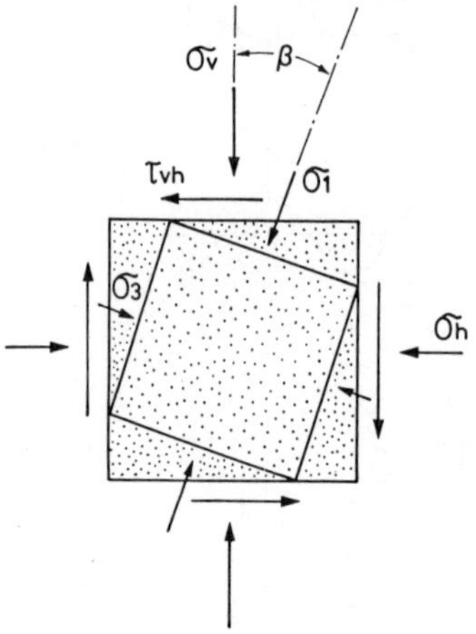

Figure 16.3 Definition of the angle of the major principal stress direction

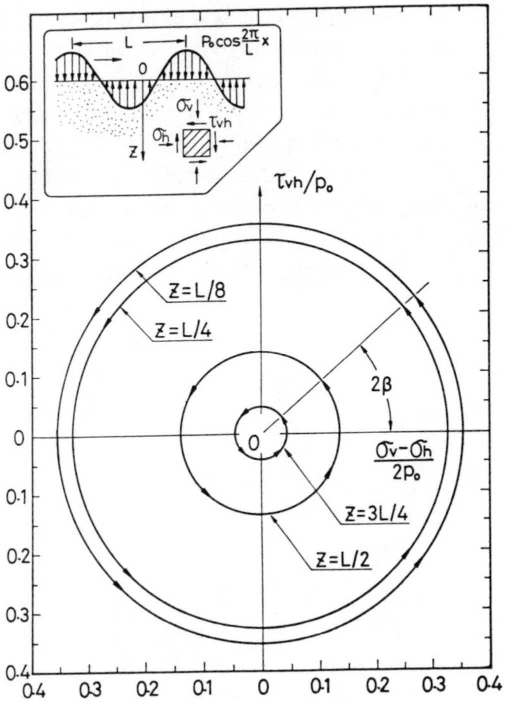

Figure 16.4 Rotation of the principal stress axes in the stress system in a semi-infinite space subjected to a moving harmonic load

state of stress as the coordinate variable, x, changes will represent the same pattern of stress changes that are induced at a fixed point in the half-space by the harmonic load moving on the surface. Viewed in this manner, the results of the analysis as expressed by equations (16.4) and (16.6) imply that the cyclic change of shear stress induced by the harmonic load moving on the surface of an elastic half-space involves a continuous rotation of the principal stress direction with the deviator stress being always maintained constant throughout the duration of loading. This characteristic feature of stress changes can be more clearly visualized when equation (16.5) is numerically presented as shown in Figure 16.4 where the shear stress is plotted versus the stress difference. It can be seen that the stress path by the harmonic loading runs around a circle in such a plot with the radius equal to the deviator stress.

16.3 TRIAXIAL TORSION SHEAR APPARATUS

The test apparatus used was a triaxial torsion shear apparatus which permits torsional shear stress to be applied to a specimen independent of vertical and horizontal stresses. A cross-section of the test device is shown in Figure 16.5. A hollow cylindrical test specimen measuring 10 cm in outer diameter, 6 cm in inner diameter, and 10.4 cm in length and encased in rubber membrane is placed in the triaxial chamber as shown in Figure 16.5. The test apparatus was designed so that the pressure inside the hollow cylindrical specimen could be varied, if necessary, independently of the pressure outside the specimen. In the present study, however, the chamber pressures both inside and outside were held at identical levels. Therefore, the three components of stress, i.e. vertical stress, lateral stress, and torsional stress were varied to produce cyclic excursion of shear stresses in the test specimen.

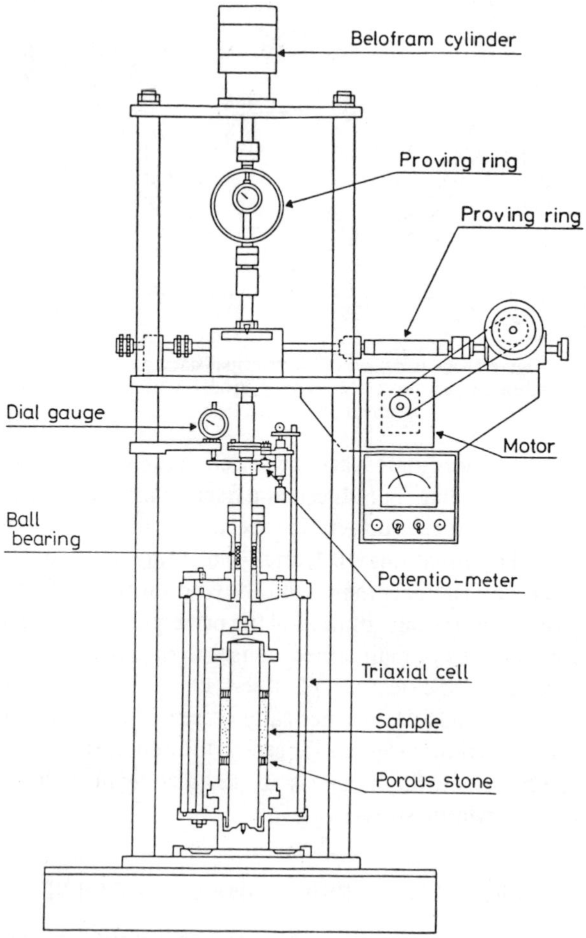

Figure 16.5 Torsional shear test apparatus

The vertical load was applied through a vertical rod connected to a Belofram cylinder which was operated by air pressure. A proving ring was used to monitor the vertical load and a dial gauge to measure the vertical displacement of the test specimen. The lateral stress was applied by a change in the chamber pressure. The torsional shear stress was applied by a rotational force transmitted to the vertical rod above the triaxial cell. The rotational force produced by a motor was monitored by a proving ring and the rotational angle measured by means of a potentiometer. Pore water pressures were monitored by an electric transducer installed at the base of the triaxial cell.

16.4 TEST MATERIAL AND PROCEDURES

The sand used in the present study was a type of Japanese standard sand called Toyoura sand. The grain size distribution curve of this sand is shown in Figure 16.6. The mean particle size, D_{50}, and uniformity coefficient, U_c, are $D_{50} = 0.17$ mm and $U_c = 2.0$, respectively. Subrounded to subangular in grain shape, the sand has maximum and minimum void ratios of $e_{max} = 0.98$ and $e_{min} = 0.60$, respectively. The specific gravity, G_s, of this sand equals 2.65.

Air-dried sand mixed with carbon dioxide gas, CO_2, was poured into the annular space between the outer and inner moulds by means of a vinyl pipe. A 10 to 13 cm height of fall was employed to spread the sand to a relative density of $D_r = 40 \sim 50$ per cent. To obtain a denser specimen with a relative density of $D_r = 80$ per cent, the sand was spread from a height of approximately 30 cm. After forming the specimen, it was saturated under a back pressure of $29.4\,kN/m^2$. With the aid of the CO_2 gas, full saturation was achieved with a b-value exceeding 0.99. The specimen was consolidated under a confining stress of $294\,kN/m^2$. The cyclic shear stress application was made statically with a shear strain rate of 0.14 per cent per minute. All the tests reported in this

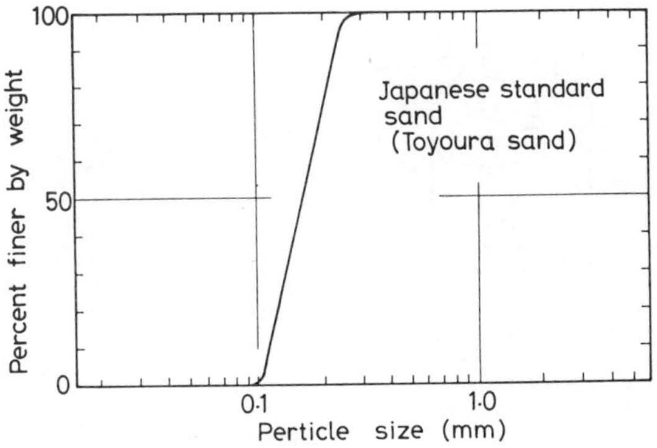

Figure 16.6 Grain size distribution curve of the sand used in the test

chapter were conducted under undrained conditions with pore water pressure measurements.

16.5 STATES OF STRESS AND LOADING SCHEME

The cyclic loading scheme involving changes in both the torsional shear stress and the axial stress difference was programmed so that the direction of the principal stress could be continuously rotated. The conventional types of cyclic torsion shear and cyclic triaxial shear tests without continuous rotation of the principal stress were also performed to provide a basis of comparison with the tests involving the rotation of the principal stress.

Consider states of stress induced on an element in the cylindrical wall as illustrated in Figure 16.7. If stresses are assumed to be uniformly distributed in the radial direction across the wall of the sample, it can be proved that the stress in the circumferential direction is equal to the cell pressure when equal pressures are applied both in the inner and outer cells of the hollow cylindrical sample. Therefore, although the horizontal stress, σ_h, can not be measured directly, its value may be taken to be equal to the applied cell pressure. If the shear stress, τ_{vh}, is applied to the sample along with other components of stress, the principal stresses in the plane of the cylindrical wall are given by

$$\begin{aligned} \sigma_1 &= \frac{\sigma_v + \sigma_h}{2} + \sqrt{\left[\left(\frac{\sigma_v - \sigma_h}{2}\right)^2 + \tau^2_{vh}\right]} \\ \sigma_3 &= \frac{\sigma_v + \sigma_h}{2} - \sqrt{\left[\left(\frac{\sigma_v - \sigma_h}{2}\right)^2 + \tau^2_{vh}\right]} \end{aligned} \tag{16.7}$$

where σ_1 and σ_3 are major and minor principal stresses. The angle between the

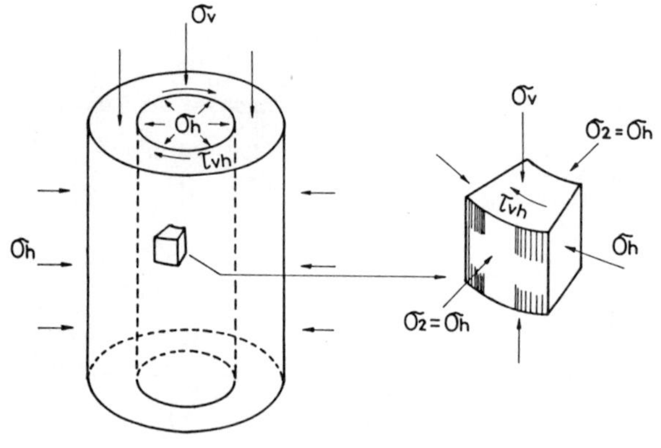

Figure 16.7 State of stress in the wall of a hollow cylindrical sample

major principal stress direction and the vertical, β, is given by

$$\tan 2\beta = \frac{2\tau_{vh}}{\sigma_v - \sigma_h} \tag{16.8}$$

Therefore, the deviator stress, defined as the difference between the major and minor principal stresses, is given by,

$$\frac{\sigma_1 - \sigma_3}{2} = \sqrt{\left[\left(\frac{\sigma_v - \sigma_h}{2}\right)^2 + \tau_{vh}^2\right]} \tag{16.9}$$

Corresponding to the stress components, it is possible to define the principal strain components as follows:

$$\left.\begin{aligned} \varepsilon_1 &= \frac{\varepsilon_v + \varepsilon_h}{2} + \tfrac{1}{2}\sqrt{[(\varepsilon_v - \varepsilon_h)^2 + \gamma_{vh}^2]} \\ \varepsilon_3 &= \frac{\varepsilon_v + \varepsilon_h}{2} - \tfrac{1}{2}\sqrt{[(\varepsilon_v - \varepsilon_h)^2 + \gamma_{vh}^2]} \end{aligned}\right\} \tag{16.10}$$

where ε_v and ε_h denote vertical and horizontal strains, respectively and γ_{vh} is the shear strain in the torsional mode of deformation. ε_1 and ε_3 are, respectively, the major and minor principal strains.

Three types of cyclic loading tests were conducted by changing the two components of shear stress either singly or in combination. The loading schemes of these tests are illustrated in the stress space shown in Figure 16.8 in which the shear stress and the stress difference are represented in a rectangular coordinate system. In the following pages, only the test with circular rotation of the principal stress axes will be described. The results of the cyclic triaxial shear test and cyclic torsion shear test will be presented elsewhere [8].

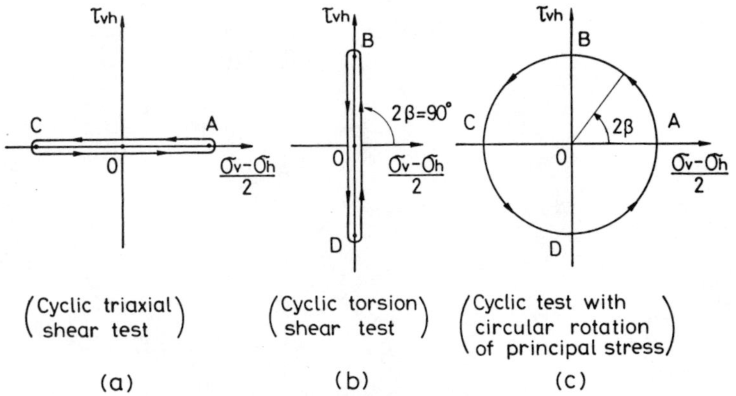

Figure 16.8 Loading schemes adopted in the tests

16.5.1 Cyclic test with circular rotation of the principal stress axes

In this type of test, both the shear stress, τ_{vh}, and the stress difference, $(\sigma_v - \sigma_h)/2$, were changed so that the deviator stress, $(\sigma_1 - \sigma_3)/2$, as defined by equation (16.9) was held constant throughout the cyclic loading test. An isotropically consolidated test specimen was first subjected to a stress difference by increasing the vertical stress, and then the shear stress, τ_{vh}, was gradually increased while reducing the vertical stress so as to maintain the deviator stress unchanged. During this phase of loading, the direction of the principal stress is changed according to equation (16.8). When the shear stress, τ_{vh}, is plotted versus the stress difference, $(\sigma_v - \sigma_h)/2$, in the rectangular coordinate system as shown in Figure 16.8(c), the stress point first moves from zero to point A during the initial application of the stress difference, and then turns counterclockwise to point B where the shear stress was increased to the maximum value with the stress difference vanishing. From then on, the vertical stress, σ_v, was reduced below the existing horizontal stress, σ_h, while decreasing the shear stress component. During this phase of loading, the stress point turns along the circle and eventually reaches point C. Loading from point C to point D involves an increase in the shear stress, τ_{vh}, in the opposite direction and a decrease in the absolute value of the stress difference, as shown in Figure 16.8(c). A reduction in the shear stress with a simultaneous increase in the stress difference constitutes the stress path from point D back to point A. It is to be noted that in the course of a round excursion of the stress change as above along the circular stress path, the direction of the principal stress rotates continuously through an angle of 180° according to equation (16.18) while the deviator stress as defined by equation (16.9) is kept constant throughout the loading process.

It is to be noted, however, that because the intermediate stress, σ_2, remains unchanged, the magnitude of the major and minor principal stresses changes relative to the intermediate principal stress during the cyclic loading. In other

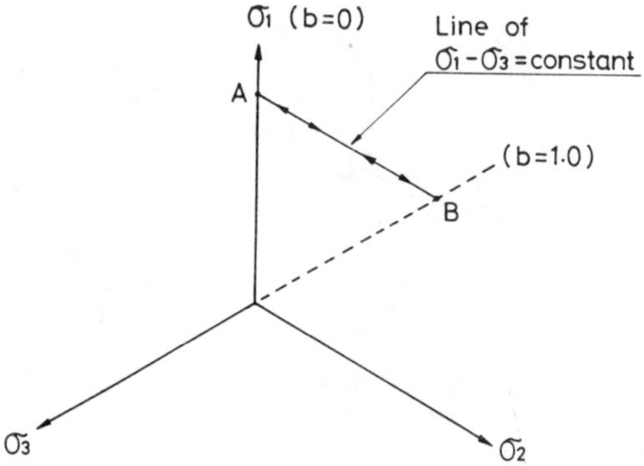

Figure 16.9 Stress path showing the change in b-value during the cyclic loading with $\sigma_1 - \sigma_3 = \text{constant}$

words, the b-value as defined by

$$b = \frac{\sigma_2 - \sigma_3}{\sigma_1 - \sigma_3} = \frac{1}{2}\left(1 - \frac{\sigma_v - \sigma_h}{\sigma_1 - \sigma_3}\right) \tag{16.11}$$

changes between 0 and 1.0 in the course of the cyclic loading. Since the changes in the b-value occurs under the condition of $\sigma_1 - \sigma_3 = \text{constant}$ in the present test scheme, the stress path on the octahedral plane of the σ_1–σ_2–σ_3 stress space consists of a back and forth movement between points A and B in Figure 16.9. Therefore, in the type of cyclic loading adopted in the present study, the cyclic changes in the deviator stress component, $\sigma_1 - \sigma_2$ and $\sigma_2 - \sigma_3$ can inevitably occur in the sample. In order to single out the nature of plastic deformation due solely to the rotation of the principal stress directions, it may be necessary to evaluate the order of magnitude of the plastic deformation caused by the cyclic changes in the other components of deviator stress. Recent studies by Yamada [6] using a true triaxial test apparatus have shown, however, that the plastic deformation during the change in b-value under the condition of approximately constant value of $\sigma_1 - \sigma_3$ is generally small. Therefore, it will be assumed in the present study that effects of the cyclic change in b-value on the shear strain and pore water pressure build-up can be neglected.

A test specimen was consolidated to a void ratio of $e = 0.811$ under the same confining pressure as in the other tests. A vertical stress of $\sigma_v = 113.8\,\text{kN/m}^2$ was first applied to produce a deviator stress of $56.9\,\text{kN/m}^2$ in the test specimen. The vertical stress was then reduced gradually while increasing the shear stress so as to maintain the initially applied deviator stress at a constant level in accordance with equation (16.9). The interchange of the shear stress, τ_{vh}, and the stress difference, $(\sigma_v - \sigma_h)/2$, was continued cyclically, thereby producing the rotation of the principal stress axes along a circular path as illustrated in Figure 16.8(c). The result of such a test is demonstrated in Figure 16.10. The pore water pressure generated during the rotation of the principal stress axes is shown in Figure 16.10(a) in terms of the decrease in the effective confining stress plotted versus the angle of the major principal stress, β, defined by equation (16.8). Figure 16.10(a) shows that, when the stress difference was initially applied by increasing the vertical stress by $\sigma_v = 113.8\,\text{kN/m}^2$, a pore water pressure of $25\,\text{kN/m}^2$ was generated, as is indicated by point 0 to A in Figure 16.10(a). In the course of the subsequent interchange of stresses from a state of triaxial compression to a state of simple shear, a pore water pressure of about $25\,\text{kN/m}^2$ was produced, as indicated by point A to point B in Figure 16.10(a). During the further rotation of the principal stress axes through 45° about $35\,\text{kN/m}^2$ of pore water pressure is shown to have developed as is represented by point B to point C. The pore water pressure response in the latter half of the one cycle of rotation is indicated by point C to D and to A_1 with the starting point C translated downwards in Figure 16.10(a). It can be seen in the figure that during the first cycle of the principal stress rotation, a pore water pressure equal to about 28 percent of the initial confining

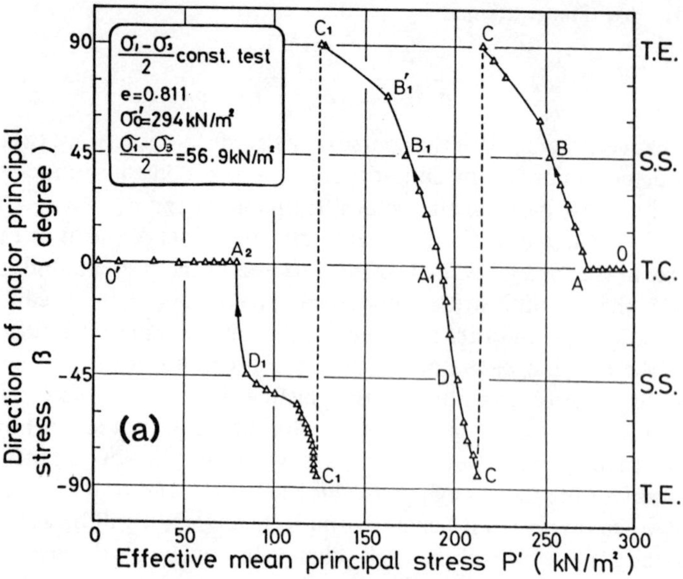

Figure 16.10 (a) Changes of effective mean principal stress due to rotation of the principal stress directions

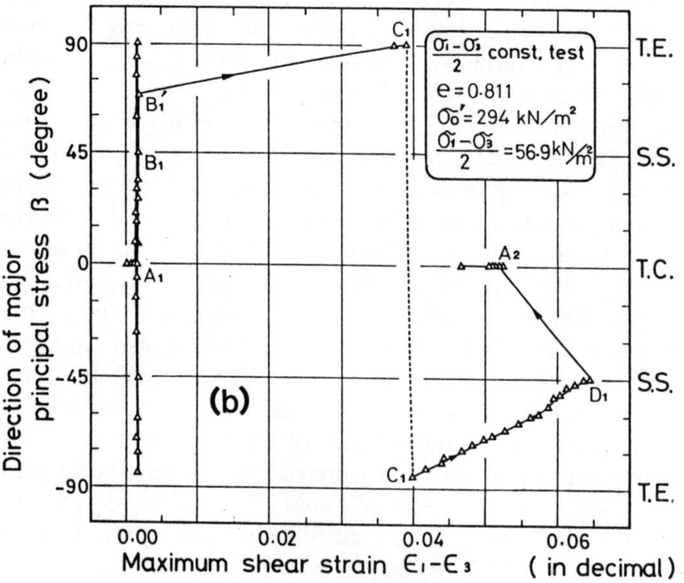

Figure 16.10 (b) Changes of deviator strain due to rotation of the principal stress directions

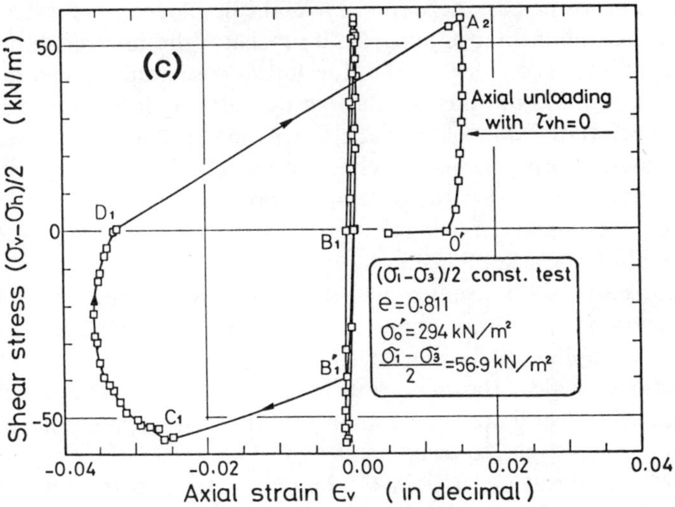

Figure 16.10 (c) Stress–strain relation in the triaxial shear mode

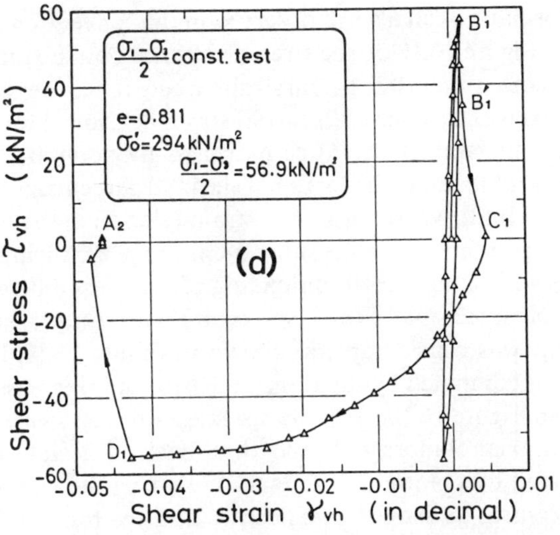

Figure 16.10 (d) Stress–strain relation in the torsional mode

stress had developed. In the second cycle, the developed pore water pressure is seen to have amounted to as much as 40 per cent of the initial confining stress. At the end of the second cycle, the developed pore water pressure amounted altogether to 220 kN/m^2, as is indicated by point A_2 in Figure 16.10(a). Upon coming back to the state of triaxial compression at point A_2, the vertical stress was removed to bring the test specimen to the initial condition of zero deviator stress. It can be seen that the pore water pressure became equal to the initial confining stress upon removing the vertical stress and a state of liquefaction was produced in the test specimen.

On the basis of the strain components measured during the cyclic test, the deviator strain, $\varepsilon_1 - \varepsilon_3$, was computed using equation (16.10). The variation of the deviator strain thus obtained in the above test is shown in Figure 16.10(b) versus the rotational angle of the major principal stress axis. It can be seen in the figure that the deviator strain remained small until the loading proceeded to about one and one-quarter cycles to reach point B_1' in Figure 16.10(b), whereupon a large deviator strain developed to point C_1. This critical state of stress is somewhere on the way towards the triaxial extension from the simple shear state of stress. In other words, the shear stress component, τ_{vh}, was reduced, while the component of the stress difference, $(\sigma_v - \sigma_h)/2$, was being increased in the region of the triaxial extension. It is generally known [5] that loose sand specimens tend to develop a large shear strain at a certain level of shear stress ratio on the side of the triaxial extension which is smaller than the corresponding stress ratio to be imposed on the triaxial compression side to produce the same amount of shear strain. It may be for this reason that the large deformation developed at the state of loading approaching the triaxial extension region. This characteristic behaviour of deformation can also be observed in the conventional types of stress–strain plots. In Figure 16.10(c), the stress difference obtained in the same cyclic loading test is plotted against the axial strain, where it may be seen that at a stage close to the triaxial extension, a large axial strain developed towards the triaxial extension side (minus axial strain) as indicated by point B_1' to C_1. However, during this phase of loading, the torsional shear strain remained small, as can be seen in Figure 16.10(d), where the stress–strain relation in the torsional mode of deformation is presented. In the subsequent stage involving loading of the torsional stress with simultaneous unloading of the stress difference, the amount of axial strain remained small as can be seen in Figure 16.10(c) but the amount of shear strain, γ_{vh}, was considerable as shown in Figure 16.10(d). Therefore, the deviator strain as computed by equation (16.10) tends to increase considerably as demonstrated in Figure 16.10(b). In the last stage, the test specimen was brought to a state of triaxial compression and then back to a state of zero deviator stress as indicated by point D_1 to A_2 and $0'$. During the last phase of loading, pore water pressures developed significantly and the test specimen was put in a state of liquefaction as can be seen in Figure 16.10(a).

16.6 CONCLUSIONS

A cyclic undrained triaxial torsion shear tests was performed in a statical manner on saturated loose specimen of sand using a triaxial torsion shear apparatus. Both the stress difference (difference between the vertical stress and horizontal stress) and the torsional shear stress were varied cyclically so that the axes of the principal stresses could be rotated continuously, while keeping the amplitude of the combined shear stress (deviator stress) constant throughout the cyclic loading test. The results of such tests indicated that the plastic deformation such as the build-up of pore water pressure can occur, even when the deviator stress is maintained unchanged, if the rotation of the principal stress axes is executed on the test specimen. In addition, it was also indicated that large shear deformation can take place under this stress condition after a couple of cycles.

ACKNOWLEDGEMENTS

The manufacture of the triaxial torsional test apparatus used in this study was sponsored by the grant-in-aid of the Ministry of Education of Japanese Government. Messrs Y. Kikuchi and S. Nakazato assisted in conducting the tests in the laboratory. Dr Kenji Mori kindly reviewed the draft with valuable comments. The authors wish to express their gratitude to the organization and individuals as cited above.

REFERENCES

1. J. R. F. Arthur, S. Bekenstein, J. T. Cermaine, and C. C. Ladd, 'Stress path tests with controlled rotation of principal stress directions', *Laboratory Shear Strength of Soil, ASTM*, STP 740, 516–540 (1981).
2. J. R. F. Arthur, K. S. Chua, T. Dunstan, and J. I. Rodoriguez, 'Principal stress rotation: a missing parameter', *Proc. ASCE*, **106**, GT. 4, 419–433 (1980).
3. B. B. Broms and A. O. Casbarian, 'Effects of Rotation of the Principal Stress Axes and of the Intermediate Principal Stress on the Shear Strength', *Proc. 6th International Conference on Soil Mechanics and Foundation Engineering*, Vol. 1, pp. 179–183, Montreal, Canada, 1965.
4. O. S. Madsen, 'Wave-induced pore pressures and effective stresses in a porous bed', *Geotechnique*, **28**, 4, 377–393 (1978).
5. Y. Yamada and K. Ishihara, 'Anisotropic deformation characteristics of sand under three dimensional stress conditions', *Soils and Foundations*, **19**, 2, 79–94 (1979).
6. Y. Yamada, Private communication (1982).
7. T. Yamamoto, 'Sea bed instability from waves', 10*th Annual off-Shore Technology Conference*, Vol. 1, pp. 1819–1842, Houston, Texas, 1978.
8. K. Ishihara and I. Towhata, 'Sand response to cyclic rotation of principal stress directions as induced by wave loads', *Soils and Foundations* Vol. 23, No. 4, pp. 11–26, (1983).

Mechanics of Engineering Materials
Edited by C. S. Desai and R. H. Gallagher

Chapter 17

A Viscoplastic Cap Model for Soils and Rock

M. G. Katona and M. A. Mulert

17.1 INTRODUCTION

17.1.1 Objective

The specific objective is to recast the inviscid cap model for rocks and soils [1 to 3] into a viscoplastic formulation based on Perzyna's elastic/viscoplastic theory [4]. Included in the objective is the development of a numerical solution algorithm at the constitutive level for determining the stress history response from an arbitrary strain loading schedule (six components). Ultimately the goal is to examine the behaviour of the viscoplastic cap model in light of experimental data.

17.1.2 Background

By definition, an inviscid plasticity model, such as the cap model, is independent of real time. Experimental evidence, however, often indicates a significant time-dependent behaviour for many soils and rocks [5 to 8]. None the less, the inviscid cap model has been successfully used in a variety of applications, and within the inviscid limitation, it appears to adequately simulate the elastic–plastic nature of soils and rocks [2, 3]. Accordingly, the cap model is adopted here to represent plastic-like behaviour of the proposed model.

The coupling of plastic behaviour with time dependency comes under the general heading of viscoplasticity. To be sure, a variety of viscoplastic formulations have been suggested in the literature. Perhaps two of the most popular formulations are the endochronic theory pioneered by Valinis [9] and the elastic/viscoplastic theory presented by Perzyna. The genesis of the former may be envisioned as a modification of classical viscoelasticity wherein plastic-like behaviour is introduced by means of an intrinsic time scale related to material deformation. Alternatively, Perzyna's theory is a modification of classical plasticity wherein viscous-like behaviour is introduced by a time-rate flow rule employing a plasticity yield function. Other viscoplastic formulations include developments by Katona [10], Bodner and Parton [11], and Phillips and Wu [12].

Motivations for adopting Perzyna's elastic/viscoplastic theory in this study are; (1) the formulation is well accepted and well used, (2) the incorporation of the inviscid cap model (or any plasticity yield function) is relatively straightforward, (3) the generality of the time-rate flow rule offers the capability of simulating time-dependent material behaviour over a wide range of loading, (4) techniques for parameter identification are feasible, and (5) the formulation is readily adaptable to a numerical algorithm suitable for a finite element procedure.

Although the last motivation is not of theoretical concern, it is certainly of great importance, i.e., it makes little sense to develop a sophisticated constitutive model if it cannot be readily incorporated into a general numerical scheme and solved to a given level of accuracy.

In 1974 Zienkiewicz and Cormeau [13] presented a finite element algorithm for a Perzyna-type viscoplastic model using an explicit forward difference time integration scheme. Computationally, the explicit scheme is simple requiring only an elastic stiffness matrix to be assembled and triangularized at the outset. Non-linearities are accommodated on the right-hand side at the element level with (if desired) a mid-interval integration and/or next-interval equilibrium correction [4]. A major drawback of the explicit method is the concern for numerical stability, limiting the time step size which, at times, compromises efficiency. Cormeau [15] presented time step limits to provide stable solutions for a restricted class of plasticity yield functions without hardening.

Hughes and Taylor [16] showed that the concerns for numerical instability for general elastic/viscoplastic models can be eliminated by using an implicit method based on a one-parameter Crank–Nicolson integration scheme. The computational penalty for the implicit scheme is the requirement to iteratively solve non-linear algebraic equations within each time step. However, various versions of the Newton–Raphson technique offer a variety of alternatives to increase efficiency of the iterative solution procedure [17, 18].

17.1.3 Approach and scope

In pursuing the stated objectives, we begin by reviewing the basic assumptions of Perzyna's elastic/viscoplastic theory followed by a description of the inviscid cap model and its associated plasticity yield functions, defined separately in three regions; cap region, failure region, and tension cut-off region. The resulting viscoplastic cap model is a non-linear set of differential equations in time relating stress and strain vectors. A numerical solution algorithm is established to compute stress responses from a prescribed strain loading history. Here a one-parameter Crank–Nicolson time integration scheme is used providing options for explicit or implicit time integration. For the implicit method, a Newton–Raphson procedure is presented to iteratively solve the equations.

The first example problem is a hypothetical uniaxial strain loading of a sand material whose plasticity parameters are given by Sandler and Rubin [1]. The intent is to examine the effects of the so-called fluidity parameter which controls the viscoplastic strain rate. Also, accuracy versus efficiency comparisons are made for various choices of the time integration parameter.

The last example presents a comparison of the viscoplastic cap model's 'fit' with experimental data for a limestone material in a triaxial stress loading where the axial stress periodically fluctuates. Model fitting procedures and limitations of the model are briefly discussed followed by recommendations for future studies.

17.2 VISCOPLASTIC FORMULATION OF CAP MODEL

In the following we consider a viscoplastic constitutive formulation utilizing standard vector notation for stress and strain inferring six independent components per vector. We begin by briefly reviewing the assumptions of Perzyna's elastic/viscoplastic theory and then specialize the theory to incorporate the inviscid cap model. Concluding this section is a numerical solution algorithm suitable for finite element applications.

17.2.1 Elastic/viscoplastic development

Two basic assumptions employed in the elastic/viscoplastic constitutive development are (1) the strain-rate vector $\dot{\varepsilon}$ is composed of elastic and viscoplastic strains, and (2) the stress-rate vector $\dot{\sigma}$ is related to the elastic strain rate via an elastic (or hypoelastic) constitutive matrix. Formally, these assumptions are written as

$$\dot{\varepsilon} = \dot{\varepsilon}_{\mathrm{e}} + \dot{\varepsilon}_{\mathrm{vp}} \tag{17.1}$$

$$\dot{\sigma} = D\dot{\varepsilon}_{\mathrm{e}} \tag{17.2}$$

where $\dot{\varepsilon}_{\mathrm{e}}$ and $\dot{\varepsilon}_{\mathrm{vp}}$ are the elastic and viscoplastic strain-rate vectors and D is an elastic constitutive matrix. A dot over any vector quantity signifies a simple time derivative.

A third and perhaps more contrived assumption is concerned with defining a viscoplastic flow rule relating $\dot{\varepsilon}_{\mathrm{vp}}$ to current values of stress and history state variables associated with work hardening. Generally this viscoplastic flow rule is expressed as

$$\dot{\varepsilon}_{\mathrm{vp}} = \gamma\phi(f)f' \tag{17.3}$$

Here, γ is a material property called the fluidity parameter (units of inverse time) which establishes the relative rate of viscoplastic straining. The scalar function $\phi(f)$ (dimensionless) is called the viscous flow function and its argument, f, is any valid plasticity yield function. The flow function ϕ, yet to be explicitly defined, monotonically increases with f when $f > 0$ and has the property $\phi(0) = 0$. Thus, $\phi(f)$ dictates the current magnitude of the viscoplastic straining rate. The current direction of $\dot{\varepsilon}_{vp}$ is given by the gradient vector f' which, similar to the associated flow rule of inviscid plasticity, is in the outward normal direction of a 'dynamic' yield surface, $f = \text{constant} > 0$.

Evidently, the choice of the plasticity yield function is of paramount importance since it influences both the magnitude and direction of $\dot{\varepsilon}_{vp}$. Specific forms for f are subsequently presented, for now, we simply denote the functional relations inferred in equation (17.3):

$$\phi(f) = \begin{cases} \phi(f), & f^L 0 \\ 0, & f \leq 0 \end{cases} = \text{viscous flow function} \tag{17.4}$$

$$f = f(\sigma, k) = \text{plasticity yield function} \tag{17.5}$$

$$f' = \frac{\partial f}{\partial \sigma} = \text{gradient (vector)} \tag{17.6}$$

In the above, the arguments of f are the current stress vector σ (usually expressed by stress invariants) and a strain hardening parameter k which in turn is defined by some hardening function of the viscoplastic strain history, i.e., $k = k(\varepsilon_{vp})$. For a given value of k, all states of stress that satisfy $f = 0$ form the current 'static' yield surface in a six-dimensional stress space (i.e., classical plasticity yield surface). Accordingly we deduce from equations (17.3) and (17.4), that the 'static' yield surface forms a boundary between elastic ($f < 0$) and viscoplastic ($f > 0$) domains. When a constant stress state is imposed such that $f > 0$, viscoplastic flow will occur and continue to occur at a constant rate if f is a non-hardening (perfectly-plastic) yield function. If f is a hardening yield function, viscoplastic flow occurs at a decreasing rate because as viscoplastic strain accumulates, $k(\varepsilon_{vp})$ changes in value such that $f(\sigma,k) \to 0$, hence $\dot{\varepsilon}_{vp} \to 0$. In other words, the static yield surface 'moves out' on a real time basis to eventually form a new static yield surface containing the imposed stress state. Once the new static yield surface has stabilized, we say the solution is steady state, $\dot{\varepsilon}_{vp} = 0$, and the resulting strains accumulated during this loading would be identical to the corresponding inviscid plasticity solution.

With the foregoing assumptions and concepts, a general viscoplastic constitutive relation is readily obtained by combining equations (17.1) and (17.2) to get

$$\dot{\sigma} = D(\dot{\varepsilon} - \dot{\varepsilon}_{vp}) \tag{17.7a}$$

Or upon replacing $\dot{\varepsilon}_{vp}$ with equation (17.3), we have

$$\dot{\sigma} = D(\dot{\varepsilon} - \gamma\phi(f)f') \tag{17.7b}$$

Note, although ε_{vp} is inferred in equation (17.7b) (i.e., $f(\sigma,k(\varepsilon_{vp}))$, we have $\varepsilon_{vp} = \varepsilon - D^{-1}\sigma$ so that equation (17.7b) implicitly provides the desired σ, ε relationship in a differential form.

17.2.2 Specialization to cap model

To specialize the foregoing viscoplastic constitutive relationship for a particular class of materials, functional forms for ϕ, f, and k must be defined along with the components of the elasticity matrix D. In this study, suitable forms for geological materials (soil and rock) are desired.

Two popular forms for the viscoplastic flow function are

$$\phi(f) = (f/f_0)^N \tag{17.8a}$$

$$\phi(f) = \exp(f/f_0)^N - 1 \tag{17.8b}$$

where N is an exponent and f_0 is a normalizing constant with the same units as f so that ϕ is dimensionless.

Although more elaborate functional forms for ϕ may be established [2], the forms given by equations (17.8a, b) appear to suffice for many geological materials [5, 19].

Specification of the plasticity yield function f is patterned after the seven parameter cap model (1) wherein J_1, the first stress invariant, and J_2' the second deviator stress invariant, are used to define the current static yield surface, as illustrated in Figure 17.1. Here, the static yield surface is divided into three regions along the J_1 axis; the tension cut-off region ($J_1 \geq T$), the failure surface region ($T > J_1 > L$), and the cap surface region ($J_1 \leq L$).

$$f(\sigma, k) = \begin{cases} f_T(J_1), & J_1 \geq T \quad \text{(TENSION)} \\ f_F(J_1, J_2'), & T > J_1 > L \quad \text{(FAILURE)} \\ f_C(J_1, J_2', k), & J_1 \leq L \quad \text{(CAP)} \end{cases}$$

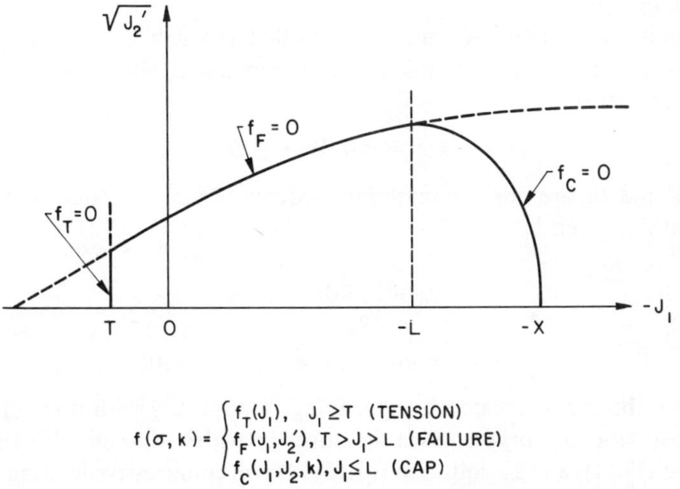

Figure 17.1 Illustration of static yield surfaces for viscoplastic cap model

17.2.2.1 *Failure surface*

The failure surface is a non-hardening, modified Drucker–Prager form with a yield function defined by

$$f_F\,(J_1, J_2') = \sqrt{J_2'} - (A - C\exp(BJ_1)) \tag{17.9}$$

where A, B, and C are positive material constants ($A \geq C$). This yield function is used to define the viscoplastic flow (equation (17.3)) whenever J_1 is in the range $T > J_1 > L$. The failure surface forms a boundary along which the cap surface can move (harden/soften).

17.2.2.2 *Cap surface*

The cap surface is a hardening surface in the shape of an ellipse quadrant when plotted in $J_1, \sqrt{(J_2')}$ space (Figure 17.1). It is defined in a 'squared' form with the normalizing constant f_0 (stress units) as

$$f_c(J_1, J_2', \bar{\varepsilon}) = (J_2' - ((X - L)^2 - (J_1 - L)^2)/R^2)/f_0 \tag{17.10}$$

The cap hardening parameters L and X are positions on the J_1 axis which locates the current cap surface ($f_c = 0$) in $J_1, \sqrt{(J_2')}$ space. The coordinate point $(L, \sqrt{(J_2^*)})$ is where the cap surface intersects the failure surface. L and X are related by the material constant R which defines the ratio of the principal ellipse radii forming the cap surface, i.e.

$$L - X = R\sqrt{(J_2^*)} \tag{17.11}$$

where $\sqrt{(J_2^*)} = A - C\exp(BL)$ is the vertical ellipse radius, and L–X is the horizontal ellipse radius.

Hardening of the cap surface is controlled through a hardening function $X = X(\bar{\varepsilon})$ where $\bar{\varepsilon}$ is an accumulation of compressive, volumetric, viscoplastic strain increments:

$$X(\bar{\varepsilon}) = \ln(\bar{\varepsilon}/W + 1)/D_0 \tag{17.12}$$

where W and D_0 are positive material hardening constants, and the hardening argument $\bar{\varepsilon}$ is given by

$$\dot{\bar{\varepsilon}} = \bar{\varepsilon}_0 + \int_0^t \bar{\varepsilon}\,\mathrm{d}t \tag{17.13}$$

$$\dot{\bar{\varepsilon}} = \min(\dot{\varepsilon}_{vp11} + \varepsilon_{vp22} + \varepsilon_{vp33}, 0) \tag{17.14}$$

Initially the cap is located by specifying X, typically with a small negative value near the J_1 origin. This in turn provides the initial values of L (equation (17.11)) and $\bar{\varepsilon}_0$ (equation (17.12)). Upon compressive loading such that $J_1 < L$ and $f_c > 0$, viscoplastic flow occurs which negatively increases $\bar{\varepsilon}$, X, and L; thereby expanding the cap as it moves in the $-J_1$ direction.

Additional hardening rules for soils have been improvised which permits the cap to retract when loading on the failure surface in order to limit excessive dilatancy. This retraction is more akin to kinematic hardening rather than strain softening [1,20].

17.2.2.3 *Tension cut-off*

The tension cut-off criterion employed in the inviscid cap model is triggered when $J_1 > T$. Here T is a material constant representing the threshold of volumetric tension stress at which abrupt stress releases occur due to tension damage. Specifically, whenever a stress state is encountered such that $J_1 \geq T$, it is assumed that all deviatoric stresses instantaneously vanish, and the volumetric stress in excess of T also vanishes. Thus, the final inviscid stress state is $\sigma_{11} = \sigma_{22} = \sigma_{33} = T/3$, all other $\sigma_{ij} = 0$.

Putting the tension cut-off criterion into a viscoplastic form infers the stresses are released at a rate controlled by a fluidity parameter γ, rather than an instantaneous release. Accordingly, it is reasonable to specify γ in the tension region (say γ_T) at a higher value than the value of γ in the failure/cap regions. Moreover, since the tension cut-off criterion treats volumetric and deviatoric stress releases independently, the viscoplastic strain rate must be independently defined in terms of volumetric and deviatoric strain-rate components.

The above assumptions are stated succinctly in the following expressions. The static yield surface for tension cut-off, $f_T = 0$, is defined by the function

$$f_T(J_1) = J_1 - T \tag{17.15}$$

where $f_T > 0$, the viscoplastic strain rate is given by

$$\dot{\varepsilon}_{\mathrm{vp}} = \gamma_T \phi(f_T) f'_T + \gamma_G \phi(f_G) f'_G \tag{17.16}$$

where $f_G = \sqrt{(J'_2)}$, $f'_G = \partial f_G / \partial \sigma$, and $f'_T = \partial f_T / \partial \sigma$.

Equation (17.16) is analogous to equation (17.3) except the first right-hand side term contains the components of the volumetric viscoplastic strain rates and the second right-hand side term contains the deviatoric viscoplastic strain rates. The two tension fluidity parameters, γ_T and γ_G, permit independent control of the volumetric and deviatoric stress release rates, respectively.

If the material is suddently strained producing an instantaneous elastic stress state such that $f_T(J_1) > 0$ then viscoplastic straining occurs (equation (17.16)), continually releasing stresses. Finally when $\dot{\varepsilon}_{\mathrm{vp}} = 0$, we have $f_T = f_G = 0$, or $J_1 = T$, and $J_2 = 0$, thereby satisfying the tension cut-off criterion in steady state.

To summarize, equation (17.7a) is the governing constitutive relationship in which $\dot{\varepsilon}_{\mathrm{vp}}$ is defined by alternative forms of the flow rule dependent on the current value J_1 (i.e., failure surface, cap surface, or tension cut-off). This is illustrated in the subsequent numerical algorithm.

17.2.3 Numerical algorithm

Using a step-by-step time integration scheme together with a Newton–Raphson iteration procedure, a numerical solution algorithm for the viscoplastic cap model is developed at the constitutive level. Here it is presumed that the strain history is specified and the objective is to determine the corresponding stress history. The extension of this algorithm to a global finite element procedure is fairly straightforward [20].

Beginning with equation (17.7a), we integrate over one time step, Δt, from time t_n to t_{n+1} to get the incremental constitutive relationship

$$\Delta\sigma = D(\Delta\varepsilon - \Delta\varepsilon_{\mathrm{vp}}) \tag{17.17}$$

where $\Delta\sigma = \sigma^{n+1} - \sigma^n$ and $\sigma^n = \sigma(t_n)$, similarly for $\Delta\varepsilon$ and $\Delta\varepsilon_{\mathrm{vp}}$.

We approximate $\Delta\varepsilon_{\mathrm{vp}}$ by a one-parameter Crank–Nicolson time integration scheme as

$$\Delta\varepsilon_{\mathrm{vp}} = \Delta t((1-\theta)\dot{\varepsilon}^n_{\mathrm{vp}} + \theta\dot{\varepsilon}^{n+1}_{\mathrm{vp}}) \tag{17.18}$$

where θ is the adjustable integration parameter in the range $0 \le \theta \le 1$. Choosing $\theta = 0$ implies the integration scheme is explicit (simple forward difference) so that $\Delta\varepsilon_{\mathrm{vp}}$ is determined directly from the known value of $\dot{\varepsilon}^n_{\mathrm{vp}}$ at the beginning of the time step. As a consequence, Δt must be restricted in size to avoid numerical instability [14, 15]. Alternatively choosing $\theta > 0$, the scheme is implicit since $\Delta\varepsilon_{\mathrm{vp}}$ is related to the unknown value $\dot{\varepsilon}^{n+1}_{\mathrm{vp}}$ at the end of the time step, requiring an iterative solution procedure within the time step. For $\theta \ge 0.5$, the implicit scheme is unconditionally stable [16], so that the choice of Δt is governed by accuracy, not stability.

Returning to equation (17.17) with $\Delta\varepsilon_{\mathrm{vp}}$ replaced by equation (17.18) and using $\Delta\sigma = \sigma^{n+1} - \sigma^n$, we rearrange equation (17.17) to get the unknown quantities on the left as

$$D^{-1}\sigma^{n+1} + \Delta t\theta\dot{\varepsilon}^{n+1}_{\mathrm{vp}} = \Delta\varepsilon - \Delta t(1-\theta)\dot{\varepsilon}^n_{\mathrm{vp}} + D^{-1}\sigma^n \tag{17.19a}$$

Or, more compactly, in a symbolic functional notation:

$$P(\sigma^{n+1}, \dot{\varepsilon}^{n+1}_{\mathrm{vp}}) = q^n \tag{17.19b}$$

where q^n are known quantities on the right side of equation (17.19a) which remain constant during the time step.

For $\theta > 0$, equation (17.19a) (or its equivalent (17.19b)) forms a coupled set of six non-linear algebraic equations for the components of σ^{n+1} with the understanding that $\dot{\varepsilon}^{n+1}_{\mathrm{vp}}$ is to be replaced by the appropriate flow rule and its associated yield function (depending on which region of the cap model is currently being activated).

To solve the above, a Newton–Raphson procedure is used by expanding the vector function P in a limited Taylor series about a stress state σ^i which is some estimate of σ^{n+1}, and $d\sigma^i$ is a first order correction to the estimate, i.e.,

$\sigma^{n+1} \simeq \sigma^i + \mathrm{d}\sigma^i$. Thus, the correction $\mathrm{d}\sigma^i$ is determined from the linear set of equations

$$P'\mathrm{d}\sigma^i = q^n - P^i$$

where $P' = \partial P^i/\partial\sigma$ is the Jacobian matrix for iteration i.

Listed below is an algorithm for the iterative procedure where it is assumed all quantities are known at time t_n as well as the applied strain at time t_{n+1}. The objective is to determine the remaining quantities at time t_{n+1}. For the first iteration ($i = 1$), P^i, P', and σ^i retain the values determined at time t_n. Thus, we proceed as follows:

(1) Compute: $\mathrm{d}\sigma^i = (P')^{-1}(q^n - P^i)$

(2) Update:

$$\sigma^{i+1} = \sigma^i + \mathrm{d}\sigma^i$$

$$\varepsilon_{\mathrm{vp}}^{i+1} = \varepsilon^{n+1} - D^{-1}\sigma^{i+1}$$

$$\left.\begin{aligned} f^{i+1} &= f(\sigma^{i+1}, \varepsilon_{\mathrm{vp}}^{i+1}) \\ \dot{\varepsilon}_{\mathrm{vp}}^{i+1} &= \gamma\phi(f^{i+1})f'^{i+1} \end{aligned}\right\} \text{dependent on } J_1$$

$$P^{i+1} = D^{-1}\sigma^{i+1} + \Delta t\theta\dot{\varepsilon}_{\mathrm{vp}}^{i+1}$$

$$P' = D^{-1} + \Delta t\theta(\partial\dot{\varepsilon}_{\mathrm{vp}}^{i+1}/\partial\sigma)$$

(3) Return to step 1 with $i = i + 1$, unless convergence is witnessed by $|\mathrm{d}\sigma| <$ tolerance.

(4) Print results and advance to next time step.

The updating procedure for f^{i+1} is dependent on the current value of J_1 which dictates which yield function of the viscoplastic cap model is currently active. Also if $J_1 > T$, $\dot{\varepsilon}_{\mathrm{vp}}^{i+1}$ is updated by equation (17.16), rather than as shown above. A detailed development of computing P' is given in ref. [20].

In passing we note that an algorithm to determine strain responses from a stress history input can be readily developed in a similar manner. However, the above algorithm is ideally suited for a displacement finite element formulation.

17.3 ILLUSTRATIVE EXAMPLES AND DISCUSSION

Two examples are presented. The first is a simulated uniaxial strain test for sand illustrating some behavioural aspects of the viscoplastic model as well as demonstrating the accuracy of the numerical algorithm. The second example presents a model fit comparison with experimental creep test data for limestone in triaxial stress loading.

17.3.1 Sand in uniaxial strain

For this example, the viscoplastic cap model employs the plasticity parameters for McCormick Ranch sand given by Sandler and Rubin [1]. Figure 17.2 defines the

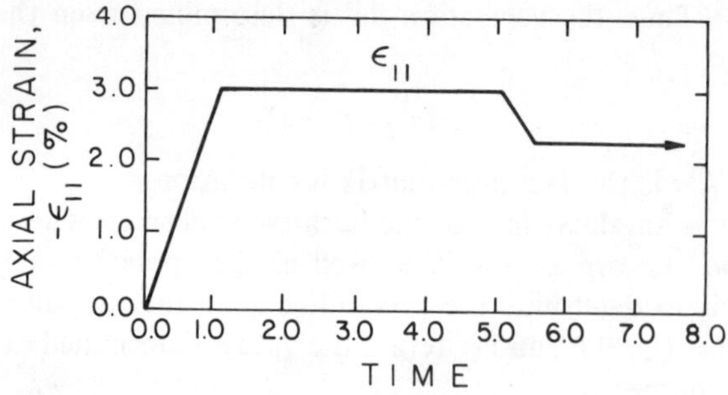

MODEL PARAMETERS

ELASTIC MODULI:
BULK = 66.7 ksi
SHEAR, = 40.0 ksi
FAILURE SURFACE:
A = 0.25 ksi, B = 0.67 ksi^{-1}, C = 0.18 ksi
CAP SURFACE:
R = 2.5, f_0 = 0.25 ksi, X_0 = -0.189 ksi
CAP HARDENING:
W = 0.066, D_0 = 0.67 ksi^{-1}
VISCOUS FLOW FUNCTION:
N = 1, f_0 = 0.25 ksi (Equation 8a)
TENSION CUTOFF NOT USED

Figure 17.2 Uniaxial strain schedule and model parameters for sand

model parameters along with a hypothetical uniaxial strain loading history. Here, ε_{11} increases in compression at a constant rate, held constant, unloaded at a constant rate, and finally held constant so that eventually the stress responses would approach steady state. Three values of the fluidity parameter ($\gamma = 0.001$, 0.01, and 0.1) are examined to illustrate its effect on the stress response, σ_{11}. Note the time units associated with γ need not be explicitly defined since they are the reciprocal of whatever time units are assumed for the strain loading history.

17.3.1.1 *Response behaviour*

Figure 17.3 shows the corresponding σ_{11} stress response for the three magnitudes of γ. For the relatively large magnitude, $\gamma = 0.1$, the response is nearly inviscid throughout the loading history, i.e., viscous straining occurs rapidly so that the response is nearly elastoplastic. Indeed, the steady state values of σ_{11} agree exactly with an inviscid plasticity solution [21], thereby providing a verification of the viscoplastic algorithm. Note if we considered $\gamma > 0.1$, the response would not change significantly, and the small peaks at time = 1.0 and 5.5 would become flattened, mimicking the inviscid solution over the entire range.

As observed in Figure 17.3, the greatest magnitude in the peak stress responses

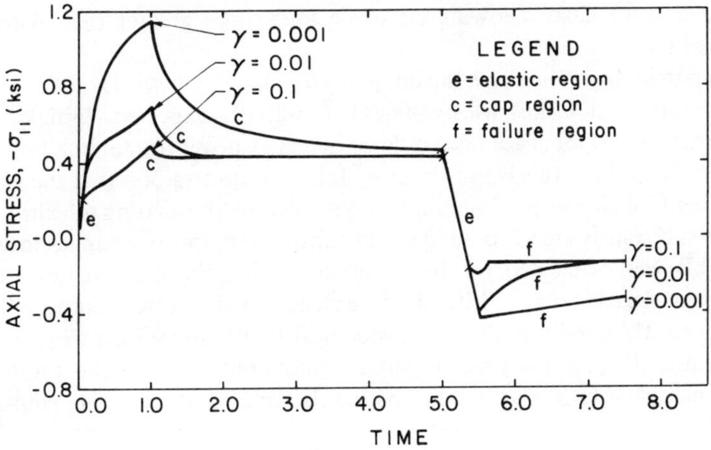

Figure 17.3 Axial stress response for three values of γ

occurs for $\gamma = 0.001$. This, of course, is because lowering γ reduces $\dot{\varepsilon}_{vp}$ implying a greater elastic strain rate and hence greater stress. In the limit as $\gamma \to 0$, the response would become purely elastic.

For reference, Figure 17.3 also indicates which cap model regions, i.e., failure surface, cap surface, or elastic domain (tension cut-off not used) are being activated during the response histories. Although not shown, the lateral stress responses ($\sigma_{22} = \sigma_{33}$) are similar in behaviour to σ_{11}.

17.3.1.2 *Solution accuracy*

Each of the three solutions were independently obtained for five choices of the integration parameter ($\theta = 0.0, 0.25, 0.5, 0.75$, and 1.0), and for each θ a sequence of solutions were obtained by successively halving the time step Δt to achieve a relative accuracy within 1 per cent. Table 17.1 shows the required values of Δt to maintain accuracy (note, although not necessary, Δt was taken uniformly throughout the loading schedule). In general $\theta = 0.5$ provided the most efficient

Table 17.1 Time step values (Δt) for 1 per cent relative accuracy

Integration parameter	Fluidity parameter $\gamma = 0.001$	$\gamma = 0.01$	$\gamma = 0.1$
0.00	0.006 25	0.012 5	0.006 25
0.25	0.006 25	0.025	0.012 5
0.50	0.025	0.025	0.012 5
0.75	0.006 25	0.012 5	0.012 5
1.00	0.006 25	0.012 5	0.006 25

scheme, in some cases allowing Δt to be four times greater than with other choices of θ.

For a particular choice of θ, the influence of γ ($0.001 \leq \gamma \leq 0.1$) on the accuracy requirement of Δt is not as significant as initially expected. Initially it was anticipated that Δt for accuracy would be inversely proportional to γ, however as shown in Table 17.1, this is not the case. It is now understood that the accuracy requirement for Δt was predominantly controlled by the loading schedule, i.e., Δt must be sufficiently small to adequately capture the abrupt changes in loading rates. This was evidenced by the observation that the greatest deviations in accuracy occurred at time = 1.0 and 5.5, i.e., where loading rates abruptly changed.

Lastly, for the conditionally stable schemes, $\theta = 0.0$ and 0.25, it was found that the Δt required for accuracy was an order of magnitude less than the stability limit for Δt. Unstable solutions are readily distinguishable from 'inaccurate' solutions by wild oscillations of the responses.

17.3.2 Limestone in triaxial stress

A rather elaborate, non-standard, triaxial test experiment on specimens of Solenhofen limestone was conducted by Robertson [7] to measure the axial strain history ε_{11} resulting from a variable axial stress loading sequence. Details of the testing apparatus and experimental programme are somewhat involved and are not repeated here. Instead, we simply identify the stress loading history (Figure 17.4) for Robertson's specimen number S-90 which is considered in this study. As shown in Figure 17.4, an initial triaxial stress state is rapidly imposed ($\sigma_1 = 96.1$ k.s.i., $\sigma_{22} = \sigma_{33} = 44.1$ k.s.i.). Thereafter, the lateral stresses are maintained constant, and the axial stress is intermittently step loaded at time = 7.2, 12.9, and 22.8 kilo seconds. After each step loading including the initial loading, σ_{11} decreases by some amount due to the nature of the hydraulic testing apparatus. Although the magnitude of these decreases were reported, their time history was not. Accordingly, the linearly decreasing functions following each jump in Figure 17.4 are approximations.

Axial strain measurements were recorded before and after each loading step, providing a data base for attempting to 'curve fit' the viscoplastic cap model. Since the experiment represents a consecutive sequence of loadings, 'curve fitting', in this case, is not a trivial exercise because the accumulated strain depends upon the entire loading history and the strain hardening parameter $\bar{\varepsilon}$ controlling the cap movement.

Figure 17.5 shows strain history data points along with a viscoplastic cap model representation producing a fairly good correlation. Since the viscoplastic model was driven by the triaxial stress loading schedule in Figure 17.4, an inverted form of the numerical algorithm is used to predict strain from stress rather than vice versa. Parameters for the viscoplastic cap model are also shown in Figure 17.4 and were largely determined by trial and error, briefly discussed next.

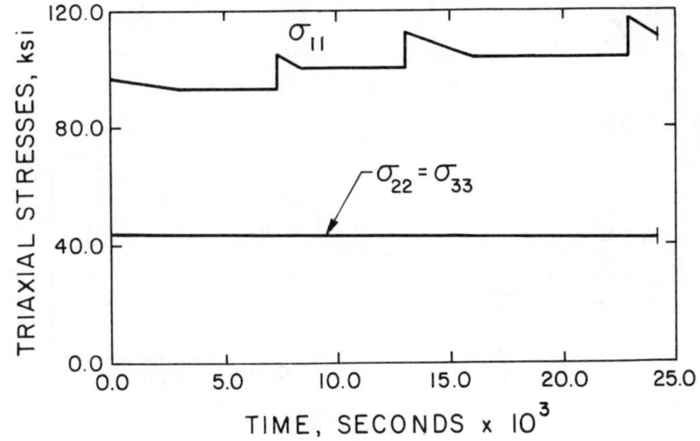

MODEL PARAMETERS FOR LIMESTONE

ELASTIC MODULI:
BULK = 3,500 ksi
SHEAR, $G(J_2') = 91(1+21\exp(-0.0012\ J_2'))$ ksi

FAILURE SURFACE:
$f_F(J_1, J_2') = \sqrt{J_2'} - (1.0 - 0.25\ J_1)$ ksi

CAP SURFACE:
$R = 2.4$, $f_0 = 1.0$ ksi, $x_0 = -212.0$ ksi

CAP HARDENING:
$W = 0.55$, $D_0 = 0.0024\ \text{ksi}^{-1}$

VISCOUS FLOW FUNCTION:
$N = 1$, $f_0 = 1.0$ ksi (Equation 17·8a)
$\gamma = 0.2 \times 10^{-9}\ \text{sec}^{-1}$

Figure 17.4 Triaxial stress loading schedule and model parameters for limestone

Isotropic elastic parameters, bulk modulus, and shear modulus, were determined by best fitting the instantaneous jump responses, i.e., no viscoplastic flow was assumed to occur during the jumps. This was best matched by a constant bulk modulus and a variable shear modulus monotonically decreasing with J_2' (Fig. 17.4). The failure surface was simplified to a standard Drucker–Prager form and the initial cap surface, shaped as a horizontal ellipse $R = 2.4$, was located well into the compression region by setting $X_0 = -212.0$ k.s.i. The motivation for this initial setting was to provide a large elastic region so that the initial jump loading did not cause excessive viscoplastic flow in accordance with observations. Also, it ensured the viscoplastic flow would be controlled by the cap surface ($J_1 \leq L$) throughout the loading schedule.

The cap hardening parameters W and D_0 as well as the fluidity parameter γ were adjusted by numerical experimentation to best match the data. No

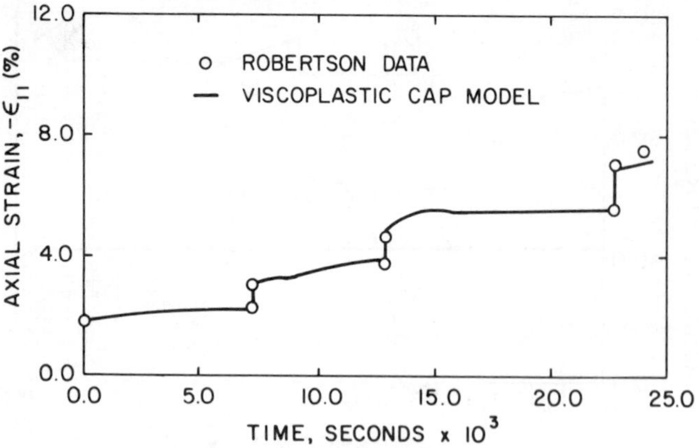

Figure 17.5 Axial strain response data and viscoplastic cap model representation

attempt was made to examine various forms of the viscous flow function and was taken in its most simple form, $\phi(f) = f/f_0$.

Certainly, it is not claimed that the model parameters chosen here are representative of the limestone material in any loading environment. Rather, we simply assert that, within the general framework presented, the viscoplastic cap model has the ability to adequately represent the time dependent behaviour of some geological materials for some loading conditions as demonstrated in Figure 17.5.

At least one shortcoming of the viscoplastic cap model is its inability to represent tertiary creep, i.e., an eventual increase in strain rate during a state of constant stress. This is because the cap surface hardens when it is loaded reducing the value of yield function f_c and hence the viscoplastic flow rate. One way to correct for this is to introduce a softening function which overrides the hardening function when the hardening parameter $\bar{\varepsilon}$ reaches a specified value. Such modifications will be explored in future investigations.

17.4 CONCLUDING REMARKS AND RECOMMENDATIONS

The viscoplastic formulation and numerical algorithm presented herein provides a general format for incorporating various plasticity models into a Perzyna-type elastic/viscoplastic constitutive relationship suitable for finite element applications. In particular, the viscoplastic cap model illustrated in this study appears to be a useful extension of the inviscid cap model, allowing a realistic, time-dependent representation of soil and rocks over a range of loading conditions.

Future work is needed in the areas of parameter identification techniques and experimental verification particularly in ground shock problems. It is anticipated that the viscoplastic cap model, as opposed to the inviscid cap model, will properly represent the strain rate effects observed from ground shock experimental data.

Useful modifications and/or refinements to the model include, (1) a more realistic criterion for tension cut-off together with an accumulative viscous-tension damage index, and (2) a more generalized hardening–softening functional representation. A motivation for the latter is to provide the capability of representing tertiary creep.

Lastly with regard to the numerical algorithm, experience to date indicates that $\theta = 0.5$ (implicit) provides the most accurate time integration scheme, as well as being unconditionally stable. However, from a computational viewpoint, the explict scheme $\theta = 0.0$ may, at times, be more efficient since no iterations within the time step are required, but at the expense of a smaller time steps to maintain accuracy/stability. Since the trade-off between implicit and explicit schemes are problem dependent, it is prudent to provide both options in an algorithm along with options for various forms of Newton–Raphson iteration, e.g., modified or quasi Newton–Raphson procedures.

ACKNOWLEDGEMENTS

This work was sponsored by the Naval Civil Engineering Laboratory, Port Hueneme, California. A deep appreciation is extended to the technical monitors, Mr John Crawford and Mr Joseph Holland for their enthusiastic support and technical guidance.

REFERENCES

1. I. S. Sandler and D. Rubin, 'An algorithm and a modular subroutine for the CAP model', *Int. Journal for Numerical and Analytical Methods in Geomechanics*, **3**, 1973–186 (1979).
2. I. S. Sandler, F. L. DiMaggio, and G. Y. Baladi, 'Generalized CAP model for geological materials), *Journal Geotechnical Engineering Division, ASCE*, **102**, No. GT7, July 1976.
3. F. L. DiMaggio and I. S. Sandler, 'Material models for granular soils', *Journal of Engineering Mechanics*, 935–950 (1971).
4. P. Perzyna, 'Fundamental problems in viscoplasticity', *Advances in Applied Mechanics*, **9**, 244–368 (1966).
5. K. Akai, T. Adachi, and K. Nishi, 'Mechanical properties of soft rocks', *IX Conference on Soil Mech. Found Eng., Tokyo*, Vol. 1, pp. 7–10, 1977.
6. F. Komamura and J. Huang, 'New rheological model for soil behavior', *Journ. of the Geotech. Eng. Div., ASCE*, **100**, 807–823 ((1974).
7. E. C. Robertson, 'Creep of Solenhofen limestone under moderate hydrostatic pressure', *Rock Deformation, Mem. Geol. Soc. Amer.*, **79**, 227–244 (1960).
8. A. Sing and J. K. Mitchell 'General stress-strain-time functions for soils), *Soil Mech. and Found. Div., ASCE*, **94**, 21–46 (1968).
9. K. C. Valanis, 'A theory of viscoplasticity without a yield surface', *Archives of Mechanics, No.* 23, 517–555 (1971).

10. M. G. Katona, 'Combo viscoplasticity: an introduction with incremental formulation', *Computers and Structures*, **11**, (3), 217–224 (1980).
11. S. R. Bodner and V. Parton, 'Constitutive equations for elastic–viscoplastic strain-hardening materials', *Journal of Applied Mech., ASME*, **42**, 385–389 (1975).
12. A. Phillips and H. C. Wu, 'A theory of viscoplasticity', *Int. Journal of Solids and Structures*, **9**, 15–30 (1973).
13. O. C. Zienkiewicz and I. C. Cormeau, 'Visco-plasticity, plasticity and creep in elastic solids—a unified numerical approach', *Int. Journal for Numer. Methods in eng.*, **8**, 821—845.
14. D. R. J. Owen and E. Hinton, *Finite Elements in Plasticity, Theory and Practice*, Pineridge Press Limited, Swansea, UK, 1980.
15. I. Cormeau, 'Numerical stability in quasi–static elasto/visco-plasticity', *Int. Journal for Numerical Methods in Eng.*, **9**, 109–127 (1975).
16. T. R. Hughes and R. L. Taylor, 'Unconditionally stable algorithms for quasi–static elasto/viscoplastic finite element analysis', *Int. Journal Numerical Methods in Engineering* (to be published).
17. J. R. Dennis and J. J. More, 'Quasi–Newton methods, motivation and theory', *SIAM Review*, **19**, 46–86 (1977).
18. H. Matthies and G. Strang, 'The solution of nonlinear finite element equations', *Int. Journal Numerical Methods in Engineering*, **14**, 1613–1626 (1979).
19. O. C. Zienkiewicz, C. Humpheson, and R. W.. Lewis, 'Associated and non-associated visco-plasticity and plasticity in soil mechanics', *Geotechnique*, **25**, (4), 671–689 (1975).
20. M. G. Katona, 'A viscoplastic algorithm for CAP75', *Report to Naval Civil Engineering Laboratory, N*68305-80-*C*-0031, Port Hueneme, CA, Sept., 1981.
21. J. Crawford, 'Capdrver: a computer program for exercising plasticity constitutive models', Naval Civil Engineering Laboratory, Port Hueneme, California, 1980.

Mechanics of Engineering Materials
Edited by C. S. Desai and R. H. Gallagher

Chapter 18

Computer Simulation of the Material Testing by Means of New Discrete Models

T. Kawai, K. Niwa, and M. Ikeda

18.1 WHAT WE CAN OBTAIN FROM THE CONVENTIONAL MATERIAL TESTING

The constitutive law of materials, more precisely the stress–strain law of materials is usually obtained by conducting conventional material testings.

The stress–strain relation to be obtained from results of these material testings only gives the recorded relation average stress and average strain of a given specimen under a given loading. It may depend upon many parameters such as shape, dimension, type of loading, boundary conditions, initial imperfection, residual stresses, and so on. Figure 18.1 shows the crushing failure modes of a concrete cylinder under uniaxial compression. It is well known that either tensile failure or shear failure mode is observed in this case depending upon the existence of frictional force at both ends of a given specimen and dimension. Such experimental evidence cannot be well explained by the existing theories of solid mechanics which is usually based on the one-dimensional stress–strain relation

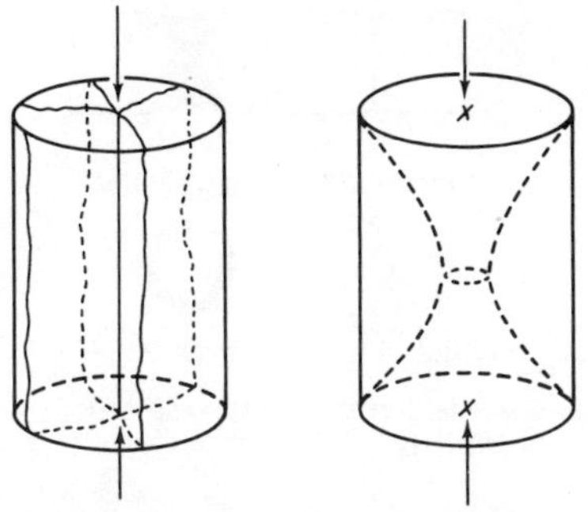

Figure 18.1 Crushing modes of a concrete cylinder under compression

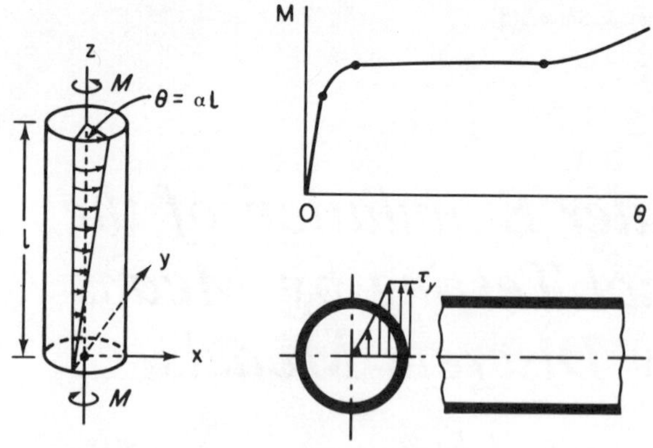

Figure 18.2 Result of mathematical analysis on the plastic torsion problem of a round steel bar by means of the existing theory of plasticity

obtained from the conventional material testing and assumes the continuity of displacement field irrespective of the deformation status [1 to 5].

Mild steel is a metallic material which shows the so-called perfect elastoplastic characteristic, and existing theory of plasticity yields the solution shown in Figure 18.2 on the plastic torsion problem of a mild steel round bar.

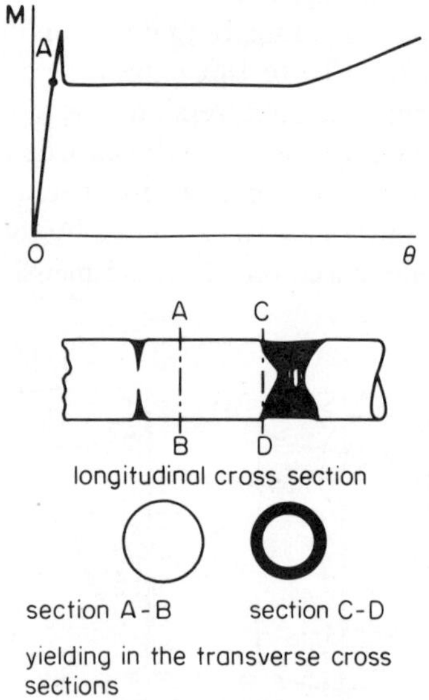

Figure 18.3 The first failure mode of round mild steel specimens in torsional testing

In actual torsion testing of mild steel round bar specimens two different moment–rotation curves, shown in Figures 18.3 and 18.4 are usually obtained.

These moment–rotation characteristics are essentially similar to the M–θ curve shown in Figure 18.2, but modes of failure are distinctively different from the calculated failure mode shown in Figure 18.2. More precisely there are two types of failure modes, i.e., yielding in the transverse cross-section as well as in the longitudinal section. In the failure on transverse cross-sections a certain portion yields completely along the cylinder axis while the other portion remains elastic as shown in Figure 18.3.

On the other hand, in the failure on the longitudinal section the yield pattern as shown in Figure 18.4 can be obtained along the specimen axis. These experimental results have drawn the keen attention of many scholars and a number of theories have been proposed to explain such peculiar failure mechanism of mild steel bars under torsion. However, none of them seems to successfully explain these points clearly. The first author believes that such a drawback of existing theory and constitutive equation may be attributed to the following two reasons.

(1) Almost all the plastic theories are based on the assumption of displacement continuity. However, slip is the essential feature of plastic deformation and it cannot be easily taken into account in the existing plasticity theory which is usually based on the assumption of displacement continuity.

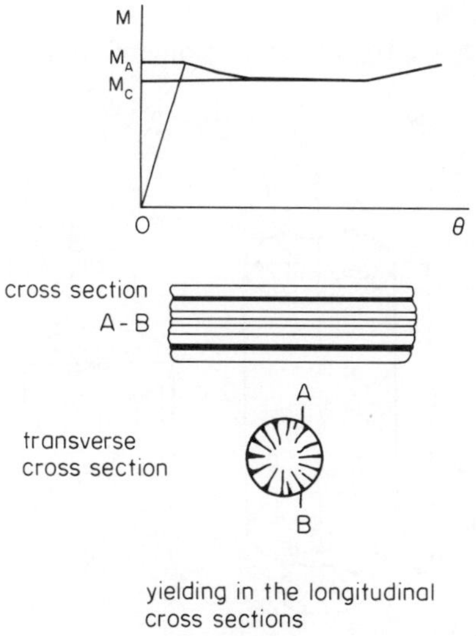

Figure 18.4 The second failure mode of round mild steel specimens in torsional testing

(2) A one-dimensional stress–strain diagram obtained from the conventional material testing is an example in which effects of many parameters such as those shown above are integrated. Therefore, it should be necessary to establish a method to obtain the real stress–strain relation of a given material, from experimental results of the material testing by eliminating influences of these parameters on the constitutive relation.

18.2 DISCRETE LIMIT ANALYSIS OF PLASTIC TORSION OF A ROUND MILD STEEL BAR

In 1976 Kawai developed a family of new discrete models entitled 'rigid bodies–spring models' (abbreviated as RBSM) [6, 7]. Through basic studies of these elements they were proved to be very useful in the limit analysis of solids and structures. In this section a method of computer simulation by using new discrete models is proposed on the plastic torsion problems. First, a method of analysis is discussed briefly and then stiffness matrices of a prism and cylindrical shell type elements are obtained.

18.2.1 Method of analysis

The total number of degrees of freedom of the 3D RBSM model is only six. In three-dimensional analysis, however, the number of elements is usually very large and therefore some consideration must be given to reduce the computing time and computer cost.

A round bar subjected to torsional moment M at both ends is shown in Figure 18.5.

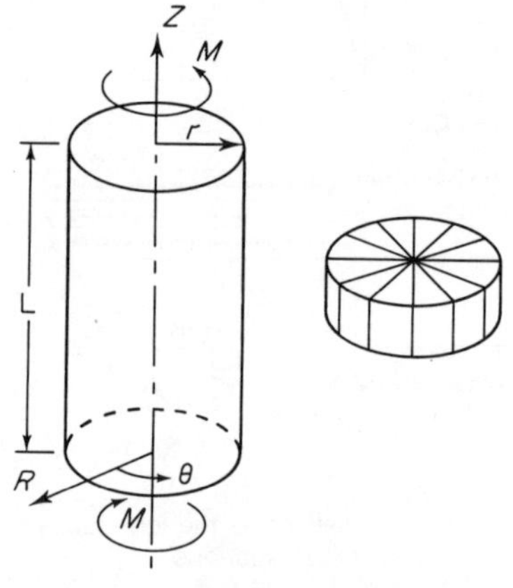

Figure 18.5 Disc and sectorial plate elements for analysis of torsion problem of a round bar

$$\mathbf{u}_I^T = (U_I, V_I, W_I, \theta_I, \phi_I, \chi_I)$$
$$\mathbf{F}_I^T = (F_{RI}, F_{\theta I}, F_{ZI}, M_{RI}, M_{\theta I}, M_{ZI})$$

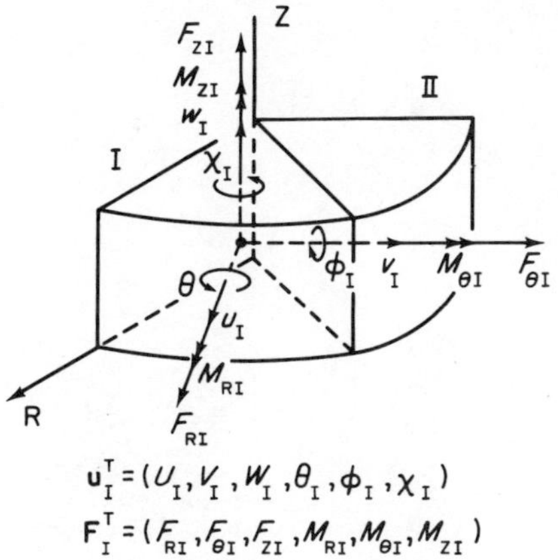

Figure 18.6 Centroidal displacement and load vector of a sectorial plate

As global coordinates, a cylindrical coordinate is considered whose z axis is taken as the axis of a given round bar. Now this round bar is sliced by a number of planes perpendicular to the z axis, and each sliced disc is divided into N sectorial plates of equal opening angle $2\pi/N$. These sectorial plates may be a superelement which consists of a number of tetrahedral RBSM elements.

When a round bar is subjected to a torsional moment, all the sectorial plate elements in the same disc may displace by the same amount. Taking two adjacent elements I and II as shown in Figure 18.6, the stiffness equation will be formulated by the method explained in Appendix I as follows:

$$\begin{bmatrix} \mathbf{K}_I & \mathbf{K}_{I,II} \\ \mathbf{K}_{I,II}^T & \mathbf{K}_{II} \end{bmatrix} \begin{Bmatrix} \mathbf{u}_I \\ \mathbf{u}_{II} \end{Bmatrix} = \begin{Bmatrix} \mathbf{F}_I \\ \mathbf{F}_{II} \end{Bmatrix} \tag{18.1}$$

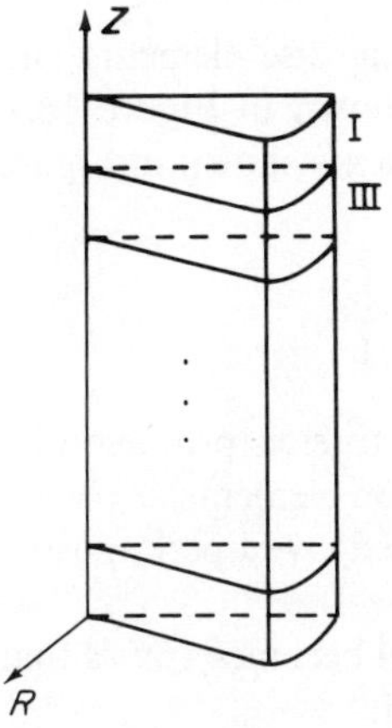

Figure 18.7 A prism substructure as composed of sectorial plate elements

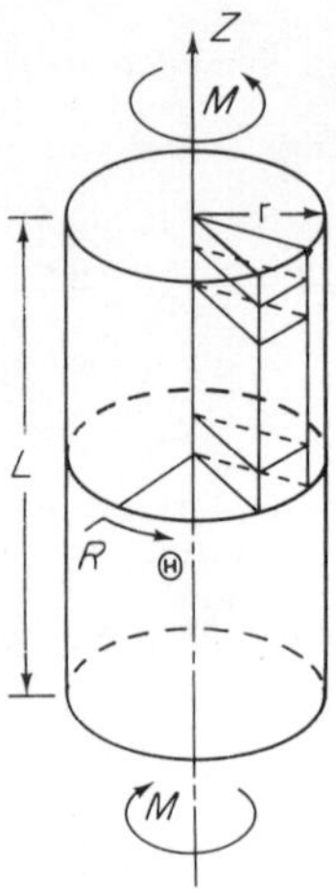

Figure 18.8 A triangular cylindrical element as an unit structure of the FEM model for plastic torsion of a round bar

where the square matrix on the left-hand side of equation (18.1) is a stiffness matrix of sectorial elements I and II. $\mathbf{u}_{\mathrm{I}}$, $\mathbf{u}_{\mathrm{II}}$; $\mathbf{F}_{\mathrm{I}}$, $\mathbf{F}_{\mathrm{II}}$ are displacement and load vectors at the centroids respectively. Their components are shown in Figure 18.6.

In the case of a torsion problem the following identity can be observed:

$$\mathbf{u}_{\mathrm{I}}=\mathbf{u}_{\mathrm{II}}, \qquad \mathbf{F}_{\mathrm{I}}=\mathbf{F}_{\mathrm{II}}, \qquad \mathbf{K}_{\mathrm{I}}=\mathbf{K}_{\mathrm{II}} \tag{18.2}$$

Therefore equation (18.1) can be condensed as follows:

$$(2\mathbf{K}_{\mathrm{I}}+\mathbf{K}_{\mathrm{I,II}}+\mathbf{K}^{\mathrm{T}}_{\mathrm{I,II}})\mathbf{u}_{\mathrm{I}}=2\mathbf{F}_{\mathrm{I}} \tag{18.3}$$

Repeating such an operation to all the elements on the same disc, the total stiffness matrix of a disc element can be condensed to that of a sectorial element as follows:

$$N(2\mathbf{K}_{\mathrm{I}}+\mathbf{K}_{\mathrm{I,II}}+\mathbf{K}^{\mathrm{T}}_{\mathrm{I,II}})\mathbf{u}_{\mathrm{I}}=N\mathbf{F}_{\mathrm{I}} \tag{18.4}$$

Similar matrix operation can be applied to the remaining disc elements and these elements correspond to a set of prism elements shown in Figure 18.7. Constructing the stiffness equation for a set of prism elements shown in Figure 18.8, the following equation can be similarly obtained:

$$N\begin{bmatrix}\mathbf{K}_{\mathrm{I}} & \mathbf{K}_{\mathrm{I,III}}\\ \mathbf{K}_{\mathrm{I,III}} & \mathbf{K}_{\mathrm{III}}\end{bmatrix}\begin{Bmatrix}\mathbf{u}_{\mathrm{I}}\\ \mathbf{u}_{\mathrm{III}}\end{Bmatrix}=N\begin{Bmatrix}\mathbf{F}_{\mathrm{I}}\\ \mathbf{F}_{\mathrm{III}}\end{Bmatrix} \tag{18.5}$$

From the above discussion, it can be concluded that the torsion problem of a round bar can be studied only by considering a set of prism elements as shown in Figure 18.8. The total stiffness matrix can be obtained by superimposing equations (18.4) and (18.5).

Therefore the number of degrees of freedom of this model becomes $1/N$ of that of the original model and the bandwidth may be reduced.

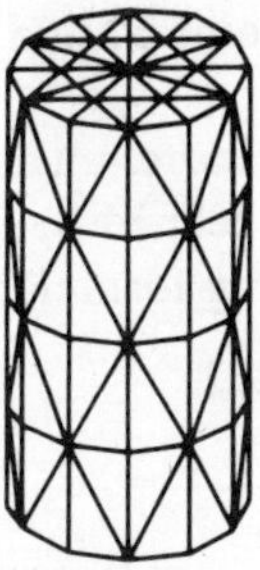

Figure 18.9 Finite element model for analysis of the torsion problem of a round bar (dodecahedral representation)

For example, in case of torsion of a dodecahedral cylinder the number of degrees of freedom of the condensed model is $288(=\frac{1}{6}\times 1728)$ and computer time is $\frac{1}{60}$ of that of one elastic analysis. It should be noted here that the model shown in Figure 18.9 is condensed to $\frac{2}{12}$.

The method of matrix condensation can be applied not only to the case of torsion of a round bar but also to torsion of a prism or hollow cylinder and tension or bending problems of a bar or beam.

18.2.2 Development of a new 3D RBSM element with respect to the cylindrical coordinates

RBSM elements consist of rigid bodies and normal as well as shear spring distributed on the interface of two adjacent elements.

Denoting the centroidal displacement of a RBSM element i by $\mathbf{u}_i$, the displacement $\mathbf{u}_i$ of an arbitrary point P of the said element can be given by using a transformation matrix $\mathbf{Q}_i$ as follows:

$$\mathbf{u}_i = \mathbf{Q}_i \mathbf{u}_i \tag{18.6}$$

In case of the cylindrical coordinates shown in Figure 18.6 $\bar{\mathbf{U}}_i$ consists of translational displacements along three axes $\bar{U}_{ri}$, $\bar{V}_{\theta i}$, $\bar{W}_{zi}$ and angular displacements around three axes θ_{ri}, $\Phi_{\theta i}$, χ_{zi}, while $\mathbf{u}_i$ consists of translational displacements along three axes u_{ri}, $v_{\theta i}$, w_{zi}. The transformation matrix $\mathbf{Q}_1$ can be derived from the corresponding matrix $\mathbf{Q}_1$ defined with respect to cartesian coordinates as follows:

$$\mathbf{Q}_i = \begin{bmatrix} 1 & 0 & 0 & 0 & (z-z_i) & -(y-y_i) \\ 0 & 1 & 0 & -(z-z_i) & 0 & (x-x_i) \\ 0 & 0 & 1 & (y-y_i) & -(x-x_i) & 0 \end{bmatrix} \tag{18.7}$$

where (x, y, z) are the coordinates of a point P, and (x_i, y_i, z_i) the centroidal coordinates of the element i.

The transformation matrices between the cartesian and cylindrical coordinates of the centroid as well as the point P are defined by $\mathbf{T}_i$ and $\mathbf{T}$ respectively.

Then $\mathbf{Q}_i$ can be obtained by the following formula:

$$\mathbf{Q}_i = \mathbf{T}^{\mathrm{T}}\mathbf{Q}_i'\mathbf{T}_i \tag{18.8}$$

where

$$\mathbf{T}_i = \begin{bmatrix} \cos\theta_i & \sin\theta_i & 0 \\ -\sin\theta_i & \cos\theta_i & 0 \\ 0 & 0 & 1 \end{bmatrix}, \qquad \mathbf{T} = \begin{bmatrix} \cos\theta & \sin\theta & 0 \\ -\sin\theta & \cos\theta & 0 \\ 0 & 0 & 1 \end{bmatrix}$$

θ_i is the θ component of the centroidal coordinate (r_i, θ_i, z_i) of a given element i, while θ is the θ component of the coordinates (r, θ, z) of a given point P.

The relative displacement $\boldsymbol{\delta}^{\mathrm{T}} = [\delta_{\mathrm{n}}, \delta_{\mathrm{t}}, \delta_{\mathrm{s}}]$ of two adjacent elements due to the centroidal rigid body displacement is now considered in which δ_{n} is the vertical, and δ_{t}, δ_{s} are tangential displacements along two orthogonal directions. Direction cosines of δ_{n}, δ_{s}, δ_{t} are defined by $(l_{\mathrm{n}}, m_{\mathrm{n}}, n_{\mathrm{n}})$, $(l_{\mathrm{s}}, m_{\mathrm{s}}, n_{\mathrm{s}})$ and $(l_{\mathrm{t}}, m_{\mathrm{t}}, n_{\mathrm{t}})$ respectively.

Based on the above preliminaries, the relative displacement vector $\boldsymbol{\delta}$ can be given by the following equation:

$$\boldsymbol{\delta} = \mathbf{R}\mathbf{u} \tag{18.9}$$

where

$$\mathbf{u}^{\mathrm{T}} = [\mathbf{u}_1^{\mathrm{T}}, \mathbf{u}_2^{\mathrm{T}}]$$

$$\mathbf{R} = \begin{bmatrix} -l_{\mathrm{n}}, & -m_{\mathrm{n}}, & -n_{\mathrm{n}}, & l_{\mathrm{n}}, & m_{\mathrm{n}}, & n^{\mathrm{n}} \\ -l_{\mathrm{s}}, & -m_{\mathrm{s}}, & -n_{\mathrm{s}}, & l_{\mathrm{s}}, & m_{\mathrm{s}}, & n_{\mathrm{s}} \\ -l_{\mathrm{t}}, & -n_{\mathrm{t}} & -n_{\mathrm{t}}, & l_{\mathrm{t}}, & m^{\mathrm{t}}, & n_{\mathrm{t}} \end{bmatrix}$$

Substituting equations (18.6) and (18.8) into equation (18.9) the following equation can be derived:

$$\boldsymbol{\delta} = \mathbf{R}\begin{bmatrix} \mathbf{T}^{\mathrm{T}}\mathbf{Q}_1'\mathbf{T}_1 & 0 \\ 0 & \mathbf{T}^{\mathrm{T}}\mathbf{Q}_2'\mathbf{T}_2 \end{bmatrix}\begin{Bmatrix} \bar{\mathbf{U}}_1 \\ \bar{\mathbf{U}}_2 \end{Bmatrix} = \mathbf{B}\bar{\mathbf{U}} \tag{18.10}$$

where

$$\mathbf{B} = \mathbf{R}\begin{bmatrix} \mathbf{T}^{\mathrm{T}}\mathbf{Q}_1'\mathbf{T}_1 & 0 \\ 0 & \mathbf{T}^{\mathrm{T}}\mathbf{Q}_2'\mathbf{T}_2 \end{bmatrix}, \qquad \mathbf{U}^{\mathrm{T}} = [\bar{\mathbf{U}}_1^{\mathrm{T}}, \bar{\mathbf{U}}_2^{\mathrm{T}}]$$

Consequently, the strain energy $\bar{V}$ to be stored in the connection spring system distributed on the interfaces of two adjacent elements and can be defined by the following equation:

$$\bar{V} = \tfrac{1}{2}\int \boldsymbol{\sigma}^{\mathrm{T}}\boldsymbol{\delta}\,\mathrm{d}s \tag{18.11}$$

where stress vector $\boldsymbol{\sigma}$ can be given by the following equation:

$$\boldsymbol{\sigma} = \mathbf{D}\boldsymbol{\delta} \tag{18.12}$$

where

$$\mathbf{D} = \begin{bmatrix} k_{\mathrm{n}} & 0 & 0 \\ 0 & k_{\mathrm{s}} & 0 \\ 0 & 0 & k_{\mathrm{t}} \end{bmatrix}$$

and

$$\left.\begin{aligned} k_{\mathrm{n}} &= \frac{1}{h}\cdot\frac{E(1+\nu)}{(1+\nu)(1-2\nu)} \\ k_{\mathrm{s}} = k_{\mathrm{t}} &= \frac{1}{h}\left(\frac{E}{1+\nu}\right) \end{aligned}\right\} \tag{18.13}$$

in which these spring constants are defined in case of the isotropic elastic materials, h is the projected length of a vector connecting centroids along the normal drawn on the tangential plane of two adjacent elements. (See Appendix II.)

Therefore the following formula of the stiffness matrix of a given RBSM element can be obtained:

$$\mathbf{K} = \int \mathbf{B}^{\mathrm{T}}\mathbf{D}\mathbf{B}\,\mathrm{d}s \tag{18.14}$$

In case of the inelastic deformation, it is assumed that the yield condition of a given material is defined as follows:

$$f(\boldsymbol{\sigma}) = 0, \qquad \boldsymbol{\sigma}^{\mathrm{T}} = [\sigma_{\mathrm{n}}, \tau_{\mathrm{s}}, \tau_{\mathrm{t}}] \tag{18.15}$$

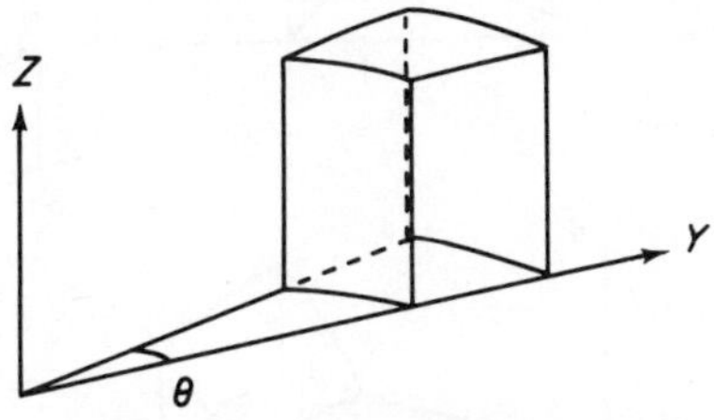

Figure 18.10 Hexahedral element

The stress–strain matrix corresponding to equation (18.15) can be derived by using Yamada's method [8] as follows:

$$K_{ij}^{(p)} = K_i^{(e)}\delta_{ij} - \frac{1}{\sum K_i^{(e)}\bar{f}_i^{(e)}}\bar{f}_i\bar{f}_j K_i^{(e)}K_j^{(e)} \tag{18.16}$$

where $\bar{f}_i = \partial f/\partial\sigma_i$, $(i = 1, 2, 3)$, $\sigma_1 = \sigma$, $\sigma_2 = \tau_s$, $\sigma_3 = \tau_t$ and $K_i^{(e)}$ are components of the elastic stiffness matrix.

In order to carry out accurate analysis of solid cylinders, hexahedral and pentahedral RBSM elements are developed whose shape functions are expressed by the linear function with respect to the cylindrical coordinates.

(a) *Hexahedral element* The following displacement fields are assumed:

$$\left.\begin{aligned} r &= \sum_{i=1}^{6} N_i r_i \\ \theta &= \sum_{i=1}^{6} N_i \theta_i \\ z &= \sum_{i=1}^{6} N_i z_i \end{aligned}\right\} \tag{18.17}$$

where (r_i, θ_i, z_i) are coordinates of the ith nodal point. N_i is the shape function given by the following equation:

$$\begin{aligned} &N_i = \tfrac{1}{8}(1+\xi_0)(1+\eta_0)(1+\zeta_0) \\ &\xi_0 = \xi\xi_i, \qquad \eta_0 = \eta\eta_i, \qquad \zeta_0 = \zeta\zeta_i \end{aligned} \tag{18.18}$$

(b) *Pentahedral element* The displacement fields assumed are given by the

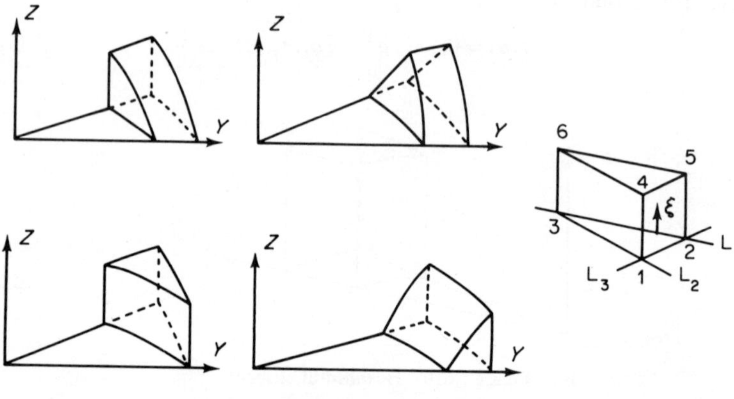

Figure 18.11 Prism element as composed by four pentahedral elements

following equation:

$$r = \sum_{i=1}^{5} N_i r_i, \qquad \theta = \sum_{i=1}^{5} N_i \theta_i, \qquad z = \sum_{i=1}^{5} N_i z_i \tag{18.19}$$

where $N_i = \frac{1}{2}(1 + \xi_0)L_i$, $\xi_0 = \xi\xi_i$, and $L_j(j = 1, 2, 3)$ are area coordinates.

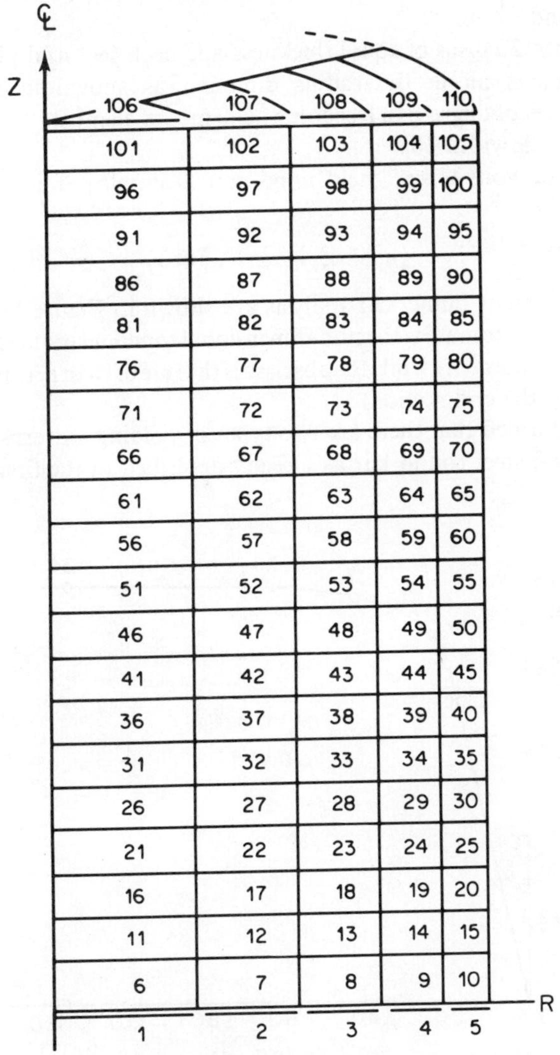

Figure 18.12 Finite element model used for the present analysis

18.3 RESULTS OF NUMERICAL ANALYSIS ON THE PLASTIC TORSION

In this section the plastic torsion problem is analysed by using the elements described in the previous section, and the ultimate strength, development of slip surfaces, failure modes, and accuracy of solutions are discussed.

The analysis model employed is a triangular prism with the opening angle 1°, length $L = 80$ mm, and its radius $r(=D/2)$ is assumed to be 0.8 mm, 2 mm, 4 mm, 8 mm, and 20 mm.

It is sliced into 20 discs of equal thickness and each sectorial plate is divided into five segments along the radius direction as shown in Figure 18.12. The total number of degrees of freedom is 660 (no. of elements 110, no. of nodes 252, and half-bandwidth 36).

The following von Mises' yield condition is employed in the numerical analysis:

$$\tfrac{1}{4}\sigma_z^2 + \tau_{xy}^2 + \tau_{yz}^2 - c_1^2 = 0 \tag{18.20}$$

Results of the present numerical analysis are shown in Figure 18.13.

In this figure the ordinate is the non-dimensional torsional moment M/M_y(M_y is the yield twisting moment), while the abscissa is the rate of twist θ/L (θ is the angular displacement of the end section).

It can be confirmed that there are two types of yielding patterns in the plastic torsion of a mild steel round bar as already described in the first section.

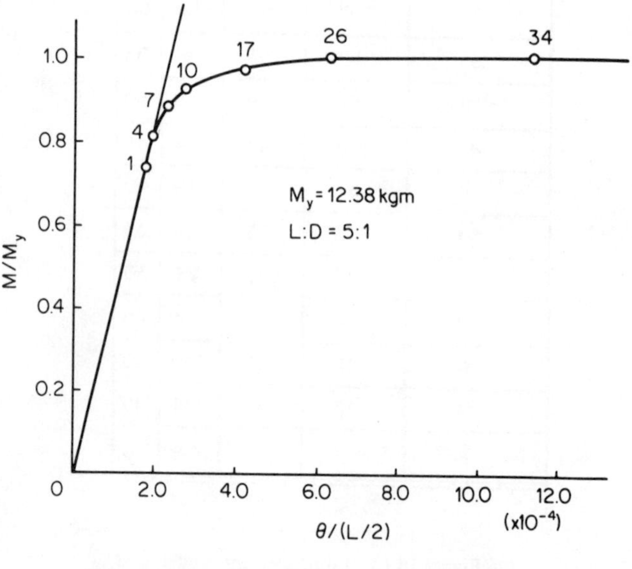

Figure 18.13 Calculated moment–rotation curve

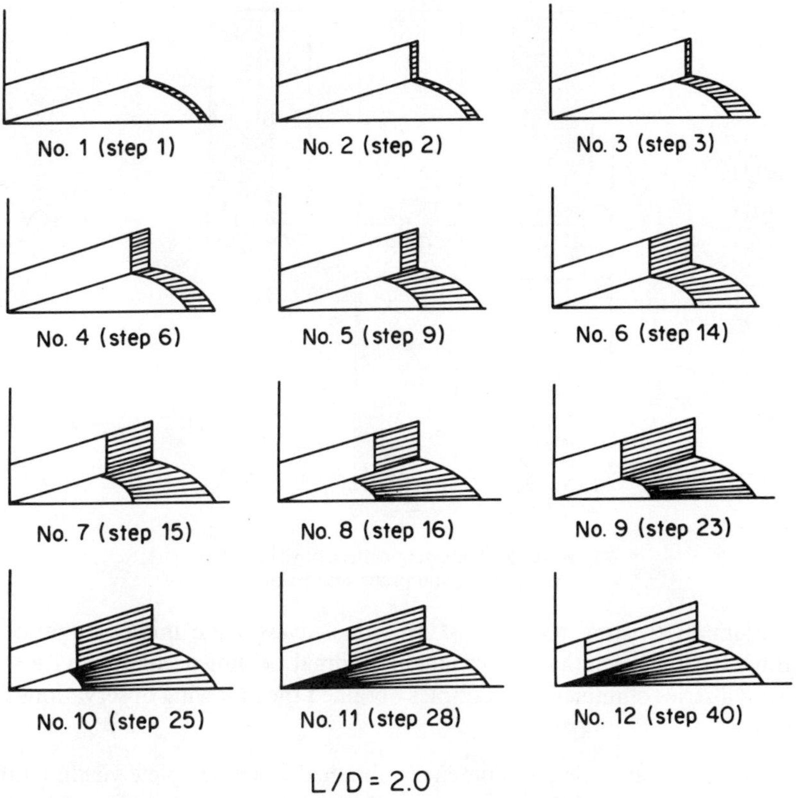

Figure 18.14 Development of slip surfaces on the middle cross-section of a twisted cylinder

Figure 18.14 shows the development of slip surfaces on the middle cross-section of a given cylinder in case of horizontal as well as vertical yielding (shown by the shaded area).

Spreading pattern of yield surfaces depends on the ratio L/D. In the case of $L/D = 2.0$ or 5.0 uniform yielding occurs, while in the other case of $L/D = 10.0$, 20.0, and 50.0 a plastic zone concentrates in the vicinity of the middle cross-section. Figure 18.15 shows the development of a plastic zone due to the transverse yielding at the ultimate twisting moment versus L/D ratio.

Such tendency can be equally observed on the plastic zone produced by the longitudinal yielding, but it may reduce rapidly as L/D increases and in the case of $L/D = 20.0$ and 50.0 no longitudinal yielding is observed. It has been reported by the previous experimental studies that yielding planes appear at almost constant $\Delta\theta$ on the cross-section of a bar in case of the longitudinal yielding. Figure 18.16 shows the result of a study on the relation between the initial yield moment versus the opening angle $\Delta\theta$, in which the ordinate implies

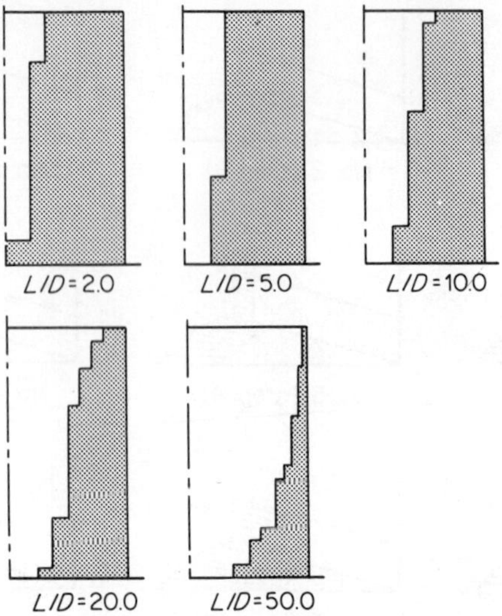

Figure 18.15 Spreading pattern of yield surfaces versus L/D ratio (transverse yielding)

the minimum twisting moment M/M_y which causes the initial longitudinal yielding. It can be seen that the initial longitudinal yielding occurs when $\Delta\theta \leq 3°$.

From a series of numerical examples obtained the following observations can be made:

(i) The ultimate twisting moment is obtained when complete yielding takes place on the middle cross-section of a homogeneous and isotropic round mild steel bar.

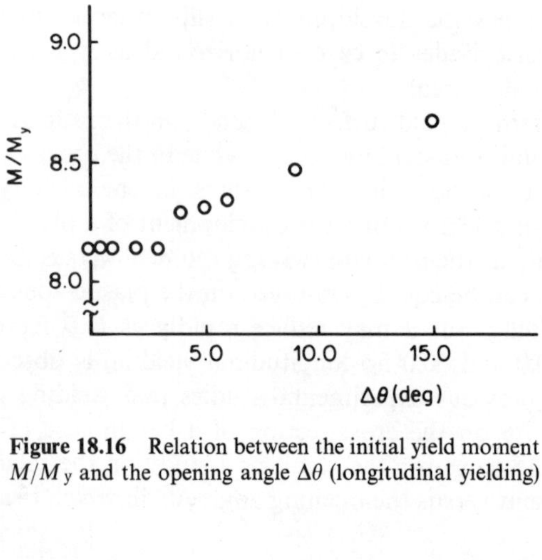

Figure 18.16 Relation between the initial yield moment M/M_y and the opening angle $\Delta\theta$ (longitudinal yielding)

(ii) Under von Mises' yield condition, transverse yielding is generally followed by the longitudinal yielding. In other words, neither purely transverse nor longitudinal yielding can be obtained in case of the homogeneous materials with von Mises' yield condition.

(iii) The moment–rotation curve obtained as shown in Figure 18.13 has not a sharp cusp as seen in Figures 18.3 or 18.4.

18.4 CONCLUSION

·om the results of computer simulation described in the previous section the llowing observations can be made. Computer simulation of the plastic stability of twisted mild steel bars by using the standard programme of the cremental procedure with RBSM elements and von Mises' yield condition as found to be unsuccessful so far.

The main reason for such failure may be attributed to lack of the function release exceeded tangential forces on the slip surfaces and to balance reaction rces of the connection spring system on the interelement boundaries in the mputer program. This process of computation, however, is unstable because leasing unbalanced tangential forces on the interelement boundaries may celerate further yielding and consequently the stiffness matrix of a given ructure may become negative.

Such considerations may suggest the possibility of successful simulation of e plastic instability to be observed in torsion testing of a mild steel bar. This ct is pointed out by Shioya [4, 5] *et al.* and they attempted to simulate this stability phenomenon by using the dislocation theory of the polycrystalline ructures.

Nakanishi [2, 3, 9, 10] proposed the failure condition of a twisted mild steel ır by the following equation:

$$\iint_s \tau r \, ds - \iint_s \tau_y r \, ds = 0$$

here τ is the shearing stress, τ_y the shearing yield stress, and s is the area to integrated depending on the transverse or longitudinal yielding.

Necessity of such programme improvement may also be supported by ample perience of instability analysis of rock foundations by means of the computer ogram of tension crack analysis developed by Kawai and Takeuchi [7].

Based on such considerations development of the efficient subroutine is now nder way. Once such an elaborate program is developed, successful computer mulation on the plastic instability of a twisted mild steel bar may be complished.

APPENDIX I COMPONENTS OF THE UNIT NORMAL DRAWN TO THE BOUNDARY SURFACE OF TWO ADJACENT ELEMENTS, $\mathbf{n} = (l_n, m_n, n_n)$

The boundary surface is assumed to be expressed by the following equations:

$$r = \sum_{i=1}^{4} N_i r_i, \qquad \theta = \sum_{i=1}^{4} N_i \theta_i, \qquad z = \sum_{i=1}^{4} N_i z_i \tag{18.21}$$

where

$$N_i = \tfrac{1}{4}(1 + \xi\xi_i)(1 + \eta\eta_i) \tag{18.22}$$

In this case $\mathbf{n}(l_n, m_n, n_n)$ can be given by:

$$\left.\begin{aligned} l_n &= \gamma \frac{\partial(\theta, z)}{\partial(\xi, \eta)} \\ m_n &= \gamma \frac{\partial(z, \gamma)}{\partial(\xi, \eta)} \\ n_n &= \gamma \frac{\partial(\gamma, \theta)}{\partial(\xi, \eta)} \end{aligned}\right\} \tag{18.23}$$

APPENDIX II CALCULATION OF h

h is given by the following vectorial formulae:

$$h = (\mathbf{n} \cdot \mathbf{g}) \tag{18.24}$$

where $\mathbf{g} = \mathbf{g}_2 - \mathbf{g}_1$ and $\mathbf{g}_i (i = 1, 2)$ are position vectors of the element centroids.

Then equation (18.24) can be given by the following equation with respect to the cylindrical coordinates:

$$\begin{aligned} h = l_n \{\gamma_2 \cos(\theta_2 - \theta) - \gamma_1 \cos(\theta_1 - \theta)\} \\ + m_n \{\gamma_2 \sin(\theta_2 - \theta) - \gamma_1 \sin(\theta_1 - \theta)\} + n_n (Z_2 - Z_1) \end{aligned}$$

REFERENCES

1. A. Nandi, *Theory of Flow and Fracture of Solids*, McGraw-Hill Book Co., 1950.
2. F. Nakanishi and K. Sato, *Strength of Materials* (in Japanese), Iwanami Zensho, 1976.
3. Y. Sato, *Strength of Materials and Plasticity* (in Japanese), No. 4 Text Series of New Mechanical Engineering, Morikita Publishing Co., 1980.
4. T. Shioya and J. Shioiri, 'Elastic-plastic analysis of the yield process in mild steel', *Journal of the Mechanics and Physics of Solids*, **24**, 187–204 (1976).
5. T. Shioya, T. Machida, R. Ishida, and Y. Fujiura, 'Scale effect of the plastic zone pattern in the yield process of mild steel', *Proc. of the 22nd Japan Congress on Materials Research*, 47–50 (1979).

6. T. Kawai, 'Some considerations on the finite element method', *International Journal for Numerical Methods in Engineering*, **16**, 81–120 (1980).
7. T. Kawai and N. Takeuchi, 'A discrete method of limit analysis with simplified elements', *Proc. of the ASCE International Conference on Computing in Civil Engineering*, New York, 1981.
8. Y. Yamada, N. Yoshimura, and T. Sakurai, 'Plastic stress–strain matrix and its application for the solution of elastic plastic problems by the finite element method', *International Journal of Mechanical Science*, **10**, 343–354 (1968).
9. F. Nakanishi and Y. Sato, 'On the stress–strain relation in plastic range', *Proc. of the 4th Japan National Congress for Applied Mechanics*, 69–74 (1954).
10. F. Nakanishi and Y. Sato, 'Strength of surface layers of beams under bending', *Proc. of the 5th Japan National Congress for Applied Mechanics*, 169–173 (1955).

Mechanics of Engineering Materials
Edited by C. S. Desai and R. H. Gallagher

Chapter 19

Viscoplasticity Based on Overstress. Experiment and Theory

E. Krempl

19.1 INTRODUCTION

Advances in electronics have introduced new capabilities in the mechanical testing of solids. Continuous strain measurement on the uniform section of a test specimen representing a macroscopically homogeneous state of deformation, a servocontrolled testing machine, and a computer to generate suitable command signals and to digitize the measured response are now available to impose a controlled displacement (strain control) or load boundary condition (load or stress control). The feedback loop then assures that the programmed conditions are actually experienced by the specimen within the small inherent error which the servosystem needs for its functioning.

The capability of imposing almost any time variation of the stress or the displacement boundary condition has not existed prior to the advent of servocontrolled testing. In hydraulically operated or screw-driven machines the speed of the crosshead can be controlled but not the strain seen by the specimen. This type of testing results in undefined boundary conditions between the extremes of strain (stiff testing machine frame) and stress control (compliant frame) without being able to reach either extreme. A deadweight loading apparatus such as that used in many yield surface investigations can impose exact constant load boundary conditions but the increase in loading is rather uncontrolled as it depends on the method of adding weights.

The capabilities of a servocontrolled testing system can give new insight into the mechanical behaviour of solids due to its capability of enforcing load or displacement boundary conditions and their variation in time. This chapter will show some examples of test results on engineering alloys obtained with a servocontrolled system. They demonstrate that some commonly held notions about the mechanical behaviour of engineering alloys are artefacts of previous testing methods and are not confirmed by experiments using defined boundary conditions. They also demonstrate that rate(time)-dependence is not negligible with these alloys at room temperature. Because of these observations an overstress theory of viscoplasticity is introduced. It considers loading-rate sensitivity, creep, and relaxation as different but equally important manifestations of rate(time)-dependence. Further, no yield surface and no separate loading and unloading conditions are employed. The theory is implemented into a finite element programme and a forward gradient scheme is used for the time integration of the resulting stiff differential equations. The viscoplastic analysis of small-scale yielding in Type 304 stainless steel at room temperature under Mode 1 loading shows significant time-dependent effects on stress distribution and crack opening displacement.

19.2 EXPERIMENTAL RESULTS

Unless otherwise noted all experiments are performed at room temperature using an MTS tension–torsion servocontrolled test system with a digital function generator or an MTS 463 data control processor with strain measured on the gauge of uniaxial specimens. Details of the experimental set-up, the materials and their heat treatment can be found in [1 to 4].

Before discussing results it is instructive to consider the theoretical restrictions imposed by the boundary conditions on the slope of the load-elongation (stress–strain†) diagram shown in Figure 19.1. It is seen that load control does not permit a load decrease during loading. This is possible during displacement (strain) control. During unloading strain can increase in load control but must decrease in strain control.

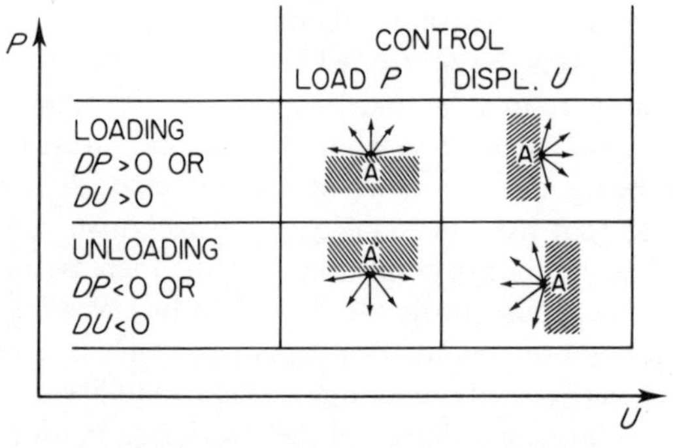

Figure 19.1 Possible slopes (indicated by arrows) of the stress–strain diagram as affected by the type of control during loading and unloading. Point A is a typical point in the first quadrant of the stress–strain space

† Engineering stress–strain or true stress–strain

19.2.1 Stress–strain diagram

The stress–strain diagram depicted schematically in Figure 19.2 can be found in almost all introductory textbooks on mechanics of solids or mechanical metallurgy, e.g. [5, 6]. The yield drop at A is mentioned to be a special property of carbon and other low alloy steels which is absent in other metals. However, the load maximum and the subsequent drop to fracture are universally reported for ductile materials when load (engineering stress) versus displacement (engineering strain) diagrams are depicted.

A comparison of Figure 19.2 with the conditions given in Figure 19.1 demonstrates that Figure 19.2 was not obtained in load control; the yield drop at A and the decrease after the maximum load are not possible for such a method of testing. It follows then that the yield drop and the decrease after maximum load are peculiar to tests with other than load boundary conditions.

Figure 19.3 shows the effect of control on the load–elongation diagram of identical specimens of SAE 1020 steel. Displacement control of the cross-head and load control were used, respectively. The yield point and the stress drop after the maximum load are absent in load control. (The symbols indicate the time intervals at which data were read.)

Figure 19.3 clearly demonstrates the importance of the boundary conditions on the appearance of the stress–strain diagram. Moreover, since a material property should be independent of the test condition, the yield drop and the decrease of load after the maximum load cannot be considered material properties.

Although the differences between load and displacement control are obvious once mentioned and demonstrated, it is interesting to note that general continuum mechanics theories have been developed which aim to model yielding [7 to 9] without considerations of boundary conditions. These attempts were criticized in [10] but not on the grounds that the yield drop is incidental and

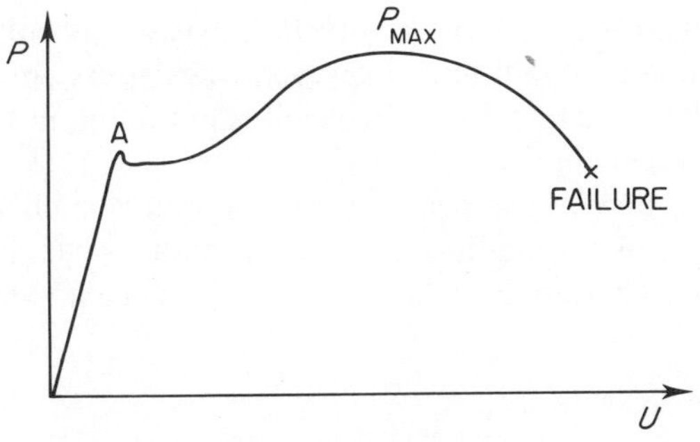

Figure 19.2 Schematic depicting an engineering stress versus engineering strain diagram with yield point and stress drop after necking

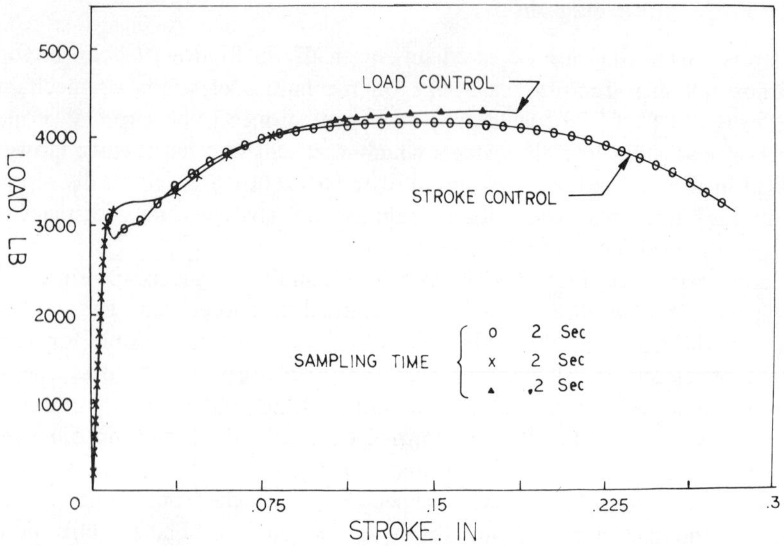

Figure 19.3 Load–elongation (stroke) diagram of SAE 1020 steel at room temperature under load and displacement (stroke) control. Note the absence of the yield point and of the descending final portion in load control. The two specimens have identical geometry

associated with displacement boundary conditions only. Also, microstructural theories aimed at explaining the yield point [11] do not mention the influence of the boundary condition at all.

As a historical note it should be mentioned that Nadai did recognize the importance of the boundary condition and demonstrated the absence of a yield drop in stress control by constructing a special testing apparatus capable of enforcing stress boundary conditions [12].

The condition $dP = 0$ is used as an indicator of the onset of necking (instability) in many analyses. It is seen from Figure 19.3 that this is only appropriate for displacement boundary conditions. However, when stress boundary conditions are necessary, $dP = 0$ cannot be used since it is not found in the corresponding basic uniaxial experiment.

The descending part of the stress–strain diagram is similarly an artefact of displacement boundary conditions which cannot occur in stress control. Constitutive equations should be able to model this behaviour.

19.2.2 Rate(time)-dependence

Plasticity theory is formulated to reproduce rate(time)-independent behaviour. It represents the common notion in mechanics that metal deformation proceeds in a time-independent fashion at room temperature. This notion is not shared in materials science where plastic deformation is viewed as a rate process. Through servocontrolled experiments the validity of the materials science viewpoint was established for AISI Type 304 stainless steel and a Ti-alloy [1 to 4]. Further experiments on A-533B pressure vessel steel, HY 80 steel, HY 130 steel, and other engineering alloys confirmed the fact that inelastic deformation is rate(time)-dependent.

In the almost linear region of the stress–strain diagram, rate(time)-independence is a good approximation for metals. However, inelastic deformation was found to be intrinsically rate(time)-dependent. The statement, 'the occurrence of plastic strains is as a rule accompanied by creep' [33], was found to be true in all the tests conducted so far.

19.2.3 Ductile fracture is time dependent

Due to the significant time dependence of plastic flow, ductile fracture was found to proceed in a time-dependent manner. In Figures 19.4(a) and 19.4(b) the creep behaviour around the ultimate load of an A533B pressure vessel steel is depicted.

Prior to the test shown in Figure 19.4(a), short-term creep tests similar to the ones depicted in Figure 19.6 were performed and a strongly decreasing creep rate, i.e. primary creep, was observed. It is prevalent at points a and b in Figures 19.4(a) and 19.4(b). Upon a slight increase in the load secondary creep is observed at c which turns into tertiary creep at d and e. (The unloading is incidental. It was necessary because of exhaustion of the extensometer travel.) The neck was observed to form at d and continued to develop in an accelerated fashion.

Neck formation and continuing development at constant load was also found in AISI Type 304 stainless steel, HY 80 steel, HY 130 steel, and the Ti-alloy. It is therefore concluded that ductile fracture is an inherently time-dependent process and should be treated as such in theoretical analyses.

19.2.4 A linear stress–strain diagram does not always imply linear elastic behaviour

The presence of significant rate(time)-dependence of the deformation implies that a nearly linear appearance of a portion of the stress–strain diagram does not necessarily indicate linear elastic behaviour. The linear appearance can be caused by viscosity if the loading rate is sufficiently high. Figure 4 of [3] demonstrates such a case for the Ti-alloy. The initial loading portion appears nearly linear at the stress rate $\dot{\sigma}_1$. When the stress rate is reduced by several orders of magnitude ($\dot{\sigma}_5$), a strong curvature is evident. Similar observations can be made when Figures 3, 5, and 7 of [3] are examined.

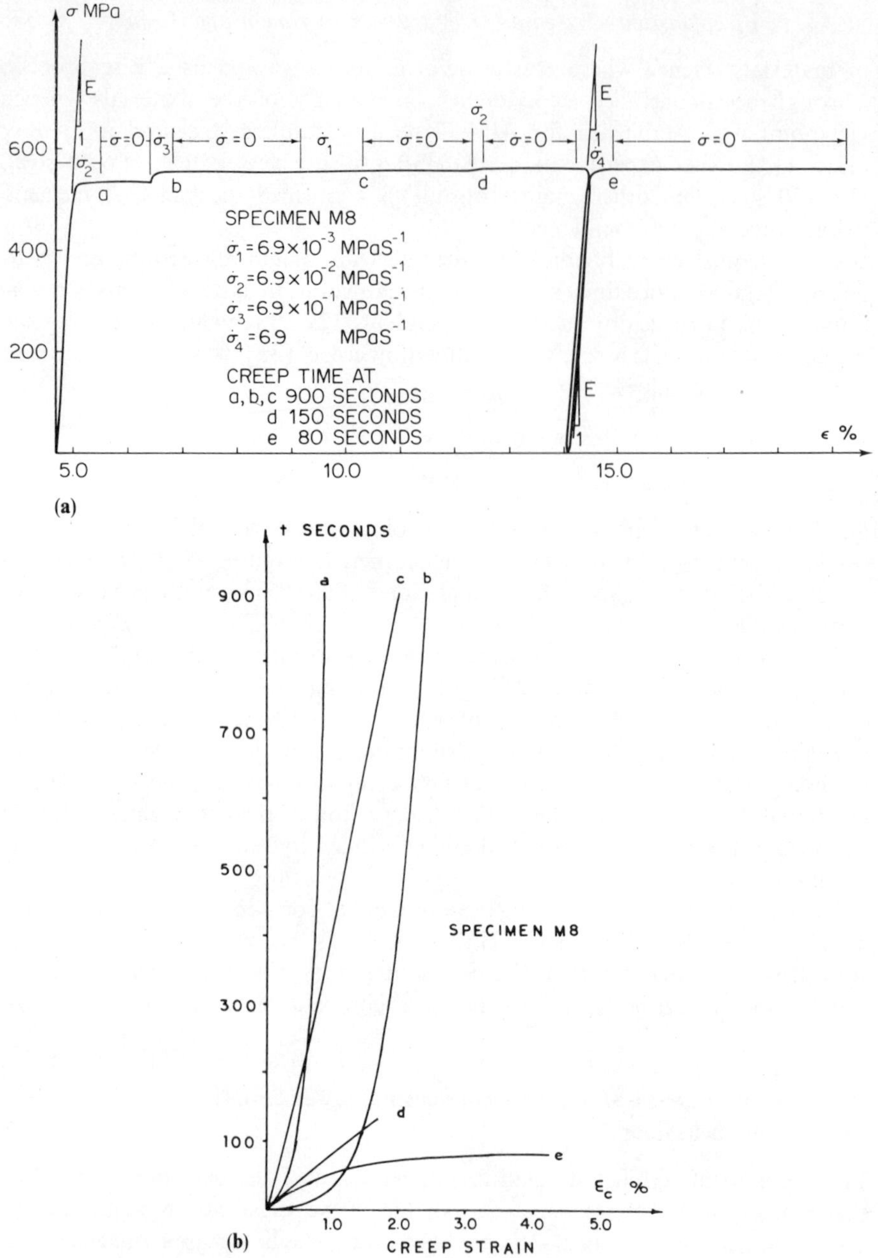

Figure 19.4 (a) Engineering stress versus engineering strain diagram for A533B Class I pressure vessel steel in load control with creep periods as indicated. Prior history: repeated loadings and unloadings involving positive stresses only. (b) Creep curves associated with the tests of (a). Note the occurrence of tertiary creep at d and e which was accompanied by necking

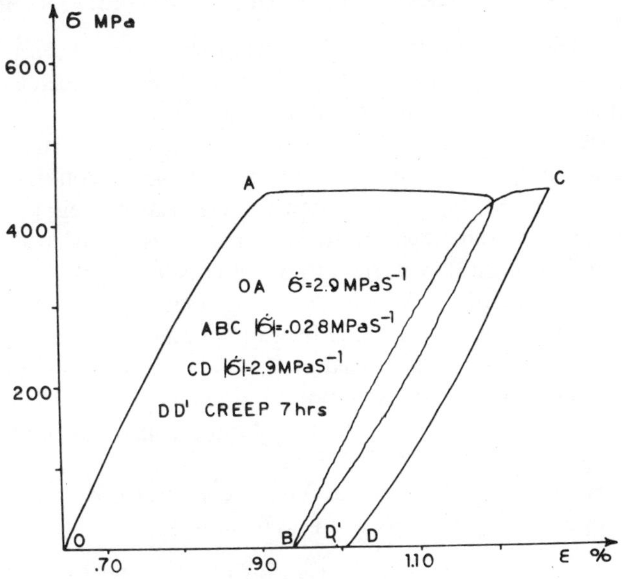

Figure 19.5 Behaviour of cold-worked AISI Type 304 stainless steel under stress control with changes in stress rate at points of unloading

When at the end of the relaxation period in Figure 5 of [3] loading resumes, a nearly elastic slope is observed. If either $\dot{\varepsilon} = 0$ or $\dot{\sigma} = 0$ were to be enforced in these linear regions, relaxation or creep would be measured.

Unloading behaviour is also very strongly influenced by rate as shown in Figure 19.5. At point C where the magnitude of the stress rate is increased a hundredfold, nearly linear unloading is observed initially. However, when the stress-rate magnitude is decreased by a factor of 100 at A, a negative slope ensues.

The above examples suffice to demonstrate that not all linear regions of a stress–strain diagram are indicative of linear elastic behaviour. When viscous deformation is present, a fast loading or unloading can cause the appearance of linear elastic behaviour.

These examples obtained from simple uniaxial tests show convincingly that servocontrolled testing can give new insight into material behaviour. These results also give rise to new ideas how to represent material behaviour in constitutive equations. In the following a new constitutive theory, the viscoplasticity theory based on overstress is introduced which considers rate(time)-dependence fundamental and which can show distinctively different behaviour in stress and in strain control as depicted in Figure 19.3.

19.3 VISCOPLASTICITY BASED ON OVERSTRESS

Overstress models in conjunction with a yield surface have been proposed for static applications (for a summary, see [13]). They have also been used for wave propagation studies [14, 15] and stress analysis.

The overstress model to be discussed shortly does not use the yield surface concept and does not need a separate loading and unloading condition. Although it shares the overstress dependence with the classical models [13 to 15], its origin lies in the fact that rate sensitivity under stress- and strain-controlled loading, creep and relaxation were considered equally important but different manifestations of rate(time)-dependence.† On the basis of the asymptotic behaviour in time of the non-linear differential equation in comparison with experiment, overstress was introduced in [16]. This overstress dependence has proven to represent creep and relaxation in a unified way and has also helped to explain phenomena which were recently found with servocontrolled testing [1 to 4].

The uniaxial form of the viscoplasticity theory based on total strain and overstress proposed in [16] in its simplest form is

$$\dot{\varepsilon} = \dot{\sigma}/E + \frac{\sigma - g[\varepsilon]}{Ek[\sigma - g[\varepsilon]]} \tag{19.1}$$

where a superposed dot represents time differentiation and square brackets following a symbol denote 'function of'. The axial component of the stress and small strain tensor are σ and ε, respectively; the positive function $k[x]$ with $dk/dx < 0$ is called the viscosity function with the dimension of time. The odd function $g[\varepsilon]$ represents the equilibrium stress–strain curve which is obtained as a solution of equation (19.1) in the limit as the loading (stress or strain) rate approaches zero. Experiments have shown that extrapolations are necessary to obtain the equilibrium curve which is significantly below the curve for a strain rate of $10^{-8}\,s^{-1}$ [3, 4, and 17].

There may be doubts whether such extrapolations are valid; however, indirect evidence will be presented which strongly suggests that the inelastic strain rate must depend in an overstress like quantity.

The theory represented by equation (19.1) might be called a deformation theory of viscoplastcity without a yield surface. It is only considered to be a valid representation of metal deformation behaviour as long as the overstress does not change sign. If this happens, a different formulation of g must be adopted which is under development.

Equation (19.1) can be rewritten as

$$\dot{\varepsilon} = \dot{\varepsilon}_{el} + \dot{\varepsilon}_{in} \tag{19.2}$$

†Classical studies [13 to 15] emphasize only loading rate dependence.

where the inelastic strain rate depends on σ and ε only through the overstress.†

For loading with constant stress or strain rate, (19.1) admits an asymptotic solution in time which is rapidly attained [4, 16]. This solution shows that the overstress is constant and that all stress–strain curves have ultimately the same slope.‡ Physically, this asymptotic solution corresponds to the attainment of fully established 'plastic flow', the flow stress.

19.3.1 Qualitative uniaxial properties of the model

Equation (19.1) was shown to exhibit the following qualitative properties observed in servocontrolled tests at room temperature an AISI Type 304 stainless steel and a Ti-alloy [1 to 4].

(1) Stress–strain curves obtained at two different strain rates differ ultimately by a constant stress and are nonlinerarly spaced. A tenfold increase of the strain rate causes a flow stress-increase which is much less than tenfold. (Flow stress is the measured stress in the plastic range obtained at constant loading rate after transient effects have died out. 'Loading rate' is used for stress and strain control.)
(2) An instantaneous large change (more than one order of magnitude) in strain rate results in an instantaneous 'elastic slope' of the stress–strain curve at every point in the plastic range.
(3) For a variable loading history involving only positive stresses the flow stress characteristic of a given strain rate is, after a transient period, always reached again. Under these loading histories the two materials forget their prior history. They do not exhibit a strain-rate history effect.§
(4) In the region where the flow stress is reached the average relaxation rate depends on the strain rate preceding the relaxation test. It increases with an increase in prior strain rate.
(5) If relaxation is started after the flow stress characteristic of a particular strain rate is reached, the amount of relaxation in a given period of time (and therefore the average relaxation rate) is independent of the stress and strain at the start of the relaxation test but depends only on the strain rate preceding the relaxation test.
(6) At a given stress level the initial creep rate increases with an increase in the prior loading rate.
(7) It is possible that two creep tests show the same initial creep rate although their stress levels are different.

† In materials science a term comparable to $\sigma - g$ is called the 'effective stress' where g is interpreted as 'internal stress'. This interpretation is difficult to reconcile with mechanics principles. A discussion of this subject is given in [4].

‡ There are some subtle differences between stress and strain control which are discussed in [4].

§ Again there are subtle differences between stress and strain control, see [18] for details.

(8) Creep rate does not necessarily increase with an increase in stress level. Even if a constant loading rate is used to change stress levels, the creep rate at a high stress level may be less than the one obtained at a low stress level.

(9) At a given stress level the creep rate following loading is higher than the creep rate at the same stress level following loading from this stress level and subsequent unloading.

The capability of the model of reproducing these features has been demonstrated in [4, 17] and it is hard to imagine how properties (5), (8), and (9) could be reproduced without an overstress dependence of inelastic strain rate.

Figure 19.6 illustrates property (8) for A533B Class 1 pressure vessel steel Although the stress level at c(e) is higher than at b(d) the accumulated creep strain in 900 s and therefore the average creep rate is higher at the low stress level than at the high stress level. The explanation of this phenomenon is given in [4]. Indeed, the tests such as shown in Figure 19.6 were conceived to confirm the overstress dependence after the properties of the model were evaluated theoretically.

Figure 19.6 also demonstrates that insignificant time dependence is observed in the elastic range (points a_0 and a_1). Plastic flow, however, is significantly time dependent for this ferritic steel.

Numerical experiments have shown no problem in strain control when the equilibrium stress–strain curve reaches a maximum and subsequently decreases.

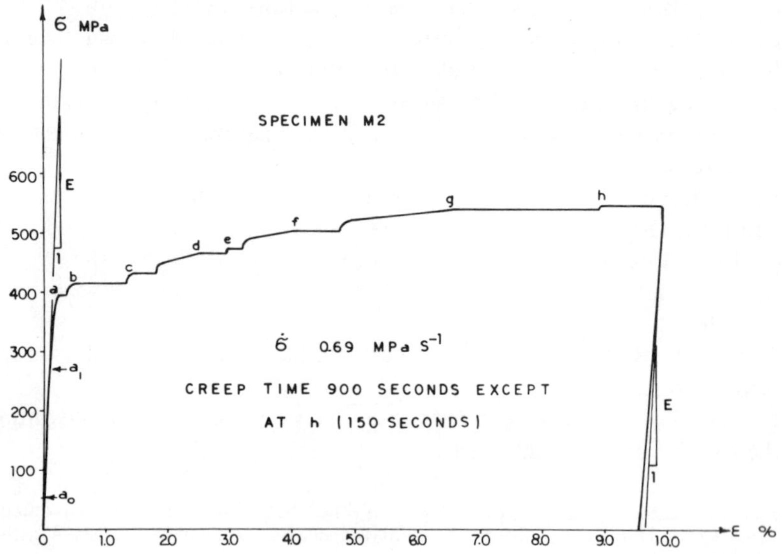

Figure 19.6 Intermittent creep tests for A533B Class 1 pressure vessel steel. Although the stress level at c (e) is higher than at b (d) the creep rate is less at c (e) than at b (d)

The asymptotic solution applies as well in this case and the stress–strain curve exhibits a negative slope. However, when the same equilibrium stress–strain curve is used together with stress control the solution becomes unstable right after the maximum of g is reached. It appears that (1) has the potential of reproducing the differences between stress and displacement control depicted in Figure 19.3.

19.3.2 Three-dimensional formulation

The theory outlined in (1) has been formulated in tensorial form for constant Poisson's ratio [19]. It is considered to be a valid representation for metals as long as no component of the overstress changes sign. Since the equilibrium curve is made to depend on strain, the present theory could be considered a deformation theory of viscoplasticity without a yield surface.

In the development of the time-independent theory of plasticity a test discriminating between deformation, flow, and physical theories was devised. It was found that a tube prestressed in tension showed always initial elastic slope when subsequently subjected to torque. Flow theory can reproduce this behaviour and it was therefore considered the appropriate theory.

The viscoplasticity theory of [19] was shown to reproduce this same behaviour [19, Appendix I] for any axial preload. It therefore exhibits a key property of metal deformation. The key repository for the modelling of this behaviour is the rate dependence which was found in metals. The demonstration given in [19] does not use an argument in which certain terms related to rate dependence can be neglected when they are small. Any amount of rate dependence no matter how small will result in the subsequent elatic slope.

The proposed constitutive equation represents a stiff differential equation which is difficult to integrate numerically. The stiffness is shared by other viscoplastic models [20–23] and is caused by the nature of viscoplastic deformation. A forward gradient scheme and an implicit operator are used for the time integration of a finite element (eight-noded quadrilateral elements with four Gauss points) formulation of the theory [24]. It is shown that the chosen method leads to stable solutions.

A special feature of the theory is the existence of an equilibrium solution which is attained under constant boundary conditions after a long time (mathematically infinite time). For a cylinder under internal pressure and an elastic, perfectly plastic, equilibrium stress–strain diagram, the equilibrium solution is shown to correspond to the elastic, perfectly plastic solution [24].

A literature survey [25] and our own experiments [26] have shown that inelastic deformation is not completely isochoric in the range of small strain. Accordingly a modification of the theory was made to account for plastic compressibility effects [27].

Often the assumption of inelastic incompressibility is justified by the Bridgman experiments [28] in which it was demonstrated that inelastic deformation was not induced in metals by pure hydrostatic pressure. If inelastic flow is then assumed to be independent of hydrostatic pressure, inelastic incompressibility follows. The flaw in such an argument is that the Bridgman results require only that inelastic deformation not be induced by pure hydrostatic pressure; they do not require that inelastic deformation be independent of hydrostatic pressure for stress states other than pure hydrostatic. It is possible to formulae a theory of inelastic flow such that the Bridgman results are reproduced, but inelastic flow is in general influenced by a superimposed hydrostatic pressure. Such is the case with the theory introduced in [19].

The three-dimensional theory corresponding to (19.1) was given in [19]. Using conventional tensor notation with summation over repeated indices implied, it is

$$\dot{\varepsilon}_{ij} = C^{\text{el}}_{ijmn}\left\{\dot{\sigma}_{mn} + \frac{\sigma_{mn} - G_{mn}[\varepsilon_{rs}]}{k[\Gamma]}\right\} \tag{19.3}$$

where ε_{ij} is the infinitesimal strain tensor, σ_{kl} is the engineering stress tensor, G_{kl} is the equilibrium stress tensor, and Γ is the overstress invariant given by

$$\Gamma = \{(\sigma_{ij} - G_{ij})(\sigma_{ij} - G_{ij})\}^{1/2} \tag{19.4}$$

and

$$C^{\text{el}}_{ijmn} = \frac{(1+\nu)}{E}\left\{\tfrac{1}{2}(\delta_{im}\delta_{jn} + \delta_{in}\delta_{jm}) - \frac{\nu}{1+\nu}\delta_{ij}\delta_{mn}\right\} \tag{19.5}$$

where ν is Poisson's ratio.

In [19] the equilibrium stress tensor was given by a particular choice. The theory is not restricted to this choice; and suitable mathematical form of **G** can be used.

The right-hand side of (19.3) is the sum of the elastic and inelastic strain-rate tensors. It can be modified to include a broader range of compressible and incompressible deformation by writing

$$\dot{\varepsilon}_{ij} = C^{\text{el}}_{ijmn}\dot{\sigma}_{mn} + C^{\text{in}}_{ijmn}\left\{\frac{\sigma_{mn} - G_{mn}[\varepsilon_{rs}]}{k[\Gamma]}\right\} \tag{19.6}$$

where C^{in}_{ijmn} is of the form (19.5) with ν replaced by α which is a compressibility factor such that $\nu \le \alpha \le 0.5$. With $\alpha = 0.5$, inelastic incompressibility is recovered.

G_{mn} can be formulated in a manner similar to commonly used power law formulations of time-independent plasticity. It is perhaps most easily recognized in its inverted form

$$\varepsilon_{ij} = \frac{(1+\nu)}{E}G'_{ij} + \frac{(1-2\nu)}{3E}G_{mm}\delta_{ij} + \beta\left\{\frac{G_e}{g_0}\right\}^{n-1}\left\{\frac{(1+\alpha)}{E}G'_{ij} + \frac{(1-2)}{3E}G_{mm}\delta_{ij}\right\} \tag{19.7}$$

where a deviatoric part is denoted by a prime.

$$G_e = \{\tfrac{3}{2}G'_{ij}G'_{ij}\}^{1/2} \tag{19.8}$$

and β, g_0, and n are constants in a uniaxial Ramberg–Osgood relation of the form

$$\varepsilon = \frac{g}{E} + \beta\frac{g}{E}\left(\frac{g}{g_0}\right)^{n-1} \tag{19.9}$$

The form of (19.7) may be recognized as that of power law plasticity as in [29] but with an extra term added to account for the influence of hydrostatic pressure on inelastic deformation. Such a choice of G_{kl} is attractive for several reasons. First, compressibility effects are added in a relatively straightforward manner. Second, elastic and inelastic parts are separate, and third, by setting $\alpha = 0.5$ the equilibrium constitutive equations are identical to those in [29] and consequently direct comparisons between equilibrium state solutions and existing power law plasticity results can be made.

A slight drawback of the formulation (19.6) is that it cannot be reduced to (19.1) exactly. When simulating a uniaxial state of stress, transverse components of the equilibrium stress develop at the transition from elastic to inelastic deformation which decay rapidly, for details see [30].

19.3.3 Application to fracture mechanics

The theory given in (19.3)–(19.9) was combined with the finite element formulation of [24] to study the small scale yielding behaviour using material property data of AISI Type 304 stainless steel [27].

A stationary crack under conditions of plane strain, Mode I loading was modelled using the eight-noded isoparametric elements with straight sides arranged in eleven circumferential layers of eight elements each. Strain singularities, with strength $1/r$, were produced following the method of [31, 32].

Following the hypothesis of small scale yielding, the elastic stress intensity factor K was prescribed on the boundary to increase with a constant rate $\dot{K}$ until K_{max} was achieved which was subsequently held constant. K_{max} was selected so that the maximum extent of the plastic zone size did not exceed 26 per cent of the radius of the finite element mesh.

A loading time of 33 seconds up to K_{max} was taken as the reference loading rate $\dot{K}_0$. Computations were made with $10^2\dot{K}_0$, $10^{-2}\dot{K}_0$, and $10^{-4}\dot{K}_0$.

It is shown that the stress level at the crack tip increases with loading rate increase and that a significant stress redistribution occurs during the hold periods at K_{max}. After one year the equilibrium stress is reached for practical purposes. At the highest loading rate and at the end of the loading ramp, the crack opening stress at the Gauss point closest to the crack tip is 1.55 its equilibrium value. After one day it is reduced to about 1.10 of its equilibrium value which is finally reached after more than one year.

A loading rate increase inhibits the development of crack opening displacement (COD). At the fastest loading rate at the end of the loading ramp, 65 per cent of the equilibrium COD is reached. (The value for the slowest loading rate is approximately 90 per cent.) During constant K_{max} COD increases and reaches 95 per cent of its equilibrium value after one day.

These examples may suffice to demonstrate the significant influence of rate(time)-dependent behaviour on the stress and displacement fields in cracked bodies. These effects may be responsible for the sometimes observed decrease of the critical stress intensity factor with increase in loading rate.

19.4 CONCLUSION

These few examples of the behaviour of engineering alloys obtained with servocontrolled testing should suffice to substantiate the claim that this type of testing will improve the understanding of the behaviour of solids. The results shown herein pertain to a simple uniaxial state of stress. Biaxial testing on thin-walled tubes is in progress and will give further results on the time(rate) and history dependent stress–strain behaviour of engineering alloys.

Servocontrolled testing in support of constitutive equation development is just starting and availability of results obtained with this method will increase in the future. It is expected that these results will give further impetus for the constitutive equation development. At the moment an initial simple version of the viscoplasticity theory based on overstress was presented which models rate sensitivity, creep, and relaxation in a unified manner. The present version, which is not applicable for cyclic loading, uses only one first-order non-linear differential equation for each tensor component to represent the viscoplastic behaviour. No auxiliary conditions are necessary. For cyclic loading a growth law for the equilibrium stress–strain curve will most probably be added.

The mechanical theory presented herein was extended to the case of coupled thermal problems such as deformation induced temperature changes and thermal cycling under constraint [34 to 36]. The qualitative and numerical solutions are encouraging. Specifically the theory captures the cyclic heating and cooling patterns found in a low-cycle fatigue test [37].

The fracture mechanics example discussed briefly shows that the implementation of the viscoplasticity theory based on overstress is feasible and can give very instructive results in explaining unsolved fracture phenomena. Since we found ductile fracture strongly time dependent in all our tests, analyses of instabilities should be investigated with a time(rate)-dependent theory.

ACKNOWLEDGEMENT

The financial support of the National Science Foundation and of the Electric Power Research Institutive is gratefully acknowledged. Messrs V. V. Kallianpur and H. Lu performed the tests.

REFERENCES

1. E. Krempl, *J. Mechanics and Physics of Solids*, **27**, 363–375 (1979).
2. D. Kujawski, V. Kallianpur, and E. Krempl, *J. Mechanics and Physics and Solids*, **28**, 129–148 (1980).
3. D. Kujawski and E. Krempl, *J. Applied Mechanics*, **48**, 55–63 (1981).
4. E. Krempl, *American Society for Testing and Materials*, *STP* 765, March 1982, 5–28.
5. E. P. Popov, *Introduction to the Mechanics of Solids*, Prentice-Hall, 1968.
6. G. E. Dieter, *Mechanical Metallurgy*, Second Edition, McGraw-Hill, 1976.
7. C. Truesdell, 'Second Order Effects in the Mechanics of Materials', *Proc. IUTAM Symp.*, specifically p. 17, Haifa, Israel, Pergamon Press, 1964.
8. A. E. Green, *J. Rational Mech. Anal.*, **5**, 725–734 (1956).
9. R. Hill, *J. Mechanics and Physics of Solids*, **7**, 209–225, footnote p. 223 (1959).
10. A. M. Freudenthal and H. Geiringer, *Hanbuch der Physik*, **VI**, specifically p. 262, Springer-Verlag, 1958.
11. A. H. Cottrell, *Dislocation and Plastic Flow in Crystals*, Oxford, 1953.
12. A. Nadai, *Theory of Flow and Fracture of Solids*, specifically p. 309, McGraw-Hill, 1950.
13. P. Perzyna, *Advances in Applied Mechanics* (Ed. C. S. Yih), **11**, 313–354. Academic Press, 1971.
14. V. V. Sokolovskii, *Doklady Akademia Nauk SSR*, **60**, 775–778 (1948).
15. L. E. Malvern, *J. Applied Mechanics*, **18**, 203–208 (1951).
16. E. P. Cernocky and E. Krempl, *Intl. J. of Non-Linear Mechanics*, **14**, 183–203 (1979).
17. M. C. M. Liu and E. Krempl, *J. Mechanics and Physics of Solids*, **27**, 377–391 (1979).
18. E. P. Cernocky, *Intl. J. of Non-Linear Mechanics*, **17**, 255–266 (1982).
19. E. P. Cernocky and E. Krempl, *Acta Mechanica*, **36**, 263–289 (1980).
20. A. K. Miller, *J. Engineering Materials and Technology*, **98**, 97–113 (1976).
21. E. W. Hart, *J. Engineering Materials and Technology*, **98**, 97–113 (1976).
22. S. R. Bodner and Y. Partom, *J. Applied Mechanics*, **42**, 385–389 (1975).
23. R. W. Rohde and J. C. Swearengen, *J. Engineering Materials and Technology*, **102**, 207–214 (1980).
24. R. M. Zirin and E. Krempl, *J. Pressure Vessel Technology*, **104**, 130–136 (1982).
25. E. Krempl and P. Hewelt, *Mechanics Research Communications*, **7**, 282–288 (1980).
26. Y. Asada, M. H. Kargarnovin, and E. Krempl, 'Inelastic volume change and Poisson's ratio in uniaxial tension and compression tests of type 304 stainless steel at room temperature', *RPI Report PRI CS* 81–3, Nov. 1981.
27. M. M. Little, E. Krempl, and C. F. Shih, American Society for Testing and Materials, STP 803, November 1983, I-615–I-636.
28. P. W. Bridgman, *Proceedings, American Academy of Arts and Sciences*, **58**, p. 166, 1923.
29. H. de Lorenzi and C. F. Shih, 'Finite element implementation of the deformation theory of plasticity', *General Electric Company Report* 80*CRD*058, April 1980.
30. M. M. Little, *A Viscoplastic Analysis of the Influences of Time and Loading Rate on Cracked Structures*, Ph.D. dissertation, Rensselaer Polytechnic Institute, Troy, NY, May 1982.
31. R. D. Henshell and K. G. Shaw, *Int'l. for Numerical Methods in Engineering*, **9**, 495–507 (1975).
32. R. S. Barsoum, *Int'l. J. for Numerical Methods in Engineering*, **11**, 85–98 (1977).
33. Yu N. Rabotnov, *Elements of Hereditary Solid Mechanics*, specifically p. 250, Mir Publishers, 1980.
34. E. P. Cernocky and E. Krempl, *International Journal of Solids and Structures*, **16**, 723–741 (1980).

35. E. P. Cernocky and E. Krempl, *J. of Thermal Stresses*, **4**, 69–82 (1981).
36. E. P. Cernocky and E. Krempl, *J. de Mecanique Appliqueé*, **5**, 293–321 (1981).
37. S. L. Adams and E. Krempl, 'Thermomechanical response of 3.5 Ni-Mo-V alloy steel and type 304 stainless steel under cyclic uniaxial inelastic deformation', *Res. Mechanica*. to appear (1984).

Mechanics of Engineering Materials
Edited by C. S. Desai and R. H. Gallagher

Chapter 20

Failure Criterion for Frictional Materials

P. V. Lade

20.1 INTRODUCTION

Failure of engineering structures, whether simple or complex, usually involves three-dimensional stress conditions in the materials. Detailed analyses of such structures therefore require knowledge of the stress–strain behaviour and the states of stress which constitute failure. However, the use of inadequate material models, especially for frictional materials, is often one of the limiting factors in the analyses procedures. In addition, the behaviour and the stress distributions are likely to be different in intact and in fractured materials. To determine the stress conditions which govern fracture of intact materials, a general three-dimensional failure criterion is required. Such a criterion is also one of the important parts of a constitutive law for frictional materials.

Frictional materials are characterized by increasing shear strengths with increasing normal stresses. Many experimental investigations have been performed, especially in recent years, to study the failure conditions for frictional materials such as sands, clays, cemented soils, concrete, mortar, rock, etc. These laboratory investigations have been characterized by utilization of increasingly improved equipment, in which the boundary conditions are well known, thus leading to reliable results of good quality. Along with these studies, formulations of failure criteria have been proposed. Some of these are aimed at practical design applications, whereas other more complex expressions have been developed for use in advanced computer codes. Most of the failure criteria proposed for three-dimensional stress states involve relatively complex expressions for which more than three material parameters are required.

These criteria have been developed to capture the experimentally determined shape of the failure surfaces as observed in the principal stress space. Several studies have shown that the failure surfaces in the principal stress space are shaped as pointed bullets with cross-sections in octahedral planes which are triangular, monotonically curved surfaces with smoothly rounded corners.

In addition to the characteristic cross-sectional shape in the octahedral plane, the three-dimensional failure surfaces for frictional materials have three independent characteristics: (1) the opening angle of the failure surface, often described by the friction angle, (2) the curvature of the failure surface in planes containing the hydrostatic axis, i.e. curved meridians, and (3) the tensile strength (which is zero for materials without effective cohesion). At least three independent parameters are necessary for description of the three separate and distinct characteristics of the failure surface for frictional materials.

A general, three-dimensional failure criterion expressed in terms of stress invariants has been developed for frictional materials. It involves three independent material parameters, each relating to one of the three characteristics mentioned above. These parameters may be determined from any type of strength tests, including the simplest possible such as uniaxial compression and triaxial compression or biaxial tests.

The procedure for determination of the three material parameters is demonstrated and comparisons between failure criterion and experimental results are presented for different frictional materials. The values of the three parameters are summarized for the various materials, and their magnitudes are compared and discussed in light of the materials to which they belong.

20.2 FRICTIONAL MATERIALS WITHOUT EFFECTIVE COHESION

Many frictional materials such as gravels, sands, silts, clays, rockfill, mine tailings, coal, feed grain, etc. do not have effective cohesion. According to Mitchell [1] 'tests over large ranges of effective stress show that the actual effective stress failure envelope is curved,' and 'that cohesion is either zero or very small, even for heavily over-consolidated clays. Thus, a significant true cohesion, if defined as strength present at zero effective stress, does not exist in the absence of chemical bonding (cementation)'. Several studies of materials without effective cohesion under two- and three-dimensional stress conditions show that these materials have many characteristics in common.

20.2.1 Evaluation of classical failure criteria

Several existing constitutive models employ the Extended Tresca criterion or the Extended von Mises (or Drucker–Prager) criterion to describe the failure condition for frictional materials. The shapes of the corresponding failure surfaces in the principal stress space are conical and they have cross-sections of a regular hexagon (Tresca) or a circle (von Mises) with their centre lines coinciding with the hydrostatic axis. According to both these criteria the material has the same strength in compression and extension for a given magnitude of mean normal stress.

In an evaluation of possible failure criteria for sands, Bishop [2] pointed out that the two failure criteria described above are in principle unable to represent

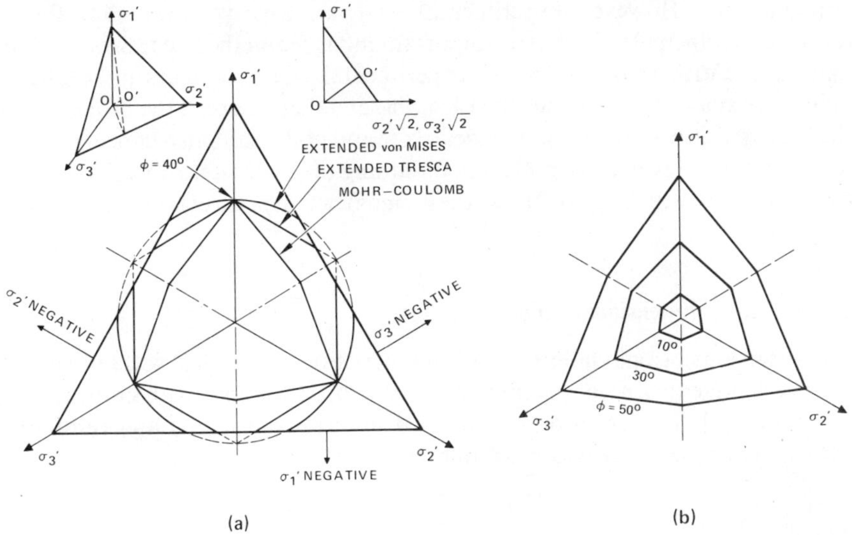

Figure 20.1 (a) Representation of classical failure criteria in principal stress space, showing boundaries of positive stress space (after Bishop [2]). (b) Cross-sections of the Mohr–Coulomb failure criterion shown for three different friction angles

the behaviour of cohesionless materials. Figure 20.1(a) indicates that the failure surfaces may extend outside the part of the principal stress space where all stresses are compressive and positive. The friction angle in triaxial compression for which both these failure surfaces are tangential to the coordinate planes is 36.9°. For higher friction angles the states of stress near triaxial extension are located in the parts of the stress space where one of the principal stresses is negative, and this is clearly unreasonable for materials without effective cohesion. Even for friction angles in triaxial compression smaller than 36.9°, the two criteria fail to model correctly the experimentally observed variation of the three-dimensional strengths of frictional materials.

The Mohr–Coulomb failure surface is also conical, and its cross-section is an irregular hexagon as indicated in Figure 20.1(a). This failure surface exhibits some of the characteristics necessary for correct modelling of failure of frictional materials. Thus, for materials without effective cohesion all principal stresses remain positive, even for very high friction angles. The shape of the cross-section of the Mohr–Coulomb failure surface resembles a regular hexagon for very small friction angles and it approaches an equilateral triangle for friction angles approaching 90°. The variation in shape for more conventional friction angles, as shown in Figure 20.1(b), is supported by experimental evidence.

The intermediate principal stress does not appear in the Mohr–Coulomb failure criterion, the failure surfaces are pointed in octahedral planes as indicated in Figure 20.1(b), and their traces in planes containing the hydrostatic axis are straight lines. However, experimental evidence clearly shows that the intermediate principal stress has an important influence on the strength of frictional materials. Furthermore, traces of experimental failure surfaces in octahedral planes are smooth throughout their lengths, and they intersect the projections of the principal stress axes at right angles. Experimental results also indicate that the failure surfaces are curved in planes containing the hydrostatic axis. The Mohr–Coulomb criterion does not model these significant aspects of failure in frictional materials.

20.2.2 Three-dimensional failure criterion

The three-dimensional failure criterion for frictional materials without effective cohesion presented here was previously developed for soils with curved failure envelopes [3]. This criterion is expressed in terms of the first and third stress invariants of the stress tensor as follows:

$$(I_1^3/I_3 - 27)\cdot(I_1/p_a)^m = \eta_1 \tag{20.1}$$

where

$$I_1 = \sigma_1 + \sigma_2 + \sigma_3 = \sigma_x + \sigma_y + \sigma_z \tag{20.2}$$

$$\begin{aligned} I_3 &= \sigma_1\cdot\sigma_2\cdot\sigma_3 \\ &= \sigma_x\cdot\sigma_y\cdot\sigma_z + \tau_{xy}\cdot\tau_{yz}\cdot\tau_{zx} + \tau_{yx}\cdot\tau_{zy}\cdot\tau_{xz} \\ &\quad - (\sigma_x\cdot\tau_{yz}\cdot\tau_{zy} + \sigma_y\cdot\tau_{zx}\cdot\tau_{xz} + \sigma_z\cdot\tau_{xy}\cdot\tau_{yx}) \end{aligned} \tag{20.3}$$

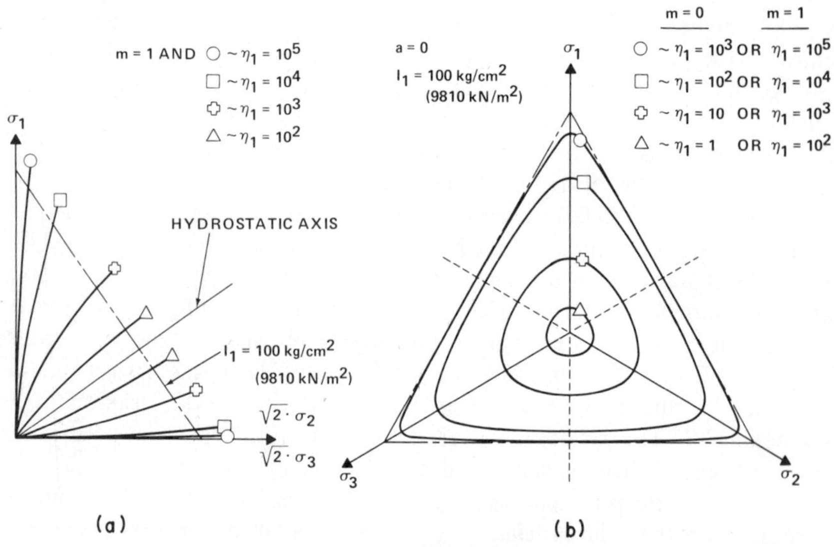

Figure 20.2 Characteristics of failure surfaces shown in principal stress space. Traces of failure surfaces shown in (a) triaxial plane, and in (b) octahedral plane

and p_a is atmospheric pressure expressed in the same units as the stresses. The value of I_1^3/I_3 is 27 at the hydrostatic axis where $\sigma_1 = \sigma_2 = \sigma_3$. The parameters η_1 and m in equation (20.1) can be determined by plotting $(I_1^3/I_3 - 27)$ versus (p_a/I_1) at failure in a log–log diagram and locate the best fitting straight line. The intercept of this line with $(p_a/I_1) = 1$ is the value of η_1 and m is the slope of the line.

In principal stress space the failure surface defined by equation (20.1) is shaped like an asymmetric bullet with the pointed apex at the origin of the stress axes as shown in Figure 20.2(a). The apex angle increases with the value of η_1. The failure surface is concave towards the hydrostatic axis, and its curvature increases with the value of m. For $m = 0$ the failure surface is straight. Figure 20.2(b) shows typical cross-sections in the octahedral plane ($I_1 =$ const.) for $m = 0$ and $\eta_1 = 1$, 10, 10^2, and 10^3. As the value of η_1 increases, the cross-sectional shape changes from circular to triangular with smoothly rounded edges in a fashion that conforms to experimental evidence. The shape of these cross-sections does not change with the value of I_1 when $m = 0$. For $m > 0$ the cross-sectional shape of the failure surface changes from triangular to become more circular with increasing value of I_1. Similar changes in cross-sectional shape are observed from experimental studies on frictional materials. The cross-sections in Figure 20.2(b) also correspond to $m = 1$ and $\eta_1 = 10^2$, 10^3, 10^4, and 10^5.

20.2.3 Comparison of failure criterion and test data

The failure criterion described above has been shown to model the experimentally determined three-dimensional strengths of sand and clay with good

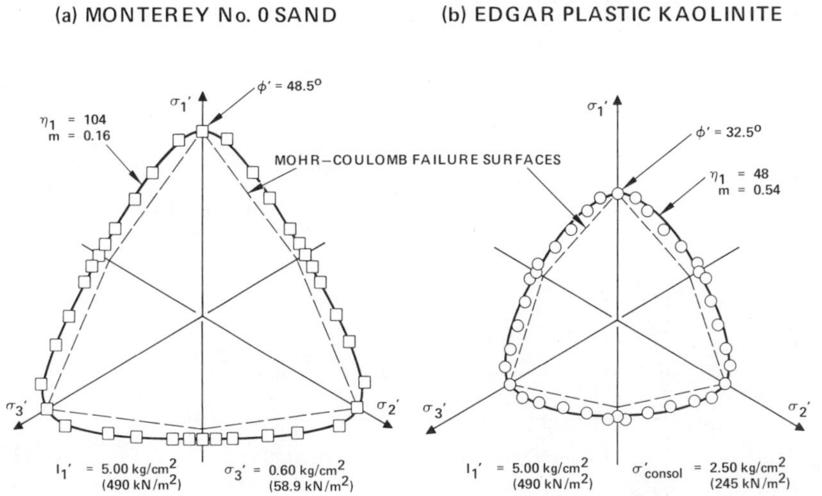

Figure 20.3 Comparison of failure criterion in octahedral planes with results of cubical triaxial tests on (a) dense Monterey No. 0 sand, and (b) normally consolidated, remoulded Edgar plastic kaolinite

accuracy in the range of stresses where the failure envelopes are concave towards the hydrostatic axis [3 to 5].

Figure 20.3 shows examples of comparisons between failure criterion and test data in terms of effective stresses for dense Monterey No. 0 sand and normally consolidated, remoulded Edgar plastic kaolinite. The values of η_1 and m suitable for description of failure in the two soils are given in Figure 20.3. The data points were projected on the common octahedral planes along curved meridians in order to provide a correct comparison between failure criterion and experimental data. It may be seen that the failure criterion models the experimentally obtained three-dimensional failure surfaces with good accuracy for both sand and clay.

20.3 FRICTIONAL MATERIALS WITH EFFECTIVE COHESION

Examples of frictional materials with effective cohesion include cemented soils (artificial and natural), frozen soils, ice, concrete, mortar, rocks, ceramics, graphite, plaster of Paris, and hydrostone. These materials exhibit tensile strengths and their shear strengths increase with increasing normal stresses.

20.3.1 General three-dimensional failure criterion for frictional materials

Because frictional materials with effective cohesion have many characteristics in common with those without effective cohesion, it has been proposed [6] that

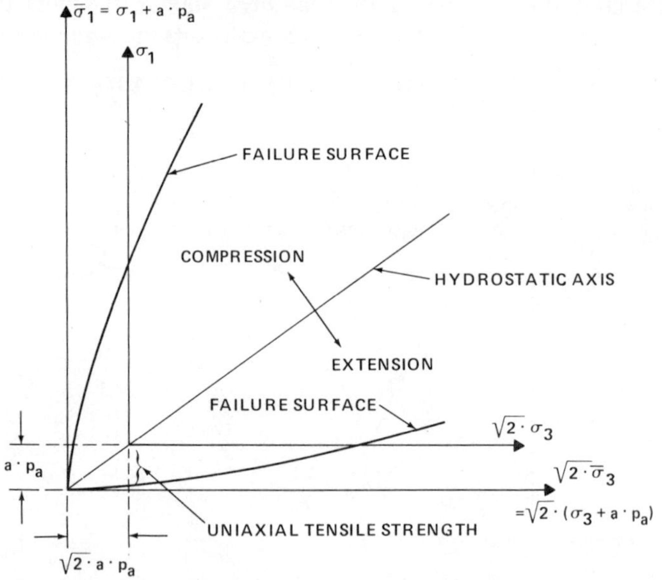

Figure 20.4 Translation of principal stress space along hydrostatic axis to include effect of tensile strength in failure criterion

their strengths can be expressed by a criterion similar to that in equation (20.1). In order to include the cohesion and the tensile strength in the failure criterion, a translation of the principal stress space along the hydrostatic axis is performed as illustrated in Figure 20.4. Thus, a constant stress $a \cdot p_a$ is added to the normal stresses before substitution in equation (20.1):

$$\bar{\sigma}_x = \sigma_x + a \cdot p_a \qquad (20.4a)$$

$$\bar{\sigma}_y = \sigma_y + a \cdot p_a \qquad (20.4b)$$

$$\bar{\sigma}_z = \sigma_z + a \cdot p_a \qquad (20.4c)$$

where 'a' is a dimensionless parameter and p_a is atmospheric pressure in the same units as σ_x, σ_y, and σ_z. The value of $a \cdot p_a$ reflects the effect of the tensile strength of the material. Although the three material parameters describe separate characteristics of the failure surface, they do interact in calculation of, for example, the uniaxial compressive strength of the material. Thus, an infinite number of combinations of 'a', η_1, and m could result in the same value of the uniaxial compressive strength.

20.3.2 Determination of material parameters

In order to determine the values of the three material parameters for a given set of experimental data, the value of 'a' is estimated and $a \cdot p_a$ is added to the normal stresses before substitution in equation (20.1). The procedure for finding η_1 and m as described above is then followed. To facilitate the estimate of 'a', advantage may be taken of the fact that $a \cdot p_a$ must be slightly greater than the uniaxial tensile strength as indicated on Figure 20.4. If tensile tests are not part of a regular testing programme, a sufficiently accurate value of the uniaxial tensile strength may be obtained from the following approximation formula.

20.3.2.1 *Uniaxial tensile strength*

The uniaxial tensile strength, σ_t, has often been expressed as a fraction of the uniaxial compressive strength, σ_c. For example, Mitchell [7] indicates that σ_t (actually given as the flexural strength which may be higher than the true value of

Table 20.1 Values of parameters T and t for various types of frictional materials

Material	T	t	Reference
Cemented soils	−0.37	0.88	Mitchell [7]
Concrete and mortar	−0.61	0.67	Wastiels [9]
Igneous rocks	−0.53	0.70	This chapter
Metamorphic rocks	−0.00082	1.6	This chapter
Sedimentary rocks	−0.22	0.75	This chapter
Ceramics	−1.0	0.73	This chapter

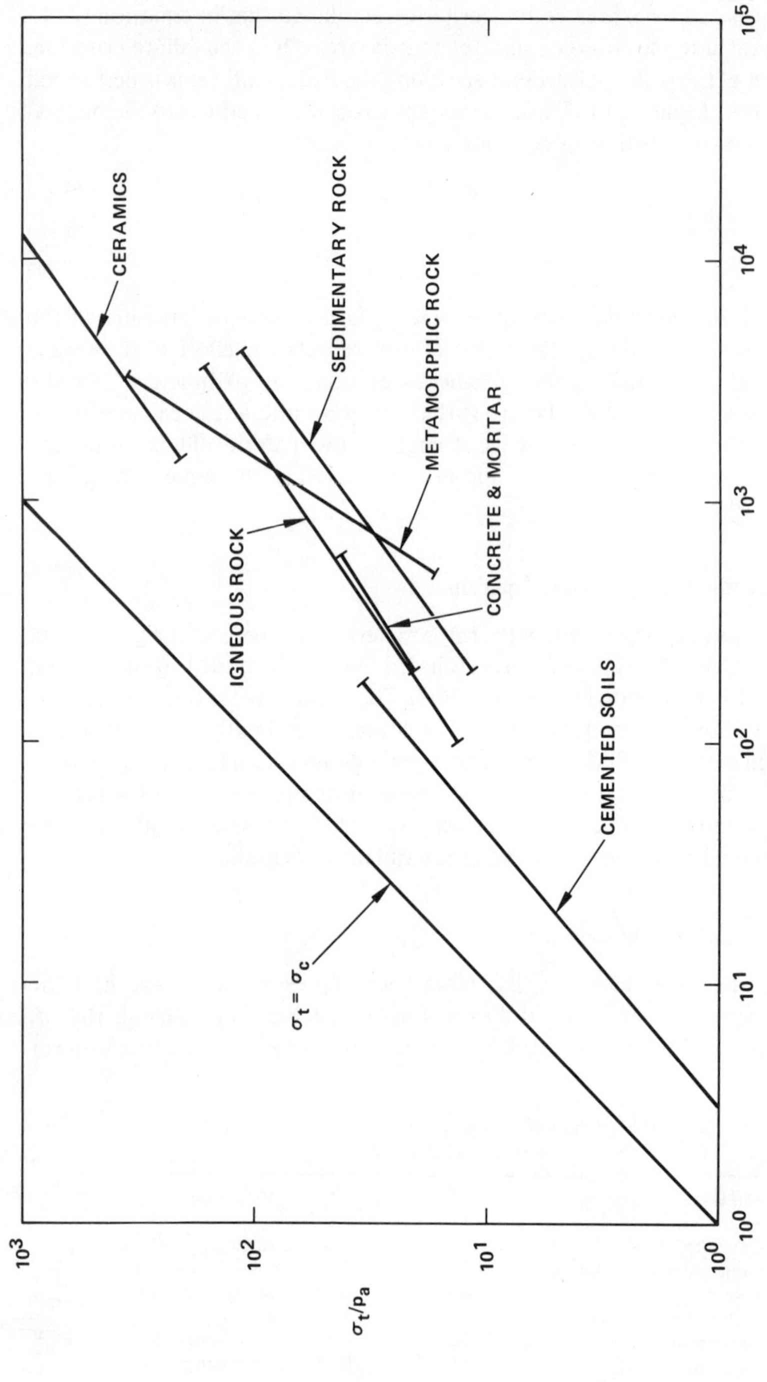

Figure 20.5 Relations between uniaxial tensile strengths and uniaxial compressive strengths for various types of frictional materials

σ_t) for cemented soils is about $\frac{1}{5}$ to $\frac{1}{3}$ of σ_c, whereas data compiled by Hannant [8] show that σ_t for concrete varies between 5 per cent and 13 per cent of σ_c. However, the values of σ_t and σ_c may be related through a power function of the following type [7, 9]:

$$\sigma_t = T \cdot p_a \cdot \left(\frac{\sigma_c}{p_a}\right)^t \tag{20.5}$$

where T and t are dimensionless numbers, and p_a is atmospheric pressure in the same units as those of σ_t and σ_c.

Values of T and t have been determined for several frictional materials and listed in Table 20.1. These values represent the best fit between experimental data and the simple expression in equation (20.5). Scatter in the test data was present for all materials, and the values of T and t given in Table 20.1 may not suitable for individual cases. The data base for ceramics was taken from only a few sources, whereas more extensive data bases were used for the other materials in Table 20.1.

Figure 20.5 shows a comparison of the relations between uniaxial tensile and uniaxial compressive strengths for the materials listed on Table 20.1. The straight lines shown on the log–log diagram span over the ranges of uniaxial compressive strengths indicated by available data. Both relatively weak and very strong frictional materials are represented in Figure 20.5. Note that the lines tend to cluster in an oblong area which slopes away from the line representing equal uniaxial tensile and uniaxial compressive strengths. Thus, the weak materials have relatively higher uniaxial tensile strengths than the strong materials.

20.3.2.2 *Regression analyses*

Because the failure criterion is expressed in terms of stress invariants, any type of test in which all stresses are measured may be used for determination of the three material parameters. However, it is advantageous to require only the simplest possible types of tests such as, for example, uniaxial compression and triaxial compression or biaxial tests for this determination, and then check whether these simple tests are sufficient for adequate characterization of the failure condition for the particular material under investigation. This may be done using various sets of data available in the literature which include both simple and more complex three-dimensional tests.

In order to obtain the overall best fitting parameters, regression analyses may be performed to determine the highest possible value of the coefficient of determination r^2. Figure 20.6 shows an example of the effect of varying the parameter 'a' on the values of r^2, η_1, and m for the tests on Mix A concrete performed by Mills and Zimmerman [10]. Only the results of the uniaxial compression and the triaxial compression tests in addition to the estimated value of the uniaxial tensile strength (from equation (20.5)) were used to

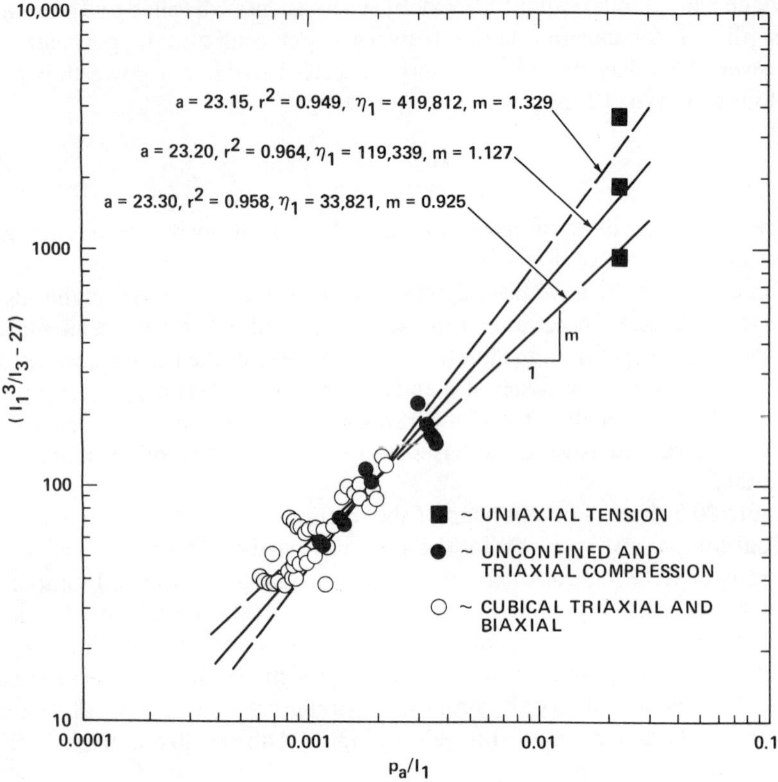

Figure 20.6 Determination of material parameters involved in failure criterion for Mix A concrete tested by Mills and Zimmerman [10]

determine the three material parameters. The uniaxial tensile strength was estimated to be $-23.1\,\text{kg/cm}^2$ ($-2266\,\text{kN/m}^2$) for Mix A concrete and the best fit value of 'a' $= 23.2$ resulted in $\eta_1 = 119{,}339$ and $m = 1.127$.

Except the three points corresponding to the uniaxial tensile strength on Figure 20.6, the points corresponding to the other tests do not translate enough on the diagram to show their separate locations. The points corresponding to the uniaxial tensile strength tend to influence the location of the best fit straight line. However, each of the three lines would describe the failure surface in the region of compressive stresses with reasonable accuracy. Thus, it is an advantage to incorporate the uniaxial tensile strength, even though it may be an estimate, in determination of the material parameters in order to stabilize the failure criterion in the region close to the origin and to describe the tensile strength for the material with reasonable accuracy.

The results of the cubical triaxial tests on Mix A concrete are shown on Figure 20.6 for comparison. It may be seen that some scatter of the data around the solid line does exist, but the material parameters selected on the basis of the simple tests appear to represent the data quite well.

20.3.3 Comparison of failure criterion and test data

In order to evaluate the capabilities of the three-dimensional failure criterion, comparisons have been made between experimental data and failure surfaces calculated from equations (20.1) and (20.4). Data were plotted on the biaxial plane, and data points were projected on the octahedral plane for all available data sets. Those data sets containing results of triaxial compression and extension tests were also shown on the triaxial plane. Examples of these comparisons are given below for concrete and rock.

Comparison of test data (points) and failure surface (solid line) is shown on the normalized biaxial plane in Figure 20.7(a) for the tests performed on concrete with $\sigma_c = 590\,\text{kg/cm}^2$ ($57\,880\,\text{kN/m}^2$) by Kupfer *et al.* [11]. All data, except those corresponding to tension–tension, were used to determine the material parameters. The failure critertion is seen to represent the test data with reasonable accuracy. In order to study the failure surface relative to the data in the tension–tension area, the data are shown on the enlarged diagram in Figure 20.7(b). It may be seen that the failure surface is smoothly rounded at the corner and that it corresponds very well with the data in this region.

A major investigation was performed by Mills and Zimmerman [10]. The results of their tests on Mix A concrete, which contained tests in triaxial, biaxial,

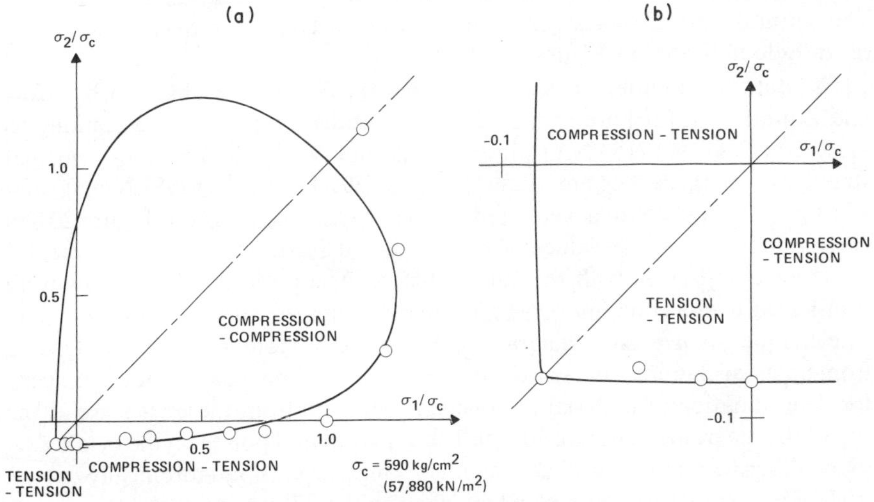

Figure 20.7 Comparison of failure criterion with results of biaxial tests performed by Kupfer, Hilsdorf, and Rusch [11] in (a) biaxial plane, and (b) enlarged tension–tension region

and octahedral planes, provided a good, coherent set of data. This set of data was used for illustration of material parameter determination in Figure 20.6. Comparisons between failure criterion and test data for Mix A concrete are shown in Figure 20.8. The material parameters for this concrete were determined on the basis of data from uniaxial compression and triaxial compression tests, and an estimated value of the uniaxial tensile strength. Thus, the good agreement shown in Figure 20.8(a) between the results of the triaxial compression tests and the failure critertion could be expected. However, the strengths obtained in triaxial extension are also well represented by the failure criterion. The data obtained in the compression–compression region of the biaxial plane are shown in Figure 20.8 (b). Although there is some scatter in the test results, the failure criterion is seen to represent the data quite well. Note that the pointed corner in the tension–tension area is actually smoothly rounded as shown in Figure 20.7(b).

Figure 20.8(a) also shows that the failure surface in extension cuts across the $\sigma_1 = 0$ plane at a very shallow angle. Therefore, any small deviation between test data and failure surface at this intersection in the triaxial plane will appear as a large deviation in the biaxial plane. Comparison of the data points for Mix A concrete indicated by arrows in Figures 20.8(a) and 20.8(b) shows that these appear to deviate somewhat from the failure surface in the biaxial plane (Figure 20.8(b)), whereas the same points in the triaxial plane are very close to the proposed failure surface. Any little amount of restraint in the testing apparatus would result in too large strength in biaxial extension, and this would show up very clearly in the biaxial plane. However, an evaluation in the triaxial plane would likely show that the test data are not that far from the actual failure surface. The natural scatter in test data could easily account for deviations of the magnitude indicated in Figure 20.8(b).

The data from cubical triaxial tests on Mix A concrete obtained by Mills and Zimmerman [10] are projected on the octahedral plane corresponding to $I_1 = 150\,\text{kg/cm}^2$ ($14\,715\,\text{kN/m}^2$) in Figure 20.8(c). Values of the minor principal stress, σ_3, of 0, $29.5\,\text{kg/cm}^2$ ($2894\,\text{kN/m}^2$), $59.11\,\text{kg/cm}^2$ ($5798\,\text{kN/m}^2$), and $88.6\,\text{kg/cm}^2$ ($8692\,\text{kN/m}^2$) were used in these tests. The points in Figure 20.8(c) corresponding to these values of σ_3 are shown separately on the octahedral plane for comparison with the failure surface. The projected data points were transferred to the common octahedral plane along the curved meridians using a technique involving the diagram in Figure 20.6. Note again that only data from uniaxial compression, triaxial compression, and uniaxial tension were used for determination of material parameters. The data from these tests are at the top of the diagrams in Figure 20.8(c). The experimental points on the octahedral plane describe a failure surface which is triangular with monotonically curved surface and smoothly rounded edges, as does the failure criterion.

A study of the three-dimensional strength of sandstone performed by Akai and Mori [12] involved uniaxial compression, triaxial compression, biaxial

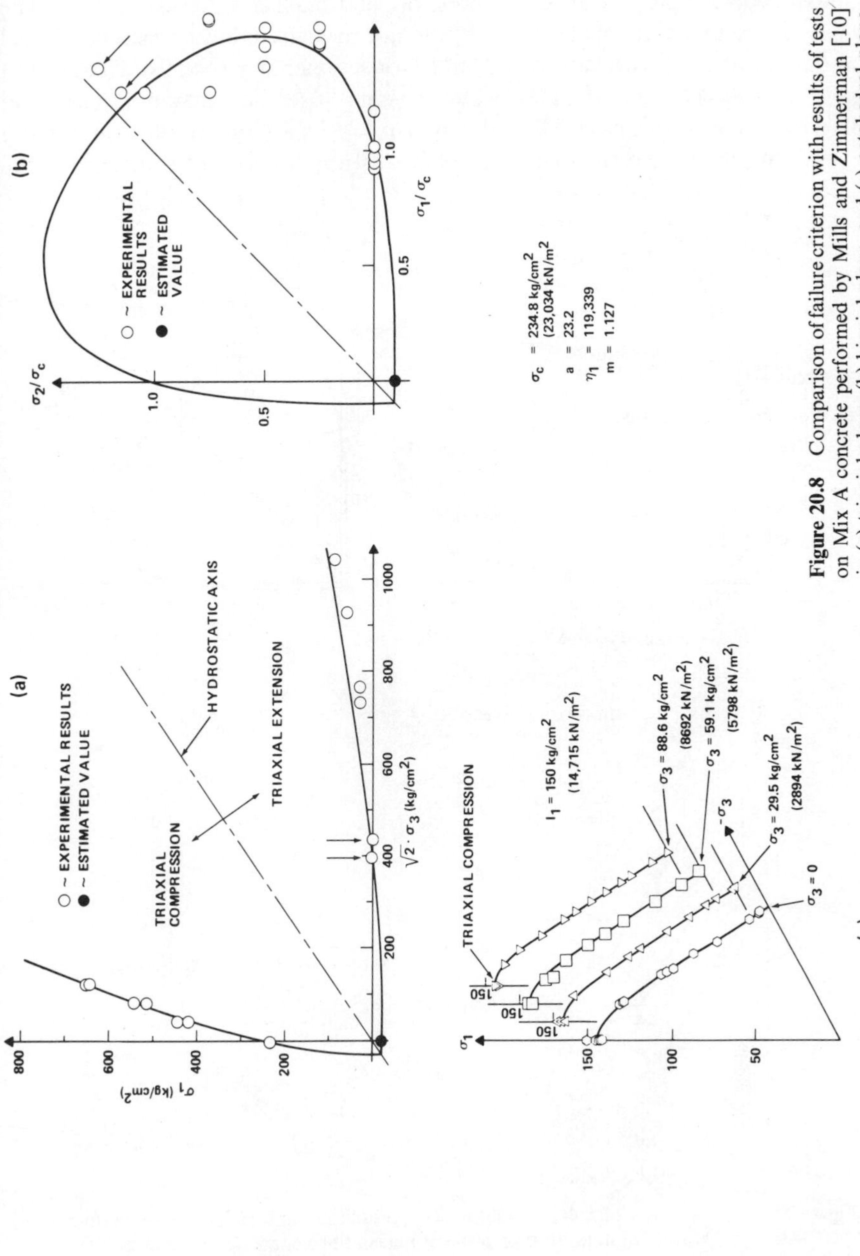

Figure 20.8 Comparison of failure criterion with results of tests on Mix A concrete performed by Mills and Zimmerman [10] in (a) triaxial plane, (b) biaxial plane, and (c) octahedral plane

compression, and cubical triaxial tests. Comparisons between failure criterion and test data for the sandstone are shown in Figure 20.9. The material parameters listed on Figure 20.9 for this rock were determined from the results of the uniaxial compression, triaxial compression, and biaxial compression tests. The value of the uniaxial tensile strength was not measured, but an estimate based on equation (20.5) with values of T and t for sedimentary rock from Table 20.1 gave $\sigma_t = -44.6$ kg/cm^2. This value was not used for material parameter determination in this case. The value of $a \cdot p_a = 37.4$ kg/cm^2 resulted in a better overall fit when the estimated value of σ_t was not employed for determination

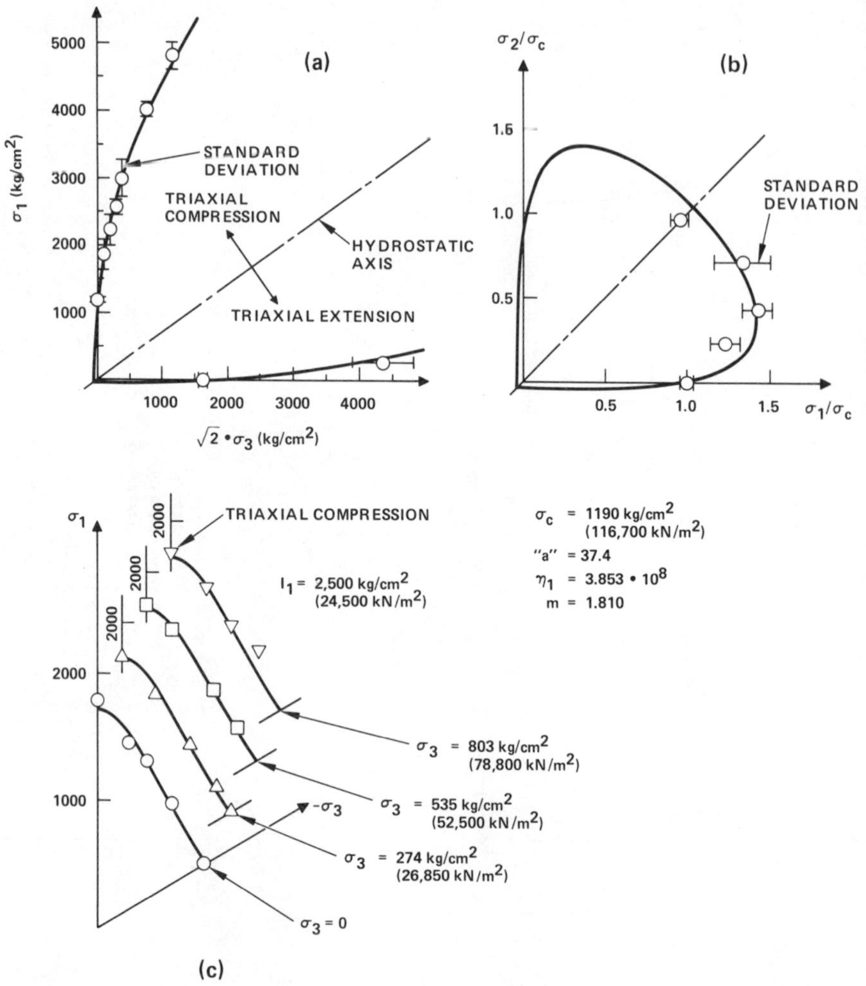

Figure 20.9 Comparison of failure criterion with results of tests on sandstone performed by Akai and Mori [12] in (a) triaxial plane, (b) biaxial plane, and (c) octahedral plane

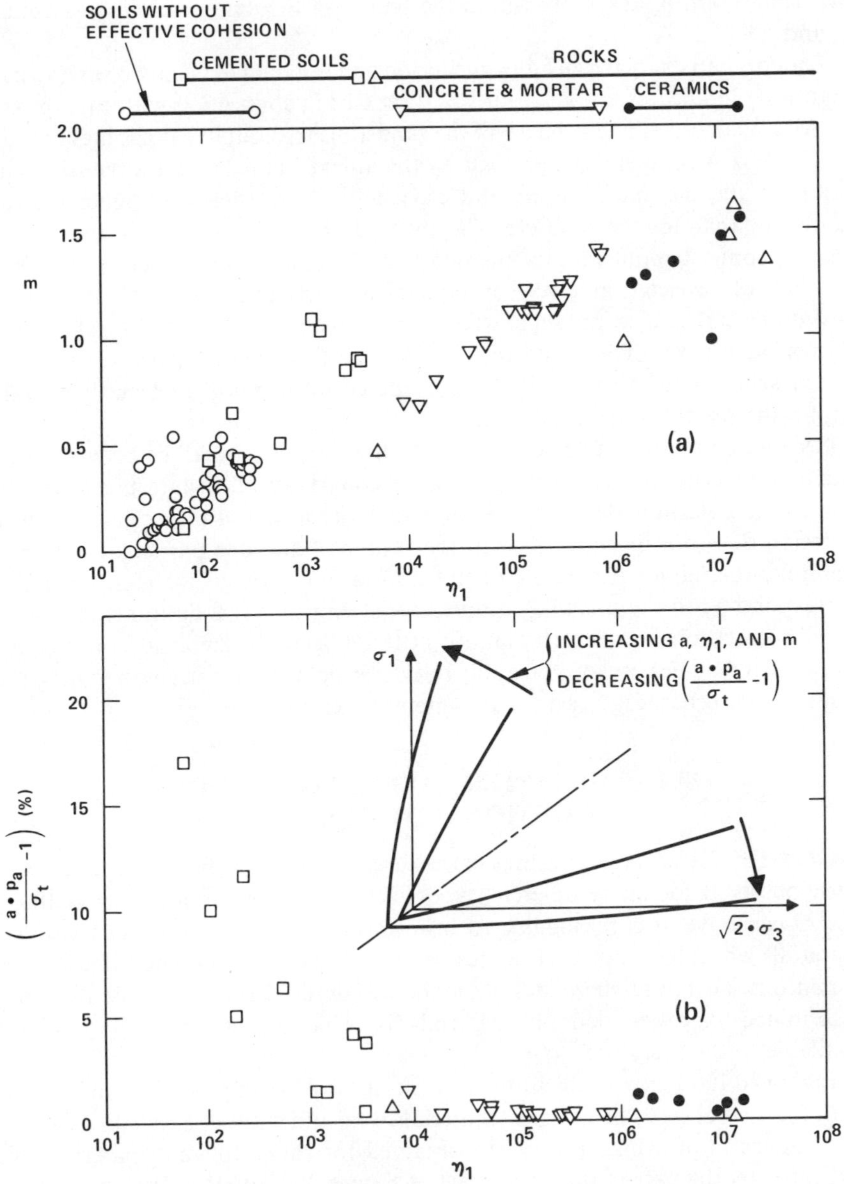

Figure 20.10 (a) Variation and ranges of the parameters η_1 and m for various frictional materials. (b) Magnitudes of the parameter 'a' relative to the uniaxial tensile strength σ_t plotted against η_1 for frictional materials with effective cohesion

of parameters. If σ_t were included, $a \cdot p_a$ would become 44.7 kg/cm^2, but these two values of $a \cdot p_a$ are easily within the scatter of the data employed to obtain T and t.

The comparison between failure criterion and test data for sandstone shown in Figure 20.9 indicates similar results as discussed in connection with Figure 20.8 for Mix A concrete. The strength of the sandstone is about five times higher than that of Mix A concrete as measured by the uniaxial compressive strengths. The values of the parameters a, m, and especially η_1 are therefore higher for the sandstone than for the concrete. The shape of the failure surface in the biaxial plane is somewhat different for the two materials, and this is correctly modelled by the failure criterion given by equations (20.1) and (20.4). The standard deviations of the major principal stresses, σ_1, as listed by Akai and Mori [12], are shown on Figures 20.9(a) and (b). The overall fit between failure surface and experimental data in Figure 20.9 is considered to be quite good and generally within the scatter of the test results.

Because emphasis was placed on obtaining the best overall fit between data and failure criterion, equal weights were placed on all test results used for parameter determination. It may therefore occur that the proposed failure criterion does not fit perfectly with the results of any one particular test. For example, the failure surface in Figure 20.7(a) does not go through the point corresponding to the uniaxial compressive strength. If it is desirable to obtain good correlation between the failure criterion and the uniaxial compressive strength, it may be necessary to apply heavier weights to this strength than to the results from other tests used for parameter determination.

20.4 PARAMETER VALUES FOR VARIOUS FRICTIONAL MATERIALS

Because the characteristic features of the three-dimensional failure surfaces have been observed for many different types of frictional materials, such as those described above, it is reasonable to believe that they exist for other frictional materials which have not yet been tested under general three-dimensional stress conditions. Therefore, the values of the three material parameters have also been determined for other frictional materials for which results of simple tests are available in the literature.

Figure 20.10(a) shows a diagram in which values of m are plotted against values of η_1 for several types of frictional materials with and without effective cohesion. The ranges of η_1 which have been obtained for these different materials are indicated by the bars in the upper part of Figure 20.10(a). It is likely that these ranges will be expanded when additional data sets for each type of material are analysed. Thus, the transition from cemented soils to rock is most likely smooth with no clear distinction between the two materials. Only a few sources of data for ceramics have been consulted for this study, and the range of the parameters for ceramics is probably larger than indicated in Figure 20.10. However, it is clear that the increased frictional strength indicated by η_1 is accompanied by an increased curvature of the failure surface as expressed by the value of m.

The parameter 'a' used for frictional materials with effective cohesion must be higher than the uniaxial tensile strength, σ_t. Values of σ_t are related to the uniaxial compressive strength, σ_c, in Figure 20.5. Thus, if uniaxial tension tests are not part of the testing programme, an approximate value of σ_t may be obtained from Figure 20.5 or from equation (20.5). Because many engineering materials are exposed to both compressive and tensile stresses in the same structure, it was proposed to include a reasonable value of the uniaxial tensile strength in the parameter determination in order to obtain an overall representative failure criterion for the material under investigation. The values of the parameter 'a' obtained in this study have been related to σ_t through the term $(a \cdot p_a - \sigma_t)/\sigma_t$ (in per cent) which expresses the percentage by which 'a' is greater than σ_t. This percentage is plotted versus η_1 in Figure 20.10(b). The high values (up to 17 per cent) of this percentage were obtained for cemented soils for which the parameter η_1 is relatively small. Values below 2 per cent were obtained for all other frictional materials with effective cohesion. In fact, a gradual decrease in percentage is obtained with increasing value of η_1.

The insert in Figure, 20.10(b) indicates the general trend for the three parameters 'a', η_1, and m. All three parameters have been found to increase with increasing strength of the frictional material.

Some materials, such as compacted, partly saturated soils, may exhibit an apparent cohesion. These materials have not been included in the present study. However, the failure conditions for such materials may be modelled by the failure criterion in equations (20.1) and (20.4) using total stressed for parameter determination and subsequent application in analysis procedures.

20.5 CONCLUSION

Failure surfaces for frictional materials with and without effective cohesion have several characteristics in common, and these are all captured by a general three-dimensional failure criterion formulated in terms of the first and the third stress invariants of the stress tensor. This failure criterion involves only three independent material parameters. Although these parameters interact with one another, each parameter corresponds to one of three failure characteristics of the frictional materials. The material parameters may be determined from simple tests such as uniaxial compression and triaxial compression or biaxial tests. For the purpose of including reasonable values of tensile strengths in the failure criterion (for frictional materials with effective cohesion), it may be necessary to include the uniaxial tensile strength in the parameter determination. A simple expression for evaluation of the uniaxial tensile strength on the basis of the uniaxial compressive strength is given. Representative sets of data for soils, concrete, and rock, for which results of three-dimensional tests were available, have been analysed, and comparisons between the failure criterion and the experimental data are made in triaxial, biaxial, and octahedral planes. Typical values of the three material parameters for various types of frictional materials have been determined, and their magnitudes are compared and discussed in light of the materials to which they belong. The ability of the general three-dimensional failure criterion to capture the characteristics of failure in frictional materials appears to be quite good with accuracies generally within the natural scatter of the test data.

ACKNOWLEDGEMENTS

Messrs E. Geiger and J. Tsai performed the tests on Edger Plastic Kaolinite, and Mr M. Kim compiled the data relating to failure of rock. Their assistance is gratefully acknowledged. Financial support for the work relating to soils has been provided under several grants from the National Science Foundation.

REFERENCES

1. J. K. Mitchell, *Fundamentals of Soils Behavior*, John Wiley & Sons, Inc., New York, NY, 1976.
2. A. W. Bishop, 'The strength of soils as engineering materials', 6th Rankine Lecture, *Geotechnique*, **16**, 2, 91–130 (1966).
3. P. V. Lade, 'Elasto-plastic stress–strain theory for cohesionless soil with curved yield surfaces', *International Journal of Solids and Structures*, **13**, 1019–1035, Pergamon Press, Inc., New York, NY, 1977.
4. P. V. Lade, 'Prediction of undrained behavior of sand', *Journal of the Geotechnical Engineering Division, ASCE*, **104**, No. GT6, Proc. Paper 13834, 721–735 (1978).
5. P. V. Lade and H. M. Musante, 'Three-dimensional behavior of remolded clay', *Journal of the Geotechnical Engineering Division, ASCE*, **104**, No. GT2, Proc. Paper 13551, 193–209 (1978).
6. P. V. Lade, 'Three-parameter failure criterion for concrete', *Journal of the Engineering Mechanics Division, ASCE*, **108**, No. EM5, Proc. Paper 17383, October 1982.
7. J. K. Mitchell, 'The properties of cement-stabilized soils', *Proc., Workshop on Materials and Methods for Low Cost Road*, pp. 365–404, Rail and Reclamation Works, Leura, Australia, September 1976.
8. D. J. Hannant, 'Nomograms for the failure of plain concrete subjected to short-term multiaxial stresses', *The Structural Engineer*, **52**, 5, 151–165 (1974).
9. J. Wastiels, 'Behaviour of concrete under multiaxial stresses—a review', *Cement and Concrete Research*, Vol. 9, pp. 35–44, Pergamon Press, 1979.
10. L. L. Mills and R. M. Zimmerman, 'Compressive strength of plain concrete under multiaxial loading conditions', *American Concrete Institute Journal*, **67**, 802–807 (1970).
11. H. Kupfer, H. K. Hilsdorf, and H. Rusch, 'Behavior of concrete under biaxial stresses', *American Concrete Institute Journal*, **66**, 656–666 (1969).
12. K. Akai and H. Mori, 'Ein Versuch über Bruchmechanismus von Sandstein under mehrachsigem Spannungszustand', *Proc. of the Second Congress of the International Society of Rock Mechanics*, Vol. II, Paper No. 3–30, Belgrade, Yugoslavia, 1970.

Mechanics of Engineering Materials
Edited by C. S. Desai and R. H. Gallagher

Chapter 21

The Constitutive Equations for High Temperatures and Their Relationship to Design

F. A. Leckie

21.1 INTRODUCTION

When load-bearing metallic components operate at temperatures in excess of approximately one-third of the melting temperature T_m, consideration must be given to the effects of creep deformations and rupture. In some components, limitations are placed on the amount of creep strain which may be accumulated. In other components such as those occurring in the process industries, less consideration need be given to deformation and the life is only terminated when rupture takes place and some form of leakage becomes evident. Rupture can be the consequence of thinning induced by large strains but it can also occur at small strains as the result of growth of damage in the metal. Metallographic inspection reveals that the damage normally occurs in the form of fissures and voids which coincide with grain boundaries. In this chapter, attention is given to the means of describing metallic creep rupture behaviour by suitable constitutive equations which can be used to predict the creep life of components with complex geometries. This branch of mechanics is often referred to as continuum creep damage mechanics because damage is observed to be smoothly distributed in contrast to classical fracture when damage appears to propagate along a well defined line. An example illustrating the continuous distribution of creep rupture damage is that studied by Leckie and Hayhurst [1] of a plate penetrated by a circular hole and subjected to constant stress σ. The damage in a copper plate was observed to vary continuously with the greatest concentration of damage at the edge of the hole. This test also revealed that the rupture life of the plate was dictated by the average tensile stress at the minimum section of the plate. For the particular plate geometry, the maximum elastic stress is 2.2σ and the maximum steady-state creep stress is 1.4σ, where σ is the average stress at the minimum cross-section. Predictions of life based on these two values would, therefore, give values which are much shorter than those observed experimentally. The implication is that the effect of stress redistribution is sufficiently large for the life of the plate to be dictated by the average stress at the minimum section. The initial stress concentrations appear to have no effect on the rupture life of the plate. In order to verify this observation further, a plate of identical size was penetrated by a slit of length equal to the diameter of the circular hole. It was found again that stress concentration effects could be neglected with the rupture life dependent on the value of the average stress at the minimum section.

The principal interest of engineers is the prediction of the magnitude of deformations and rupture time of load-bearing components while little attention is given to describing the physical processes occurring in the material which is the special interest of the metallurgist. It is not surprising, therefore, that the procedures developed to suit the needs of the two groups appear to be at odds. Because of space limitations, no substantial attempt is made in this paper to illustrate that the two approaches are in many respects quite similar and that only modest modifications are required to bring them into agreement. Since damage generally grows in an anisotropic manner, the constitutive equations which describe the material behaviour are rather complex and require the results of test which are difficult and time-consuming. Furthermore, the calculations involving the use of these constitutive equations in finite element procedures prove to be so demanding that their use is likely to be limited to the very final stages of the design process. There is a need therefore to develop constitutive equations which faithfully represent the most important aspects of material behaviour but retain sufficient simplicity for mathematical manipulation to be possible. Using this approach, it is possible to obtain results which result in procedures useful at the early stages of the design process. Some of the effort to develop these procedures are discussed.

21.2 PRELIMINARY DISCUSSION ON CONSTITUTIVE EQUATIONS

The engineering approach is to develop constitutive equations which describe the macroscopic behaviour of the material. It is the constitutive equations in conjuction with the laws of continuum mechanics which determines component behaviour. In the case of creep rupture, attempts are made to calculate the time at which cracks first appear, the growth of damage fronts and time for complete rupture. In order to describe the macroscopic behaviour of the materials, it is necessary to introduce internal state variables which, in some sense, are a measure of the physical state of the material. Since engineers do not attempt to give a precise physical description of the state variable, this approach may on the face of it, appear to be at variance with the approach of the metallurgist who wishes to give an accurate description of the mechanisms which take place. This, however, is not so and there are close similarities in the approaches adopted.

If, for convenience of discussion, the primary portion of the creep curve is neglected, then the strain/time curve in a constant stress test will have the form shown in Figure 21.1. In fact, it will be assumed throughout that transient effects brought about by change in dislocation structure will be neglected. The creep rate increases from the initial steady-state value as damage causes an advance into the tertiary portion of the curve. The steady-state creep rate can be expressed as a function of the applied stress σ alone. In order to account for the increase in strain rate during a constant stress test, it is necessary to introduce a new variable into the strain rate equation. Since the increase in strain rate is the result of a damage process, the variable ω introduced is referred to as the damage state variable. The strain rate equation then takes the form

$$\mathrm{d}\varepsilon/\mathrm{d}t = f(\sigma, \omega) \tag{21.1}$$

where f is a function which has yet to be defined. However, in the undamaged state when $\omega = 0$, the equation must reduce to the form observed for steady-state conditions. To define the strain rate, it is necessary that the value of ω as well as the stress σ be known and consequently, an equation must be introduced which defines the growth of the damage state variable. Assuming that the damage rate depends both on the current state of stress and damage gives the equation

$$\mathrm{d}\omega/\mathrm{d}t = g(\sigma, \omega) \tag{21.2}$$

Failure occurs when

$$\omega = 1 \tag{21.3}$$

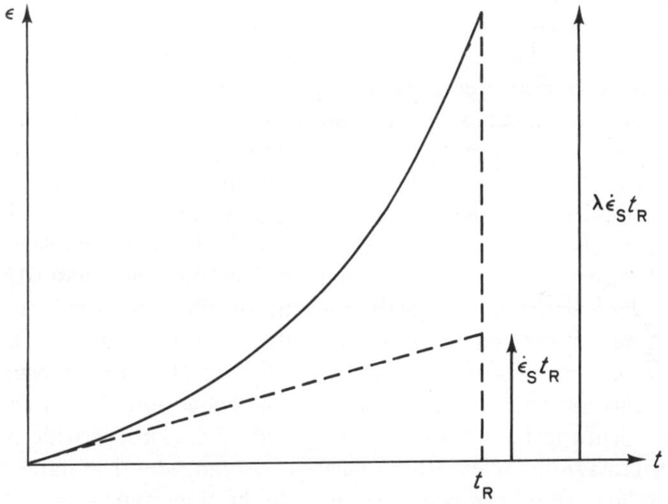

Figure 21.1 Tertiary creep curve

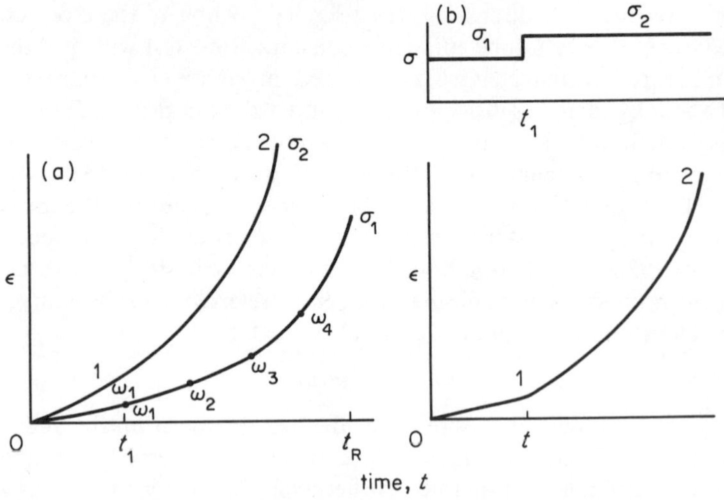

Figure 21.2 State variable tests

The problem is how to find the function f and g in a systematic manner and this has been tackled in a number of ways. In metallurgical studies, the growth of damage may be measured directly by microscopic examination. Instead of using direct observation, the equations may also be deduced from a study of the micromechanics of the basic mechanisms which occur. The efforts in this direction have been recently presented by Cocks and Ashby [2]. The engineering approach, which is to use the macroscopic observations of mechanical test, was outlined by Leckie and Hayhurst [3] and uses the result of step tests. Neglecting the primary portion of the creep curve of the strain/time graphs for two constant stress tests carried out at stresses σ_1 and σ_2 would have the form shown in Figure 21.2(a). Suppose in constants σ_1 test that ω varies between zero at time $t = 0$ and unity at the rupture time t_R. Select an arbitrary variation of ω along the σ_1 curve (Figure 21.2(a)). Now perform a test in which the stress σ is applied for time t_1 when the stress is increased to σ_2 and maintained constant at this value to give the creep curve shown in Figure 21.2(b). Since a single state variable is sufficient to define the creep curves it follows that Section 12 of the curve 012 must be identical in shape with a portion of the constant stress creep curve σ_2. In this way, the state ω_1 may be defined on the σ_2 curve. By performing similar tests constant ω contours may be constructed as shown in Figure 21.3. This information can then be used to construct the functions f and g of equation (21.2). In contrast to the metallurgical approach, the engineering method does not provide an absolute measure of damage but gives rather a means of defining identical states. However, the two methods should in principle, provide the same type of growth laws of equation (21.2). Complete match in metallurgical and engineering studies has yet

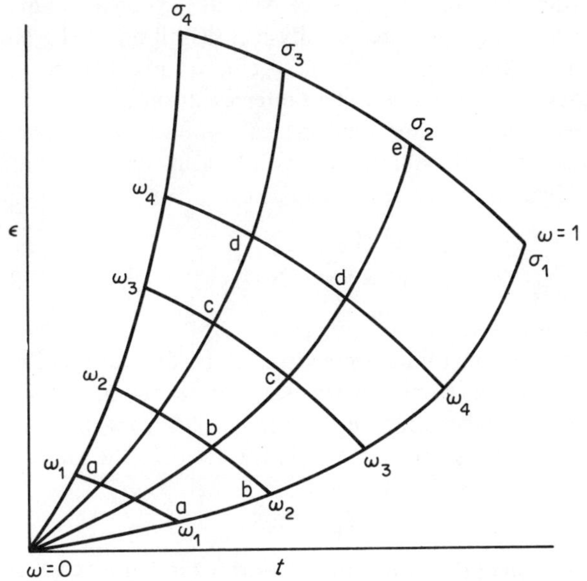

Figure 21.3 Constant damage contours

to be achieved but the mismatch is, at least in the opinion of the author, more a matter of detail rather than of substance.

21.3 CONSTITUTIVE EQUATIONS FOR PROPORTIONAL LOADING

Unfortunately, it is difficult and time consuming to conduct experiments described in the previous section and the variable stress experiments which have been performed are limited to the prediction of rupture life under cyclic conditions of loading. Until recently, multiaxial tests have been limited to constant loading so that it was not possible to investigate the tensorial nature of damage. This point will be picked up again later and in the meantime constitutive equations are developed which are suitable to describe the effects of proportional loading. Rabotnov [4] proposed modifications of constitutive equations first suggested by Kachanov [5] which described existing experimental results. For uniaxial stress tests, the growth laws are assumed to have the simple form

$$\dot{\varepsilon}/\dot{\varepsilon}_0 = (\sigma/\sigma_0)^n/(1-\omega)^m \tag{21.4a}$$

$$\dot{\omega}/\dot{\omega}_0 = (\sigma/\sigma_0)^\nu/(1-\omega)^\eta \tag{21.4b}$$

It is assumed that $\omega = 0$ when the material is in its undamaged state and that $w = 1$ at rupture. In the equations, $n, m, \nu, \dot{e}_0, \dot{\varepsilon}_0$, and σ_0 are constants to be defined and η is defined in equation (21.4c). It should be noted that in common with the spirit of Section 21.2, no precise physical definition of the parameter ω is attempted. In fact, this point has been specifically made by Rabotnov but for some reason there has been a persistent interpretation of the term $(1 - \omega)$ as the current cross-sectional area and the term $\sigma/(1 - \omega)$ as the effective stress. By the use of algebra and the introduction of two physical constants λ and t_0, it is not difficult to produce the constitutive equations

$$\varepsilon/\dot{\varepsilon}_0 = \{\sigma/(1 - \omega)\sigma_0\}^n \tag{21.4c}$$
$$\dot{\omega} = (\lambda - 1)(\sigma/\sigma_0)^\nu/\lambda n t_0(1 - \omega)^\eta$$

where $\eta = \lambda n/(\lambda - 1) - 1$.

The constants are easy to determine from tests. The values of n and ν are the creep and rupture indices of customary usage. In a constant stress test conducted at stress σ_0, $\dot{\varepsilon}_0$ is the steady state creep rate and t_0 is the rupture time. The constant λ is obtained from the same test and is the ratio of the rupture strain ε_0^R to $\dot{\varepsilon}_0 t_0$; therefore

$$\lambda = \varepsilon_0^R/\dot{\varepsilon}_0 t_0 \tag{21.5}$$

The determination of this constant is illustrated in Figure 21.1. It is sometimes referred to as the creep ductility since (as can be readily shown) it is this value which determines the ability of the material to resist local cracking at points of high stress concentration.

The effect of multiaxial states of stress on rupture are normally expressed in terms of isochronous surfaces. These are surfaces in stress space for which rupture times are constant and equal to t_0. Hayhurst [6], after an extensive review of available data, concluded that isochronous surfaces for rupture time t_0 can be represented by the equation

$$\Delta(\sigma_{ij}/\sigma_0) = [\alpha(\sigma_1/\sigma_0) + \beta(\bar{\sigma}/\sigma_0) + \gamma(J_1/\sigma_0)] = 1 \tag{21.6a}$$

In this expression α, β, and γ are constants with $\alpha + \beta + \gamma = 1$, σ_1 is the maximum stress, $\bar{\sigma}$ the effective stress, and $J_1(= \sigma_1 + \sigma_2 + \sigma_3)$ is the first stress invariant. The constant γ is introduced to describe the reduction in the value sometimes observed when metals are subjected to equal biaxial states of stress. It appears, however, that γ is usually small [6] and when neglected (equation (21.6a)) simplifies to the form

$$\Delta(\sigma_{ij}/\sigma_0) = [\alpha(\sigma_1/\sigma_0) + (1 - \alpha)(\bar{\sigma}/\sigma_0)] = 1 \tag{21.6b}$$

The form of the generalization for multiaxial strain rates was influenced by the experimental results of Johnson *et al.* [7] which indicate that the strain rates are dependent on the effective stress $\bar{\sigma}$ and that the components of strain rate $\dot{\varepsilon}_{ij}$ are proportional to the deviatoric stress s_{ij}. These results when expressed mathematically take form

$$\dot{\varepsilon}_{ij}/\dot{\varepsilon}_0 = (3/2)(\bar{\sigma}/\sigma_0)^{n-1} s_{ij}/\sigma_0 \tag{21.7}$$

with n, $\dot{\varepsilon}_0$, and σ_0 having the same significance as in equation (21.4). Investigation of the torsion–tension experimental results of Johnson *et al.*, on copper and aluminium show in both cases that when deterioration is taking place in the tertiary region, the ratio of the strain rate components remains sensibly constant and equal to the value in the steady-state condition. Consequently, the strain rates in the tertiary region must be represented by constitutive equations with the form

$$\dot{\varepsilon}_{ij}/\dot{\varepsilon}_0 = (3/2)k(t)(\bar{\sigma}/\sigma_0)^{n-1}s_{ij}/\sigma_0 \qquad (21.8)$$

where $k(t)$ is a scalar quantity increasing monotonically with time. Johnson *et al.*, also observed that the growth of k is dependent on the multiaxial form of the stress field. In copper, it is maximum stress dependent and in aluminium shear stress dependent. In fact, the stress state affecting the growth of k appears to be the same as that dictating the form of the isochronous surface.

Comparing the form of equation (21.8) with the uniaxial relation (21.4a) suggests the following constitutive equation for the strain rates:

$$\dot{\varepsilon}_{ij}/\dot{\varepsilon}_0 = (3/2)(\bar{\sigma}/\sigma_0)^{n-1}(s_{ij}/\sigma_0)/(1-\omega)^n \qquad (21.9a)$$

The proposed damage rate equation following equation (21.4b) has the form

$$\dot{\omega} = \{(\lambda-1)/\lambda n t_0\}\Delta^{\nu}(\sigma_{ij}/\sigma_0)/(1-\omega)^{\eta} \qquad (21.9b)$$

where $\eta = \{\lambda n/(\lambda-1)\} - 1$, and t_0 and ν have the same significance as in equations (21.3) and $\Delta(\sigma_{ij}/\sigma_0) = 1$ is the isochronous surface for failure time t_0.

An important prediction of the constitutive equations is that the strain to failure can be greatly dependent on the form of the applied stress and on the isochronous surface Δ. The equations are simpler to integrate if it is assumed that $n = \nu$. With this assumption, it is not difficult to obtain the result

$$\varepsilon_{\Delta}/\sigma_0 = [(\bar{\sigma}/\sigma_0)/\Delta]^n \qquad (21.10)$$

where ε_{Δ} is the strain at rupture when the multiaxial stress is applied and σ_0 is the uniaxial rupture strain.

When $\Delta = \bar{\sigma}/\sigma_0$, as it is for aluminium, then the rupture strain is independent of stress state.

For copper, however, the strain at failure is

$$\varepsilon_{\Delta}/\varepsilon_0 = [1/(0.83(\sigma_1/\bar{\sigma}) + 0.17)]^{5.9} \qquad (21.11)$$

Leckie [8] compared the predictions of this formula with the experimental results of a variety of multiaxial stress states and found the comparison to be satisfactory. The important consequence of the formula, equation (21.11) is that for conditions of high constraint when $\sigma_1/\bar{\sigma}$ is large, the strain at failure can be very small indeed.

The constitutive equations (21.9) which fit the results of proportional loading are clearly unsuitable for non-proportional loading when the anisotropic effects of damage will become pronounced. The topic of damage anisotropy will be discussed in a later section but in the meantime, the potential of the constitutive equations will be investigated further.

21.4 BOUNDING THEOREMS FOR RUPTURE TIME

By using the constitutive equations (21.9) in conjunction with the continuum conditions of equilibrium and compatibility, it is possible to calculate the time-dependent stress and strain fields in load bearing components. It is also possible to calculate the growth of damage so that the time to first local failure and final rupture may be determined. Apart from problems of especially simple geometry, it is normally necessary to resort to the use of the finite element method. An example illustrating the procedure [9] is that of the plate under tension penetrated by a hole. The calculations indicate that this is not a simple task however, and demands very considerable skill. Another approach which has been attempted is to determine bounds on the rupture time.

Upper bound calculations of rupture life may be obtained by making the assumption that the isochronous surface $\Delta(\sigma_{ij}/\sigma_0) = 1$ is convex. The experimental evidence presented [6] suggests that in most circumstances this is likely to be a justified assumption. The assumption appears always to be valid when the compressive stress is smaller in magnitude than the tensile stress. In aluminium, the isochronous surface is effective stress dependent even when the stresses are compressive so that the surface is always convex. However, in copper, the isochronous surface is concave when the magnitude of the compressive stress is somewhat greater than the tensile stress.

Suppose a structure of total volume V is subjected to constant loads P_i. An upper bound t_u on the rupture life t_R of the structure is given by the following expression [10]:

$$t_R/t_0 < t_u/t_0 = V \int_V \Delta^\nu(\sigma_{ij}^s/\sigma_0)\,dv \tag{21.12}$$

In this expression, σ_{ij}^s is the stress distribution found by performing the analysis on a structure of identical geometry and load to that under consideration but with a creep law of the form

$$\dot{\varepsilon}_{ij}/\dot{\varepsilon}_0 = \Delta^{\nu-1}(\sigma_{ij}/\sigma_0)\partial\Delta/\partial(\sigma_{ij}/\sigma_0) \tag{21.13}$$

This calculation is equivalent to the standard steady-state analysis but with a different creep law. The substitution of the stress field σ_{ij}^s in equation (21.12) is a simple calculation.

It is found convenient in practice to express the result (21.12) in terms of the so-called reference rupture stress. If a uniaxial specimen is subjected to a constant stress σ_u so that the rupture life is t_u, then $t_u/t_0 = (\sigma_0/\sigma_u)^\nu$. Equating this to the time t_u calculated for the component gives an expression for σ_u according to the equation

$$(\sigma_u/\sigma_0) = \left\{ \int_V \Delta^\nu(\sigma_{ij}^s/\sigma_0)\,\mathrm{d}V/V \right\}^{1/\nu} \tag{21.14}$$

If this value σ_u is used in conjunction with the uniaxial creep rupture data then an upper bound is obtained on the time to rupture of the structure. This is a convenient way of relating component performance directly to uniaxial data and avoids data fitting procedures.

Another upper bound on rupture time has been developed by Goodall *et al.* [11]. In their calculation, the limit load P_0 of a component of identical geometry is determined for a material with a yield stress σ_0 and a yield surface $\Delta(\sigma_{ij}/\sigma_0) = 1$, which is identical in shape to the isochronous surface. Then the representative rupture stress σ_u is given by

$$\sigma_u/\sigma_0 = P_i/P_0 \tag{21.15}$$

and once again when used in conjuction with the uniaxial creep rupture data gives an over-estimate of the rupture life of the component. This expression gives a better bound than that of equation (21.14) but the difference is not great and the selection of the procedure will normally be related to convenience and local computing expertise.

The expression (21.14) illustrates that the reference rupture stress is related to a weighted average stress over the volume which is in accord with the experimental observations discussed in the introduction. While the above procedures do give upper and therefore unsafe bounds on rupture life, experience from experiments reported in literature [11] suggest that the bounds are often good enough for practical design purposes and are easy to apply.

It has proved difficult to find a procedure which gives a lower bound on rupture life. Some limited progress has been made for components which are kinematically determinate. For materials for which the isochronous surface is given by the Mises condition $\Delta(\sigma_{ij}/\sigma_0) = (\bar{\sigma}/\sigma_0)$ then the reference rupture stress σ_L [1] giving a lower bound on time is

$$\sigma_L/\sigma_0 = \left\{ \int_V (\bar{\sigma}_s/\sigma_0)^{n+1+\nu}\,\mathrm{d}v \Bigg/ \int_V (\bar{\sigma}_s/\sigma_0)^{n+1}\,\mathrm{d}v \right\}^{1/\nu} \tag{21.16}$$

where $\bar{\sigma}_s$ is the steady-state effective stress distribution.

For components in a state of plane stress, the reference rupture stress is

$$\sigma_L/\sigma_0 = \frac{\int_V (\bar{\sigma}_s/\sigma_0)^{n+1}\Delta^\nu(\sigma_{ij}^s/\sigma_0)\,\mathrm{d}v}{\int_V (\bar{\sigma}/\sigma_0)^{n+1}\,\mathrm{d}v} \tag{21.17}$$

where σ_{ij}^s is the steady-state stress distribution.

Again, it can be observed that this reference stress which gives a lower or conservative estimate of rupture life is a weighted average of stresses over the component volume and can be calculated using steady-state stress distributions.

21.5 REPRESENTATION OF ANISOTROPIC DAMAGE

Visual inspection of creep damage readily confirms that in certain metals such as copper and nickel-based alloys, damage is isotropic. In an effort to form a clearer picture of the anisotropic growth of damage copper and aluminium have been subjected to nonproportional histories of loading [12]. In these tests, the value of the maximum shear and normal stresses remained constant so that the value of $\Delta(\sigma_{ij}/\sigma_u)$ remained constant (equation (21.6a)) but the directions of the stress fields rotated through 33.7 degrees. The behaviour of the aluminium and copper is quite different. The damage growth in aluminium is independent of the stress direction while in copper the damage growth in the two stress directions apparently grows independently of one another.

The multiaxial generalization of equations (21.4) must take into consideration the effects of the anisotropic growth of the grain boundary damage. Leckie and Onat [13] showed that by applying rotation invariance requirements, the damage may be expressed in terms of a number of irreducible tensors of even rank. The result is general and it is in keeping with the theory of Krachinovitz [14] who showed that the physical representation of brittle damage may be described by a second order tensor. In the case of creep damage, the physical interpretation of the damage tensors may be given in terms of the moments of the voids on the grain boundaries. If the volume distribution of damage on a unit sphere in the direction of the normal $\bar{n}$ is $V(\bar{n})$, it may be demonstrated [15] that

$$V(\bar{n}) = V_0 + V_{ij}f_{ij} + V_{ijkl}f_{ijkl} + \dots \tag{21.18}$$

where $f_{ij}(\bar{n}) = n_i n_j - (\frac{1}{3})\delta_{ij}$

$$f_{ijkl}(\bar{n}) = n_i n_j n_k n_l - (\tfrac{1}{7})(\delta_{ij}n_k n_l + \delta_{ik}n_j n_l + \delta_{il}n_j n_k + \delta_{jk}n_i n_l + \delta_{jl}n_i n_k + \delta_{kl}n_i n_j) + (\tfrac{1}{35})(\delta_{ij}\delta_{kl} + \delta_{ik}\delta_{jl} + \delta_{ik}\delta_{jl} + \delta_{il}\delta_{jk})$$

The expressions for the damage tensors are

$$\begin{aligned} V_0 &= (1/4\pi)\int_A V(\bar{n})\,\mathrm{d}A \\ V_{ij} &= (1/4\pi)C_{ij}\int_A V(\bar{n})f_{ij}\,\mathrm{d}A \end{aligned} \tag{21.19}$$

with $C_{ij} = 5$ when $i = j$, $\frac{15}{2}$ when $i \neq j$.

Hence, by means of these results, it is possible to convert physical measurement of creep damage into its tensorial components which follow the appropriate transformation rules. Similarly, at the conclusion of a finite element calculation, the tensorial components can be used to construct the physical damage according to equation (21.18).

The non-proportional tests on aluminium referred to earlier suggest that constitutive equations of the following form describe the behaviour

$$\dot{\varepsilon} = f(J_2 v_0) s_{ij} + C_{ijkl} \dot{\sigma}_{kl}$$
$$\mathrm{d}v_0/\mathrm{d}t = g(J_2, v_0), \qquad (21.20)$$

and for rupture, $v_0 J_2 = \text{constant}$.

Only the first term in the damage expansion of equation (21.18) is required. The same is not true to copper, however, which shows such strongly anisotropic behaviour that the distributions of damage are very nearly peaked. It is noted from the experiments that damage during a given loading occurs on faces whose normals are in the direction of the current maximum stress. When the stress field is rotated, damage again accumulates on the face with maximum normal tensile stress and damage on the other faces appears to stop. However, the strain rates continue to grow and do not appear to reflect the anisotropic behaviour characteristics of damage. If *i* faces suffer damage during the stress history, then the constitutive equations for damage on the face with normal $\bar{n}_1$ appear to have the form

$$\mathrm{d}v_i/\mathrm{d}t = f(v_i \sigma^{\bar{n}_i}) \qquad (21.21a)$$

where $\sigma^{\bar{n}_i}$ is the normal stress corresponding to the $\bar{n}_i$ direction and v_i is the volume density on the face whose normal is $\bar{n}_i$.

The strain rate is given by

$$\dot{\varepsilon}_{ij} = f(J_2 \Sigma v_i) s_{ij} + C_{ijkl} \dot{\sigma}_{kl} \qquad (21.21b)$$

and for rupture,

$$\max v_i \sigma^{\bar{n}_i} = \text{constant} \qquad (21.21c)$$

Further attempts are necessary to develop a representation which describes this type of behaviour because, in its present form, the computational effort required to keep track of all the damaged faces might be considerable.

ACKNOWLEDGEMENTS

The author acknowledges a grant from the National Science Foundation.

REFERENCES

1. F. A. Leckie and D. R. Hayhurst, ‘Creep rupture of structures’, *Proc., Roy. Soc., London*, *A*240, p. 323.
2. A. C. F. Cocks and M. F. Ashby, ‘On creep fracture by void growth’, *Progress in Materials*, pp. 1–56, Pergamon Press, 1981.

3. F. A. Leckie and D. R. Hayhurst, 'The damage concept in creep mechanics', *Mech. Res. Commun.*, **2**, (1) (1975).
4. Y. N. Rabotnov, *Creep Problems in Structural Members*, Amsterdam, North-Holland Pub. Co., 1969.
5. L. M. Kachanov, 'Time of the fracture process under creep conditions', *Izv. Akad. Nauk., SSR OTN*, *Tekh. Nauk.*, Vol. 8, No. 26 (1958).
6. D. R. Hayhurst, 'Creep rupture under multiaxial states of stress', *J. Mech. Phys. Solids*, **20**, 381 (1972).
7. A. E. Johnson, J. Henderson, and B. Khan, 'Complex Stress and creep relaxation and Fracture of Metallic Alloys', Edinburgh, HMSO, 1962.
8. F. A. Leckie, 'The constitutive equations of continuum creep damage mechanics', *Phil. Trans. Roy. Soc.*, London A288, p. 27 (1978).
9. D. R. Hayhurst, P. R. Dimmer, and M. W. Chernuka, 'Estimates of the creep rupture lifetime of structures using the finite element method', *J. Mech. Phys. Solids*, **23**, 335–355 (1975).
10. F. A. Leckie and W. Wojewodzki, 'Estimates of rupture life—constant load', *Int. J. Solids and Struct.*, **11**, 1357 (1975).
11. I. N. Goodall, R. D. H. Cockroft, and E. J. Chubb, 'An approximate description of the creep rupture of structures', *Int. J. Mech. Sci.*, **17**, 351 (1975).
12. W. A. Trampczynski, D. R. Hayhurst, and F. A. Leckie, 'Creep rupture of copper and aluminum under Non-Proportional Loading', *J. Mech. Phys. Solids*, **29**, 353 (1981).
13. F. A. Leckie and E. T. Onat, 'Tensorial nature of damage measuring internal variables', *U.I.T.A.M. Symp., Physical Non-Linearities in Structural Analysis*, Springer-Verlag, 1980.
14. D. Krachinovitz and G. V. Fonseka (1982), 'The continuous damage theory of British materials—general theory', to appear in *J. Appl. Mech.*
15. E. T. Onat, Private Communication, 1982.

Mechanics of Engineering Materials
Edited by C. S. Desai and R. H. Gallagher

Chapter 22

Uniform Formulation of Constitutive Equations for Clays and Sands

Z. Mróz and O. C. Zienkiewicz

22.1 INTRODUCTION

The chapter consists of two parts. The first part (Sections 22.2 and 22.3) is devoted to a general discussion of rate, time-independent constitutive equations, and the associated various forms of loading, unloading, and reloading conditions. The important question of admissibility conditions which should be imposed on constitutive equations is briefly discussed and the violation of these conditions by some formulations is indicated. Further, the formulations within non-linear elasticity, hypoelasticity, and plasticity are critically reviewed. In particular, it is shown that when unloading and continuity conditions are incorporated into the model, there is no difference between hypoelasticity and plasticity models and both types of models can be described in one uniform formulation.

The second part of the chapter is devoted to description of inelastic clays' and sands' behaviour within the same set of fundamental assumptions. Such formulation has a definite advantage since the mechanical response of both materials can be treated within one model and only particular material parameters are varied when passing from one to the other material. In Sections 22.4 and 22.5 both the details of description and also some applications will be presented. An extensive discussion of model can also be found in [35, 37].

22.2 NON-LINEAR ELASTICITY AND HYPOELASTICITY MODELS

The application of *linear* elasticity in soil mechanics provides only the first approximation valid for stress states lying far below the limit state. In fact, the linear elastic analysis does not describe the considerable stress redistribution occurring in the non-linear range prior to failure and underestimates deformations of soil or settlement of structures. A more realistic description is therefore obtained by assuming *non-linear* response under compression or shear and a non-linear relation between volumetric strain and applied pressure. For monotonically increasing stress components the material can be assumed as non-linearly elastic, so there is no need to decompose strains into elastic and plastic portions. Under load reversal, the elastic stress–strain response, is the same and there is no residual deformation after unloading, Figure 22.1(a).

The total stress state $\boldsymbol{\sigma}^{\mathrm{t}}$ is decomposed, as usually, into the effective stress $\boldsymbol{\sigma}$ and the pore pressure p_{w}, thus

$$\boldsymbol{\sigma}^{\mathrm{t}} = \boldsymbol{\sigma} - p_{\mathrm{w}}\boldsymbol{\delta} \tag{22.1}$$

where $\boldsymbol{\delta}$ denotes the unit tensor (or Kronecker delta). The compressibility modulus of fluid is denoted by K_{f} and that of soil skeleton by K_{s}. For soils, it is usually assumed that $K_{\mathrm{f}} \rightarrow \infty$ and $K_{\mathrm{s}} \rightarrow \infty$ and the compressibility of soil is only due to void closure or growth. The constitutive equations of soils are then expressed in terms of effective stresses. The effective stress and strain deviators $s_{ij} = \sigma_{ij} - \frac{1}{3}\sigma_{kk}\delta_{ij}$, $e_{ij} = \varepsilon_{ij} - \frac{1}{3}\varepsilon_{kk}\delta_{ij}$ provide the invariants

$$\sigma_{\mathrm{e}} = (\tfrac{3}{2}\mathbf{s}\cdot\mathbf{s})^{1/2} = 3J_2' \qquad \varepsilon_{\mathrm{e}} = (\tfrac{2}{3}\mathbf{e}\cdot\mathbf{e})^{1/2} = (3I_2')^{1/2} \tag{22.2}$$

where J_2' and I_2' are the second invariants of s_{ij} and e_{ij} and the dot between two vectors or tensors of the same order denotes their scalar product. Similarly,

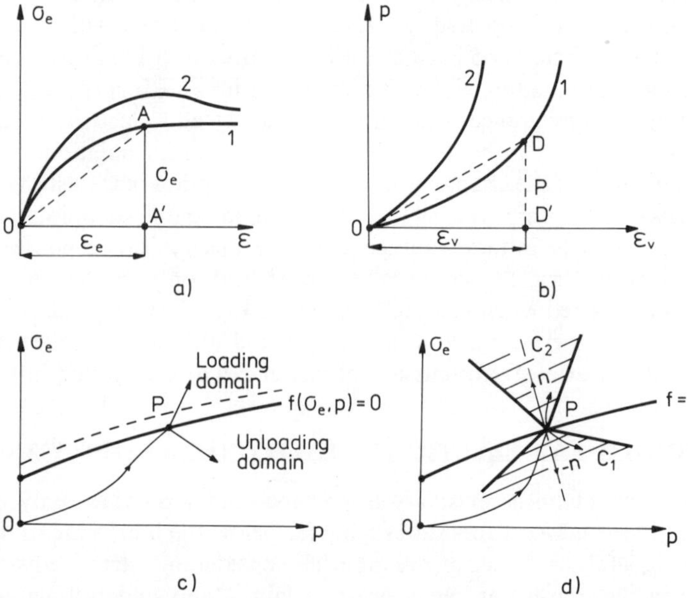

Figure 22.1 Non-linear response of a normally consolidated (curves 1) and over-consolidated (curves 2) soil; (a) shear (or deviatroic) stress–strain response, (b) isotropic compression curves, (c) loading–unloading domains in the σ_{e}, p-plane, (d) cones of admissible loading paths satisfying convexity condition

denote

$$p = -\tfrac{1}{3}J_1 \qquad \varepsilon_v = -I_1 \tag{22.3}$$

where $J_1 = \sigma_{kk}$ and $I_1 = \varepsilon_{kk}$ are the first invariants of the stress and strain tensors.

Assume that the non-linear relations between stress and strain have the same form as linear relations for an isotropic material, that is

$$s_{ij} = 2G_s e_{ij}, \qquad p = K_s \varepsilon_v \tag{22.4}$$

where

$$G_s = \frac{\sigma_e}{3\varepsilon_e} = G_s(\varepsilon_e, \varepsilon_v), \qquad K_s = \frac{p}{\varepsilon_v} = K_s(\varepsilon_e, \varepsilon_v), \tag{22.5}$$

are the *secant moduli* for deviatoric and spherical components of the stress and strain tensors. These moduli can easily be identified from the non-linear response curves between σ_e, ε_e, and p, ε_v, Figures 22.1(a), (b).

Requiring the relations (22.4) to represent a non-linear elastic material it should be demonstrated that they follow from the elastic strain or stress potentials $U(\varepsilon_{ij})$ or $W(\sigma_{ij})$, namely

$$\sigma_{ij} = \frac{\partial U}{\partial \varepsilon_{ij}}, \qquad \varepsilon_{ij} = \frac{\partial W}{\partial \sigma_{ij}} \tag{22.6}$$

For an isotropic material there is $U = U(\varepsilon_e, \varepsilon_v)$ and

$$s_{ij} = \frac{\partial U}{\partial \varepsilon_e}\frac{\partial \varepsilon_e}{\partial e_{ij}} = \frac{\partial U}{\partial \varepsilon_e}\frac{2}{3}\frac{e_{ij}}{\varepsilon_e}, \qquad p = \frac{\partial U}{\partial \varepsilon_v} \tag{22.7}$$

The secant moduli can now be related to the specific strain energy

$$G_s = \frac{1}{3}\frac{\partial U}{\partial \varepsilon_e}\frac{1}{\varepsilon_e}, \qquad K_s = \frac{\partial U}{\partial \varepsilon_v}\frac{1}{\varepsilon_v} \tag{22.8}$$

and from (22.8) it follows that

$$\frac{1}{\varepsilon_v}\frac{\partial G_s}{\partial \varepsilon_v} = \frac{1}{3\varepsilon_e}\frac{\partial K_s}{\partial \varepsilon_e}. \tag{22.9}$$

The potentiality relations (22.9) impose the restrictions on the forms (22.5), so the shear moduli cannot be the arbitrary functions of ε_e and ε_v. In particular when

$$G_s = G_s(\varepsilon_e), \qquad K_s = K_s(\varepsilon_v) \tag{22.10}$$

the conditions (22.9) are obviously satisfied. The elastic strain energy is then a sum of deviatoric and volumetric strain energies.

In order to determine the tangent stiffness matrix, let us differentiate the finite relations (22.4) and obtain

$$\begin{bmatrix} \dot{s}_{ij} \\ \dot{p} \end{bmatrix} = \begin{bmatrix} A_{dd} & A_{ds} \\ A_{sd} & A_{ss} \end{bmatrix} \begin{bmatrix} \dot{e}_{kl} \\ \dot{\varepsilon}_v \end{bmatrix} \tag{22.11}$$

where

$$A_{\mathrm{dd}} = 2G_{\mathrm{s}}\delta_{ij} + \frac{4}{3}\frac{\partial G_{\mathrm{s}}}{\partial \varepsilon_{\mathrm{e}}} e_{ij}\frac{e_{\mathrm{kl}}}{\varepsilon_{\mathrm{e}}}, \qquad A_{\mathrm{ss}} = K_{\mathrm{s}} + \frac{\partial K_{\mathrm{s}}}{\partial \varepsilon_{\mathrm{v}}}\varepsilon_{\mathrm{v}}, \tag{22.12}$$

$$A_{\mathrm{ds}} = 2\frac{\partial G_{\mathrm{s}}}{\partial \varepsilon_{\mathrm{v}}} e_{ij}, \qquad A_{\mathrm{sd}} = \frac{2}{3}\frac{\partial K_{\mathrm{s}}}{\partial \varepsilon_{\mathrm{e}}}\varepsilon_{\mathrm{v}}\frac{e_{\mathrm{kl}}}{\varepsilon_{\mathrm{e}}}$$

The off-diagonal terms A_{sd} and A_{ds} representing the coupling between deviatoric and volumetric stress and strain changes are not equal in general, that is $A_{\mathrm{sd}} \neq A_{\mathrm{ds}}$. The equality $A_{\mathrm{sd}} = A_{\mathrm{ds}}$ occurs only when the potentially relations (22.9) are satisfied. In particular, when (22.10) occurs, the off-diagonal terms vanish and we have

$$\begin{bmatrix} \dot{s}_{ij} \\ \dot{p} \end{bmatrix} = \begin{bmatrix} A_{dd} & 0 \\ 0 & A_{ss} \end{bmatrix} \begin{bmatrix} \dot{e}_{ij} \\ \dot{\varepsilon}_{\mathrm{v}} \end{bmatrix} \tag{22.13}$$

or simply

$$\dot{s}_{ij} = 2G_{\mathrm{t}}\dot{e}_{ij}, \qquad \dot{p} = K_{\mathrm{t}}\dot{\varepsilon}_{\mathrm{v}} \tag{22.14}$$

where $G_{\mathrm{t}} = \frac{1}{2}A_{dd}$, $K_{\mathrm{t}} = A_{ss}$ are the *tangent* shear and bulk moduli. The relations (22.14) represent the linear isotropic response in rates (or increments) of stress and strain whereas the relations (22.11) represent an anisotropic behaviour with anisotropy induced by the strain tensor. As the relations (22.12) are linear and homogeneous of order one in stress and strain rates, not necessarily following from the elastic potential, they can be regarded as *hypoelastic* constitutive relations. In fact, we could start from the rate form (22.11) without any recourse to the finite relations (22.4) and identify the constitutive matrix coefficients A_{dd}, A_{ss}, A_{ds}, A_{sd} from experimental data. In this case, these coefficients can be assumed as either strain or stress dependent.

A particular form of (22.14) was elaborated by Kondner and Zelasko [22, 23] and later by Duncan and Chang [17] who assumed a hyperbolic relation for a form of the compression curve in a triaxial test. For a general stress state, this form can be expressed in terms of σ_{e} and ε_{e}, namely

$$\sigma_{\mathrm{e}} = \frac{\varepsilon_{\mathrm{e}}}{A + B\varepsilon_{\mathrm{e}}} \tag{22.15}$$

where A and B are the material parameters. The tangent elastic bulk modulus can easily be identified from the isotropic compression test and it is usually assumed to depend on the mean hydrostatic stress,

$$K_{\mathrm{t}} = k_0 p_{\mathrm{a}} \left(\frac{p}{p_{\mathrm{a}}}\right)^m \tag{22.16}$$

where k_0 and m are the material parameters and p_{a} denotes the atmospheric pressure. However, in order to account for the dilatancy effect, it can be assumed that there is an additional volumetric strain rate $\dot{\varepsilon}_v^d$ related directly to the deviatoric strain rate, $\dot{\varepsilon}_v^d = D_d\dot{\varepsilon}_e$, where D_d denotes the dilatancy parameter, cf. Byrne and Eldridge [5].

Writing

$$\dot{p} = K_t(\dot{\varepsilon}_v - D_d\dot{\varepsilon}_v^d) = K_t\dot{\varepsilon}_v - K_t D_d \frac{2}{3}\frac{e_{kl}\dot{e}_{kl}}{\varepsilon_e} \tag{22.17}$$

the form (22.11) of rate equations is obtained for which

$$A_{ds} = 0, \qquad A_{sd} = -K_t D_d \frac{2}{3}\frac{e_{kl}}{\varepsilon_e}. \tag{22.18}$$

So far, the rate equations (22.11) or (22.14) are assumed to apply for monotonic loading though the criterion of unloading has not been formulated. For instance, it can be assumed that these equations apply only when $\dot{f} = \dot{\sigma}_e - \alpha\dot{p} > 0$ whereas for $\dot{f} < 0$, different rate equations are to be formulated, Figure 27.1(c). This kind of approach was used by Baron and Nelson [2] and recently by Dickin and King [9]. For instance, in [2] it was assumed that

$$\begin{aligned} s_{ij} &= 2G_t\dot{e}_{ij} \quad \text{for } \dot{\sigma}_e > 0, \qquad \dot{p} = K_t\dot{\varepsilon}_v \quad \text{for } \dot{p} > 0 \\ \dot{s}_{ij} &= 2G_u\dot{e}_{ij} \quad \text{for } \dot{\sigma}_e < 0, \qquad \dot{p} = K_u\dot{\varepsilon}_v \quad \text{for } \dot{p} < 0 \end{aligned} \tag{22.19}$$

where G_u and K_u are the new tangent moduli for the unloading paths. In [9], only one loading condition was applied following the Coulomb yield condition.

Such formulations of loading–unloading conditions associated with linear relations (22.11) or (22.14) possess unfortunately the deficiency, namely they violate the continuity condition for neutral stress paths satisfying the equality $\dot{f} = 0$. Thus, for instance, for stress paths satisfying $\dot{\sigma}_e = 0$, the relations (22.18) provide

$$\dot{\mathbf{e}}_1 = \frac{1}{2G_t}\dot{\mathbf{s}}, \qquad \dot{\mathbf{e}}_2 = \frac{1}{2G_t}\dot{\mathbf{s}}$$

that is different strain rates for the same deviatoric stress rate. This local non-uniqueness introduces discontinuity in the material response and may result in the global non-uniqueness of solution of boundary value problems.

It can be argued that the proposed rate equations, different in loading and unloading domains and violating the continuity condition on the separating boundary, are applicable only in some conical sub-domains C_1 and C_2 in the stress space, such that the admissible stress path cannot lie on the surface $f = 0$. Such admissibility conditions were discussed by Mróz [28] by starting from the condition of local *convexity of transformation* between rates of stress and strain, namely

$$(\dot{\boldsymbol{\sigma}}_2 - \dot{\boldsymbol{\sigma}}_1)\cdot(\dot{\boldsymbol{\varepsilon}}_2 - \dot{\boldsymbol{\varepsilon}}_1) > 0 \tag{22.20}$$

for any two pairs of rates $\dot{\sigma}_1, \dot{\varepsilon}_1$ and $\dot{\sigma}_2, \dot{\varepsilon}_2$ interconnected by the constitutive relation. Inequality (22.20) provides the sufficient condition for *uniqueness* of the boundary-value problems. It was shown in [28] that (22.20) implies the continuity of material response on separating surfaces $f = 0$. However, when this continuity condition is not satisfied, then (22.20) provides the *admissible domains* C_1 and C_2 in the stress space, such as those shown in Figure 22.1(d). The applicability of the rate relations is then limited to such loading paths that belong to the admissible domains. In the next section, we shall discuss the loading–unloading conditions in terms of the plasticity theory.

22.3 RATE RELATIONS, LOADING–UNLOADING CONDITIONS, AND MATERIAL MEMORY FOR ELASTIC–PLASTIC BEHAVIOUR

22.3.1 Uniaxial stress–strain response

The rate relations discussed in the previous section were referred to total strain rates and there was no distinction between elastic and irreversible portions.

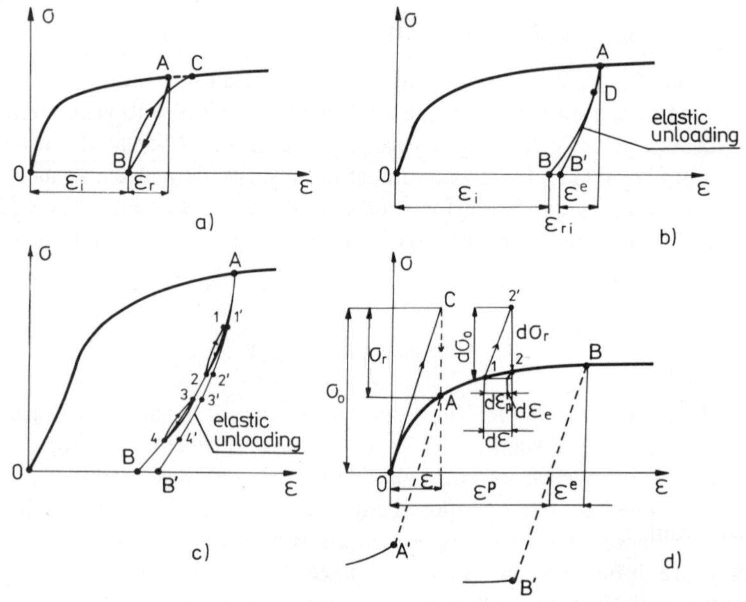

Figure 22.2 (a) Typical triaxial compression or shear curve for loading, unloading, and reloading, (b) determination of the yield point D on the unloading curve (departure from the elastic rebound curve AB′), (c) determination of the elastic unloading curve by executing small cycles 1–2 or 3–4 at various stress levels, (d) idealized model: irreversible strains beneath the prestress level are neglected and the linear response is elastic

However, such distinction is necessary when both the loading and unloading process are to be incorporated into one formulation.

Referring to Figure 22.2, presenting the uniaxial compression curve, it is seen that when the stress is removed after loading to A, the recoverable strain is ε_r and irreversible strain equals ε_i. When elastic behaviour is linear and the unloading curve AB does not depart from linearity, then unloading is elastic and the irreversible strain equals plastic strain, $\varepsilon_i = \varepsilon^p$. However, when the unloading curve is not linear and during reloading there is a hysteresis loop (ABC in Figure 22.2(a)), we conclude that reverse plastic strains also develop on the unloading portion AB. To determine the elastic unloading curve, small loading–unloading cycles should be executed at consecutive points on AB. Assuming that the initial stiffness modulus on the unloading or reloading branches represents elastic stiffness, the elastic unloading curve such as AB′ can be constructed, Figure 22.2(c). Now the total strain is composed of three portions, namely

$$\varepsilon = \varepsilon_i + \varepsilon_{ri} + \varepsilon^e \tag{22.21}$$

when ε_i is the irreversible strain after unloading, ε_{ri} is the reverse plastic stain and ε^e denotes the elastic strain. When the unloading branch ADB departs from the elastic curve AB′ at D, this point can be asumed as the yield limit during unloading. In actuality, the determination of D depends on the conventional offset limit and for some idealizations the plastic strains developed during unloading and reloading are neglected. Such is the case presented in Figure 22.2(d) where the line AA′ is assumed as elastic and during reverse loading the plastic strain increments are assumed to develop at A′.

It is thus seen that the size of the elastic domain after initial loading much depends on the definition of the yield point, that is on the value of plastic strain defining the onset of yielding. For small offset values, the point D will be very close to A or coincide with it, whereas for large offset values the elastic domain will be large such as that in Figure 22.2(d). For cyclic loading problems, when the description of hysteresis loops is important, more accurate constitutive models should be used and the elastic domain is then usually assumed as small or vanishing. The reverse plastic strain then occurs on the whole unloading path AB.

Instead of decomposing the total strain into elastic and plastic portions, an alternative decomposition can be carried out for stress corresponding to the prescribed strain, namely

$$\dot{\sigma} = \dot{\sigma}^0 - \dot{\sigma}^r \tag{22.22}$$

when σ^0 denotes the elastic stress corresponding to the prescribed strain and σ^r is the relaxation stress, Figure 22.2(d). In the case of linear elastic behaviour, the elastic stress σ^0 is calculated from Hooke's law, whereas the relaxation stress σ^r represents the departure from non-linearity and reduces the elastic stress to the actual stress. Similar decomposition can also be carried out for the unloading curve. The yield point can therefore be also defined by the magnitude of the relaxation stress instead of permanent strain. For increments, or rates, of stress and strain the similar decomposition occurs, namely

$$d\varepsilon = d\varepsilon^e + d\varepsilon^p, \qquad d\sigma = d\sigma^0 - d\sigma^r \tag{22.23}$$

Let us first describe the material response presented in Figure 22.2 for the uniaxial case. Considering the loading and unloading at point A on the *stable* (hardening) portion of stress–strain curve, Figure 22.2(a), we can write

$$\begin{aligned} d\varepsilon &= L_l\, d\sigma \qquad \text{for } d\sigma > 0 \\ d\varepsilon &= L_u\, d\sigma \qquad \text{for } d\sigma < 0 \end{aligned} \tag{22.24}$$

where L_l and L_u denote the loading and unloading tangent compliance moduli. Note that the relations (22.24) are linear with respect to increments both for loading and unloading. An alternative way is to replace (22.24) by a single *non-linear* relation

$$d\varepsilon = B\, d\sigma + B_1 |d\sigma| \tag{22.25}$$

in which

$$B = \tfrac{1}{2}(L_l + L_u), \qquad B_1 = \tfrac{1}{2}(L_l - L_u) \tag{22.26}$$

In fact, the relation (22.25) has the forms

$$\begin{aligned} d\varepsilon &= (B + B_1)\, d\sigma \quad \text{for } d\sigma > 0, \\ d\varepsilon &= (B - B_1)\, d\sigma \quad \text{for } d\sigma < 0 \end{aligned} \tag{22.27}$$

and in view of (22.26), the relations (22.24) are obtained. Alternatively, it can be written

$$d\sigma = E\, d\varepsilon + E_1 |d\varepsilon| \tag{22.28}$$

and again, we have

$$\begin{aligned} d\sigma &= (E + E_1)\, d\varepsilon \quad \text{for } d\varepsilon > 0, \\ d\sigma &= (E - E_1)\, d\varepsilon \quad \text{for } d\varepsilon < 0 \end{aligned} \tag{22.29}$$

Note that the relations (22.28) or (22.29) apply for both *hardening and softening* portions of the stress–strain curve. Requiring

$$\begin{aligned} E &= \frac{1}{2}\left(\frac{1}{L_l} + \frac{1}{L_u}\right) = \tfrac{1}{2}(K_l + K_u), \\ E_1 &= \frac{1}{2}\left(\frac{1}{L_l} - \frac{1}{L_u}\right) = \tfrac{1}{2}(K_l - K_u) \end{aligned} \tag{22.30}$$

where $K_l = L_l^{-1}$ and $K_u = L_u^{-1}$ denote the tangent stiffness and unloading moduli, we obtain the relations

$$\begin{aligned} d\sigma &= K_l\, d\varepsilon \qquad \text{for } d\varepsilon > 0 \\ d\sigma &= K_u\, d\varepsilon \qquad \text{for } d\varepsilon < 0 \end{aligned} \tag{22.31}$$

which is the inverse form of (22.24). Note that $E_1 < 0$, since $K_1 < K_u$, Figure 22.3. It is thus seen that the local loading–unloading behaviour can be described either by two linear relations accompanied by inequality conditions or by single non-linear relations in stress or strain increments. These relations are prototypes for more general plasticity and endochronic models.

A more familiar form can be ascribed to relations (22.25) or (22.28) by assuming that their first terms represent the elastic behaviour. Thus, (22.25) takes the form $d\varepsilon = d\varepsilon^e + d\varepsilon^p$ where $d\varepsilon^e = B\,d\sigma$ $d\varepsilon^p = B_1|d\sigma|$ and B denotes the elastic compliance modulus. Similarly, relation (22.28) takes the form $d\sigma = d\sigma^0 - d\sigma^r$ where $d\sigma^0$ is the elastic stress increment and $d\sigma^r$ denotes the relaxation stress increment and E is the elastic stiffness modulus. Thus, we have

$$\begin{aligned} &d\varepsilon^e = B\,d\sigma, \qquad d\varepsilon^p = B_1|d\sigma|, \qquad d\varepsilon = d\varepsilon^e + d\varepsilon^p, \\ &d\sigma^0 = E\,d\varepsilon, \qquad d\sigma^r = -E_1|d\varepsilon|, \qquad d\sigma = d\sigma^0 - d\sigma^r \end{aligned} \tag{22.32}$$

and then it follows immediately that the formulations (22.25) and (22.28) imply the larger unloading stiffness modulus than the elastic modulus. In fact, since $E_1 < 0$, the unloading stiffness modulus equals $E - E_1$ and is greater than E, Figure 22.3(b).

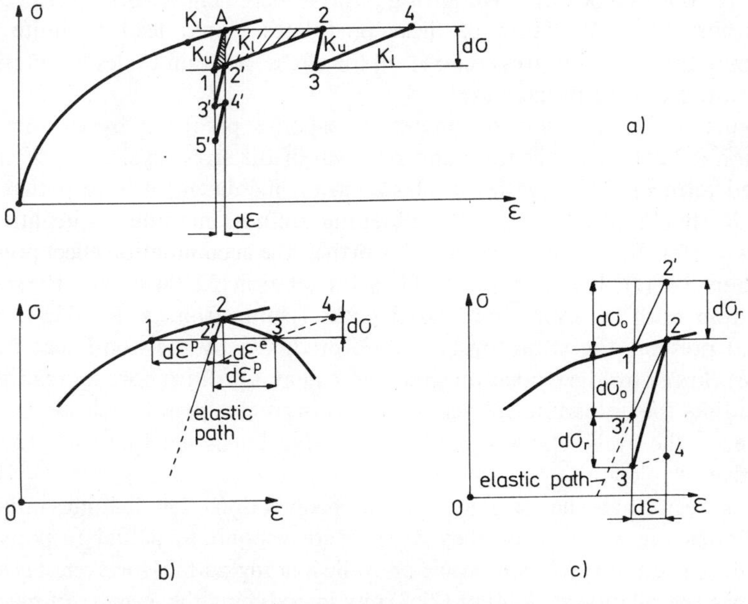

Figure 22.3 Loading–unloading conditions corresponding to linear and nonlinear incremental relations: (a) (24), (b) (25), and (c) (28). Progressive ratchetting and relaxation phenomena for infinitesimal cycles

The predictions of the presented relations for small stress and strain cycles are illustrated in Figure 22.3. In Figure 22.3(a) small stress or strain cycles are applied from the state A and the prediction of (22.24) or (22.31) is presented. Since the unloading stiffness modulus is greater than the loading modulus, that is $K_u > K_l$ or $L_u < L_l$, after each cycle the permanent strain A–z, z–4, etc. accumulates. Similarly, for the prescribed stress cycle, the progressive stress relaxation A–z', z–$4'$, etc. develops. Identifying the unloading stiffness modulus with the elastic modulus, it is seen that for each cycle the incremental work is negative, that is

$$\oint_{\sigma} (\sigma - \sigma_A)\,\mathrm{d}\varepsilon = \tfrac{1}{2}\mathrm{d}\sigma \cdot \mathrm{d}\varepsilon^p = \tfrac{1}{2}(\mathrm{d}\sigma)^2(L_u - L_l)$$

$$= \tfrac{1}{2}(\mathrm{d}\sigma)^2\left(\frac{1}{K_u} - \frac{1}{K_l}\right) < 0 \tag{22.33}$$

and similarly, each strain cycle is associated with the negative incremental work

$$\int_{\varepsilon} (\sigma - \sigma_A)\,\mathrm{d}\varepsilon = \tfrac{1}{2}\mathrm{d}\varepsilon \cdot \mathrm{d}\sigma^r = -\tfrac{1}{2}(\mathrm{d}\varepsilon)^2(K_u - K_i) < 0 \tag{22.34}$$

These inequalities implicate *cyclic instability* of the behaviour in the sense of Drucker or Ilyushin postulates. In fact, (22.23) corresponds to Drucker and (22.34) to Ilyushin postulates requiring positive incremental work on closed stress or strain cycles. Violation of these postulates would lead to finite strain accumulation or finite stress relaxation for stress or strain cycles infinitesimally departing from the initial state.

Figure 22.3(b) presents the material response predicted by the nonlinear relation (22.25) for application and removal of the stress-cycle. Identifying the second term of (22.25) with the plastic strain increment, it is seen that when $B_1 > B$, that is $|\mathrm{d}\varepsilon_p| > |\mathrm{d}\varepsilon_e|$, the unloading stiffness modulus is negative and $\mathrm{d}\sigma \cdot \mathrm{d}\varepsilon < 0$ for the unloading path 2–3. Further, the accumulation effect presented in Figure 22.3(a) obviously occurs. Thus, the relation (22.25) violates the stability condition both for cyclic and radial paths issuing from A. Similarly, Figure 22.3(d) presents the strain cycle 1–2–3, predicted by the non-linear relation (22.28). Since now the relaxation stress increment is positive both for loading and unloading, the unloading stiffness modulus is greater than the elastic modulus. However, the cyclic stability condition of Ilyushin is violated since (22.34) is negative.

It is seen that the presented three propositions for loading–unloading conditions are *defective* as they do not correspond to actual response and introduce instability for monotonic or cyclic loading paths. However, it is easy to improve the relations (22.24) or (22.31) by introducing the *memory* of particular loading events. Consider the process of loading to the prescribed stress A, unloading to B, and subsequent reloading to A. It can be required that

$$
\begin{aligned}
d\varepsilon &= L_l\, d\sigma && \text{for } d\sigma > 0\text{—loading to } A \\
d\varepsilon &= L_u\, d\sigma && \text{for } d\sigma < 0, \sigma < \sigma_A\text{—unloading from } A \\
d\varepsilon &= L_r\, d\sigma && \text{for } d\sigma > 0, \sigma > \sigma_B, \sigma < \sigma_A\text{—reloading from } B \\
d\varepsilon &= L_l\, d\sigma && \text{for } d\sigma > 0, \sigma > \sigma_A\text{—loading from } A
\end{aligned}
\tag{22.35}
$$

where the compliance modulus L_r may be equal to the unloading modulus K_u or different. In this way, not only the sign of $d\sigma$ defines the loading–unloading events but also the memory of points where stress increments change their sign. We shall call σ_A the *maximum prestress* and σ_B *the stress reversal* in our loading history.

The non-linear relations (22.25) and (22.28) do not require this definition of particular loading events and are much simpler. On the other hand, they possess disadvantages discussed in this section. In particular, (22.28) is a uniaxial prototype of endochronic theories [3, 48] where the 'intrinsic time' measure is expressed as the absolute value of the total strain, and of non-linear models in which the stress rate is a non-linear function of the strain rate [6]. In what follows, we shall extend our discussion to the multiaxial case and discuss loading–unloading conditions expressed in terms of stress or strain increments.

22.3.2 Multiaxial stress–strain response

There are numerous possibilities to generalize the uniaxial conditions to a general multiaxial case. Figure 22.4 illustrates some of these possibilities. Consider the stress path OP in the stress space and associate with it a *loading* surface $f_1 = 0$ such that the stress point remains on this surface. Thus for a hardening material any stress path directed in the exterior of this surface corresponds to loading process whereas the stress path directed into the interior of $f_1 = 0$ corresponds to *unloading process* (not necessarily elastic). Different rate relations apply for loading and unloading, so we have

$$
\begin{aligned}
\dot{\boldsymbol{\varepsilon}} &= \mathbf{C}_1 \dot{\boldsymbol{\sigma}} && \text{for } \dot{\boldsymbol{\sigma}} \cdot \mathbf{n}_l > 0 \\
\dot{\boldsymbol{\varepsilon}} &= \mathbf{C}_2 \dot{\boldsymbol{\sigma}} && \text{for } \dot{\boldsymbol{\sigma}} \cdot \mathbf{n}_l < 0
\end{aligned}
\tag{22.36}
$$

where $\mathbf{C}_1$ and $\mathbf{C}_2$ are the compliance matrices depending on the stress state and on a set of state parameters, but not on the stress rate. The relations (22.18) are therefore *linear* in respective semi-spaces $\dot{\boldsymbol{\sigma}} \cdot \mathbf{n}_l > \mathbf{0}$ and $\dot{\boldsymbol{\sigma}} \cdot \mathbf{n}_l < 0$. Here $\mathbf{n}_l$ denotes the unit vector (or normalized tensor) normal to the loading surface $f_1 = 0$, that is

$$
\mathbf{n}_l = \frac{\partial f_1 / \partial \boldsymbol{\sigma}}{\left[\dfrac{\partial f_1}{\partial \boldsymbol{\sigma}} \cdot \dfrac{\partial f_1}{\partial \boldsymbol{\sigma}} \right]^{1/2}}
\tag{22.37}
$$

To define the unloading or reloading processes occurring within the loading surface $f_l = 0$, let us introduce the *unloading surface* $f_u = 0$ lying within the domain enclosed by $f_l = 0$. The unloading process will continue when the stress path is directed into the interior of the unloading surface $f_u = 0$ and reloading occurs for the stress trajectory directed into the exterior of $f_u = 0$, thus

$$\begin{aligned} &\text{unloading:} \quad \dot{\boldsymbol{\varepsilon}} = \mathbf{C}_2\dot{\boldsymbol{\sigma}} \quad \text{for } \dot{\boldsymbol{\sigma}}\cdot\mathbf{n}_u < 0, \quad f_l < 0 \\ &\text{reloading:} \quad \dot{\boldsymbol{\varepsilon}} = \mathbf{C}_3\dot{\boldsymbol{\sigma}} \quad \text{for } \dot{\boldsymbol{\sigma}}\cdot\mathbf{n}_u > 0, \quad f_l < 0 \end{aligned} \tag{22.38}$$

where $\mathbf{n}_u$ is the unit normal vector directed into the exterior of $f_u = 0$. Figures 22.3(a) and 22.3(b) present two cases of unloading surfaces: whereas in Figure 22.3(a) the unloading surface shrinks and expands with respect to a fixed point C which can be its symmetry centre, the unloading surface in Figure 22.3(b) possesses its transformation centre at A so it may translate, shrink, and expand during the loading process. More generally, it can be assumed that the unloading surface may translate and shrink in a more complex way, depending on the plastic strain.

As seen from Figures 22.3(a), (b), for stress paths, such as PS emanating from P, at some point S the path passes into the exterior of $f_u = 0$ and then the *reloading event* commences for which the surface $f_u = 0$ expands with the stress point. Thus

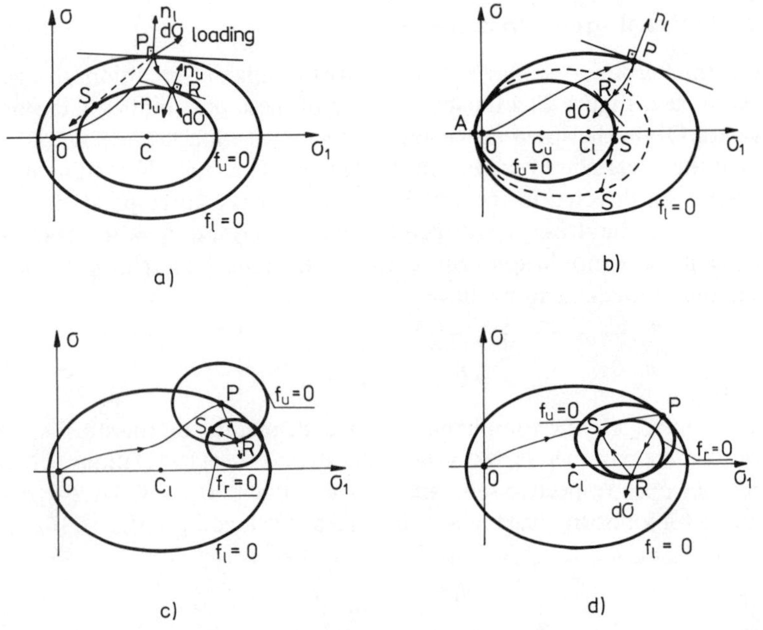

Figure 22.4 Loading, unloading, or reloading surfaces in the stress-space. (a) Unloading surface shrinking toward fixed centre C, (b) unloading surface shrinking toward fixed point A, (c) unloading surface expanding from the stress reversal point P, (d) unloading surface tangential at P to the maximal loading surface

the same surface is used to define both unloading and reloading events. This definition has the disadvantage as there is no memory of past loading events and only three kinds of loading processes exist: loading when the stress point is on the surface $f_1 = 0$, unloading or reloading when the stress point is on the surface $f_u = 0$. On the other hand, the memory of maximal prestress is incorporated into the model by specifying the maximal loading surface $f_1 = 0$.

Figures 22.3(c) and 22.3(d) present two other possibilities of defining the unloading process. The new unloading surface $f_u = 0$ grows in size and the unloading process terminates when the stress point reaches the surface $f_1 = 0$. We have therefore

$$\begin{aligned} &\text{unloading:} \quad && \dot{\boldsymbol{\varepsilon}} = \mathbf{C}_2 \dot{\boldsymbol{\sigma}} \quad && \text{for } \dot{\boldsymbol{\sigma}} \cdot \mathbf{n}_u > 0, \quad f_1 < 0 \\ &\text{reloading:} \quad && \dot{\boldsymbol{\varepsilon}} = \mathbf{C}_3 \dot{\boldsymbol{\sigma}} \quad && \text{for } \dot{\boldsymbol{\sigma}} \cdot \mathbf{n}_u > 0, \quad f_1 < 0 \end{aligned} \tag{22.39}$$

where $\mathbf{n}_u$ is the unit normal vector directed into the exterior of $f_u = 0$. In Figure 22.3(c) the unloading surface is centred at P and intersects the surface $f_1 = 0$, whereas in Figure 22.3(d) the unloading surface is tangential at P to the initial loading surface. When the reloading occurs, a new stress reversal surface develops from the stress reversal point R, Let us note that not only the particular loading events are defined, but the model possesses the memory of stress reversal points, such as P and R. This memory is erased by subsequent loading events of greater intensity, that is when the stress point reaches the previous loading or stress reversal surface.

One of important constraints imposed on compliance matrices $\mathbf{C}_1$, $\mathbf{C}_2$, and $\mathbf{C}_3$ is the *continuity condition.* Namely, for neutral *stress paths* lying on the loading surface, the strain rates predicted by two constitutive relations should be the same, namely

$$\mathbf{C}_1 \dot{\boldsymbol{\sigma}} = \mathbf{C}_2 \dot{\boldsymbol{\sigma}} \qquad \text{for } \dot{\boldsymbol{\sigma}} \cdot \mathbf{n}_l = 0 \tag{22.40}$$

and similar continuity conditions should occur on the surface separating unloading and reloading domains. Violation of this condition implies discontinuity of material response and violates the convexity condition (22.20) and hence implies non-uniqueness of solution of boundary value problems. The condition (22.20) is only satisfied in conical subdomains shown in Figure 22.1(d) and they can be regarded as *domains of applicability* of discontinuous rate equations. Determination of these domains for the deformation theories of plasticity was discussed by Mróz in [28].

It was shown in [28] that for regular loading surfaces, the matrices $\mathbf{C}_1$ and $\mathbf{C}_2$ satisfying the continuity condition (22.40) are interrelated, namely

$$\dot{\boldsymbol{\varepsilon}}_1 = \mathbf{C}_1 \dot{\boldsymbol{\sigma}} = \mathbf{C}_2 \dot{\boldsymbol{\sigma}} + \mathbf{g}\mathbf{n}_l \cdot \dot{\boldsymbol{\sigma}}, \qquad \mathbf{C}_1 = \mathbf{C}_2 + \mathbf{g}\mathbf{n}_l \tag{22.41}$$

where $\mathbf{g}$ is an arbitrary tensor. In fact, for neutral loading paths there is $\dot{\boldsymbol{\sigma}} \cdot \mathbf{n}_l = 0$ and $\mathbf{C}_1 \dot{\boldsymbol{\sigma}} = \mathbf{C}_2 \dot{\boldsymbol{\sigma}}$. In the particular case, the matrix $\mathbf{C}_2$ can be identified with the elastic compliance matrix (elastic unloading). Then

$$\begin{aligned} \dot{\boldsymbol{\varepsilon}}_2 &= \dot{\boldsymbol{\varepsilon}}_2^e = \mathbf{C}_2\dot{\boldsymbol{\sigma}} \\ \dot{\boldsymbol{\varepsilon}}_1 &= \dot{\boldsymbol{\varepsilon}}^e + \dot{\boldsymbol{\varepsilon}}^p = \mathbf{C}_2\dot{\boldsymbol{\sigma}} + \frac{1}{K}\mathbf{n}_g\mathbf{n}_l\cdot\dot{\boldsymbol{\sigma}} \end{aligned} \tag{22.42}$$

where $\mathbf{n}_g$ is the normalized tensor, $\mathbf{n}_g\cdot\mathbf{n}_g = 1$, and the scalar K will be called the hardening modulus. In fact, K is proportional to the hardening modulus of the σ–ε_p-curve in the uniaxial case. From (22.42) it follows that

$$K = \frac{\mathbf{n}_1\cdot\dot{\boldsymbol{\sigma}}}{(\dot{\boldsymbol{\varepsilon}}^p\cdot\dot{\boldsymbol{\varepsilon}}^p)^{1/2}} \tag{22.43}$$

Since in general $\mathbf{n}_g \neq \mathbf{n}_l$, the plastic strain-rate vector departs from the direction of the exterior normal $\mathbf{n}_l$ to the loading surface. The case of the *associated flow rule* is obtained by postulating $\mathbf{n}_g = \mathbf{n}_l$ and then

$$\dot{\boldsymbol{\varepsilon}}^p = \frac{1}{K}\mathbf{n}_l(\mathbf{n}_l\cdot\dot{\boldsymbol{\sigma}}) \tag{22.44}$$

Similar relations can be derived for strain controlled processes. Consider the loading and unloading surfaces $\phi_l = 0$, $\phi_u = 0$ (Figure 22.5) and write the inverse relations

$$\begin{aligned} \dot{\boldsymbol{\sigma}} &= \mathbf{D}_1\dot{\boldsymbol{\varepsilon}} \qquad \text{for } \mathbf{n}_\phi\cdot\dot{\boldsymbol{\varepsilon}} > 0\text{—loading} \\ \dot{\boldsymbol{\sigma}} &= \mathbf{D}_2\dot{\boldsymbol{\varepsilon}} \qquad \text{for } \mathbf{n}_\phi\cdot\dot{\boldsymbol{\varepsilon}} < 0\text{—unloading} \end{aligned} \tag{22.45}$$

where $\mathbf{n}_\phi$ is the normalized gradient tensor of the loading surface, $\mathbf{n}_\phi\cdot\mathbf{n}_\phi = 1$.

Note that the loading–unloading conditions (22.45) are now valid for both *hardening* and *softening* response.

Again the continuity condition

$$\mathbf{D}_2\dot{\boldsymbol{\varepsilon}} = \mathbf{D}_1\dot{\boldsymbol{\varepsilon}} \qquad \text{for } \phi = 0, \quad \mathbf{n}_\phi\cdot\dot{\boldsymbol{\varepsilon}} = 0 \tag{22.46}$$

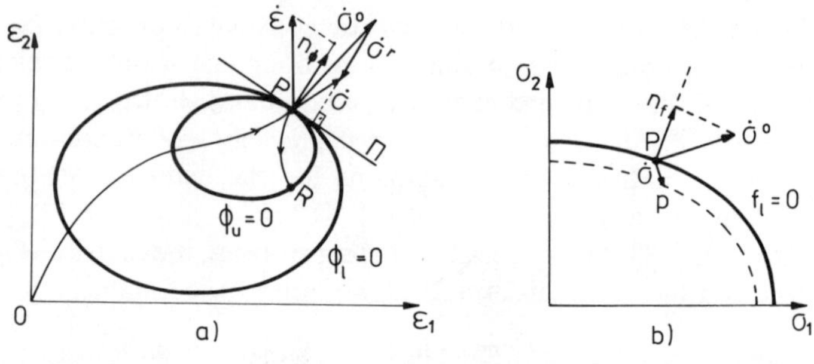

Figure 22.5. (a) Loading and unloading surfaces in the strain space, (b) loading–unloading conditions in the stress space expressed in terms of the elastic stress rate $\dot{\sigma}^0$

provides the interrelation between the stiffness matrices $\mathbf{D}_1$ and $\mathbf{D}_2$, namely

$$\dot{\boldsymbol{\sigma}} = \mathbf{D}_1\dot{\boldsymbol{\varepsilon}} = (\mathbf{D}_2 - M\mathbf{n}_h\mathbf{n}_\phi)\cdot\dot{\boldsymbol{\varepsilon}} \qquad \text{for } \mathbf{n}_\phi\cdot\dot{\boldsymbol{\varepsilon}} > 0 \tag{22.47}$$

where $\mathbf{n}_h$ is an arbitrary normalized tensor and M is the scalar parameter. Identifying $\mathbf{D}_2$ with the elastic stiffness matrix, the relation (22.47) can be presented as follows

$$\dot{\boldsymbol{\sigma}} = \dot{\boldsymbol{\sigma}}^0 - \dot{\boldsymbol{\sigma}}^r = \mathbf{D}_2\boldsymbol{\varepsilon} - M\mathbf{n}_h\mathbf{n}_\phi\cdot\dot{\boldsymbol{\varepsilon}} \tag{22.48}$$

where $\dot{\boldsymbol{\sigma}}^0$ and $\dot{\boldsymbol{\sigma}}^r$ are the elastic and the relaxation stress rates discussed already in the uniaxial case. The scalar M will be called the *relaxation modulus* and in view of (22.48) it is defined as follows

$$M = \frac{(\dot{\boldsymbol{\sigma}}^r\cdot\dot{\boldsymbol{\sigma}}^r)^{1/2}}{(\mathbf{n}_\phi\cdot\dot{\boldsymbol{\varepsilon}})} \tag{22.49}$$

that is as the ratio of modulus of $\dot{\boldsymbol{\sigma}}^r$ and the projection of $\dot{\boldsymbol{\varepsilon}}$ on normal direction $\mathbf{n}_\phi$, Figure 22.5. Since in general $\mathbf{n}_\phi \neq \mathbf{n}_h$, the direction of the relaxation stress does not coincide with that of $\mathbf{n}_\phi$. The case of *associated relaxation rule* occurs when $\mathbf{n}_h = \mathbf{n}_\phi$. The hardening and relaxation moduli are interrelated and in [34] it was shown that.

$$M = \frac{1}{K' + \mathbf{n}_\phi\cdot\mathbf{C}_2\cdot\mathbf{n}_\phi}, \qquad K' = \frac{K}{(\mathbf{D}_2\mathbf{n}_f)\cdot(\mathbf{D}_2\mathbf{n}_f)}, \qquad \mathbf{C}_2 = \mathbf{D}_2^{-1} \tag{22.50}$$

The presented rate relations for loading and unloading are general and can be referred to plasticity or hypoelasticity formulations of constitutive equations. It is shown that the structure of constitutive relations is the *same* in both cases and only different interpretations of stiffness and compliance matrices can be ascribed in particular formulations. The unloading behaviour need not to be elastic and the matrices $\mathbf{C}_1$ and $\mathbf{D}_1$ may represent the inelastic deformation as well. In particular, when the matrices $\mathbf{C}$ and $\mathbf{D}$ are symmetric, there is normality of the plastic strain rate in the stress space and of the relaxation stress in the strain space to respective loading surfaces.

Let us note that the loading–unloading conditions (22.45) expressed in terms of total strain rates can also be retransformed into the stress space. In fact, since $f(\boldsymbol{\sigma}, \boldsymbol{\alpha}) = f[\mathbf{D}^e(\boldsymbol{\varepsilon} - \boldsymbol{\varepsilon}^p), \boldsymbol{\alpha}] = \phi(\boldsymbol{\varepsilon}, \boldsymbol{\alpha}) = 0$ there is

$$\mathbf{D}^e\frac{\partial f}{\partial \boldsymbol{\sigma}} = \frac{\partial \phi}{\partial \boldsymbol{\varepsilon}}, \qquad \frac{\partial \phi}{\partial \boldsymbol{\varepsilon}}\cdot\dot{\boldsymbol{\varepsilon}} = \frac{\partial f}{\partial \boldsymbol{\sigma}}\cdot\dot{\boldsymbol{\sigma}}^e \tag{22.51}$$

and the loading–unloading conditions in the stress space are

$$f_1 = 0, \qquad \mathbf{n}_f\cdot\dot{\boldsymbol{\sigma}}^0 > 0, \qquad f_1 = 0, \qquad \mathbf{n}_f\cdot\dot{\boldsymbol{\sigma}}^0 < 0 \tag{22.52}$$

where $\dot{\boldsymbol{\sigma}}^0 = \mathbf{D}^e\dot{\boldsymbol{\varepsilon}}$ is the elastic stress rate defined by (22.48). Thus, the loading surface in the stress surface may already shrink, but the vector $\dot{\boldsymbol{\sigma}}^0$ is always directed into its exterior for a loading process, Figure 22.5(b).

The loading–unloading conditions illustrated in Figures 22.3(a)–(d) may now be referred to various proposals of constitutive models. Thus, Figures 22.3(a), (b) may be referred to loading–unloading conditions used in hypoelastic models by Romano [44] or Davis and Mullenger [16] who assumed the sign of work rate as specifying loading and unloading, namely

$$\begin{aligned} &\text{loading:} && \dot{\boldsymbol{\varepsilon}} = \mathbf{C}_1\dot{\boldsymbol{\sigma}} && \text{for } \boldsymbol{\sigma}\cdot\dot{\boldsymbol{\varepsilon}} > 0 \\ &\text{unlodaing:} && \dot{\boldsymbol{\varepsilon}} = \mathbf{C}_2\dot{\boldsymbol{\sigma}} && \text{for } \boldsymbol{\sigma}\cdot\boldsymbol{\varepsilon} < 0 \end{aligned} \tag{22.53}$$

It was shown in [29] that these loading–unloading conditions can be geometrically interpreted by introducing the unloading surface $f_u = 0$ and the unloading process terminates when the stress path passes in the exterior of this surface. The typical response in uniaxial tension or compression is that presented in Figures 22.3(a), (b); namely, for small stress cycles there is a continuing ratchetting effect and for small strain cycles the continuing relaxation effect occurs. Similarly as for non-linear relations (22.25) or (22.28) the present loading–unloading conditions induce cyclic instability and violate both Drucker and Ilyushin postulates and also the convexity condition (22.20).

The unloading surface of type shown in Figure 22.3(b) was recently applied by Hashiguchi [20], and Dafalias and Herrmann [13]. In fact, their formulation suffers from the same disadvantage that any small cyclic variation of stress or strain produces finite changes in total strain or stress. The unloading surface shown in Figure 22.3(c) was applied by Hueckel and Nova [21] and by Bazant and Kim [4] in modelling hysteric behaviour of rock and concrete. The disadvantage of their formulation lies in violation of the continuity condition for neutral paths $f_1 = 0$ or $f_u = 0$ and hence violation of the convexity (or uniqueness) condition (22.20).

The unloading surface which is tangential to the initial loading surface $f_l = 0$ at the prestress point P, Figure 22.3(d), was applied by Mróz *et al.* [31, 35, 37] in elaborating their multisurface model and its simplified versions. This model will be discussed further in subsequent sections.

Finally, we should mention of loading–unloading conditions specified in particular subdomains of the stress space. For instance, Darve, Boulon, and Chambon [6] used piecewise linear rate relations applicable in subdomains corresponding to positive or negative signs of the principal stress rates. More generally, instead of a single loading–unloading condition, several subdomains in the stress space can be introduced corresponding to various signs of linear stress functions $f_1, f_2, \ldots$, that is (cf. Figure 22.6)

$$\begin{aligned} &1. && \dot{\boldsymbol{\varepsilon}} = \mathbf{C}_1\dot{\boldsymbol{\sigma}} && \text{for } \dot{f}_1 > 0, \quad \dot{f}_2 > 0 \\ &2. && \dot{\boldsymbol{\varepsilon}} = \mathbf{C}_2\dot{\boldsymbol{\sigma}} && \text{for } \dot{f}_1 > 0, \quad \dot{f}_2 < 0 \\ &3. && \dot{\boldsymbol{\varepsilon}} = \mathbf{C}_3\dot{\boldsymbol{\sigma}} && \text{for } \dot{f}_1 < 0, \quad \dot{f}_2 < 0 \\ &4. && \dot{\boldsymbol{\varepsilon}} = \mathbf{C}_4\dot{\boldsymbol{\sigma}} && \text{for } \dot{f}_1 < 0, \quad \dot{f}_2 > 0 \end{aligned} \tag{22.54}$$

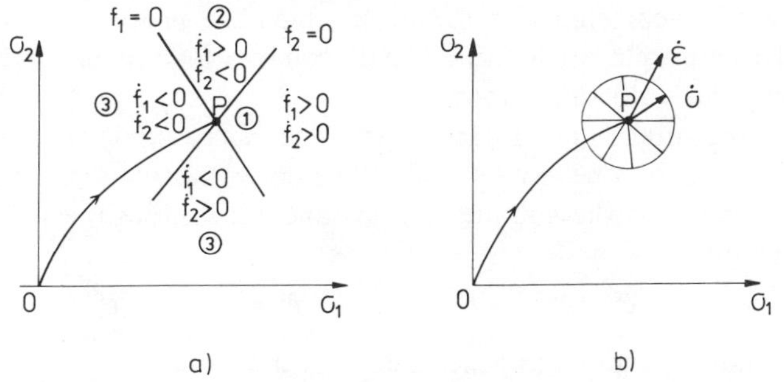

Figure 22.6 (a) Singular loading surfaces and (b) non-linear rate relations as a limiting case of singular loading conditions

with proper continuity conditions between the subdomains 1, 2, 3, 4. When the number of subdomains tends to infinity, the piecewise linear relations (22.54) become *non-linear* relations, thus

$$\dot{\boldsymbol{\sigma}} = \mathbf{C}(\dot{\boldsymbol{\sigma}})\dot{\boldsymbol{\sigma}}, \qquad \dot{\boldsymbol{\varepsilon}} = \mathbf{D}(\dot{\boldsymbol{\varepsilon}})\dot{\boldsymbol{\varepsilon}} \tag{22.55}$$

where $\mathbf{C}(\dot{\boldsymbol{\sigma}})$ and $\mathbf{D}(\dot{\boldsymbol{\varepsilon}})$ are homogeneous functions of stress and strain rates, respectively of order zero. Such non-linear rate relations were recently considered by Kolymbas [45] and Benedetto and Darve [49]. For instance, a simple generalization of (22.28) can be presented in a form

$$\dot{\boldsymbol{\sigma}} = \mathbf{D}^{e}\dot{\boldsymbol{\varepsilon}} - \mathbf{d}(\dot{\boldsymbol{\varepsilon}} \cdot \dot{\boldsymbol{\varepsilon}})^{1/2} = \dot{\boldsymbol{\sigma}}^{0} - \dot{\boldsymbol{\sigma}}^{r} \tag{22.56}$$

where $\mathbf{D}^{e}$ is a constant elastic stiffness matrix and $\mathbf{d}$ is a second order tensor, not depending on the strain rate. The relation (22.56) can be identified with a familiar decomposition of the stress rate into elastic and relaxation portions. Similar structures to (22.56) can be ascribed to endochronic theories of plasticity [3, 48]. The disadvantage of such formulations lies in violating the uniqueness (or convexity) and cyclic stability conditions, discussed in the uniaxial case, (cf. also Sandler [46] and Rivlin [47] articles concerned with endochronic theories of plasticity). A new version of endochronic theory discussed by Valanis and Read [48] uses the plastic strain rate in defining the monotonically growing endochronic time, thus resembling the classical plasticity formulation.

22.4 ANISOTROPIC HARDENING MODEL FOR CLAY

The anisotropic hardening rule constitutes a generalization and extension of isotropic hardening models, formulated previously for clays and sands [10, 15, 18, 26, 27, 36, 50, 51, 52]. In these models, it is assumed that the state of material depends on a scalar parameter varying with plastic deformation and during unloading, only elastic changes occur. The applicability of such models is therefore limited to monotonically growing loads, a situation which is rather rare in geotechnical systems, in which the processes of excavation, erecting a structure, etc. are associated with complex loading histories involving both loading and unloading events.

Assuming the solid to be incompressible, the macroscopic volume variation is due to closing or opening of voids. Denoting the volume void ratio by $e = V_v/V_m$ where V_v is the volume of voids and V_m denotes the volume of soil skeleton, there is

$$\dot{e} = \dot{e}^e + \dot{e}^p = -(1+e)\dot{\varepsilon}_v, \qquad \dot{e}^p = -(1+e)\dot{\varepsilon}_v^p \tag{22.57}$$

The yield condition can be in general expressed as follows

$$f(\boldsymbol{\sigma}, \boldsymbol{\alpha}) = 0 \tag{22.58}$$

where $\boldsymbol{\alpha}$ denotes collectively the *state* or *hardening* parameters which may be a set of scalars or tensors. Depending on the selection of hardening parameters and their evolution rules, different constitutive models can be proposed.

A simple assumption for clays is that the maximal consolidation pressure or irreversible void ratio is the only hardening parameter, thus

$$f(\boldsymbol{\sigma}, e^p) = 0 \quad \text{or} \quad f(\boldsymbol{\sigma}, p_c) = 0 \tag{27.59}$$

Assuming the associated flow rule (22.44) and using the consistency condition

$$\frac{\partial f}{\partial \boldsymbol{\sigma}} \cdot \dot{\boldsymbol{\sigma}} + \frac{\partial f}{\partial e^p} \dot{e}^p = 0 \tag{22.60}$$

the hardening modulus can be expressed in form

$$K = \left(-\frac{\partial f}{\partial e^p} \right) (1+e) \operatorname{tr} \mathbf{n} \left[\frac{\partial f}{\partial \boldsymbol{\sigma}} \cdot \frac{\partial f}{\partial \boldsymbol{\sigma}} \right]^{-1/2} \tag{22.61}$$

where $\operatorname{tr} \mathbf{n} = n_{kk}$ and $\mathbf{n}$ is the unit normal vector to the yield surface. Figure 22.7(a) presents a typical yield condition in the plane J_1, J_2'. The critical state lines (c.s.l.) OB and OB′ correspond to $K = \operatorname{tr} \mathbf{n} = 0$ and separate the domain of consolidation OBB′A ($\operatorname{tr} \mathbf{n} < 0$) and the domain of dilatancy or softening OBB′C ($\operatorname{tr} \mathbf{n} > 0$). This hardening rule generally known as the critical state or density hardening model is applicable to clays under monotonic loading. Its disadvantage lies in the inability to simulate properly the softening response and dilatancy as well as pore pressure variation is undrained tests for over-consolidated clays. For a detailed exposition of this model, cf. [15, 50].

As over-consolidated clays and dense sands exhibit stable behaviour despite dilatancy, it may be expected that besides density hardening, there is also an additional hardening effect due to shear action. Let us introduce the combined hardening parameter $\mathscr{H}$ whose rate is expressed as follows

$$\dot{\mathscr{H}} = \beta(\dot{\mathbf{e}}^p \cdot \dot{\mathbf{e}}^p)^{1/2} - \dot{e}^p = \beta(\dot{e}_{ij}^p \dot{e}_{ij}^p)^{1/2} + (1+e)\dot{\varepsilon}_v^p \tag{22.62}$$

where β is a material constant. Using the flow rule (22.44) and satisfying the

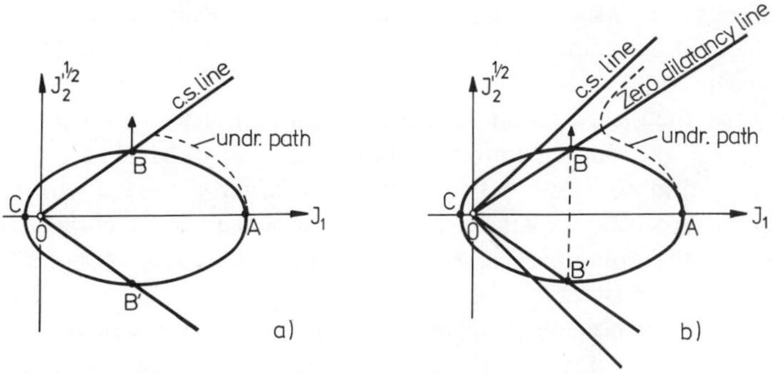

Figure 22.7 Isotropic hardening rules: (a) density (or volumetric) hardening model, (b) combined hardening model

consistency condition (22.60), it is obtained

$$K = \left(-\frac{\partial f}{\partial \mathscr{H}}\right)[\beta(\operatorname{dev}\mathbf{n}\cdot\operatorname{dev}\mathbf{n})^{1/2} - (1+e)\operatorname{tr}\mathbf{n}]\left[\frac{\partial f}{\partial \boldsymbol{\sigma}}\cdot\frac{\partial f}{\partial \boldsymbol{\sigma}}\right]^{1/2} \tag{22.63}$$

where dev **n** denotes the deviatoric portion of **n**. As $\partial f/\partial \mathscr{H} < 0$, the hardening modulus is positive when

$$\operatorname{tr}\mathbf{n} \leq \frac{\beta}{1+e}(\operatorname{dev}\mathbf{n}\cdot\operatorname{dev}\mathbf{n})^{1/2} \tag{22.64}$$

and the failure line OD satisfying the condition $K=0$ separates the domains of hardening and softening, Figure 22.7(b), whereas the zero-dilatancy line lies beneath the failure line. Such a combined hardening parameter was applied by Nova and Wood [36] and by Wilde [51] in describing inelastic behaviour of sands.

Let us now discuss the anisotropic hardening model for clays, described in detail in [38, 39] and constituting an extension of a class of multisurface hardening rules elaborated in [31 to 34]. A detailed review of these models is presented by Mróz and Norris [34] and here we only discuss one version of these rules applicable both for clays and sands.

Consider first the case of a triaxial test when the two principal effective stresses are equal, $\sigma_2 = \sigma_3$. The commonly used stress and strain components then are

$$\begin{aligned} p &= -\tfrac{1}{3}(\sigma_1 + 2\sigma_2), \qquad & q &= \sigma_2 - \sigma_1, \\ \varepsilon_v &= -(\varepsilon_1 + 2\varepsilon_2), \qquad & \varepsilon_q &= \tfrac{2}{3}(\varepsilon_2 - \varepsilon_1). \end{aligned} \tag{2.65}$$

Assume that the degree of consolidation of soil is represented by the consolidation surface $F_c = 0$. For clays, this surface is constituted by the initial consolidation process either in natural deposit or in laboratory and it preserves the memory of maximal prestress from the loading history. The yield surface

$f_0 = 0$ encloses the elastic domain enclosed by the consolidation surface $F_c = 0$. Usually the elastic domain is very small and it may be assumed that this domain shrinks to a point.

In this case, the plastic strain increment occurs for any stress increment, both inside and on the consolidation surface. In view of this assumption, the distinction should be made between the *consolidation process* with the stress point remaining on the consolidation surface $F_c = 0$ and *unloading or reloading process* within the domain enclosed by $F_c = 0$. Let us discuss consecutively the description of these two types of processes.

Consider first the consolidation process. The stress path is then directed into the exterior of the domain $F_c \leq 0$ and the stress point remains on the consolidation surface. Assume that this surface is represented in the p, q-plane by the portions of two ellipses rotated with respect to the p, q-axes, thus

$$F_c = p^2 + \frac{2(\zeta - 1)}{m_c} pq + \frac{(1 - 2\zeta)^2}{m_c^2} q^2 - 2a_c p - \frac{4(\zeta - 1)}{m_c} a_c q = 0, \qquad q > 0,$$

and (22.66)

$$F_e = p^2 + \frac{2(1 - \zeta)}{m_e} pq + \frac{(1 - 2\zeta)^2}{m_c^2} q^2 - 2a_c p - \frac{4(1 - \zeta)}{m_e} a_c q = 0, \qquad q < 0$$

where $\zeta = a_c/c$, $m_c = \tan \omega_c$, $m_e = \tan \omega_e$ and ω_c, ω_e are the angles of inclination

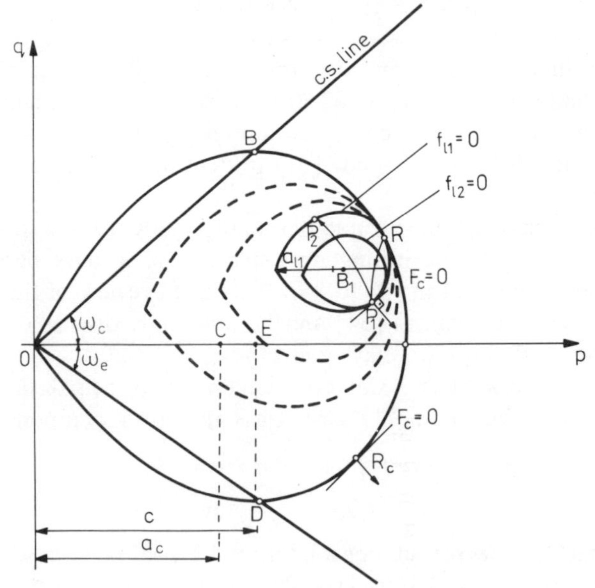

Figure 22.8 Consolidation and loading surfaces in the p, q-plane for clay

of the critical state lines to the p-axis. The angles ω_c and ω_e satisfy the relations

$$m_c = \frac{q_R}{c} = \frac{6 \sin \varphi}{3 - \sin \varphi}, \qquad m_e = -\frac{q_D}{c} = \frac{6 \sin \varphi}{3 + \sin \varphi} \tag{22.67}$$

following from the Coulomb yield condition, where φ denotes the angle of internal friction at failure, Figure 22.8. Let us note that for $\zeta = 1$ there is $a_c = c$ and the principal axis of each ellipse coincides with the p-axis, thus

$$F_c = (p - a_c)^2 + \frac{q^2}{m_c^2} - a_c^2 = 0 \tag{22.68}$$

The value of ζ will be identified from the K_0-consolidation test as it was shown in [32] that the use of (22.68) does not provide accurate values of lateral pressures in the uniaxial consolidation. Further, it will be assumed that the consolidation surface represents the isotropic hardening of the material and the surface diameter varies with the irreversible void ratio, thus $F_c(p, q, e^p) = 0$. The parameters m_c, m_e, and ζ will be assumed as constant and $a_c = a_c(e^p)$.

Writing the associated flow rule

$$d\varepsilon_v^p = \frac{1}{K_p} n_p (d\boldsymbol{\sigma} \cdot \mathbf{n}), \qquad d\varepsilon_q^p = \frac{1}{K_p} n_q (d\boldsymbol{\sigma} \cdot \mathbf{n}) \tag{22.69}$$

the hardening modulus K_p is expressed as follows

$$K_p = H_c \left[\left(\frac{\partial E_c}{\partial p} \right)^2 + \left(\frac{\partial F_c}{\partial q} \right)^2 \right]^{1/2}, \tag{22.70}$$

$$H_c = (1 + e) \frac{\partial F_c}{\partial e^p} \frac{\partial F_c}{\partial p} = -4(1 + e) \frac{da_c}{de^p} \left(p + \frac{2(\zeta - 1)}{m_c} q \right) \left(p + \frac{\zeta - 1}{m_c} q - a_c \right)$$

Now, let us discuss the reverse active loading process occurring within the domain $F_c \leq 0$. Consider the situation presented in Figure 22.8 where the initial consolidation process is terminated at the point R with the subsequent stress path RP_1 directed into the domain $F_c < 0$. When elastic domain vanishes, the reverse plastic strain occurs along the whole path RP_1 otherwise the yield surface $f_0 = 0$ translates with the point P_1.

Let us introduce the concepts of *active loading* and *stress reversal surfaces* that will enable us to make distinction between particular loading events. Assume that after reaching the maximal prestress point R all reverse loading surfaces are tangential at R to the consolidation surface. For the loading programme RP_1, the stress point always remains on the active loading surface $f_{l1} = 0$ translating and expanding so that it always remains tangential to $F_c = 0$ at R. The plastic response at P_1 is described by the flow rule (22.69) applied to $f_{l1} = 0$ with the plastic modulus K_p governed, for instance, by the relation

$$K_p = K_{pc} + (K_y - K_{pc}) R_1^r \tag{22.71}$$

where

$$\mathrm{R}_1 = \frac{a_c - a_{11}}{a_c - a_0} \tag{22.72}$$

where a_c, a_{11}, a_0 are size parameters of the consolidation, active loading and yield surfaces, K_y is the initial modulus when the stress point reaches the yield surface, that is $a_{11} = a_0$ and K_{pc} is the value of the hardening modulus at the associated point R_c' on the consolidation surface for which the normal vector has the same direction as that at P_1. When $a_{11} = a_c$, that is the stress point reaches the consolidation surface, there is $K_p = K_{pc}$. Further, setting $a_0 = 0$ in (22.72), we reduce the yield surface to a point, that is neglect the elastic domain.

The first reverse loading (or unloading) programme continues when $\dot{a}_{11}$ increases, thus $\dot{a}_{11} > 0$. However, when at P_1 the stress path reverses and is directed into the interior of $f_{11} = 0$, then the second reverse loading (or reloading) event commences and the active loading surface $f_{12} = 0$ for this programme is tangential at P_1 to $f_{11} = 0$ and passes through the stress reversal point. The previously active loading surface $f_{11} = 0$ now becomes the *stress reversal surface* and preserves the memory of the stress reversal point P_1. Denoting the size parameter of $f_{12} = 0$ by a_{12}, the rule (22.71) for the hardening modulus variation is still used with R_1 replaced by R_2, where

$$\mathrm{R}_2 = \frac{a_c - a_{12}}{a_c - a_0} \tag{22.73}$$

The second reverse loading process continues provided $\dot{a}_{12} > 0$ and $a_{12} < a_{11}$. When at P_2 the surfaces $f_{12} = 0$ and $f_{11} = 0$ coincide and the stress path is directed in the exterior of $f_{11} = 0$, the second loading event and the stress reversal point P_1 are erased from the material memory. If the stress path $\mathrm{P}_1\mathrm{P}_2$ continues and moves beyond the domain enclosed by the consolidation surface, the consolidation process starts again. The memory of R is erased and a new maximal prestress point is created.

The yield and active loading surfaces are similar to the consolidation surface and are composed of portions of two ellipses

$$f_{1c} = (p - \alpha_p)^2 + \frac{2(\zeta - 1)}{m_c}(q - \alpha_q)(p - \alpha_p) + \frac{(1 - 2\zeta)^2}{m_c^2}(q - \alpha_q)^2 + 2a_1(p - \alpha_p)\frac{1 - \zeta}{\zeta}$$

$$+ \frac{2(\zeta - 1)(1 - 2\zeta)}{m_c \zeta} a_1(q - \alpha_q) + a_1^2 \frac{1 - 2\zeta}{\zeta^2} = 0; \qquad q - \alpha_q > 0 \tag{22.74a}$$

$$f_{1e} = (p - \alpha_p)^2 + \frac{2(1 - \zeta)}{m_e}(q - \alpha_q)(p - \alpha_p) + \frac{(1 - 2\zeta)^2}{m_e^2}(q - \alpha_q)^2 + 2a_1(p - \alpha_p)\frac{1 - \zeta}{\zeta}$$

$$+ \frac{2(1 - \zeta)(1 - 2\zeta)}{m_e \zeta} a_1(q - \alpha_q) + a_1^2 \frac{1 - 2\zeta}{\zeta^2} = 0; \qquad q - \alpha_q < 0 \tag{22.74b}$$

where α_p, α_q define the position of the centre point B_1, and the parameters ζ, m_c, m_e are the same as those for the consolidation surface. The function $a_c = a_c(e^p)$ is assumed in the usual exponential form

$$a = a_i \exp\left(\frac{e_i^p - e^p}{\lambda - k}\right) \tag{22.75}$$

and the curves of isotropic consolidation and unloading are specified as follows

$$p = p_i \exp\left(\frac{e_i - e}{\lambda}\right), \qquad p = p_0 \exp\left(\frac{e_0^e - e^e}{k}\right) \tag{22.76}$$

where p_i, e_i are the initial values of p and e in the consolidation process whereas p_0, e_0 are the initial values for the unloading process; k and λ are the material parameters. The elastic strain increments are specified by the relations

$$d\varepsilon_v^e = \frac{dp}{K}, \qquad d\varepsilon_q^e = \frac{dq}{3G} \tag{22.77}$$

For any position P_1 of the stress point on the active loading surface, there exists an associated point R_c on the consolidation surface having the same direction of the normal vector. Thus, we have

$$p_c - c = \frac{a_c}{a_{11}}(p_p - \alpha_p), \qquad q_c = \frac{a_c}{a_{11}}(q_p - \alpha_q) \tag{22.78}$$

where p_c, q_c denote the stress components at the associated point R_c and p_p, q_p denote the components of the loading point P.

The present formulation can formally be deduced from the multisurface hardening model, discussed in detail in [30, 31] and [41]. In fact, assuming the field of hardening moduli to be represented by an infinite number of loading surfaces, the active reverse loading surfaces can easily be identified and the interpolation hardening rule (22.71) follows from the corresponding rule for a finite number of loading surfaces. On the other hand, a simplified model using only yield and consolidation surfaces was discussed and applied in [31 to 33, 38] to study clay behaviour under a variety of loading conditions. This model constituted an extension of a previously proposed description of the hardening of metals by Dafalias and Popov [11] and Krieg [25]. However, it turned out that the two-surface model does not possess sufficient memory to make a distinction between particular loading events and difficulties occur in its application to cyclic loading conditions, cf. [33, 38]. We therefore confine ourselves to discussing the present version based on the concept of an infinite number of loading surfaces.

Let us discuss first the K_0-*consolidation process* from which we want to identify the parameter ζ. When $\varepsilon_2 = \varepsilon_3 = 0$ and ε_1 is the only non-vanishing strain component, then $d\varepsilon_q = \frac{2}{3} d\varepsilon_v$. Neglecting elastic strains, the uniaxial consolidation is described by the relations

$$\frac{\partial F_c}{\partial q} - \frac{2}{3}\frac{\partial F_c}{\partial p} = 0 \tag{22.79}$$

and

$$S_{\mathrm{ko}} = \frac{q}{p}$$

$$= \frac{2m_c^2}{\{[4(\zeta-1)(3\zeta-m_c)+3]^2+4m_c[(1-2\zeta)^2(m_c+3\zeta-3)-4m_c(\zeta-1)^2]\}^{1/2}+4(\zeta-1)(3\zeta-m_c)+3} \tag{22.80}$$

The value of K_0 is now expressed as follows

$$K_0 = \frac{\sigma_2}{\sigma_1} = \frac{3 - S_{\mathrm{ko}}}{3 + 2S_{\mathrm{ko}}} \tag{22.81}$$

Thus, for a rigid plastic material the consolidation path will be a straight line in

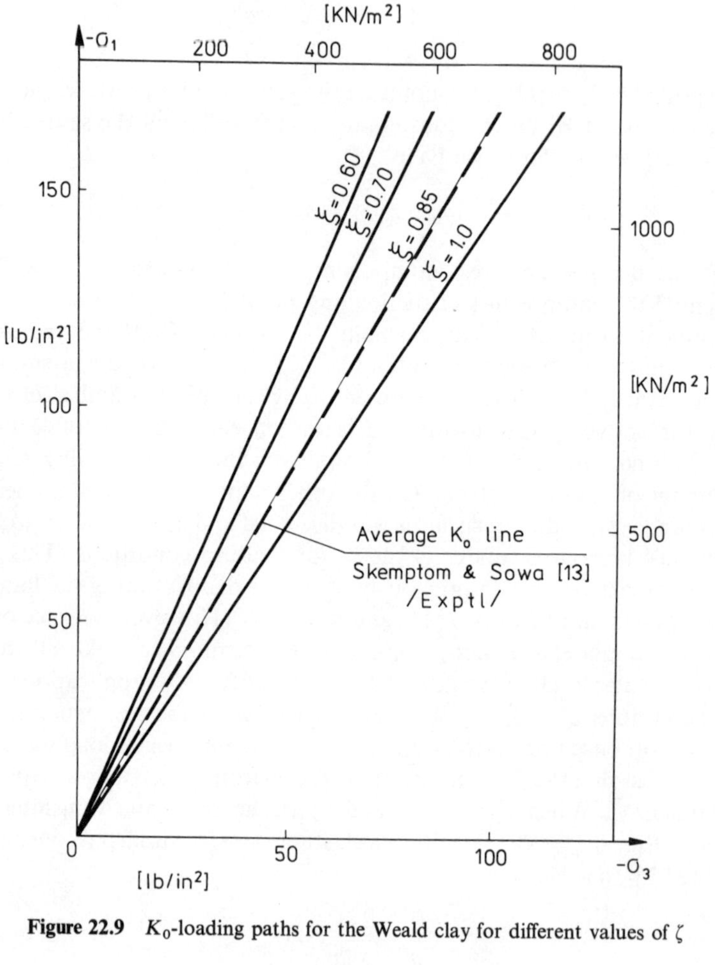

Figure 22.9 K_0-loading paths for the Weald clay for different values of ζ

the p, q-plane, Figure 22.9. Equation (22.80) can be used to identify the parameter ζ by comparing the predicted and measured K_0 values. Thus, for the Weald clay, the value $\zeta = 0.85$ is selected to simulate the consolidation path obtained experimentally by Skempton and Sowa [53]. Further examples of application of this model to study undrained compression and extension after anisotropic consolidation can be found in [37, 38].

The derivation of constitutive equations for a general stress state can be performed by introducing three invariants $J_{\text{m}}, \bar{\sigma}$, and J_3 of the 'translated' effective stress $\sigma_{ij} - \alpha_{ij}$, defined as follows

$$J_{\text{m}} = \tfrac{1}{3}(\sigma_{ii} - \alpha_{ii}), \quad \bar{\sigma} = [\tfrac{1}{2}(s_{ij} - \bar{\sigma}_{ij})(s_{ij} - \bar{\alpha}_{ij})]^{1/2},$$
$$J_3 = \tfrac{1}{3}(s_{ij} - \bar{\alpha}_{ij})(s_{ki} - \bar{\alpha}_{ki})(s_{kj} - \bar{\alpha}_{kj}) \tag{22.82}$$

where $\bar{\alpha}_{ij}$ and α_{ij} are the deviatoric and spherical components of α_{ij}. The angle measure of the third invariant is specified as follows

$$\theta = \tfrac{1}{3} \operatorname{arc\,sin}\left(- \frac{3\sqrt{3}}{2} \frac{J_3}{\bar{\sigma}^3} \right), \qquad -\frac{\pi}{\sigma} \leq \theta \leq \frac{\pi}{\sigma} \tag{22.83}$$

and this angle can be identified on the octahedral plane. The principal stresses are now expressed as follows

$$\sigma_1 = \sigma_{\text{m}} + \frac{2}{\sqrt{3}} \bar{\sigma} \sin(\theta + \tfrac{2}{3}\pi), \qquad \sigma_2 = \sigma_{\text{m}} + \frac{2}{\sqrt{3}} \bar{\sigma} \sin \theta,$$
$$\sigma_3 = \sigma_{\text{m}} + \frac{2}{\sqrt{3}} \sin(\theta + \tfrac{4}{3}\pi) \tag{22.84}$$

Assume now that the Π-plane section of the loading surface is described as follows

$$\bar{\sigma} = \bar{\sigma}_+ g(\theta) \tag{22.85}$$

where $\bar{\sigma}_+$ denotes the values of $\bar{\sigma}$ for $\theta = \pi/6$, $\sigma_2 = \sigma_3$ and $g(\theta)$ is assumed in the form

$$g(\theta) = \frac{2k}{1 + k - (1 - k) \sin 3\theta}, \qquad k = \frac{3 - \sin \phi}{3 + \sin \phi} \tag{22.86}$$

so that $I(\pi/6) = 1$. The equation of the loading surface (22.74) can now be rewritten in the form

$$f_1 = 3 \sin^2 \phi (\sigma_{\text{m}} - \tfrac{1}{3}\alpha)^2 + 6 \sin^2 \phi \frac{1 - \zeta}{\zeta} a_1 (\sigma_{\text{m}} - \tfrac{1}{3}\alpha)$$
$$- \sqrt{3}(\zeta - 1) \sin \phi \, (3 - \sin \phi \sin 3\theta) \left[(\sigma_{\text{m}} - \tfrac{1}{3}\alpha) - \frac{1 - 2\zeta}{\zeta} a_1 \right] \bar{\sigma}$$
$$+ [\tfrac{1}{2}(1 - 2\zeta)(3 - \sin \varphi \sin 3\theta)]^2 \bar{\sigma}^2 + 3a_1^2 \sin^2 \varphi \frac{1 - 2\zeta}{\zeta^2} = 0 \tag{22.87}$$

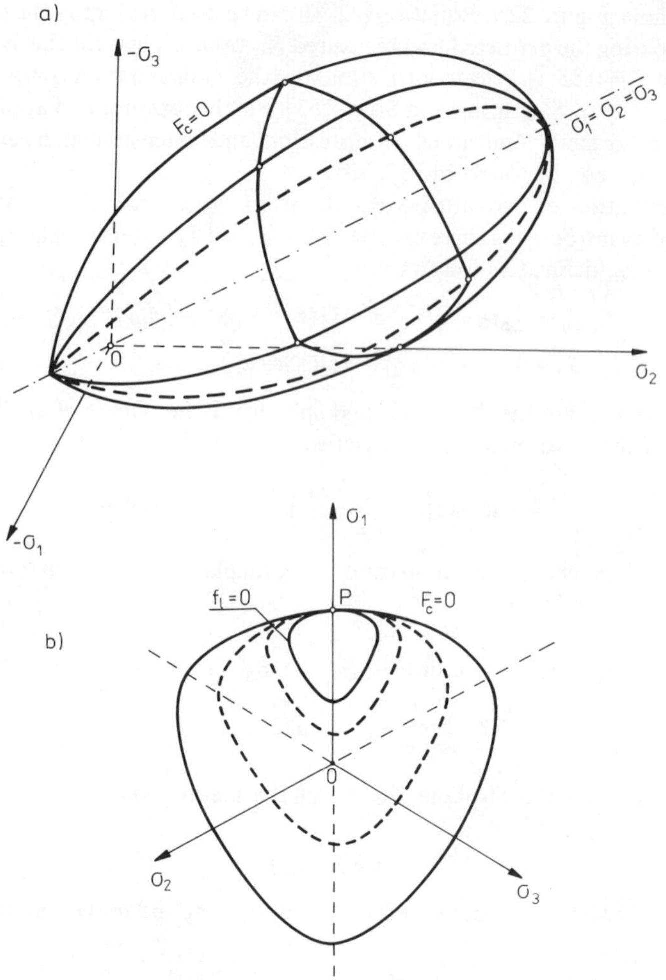

Figure 22.10 Consolidation and loading surfaces in the principal stress space

and is geometrically represented by a surface with similar cross-sectional shapes in all π-planes, Fig. 22.10. Setting $\bar{\alpha}_{ij}=0$, $\alpha_{ii}=-3(a_c/\zeta)$, $a_1=a_c$, from (22.81) we obtain the equation of the consolidation surface $F_c=0$.

Let us apply this model to investigate the mechanical response of clay specimens trimmed at various angles to the major preconsolidation stress in K_0-consolidation. It is known that the clay deposit consolidated one-dimensionally exhibits anisotropic behaviour and in particular, the undrained failure strength varies with the direction of subsequent loading. To study this effect, consider the deformation programme in which after K_0-consolidation the material follows

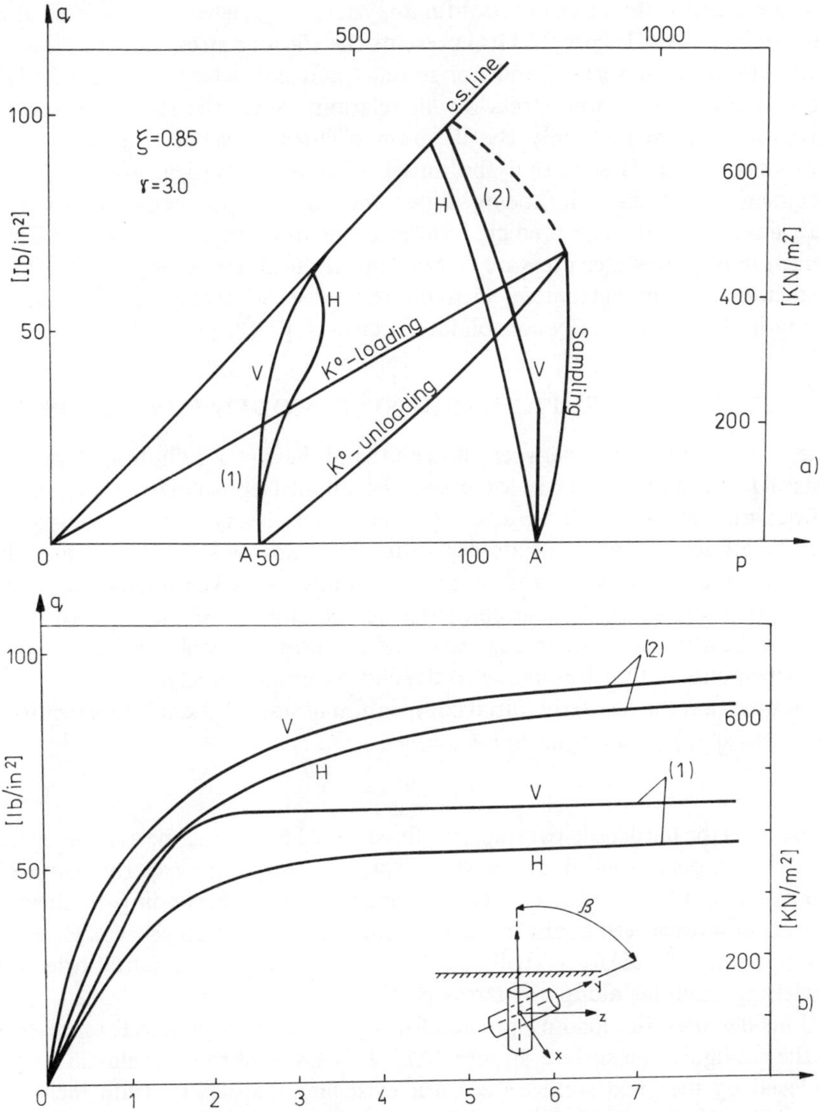

Figure 22.11 (a) Undrained stress paths for specimens oriented horizontally (*H*) and vertically (*V*) with respect to the direction of initial consolidation, (b) shear stress versus axial strain for stress paths 1 and 2 ($K_y = 0.8 \cdot 10^5$ lb/in^2 = $5.515 \cdot 10^8$ KN/m^2)

K_0-unloading or undrained unloding ($d\varepsilon_v = 0$). Subsequently, starting from some hydrostatic stress state, the undrained compression process is performed for specimens differently oriented with respect to the direction of initial consolidation. Note that now the deformation state is not axisymmetric and should be represented in the cartesian coordinate system x, y, z, where y is directed along the specimen axis. Figure 22.11(a) presents the effective stress paths in the p, q-plane obtained for vertical and horizontal specimens, whereas Figure 22.11(b) shows the corresponding stress–strain relations. Since the strain state is not axisymmetric, we plot only the diagram of stress q versus the axial strain component ε_y. It is seen that the initial stiffness for vertical and horizontal specimens differ markedly, being higher for vertical specimens and so is the undrained strength. The predicted variation of strength is a result of different evolution of pore water pressure rather than textural anisotropy of clay. These predictions are in agreement with numerous experimental observations on strength of anisotropically consolidated clays, cf. [37, 38].

22.5 ANISOTROPIC HARDENING MODEL FOR SANDS

One of major differences between clays and sands lies in the definition of an initial consolidated state. Whereas for clays, the applied consolidation pressure p_c defines uniquely the void ratio e, for granular materials any degree of compaction may be attained by appropriate deposition of grains under gravity forces. The required consolidation pressure p_c corresponding to a given relative density or void ratio may be very high as compared to actual values of mean pressure and may not be attained at all during tests. The compressibility of sand, as well as its shear response is therefore made to depend on initial density.

Assume that the degree of initial compaction of sand element is represented in the stress space by a *configuration surface*

$$F_c(\boldsymbol{\sigma}, \mathscr{H}) = 0 \tag{22.88}$$

where $\mathscr{H}$ is the hardening parameter defined by (22.62). Using the associated flow rule, the hardening modulus for stress states corresponding to this surface is expressed by (22.63). In actuality, the stress state corresponding to the configuration surface need not be reached during the deformation process. However, the expression (22.63) for K_c will be used in formulating the variation rule of the hardening modulus along any stress path.

Consider now the loading process for a given state of material represented by the configuration surface, Figure 22.12. It is assumed that the elastic domain enclosed by the yield surface does not exist and the plastic strain increment occurs for any stress increment. The concept of active loading and stress reversal surfaces discussed for clays now still applies, so the actual stress state corresponds always to the active loading surface $f_1^{(i)} = 0$ for the ith loading event whereas the stress reversal surface $f_r^{(i-1)} = 0$ represents the maximal prestress from the past loading history. The plastic strain rate along the stress path is generated by the flow rule (22.69) associated with the active loading surface. The hardening modulus K varies between its initial high value $K = K_y$ for vanishing diameter or other size parameter a_1 of the loading surface and

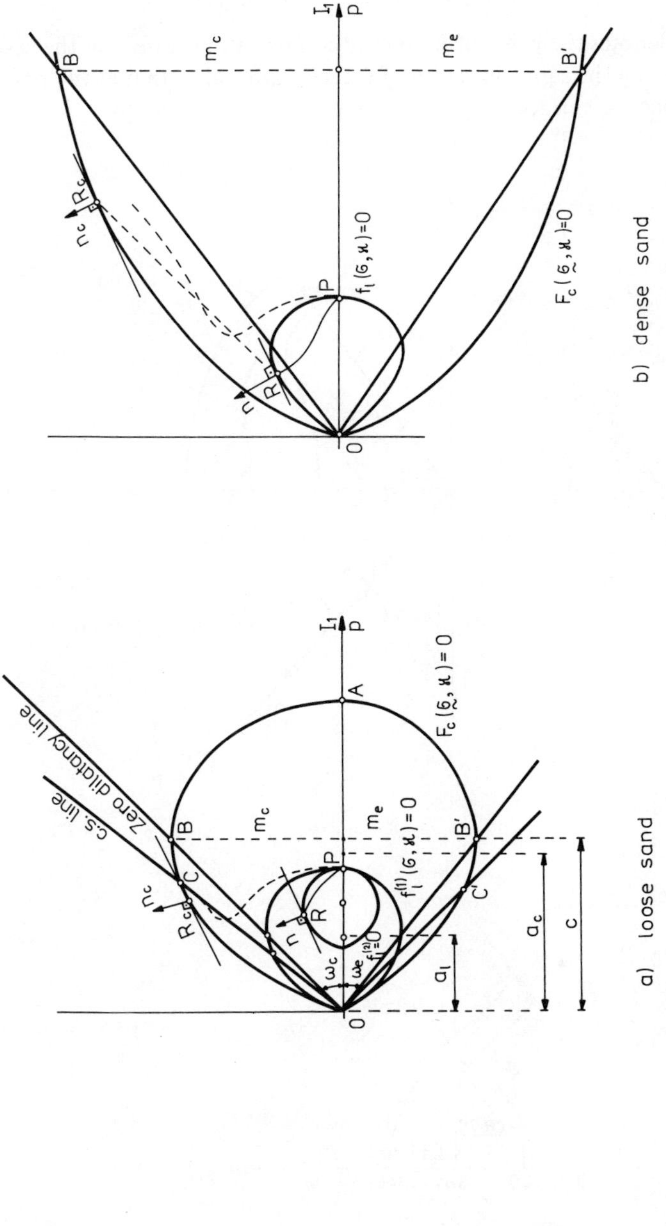

Figure 22.12 Configuration and loading sufaces for: (a) loose sand, (b) dense sand

the value $K = K_c$ corresponding to the configuration surface when $a_1 = a_c$. The variation of hardening modulus can be described by (22.71) or by a similar relation

$$K_p = K_y + (K_{pc} - K_y)\left(\frac{a_1}{a_c}\right) \tag{22.89}$$

where K_{pc} denotes the value of K_p on the configuration surface at the associated point R_c having the same direction of exterior normal as that at the stress point R on the loading surface.

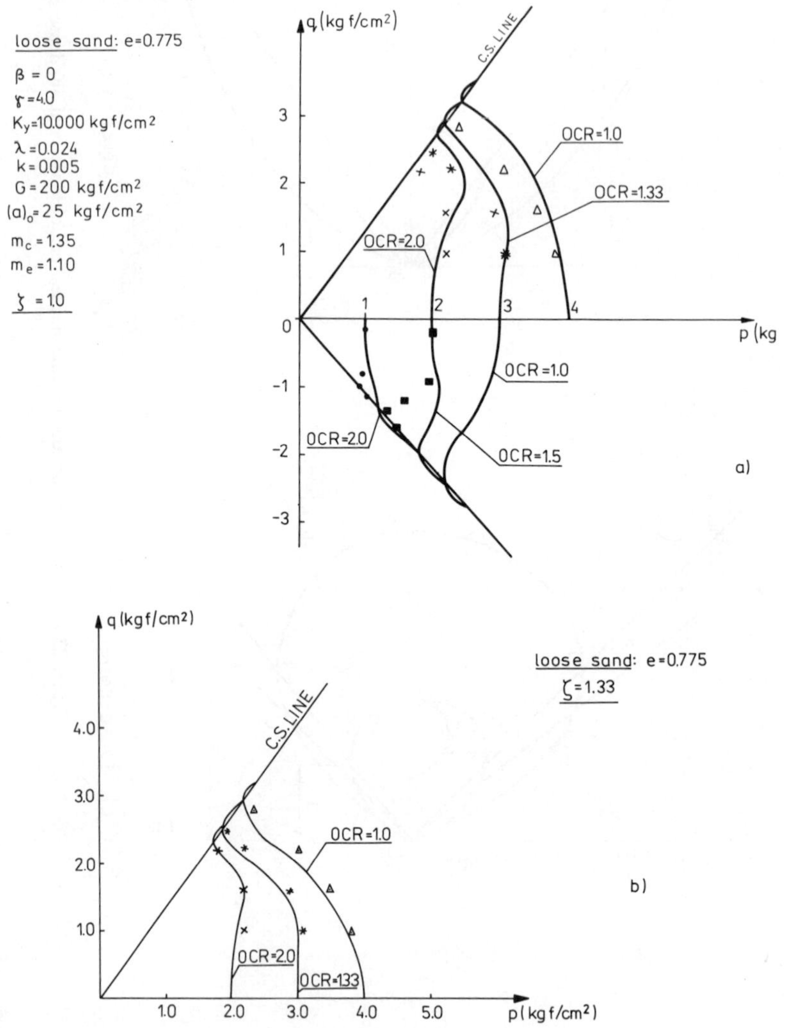

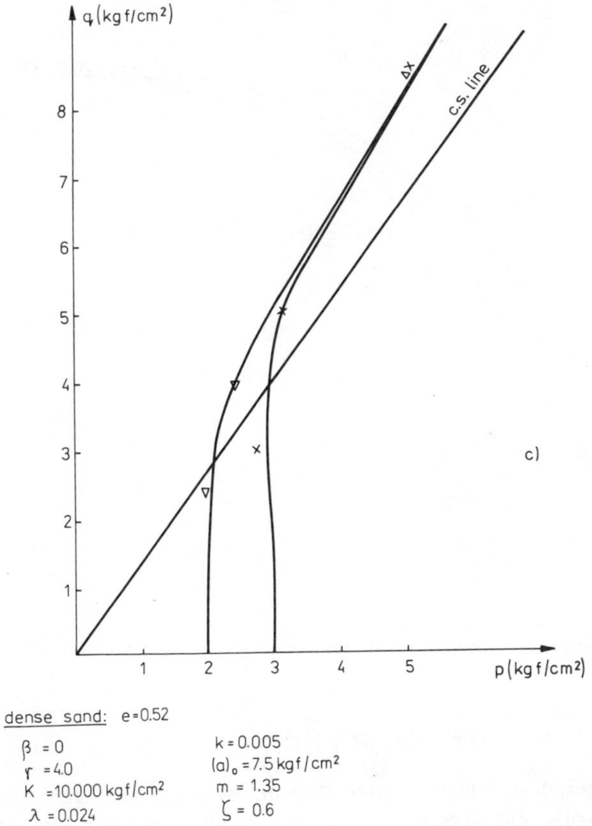

Figure 22.13 (a) Predicted and experimental undrained paths for a loose sand for $\zeta = 1.0$, (b) predicted and experimental undrained paths for a loose sand for $\zeta = 1.3$, (c) undrained paths for a dense sand ($\zeta = 0.6$)

Further development parallels closely that presented for clays. In particular, the loading and configuration surfaces are described by (22.74) and the hardening parameter $\mathscr{H}$ is specified by (22.62). The difference between initially loose and dense sand deposits lies in different sizes of the configuration surface and different values of ζ, Figure 22.12. Studying undrained paths in compression and extension it was found that the best description was obtained for $\zeta = 1.3$ for loose sand ($e_0 = 0.775$) and $\zeta = 0.6$ for dense sand. Figures 22.13(a), (b), (c) illustrate the undrained paths for loose and dense sand predicted by the present model and compared with the experimental data of Ishihara and Okada [54]. Figure 22.14 presents the predicted and experimental curves q/p versus ε_q for sand initially precompressed and subsequently subjected to undrained compression.

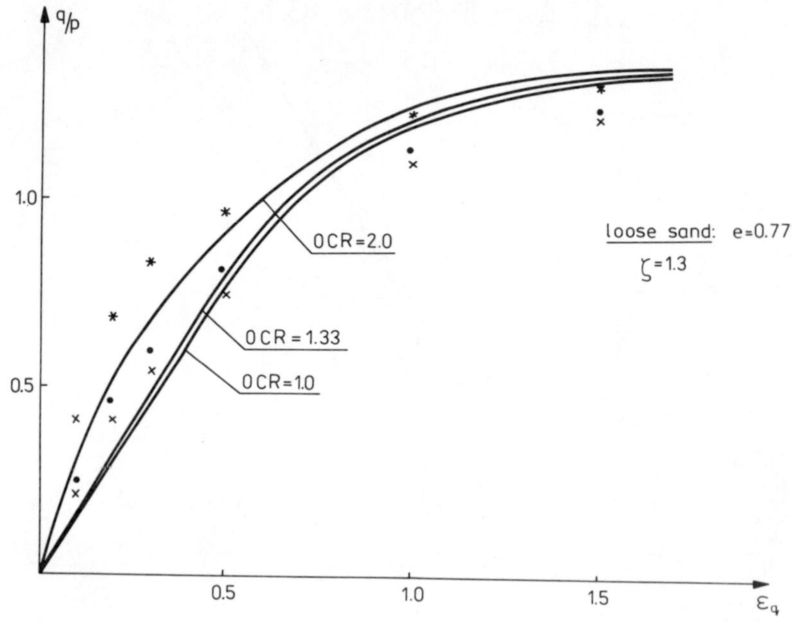

Figure 22.14 Predicted and experimental $q/p - \varepsilon_q$ curves for a loose sand

22.6 CONCLUDING REMARKS

The present paper discusses a broad class of formulations of generalized plasticity models for soils and presents a critical review of some formulations. Next a uniform presentation of an anisotropic hardening rule is given and some of its applications are presented both for clays and sands. Further development and improvement of this rule is needed, especially for the case of cyclic loading. In particular, to simulate pore pressure build-up and consecutive liquefaction in sands, an additional translation of the loading surface toward the origin should be specified in terms of cyclic strain amplitude, similarly as was done for clays in [33].

A critical review of various constitutive models of soils from the viewpoint of their ability to reproduce real soil behaviour presented recently by Pande and Pietruszczak [43] indicates the necessity for further improvements to be introduced in existing model formulations. The use of non-associated flow rules was advocated for some descriptions of [7, 8, 10] in order to increase accuracy of simulation of dilatancy phenomena. Although the formulation discussed in Sections 22.4 and 22.5 was restricted to the associated flow rules, there is no essential difficulty in introducing the departure from the normality rule.

REFERENCES

1. J. H. Atkinson and P. L. Bransby, *The Mechanics of Soils. An Introduction to Critical State Soil Mechanics*, McGraw-Hill Publ. Co., 1979.
2. M. I. Baron and I. Nelson, 'Application of variable moduli models to the soil behaviour', *Int. J. Solids Struct.*, **7**, 389–417 (1971).
3. Z. P. Bažant, 'Endochronic inelasticity and incremental plasticity', *Int. J. Sol. Struct.*, **14**, 691–714 (1978).
4. Z. P. Bažant and S. Kim, 'Plastic-fracturing theory for concrete', *J. Eng. Mech. Div., ASCE*, **105**, No EM 3, 407–428 (1979).
5. P. M. Byrne and T. L. Eldridge, 'A three parameter dilatant elastic stress–strain model for sand', *Proc. Int. Symp. on Numerical Models in Geomechanics*, pp. 73–80, Zurich, 1982.
6. F. Darve, M. Boulon, and R. Chambon, 'Loi rheologique incrementale des sols', *J. de Mécanique*, **17**, 679–716 (1978).
7. C. S. Desai, 'A general basis for yield, failure and potential functions in plasticity', *Int. J. Num. Anal. Meth. Geom.*, **4**, 361–375 (1980).
8. C. S. Desai and H. J. Siriwardane, 'A concept of correction functions to account for non-associative characteristics of geological media', *Int. J. Num. Anal. Meth. in Geom.*, **4**, 377–387 (1980).
9. E. A. Dickin and G. J. W. King, 'The behaviour of hyperbolic stress–strain models in triaxial and plane-strain compression', *Proc. Intern. Symp. on Numerical Models in Geomechanics*, pp. 303–311, A. A. Balkema Publ., Zurich, 1982.
10. P. K. Banerjee and A. S. Stipho, 'Associated and non-associated constitutive relations for undrained behaviour of isotropic soft clays', *Int. J. Num. Anal. Meth. Geom.*, **2**, 35–56 (1978).
11. Y. F. Dafalias and E. P. Popov, 'Plastic internal variables formalism of cyclic plasticity', *ASME J. Appl. Mech.*, **98**, 645–650 (1976).
12. Y. F. Dafalias and L. R. Herrmann, 'A bounding surface soil plasticity model', *Proc. Int. Symp. Soils Under Cyclic and Transient Loading* (Eds. G. N. Pande and O. C. Zienkiewicz), pp. 335–345 A. A. Balkema Publ., 1980.
13. Y. F. Dafalias and L. R. Herrmann, 'Bounding surface formulation of soil plasticity', in *Soil Mechanics—Transient and Cyclic Loads* (Ed. G. N. Pande and O. C. Zienkiewicz).
14. D. C. Drucker and W. Prager, 'Soil mechanics and plastic analysis or limit design', *Quart. Appl. Math.*, **10**, 157–165 (1952).
15. D. C. Drucker, R. E. Gibson, and D. J. Henkel, 'Soil mechanics and work-hardening theories of plasticity', *Trans. ASCE*, **122**, 338–346 (1957).
16. R. O. Davis and G. Mullenger, 'A rate type constitutive model for soil with a critical state', *Int. J. Num. Anal. Meth. Geom.*, **2**, 255–283 (1978).
17. J. M. Duncan and C. Y. Chang, 'Non-linear analysis of stress and strain in soils', *J. Soil Mech. Found, Div. ASCE*, **96**, No. SM 5, 1629–1651 (1970).
18. G. Gudehus, 'Elastoplastische Stoffgleichungen für trockener Sand', *Ing. Arch.*, **42** (1973).
19. K. Hashiguchi, 'Constitutive equations of granular media with an anisotropic hardening', *Third Int. Conf. Num. Meth. Geom.* Aachen, Germany, Vol. 4, (1979).
20. K. Hashiguchi, 'Constitutive equations of elasto-plastic materials with elastic-plastic transition', *ASME J. Appl. Mech.*, **47**, 266–272 (1980).
21. T. Hueckel and R. Nova, 'Some hysteric effects of the behaviour of geologic media', *Int. J. Solids Struct.*, **15**, 625–642 (1979).

22. R. L. Kondner and I. S. Zelasko, 'Hyberbolic stress–strain response: cohesive soils', *J. Soils Mech. Found. Div. ASCE*, **89**, No. SM 1, 115–143 (1963).
23. R. L. Kondner and J. S. Zelasko, 'A hyperbolic stress–strain formulation for sands', *Proc. 2nd Pan-Am. Conf. Soil Mech. Found. Eng.* Brazil, **1**, 289–324 (1963).
24. W. D. Iwan, 'On a class of models for the yielding behaviour of continuous and composite systems', *J. Appl. Mech.*, **34**, 612–617 (1967).
25. R. D. Krieg, 'A practical two-surface plasticity theory', *J. Appl. Mech. Trans. ASME*, E 42, **97**, 641–646 (1975).
26. P. V. Lade and J. M. Duncan, 'Elastoplastic stress–strain theory for cohesion-less soil', *J. Geotechn. Eng. Div. ASCE*, **101**, No. GT 10, 1037–1053 (1975).
27. P. V. Lade, 'Elastoplastic stress–strain theory for cohesion-less soil with curved yield surfaces', *Int. J. Solids and Structures*, **13**, 1019–1035 (1975).
28. Z. Mróz, 'On forms of constitutive laws for elastic-plastic solids', *Arch. Mech. Stos.*, **18**, 3–35 (1966).
29. Z. Mróz, 'On hypoelasticity and plasticity approaches to constitutive modelling of inelastic behaviour of soils', *Int. J. Num. Anal. Meth. Geom.*, **4**, 45–55 (1980).
30. Z. Mróz, 'On the description of anisotropic work-hardening', *J. Mech. Phys. Solids*, **15**, 163–175 (1967).
31. Z. Mróz, V. A. Norris, and O. C. Zienkiewicz, 'An anisotropic hardening model for soils and its application to cyclic loading', *Int. J. Num. Anal. Meth. Geom.*, **2**, 203–221 (1978).
32. Z. Mróz, V. A. Norris, and O. C. Zienkiewicz, 'Application of an anisotropic hardening model in the analysis of elastoplastic deformation of soils', *Geotechnique*, **29**, 1–34 (1979).
33. Z. Mróz, V. A. Norris, and O. C. Zienkiewicz, 'An anisotropic critical state model for soils subject to cyclic loading, *Geotechnique*, **31**, 451–469 (1981).
34. Z. Mróz and V. A. Norris, 'Elastoplastic and viscoplastic constitutive models for soils with application to cyclic loading', in *Soil Mechanics—Transient and Cyclic Loads* (Ed. G. N. Pande and O. C. Zienkiewicz), Chapter 8, pp. 173–217, J. Wiley and Sons, 1982.
35. Z. Mróz and S. Pietruszczak, 'A constitutive model for sand with anisotropic hardening rule', *Int. J. Num. Anal. Math. Geom.*, **7**, 19–38, (1983).
36. R. Nova and D. M. Wood, 'A constitutive model for sand in triaxial compression', *Int. J. Num. Anal. Meth. Geom.*, **3**, 255–278 (1979).
37. S. Pietruszczak and Z. Mróz, 'On hardening anisotropy of K_0-consolidated clays', *Int. J. Num. Anal. Meth. Geom.*, **7**, 19–38, (1983).
38. S. Pietruszczak and Z. Mróz, 'Description of anisotropic consolidation of clays', *Proc. Symp. CRNS, Comportement Mécanique des Solides Anisotropes*, pp. 399–623, Noordhoff Sc. Publ., 1982.
39. S. Pietruszczak and Z. Mróz, 'Finite element analysis of deformation of strain softening materials', *Int. J. Num. Meth. Eng.*, **17**, 327–334 (1981).
40. J. H. Prevost, 'Mathematical modelling of monotonic and cyclic undrained clay behaviour', *Int. J. Num. Anal. Meth. Geom.*, **1**, 195–216 (1977).
41. J. H. Prevost, 'Plasticity theory for soil stress–strain behaviour', *J. Eng. Mech. Div. ASCE*, **104**, No. EM5, 1177–1194, 1978.
42. J. H. Prevost and K. Hoeg, 'Effective stress–strain strength model for soils', *Proc. ASCE*, **101**, GT 3, 259–278 (1975).
43. G. Pande and S. Pietruszczak, 'A sideways look at numerical models of soils', *Univ. Coll. Swansea, Civ. Eng. Dep.* Report C/R/433/82, Dec. 1982.
44. M. Romano, 'A continuum theory for granular media with a critical state', *Arch. Mech. Stos.*, **26**, 1011–1028 (1974).

45. D. A. Kolymbas, 'A rate-dependent constitutive equation for soils', *Mech. Res. Comm*, **4**, 367–372 (1977).
46. I. S. Sandler, 'On the uniqueness and stability of endochronic theories of material behaviour', *Trans. ASME, J. Appl. Mech.*, **45**, 263–266 (1978).
47. R. S. Rivlin, 'Some comments on the endochronic theory of plasticity', *Int. J. Sol. Struct.* 1981.
48. K. C. Valanis and H. E. Read, 'A new endochronic plasticity model for soils', in *Soil Mechanics—Transient and Cyclic Loads* (Eds. G. N. Pande and O. C. Zienkiewicz), pp. 375–417, 1982, John Wiley and Sons.
49. H. Di Benedetto and F. Darve, 'Comparaison de lois rheologiques en cinematique rotationelle', *J. de Mec. Theor. Appl.*, 1983 (in print).
50. A. N. Schofield and P. Wroth, *Critical State Soil Mechanics*, McGraw-Hill, 1968.
51. P. Wilde, 'Two invariant depending models of granular media', *Arch. Mech. Stos.*, **29**, 199–209, 1977.
52. S. Nemat-Nasser and A. Shokooh, 'On finite plastic flows of compressible materials with internal friction', *Int. J. Sol. Struct.*, **16**, 495–514 (1980).
53. A. W. Skempton and V. A. Sowa, 'The behaviour of saturated clays during sampling and testing', *Geotechnique*, **13**, 269–280 (1963).
54. K. Ishihara and S. Okada. 'Yielding of overconsolidated sand and liquefaction model under cyclic stresses', *Soils and Found.*, **18**, 57–72 (1978).

Mechanics of Engineering Materials
Edited by C. S. Desai and R. H. Gallagher

Chapter 23

Micromechanically Based Rate Constitutive Descriptions for Granular Materials

S. Nemat-Nasser and M. M. Mehrabadi

23.1 INTRODUCTION

For a granular mass that supports the overall applied loads through contact friction, one may seek to describe the overall mechanical response on the basis of simple micromechanical models. Broadly speaking, this requires the description of overall stress, characterization of fabric, representation of kinematics, development of local rate constitutive relations, and evaluation of the overall rate constitutive relations in terms of the local quantities.

The representation of the overall stress tensor in terms of the volume average of the contact forces and 'branches'† was first given by Christoffersen, Mehrabadi, and Nemat-Nasser [1], using the principle of virtual work. The relation between such *volume average* description of stress and an alternative representation in terms of average tractions transmitted over three mutually orthogonal planes, has been discussed by Mehrabadi, Nemat-Nasser, and Oda [3], Oda, Nemat-Nasser, and Mehrabadi [4], and Nemat-Nasser [5]; representation of stress in terms of surface averages has been examined by Drescher and de Josselin de Jong [6], Oda [7], Satake [2], and Konishi [8]. In the present work the results of Christoffersen, Mehrabadi, and Nemat-Nasser [1] are used to relate the overall stress to the measures of granular fabric (or the microstructure).

The term fabric refers to the microscopic mutual arrangement of the particles and the associated voids, which gives rise to the overall anisotropic response; Horne [9, 10], Oda [7, 11], Konishi [8], Oda, Konishi, and Nemat-Nasser [12], Oda, Nemat-Nasser, and Mehrabadi [4], and Nemat-Nasser and Mehrabadi [13]. In the present work the fabric is identified with the distribution of contact unit normals and unit branches which are unit vectors along the branches. Fabric

† Branches are vectors which connect the centroids of mutually contacting granules; Satake [2].

measures involving both unit vectors emerge in kinematics, dynamics, and rate constitutive relations of granular materials.

When a granular mass consists of, say, rigid (idealized) granules, the change in overall stress is accompanied by a change in microstructure or fabric. The particle movements by relative sliding and rolling change the distribution of contact normals and unit branches. The local deformation rate and spin, therefore, may be attributed to two accompanying processes: (1) a 'plastic' or 'inelastic' part *which leaves fabric unchanged*; and (2) an accommodating part which produces a change in fabric and results in a change in stress. This decomposition is analogous to the description of kinematics of crystalline solids, where the velocity gradient in each crystal is regarded to stem from a plastic contribution due to slip on crystallographic planes, accompanied by an elastic part due to lattice distortion which produces stress changes and renders the total velocity gradient compatible; see Mandel [14], Hill [15], Hill and Rice [16], Asaro [17], Nemat-Nasser, Mehrabadi, and Iwakuma [18], Havner [19], and Nemat-Nasser [20], who gives additional relevant references.

Based on the observation that fabric change induces change in contact forces, simple constitutive relations are assumed at the microlevel. These relate an objective local stress rate, associated with the contact force, to the fabric deformation rate.† To proceed further, the average of the rate quantities must be related to the rate of average quantities. In finite deformations, the average of the velocity gradient and that of the nominal stress rate are suitable kinematic and dynamic variables; Hill [21]. Using these, we relate the overall nominal stress rate to the overall velocity gradient and calculate the overall instantaneous moduli in terms of the corresponding local ones. The completely 'self-consistent calculation' (Hill [22]) is rather complicated, especially in the general setting considered here (Iwakuma and Nemat-Nasser [23] and Nemat-Nasser and Iwakuma [24]), where the dominant moduli are of the order of pressure and vanish at zero pressure. A simpler alternative procedure is, therefore, outlined.

An interesting feature of the theory is that the overall, as well as the local, constitutive relations are highly anisotropic even when the simplest isotropic form is used to relate the local stress rate and fabric strain rate, and a simple frictional law is used to express the local flow rule.

23.2 STRESS

Let $\bar{\sigma}_{ij}$ stand for the components of the overall stress, $\bar{\boldsymbol{\sigma}}$; throughout, a fixed rectangular Cartesian coordinate system is used. Christoffersen, Mehrabadi, and Nemat-Nasser [1] have shown that

$$\bar{\sigma}_{ij} = \frac{1}{N} \sum_{\alpha=1}^{N} t_{ij}^{\alpha} \equiv \langle t_{ij} \rangle \tag{23.1}$$

† The term 'fabric deformation rate' (or 'fabric spin', etc.) refers to the deformation rate which stems from fabric changes.

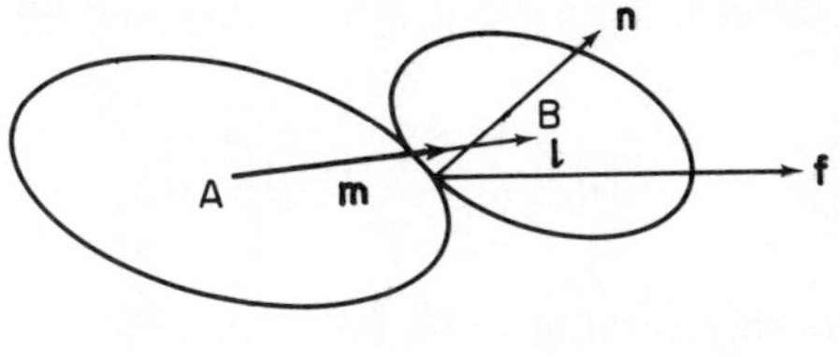

Figure 23.1

where

$$t_{ij}^{\alpha} = Nl^{\alpha} f_i^{\alpha} m_j^{\alpha}, \qquad l_j^{\alpha} = l^{\alpha} m_j^{\alpha} \qquad \text{(no sum on } \alpha) \tag{23.2}$$

Here, N is the number of contacts per unit volume of a typical sample of the granular mass, f_i^{α} is the contact force at contact α, and l_j^{α} is the corresponding 'branch vector', i.e. a vector connecting the centroids of two granules which are in contact at α; l^{α} is the 'branch length' and m_j^{α} is the 'unit branch' (Figure 23.1). In the sequel, averages will be denoted by the angular brackets, as in (23.1). Also, the superimposed α will be omitted whenever it is felt that no confusion will result. For example, instead of (23.2), we write $t_{ij} \equiv Nlf_i m_j$. Note that, for a sample of volume V containing M contacts, $N = M/V$.

There are several ways that the contact force f_i can be expressed in terms of some 'local stress' measure. In [1, 3], f_i is decomposed along the contact normal, the sliding direction, and a direction normal to these, whereas in [25] the quantity Nlf_i is written as $2A_{ik}^{r} m_k$ where A_{ik}^{r} is a local stress. In [13], on the other hand, the contact force is expressed as

$$f_i = a\tau_{ij} n_j \tag{23.3}$$

where a is an 'effective' contact area, τ_{ij} is a local stress, and **n** (with components n_j) is the unit contact normal. While these and other representations have their own appeal, here we shall use (23.3). Then (23.2) becomes

$$t_{ij} = M\frac{al}{V}\tau_{ik} n_k m_j = \tau_{ik} h_{kj} \tag{23.4}$$

where

$$h_{kj} \equiv \varepsilon n_k m_j, \qquad \varepsilon \equiv M\frac{al}{V} = Nal \tag{23.5}$$

It is reasonable to regard the local stress tensor τ_{ij} independent of the orientations n_k and m_j, as well as independent of ε, and from (23.1) and (23.4) obtain

$$\begin{aligned} \bar{\sigma}_{ij} &= \langle \tau_{ik} h_{kj} \rangle = \langle \tau_{ik} \rangle \langle h_{kj} \rangle \\ &= T_{ik} H_{kj} = \tfrac{1}{2}(T_{ik} H_{kj} + T_{jk} H_{ki}) \end{aligned} \tag{23.6}$$

where the overall stress is regarded to be symmetric,

$$T_{ij} = \langle \tau_{ij} \rangle \qquad \text{and} \qquad H_{ij} = \langle h_{ij} \rangle \tag{23.7}$$

In [13] the fabric tensor $(23.7)_2$ is denoted by $\hat{H}_{ij}$.

For composite materials, the average stress is defined by

$$\bar{\sigma}_{ij} = \frac{1}{V} \int_V \sigma_{ij}(\mathbf{x}) \, \mathrm{d}V \tag{23.8}$$

where $\sigma_{ij}(\mathbf{x})$ is a smoothly varying stress field in equilibrium with the uniform applied surface loads. While definition (23.8) is used in deriving (23.1) (see [1]), in (23.3) and (23.6) we do not identify τ_{ij} with $\sigma_{ij}(\mathbf{x})$. The local stress τ_{ij} characterizes the local contact force which acts on a *small* contact area. Hence, unlike $\sigma_{ij}(\mathbf{x})$, τ_{ij} is expected to have sharp variations from one contact neighbourhood to another. Furthermore, τ_{ij} and hence its average T_{ij} need not be symmetric, whereas $\sigma_{ij}(x)$ in (23.8) is usually assumed to be symmetric, and this is in line with the symmetry of the overall stress $\bar{\boldsymbol{\sigma}}$. Observe that $\mathbf{H}$, in general, is *not* symmetric. For spherical granules (circular in two dimensions), however, the unit normal $\mathbf{n}$ coincides with the unit branch $\mathbf{m}$, and $\mathbf{H}$ becomes symmetric. In this case we assume that τ_{ij} (and hence T_{ij}) is also symmetric, and from the symmetry of the overall stress, $\bar{\boldsymbol{\sigma}}$, conclude that all three tensors, $\bar{\boldsymbol{\sigma}}$, $\mathbf{T}$ and $\mathbf{H}$, are coaxial (have the same principal directions). This immediately leads to the following *exact* stress–fabric relation (Nemat-Nasser and Mehrabadi [13]):

$$\bar{\sigma}_{ij} = A_0 \delta_{ij} + A_1 H_{ij} + A_2 H_{ik} H_{kj} \tag{23.9}$$

where A_0, A_1, and A_2 are, in general, functions of the basic invariants of $\bar{\boldsymbol{\sigma}}$ and $\mathbf{H}$.

Representation (23.9) is exact whenever the three tensors $\bar{\boldsymbol{\sigma}}$, $\mathbf{T}$, and $\mathbf{H}$ are symmetric. Experimental results on biaxial deformation of oval cross-sectional photoelastic rods seem to support the representation (23.9) even for non-circular granules; see Oda, Konishi, and Nemat-Nasser [26].

Note that when neither $\mathbf{T}$ nor $\mathbf{H}$ is symmetric, the symmetry of $\bar{\boldsymbol{\sigma}}$ requires

$$T_{ik} H_{kj} - T_{jk} \mathbf{H}_{ki} = 0 \tag{23.10}$$

23.3 FABRIC

Fabric refers to the microstructure of the granular mass; see [4, 12, 13] for discussion and references. Various measures may be used to quantify fabric. One such measure is $\mathbf{H}$ defined by (23.5) and $(23.7)_2$. Another measure, introduced by Satake [2], is

$$J_{ij} = \langle n_i n_j \rangle \tag{23.11}$$

Other measures may be used, depending on the particular context, but, as pointed out in [13] and elsewhere, measures like (23.11) must be even order tensors, the lowest order being a scalar which may be identified with the void ratio or the average coordination number.

In this work we view the fabric as being characterized by the average coordination number and the distribution of contact unit normals and unit branches. Note that the fabric tensor $H_{ij} = \langle \varepsilon n_i m_j \rangle$ includes information about the average number of contacts and the distribution of branch lengths and contact areas (through $\varepsilon = Nal$), as well as the distribution of contact normals and unit branches. H_{ij}, therefore, is an effective measure of the microstructure; see [13].

In Section 23.5 we shall assume spherical granules and develop local rate constitutive relations associated with a typical *active* contact. These relations include the following fabric measures:

$$h_{ij} = \varepsilon n_i n_j, \qquad r_{ij} = \tfrac{1}{2}(s_i n_j + s_j n_i), \qquad \omega_{ij} = \tfrac{1}{2}(s_i n_j - s_j n_i) \tag{23.12}$$

where **s**, with components s_i, is the unit vector in the 'sliding' direction at the considered active contact.

With reference to the representation of the overall stress by (23.6), one may use the following alternative,

$$\bar{\sigma}_{ij} = \langle \varepsilon \tau_{ik} n_k m_j \rangle = \langle \varepsilon \tau_{ik} \rangle \langle n_k m_j \rangle$$

and view $\langle n_k m_j \rangle$ as the fabric measure and $\langle \varepsilon \tau_{ik} \rangle$ as a local stress measure. Since the average number of contacts and the distribution of contact areas and branch lengths are all relevant measures of fabric, the representation (23.6) is preferred in the present work.

23.4 KINEMATICS

In the rate-independent response range, the velocity gradient for a single *crystal* is often viewed to stem from two accompanying microprocesses: (1) slip over active crystallographic slip planes; and (2) an accommodating elastic lattice distortion; see Mandel [14], Hill [15], Kocks [27], Hill and Rice [16], Asaro [17], Havner [19], Nemat-Nasser, Mehrabadi, and Iwakuma [18], and Nemat-Nasser [20]. An objective stress rate hence is related to the elastic lattice distortion rate by a suitable elastic constitutive relation.

For a granular mass which consists of 'rigid' (idealized) granules, the 'local' velocity gradient, l_{ij}^{α}, in part stems from the (inelastic) relative slip and rolling of granules, but since no elastic distortion exists, *the total velocity gradient is rendered compatible by an accompanying distortion associated with the 'fabric change'*. It thus follows that

$$l_{ij} = l_{ij}^{*} + l_{ij}^{**} \tag{23.13}$$

where l_{ij}^{**} *is the inelastic part of the velocity gradient, which leaves the fabric unchanged, and* l_{ij}^{*} *is the accompanying part that stems from the fabric change.* The quantities l_{ij}, l_{ij}^{*}, and l_{ij}^{**} are associated with a typical contact α, but for simplicity the superimposed α is omitted in (23.13) and in the sequel.

For a single crystal, $l_{ij}^{**} = \dot{\gamma} s_i n_j$, where the unit vector s_i defines the slip direction, and $\dot{\gamma}$ is the corresponding shear rate. Such a slip-induced rate of deformation does not affect the existing lattice distortion, though, in general, it violates compatibility. In line with this, we write for the inelastic part of the local velocity gradient associated with contact α,

$$l_{ij}^{**} = (s_i n_j + \zeta n_i n_j)\dot{\gamma} \tag{23.14}$$

where $\dot{\gamma}$ is the magnitude of the inelastic shearing distortion, ζ is the local rate of inelastic volumetric strain rate per unit shearing, and **s** is a unit vector normal to **n**, in the sliding direction. A representation of this kind has been given by Nemat-Nasser, Mehrabadi, and Iwakuma [18] for pressure sensitive materials. From (23.14),

$$l_{ij}^{**} = d_{ij}^{**} + w_{ij}^{**} \tag{23.15}$$

where

$$d_{ij}^{**} = p_{ij}\dot{\gamma} \quad \text{and} \quad w_{ij}^{**} = \omega_{ij}\dot{\gamma} \tag{23.16}$$

respectively, are the local inelastic deformation rate and spin; here

$$p_{ij} = \tfrac{1}{2}(s_i n_j + s_j n_i) + \zeta n_i n_j, \qquad \omega_{ij} = \tfrac{1}{2}(s_i n_j - s_j n_i) \tag{23.17}$$

Similarly, the accommodating velocity gradient, l_{ij}^*, associated with the fabric change, is written as

$$l_{ij}^* = d_{ij}^* + w_{ij}^* \tag{23.18}$$

where

$$d_{ij}^* = \tfrac{1}{2}(l_{ij}^* + l_{ji}^*) \quad \text{and} \quad w_{ij}^* = \tfrac{1}{2}(l_{ij}^* - l_{ji}^*) \tag{23.19}$$

are the corresponding deformation rate and spin tensors. Locally, therefore, we have

$$d_{ij} = d_{ij}^* + p_{ij}\dot{\gamma}, \qquad w_{ij} = w_{ij}^* + \omega_{ij}\dot{\gamma} \tag{23.20}$$

for the deformation rate and spin.

Consider now a representative sample of a granular mass of volume V and surface S and let it be subjected to a boundary velocity field compatible with the uniform overall velocity gradient L_{ij}; i.e., let it be subjected to

$$v_i = L_{ij} x_j \qquad \text{on } S \tag{23.21}$$

where L_{ij} is constant. (Note that this is what one often attempts to achieve in experiments.) Within the sample, the velocity field, $v_i = v_i(\mathbf{x})$, will not generally be linear. However, with the aid of the divergence theorem,

$$\frac{1}{V}\int_S v_i \nu_j \,\mathrm{d}S = \frac{1}{V}\int_V \frac{\partial v_i}{\partial x_j}\mathrm{d}V$$

so that

$$L_{ij} = \left\langle \frac{\partial v_i}{\partial x_j} \right\rangle \equiv \langle l_{ij} \rangle \tag{23.22}$$

where $\partial v_i/\partial x_j$ is the variable velocity gradient within the sample, and $\boldsymbol{\nu}$ is the exterior unit normal on S.

Define the overall deformation rate and spin, respectively, by

$$D_{ij} = \langle d_{ij} \rangle = \tfrac{1}{2}(L_{ij} + L_{ji}) \tag{23.23}$$

$$W_{ij} = \langle w_{ij} \rangle = \tfrac{1}{2}(L_{ij} - L_{ji}) \tag{23.24}$$

and observe that the corresponding 'inelastic' and 'fabric-induced' parts are *not*, in general, equal to the averages of the associated local quantities.

23.5 LOCAL RATE CONSTITUTIVE RELATIONS

Henceforth attention is focused on spherical granules only.Then $\mathbf{n} \equiv \mathbf{m}$ and $h_{ij} = \varepsilon n_i n_j$. Furthermore, the stress–fabric relation (23.9) holds exactly.

Let superimposed ∇ denote the time rate of change *corotational with fabric*. Since the contact normal $\mathbf{n}$ (and hence $\mathbf{s}$) characterizes the local fabric, we require that

$$\overset{\nabla}{n}_i = \dot{n}_i - w^*_{ik} n_k = 0, \qquad \overset{\nabla}{s}_i = \dot{s}_i - w^*_{ik} s_k = 0 \tag{23.25}$$

Moreover, from the basic assumption that the microscopic inelastic contribution to the total *local* velocity gradient does *not* affect the fabric, it follows that the change in $\varepsilon = Nal$ stems from fabric changes only. Since ε represents an 'effective' fractional number of contacts associated with the small volume al, it is reasonable to attribute the rate of change of ε to the volumetric strain rate produced by the rate of change of fabric, i.e. to d^*_{kk}, and postulate the following 'conservation' relation:

$$\frac{\dot{\varepsilon}}{\varepsilon} + d^*_{kk} = 0 \tag{23.26}$$

Note the correspondence with the conservation of mass.

From (23.25) and (23.26) we have

$$\overset{\nabla}{h}_{ij} + d^*_{kk} h_{ij} = 0 \tag{23.27}$$

which defines the evolution of the local fabric tensor, h_{ij}.

Now consider (23.4), i.e. $t_{ij} = \tau_{ik} h_{kj}$, and in view of (23.25) to (23.27) obtain†

$$\overset{\nabla}{t}_{ij} + d^*_{kk} t_{ij} = \overset{\nabla}{\tau}_{ik} h_{kj} \tag{23.28}$$

The local stress τ_{ij} directly relates to the contact forces f_i; i.e., $1/(a) f_i = \tau_{ik} n_k$, equation (23.3). One of our basic assumptions is that changes in stress, and hence in contact force, arise from the fabric change. It is therefore reasonable to relate

† Here and in the sequel we regard t_{ij} and its rates to be symmetric, given by the symmetric part of the right-hand side, although this is not made explicit.

$(f_i/a)^\nabla = \overset{\nabla}{\tau}_{ik} n_k$ directly to the fabric strain rate, d^*_{ij}. To this end, let

$$\overset{\nabla}{\tau}_{ij} = L_{ijkl} d^*_{kl} \tag{23.29}$$

where $L_{ijkl} = L_{jikl} = L_{ijlk}$ is independent of d^*_{ij}. For the present application, L_{ijkl} may be regarded as a constant tensor. In fact, from (23.28),

$$\overset{\nabla}{t}_{ij} + d^*_{kk} t_{ij} = L_{imkl} h_{mj} d^*_{kl} \tag{23.30}$$

which shows that even an isotropic tensor,

$$L_{ijkl} = \alpha_0 \delta_{ij} \delta_{kl} + \tfrac{1}{2} \beta_0 (\delta_{ik}\delta_{jl} + \delta_{jk}\delta_{il}) \tag{23.31}$$

yields an anisotropic stress rate–fabric strain rate relation; i.e.,

$$\overset{\nabla}{t}_{ij} + d^*_{kk} t_{ij} = \alpha_0 h_{ij} d^*_{kk} + \beta_0 d^*_{ik} h_{kj}$$

At an active contact, the local friction law, $\tau + \mu\sigma = 0$, is assumed to hold,

$$\dot{\tau} + \mu\dot{\sigma} = 0, \qquad \tau = \left(\frac{1}{a} f_i\right) s_i, \qquad \sigma = \left(\frac{1}{a} f_i\right) n_i \tag{23.32}$$

where μ is the coefficient of friction (assumed constant). Because of (23.25), equation $(23.32)_1$ becomes

$$\dot{\tau} + \mu\dot{\sigma} = \overset{\nabla}{\tau}_{ij} q_{ij} = 0 \tag{23.33}$$

where

$$q_{ij} = \tfrac{1}{2}(s_i n_j + s_j n_i) + \mu n_i n_j \tag{23.34}$$

From (23.29), (23.33), and (23.20), it follows that

$$\dot{\gamma} = K q_{ij} L_{ijkl} d_{kl}, \qquad K = (q_{ij} L_{ijkl} p_{kl})^{-1} \tag{23.35}$$

Define the Jaumann rate of the local stress t_{ij}, corotational with the *total* local spin, w_{ij}, by

$$\mathring{t}_{ij} = \dot{t}_{ij} - w_{ik} t_{kj} - w_{jk} t_{ki} \tag{23.36}$$

and from (23.30) and (23.35) obtain

$$\mathring{t}_{ij} + d_{kk} t_{ij} = \{L_{imkl} h_{mj} - [L_{iqrs} h_{qj} p_{rs} + \omega_{ik} t_{kj} + \omega_{jk} t_{ki} - \zeta t_{ij}]\ K q_{mn} L_{mnkl}\} d_{kl} \tag{23.37}$$

where p_{ij}, ω_{ij}, and q_{ij} are given by (23.17) and (23.34).

Consider now a representative sample of volume V and surface S, subjected on its boundary to uniform tractions, T^0_i, measured per unit current area. Let these tractions be changed at the rate $\dot{T}^0_i$ (uniform on S), again measured per unit current area. With $\dot{n}_{ij}$ denoting the nominal stress rate (non-symmetric), it follows that

$$\frac{\partial \dot{n}_{ij}}{\partial x_i} = 0 \quad \text{in } V \qquad \text{and} \qquad \dot{n}_{ij} v_i = \dot{T}^0_j \quad \text{on} \quad S \tag{23.38}$$

The nominal stress rate will generally be non-uniform within V. However, with the aid of the divergence theorem,

$$\frac{1}{V}\int_S x_i \dot{T}_j^0 \, dS = \frac{1}{V}\int_V \dot{n}_{ij} dV \equiv \langle \dot{n}_{ij} \rangle$$

it is seen that the overall *uniform* nominal stress rate

$$\dot{N}_{ij} \equiv \langle \dot{n}_{ij} \rangle \tag{23.39}$$

is compatible with the uniform surface traction rates, $(23.38)_2$, i.e., $\dot{T}_j^0 = \dot{N}_{ij} v_i$ on S. Note that, in general, other stress rates, e.g. the Cauchy stress rate, do not yield such a simple self-consistent averaging result; Hill [21].

We use (23.39) as the definition of the overall nominal stress rate, and (23.22) as the corresponding definition of the overall velocity gradient. Since

$$\dot{n}_{ij} = \dot{t}_{ij} + d_{kk} t_{ij} - d_{ik} t_{kj} + w_{jk} t_{ki} \tag{23.40}$$

the local constitutive relations (23.37) may be expressed as

$$\dot{n}_{ij} = \mathscr{F}^{\alpha}_{ijkl} l_{kl} \tag{23.41}$$

where

$$\begin{aligned}\mathscr{F}^{\alpha}_{ijkl} = {} & L_{imkl} h_{mj} - \tfrac{1}{2}(\delta_{ik}\delta_{lm} + \delta_{il}\delta_{km}) t_{mj} + \tfrac{1}{2}(\delta_{jk}\delta_{lm} - \delta_{jl}\delta_{km}) t_{mi} \\ & - (L_{iqrs} h_{qj} p_{rs} - \zeta t_{ij} + \omega_{ir} t_{rj} + \omega_{jr} t_{ri}) K q_{mn} L_{mnkl}\end{aligned} \tag{23.42}$$

Here, superimposed α emphasizes that the moduli pertain to the contact α, and hence vary from contact to contact. Except for the explicit inclusion of the fabric tensor h_{ij}, equation (23.42) is similar to that of Iwakuma and Nemat-Nasser [23], who seek to model dilatant, pressure sensitive single crystals or jointed rocks. Observe that, when the symmetry of the local Cauchy stress is rendered explicit, (23.42) reads

$$\begin{aligned}\mathscr{F}^{\alpha}_{ijkl} = {} & \tfrac{1}{2}(L_{imkl} h_{mj} + L_{jmkl} h_{mi}) - \tfrac{1}{2}(\delta_{ik}\delta_{lm} + \delta_{il}\delta_{km}) t_{mj} \\ & + \tfrac{1}{2}(\delta_{jk}\delta_{lm} - \delta_{jl}\delta_{km}) t_{mi} - [\tfrac{1}{2}(L_{iqrs} h_{qj} + L_{jqrs} h_{qi}) p_{rs} \\ & - \zeta t_{ij} + \omega_{ir} t_{rj} + \omega_{jr} t_{ri}] K q_{mn} L_{mnkl}\end{aligned} \tag{23.43}$$

23.6 OVERALL RATE CONSTITUTIVE RELATIONS

We follow here the procedure developed by Iwakuma and Nemat-Nasser [23] and Nemat-Nasser and Iwakuma [24] for elastic–plastic composites and polycrystals at finite strains. Guided by the general results of Hill [21, 22], these authors formulate and actually calculate the overall instantaneous moduli at finite strains, using the nominal stress rate and the velocity gradient in a self-consistent manner. In this approach, the overall nominal stress rate, (23.39), is related to the overall velocity gradient, (23.22), by

$$\dot{N}_{ij} = \mathscr{F}_{ijkl} L_{kl} \tag{23.44}$$

and the objective is to estimate $\mathscr{F}_{ijkl}$ in terms of $\mathscr{F}^{\alpha}_{ijkl}$, for a representative sample of the granular mass.

To this end, it is necessary to express the local velocity gradient, l_{ij}, in terms of the global one, L_{ij}. Let $\mathscr{A}^{\alpha}_{ijkl}$ be the 'concentration' tensor such that

$$l_{ij} = \mathscr{A}^{\alpha}_{ijkl} L_{kl} \tag{23.45}$$

for the αth contact. Then from (23.41), (23.22), (23.39), and (23.44), it follows that

$$\mathscr{F}_{ijkl} = \langle \mathscr{F}^{\alpha}_{ijmn} \mathscr{A}^{\alpha}_{mnkl} \rangle \tag{23.46}$$

where the average is taken over all active contacts. Moreover, averaging (23.45), it is seen that the self-consistency requires

$$\langle \mathscr{A}^{\alpha}_{ijkl} \rangle = I_{ijkl} \tag{23.47}$$

the identity tensor.

To complete this solution, $\mathscr{A}^{\alpha}_{ijkl}$ must be expressed in terms of quantities which can be calculated. One approach is to employ Hill's self-consistent method, as applied by Iwakuma and Nemat-Nasser [23] to polycrystals and composites at finite strains. The basic procedure is to embed a representative small region, associated with a typical contact α, in a homogeneous body of instantaneous uniform moduli $\mathscr{F}_{ijkl}$ (as yet unknown), and then estimate the local velocity gradient, l_{ij}, in terms of the overall one, L_{ij}. When the operator

$$\mathscr{F}_{ijkl} \frac{\partial^2}{\partial x_i \partial x_l} \tag{23.48}$$

is elliptic, the method works. On the other hand, as soon as (23.48) ceases to be elliptic, then the assumed overall uniform deformation rate cannot be maintained, leading to instability by possible localized deformations; Iwakuma and Nemat-Nasser [23] and Nemat-Nasser and Iwakuma [24].

It is convenient to associate a spherical region, Ω, of unit radius with the contact α, and follow the procedure of Iwakuma and Nemat-Nasser [23] to obtain

$$\mathscr{A}^{\alpha}_{ijkl} = [I_{ijkl} + J_{ijmn}(\mathscr{F}_{mnkl} - \mathscr{F}^{\alpha}_{mnkl})]^{-1} \tag{23.49}$$

where

$$
\begin{aligned}
J_{ijmn} &= \tfrac{1}{2} \int_{-1}^{1} H_{ijmn}(\xi)\, \mathrm{d}\xi \\
H_{jkmn} &= \frac{-1}{2\pi i} \int_{\gamma} \left[\frac{N_{nj}(\boldsymbol{\xi}) \xi_k \xi_m}{D(\boldsymbol{\xi}) z} \right] \mathrm{d}z, \qquad z = \mathrm{e}^{-\mathrm{i}\theta}, i = \sqrt{(-1)} \\
N_{ij} &= \text{cofactor } (\mathscr{F}_{kijl} \xi_k \xi_l), \qquad D(\boldsymbol{\xi}) = \det(\mathscr{F}_{kijl} \xi_k \xi_l)
\end{aligned}
\tag{23.50}
$$

Here γ is a unit circle in the complex z-plane, and the unit vector $\boldsymbol{\xi}$ is expressed as

$$\xi_1 = (1 - \xi^2)^{1/2} \cos\theta, \qquad \xi_2 = (1 - \xi^2)^{1/2} \sin\theta, \qquad \xi_3 = \xi \tag{23.51}$$

Iwakuma and Nemat-Nasser [23] show that as long as the roots of $D(\xi)=0$ are complex, the operator (23.48) remains elliptic, and (23.50) and (23.49) yield the required concentration tensor. Observe that J_{ijmn} depends on $\mathscr{F}_{ijkl}$ which is not known until $\mathscr{A}^{\alpha}_{ijkl}$ is obtained and used in (23.46). Therefore, a numerical iterative procedure is required.

The method simplifies considerably when one is able to estimate the concentration tensor $\mathscr{A}^{\alpha}_{ijkl}$, in a manner that does not involve the unknown moduli $\mathscr{F}_{ijkl}$. For polycrystals, for example, the elastic modulus tensor may be used in place of $\mathscr{F}_{ijkl}$ in (23.49) and (23.50). This then will correspond to a modified, finite deformation version of models considered by Budiansky and Wu [28] and Hutchinson [29]. As shown by Hutchinson [30] for polycrystals at small strains, models of this kind yield stiffer response.

For the present problem, an interesting possibility is to employ

$$\bar{\mathscr{F}}^{*}_{ijkl}=\tfrac{1}{2}(L_{imkl}H_{mj}+L_{jmkl}H_{mi})-\tfrac{1}{2}(\delta_{ik}\delta_{lm}+\delta_{il}\delta_{km})\bar{\sigma}_{mj}$$
$$+\tfrac{1}{2}(\delta_{jk}\delta_{lm}-\delta_{jl}\delta_{km})\bar{\sigma}_{mi} \qquad (23.52)$$

in place of $\mathscr{F}_{ijkl}$ in (23.49) and (23.50). The concentration tensor will then be affected by the current overall fabric and stress only, and the resulting overall moduli will probably correspond to a stiffer material relative to those estimated by the complete self-consistent method outlined above.

Since the average of the local Cauchy stress rate does not equal the rate of the overall Cauchy stress, this latter quantity must be defined in terms of the nominal stress rate $\dot{N}_{ij}$. Here we set

$$\dot{\bar{\sigma}}_{ij}+D_{kk}\bar{\sigma}_{ij}\equiv\tfrac{1}{2}(\dot{N}_{ij}+\dot{N}_{ji}+L_{ik}\bar{\sigma}_{kj}+L_{jk}\bar{\sigma}_{ki}) \qquad (23.53)$$

and calculate the overall Cauchy stress at time $t+\Delta t$ by

$$\bar{\boldsymbol{\sigma}}(t+\Delta t)=\bar{\boldsymbol{\sigma}}(t)+\Delta t\dot{\bar{\sigma}}(t)+0(\Delta t^2) \qquad (23.54)$$

ACKNOWLEDGEMENTS

This work was supported by the US Air Force Office of Scientific Research, Grant No. AFOSR-80-0017 to Northwestern University.

REFERENCES

1. J. Christoffersen, M. M. Mehrabadi, and S. Nemat-Nasser, ‘A micromechanical description of granular material behavior’, *J. Appl. Mech.*, **48**, 339–344 (1981).
2. M. Satake, ‘Constitution of mechanics of granular materials through the graph theory’, in *Proc. U.S.–Japan Seminar on Continuum-Mechanical and Statistical Approaches in the Mechanics of Granular Materials* (Eds. S. C. Cowin and M. Satake), pp. 47–62, Gakujutsu Bunken Fukyukai, Tokyo, 1978.
3. M. M. Mehrabadi, S. Nemat-Nasser, and M. Oda, ‘On statistical description of stress and fabric in granular materials’, *Int. J. Numer. Anal. Methods Geomech.*, **6**, 95–108 (1982).

4. M. Oda, S. Nemat-Nasser, and M. M. Mehrabadi, 'A statistical study of fabric in a random assembly of spherical granules', *Int. J. Numer. Anal. Methods Geomech.*, **6**, 77–94 (1982).
5. S. Nemat-Nasser, 'Fabric and its influence on mechanical behavior of granular materials', in *Deformation and Failure of Granular Materials* (Eds. P. A. Vermeer and H. J. Luger), pp. 37–42, A. A. Balkema, Rotterdam, 1982.
6. A. Drescher and G. de Josselin de Jong, 'Photoelastic verification of a mechanical model for the flow of a granular material', *J. Mech. Phys. Solids*, **20**, 337–351 (1972).
7. M. Oda, 'The mechanism of fabric changes during compressional deformation of sand', *Soils and Foundations*, **12**, 1–18 (1972).
8. J. Konishi, 'Microscopic model studies on the mechanical behavior of granular materials', in *Proc. U.S.–Japan Seminar on Continuum-Mechanical and Statistical Approaches in the Mechanics of Granular Materials* (Eds. S. C. Cowin and M. Satake), pp. 27–45, Gakujutsu Bunken Fukyukai, Tokyo, 1978.
9. M. R. Horne, 'The behaviour of an assembly of rotund, rigid, cohesionless particles (I and II)', *Proc. Roy. Soc. London*, **A286**, 62–97 (1965).
10. M. R. Horne, 'The behaviour of an assembly of rotund, rigid, cohesionless particles (III)', *Proc. Roy. Soc. London*, **A310**, 21–34 (1969).
11. M. Oda, 'Significance of fabric in granular mechanics', in *Proc. U.S.–Japan Seminar on Continuum-Mechanical and Statistical Approaches in the Mechanics of Granular Materials* (Eds. S. C. Cowin and M. Satake), pp. 7–26, Gakujutsu Bunken Fukyukai, Tokyo, 1978.
12. M. Oda, J. Konishi, and S. Nemat-Nasser, 'Some experimentally based fundamental results on the mechanical behaviour of granular materials', *Géotechnique*, **30**, 479–495 (1980).
13. S. Nemat-Nasser and M. M. Mehrabadi, 'Stress and fabric in granular masses', in *Mechanics of Granular Materials: New Models and Constitutive Relations* (Eds. J. T. Jenkins and M. Satake), pp. 1–8, Elsevier Sci. Pub., 1983.
14. J. Mandel, 'Généralisation de la théorie de plasticité de W. T. Koiter', *Int. J. Solids Structures*, **1**, 273–295 (1965).
15. R. Hill, 'Generalized constitutive relations for incremental deformation of metal crystals by multislip', *J. Mech. Phys. Solids*, **14**, 95–102 (1966).
16. R. Hill and J. R. Rice, 'Constitutive analysis of elastic-plastic crystals at arbitrary strain', *J. Mech. Phys. Solids*, **20**, 401–413 (1972).
17. R. J. Asaro, 'Geometrical effects in the inhomogeneous deformation of ductile single crystals', *Acta Met.*, **27**, 445–453 (1979).
18. S. Nemat-Nasser, M. M. Mehrabadi, and T. Iwakuma, 'On certain macroscopic and microscopic aspects of plastic flow of ductile materials', in *Three-Dimensional Constitutive Relations and Ductile Fracture*, (Ed. S. Nemat-Nasser), pp. 157–172, North-Holland, Amsterdam, 1980.
19. K. S. Havner, 'The theory of finite plastic deformation of crystalline solids', in *Mechanics of Solids, The Rodney Hill 60th Anniversary Volume* (Eds. H. G. Hopkins and M. J. Sewell), pp. 265–302, Pergamon Press, Oxford, 1982.
20. S. Nemat-Nasser, 'On finite deformation elasto-plasticity', *Int. J. Solids Structures*, **18**, 857–872 (1982).
21. R. Hill, 'On constitutive macro-variables for heterogeneous solids at finite strain', *Proc. Roy. Soc. London* **A326**, 131–147 (1972).
22. R. Hill, 'Continuum micro-mechanics of elastoplastic polycrystals', *J. Mech. Phys. Solids*, **13**, 89–101 (1965).
23. T. Iwakuma and S. Nemat-Nasser, 'Finite elastic-plastic deformation of polycrystalline metals', *Proc. Roy. Soc. London* (1984) in press.

24. S. Nemat-Nasser and T. Iwakuma, 'Finite elastic-plastic deformation of composites', in *Mechanics of Composite Materials—Recent Advances* (Eds. Z. Hashin and C. T. Herakovich), pp. 47–55, Pergamon, 1983.
25. M. M. Mehrabadi and S. Nemat-Nasser, 'Stress, dilatancy, and fabric in granular materials', *Mechanics of Materials*, **2**, 155–161 (1983).
26. M. Oda, J. Konishi, and S. Nemat-Nasser, 'Experimental micromechanical evaluation of strength of granular materials: Effects of particle rolling', *Mechanics of Materials*, **1**, 269–283 (1982).
27. U. F. Kocks, 'The relation between polycrystal deformation and single-crystal deformation', *Met Trans.*, **1**, 1121–1143 (1970).
28. B. Budiansky and T. T. Wu, 'Theoretical prediction of plastic strains of polycrystals', in *Proc. U.S. Nat. Congr. Appl. Mech.*, pp. 1175–1185, 1962.
29. J. W. Hutchinson, 'Plastic stress–strain relations of F.C.C. polycrystalline metals hardening according to Taylor's rule' and 'Plastic deformation of B.B.C. polycrystals', *J. Mech. Phys. Solids*, **12**, 11–24 and 25–33 (1964).
30. J. W. Hutchinson, 'Elastic–plastic behaviour of polycrystalline metals and composites', *Proc. Roy. Soc. London*, **A319**, 247–272 (1970).

Mechanics of Engineering Materials
Edited by C. S. Desai and R. H. Gallagher

Chapter 24

Experimental Foundations of Thermoplasticity and Viscoplasticity

A. Phillips

24.1 INTRODUCTION. THE YIELD SURFACE AND THE LOADING SURFACE

The purpose of this chapter is to discuss the present state of the experimental foundations of thermoplasticity and viscoplasticity with special emphasis on the relevant research done at Yale University by the author and his students during the past 20 years. Early versions of this research have been presented in [1, 2, 3] and we shall emphasize here the research done since the time our previous

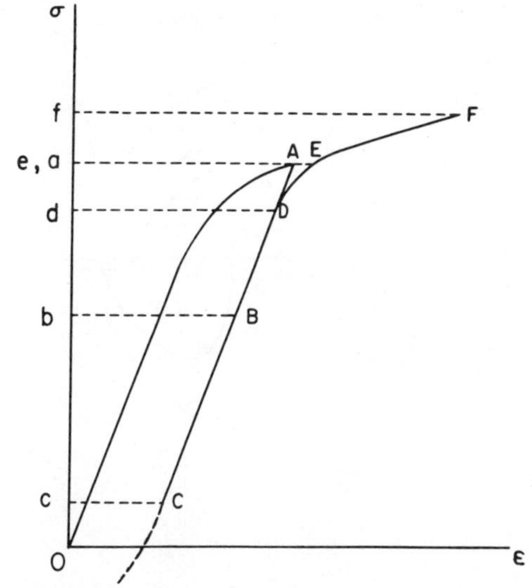

Figure 24.1 Typical stress–strain relation with loading OA, unloading AB, and reloading BDEF. The corresponding stress space path will be Oa, ab, bdef

presentations were prepared. In this paper no attempt is made to incorporate the very valuable experimental research of others on the same subject. Most of our experiments were done on pure aluminium tubes, but a number of the experiments were performed on copper and brass tubes. All the test results presented in this chapter are with pure aluminium tubes.

The research to be discussed has been initiated with our 1965 paper [4] in which, to our knowledge, the first presentation of the concept of two yield surfaces (instead of a single yield surface) in stress space was presented. The one surface encloses the other one and the inner surface was called 'yield surface' while the outside surface was called 'loading surface'. To explain how these two surfaces move in stress space as loading, unloading, and reloading proceeds we shall first refer to the stress–strain diagram, Figure 24.1. In this figure we see the cases of loading OA, unloading AB, and reloading BDF The point A corresponds to the highest stress reached during loading before unloading AB starts. It is seen that during reloading the proportional limit D lies below A.

It is also assumed that unloading below A produces only elastic strains but that reloading will produce plastic strain if carried beyond the level of stress represented by the point D. In the classical theory, on the other hand, the two points A and D coincide. To these two points correspond in stress space,

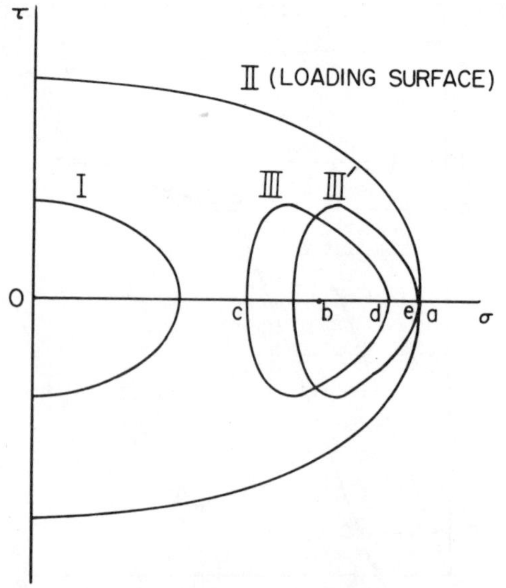

Figure 24.2 The loading surface II and the yield surfaces III and III′. When the stress remains stationary at 'a' the yield surface moves slowly from position III to position III′

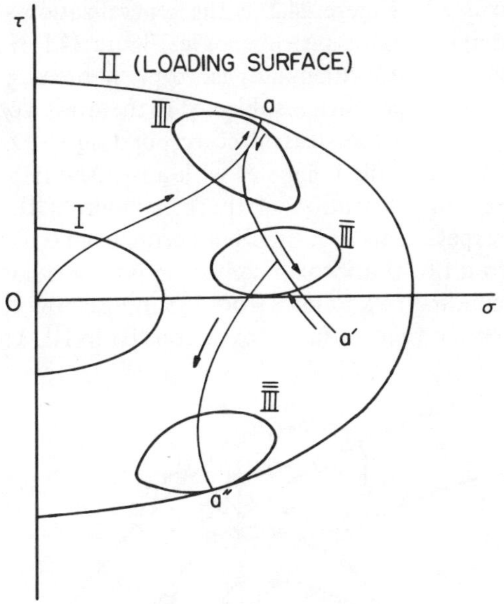

Figure 24.3 Motion of the yield surface within the loading surface. The path Oaa′ a″ generates a motion of the yield surface illustrated by the sequence of positions I, III, $\overline{\text{III}}$, $\overline{\overline{\text{III}}}$. Observe that $\overline{\overline{\text{III}}}$ is tangential to the loading surface

Figure 24.2, the yield surface III (point D, d) and the loading surface II (point A, a).

The point D corresponds to the stress where the first indication of plastic deformation during reloading appears. This stress is obtained by means of a well-defined operational procedure. This procedure consists of making small excursions of the stress point into the plastic region by amounts of plastic strain agreed upon in advance (usually 2 μin/in but in some cases up to 5 μin/in) and then performing a backward extrapolation to the elastic line. By obtaining the point D (yield point) by means of the above-mentioned operational procedure we ensure that the region enclosed by the yield surface in stress space is purely elastic and that the entire purely elastic region is enclosed by the yield surface. Therefore, the yield surface does not enclose some part of the plastic region.[†]

The stress space, Figure 24.2, and on the stress axis, Figure 24.1, the points corresponding to A, B, C, D, and E are a, b, c, d, and e, respectively.

[†] The backward extrapolation introduced here is radically different conceptually and in its effects from the ones used by Taylor and Quinney [5], Lode [6], and Mair and Pugh [7]. These authors used penetrations into the plastic region which are of the order of 1000 μin/in or more. In these cases the region enclosed by the resulting yield surface included to a substantial degree a plastic region.

The loading surface II, Figure 24.2, is the generalization stress space of the stress point A (and a) in the stress–strain curve, Figure 24.1. It can be considered as being generated by an approximately isotropic hardening process from the initial yield surface. If the specimen is subjected to the stress a corresponding to A for a sufficiently long time the stress d corresponding to D, moves gradually upwards towards a and finally coincides with a (surface III′, Figure 24.2).

Figure 24.3 gives an illustration of the behaviour of the two surfaces as evidenced by our experiments [3]. Loading occurs from 0 to a and the loading surface expands from I to II approximately as in isotropic hardening. The yield surface III corresponding to a follows from I. Suppose now that we move from inside III to a′; then the yield surface moves from III to $\overline{\text{III}}$. Moving from inside

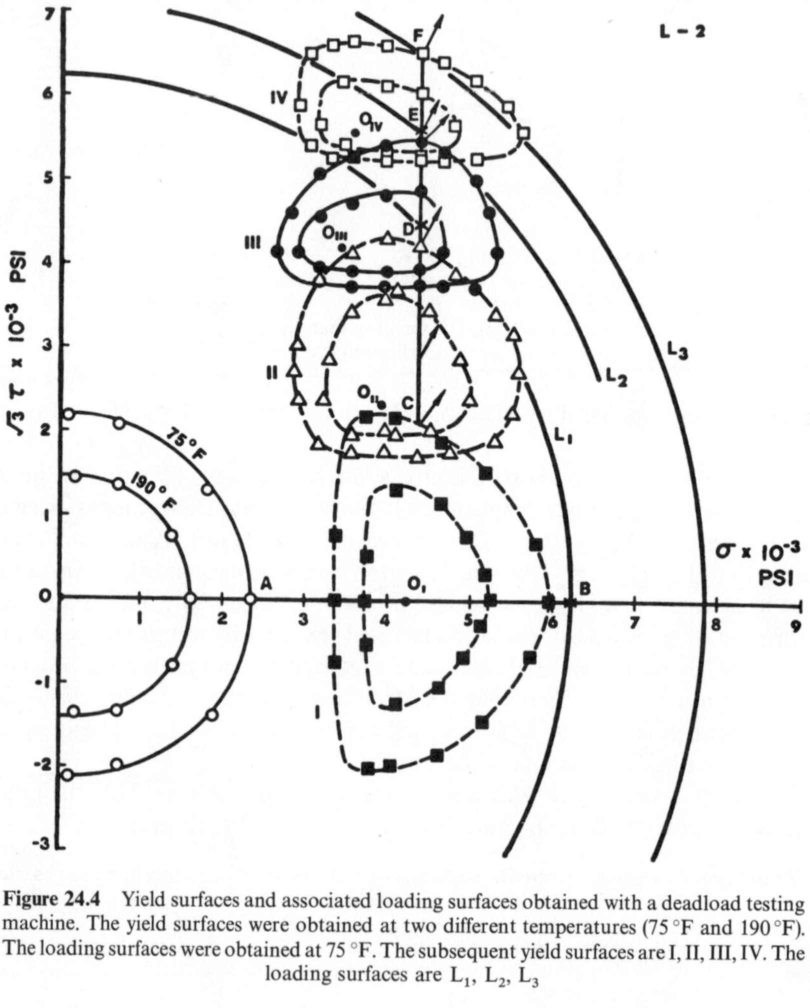

Figure 24.4 Yield surfaces and associated loading surfaces obtained with a deadload testing machine. The yield surfaces were obtained at two different temperatures (75 °F and 190 °F). The loading surfaces were obtained at 75 °F. The subsequent yield surfaces are I, II, III, IV. The loading surfaces are L_1, L_2, L_3

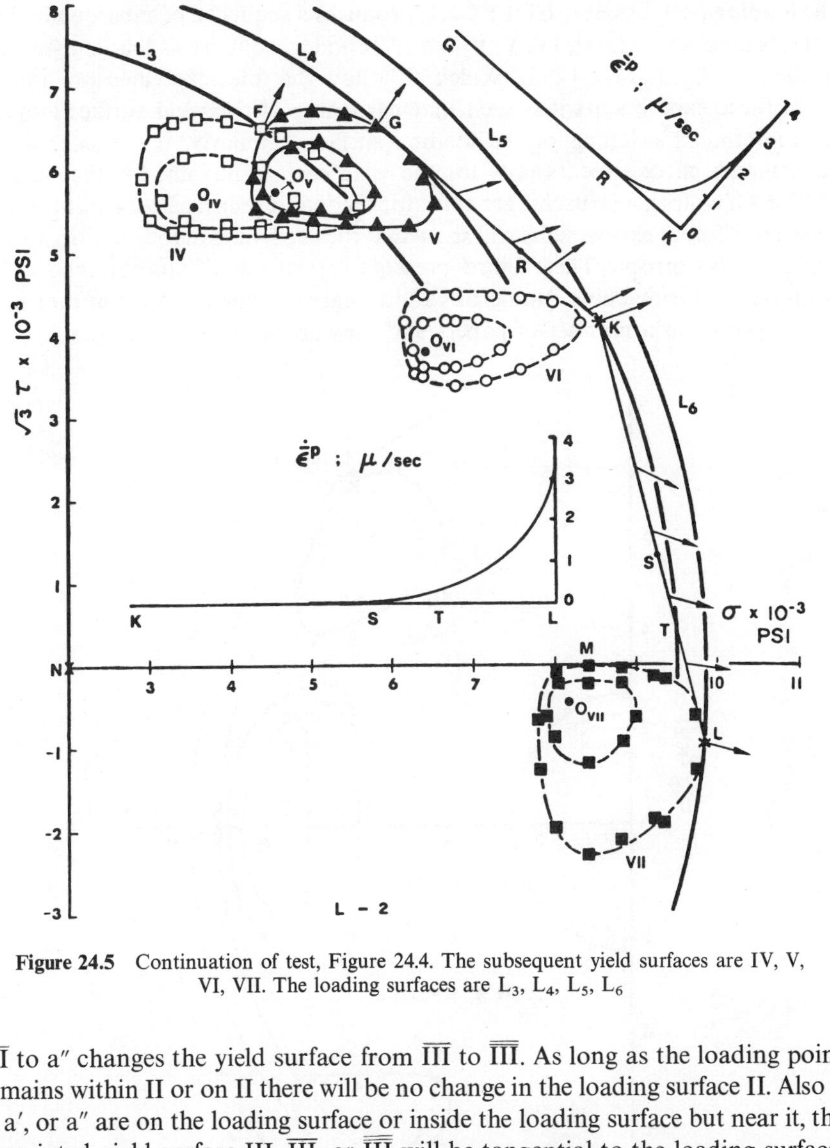

Figure 24.5 Continuation of test, Figure 24.4. The subsequent yield surfaces are IV, V, VI, VII. The loading surfaces are L_3, L_4, L_5, L_6

$\overline{\text{III}}$ to a″ changes the yield surface from $\overline{\text{III}}$ to $\overline{\overline{\text{III}}}$. As long as the loading point remains within II or on II there will be no change in the loading surface II. Also if a, a′, or a″ are on the loading surface or inside the loading surface but near it, the associated yield surface III, $\overline{\text{III}}$, or $\overline{\overline{\text{III}}}$ will be tangential to the loading surface. When the loading point moves beyond II then both the loading and yield surfaces will change according to their respective laws of hardening.

The experimental evidence upon which the above model is based is presented in Figure 24.4 and 24.5 from [3]. The experiments in Figure 24.4 and 24.5, in which the yield surfaces were obtained at two different temperatures (75 °F and 190 °F) in each instance, show that the behaviour outlined above is indeed valid.

The loading path $OABO_1$ CDEFGKL produces a sequence of subsequent yield surfaces labelled I, II, III, IV, V, VI, and VII and a sequence of loading surfaces labelled L_1, L_2, L_3, L_4, L_5, L_6 which all follow the rules explained previously. From these experiments it is seen that hardening of the yield surface is quite different than hardening of the loading surface. Similarly, it seems that the Bauschinger effect appears only for the yield surface but not for the loading surface. Of course, a Bauschinger effect for the loading surface may still become apparent if future experiments will show that the hardening of the loading surface is not strictly isotropic. These figures present experimental results obtained with a deadload testing machine. Numerous additional experiments by the author have been reported in a number of papers and they are summarized in [2].

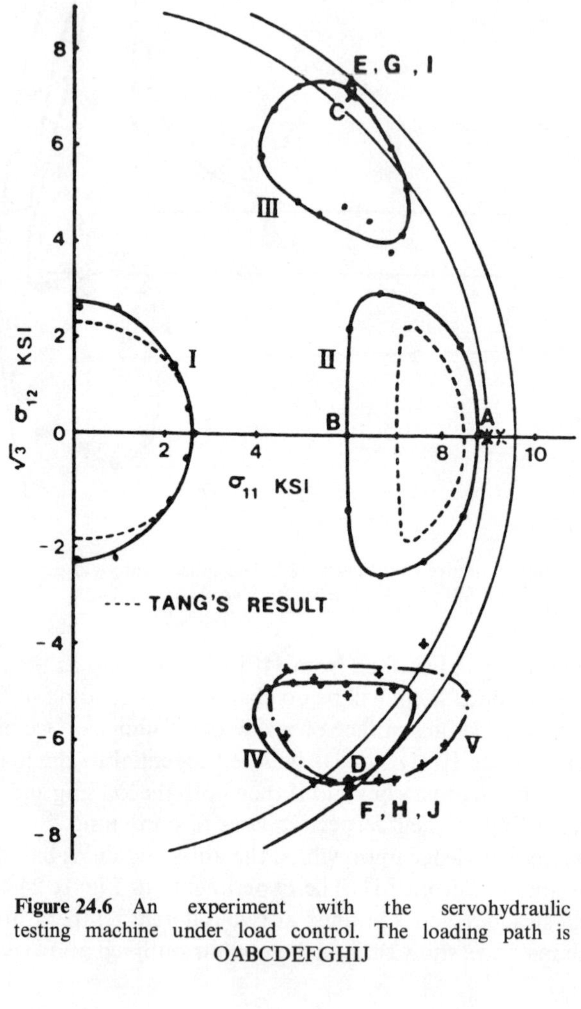

Figure 24.6 An experiment with the servohydraulic testing machine under load control. The loading path is OABCDEFGHIJ

Such experiments with deadload testing machines are very laborious to perform but the results are very accurate. During the past few years we installed in our laboratory a computer controlled closed loop servohydraulic combined stress (axial–torsional–temperature and axial–torsional–internal pressure) testing machine and one of our first tasks was to obtain yield surfaces with this new testing equipment and compare them with those obtained previously. The versatility and speed of experimentation possible with the new equipment is of a much higher order of magnitude than those previously existing.

Experiments in which the testing machine is under load control and the strains are measured provide a case nearest to the earlier experiments with the deadload machine. In this case creep can occur but relaxation cannot occur. Figure 24.6 gives some of the results. After first obtaining the initial yield surface I we prestressed to A and obtained yield surface II. The initial and first subsequent yield surfaces obtained with the deadload machine are illustrated with the broken lines. They are smaller than the new ones and the reason for this difference in size is that with the automatic equipment deeper penetration into the plastic region was needed in order to obtain the yield surfaces than was the case with the deadload equipment; consequently the new yield surfaces are larger in size.

We continued the experiment by prestressing from B to C, thus obtaining yield

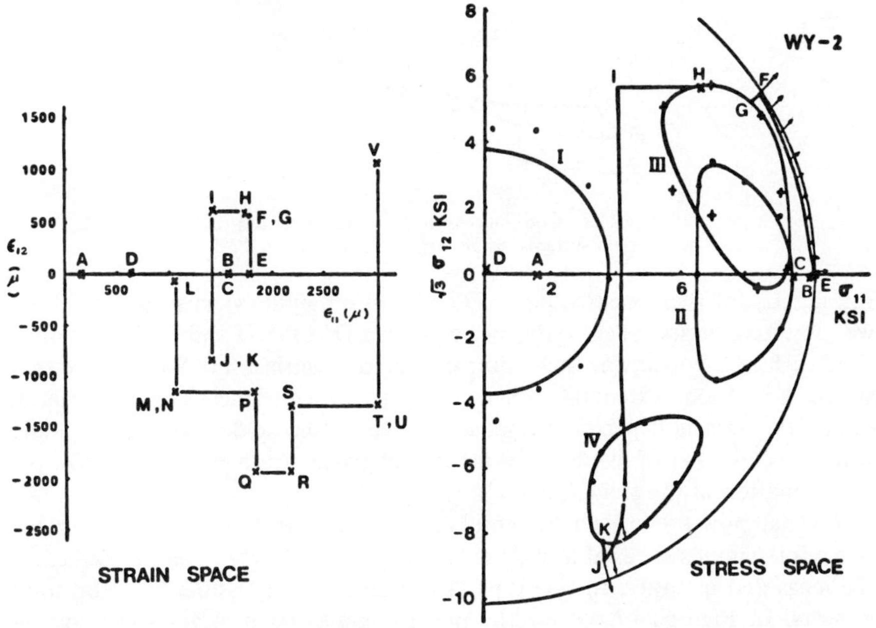

Figure 24.7 An experiment with the servohydraulic testing machine under strain control. The straining path is on the left side and the first part of the stress path as well as the yield surfaces I to IV are on the right side

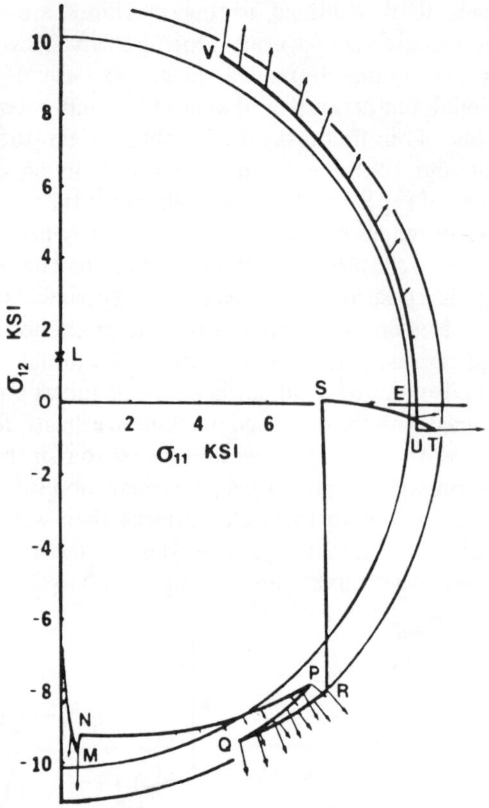

Figure 24.8 Continuation of the stress path and the yield surfaces of the experiment, Figure 24.7

surface III, and then prestressing to D and thus obtaining yield surface IV. Next we subjected the specimen to the cyclic loading DEFGHIJ and finally obtained yield surface V which was then compared to yield surface IV. We observe that surface V is to some extent different from surface IV which means that the interim cyclic development of plastic strain had an effect on the yield surface. The loading surface seems also to depend on the amount of plastic strain developed before the determination of the yield surface.

We shall now present experimental results under strain control in which the strain is the imposed variable and the load experienced by the loading element is the measured quantity. In this type of test creep is impossible but relaxation appears. In Figure 24.7 we see the prestraining path in strain space and in Figures 24.7 and 24.8 we see the stress path and the yield surfaces in stress space. We first obtained the initial yield surface I, Figure 24.7 and prestrained from A to B. The strain remained at B for 14 hours and stress relaxation took place at the end of which the stress was at C and the first subsequent yield surface II was obtained. The specimen was then unloaded to D and then prestrained according to the path DEF. The strain remained at F for 16 hours and during this time the stress was relaxed to G. The yield surface III was then obtained. The third prestrain follows the path HIJ and the corresponding stress path is shown again as HIJ. Then while the specimen remained at J for 19 hours the stress relaxed to K and the third subsequent yield surface IV was obtained.

After the determination of the third subsequent yield surface the specimen was accidentally strained to L. The specimen was then strained in shear LM. At M the strain was kept constant for 10 hours and stress relaxation took place. The corresponding stress path LMN and the vectors of the plastic strain increments are indicated in Figure 24.8. The specimen was then strained in a zigzag path NPQR and the corresponding stress path as well as the vectors of the plastic strain rate are shown. Finally the remaining straining path was RSTUV where the prestraining was interrupted at T for $3\frac{1}{2}$ hours while the stress relaxed to U. We observe the plastic strain increments in the path STUV and it is obvious that they are very nearly normal to the loading surface. The same observation can be made for the plastic strain increments in the paths NPQR. More information concerning these experiments under strain control can be obtained from [8].

The question of the development of plastic strain and the question of normality are two items which are only slightly touched here, but which have been extensively dealt with in our previous papers, as for example in [2].

24.2 YIELD SURFACES IN THE STRESS–TEMPERATURE SPACE. THERMAL LOADING

In the previous section we discussed yield surfaces in the stress space. In this section we shall consider yield surfaces in the stress–temperature space. In 1972 we published a paper [9] in which a yield surface in the stress–temperature space was obtained, by means of a series of isothermal yield curves. This so obtained yield surface was also verified at selected points by means of pure thermal loading. Recently, we obtained [10] complete yield surfaces in the stress–temperature space by means of pure thermal loading. The form of a yield surface in the stress–temperature space $(\sigma, \sqrt{(3)}\tau)$ has the general form of Figure 24.9. The effects of thermal prestressing on the yield surface have been considered in two of our papers [10, 11] in which the following experimental procedure was adopted.

After a yield curve at room temperature was determined, Figure 24.9, from a location A on the yield curve, or very near inside it, we introduced a pure thermal loading path AB by simply increasing the temperature. The increase in the temperature means that the thermal loading path AB will penetrate more or less deeply the plastic region and it will therefore produce plastic strains as well as displace and deform the yield surface. Then, we decreased the temperature to its initial value and we determined the displaced yield surface by obtaining

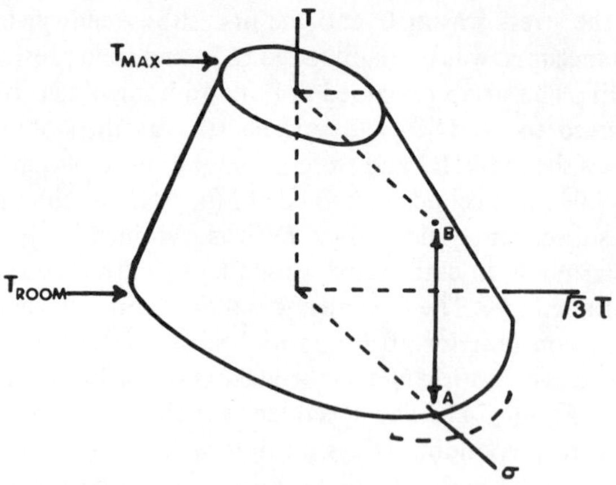

Figure 24.9 Typical yield surface in the (σ, $\sqrt{(3)}\tau$, T) space. ABA represents a typical thermal loading cycle. The yield surface is illustrated only in the temperature region between room temperature and maximum testing temperature

isothermal yield curves at several temperatures. Thus, we applied a thermal loading cycle ABA, Figure 24.10 (this is the type of thermal loading cycle used in [10]). Figure 24.11 from [10] gives an example in which both the initial yield surface and the first subsequent yield surface are shown. The thermal loading cycle occurred at stress point A. We observe that the subsequent yield surface is similar to yield surfaces obtained by isothermal prestressing. It does not enclose the origin and lack of cross effect is evident.

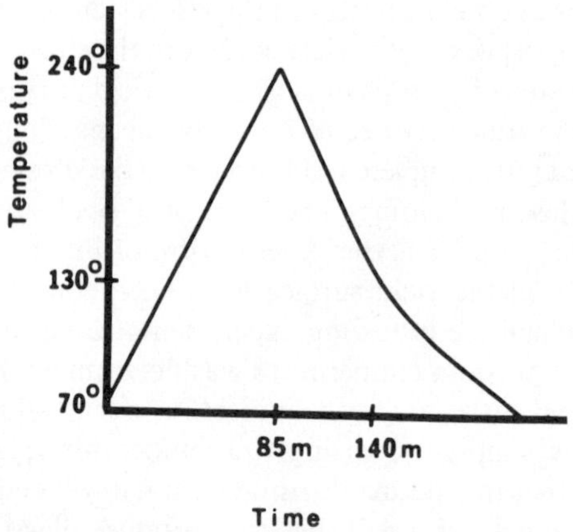

Figure 24.10 The temperature–time relation during the thermal loading cycle

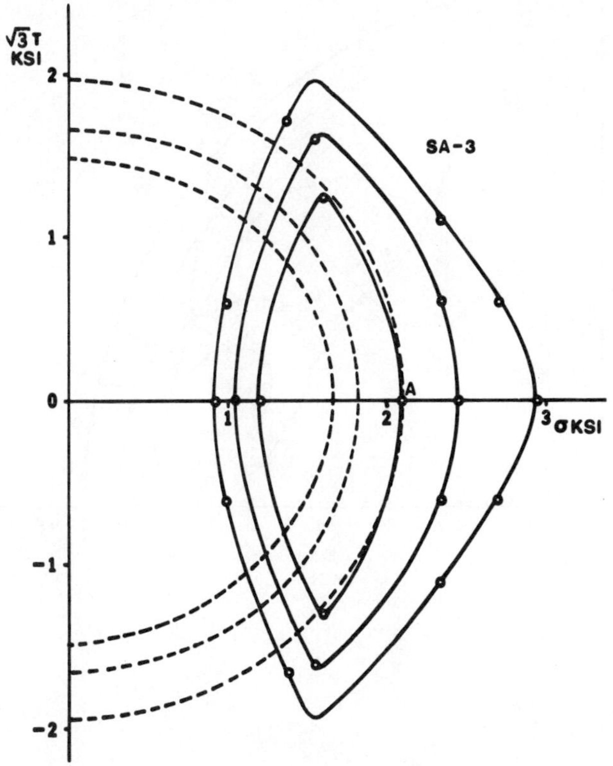

Figure 24.11 A first subsequent yield surface obtained by a thermal loading cycle. The first subsequent yield surface (solid line) is compared to the initial yield surface (broken lines). The solid lines are the three isothermals at 70°, 147°, and 220 °F. The thermal loading cycle was to 240°F at A

The question can now be raised about the existence of loading surfaces at elevated temperatures. Indeed, there is no reason why the concept of the loading surface should be valid only at room temperature. Our experiments have shown that once a loading surface is established at room temperature, there exists a family of loading surfaces for all temperatures. If at room temperature the yield surface is tangential to the loading surface then at an elevated temperature, the elevated temperature yield surface will be tangent to the corresponding elevated temperature loading surface. Figure 24.12 illustrates this finding. In this figure we see two consecutive yield surfaces. We observe that the yield curves at room temperature are tangential to the same Mises curve B′M′ (loading surface at room temperature). Also, we observe that the two yield curves at elevated temperature 190 °F) are tangential to another Mises curve $B''M''$ (loading surface at 190 °F).

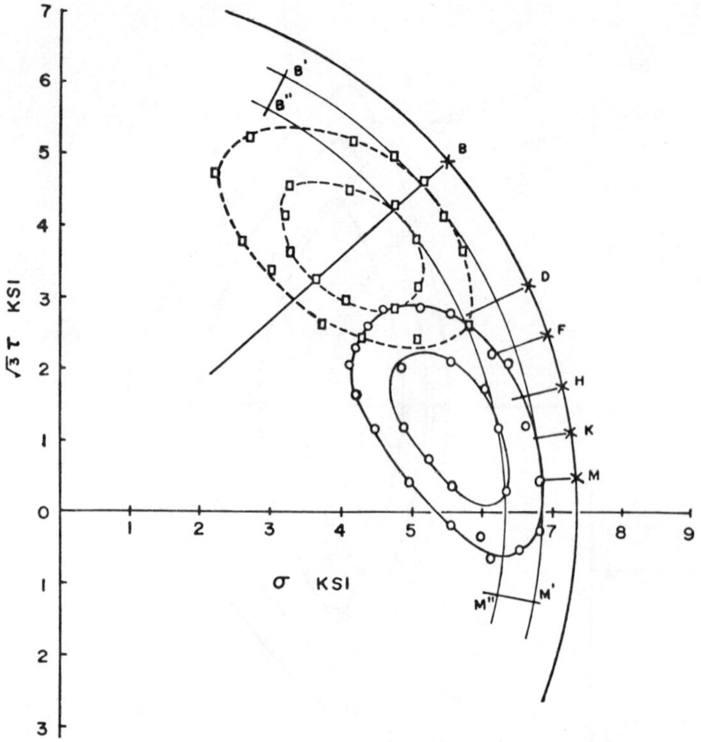

Figure 24.12 The change of the loading surface with the temperature. B′M′ and B″M″ are loading surfaces at two different temperatures

In Figure 24.13 the yield surface is depicted by means of a series of yield curves at increasing temperature levels. This yield surface could be either the initial one or one due to prestressing at constant temperature, or one due to thermal prestressing. We shall consider now the relationship between the isothermal yield curves as the temperature increases. We shall draw a series of arbitrary parallel strainght lines $A\bar{A}$, $B\bar{B}$, $C\bar{C}$, $D\bar{D}$, $E\bar{E}$, which could be parallel to the prestressing direction. The intersections of these lines with the yield curves are plotted in a stress–temperature diagram as shown in the lower part of the figure. We observe that the intersections produced pairs of straight lines ($\bar{A}A'$, AA'), ($\bar{B}B'$, BB'),... and that the intersections B′, C′, and D′, lie on a straight line $\bar{Q}Q$ which represents a high temperature limit of the sequence of yield curves. This fact suggests the validity of an anisotropic sandhill analogy. The slope of the sandhill (maximum temperature gradient) varies with the orientation of the normal to the isothermal surfaces with the prestressing direction; it is minimum in the direction of prestressing, and maximum in the opposite direction. In Figure 24.13, lines $\bar{B}B'$,

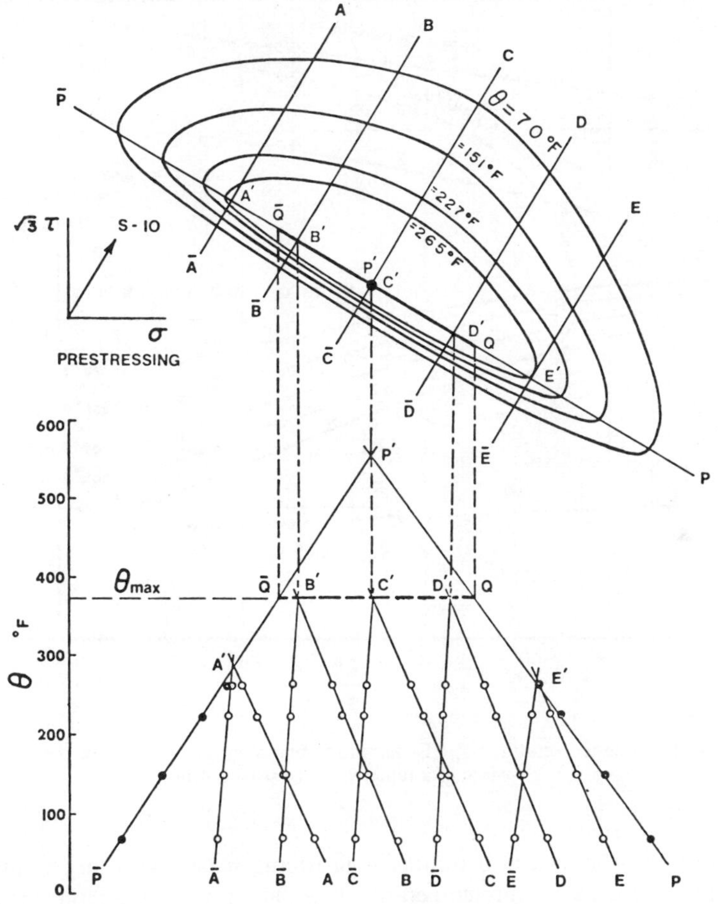

Figure 24.13 At an appropriate temperature the yield curve degenerates into a straight ridge line $\bar{Q}Q$

$\bar{C}C'$, and $\bar{D}D'$ have approximately the same slope. The slopes of the lines of the front side of the surface markedly smaller. The straight line $\bar{Q}Q$ to which the yield curves degenerate is the familiar ridge line of the sandhill analogy. For the initial yield surface one might expect that the sandhill would be more or less isotropic and for the Mises surface we would expect that the ridge line would shrink to a point at the apex of a circular cylindrical core. Our experiments indicate that for monotonic radial loading the length of the ridge line increases with plastic strain developed during prestressing. We also observe that at each prestressing there exists a straight line perpendicular to the temperature axis: this straight line is the limiting yield curve and it occurs at a limiting temperature.

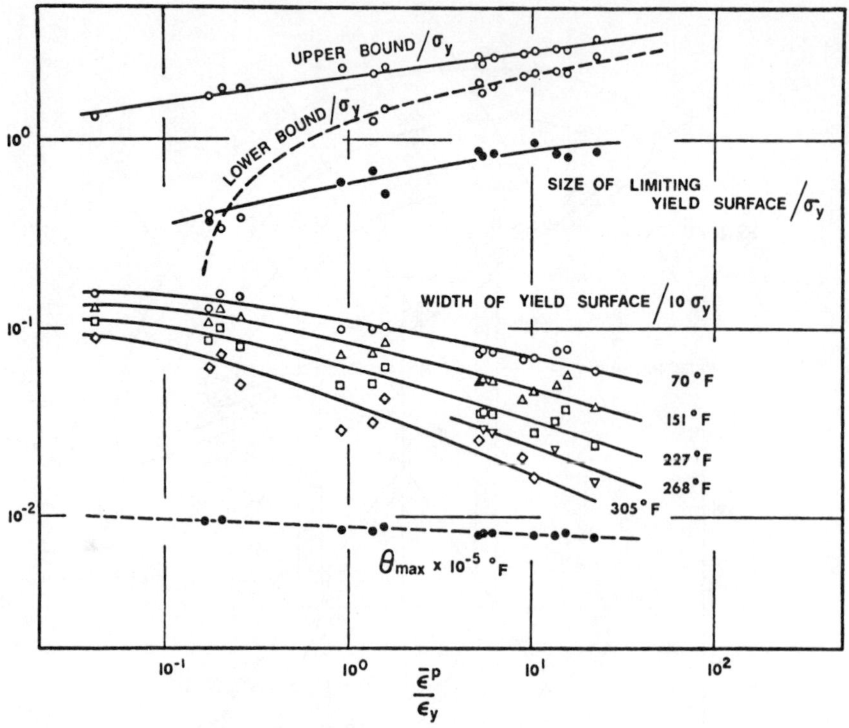

Figure 24.14 The temperature θ_{max}(°F) versus the plastic strain. The size of the limiting yield surface as a function of the plastic strain

Figure 24.14 shows in a double logarithmic scale, for a larger number of experiments, the maximum temperature θ_{max}(°F), at which the ridge line appears, as a function of the plastic strain; the relationship, in this presentation, is a linear one. The size of the limiting yield surface, that is, the length of the straight line to which the yield surface degenerates at θ_{max}, increases linearly (in double logarithmic scale) with the plastic strain. Additional information concerning these results is given in [12].

24.3 THE EQUILIBRIUM STRESS–STRAIN CURVE

Intimately related to the concept of the yield surface is the concept of the equilibrium stress–strain curve which was introduced by the author in 1972 [1]. Figure 24.15 shows the equilibrium stress–strain curve AC corresponding to a zero stress rate. It is the sequence of equilibrium positions due to successively larger values of applied stress, that is, each increment of stress is applied only after the permanent strain due to the previous strain increments have developed fully.

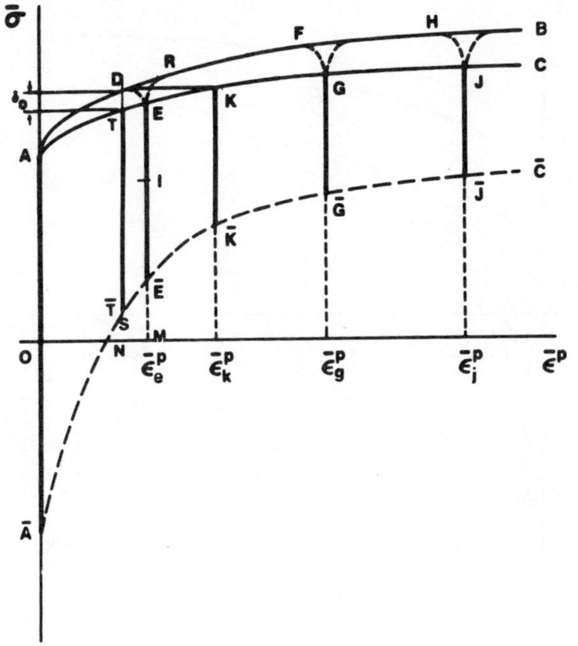

Figure 24.15 The equilibrium stress–strain lines AC and $\bar{A}\bar{C}$

The equilibrium stress–curve can be obtained in principle by successively unloading to within the elastic range, at different values of strain, and then reloading very slowly. We find that we are within the elastic range when no plastic strain appears while unloading, and we find that we have reached the equilibrium stress–strain line while reloading when we observe that we reached the proportional limit. By repeating this procedure in succession at several values of strain, we can obtain the equilibrium stress–strain curve.

The yield curves at room temperature at different values of permanent strain are shown in Figure 24.15 by means of the straight lines $A\bar{A}$, $T\bar{T}$, $E\bar{E}$, $K\bar{K}$, $G\bar{G}$, $J\bar{J}$. Thus, the equilibrium stress–strain curve AC represents the forward limits of the room temperature yield curves at different values of permanent strain (or prestressing). The backwards limits of the room temperature yield curves are given by the line $\bar{A}\bar{C}$. Since both lines as a pair are valid for an increasing permanent strain we shall call them *ascending* equilibrium stress–strain lines. At a higher temperature the corresponding yield curve lies inside the room temperature yield curve and, consequently, at a higher temperature the ascending equilibrium stress–strain lines have the general form shown in Figure 24.16. Because of the existence of the ridge lines of the yield surface, discussed earlier, it follows that each two ascending equilibrium stress–strain curves (of the same

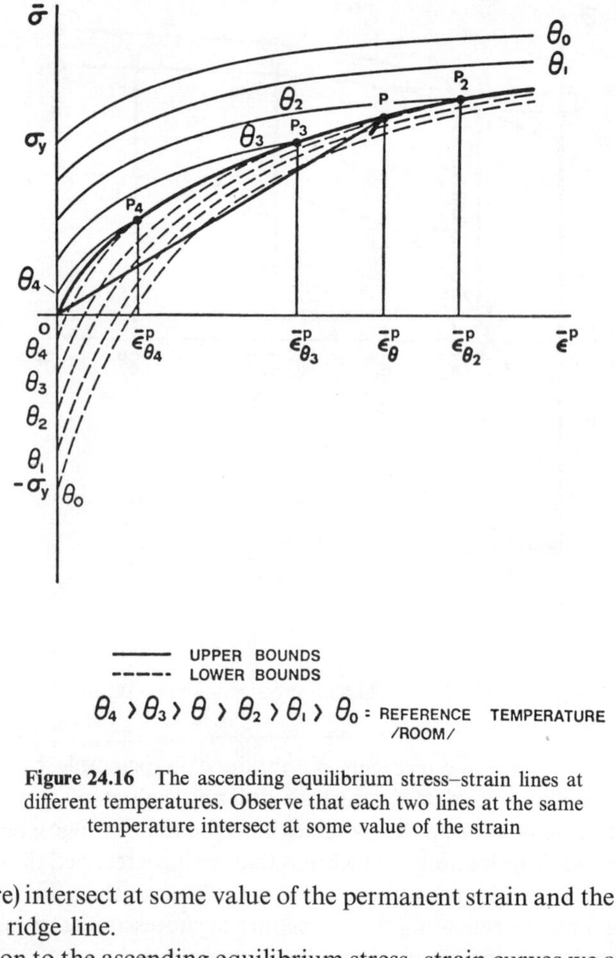

Figure 24.16 The ascending equilibrium stress–strain lines at different temperatures. Observe that each two lines at the same temperature intersect at some value of the strain

temperature) intersect at some value of the permanent strain and the intersection depicts the ridge line.

In addition to the ascending equilibrium stress–strain curves we also have, for decreasing permanent strain *descending* equilibrium stress–strain curves. Figure 24.17 presents an example from an experiment reported in [11]. In this figure AA′ represents the yield surface at the highest ascending position. CC′ represents the yield surface at the furthest descending position and BB′ represents the yield surface at an intermediate descending position. We see that A′BC is the descending lower equilibrium stress–strain curve and AB′C′ is the descending upper equilibrium stress–strain curve. The descending upper equilibrium stress–strain curve is now the locus of the backwards limits of the room temperature yield curves, while the descending lower equilibrium strain–strain curve represents the forward limits of the room temperature yield curves at different values of prestressing. Additional information concerning the ascending and descending equilibrium stress–stress curves can be obtained for [11, 12].

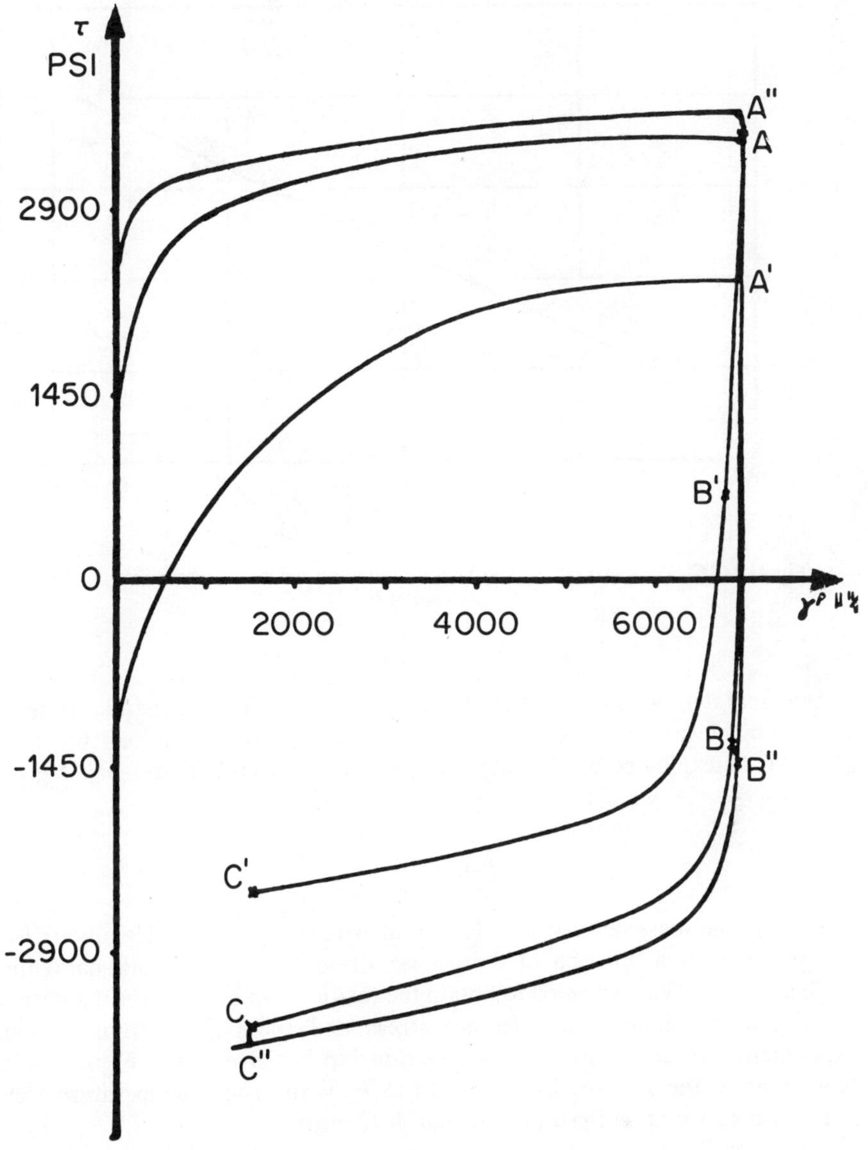

Figure 24.17 The descending equilibrium stress–strain lines

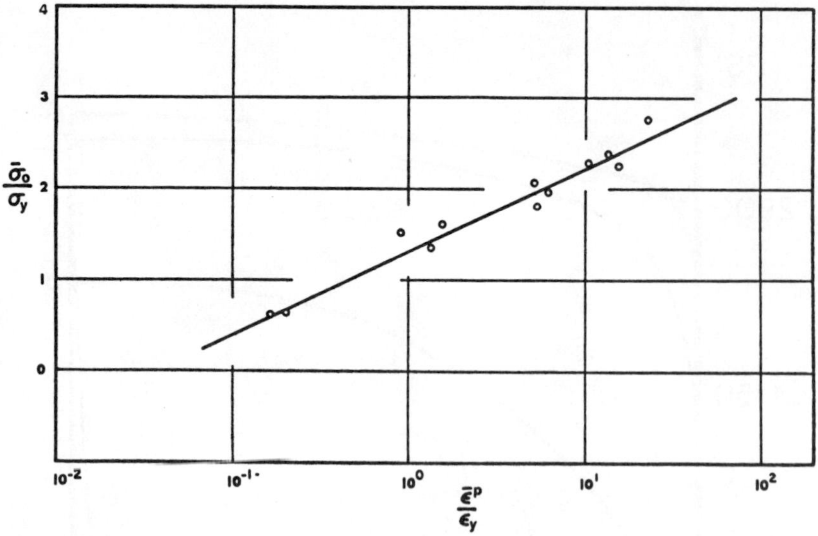

Figure 24.18 The linear relationship between thermodynamic reference stress $\bar{\sigma}_0$ and plastic strain $\bar{\varepsilon}^p$

As a final item we shall consider the concept of the thermodynamic reference stress which was introduced from thermodynamic considerations by us in [13, 14]. As explained in [12] the thermodynamic reference stress σ^0_{kl} is given by

$$\sigma^0_{kl} = \frac{d\psi''}{d\varepsilon^0_{kl}}$$

where ρ is the mass density, and ψ'' is that part of the specific Helmholtz free energy which is a function of the plastic strain and of the work hardening coefficient only. We succeeded to obtain the relationship between the magnitude $\bar{\sigma}_0$ of the thermodynamic reference stress and the plastic strain $\bar{\varepsilon}^p$. Our experimental results show that this relationship is a linear one, Figure 24.18. Note that in the Figures 24.14 and 24.18 σ_y is the room temperature yield stress and $\varepsilon_y = \sigma_y/E$ is the strain at that yield stress.

24.4 CONCLUSIONS

The concepts discussed in the previous sections are at the foundation of such theories of thermoplasticity and viscoplasticity which are experimentally developed and verifiable. In [2] we introduced the concept of building and verifying a theory by means of the axiomatic method. According to this concept the theory is presented on the basis of several fundamental assumptions which are selected in such a way that they can be verified experimentally independently of one another. Once each assumption of the theory is verified independently of the other assumptions, then it follows that the entire theory is correct, and depending on the range of validity of the assumptions, the range of validity of the theory becomes known without ambiguity.

If one of the assumptions of the theory is found to be invalid it can always be modified so that the new version of this assumption will agree with the experimental results. Thus, it is always possible to change the theory rationally so that its fundamental assumptions will agree with the experiments. More generally a way is thus provided for theory and experiment to interact in a fruitful manner and thus generate new questions the answer to which provides for the advancement of understanding of both the theoretical framework and the experimental results. We hope that this contribution may help in the development of correct theories of viscoplasticity and thermoplasticity.

ACKNOWLEDGEMENTS

The support of this research by the National Science Foundation is gratefully acknowledged. The author wishes to thank Mr S. Murthy and Mr Y. Macheret, graduate students at Yale University, for their help in the timely completion of this paper.

REFERENCES

1. A. Phillips, 'Experimental plasticity. Some thoughts on its present status and possible future trends', in *Problems of Plasticity*, Proc. Int. Symp. Foundations of Plasticity, (Ed. A. Sawczuk), Vol. II, pp. 193–234, Warsaw, August 30–September 2, 1972, Noordhoff, Leyden, 1974.
2. A. Phillips, 'The foundations of plasticity', in Ch. Massonnet, W. Olszak, and A. Phillips, *Plasticity in Structural Engineering*, pp. 191–272, CISM Courses and Lectures No. 241, Springer-Verlag, Wien–New York, 1980.
3. A. Phillips and C. W. Lee, 'Yield surfaces and loading surfaces. Experiments and recommendations', *International Journal Solids and Structures*, **15**, 715–729 (1979).
4. A. Phillips and R. L. Sierakowski, 'On the concept of the yield surface', *Acta Mechanica*, **1**, 29–32 (1965).
5. G. I. Taylor and H. Quinney, 'The plastic distortion of metals', *Phil. Transactions Royal Society*, London, **Ser. A**, **230**, 323–362 (1931).
6. W. Lode, *Z. Physik*, **36**, 913–939 (1926).
7. W. M. Mair and H. Ll. D. Pugh, 'Effects of prestrain on yield surfaces in copper', *Journal Mech. Eng. Sci.*, **6**, 150–163 (1964).
8. A. Phillips and W. Y. Lu, 'An Experimental Investigation of Yield Surfaces and Loading Surfaces of Pure Aluminum with Stress-controlled and Strain-controlled Paths of Loading', (submitted for publication) 1984.
9. A. Phillips, K. Liu, and W. J. Justusson, 'An experimental investigation of yield Surfaces at elevated temperatures', *Acta Mechanica*, **14**, 119–146 (1972).
10. A. Phillips and S. Murthy, 'Yield surfaces in the stress–temperature space generated by thermal loading', to be published in the *D. C. Drucker Anniversary Volume*, 1984 (in press).

11. A. Phillips and W. A. Kawahara, 'The Effect of Thermal Loading on the Yield Surface of Aluminum. An Experimental Investigation. Acta Mechanica 1984 (in press).
12. M. A. Eisenberg, C. W. Lee, and A. Phillips, 'Observations on the theoretical and experimental foundation of thermoplasticity', *Intern. Journal Solids Structures*, **13**, 1239–1255 (1977).
13. A. Phillips and M. Eisenberg, 'Observations on certain inequality conditions in plasticity', *Int. Journal Nonlinear Mechanics*, **1**, 247–256 (1966).
14. A. Phillips, 'The foundations of thermoplasticity, experiments and theory', in *Topics in Applied Continuum Mechanics* (Eds. J. L. Zeman and F. Ziegler), H. Parkus Anniversary Volume, pp. 1–21, Springer-Verlag, Wien–New York, 1974.

Mechanics of Engineering Materials
Edited by C. S. Desai and R. H. Gallagher

Chapter 25

Formulation and Numerical Integration of Elastoplastic and Elasto-viscoplastic Rate Constitutive Equations

P. M. Pinsky, K. S. Pister, and R. L. Taylor

25.1 INTRODUCTION

Constitutive equations appropriate for the finite deformation analysis of elastoplastic and elasto-viscoplastic materials are most frequently expressed in a spatial rate form. A thermodynamic framework for the development of such equations can be provided by characterizing irreversible processes through the use of internal variables [1 to 4]. In this case, spatial rate constitutive equations, expressed in terms of objective rates, will be required for the stress tensor and a set of internal variables. In any numerical scheme employed for the analysis of elastoplastic or elasto-viscoplastic problems it will be necessary to integrate these rate constitutive equations for the stress and internal variables. It is the object of this chapter to derive adequate spatial rate constitutive equations for a limited class of elastoplastic and elasto-viscoplastic materials undergoing finite deformation and to propose a numerical algorithm for their integration.

The correct choice of objective rate appearing in spatial rate constitutive equations has been the subject of considerable conjecture [5 to 8] since the principle of objectivity alone does not uniquely determine this choice. It is noted, however, that the principles of mechanics must be invariant with respect to the choice of reference configuration [9]. When this principle is invoked together with the assumed existence of a free energy density, it is shown that the indeterminacy in the choice of objective rates for the spatial stress tensor and the spatial internal variables is removed.

A number of theories of plasticity have been based on different kinematic assumptions regarding the elastic–plastic split of the deformation [10 to 12]. A requirement for the success of the numerical integration algorithm proposed in this paper is that the spatial rate of deformation tensor admit an additive decomposition into an 'elastic' and a 'plastic' part. A thermodynamic argument for such a decomposition is provided within the framework of the internal variable theory.

The discussion of constitutive equations is concluded by the presentation of specific examples which are general enough to accommodate perfect and hardening viscoplasticity as well as perfect and hardening plasticity. In this chapter, inviscid or rate-independent plasticity is treated as the limiting case of viscoplasticity as the viscosity of the material tends to zero or, alternatively, as an infinite length of time is allowed for the stress and internal variables to relax to their asymptotic values.

Following the development of the rate constitutive equations, a numerical algorithm is presented for their integration. As a consequence of the additive decomposition of the spatial rate of deformation tensor, the complete set of spatial rate constitutive equations also exhibits and additive decomposition into elastic and plastic parts. This 'operator split' into component parts suggests the application of the product formula techniques for the construction of an efficient algorithm. Operator split methods have recently been successfully applied to the finite element analysis of problems in a number of areas. For example, the heat conduction problem [13], the structural dynamics problem [14] and in plasticity [15, 16]. However, as will be made clear below, the application of the operator split method for plasticity reported in [15, 16] is quite different to the approach being pursued in the present chapter. After providing a brief general overview of product formula techniques, their application to the integration of rate constitutive equations is considered in detail. It is shown that the product algorithm consists of first integrating the elastic rate constitutive equations, ignoring the plasticity of the material. The stresses resulting from this operation are then allowed to relax towards the elastic domain, which itself is evolving according to the internal variable rate equations.

From a numerical point of view, an algorithm for the integration of rate constitutive equations should satisfy three requirements:

(i) Consistency with the constitutive equations.
(ii) Numerical stability.
(iii) Incremental objectivity.

Conditions (i) and (ii) are required for the convergence of the integration scheme [17]. Condition (iii) results from the physical requirement that the algorithm be invariant with respect to superimposed rigid body motion [18, 19]. It is shown that the proposed product algorithm provides a basis for demonstrating the consistency and numerical stability of the resulting algorithm. The requirements of incremental objectivity are also considered in detail.

Although the development of global algorithms for the solution of the boundary value problem of linear momentum balance is outside the scope of the present chapter, such algorithms provide the motivation for developing numerical schemes for the integration of the rate constitutive equations. Indeed, the two are strongly interdependent and cannot be entirely separated in any reasonably complete discussion of either algorithm. Accordingly, a brief description of global algorithms is presented. It is noted that 'implicit' global algorithms which employ an elastoplastic (or elasto-viscoplastic) tangent modulus tensor suffer from certain computational disadvantages. These disadvantages lead to the development of alternative algorithms based on the operator split of the momentum balance equation (16). Unfortunately, the error introduced into the product algorithm through the operator split tends to dominate at 'practical' time step sizes. Refining the time step size for the global algorithm is very costly. In response, an implicit global algorithm is proposed in conjunction with the product formula algorithm for the constitutive equations which eliminates the disadvantages of the global implicit method resulting in an efficient and very accurate solution scheme.

In order to further improve the accuracy of the proposed algorithm, a method for refining the time step size in the product algorithm for the constitutive equations is presented. The time step size for the global algorithm is unaffected. This has the desirable effect of improving the accuracy (by reducing the error introduced by the operator split) of the overall scheme without incurring considerable cost, since the constitutive equations are integrated locally.

An interpretation of these algorithms for finite element analysis is considered throughout the development.

Finally, numerical examples are presented to demonstrate the effectiveness of the proposed algorithms.

25.2 FIELD EQUATIONS FOR FINITE DEFORMATION ELASTOPLASTICITY

25.2.1 Preliminaries

A motion of a deformable body in the ambient space R^N, relative to a reference configuration B, is given by a time dependent mapping $\boldsymbol{\phi}_t(\mathbf{X})\colon B \to R^N$, $t > 0$. Here, $\mathbf{X}$ denotes a set of material coordinates defined on the reference configuration.

The material velocity of the motion $\boldsymbol{\phi}_t$ is defined as a vector field $\mathbf{V}$ over the reference configuration, such that $\mathbf{V} = (\mathrm{d}/\mathrm{d}t)\boldsymbol{\phi}_t$. The spatial velocity field $\mathbf{v}$ is defined by $\mathbf{v} = \mathbf{V} \circ \boldsymbol{\phi}_t^{-1}$. Note that the spatial velocity field $\mathbf{v}(\mathbf{x}, t)$ is dependent on a set of spatial coordinates denoted $\mathbf{x}$.

The deformation gradient is defined by $\mathbf{F} = \partial\boldsymbol{\phi}/\partial\mathbf{X}$ with components, $F^a_A = \partial\phi^a/\partial X^A$. The polar decomposition of the deformation gradient is given by $\mathbf{F} = \mathbf{R}\cdot\mathbf{U} = \mathbf{V}\cdot\mathbf{R}$, where $\mathbf{R}$ is an orthogonal rotation tensor and $\mathbf{U}$ and $\mathbf{V}$ are respectively positive definite and symmetric right and left stretch tensors. From $\mathbf{F}$ one can obtain the Jacobian of the motion $J = \det\mathbf{F}$ and the right Cauchy–

Green deformation tensor $\mathbf{C}$ which is related to the deformation gradient by $\mathbf{C} = \mathbf{F}^T \cdot \mathbf{F}$.

The spatial velocity gradient tensor $\mathbf{l}$ is given by $\mathbf{l} = \nabla \mathbf{v}$ where ∇ denotes the gradient with respect to the spatial coordinates $\mathbf{x}$. The symmetric part of $\mathbf{l}, \mathbf{d} \equiv \nabla^S \mathbf{v}$ is the spatial rate of deformation tensor, and the skew-symmetric part $\omega \equiv \nabla^A \mathbf{v}$ is the spin rate or vorticity tensor.

If γ is a tensor field defined on the deformed configuration $\boldsymbol{\phi}_t(B)$, the pull-back of γ through the motion $\boldsymbol{\phi}_t$ defines a tensor field $\boldsymbol{\Gamma}$ on B denoted by $\boldsymbol{\Gamma} = \boldsymbol{\phi}_t^*(\gamma)$ [9]. For example, in the case of a second order contravariant tensor the pull-back operation takes the form

$$\Gamma^{AB} = (F^{-1})^A_a (F^{-1})^B_b (\gamma^{ab} \circ \boldsymbol{\phi}_t)$$

This definition may be readily generalized to spatial tensor fields of any order.

Likewise, if $\boldsymbol{\Gamma}$ is a material tensor field defined on B, the push forward of $\boldsymbol{\Gamma}$ through the motion $\boldsymbol{\phi}_t$ defines a spatial tensor field γ on $\boldsymbol{\phi}_t(B)$ denoted by $\gamma = \boldsymbol{\phi}_{t*}(\boldsymbol{\Gamma})$. In this case, for the example used above, the push forward operation takes the form

$$\gamma^{ab} = F^a_A F^b_B (\Gamma^{AB} \circ \boldsymbol{\phi}_t^{-1})$$

A related concept associated with the push forward of material rates of material tensors is that of the Lie derivative of a spatial tensor with respect to the spatial velocity field. The Lie derivative entails pulling back the spatial tensor to the reference configuration, taking the material time derivative of the resulting material tensor and pushing forward the result into the current configuration. Formally, for a spatial tensor γ,

$$L_v(\gamma) \equiv \boldsymbol{\phi}_{t*}\left(\frac{\mathrm{d}}{\mathrm{d}t} \boldsymbol{\phi}_t^*(\gamma)\right) \tag{25.1}$$

This notation allows for a compact expression of many relations in continuum mechanics. For example, the Cauchy stress tensor $\boldsymbol{\sigma}$ defined on the current configuration and the second Piola–Kirchhoff stress tensor $\mathbf{S}$ associated with the reference configuration are related by

$$\mathbf{S} = J\boldsymbol{\phi}_t^*(\boldsymbol{\sigma}) \quad \text{or} \quad \boldsymbol{\sigma} = \boldsymbol{\phi}_{t*}(J^{-1}\mathbf{S})$$

These relations, involving J, are called Piola transformations.

The forward Piola transformation of the material time derivative of $\mathbf{S}$, denoted $\boldsymbol{\sigma}$, is

$$\boldsymbol{\sigma} = \phi_{t*}(J^{-1}\dot{\mathbf{S}}) \tag{25.2}$$

to so-called Truesdell rate of Cauchy stress. For contravariant components (25.2) has the form

$$\boldsymbol{\sigma} = \dot{\boldsymbol{\sigma}} - \mathbf{l} \cdot \boldsymbol{\sigma} - \boldsymbol{\sigma} \cdot \mathbf{l}^T + \boldsymbol{\sigma}\, \mathrm{tr}\,(\mathbf{d}) \tag{25.3}$$

where $\dot{\boldsymbol{\sigma}}$ denotes the material time derivative of $\boldsymbol{\sigma}$ given by $\dot{\boldsymbol{\sigma}} = \partial\boldsymbol{\sigma}/(\partial t) + \nabla\boldsymbol{\sigma} \cdot \mathbf{v}$.

Alternatively, in terms of the Lie derivative, (25.2) can be written

$$\mathring{\sigma} = J^{-1}\boldsymbol{\phi}_{t*}\left(\frac{\mathrm{d}}{\mathrm{d}t}\boldsymbol{\phi}_t^*(J\boldsymbol{\sigma})\right) = J^{-1}L_{\mathrm{v}}(\boldsymbol{\tau}) \tag{25.4}$$

where $\boldsymbol{\tau} \equiv (J \circ \boldsymbol{\phi}_t^{-1})\boldsymbol{\sigma}$ is the Kirchhoff stress tensor.

The principle of objectivity requires that intrinsic physical properties of a body be independent of the body's location or orientation in space. This principle is embodied in constitutive theory by requiring that constitutive equations contain only objective tensor fields. Consequently, since the Truesdell rate is objective [9, 20] it may be considered as a candidate for use in spatial rate constitutive equations.

Many other objective rates have been proposed within the context of constitutive theory. One that frequently arises is the Jaumann, or co-rotational rate of Kirchhoff stress,

$$\overset{\nabla}{\boldsymbol{\tau}} = \dot{\boldsymbol{\tau}} - \boldsymbol{\omega}\cdot\boldsymbol{\tau} + \boldsymbol{\tau}\cdot\boldsymbol{\omega} \tag{25.5}$$

The co-rotational rate can be related to the Lie derivative. To explicate this relationship, pull back and push forward operations associated with the rotational part of the deformation gradient, or co-rotational pull back and push forward operations, are introduced as follows.

Given any spatial tensor γ, the co-rotational pull back of γ, $\boldsymbol{\phi}_t^{R*}(\gamma)$, is defined by formally replacing the deformation gradient appearing in the pull back operation by its rotational component tensor $\mathbf{R}$. The co-rotational push forward operation is similarly defined. With this notation, the Jaumann or co-rotational rate of Kirchhoff stress is given by

$$\overset{\nabla}{\boldsymbol{\tau}} = \boldsymbol{\phi}_{t*}^{R}\left(\frac{\mathrm{d}}{\mathrm{d}t}\boldsymbol{\phi}_t^{R*}(\boldsymbol{\tau})\right) \tag{25.6}$$

Comparing this expression with (25.1) it is noted that the Jaumann rate coincides with the Lie derivative under the assumption that the rate of deformation tensor vanishes, that is

$$\overset{\nabla}{\boldsymbol{\tau}} = L_{\mathrm{v}}(\boldsymbol{\tau})|_{\mathbf{d}=0} \tag{25.7}$$

Noting that $\dot{\mathbf{F}} = \boldsymbol{\omega}\cdot\mathbf{F}$ when $\mathbf{d} = 0$, where $\boldsymbol{\omega}$ is the spin rate tensor, it follows that

$$L_{\mathrm{v}}(\boldsymbol{\tau})|_{\mathbf{d}=0} = \dot{\boldsymbol{\tau}} - \boldsymbol{\omega}\cdot\boldsymbol{\tau} - \boldsymbol{\tau}\cdot\boldsymbol{\omega}^T = \overset{\nabla}{\boldsymbol{\tau}}$$

for contravariant components of $\boldsymbol{\tau}$.

Finally, the local form of linear momentum balance together with traction and kinematic boundary conditions can be expressed as

$$\begin{aligned} \rho\dot{\mathbf{v}} &= \boldsymbol{\nabla}\cdot\boldsymbol{\sigma} + \rho\mathbf{b} && \mathbf{x}\in\boldsymbol{\phi}_t(B) \\ \boldsymbol{\sigma}\cdot\mathbf{n} &= \bar{\mathbf{t}} && \mathbf{x}\in\partial_\sigma\boldsymbol{\phi}_t(B) \\ \boldsymbol{\phi} &= \bar{\boldsymbol{\phi}} && \mathbf{x}\in\partial_u\boldsymbol{\phi}_t(B) \end{aligned} \tag{25.8}$$

where ρ is the mass density in $\boldsymbol{\phi}_t(B)$, $\mathbf{b}$ is a spatial body force field, and $\bar{\mathbf{t}}$ and $\bar{\boldsymbol{\phi}}$ are the prescribed tractions and motion over the traction and kinematic boundaries $\partial_\sigma \boldsymbol{\phi}_t(B)$ and $\partial_u \boldsymbol{\phi}_t(B)$, respectively.

Suitable constitutive equations need to be introduced in order to complete the specification of an initial boundary value problem. The form of these equations for a limited class of elastoplastic and elasto-viscoplastic materials is the subject of the next section.

25.2.2 Rate constitutive equations for finite deformation elastoplasticity and elasto-viscoplasticity

Constitutive equations appropriate for the finite deformation analysis of elastoplastic and elasto-viscoplastic materials are most frequently expressed in a spatial rate form. A thermodynamic framework for the development of such equations can be provided by characterizing irreversible processes through the use of internal variables [1 to 4]. In this case, spatial rate constitutive equations will be required for the stress tensor and a set of internal variables.

Rate constitutive equations can be alternatively formulated in a material or a spatial setting. The former case involves rates of material tensors which are always objective. In a spatial formulation, however, material rates of objective tensors are not objective and objective stress rates, such as the Truesdell or Jaumann rates, must be introduced. The approach taken here is based on a thermodynamic formulation in a material setting, the spatial representation of which is then consistently derived. As noted in Section 25.1, the correct choice of objective rate appearing in spatial rate constitutive equations has been the subject of considerable conjecture [5 to 8] since the principle of objectivity alone does not uniquely determine this choice. It is demonstrated below, however, that when the invariance of constitutive equations to the choice of reference configuration is invoked together with the assumed existence of a free energy density, the indeterminacy in the choice of objective rates for the spatial stress tensor and the spatial internal variables is removed.

A number of theories of plasticity have been based on different kinematic assumptions regarding the elastic–plastic split of the deformation [10 to 12]. A requirement for the success of the numerical integration algorithm proposed in this paper is that the spatial rate of deformation tensor admit an additive decomposition into an 'elastic' and a 'plastic' part. A thermodynamic argument for such a decomposition is provided within the framework of the internal variable theory.

The existence of a complementary free energy potential per unit mass of $\boldsymbol{B}$, denoted $\chi(\mathbf{S}, \mathbf{Q}_1, \mathbf{Q}_2, \ldots, \mathbf{Q}_{niv})$ is assumed, Here, $\mathbf{S}$ is the second Piola–Kirchhoff stress tensor and $\{\mathbf{Q}_\alpha\}$, $\alpha = 1, \ldots, niv$, is a set of internal variables, the members of which may be scalars or tensors of any order, defined on the reference configuration B. The justification for such an assumption is argued in [21].

Assuming only mild restrictions on the structure of the internal variable rate constitutive equations [21], it can be shown from the Clausius–Duhem inequality [21, 23] that the complementary free energy density is a potential for the right Cauchy–Green deformation tensor **C**, i.e.,

$$\mathbf{C} = 2\rho_0 \frac{\partial \chi}{\partial \mathbf{S}} \tag{25.9}$$

where ρ_0 denotes the reference mass density. A rate form of (25.9) is obtained by taking the material time derivative, resulting in

$$\dot{\mathbf{C}} = \mathbf{M}:\dot{\mathbf{S}} + \sum_{\alpha=1}^{niv} \mathbf{N}_\alpha \cdot \dot{\mathbf{Q}}_\alpha \tag{25.10}$$

where **M** is the elastic compliance tensor defined by

$$\mathbf{M} = 2\rho_0 \frac{\partial^2 \chi}{\partial \mathbf{S}^2} \tag{25.11}$$

and $\mathbf{N}_\alpha$ is an inelastic compliance tensor defined by

$$\mathbf{N}_\alpha = 2\rho_0 \frac{\partial^2 \chi}{\partial \mathbf{Q}_\alpha \partial \mathbf{S}} \tag{25.12}$$

In (25.10), the contraction $\mathbf{N}_\alpha \cdot \dot{\mathbf{Q}}_\alpha$ is to be interpreted according to the order of $\mathbf{Q}_\alpha$. Since material time derivatives of material tensors are objective, constitutive equation (25.10) is also objective.

A spatial form of constitutive equation (25.10) can be obtained as follows. Noting that

$$\dot{\mathbf{C}} = 2\boldsymbol{\phi}_t^*(\mathbf{d}) \tag{25.13}$$

recalling (25.2) and introducing spatial internal variables $\mathbf{q}_\alpha = \boldsymbol{\phi}_{t*}(J^{-1}\mathbf{Q}_\alpha)$, the Truesdell rate of which is given by $\mathring{\mathbf{q}}_\alpha = \boldsymbol{\phi}_{t*}(J^{-1}\dot{\mathbf{Q}}_\alpha)$, equation (25.10) has the alternative form

$$2\boldsymbol{\phi}_t^*(\mathbf{d}) = J\mathbf{M}:\boldsymbol{\phi}_t^*(\mathring{\boldsymbol{\sigma}}) + \sum_{\alpha=1}^{niv} J\mathbf{N}_\alpha \cdot \boldsymbol{\phi}_t^*(\mathring{\mathbf{q}}_\alpha) \tag{25.14}$$

The push forward of (25.14) reads

$$\mathbf{d} = \mathbf{m}:\mathring{\boldsymbol{\sigma}} + \sum_{\alpha=1}^{niv} \mathbf{n}_\alpha \cdot \mathring{\mathbf{q}}_\alpha \tag{25.15}$$

where $\mathbf{m} = \frac{1}{2}J\boldsymbol{\phi}_{t*}(\mathbf{M})$ is the spatial elastic compliance tensor and $\mathbf{n}_\alpha = \frac{1}{2}J\boldsymbol{\phi}_{t*}(\mathbf{N}_\alpha)$ is a spatial inelastic compliance tensor. The material and spatial forms, (25.10) and (25.15) respectively, must be equivalent and simply represent alternative expressions of the same constitutive hypothesis. This equivalence is required by the principle that the laws of continuum mechanics must be invariant with respect to the choice of reference configuration. It is interesting to note that this invariance principle uniquely determines the form of the spatial rate constitutive equations. In particular, the Truesdell rate of Cauchy stress appears as a natural choice of objective spatial stress rate consistent with the material formulation.

Furthermore, equation (25.15) has the interpretation that the rate of deformation tensor $\mathbf{d}$ has an additive decomposition into an 'elastic' part $\mathbf{d}^e = \mathbf{m}:\mathring{\boldsymbol{\sigma}}$ and an 'inelastic' or 'plastic' part $\mathbf{d}^p = \sum_{\alpha=1}^{niv} \mathbf{n}_\alpha \cdot \mathring{\mathbf{q}}_\alpha$, that is

$$\mathbf{d} = \mathbf{d}^e + \mathbf{d}^p \tag{25.16}$$

This decomposition has been obtained independently of any kinematic considerations. Other theories of plasticity based on specific kinematic assumptions [10 to 12] can be brought into correspondence with (25.16) by appropriate definitions of the kinematic variables. Combining (25.15) and (25.16), it follows that

$$\begin{aligned} \mathring{\sigma} &= \mathbf{a}:(\mathbf{d} - \mathbf{d}^p) \\ \mathbf{d}^p &= \sum_{\alpha=1}^{niv} \mathbf{n}_\alpha \cdot \mathring{\mathbf{q}}_\alpha \end{aligned} \tag{25.17}$$

where $\mathbf{a} = \mathbf{m}^{-1}$ is the spatial elastic modulus tensor.

It is noted that (25.17) is expressible in terms of other stress rates if the difference between these rates and the Truesdel rate of Cauchy stress is absorbed in the definition of the elastic modulus tensor. In this case, the modified elastic modulus tensor will, in general, be a function of the deformation and the spatial stress tensor [19, 20].

In order to have a complete set of constitutive equations one has to supplement (25.17) with constitutive relations for $\mathring{\mathbf{q}}_\alpha$ as well as supplying the functional form of the spatial compliances $\mathbf{m}$ and $\mathbf{n}_\alpha$. The 'rate-dependent' or 'rate-independent' characterization of plasticity will reside in the structure of the constitutive equations for $\mathring{\mathbf{q}}_\alpha$ [21]. If these rate constitutive equations are homogeneous (of degree one) in some measure of time then rate-independent (inviscid plastic) behaviour is obtained, otherwise rate-dependent (viscoplastic) behaviour results. The constitutive equations for $\mathring{\mathbf{q}}_\alpha$ cannot be specified arbitrarily but must satisfy

$$\sum_{\alpha=1}^{niv} J\phi_{t*}\left(\frac{\partial \chi}{\partial \mathbf{Q}_\alpha}\right):\mathring{\mathbf{q}}_\alpha \geq 0$$

which is a representation of the Clausius–Planck dissipation inequality [21, 23].

For the present purpose it will suffice to assume that $\mathbf{d}^p$ can be expressed as a function of the spatial stress and the spatial internal variables

$$\mathbf{d}^p = \mathbf{T}(\boldsymbol{\sigma}, \mathbf{q}_1, \ldots, \mathbf{q}_{niv}) \tag{25.18}$$

where the internal variables $\mathbf{q}_\alpha$ may for example represent some invariant of the yield stress for an isotropic hardening model or the translation of the elastic domain for a kinematic hardening model.

We introduce here two particularly simple examples of the constitutive mapping $\mathbf{T}$ appearing in (25.18).

25.2.2.1 *Perfect viscoplasticity*

For a perfectly viscoplastic material we first introduce a closed convex elastic domain C in stress space $S \equiv R^6$ which contains the origin and has a smooth boundary ∂C. Then, for every point $\boldsymbol{\sigma} \in S$, we assume constitutive equation (25.18) has the form

$$\mathbf{T}(\boldsymbol{\sigma}) = \begin{cases} \lambda \mathbf{n}_\sigma & \text{if } \boldsymbol{\sigma} \in S - \text{Int}(C) \\ 0 & \text{if } \boldsymbol{\sigma} \in \text{Int}(C) \end{cases} \tag{25.19}$$

where $\mathbf{n}_\sigma$ is the outward normal to ∂C at the (unique) point on ∂C which is closest to $\boldsymbol{\sigma}$, $\lambda \geq 0$ and $\text{Int}(C) = C - \partial C$. Note that no internal variables are required for this model. Equation (25.19) may be viewed as a generalization of the usual normality assumption of infinitesimal plasticity.

A specific example of such a constitutive equation can be constructed as follows. Given any point $\boldsymbol{\sigma} \in S$, then from the assumed convexity of C, there is always a unique point $\mathbf{P}_C \boldsymbol{\sigma}$ in C which is closest to $\boldsymbol{\sigma}$. The mapping $\mathbf{P}_C$ is called the closest point mapping (relative to C). Clearly, if $\boldsymbol{\sigma} \in C$ then $\mathbf{P}_C \boldsymbol{\sigma} = \boldsymbol{\sigma}$. If, on the other hand, $\boldsymbol{\sigma} \in S - C$ then $\boldsymbol{\sigma} - \mathbf{P}_C \boldsymbol{\sigma} = \mu \mathbf{n}_\sigma$ with $\mu > 0$. This suggests taking the viscoplastic constitutive mapping (25.19) in the form

$$\mathbf{T}(\boldsymbol{\sigma}) = \frac{\boldsymbol{\sigma} - \mathbf{P}_C \boldsymbol{\sigma}}{\eta} \tag{25.20}$$

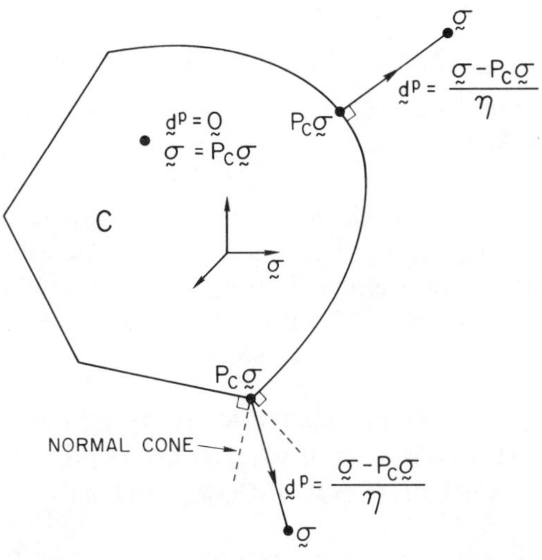

Figure 25.1 Definition of viscoplastic constitutive mapping

The parameter η is the viscosity of the material. If $\boldsymbol{\sigma}$ belongs to C then $\mathbf{P}_C\boldsymbol{\sigma}=\boldsymbol{\sigma}$ and $\mathbf{d}^p=\mathbf{T}(\boldsymbol{\sigma})=0$. If, on the other hand, $\boldsymbol{\sigma}$ does not belong to C then $\mathbf{d}^p$ is directed along the vector that joins $\boldsymbol{\sigma}$ and its closest point in C and it points outside the elastic region. The magnitude of $\mathbf{d}^p$ is proportional to the distance from $\boldsymbol{\sigma}$ to C, the proportionality constant being $1/\eta$, Figure 25.1. It clear that (25.20) is also well behaved (single valued) when ∂C is not smooth but exhibits corners, Figure 25.1. Examples employing the von Mises yield criterion in the definition of the elastic domain C are given in Section 25.3.

In this paper, inviscid or rate-independent plasticity is treated as the limiting case of viscoplasticity as the viscosity η of the material tends to zero. Alternatively, one may think of this limiting process as the result of allowing an infinite period of time to elapse for the relaxation of the stresses towards the elastic region. Examples of the process are considered in Section 25.3.

25.2.2.2 *Hardening viscoplasticity*

In the case of hardening viscoplasticity, a set of spatial internal variables $\mathbf{q}\equiv\{\mathbf{q}_\alpha\}, \alpha=1,\ldots,niv$ is introduced such that the elastic domain $C(\mathbf{q})$ now depends on the current values of the internal variables. The plastic constitutive mapping (25.20) can be generalized for this case by taking

$$\mathbf{T}(\boldsymbol{\sigma},\mathbf{q})=\frac{\boldsymbol{\sigma}-\mathbf{P}_{C(\mathbf{q})}\boldsymbol{\sigma}}{\eta} \tag{25.21}$$

where $\mathbf{P}_{C(\mathbf{q})}$ denotes the closest point mapping relative to $C(\mathbf{q})$.

It is assumed in this hardening case that the evolution of the internal variables is governed by evolutionary equations of the form

$$\mathring{\mathbf{q}}_\alpha=\mathbf{f}_\alpha(\boldsymbol{\sigma},\mathbf{q}_1,\ldots,\mathbf{q}_{niv}), \qquad \alpha=1,\ldots,niv \tag{25.22}$$

Note that the use of the Truesdell rate in the left-hand side of (25.22) makes these equations objective and consistent with a set of material kinetic equations given by $\dot{\mathbf{Q}}_\alpha=\mathbf{H}_\alpha(\mathbf{S},\mathbf{Q}_1,\ldots,\mathbf{Q}_{niv})$ where $\mathbf{f}_\alpha=\boldsymbol{\phi}_{t*}(J^{-1}\mathbf{H}_\alpha)$.

As for the case of perfect (inviscid) plasticity, hardening plasticity will be treated as the limiting case of hardening viscoplasticity as the viscosity η of the material tends to zero or, alternatively, as an infinite period of time is allowed for the stress and internal variables to relax to their asymptotic values. Examples of this process are considered in the following section.

25.3 NUMERICAL INTEGRATION OF RATE CONSTITUTIVE EQUATIONS FOR ELASTOPLASTICITY AND ELASTO-VISCOPLASTICITY

25.3.1 Introduction

As noted in the Introduction, any numerical scheme for the solution of the boundary value problem of linear momentum balance for elastoplastic or elasto-viscoplastic materials will require an algorithm for the integration of the rate constitutive equations. This section addresses the numerical integration of rate constitutive equations for the stress and internal variables introduced in Section 25.2 and summarized from (25.17), (25.18) and (25.22) as

$$\begin{aligned} \mathring{\boldsymbol{\sigma}} &= \mathbf{a}:(\mathbf{d} - \mathbf{T}(\boldsymbol{\sigma}, \mathbf{q}_1, \ldots, \mathbf{q}_{niv})) \\ \mathring{\mathbf{q}}_\alpha &= \mathbf{f}_\alpha(\boldsymbol{\sigma}, \mathbf{q}_1, \ldots, \mathbf{q}_{niv}), \qquad \alpha = 1, \ldots, niv \end{aligned} \tag{25.23}$$

As a consequence of the additive decomposition of the spatial rate of deformation tensor (25.16), constitutive equations (25.23) also exhibit an additive decomposition into an elastic part

$$\begin{aligned} \mathring{\boldsymbol{\sigma}} &= \mathbf{a}:\mathbf{d} \\ \partial \mathbf{q}_\alpha / \partial t &= 0, \qquad \alpha = 1, \ldots, niv \end{aligned} \tag{25.24}$$

and a plastic part

$$\begin{aligned} \partial \boldsymbol{\sigma} / \partial t &= -\mathbf{a}:\mathbf{T}(\boldsymbol{\sigma}, \mathbf{q}_1, \ldots, \mathbf{q}_{niv}) \\ \mathring{\mathbf{q}}_\alpha &= \mathbf{f}_\alpha(\boldsymbol{\sigma}, \mathbf{q}_1, \ldots, \mathbf{q}_{niv}), \qquad \alpha = 1, \ldots, niv \end{aligned} \tag{25.25}$$

From a numerical point of view, an algorithm for the integration of (25.23) should satisfy three requirements:

(i) Consistency with the constitutive equations.
(ii) Numerical stability.
(iii) Incremental objectivity.

Conditions (i) and (ii) are required for the convergence of the numerical integration scheme [17]. Condition (iii) results from the physical requirement that the algorithm must be invariant with respect to superimposed rigid body motions. This idea is considered in detail in subsequent sections. Few algorithms reported in the literature seem to satisfy all these requirements.

Equations (25.24) and (25.25) suggest the possibility of using product formula techniques for constructing efficient solution algorithms for (25.23) which will also provide a basis for demonstrating the consistency and numerical stability of the resulting algorithms. Before presenting the details of such an algorithm a brief discussion of general product algorithms will be useful.

25.3.2 Operator splits and product algorithms

The operator split method has recently been applied to the finite element analysis of the heat conduction problem [13], to the structural dynamics problem [14] and also, as discussed above, to the finite deformation elastoplastic dynamic problem [16]. A collection of results regarding operator split methods and product formula algorithms for general non-linear equations of evolution is presented in [13, 14]. These results illustrate the point that product formulas can be advantageously applied to any set of equations of evolution where the evolutionary operator has an additive decomposition (operator split) into component operators. The basic idea underlying product formulas is that of treating each one of the component operators independently. In a typical integration process, one applies an algorithm to the solution vector that is consistent with the first component operator, the result of which is then operated upon with an algorithm which is consistent with the second component operator, and so on.

Consider the following general evolution equation

$$\mathbf{A}\dot{\mathbf{x}} + \mathbf{B}(\mathbf{x}) = \mathbf{f}; \qquad \mathbf{x}(0) = \mathbf{x}_0 \tag{25.26}$$

where $\mathbf{x}$ is an s-dimensional vector, $\mathbf{A}$ is a positive definite symmetric matrix, and $\mathbf{B}$ is a non-linear function from R^s into R^s. We endow R^s with the 'energy' inner product $\langle \mathbf{x}, \mathbf{y} \rangle = \mathbf{x}^T\mathbf{A}\mathbf{y}$ for every $\mathbf{x}, \mathbf{y} \in R^s$, with its associated norm $\|\mathbf{x}\|^2 = \langle \mathbf{x}, \mathbf{x} \rangle$. In this context, an unconditionally stable algorithm for equation (25.26) is a one-parameter family of (non-linear) functions $\mathbf{F}(h): R^s \rightarrow R^s$, $h > 0$, satisfying

(1) Consistency:

$$\lim_{h \rightarrow 0+} \mathbf{A}\frac{\mathbf{F}(h)\mathbf{x} - \mathbf{x}}{h} = \mathbf{B}(\mathbf{x}) + \mathbf{f} \qquad \text{for every } \mathbf{x} \in R^s \tag{25.27}$$

(2) Unconditional stability:

$$\|\mathbf{F}(h)\mathbf{x} - \mathbf{F}(h)\mathbf{y}\| \leq \|\mathbf{x} - \mathbf{y}\| \qquad \text{for every } \mathbf{x}, \mathbf{y} \in R^s, \quad h > 0 \tag{25.28}$$

If $\mathbf{F}(h)$ is a consistent and stable algorithm for (25.26) in the sense of (25.27) and (25.28) then convergence is guaranteed under mild conditions on $\mathbf{B}$ [17].

Suppose the evolutionary operator $\mathbf{B}$ and the forcing term $\mathbf{f}$ admit an additive decomposition

$$\mathbf{B} = \sum_{i=1}^{N} \mathbf{B}_i; \qquad \mathbf{f} = \sum_{i=1}^{N} \mathbf{f}_i \tag{25.29}$$

and let $\mathbf{F}_i(h), i = 1, 2, \ldots, N$ denote stable algorithms consistent with

$$\mathbf{A}\dot{\mathbf{x}} + \mathbf{B}_i(\mathbf{x}) = \mathbf{f}_i$$

Then the corresponding global product algorithm takes the form

$$\mathbf{F}(h) = \mathbf{F}_N(h)\mathbf{F}_{N-1}(h) \cdots \mathbf{F}_1(h) \equiv \prod_{i=1}^{N} \mathbf{F}_i(h) \tag{25.30}$$

In other words, the algorithm $\mathbf{F}(h)$ amounts to applying the individual algorithms $\mathbf{F}_i(h)$ consecutively to the solution vector, taking the result from each one of these applications as the initial conditions for the next algorithm. The global algorithm is complete for a given time step when all the individual algorithms have been applied.

It can be shown that if all the individual algorithms $\mathbf{F}_i(h)$ are consistent with $\mathbf{A}$, $\mathbf{B}_i$, and $\mathbf{f}_i$ in the sense of (25.27), then the global product algorithm $\mathbf{F}(h)$ given by (25.30) is consistent with $\mathbf{A}$, $\mathbf{B}$ and $\mathbf{f}$ [14]. It can also be shown that if all the individual algorithms $\mathbf{F}_i(h)$ are unconditionally stable in the sense of (25.28), then the global product algorithm $\mathbf{F}(h)$ is also unconditionally stable [14]. In other words the norm stability of the individual algorithms, in the sense of (25.28), is preserved by the product formula (25.30). A general discussion of these and other related issues can be found in [13, 14].

25.3.3 A product algorithm for rate constitutive equations

A product algorithm for the integration of the rate constitutive equations (25.23) relative to the additive decomposition given by (25.24) and (25.25) can be constructed as follows. Consider two unconditionally stable algorithms $\mathbf{F}^{\mathrm{el}}(h)$ and $\mathbf{F}^{\mathrm{pl}}(h)$ which are consistent, in the sense of (25.27), with (25.24) and (25.25) respectively. Then an unconditionally stable algorithm $\mathbf{F}(h)$ which is consistent with the complete set of constitutive equations (25.23) is given, in analogy with (25.30), by

$$\mathbf{F}(h) = \mathbf{F}^{\mathrm{pl}}(h)\mathbf{F}^{\mathrm{el}}(h) \tag{25.31}$$

such that

$$\mathbf{F}(h)\begin{Bmatrix} \boldsymbol{\sigma} \\ \mathbf{q}_\alpha \end{Bmatrix} = \mathbf{F}^{\mathrm{pl}}(h)\left[\mathbf{F}^{\mathrm{el}}(h)\begin{Bmatrix} \boldsymbol{\sigma} \\ \mathbf{q}_\alpha \end{Bmatrix}\right] = \begin{Bmatrix} \boldsymbol{\sigma}(h) \\ \mathbf{q}_\alpha(h) \end{Bmatrix}, \qquad \alpha = 1, \ldots, niv$$

The product formula (25.31) states that a solution algorithm is obtained by first integrating the elastic constitutive equations and then applying to the solution vector so obtained a plastic algorithm operating on the stress and internal variables reflecting the effect of the plastic part of the constitutive equations. The remainder of this section is concerned with the development of the $\mathbf{F}^{\mathrm{el}}(h)$ and $\mathbf{F}^{\mathrm{pl}}(h)$ algorithms.

25.3.3.1 *Elastic algorithm*

The desired algorithm $\mathbf{F}^{\mathrm{el}}(h)$ must be unconditionally stable and consistent with (25.24) in the sense of (25.27). It is observed that the evolution equations (25.24) effect only the stresses with the internal variables $\mathbf{q}_\alpha$ remaining constant, thus the algorithm can be expressed as

$$\mathbf{F}^{\mathrm{el}}(h)\begin{Bmatrix} \boldsymbol{\sigma} \\ \mathbf{q}_\alpha \end{Bmatrix} = \begin{Bmatrix} \boldsymbol{\sigma}(h) \\ \mathbf{q}_\alpha \end{Bmatrix}, \qquad \alpha = 1, \ldots, niv \tag{25.32}$$

Objective rates appearing in spatial constitutive equations have the effect of introducing some complications in the development of numerical integration algorithms and have motivated research on this problem [18, 24, 25]. However, few algorithms presented in the computational literature appear to be consistent with the constitutive equations which they are purporting to integrate. A family of algorithms, appropriate for the integration of (25.24), which does satisfy the requirements of consistency, numerical stability and incremental objectivity, has recently been proposed in [19]. The essential ideas are as follows. We start by noting that from a mathematical point of view, the usual linear space operations such as addition and scalar multiplication can only be rigorously applied to relate tensor fields associated with a common configuration [19]. It is thus natural to use the idea of pulling back spatial quantities to a common reference configuration in order to define difference operators to be used in numerical algorithms. This suggests defining algorithms for the integration of (25.24a) based upon difference operators employing the second Piola–Kirchhoff stress tensor. A generalized midpoint rule algorithm can be introduced as follows:

$$\mathbf{S}_{n+1} - \mathbf{S}_n = h\dot{\mathbf{S}}_{n+\alpha} 0 \leq \alpha \leq 1 \tag{25.33}$$

where subscripts refer to the time step, the time step size $h = t_{n+1} - t_n$ and $\dot{\mathbf{S}}_{n+\alpha}$ is to be evaluated on an intermediate configuration defined by the mapping $\boldsymbol{\phi}_{n+\alpha}(\mathbf{X}): B \rightarrow R^N$ with

$$\boldsymbol{\phi}_{n+\alpha} = \alpha\boldsymbol{\phi}_{n+1} + (1-\alpha)\boldsymbol{\phi}_n 0 \leq \alpha \leq 1 \tag{25.34}$$

Using the inverse of (25.2) and noting again that $\mathbf{S} = J\boldsymbol{\phi}_t^*(\boldsymbol{\sigma})$, (25.33) has the alternative representation

$$\boldsymbol{\phi}_{n+1}^*(J\boldsymbol{\sigma}) - \boldsymbol{\phi}_n^*(J\boldsymbol{\sigma}) = h\boldsymbol{\phi}_{n+\alpha}^*(J\mathring{\boldsymbol{\sigma}}) \tag{25.35}$$

To simplify (25.35) the reference configuration B is selected to coincide instantaneously with the configuration at time t_{n+1} such that $\boldsymbol{\phi}_{n+1} = \mathbf{I}$, in which case (25.35) reduces to

$$\boldsymbol{\sigma}_{n+1} - \boldsymbol{\phi}_n^*(J\boldsymbol{\sigma}) = h\boldsymbol{\phi}_{n+\alpha}^*(J\mathring{\boldsymbol{\sigma}}) \tag{25.36}$$

Defining the deformation gradients

$$\boldsymbol{\Lambda}_{n+\alpha} = \left(\frac{\partial\boldsymbol{\phi}_{n+\alpha}}{\partial\mathbf{x}_{n+1}}\right)^{-1} 0 \leq \alpha \leq 1 \tag{25.37}$$

and Jacobians

$$J_{n+\alpha} = \det(\boldsymbol{\Lambda}_{n+\alpha}) 0 \leq \alpha \leq 1 \tag{25.38}$$

then, for contravariant components of stress, (25.36) has the form

$$\boldsymbol{\sigma}_{n+1} - J_n^{-1}\boldsymbol{\Lambda}_n\cdot\boldsymbol{\sigma}_n\cdot\boldsymbol{\Lambda}_n^T = hJ_{n+\alpha}^{-1}\boldsymbol{\Lambda}_{n+\alpha}\cdot\mathring{\boldsymbol{\sigma}}_{n+\alpha}\cdot\boldsymbol{\Lambda}_{n+\alpha}^T \tag{25.39}$$

This equation is completed by introducing the rate constitutive equation for $\mathring{\boldsymbol{\sigma}}_{n+\alpha}$. For example, for constitutive equation (25.24a), equation (25.39) is expressed by

$$\boldsymbol{\sigma}_{n+1} - J_n^{-1}\boldsymbol{\Lambda}_n\cdot\boldsymbol{\sigma}_n\cdot\boldsymbol{\Lambda}_n^T = hJ_{n+\alpha}^{-1}\boldsymbol{\Lambda}_{n+\alpha}\cdot(\mathbf{a}:\mathbf{d})|_{n+\alpha}\cdot\boldsymbol{\Lambda}_{n+\alpha}^T \tag{25.40}$$

This form requires evaluation of the quantities $\boldsymbol{\Lambda}_{n+\alpha}$, $\mathbf{a}_{n+\alpha}$, and $\mathbf{d}_{n+\alpha}$. Assuming $\boldsymbol{\Lambda}_n$ to be a known quantity (representing the incremental motion for which the corresponding stresses are desired), it is shown in [19] that $\boldsymbol{\Lambda}_{n+\alpha}$ is given by

$$\boldsymbol{\Lambda}_{n+\alpha} = [(1-\alpha)\mathbf{I} + \alpha\boldsymbol{\Lambda}_n]^{-1}\cdot\boldsymbol{\Lambda}_n \tag{25.41}$$

and that $\mathbf{d}_{n+\alpha}$ is consistently approximated by

$$\mathbf{d}_{n+\alpha} = \frac{1}{h}[\{(1-\alpha)\mathbf{I} + \alpha\mathbf{\Lambda}_n\}^{-1}\{\mathbf{\Lambda}_n - \mathbf{I}\}]^S \tag{25.42}$$

As noted above, the algorithm (25.40)–(25.42) must to satisfy the three requirements of consistency with the spatial rate constitutive equations, numerical stability and incremental objectivity. It is demonstrated in [19] that the above algorithm is consistent with the rate constitutive equation and that it is unconditionally stable for $\alpha \geq 0.5$, moreover, it is second order accurate for $\alpha = 0.5$.

The condition of incremental objectivity is a physical requirement expressing the fact that the algorithm has to be invariant with respect to superimposed rigid body motions occurring over the time step and has the effect of restricting the admissible values of α. This idea was first expressed in an algorithmic context in [18] and further considered in [18, 19]. Formally, let $\mathbf{R}$ be the group of all orthogonal second order tensors and $\mathbf{M}$ the group of all positive definite symmetric second order tensors. The algorithm defined by (25.40)–(25.42) is incrementally objective if and only if

(a) $$\mathbf{\Lambda}_n \in \mathbf{R} \Leftrightarrow \mathbf{d}_{n+\alpha} = 0$$

(b) $$\mathbf{\Lambda}_n \in \mathbf{M} \Leftrightarrow \omega_{n+\alpha} = 0$$

where $\boldsymbol{\omega}_{n+\alpha}$ is the spin rate tensor determined by replacing the symmetric part on the right-hand side of (25.42) by the skew-symmetric part. Condition (a) ensures that the integration algorithm reduces to $\boldsymbol{\sigma}_{n+1} = \mathbf{\Lambda}_n \cdot \boldsymbol{\sigma}_n \cdot \mathbf{\Lambda}_n^T$ in the event that $\mathbf{\Lambda}_n \in \mathbf{R}$. It is demonstrated in [19] that the algorithm (25.40)–(25.42) is incrementally objective if and only if $\alpha = 0.5$.

Equation (25.40) may easily be generalized to accommodate choices of objecting stress rate other than the Truesdell rate of Cauchy stress by embedding the difference between the stress rate definitions in the elastic modulus tensor $\mathbf{a}$. In this case (25.40) will, in general, become implicit in $\boldsymbol{\sigma}_{n+1}$ and may be solved by means of an iterative solution produre [19]. The algorithm (25.40)–(25.42) fits naturally into a finite element implementation since it employs quantities that are readily available from standard shape function routines [20].

It is noted finally that the incrementally objective algorithm introduced in [18] can be related to the present algorithm (25.40)–(25.42) by replacing the pull-back operations in (25.35) by their co-rotational counterparts and using the fact that the Lie derivative and the Jaumann rate coincide under the assumption that the rate of deformation tensor vanishes (see Section 25.2.1) [20]. This relation can be used to draw conclusions about the applicability of the algorithm presented in [18].

25.3.3.2 *Plastic/viscoplastic algorithm*

The desired algorithm $\mathbf{F}^{\mathrm{pl}}(h)$, which can be expressed by

$$\mathbf{F}^{\mathrm{pl}}(h)\begin{Bmatrix}\boldsymbol{\sigma}\\ \mathbf{q}_\alpha\end{Bmatrix}=\begin{Bmatrix}\boldsymbol{\sigma}(h)\\ \mathbf{q}_\alpha(h)\end{Bmatrix}, \qquad \alpha=1,\ldots,niv \tag{25.43}$$

must be unconditionally stable and consistent with the plastic part of the constitutive equations (25.25).

The constitutive equations (25.25) may admit closed form solutions for particular forms of $\mathbf{T}(\boldsymbol{\sigma},\mathbf{q}_1,\ldots,\mathbf{q}_{niv})$ and $\mathbf{f}_\alpha(\boldsymbol{\sigma},\mathbf{q}_1,\ldots,\mathbf{q}_{niv})$ appearing in (25.25). This introduces the possibility of using the closed form solutions to (25.25) as the algorithm $\mathbf{F}^{\mathrm{pl}}(h)$ and is the approach used here. It should be emphasized, however, that the plastic relaxation equations (25.25) will not admit closed form solutions in general. That closed form solutions may be found in the present case results from the simplicity of the assumed structure of the constitutive mappings. In general, as in the case of $\mathbf{F}^{\mathrm{el}}(h)$ given above, numerical solution schemes for (25.25) will have to be resorted to with the usual considerations for consistency, numerical stability, and incremental objectivity.

Four examples of the constitutive mappings $\mathbf{T}(\boldsymbol{\sigma},\mathbf{q}_\alpha)$ and $\mathbf{f}_\alpha(\boldsymbol{\sigma},\mathbf{q}_\alpha)$ appearing in (25.25) are given according to the discussion in Section 25.2.2. These examples include perfect viscoplasticity and perfect plasticity (limiting case of viscoplasticity, see Section 25.2.2.1), hardening viscoplasticity, and hardening plasticity (see Section 25.2.2.2). Closed from solutions for (25.25) corresponding to these constitutive equations have been found in [15, 16] for use in a different context and will be utilized here. These solutions are summarized as follows.

25.3.3.2.1 *Perfect viscoplasticity*

Noting again that no internal variables are needed for this model, the algorithm (25.43) becomes

$$\mathbf{F}^{\mathrm{pl}}(h)(\boldsymbol{\sigma})=\boldsymbol{\sigma}(h) \tag{25.44}$$

Using (25.20), the relaxation equations (25.25) take the form

$$\partial\boldsymbol{\sigma}/\partial t=-\mathbf{a}:\mathbf{T}(\boldsymbol{\sigma})=-\mathbf{a}:\frac{\boldsymbol{\sigma}-\mathbf{P}_C\boldsymbol{\sigma}}{\eta} \tag{25.45}$$

where it is recalled from Section 25.2.2.1 that C denotes the convex elastic domain in stress space, $\mathbf{P}_C$ denotes the closest point mapping relative to C and the parameter η is the viscosity of the material. Equation (25.45) represents a system of ordinary differential equations whose solution is

$$\boldsymbol{\sigma}(t)=\begin{cases}\boldsymbol{\sigma}_0 & \text{if } \boldsymbol{\sigma}_0\in C\\ \exp(-\mathbf{a}t/\eta):\boldsymbol{\sigma}_0+[\mathbf{I}-\exp(-\mathbf{a}t/\eta)]:\mathbf{P}_C\boldsymbol{\sigma}_0 & \text{otherwise}\end{cases} \tag{25.46}$$

For the case of isochoric plasticity in which C is a cylinder oriented along the hydrostatic axis and for isotropic elasticity, equation (25.46b) simplifies to

$$\begin{aligned}\boldsymbol{\sigma}(t) &= \mathrm{e}^{-t/\tau}\boldsymbol{\sigma}_0 + (1 - \mathrm{e}^{-t/\tau})\mathbf{P}_C\boldsymbol{\sigma}_0 \\ &= p_0\mathbf{I} + \mathrm{e}^{-t/\tau}\mathbf{s}_0 + (1 - \mathrm{e}^{-t/\tau})\mathbf{P}_C\mathbf{s}_0 \end{aligned} \tag{25.47}$$

where $p_0 = \boldsymbol{\sigma}_0 : \mathbf{I}$ is the initial hydrostatic pressure, $\mathbf{s}_0$ is the deviatoric part of $\boldsymbol{\sigma}_0$, and $\tau = \eta/G$ is the relaxation time of the process, given in terms of the shear modulus of the material G.

For the von Mises yield criterion, the elastic domain C is the set $\{\boldsymbol{\sigma} \in S$ such that $J_2 \leq k^2\}$, where k is the shear yield stress, $J_2 = 1/2\,\mathbf{s}:\mathbf{s}$ and $\mathbf{s}$ is the deviatoric part of $\boldsymbol{\sigma}$. In this case, equation (25.47) reduces to

$$\boldsymbol{\sigma}(t) = p_0\mathbf{I} + \mathrm{e}^{-t/\tau}\mathbf{s}_0 + (1 - \mathrm{e}^{-t/\tau})\frac{k}{r_0}\mathbf{s}_0 \tag{25.48}$$

where $r_0 = (1/2\,\mathbf{s}_0 : \mathbf{s}_0)^{1/2}$.

As noted above, these closed form solutions can be used for the $\mathbf{F}^{\mathrm{pl}}(h)$ algorithm. While this algorithm is obviously consistent with (25.25) specialized for the given constitutive equations, the stability of the algorithm is not automatic. A general discussion of the requirements for unconditional stability of $\mathbf{F}^{\mathrm{pl}}(h)$ can be found in [15, 16] where it is also demonstrated that the solutions presented above satisfy such requirements.

Finally, the product algorithm (25.31) consists of first integrating the elastic constitutive equations, with time step h, ignoring the plasticity of the material. The stresses resulting from this operation are then allowed to relax according to (25.46) for a period of time h, Figure 25.2. Clearly, the stresses resulting from the application of the elastic algorithm that lie inside the elastic region are unaffected by this relaxation process. The unconditional stability of the product algorithm follows from the unconditional stability of the component algorithms $\mathbf{F}^{\mathrm{el}}(h)$ and $\mathbf{F}^{\mathrm{pl}}(h)$.

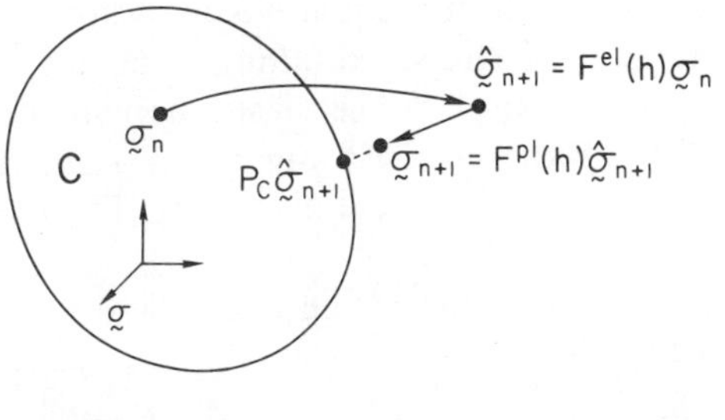

Figure 25.2 Schematic representation of product algorithm

25.3.3.2.2 *Perfect plasticity* As discussed in Section 25.2.2.1, perfect plasticity is considered as the limiting case of viscoplasticity as the viscosity η tends to zero or as an infinite period of time is allowed to elapse permitting the relaxation of the stress towards the elastic domain C. Taking this limit on the viscoplastic algorithm (25.44), the detailed form of which is given by (25.46), an algorithm for the case of perfect plasticity can be expressed as

$$\mathbf{F}^{\mathrm{pl}}(h)(\boldsymbol{\sigma}) = \boldsymbol{\sigma}(\infty) = \mathbf{P}_C\boldsymbol{\sigma} \tag{25.49}$$

where $\mathbf{P}_C\boldsymbol{\sigma}$ again denotes the closest point projection of $\boldsymbol{\sigma}$ on to the elastic domain C. If $\boldsymbol{\sigma} \in C$ then $\mathbf{P}_C\boldsymbol{\sigma} = \boldsymbol{\sigma}$ and the stresses are unaffected by the plastic algorithm. It is noted that (25.49) is independent of the time step size h, reflecting the rate-independent character of this algorithm.

Applying this limiting condition to (25.48), an algorithm for perfect plasticity with the von Mises yield condition results and is given by

$$\mathbf{F}^{\mathrm{pl}}(h)(\boldsymbol{\sigma}) = \boldsymbol{\sigma}(\infty) = \mathbf{P}_C\boldsymbol{\sigma} = \begin{cases} \boldsymbol{\sigma} & \text{if } r \le k \\ p_0\mathbf{I} + \dfrac{k}{r_0}\mathbf{s}_0 & \text{otherwise} \end{cases} \tag{25.50}$$

A more rigorous mathematical treatment of the viscoplastic approximation to inviscid plasticity can be found in [26].

The closest point mapping algorithm (25.49) may be viewed as a 'return mapping' algorithm, other examples are discussed in [27 to 30]. Certainly, the closest point mapping does not exhaust all the possible choices of return mapping that can be used to project stresses back to the elastic domain. If consistent numerical schemes are introduced for the integration of the plastic constitutive equations, rather than the closed form solution adopted above, then these algorithms will result in various return mappings of the stress. A set of conditions on the return mapping has been presented in [26] that guarantees the consistency and numerical stability of the resulting plastic algorithm. In particular, it is demonstrated in [26] that the algorithm (25.50) satisfies the conditions for unconditional stability.

Finally, for the case of perfect plasticity, the product algorithm (25.31) consists of integrating the elastic constitutive equations with time step h, ignoring the plasticity of the material. The stresses resulting from this operation are then projected on to the closest point of the elastic domain C, Figure 25.2. The unconditional stability of the product algorithm again follows from the unconditional stability of the component algorithms $\mathbf{F}^{\mathrm{el}}(h)$ and $\mathbf{F}^{\mathrm{pl}}(h)$.

25.3.3.2.3 *Hardening viscoplasticity* Using (25.21) and (25.22), the relaxation equations (25.25) take the form

$$\partial\boldsymbol{\sigma}/\partial t = -\mathbf{a}:\mathbf{T}(\boldsymbol{\sigma},\mathbf{q}) = -\mathbf{a}:\frac{\boldsymbol{\sigma} - \mathbf{P}_{C(\mathbf{q})}\boldsymbol{\sigma}}{\eta}$$

$$\mathring{\mathbf{q}}_\alpha = \mathbf{f}_\alpha(\boldsymbol{\sigma},\mathbf{q}), \qquad \alpha = 1,\ldots,niv \tag{25.51}$$

where $\mathbf{q}$ represents the set of spatial internal variables $\{\mathbf{q}_\alpha\}$, $\alpha = 1, \ldots, niv$.

For the case of isotropic hardening we take $niv = 1$ and identify $\mathbf{q}_1 = k$. Noting (25.3), it follows that

$$\overset{\circ}{k} = \partial k/\partial t + \partial k/\partial \mathbf{x} \cdot \mathbf{v} + k \operatorname{tr}(\mathbf{d}). \tag{25.52}$$

Assuming a von Mises yield criterion with isotropic bilinear hardening and isotropic elasticity, (25.51) reduces to

$$\begin{aligned} \partial \boldsymbol{\sigma}/\partial t &= -2G\mathbf{d}^p = \frac{-G\langle\sqrt{(J_2 - k)}\rangle}{\eta}\frac{\mathbf{s}}{\sqrt{(J_2)}} \\ \partial k/\partial t &= 2H(\tfrac{1}{2}\mathbf{d}^p : \mathbf{d}^p)^{1/2} = H\frac{\langle\sqrt{(J_2)} - k\rangle}{\eta} \end{aligned} \tag{25.53}$$

where H denotes the shear plastic modulus. Note that the last two terms on the right-hand side of (25.52) have been neglected. The solution of (25.53) is found to be [15, 16]

$$\boldsymbol{\sigma}(t) = \boldsymbol{\sigma}_0, \qquad k(t) = k_0, \qquad \text{if } r_0 \leq k_0$$

$$\boldsymbol{\sigma}(t) = \boldsymbol{\sigma}_0 - \frac{\tau}{\tau_s}(r_0 - k_0)(1 - \mathrm{e}^{-t/\tau})\frac{\mathbf{s}_0}{r_0} \tag{25.54}$$

otherwise

$$k(t) = k_0 + \frac{\tau}{\tau_q}(r_0 - k_0)(1 - \mathrm{e}^{-t/\tau})$$

where

$$\tau_s = \frac{\eta}{G}; \qquad \tau_q = \frac{\eta}{H}; \qquad \tau = \frac{\tau_s \tau_q}{\tau_s + \tau_q} \tag{25.55}$$

and relaxation times for the process.

It is seen from (25.54) that during a relaxation process corresponding to initial stresses outside the elastic domain, the stresses steadily approach the elastic domain, which at the same time expands towards the stress point. As in the case of perfect viscoplasticity, the closed form solution (25.54) can be utilized to define the plastic algorithm (25.43). Requirements for the unconditional stability of (25.54) are considered in [15] where a similar algorithm is presented for the case of kinematic hardening.

25.3.3.2.4 *Hardening plasticity* As in the case of perfect plasticity, plastic algorithms for the hardening case can be obtained as the limiting case of (25.43) as the viscosity η tends to zero or as an infinite length of time is allowed for the relaxation of the stress and internal variables and is expressed by

$$\mathbf{F}^{\mathrm{pl}}(h)\begin{Bmatrix} \boldsymbol{\sigma} \\ \mathbf{q}_\alpha \end{Bmatrix} = \begin{Bmatrix} \boldsymbol{\sigma}(\infty) \\ \mathbf{q}_\alpha(\infty) \end{Bmatrix}, \qquad \alpha = 1, \ldots, niv \tag{25.56}$$

For the case of bilinear isotropic hardening and a von Mises yield criterion, the asymptotic values in (25.56) can be obtained from (25.54) as

$$\boldsymbol{\sigma}(\infty)=\boldsymbol{\sigma}_0, \qquad k(\infty)=k_0 \qquad \text{if } r_0 \leq k_0$$

$$\boldsymbol{\sigma}(\infty)=\boldsymbol{\sigma}_0-\frac{G}{G+H}(r_0-k_0)\frac{\mathbf{s}_0}{r_0} \tag{25.57}$$

otherwise

$$k(\infty)=k_0+\frac{H}{G+H}(r_0-k_0)$$

Note that these limiting values are independent of η and the time step size h.

Equation (25.57) yields a suitable 'return mapping' for the isotropic hardening rule. It is seen from (25.57) that the stress point and the yield surface meet at some intermediate point on the segment joining their initial values, the distances from these being proportional to G and H, respectively. A similar geometric interpretation can be derived for the return mapping corresponding to the kinematic hardening rule [15]. Note that the perfectly plastic case is recovered by setting $H=0$.

Finally, the unconditional stability of the plastic algorithm induced by this return mapping follows from that of the corresponding viscoplastic case. Consequently, the unconditional stability of the product algorithm (25.31) follows from the unconditional stability of the component algorithms $\mathbf{F}^{el}(h)$ and $\mathbf{F}^{pl}(h)$.

25.4 GLOBAL ALGORITHMS FOR THE FINITE DEFORMATION DYNAMIC PROBLEM

25.4.1 Introduction

As noted in Section 25.1, the development of global algorithms for the solution of the boundary value problem of linear momentum balance is outside the scope of the present paper. However, the global solution algorithm and the algorithm for the integration of the rate constitutive equations will be interdependent and cannot be entirely separated in any reasonably complete discussion of either algorithm. Accordingly, a brief description of global algorithms is presented, including 'implicit' algorithms which employ an elastoplastic (or elasto-viscoplastic) tangent modulus tensor and a product algorithm based on an operator split of the momentum balance equation. It is noted that both these global algorithms suffer from computational disadvantages that make their use very costly in general. An alternative global algorithm is therefore proposed.

25.4.2 Boundary value problem of linear momentum balance

The boundary value problem of linear momentum balance for elastoplastic or elasto-viscoplastic materials introduced in Section 25.2 is summarized as

$$
\begin{aligned}
\dot{\boldsymbol{\phi}}_t &= \mathbf{v}\circ\boldsymbol{\phi}_t \\
\rho\dot{\mathbf{v}} &= \nabla\cdot\boldsymbol{\sigma} + \rho\mathbf{b} \\
\mathring{\boldsymbol{\sigma}} &= \mathbf{a}:(\mathbf{d} - \mathbf{T}(\boldsymbol{\sigma}, \mathbf{q}_1, \ldots, \mathbf{q}_{niv})) \\
\breve{\mathbf{q}}_\alpha &= \mathbf{f}_\alpha(\boldsymbol{\sigma}, \mathbf{q}_1, \ldots, \mathbf{q}_{niv}) \quad \alpha = 1, \ldots, niv \\
\boldsymbol{\sigma}\cdot\mathbf{n} &= \bar{\mathbf{t}} \qquad \mathbf{x}\in\partial_\sigma\boldsymbol{\phi}_t(B) \\
\boldsymbol{\phi} &= \bar{\boldsymbol{\phi}} \qquad \mathbf{x}\in\partial_u\boldsymbol{\phi}_t(B)
\end{aligned}
\tag{25.58}
$$

A variety of techniques has been proposed for the solution of (25.58) (although many of these have been based on the equivalent rate of momentum balance problem derived from time differentiation of (25.58b) together with appropriate traction and kinematic rate boundary conditions). Implicit methods which employ an elastoplastic 'tangent' modulus, e.g. [22, 24, 25], suffer from numerical difficulties associated with enforcing the consistency condition of plasticity which requires that the stress trajectory be confined to the elastic domain. Frequently, projection techniques have been introduced to restore consistency. Such methods also require elaborate schemes for making the transition from the elastic to the plastic regimes and frequently require truncating or discarding of time steps. Although these methods are potentially quite accurate, they can be very costly in practice.

The limitations of the tangent modulus methods motivated a search for alternative methods of solution. One such alternative method, originally proposed by Mendelson [30] for the case of infinitesimal plasticity, employs the concept of a 'return mapping' algorithm which automatically ensures satisfaction of the plastic consistency condition. The accuracy of this method has also been considered for a limited class of material models [27 to 29].

In order to be convergent, a numerical solution scheme must satisfy the requirements of consistency and numerical stability. A formal study of the consistency and numerical stability properties of global solution schemes arising from the use of return mapping algorithms has recently been considered within the framework of operator split methods [15, 16, 26]. It is noted in [16] that the field equations (25.58) exhibit an additive decomposition into an 'elastic' part (which defines an elastodynamic boundary value problem in which only the motion and stress tensor are involved) and a 'plastic' part (which leaves the configuration unchanges and defines a pointwise relaxation process for the stress tensor and internal variables). This suggests using the product formula techniques discussed in Section 25.3.2 to construct a solution algorithm for (25.58).

In the present context, a product algorithm relative to the elastoplastic additive decomposition of the equations of motion takes the following meaning [16]. Consider two algorithms $\mathbf{G}^{el}(h)$ and $\mathbf{G}^{pl}(h)$ which are consistent in the sense of (25.27) with the 'elastic' and 'plastic' parts of (25.58) respectively and which are unconditionally stable in the sense of (25.28). Then an unconditionally stable algorithm $\mathbf{G}(h)$ consistent with the full equations of motion (25.58) can be obtained by means of the general product formula (25.30) such that

$$\mathbf{G}(h) = \mathbf{G}^{pl}(h)\mathbf{G}^{pl}(h) \tag{25.59}$$

The product formula (25.59) simply states that a solution algorithm for the elastoplastic problem can be obtained by solving for each time step an elastic dynamic problem first, and then applying to the solution vector so obtained a plastic algorithm operating on the stresses and internal variables bringing in the effect of the plastic constitutive equations. It is interesting to note that all the boundary value aspects of the elastoplastic dynamic problem are included in the 'elastic' part of (25.58) and are taken care of by the elastic algorithm $\mathbf{G}^{el}(h)$ [16].

In practice, the above equations are solved by means of some spatial discretization technique such as the finite element method. In this case, the 'plastic' part of the equations of motion will correspond to a set of relaxation equations expressed at the integration points within the elements and the plastic algorithm $\mathbf{G}^{pl}(h)$ is accordingly applied integration point by integration point. The construction of the global algorithm given by (25.59) is discussed in detail in [16].

Although the unconditionally stable global product algorithm described above is consistent with the field equations it is observed in [16] that the accuracy of the global product algorithm deteriorates as the size of the time step is increased. The effectiveness of the operator split method depends strongly on the error introduced by the splitting. Experience in a number of areas of application indicate that the splitting error dominates beyond certain critical time step sizes [14 to 16]. For some problems these critical time step sizes are small enough to eliminate the possible benefits of the method.

The fully implicit methods employing the elastoplastic tangent modulus are potentially more accurate than the global product algorithm described above but suffer from the difficulties also noted above. In this paper an implicit global solution scheme based on the elastoplastic tangent modulus is proposed in combination with a product algorithm applied to the integration of the constitutive equations. It is demonstrated below that such a scheme eliminates the difficulties associated with the implicit methods and results in an accurate and efficient solution procedure.

25.4.3 Alternative implicit global algorithm

The global algorithm is an iterative scheme based upon the consistent linearization of a weak form of the boundary value problem of linear momentum balance (25.58). The construction of the weak form as well as the mathematical ideas necessary for the consistent linearization of the weak form will not be discussed in detail here. Similar notions have been considered in [16, 20]. The construction of this implicit global algorithm entails the following steps:

(*i*) *Spatially discretized weak form.* A weak form of (25.58) is expressed by

$$\int_{\varphi_t(B)} [\rho(\dot{\mathbf{v}} - \mathbf{b})\cdot\boldsymbol{\eta} + \boldsymbol{\sigma}:\nabla\boldsymbol{\eta}]\,\mathrm{d}v = \int_{\partial_\alpha \Phi_t(B)} \bar{\mathbf{t}}\cdot\boldsymbol{\eta}\,\mathrm{d}a \tag{25.60}$$

for all weighting functions η which satisfy the homogeneous boundary conditions on $\partial_u\boldsymbol{\phi}_t(B)$. Spatial discretization of (25.60) can be accomplished using the finite element method in which a set of global finite element interpolation functions is introduced for the nodal values of the motion $\boldsymbol{\phi}_t$ and the weighting functions $\boldsymbol{\eta}$. In the present 'displacement' method the stress is not independently interpolated but is thought of as being a function of the motion. In this case the spatially discretized weak form (25.60) can be expressed as

$$\mathbf{M}\cdot\dot{\mathbf{v}} + \mathbf{P}(\boldsymbol{\sigma}, t) = \mathbf{F}(t) \tag{25.61}$$

where $\mathbf{M}$ is the mass matrix, $\mathbf{v}$ is the vector of nodal velocities, $\mathbf{P}(\boldsymbol{\sigma}, t)$ is the 'internal force' vector, and $\mathbf{F}(t)$ is the global force vector.

(*ii*) *Temporal integration of the weak form.* The spatially discretized weak form (25.61) can be numerically integrated in time by the application of an algorithm such as the implicit Newmark algorithm defined by

$$\mathbf{M}\cdot\dot{\mathbf{v}}_{n+1} + \mathbf{P}_{n+1}(\boldsymbol{\sigma}_{n+1}) = \mathbf{F}_{n+1} \tag{25.62}$$

$$\boldsymbol{\phi}_{n+1} = \boldsymbol{\phi}_n + h\mathbf{v}_n + h^2[(\tfrac{1}{2} - \beta)\dot{\mathbf{v}}_n + \beta\dot{\mathbf{v}}_{n+1}] \tag{26.63}$$

$$\mathbf{v}_{n+1} = \mathbf{v}_n + h[(1-\gamma)\dot{\mathbf{v}}_n + \gamma\dot{\mathbf{v}}_{n+1}] \tag{25.64}$$

where subscripts n and $n+1$ denote variables evaluated at the nth and $n+1$th time steps, $h = t_{n+1} - t_n$ is the time step size and β and γ are the Newmark parameters. Substituting (25.63) and (25.64) into (25.62) results in

$$G(\boldsymbol{\phi}_{n+1}) \equiv \frac{1}{\beta h^2}\mathbf{M}\cdot\boldsymbol{\phi}_{n+1} + \mathbf{P}_{n+1}(\boldsymbol{\sigma}_{n+1}) - \hat{\mathbf{P}}_{n+1} = 0 \tag{25.65}$$

where $\hat{\mathbf{P}}_{n+1} = \mathbf{F}_{n+1} - \mathbf{M}\cdot\left[\left(1 - \frac{1}{2\beta}\right)\dot{\mathbf{v}}_n - \frac{1}{\beta h}\mathbf{v}_n - \frac{1}{\beta h^2}\boldsymbol{\phi}_n\right]$. For this formulation $\beta \neq 0$.

(*iii*) *Linearization.* Formally, it is found that a consistent linearization procedure may be based on Taylor's formula for C^1 functions [9, 31]. Without entering into mathematical detail, the linearization of (25.65) about the motion $\boldsymbol{\phi}_{n+1}$ can be expressed as

$$L[G(\boldsymbol{\phi}_{n+1})] = G(\boldsymbol{\phi}_{n+1}) + \mathbf{D}G(\boldsymbol{\phi}_{n+1})\cdot\mathbf{u}_{n+1} = 0 \tag{25.66}$$

where

$$\mathbf{D}G(\boldsymbol{\phi}_{n+1})\cdot\mathbf{u}_{n+1} = \frac{1}{\beta h^2}\mathbf{M}\cdot\mathbf{u}_{n+1} + \mathbf{D}\mathbf{P}_{n+1}\cdot\mathbf{u}_{n+1}. \tag{25.67}$$

is the directional derivative of $G(\boldsymbol{\phi}_{n+1})$ in the direction of the incremental motion $\mathbf{u}_{n+1}$ [16]. Similarly, in (25.67) $\mathbf{DP}_{n+1}\cdot\mathbf{u}_{n+1}$ is the directional derivative of the internal force vector. This last directional derivative involves finding the directional derivative of the spatial stress tensor. This can be accomplished by using results from differential geometry. The consistent approximation of a certain derivative appearing in the development allows the introduction of the rate constitutive equations for the stress given by (25.58c). Details may be found in [16].

Combining the above concepts with the product algorithm (25.31) for the integration of the rate constitutive equations, an iterative Newton–Raphson solution scheme can be expressed as

$$\begin{aligned}
&\text{(i)}\ \ \mathbf{D}G(\boldsymbol{\phi}^i_{n+1})\cdot\mathbf{u}^i_{n+1} \equiv \mathbf{K}^i_{n+1}\cdot\mathbf{u}^i_{n+1}\\
&\text{(ii)}\ \ \mathbf{u}^i_{n+1} = -(\mathbf{K}^i_{n+1})^{-1}\cdot G(\boldsymbol{\phi}^i_{n+1})\\
&\text{(iii)}\ \ \boldsymbol{\phi}^{i+1}_{n+1} = \boldsymbol{\phi}^i_{n+1} + \mathbf{u}^i_{n+1}\\
&\text{(iv)}\ \ \begin{Bmatrix}\boldsymbol{\sigma}\\ \mathbf{q}_\alpha\end{Bmatrix}^{i+1}_{n+1} = \mathbf{F}^{\text{pl}}(h)\mathbf{F}^{\text{el}}(h)\begin{Bmatrix}\boldsymbol{\sigma}\\ \mathbf{q}_\alpha\end{Bmatrix}_n\\
&\text{(v)}\ \ i \leftarrow i+1, \quad \text{go to (i)}
\end{aligned} \tag{25.68}$$

where the superscript i is the iteration counter within the time step $h = t_{n+1} - t_n$. The quantity with no superscript appearing on the right-hand side of (iv) implies that the converged values at the indicated time are to be used.

The solution procedure is illustrated by means of an example in Figure 25.3. The problem concerns an elastoplastic material with isotropic hardening subjected to a cycle of uniaxial tension while in a state of plane strain. Figure 25.3, which depicts axial stress versus axial stretch, illustrates two solutions. One of these corresponds to the global product algorithm (25.59) reported in [16] and discussed in Section 25.4.2 and the other corresponds to a solution obtained from the algorithm given by (25.68). The solution obtained by use of (25.59) employed 400 time steps. In contrast, the solution obtained by use of (25.68) employed a total of four time steps. With reference to the algorithm (25.68), the following comments apply:

(a) In the first iteration of each time step ($i = 1$) the 'tangent modulus matrix' $\mathbf{K}^1$ is assumed to be only elastic and the product algorithm (iv) employs only the elastic algorithm $\mathbf{F}^{\text{el}}(h)$.

(b) If the resulting stress lies inside the elastic domain the iterative procedure progresses by ignoring the plasticity of the material (although in each subsequent iteration the stresses are always checked to ensure that their

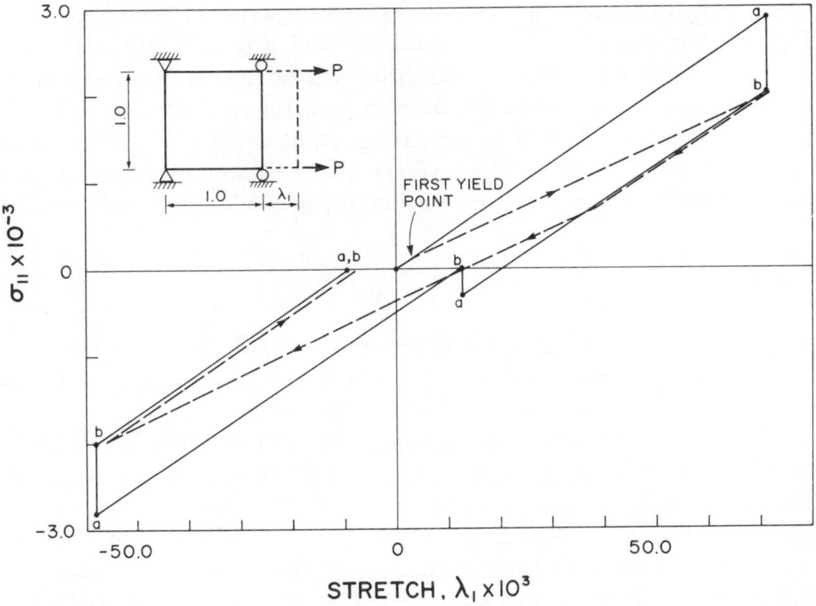

Figure 25.3 Cycle of axial stress versus stretch for plane strain extension of an elastoplastic material with von Mises' yield criterion and isotropic hardening. The dashed lines correspond to the global product algorithm (25.59) with 400 time steps. The solid lines correspond to the algorithm (25.68) with four time steps. Point *a* represents the 'elastic' solution and point *b* the final solution for each time step

trajectory lies inside the elastic domain). However, if the stress lies outside the elastic domain the tangent modulus matrix $\mathbf{K}^i$ is modified to include the plastic constitutive equations and the full product algorithm (iv) is employed.

In this way the transition from elastic to plastic loading is handled automatically. Figure 25.3 shows the results of step (iv) in the last iteration of each of the four time steps. The points marked *a* and *b* correspond to $\mathbf{F}^{\text{el}}(h)\begin{Bmatrix} \boldsymbol{\sigma} \\ \mathbf{q}_\alpha \end{Bmatrix}$ and $\mathbf{F}^{\text{pl}}(h)\mathbf{F}^{\text{el}}(h)\begin{Bmatrix} \boldsymbol{\sigma} \\ \mathbf{q}_\alpha \end{Bmatrix}$ respectively. It is evident that the algorithm (25.68) employing four time steps is close to the solution given by (25.59) which employed 400 time steps. The solution obtained from (25.59) with four time steps is extremely inaccurate, with stresses in error by almost 50 per cent (this solution is not shown in Figure 25.3).

25.5 IMPROVING ACCURACY BY TIME STEP REFINEMENT

The operator split of (25.23) into (25.24) and (25.25) will introduce an error into the associated product formula (25.31) which is step (iv) in (25.68). For viscoplasticity (plasticity) the spatial stress tensor plays a dominant role in the definition of the tangent modulus matrix **K** and it is thus important for the convergence rate of the Newton–Raphson scheme (25.68) that this tensor be evaluated as accurately as possible from the integration of the rate constitutive equations. This suggests that by improving the accuracy of (25.31) the overall scheme will enjoy an improved convergence rate without changing the global time size. This may be accomplished by modifying (25.31) to the following form

$$\mathbf{F}(h) = \sum_{i=1}^{N} \mathbf{F}_i^{\mathrm{pl}}\left(\frac{h}{N}\right)\mathbf{F}_i^{\mathrm{pl}}\left(\frac{h}{N}\right) \tag{25.69}$$

The algorithm defined by (25.69) involves subdividing the time step into N parts and applying the product algorithm N times. The algorithms $\mathbf{F}_i^{\mathrm{el}}$ and $\mathbf{F}_i^{\mathrm{pl}}$ have the form described in Sections 25.3.3.1 and 25.3.3.2 respectively.

This algorithm will satisfy the requirements of consistency in the sense of (25.27) and stability in the sense of (25.28). However, the requirement of incremental objectivity places a restriction on the interpretation of $\mathbf{F}^{\mathrm{el}}(h/N)$. Recalling equations (25.40)–(25.42), which are the detailed form of $\mathbf{F}^{\mathrm{el}}(h)$, it may be shown that incremental objectivity is satisfied in (25.69) only if $\mathbf{d}_{n+\alpha}$ appearing in (25.42) is replaced by $\mathbf{d}_{n+\alpha}/N$. Thus the deformation gradient over

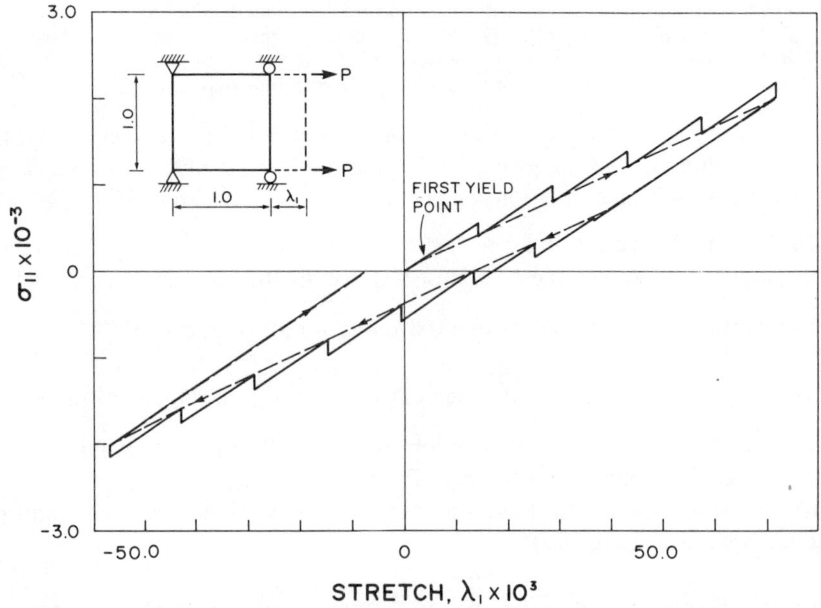

Figure 25.4 Cycle of axial extension for material described in Figure 25.3. The dashed lines correspond to the global product algorithm (25.59) with 400 time steps. The solid lines correspond to algorithm (25.68)–(25.69) with four time steps and $N = 5$ in (25.69)

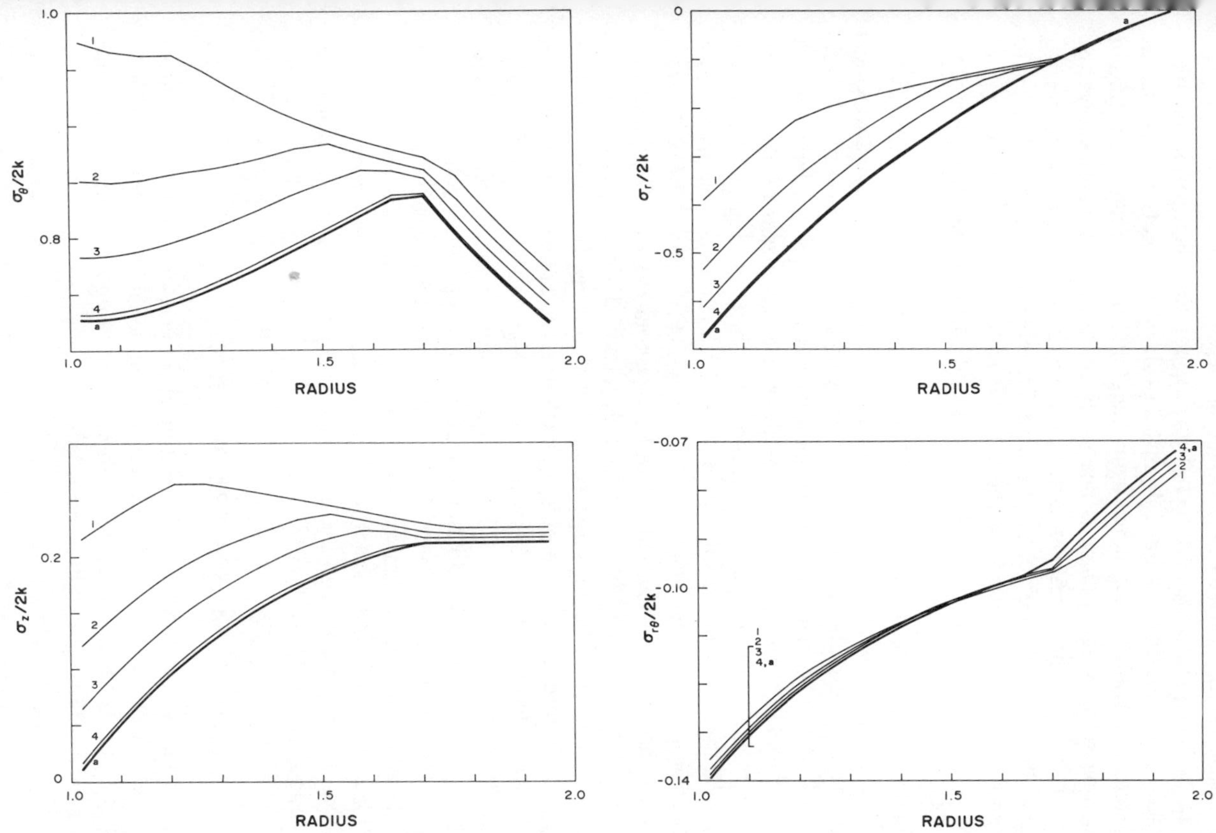

Figure 25.5 Stress distribution through the wall thickness of an infinitely long internally pressurized thick walled cylinder. The material is isotropic hardening plastically with a plastic hardening modulus which is 10 per cent of the Young's modulus. Curves 1–4 correspond to algorithm (25.59) and were obtained with 2, 4, 8, and 100 time steps respectively. Curve a corresponds to the algorithm (25.68)–(25.69) with $N = 2$ and was obtained with 2, 4, and 8 time steps (the solutions were coincident)

the interval h is subdivided rather than the deformation itself. The resulting algorithm appears to display excellent accuracy for N in the range of 2 to 5 for numerical examples which have been considered.

As an example, the cycle of uniaxial loading for the plane strain problem discussed above is again considered. Figure 25.4 shows two plots of stress versus axial stretch. The solution obtained from the global product algorithm (25.59) in 400 time steps is again shown. The other curve in Figure 25.4 shows the results for the algorithm (25.68) modified by using (25.69) for step (iv). As before, four time steps are taken globally but N in (25.69) is selected as 5. The curve again shows the stress plotted for the solution at the last iteration in each of the four time steps. The increase in accuracy is evident by comparing Figures 25.3 and 25.4.

A further example of the accuracy of the algorithm given by (25.68)–(25.69) compared to the global product algorithm (25.59) is shown in Figure 25.5 which depicts the stress distribution in an infinitely long internally pressurized thick walled cylinder. The material is elastoplastic with isotropic hardening and a von Mises' yield criterion. The tube was discretized by means of plane strain quadrilateral elements with 16 elements across the thickness. Figure 25.5 depicts the stress components in the tube for an internal pressure corresponding to a plastification of 70 per cent of the wall thickness. Five curves for each stress component are plotted. The curves 1–4 correspond to the global product algorithm (25.59) and were obtained with 2, 4, 8, and 100 time steps respectively. These curve illustrate the convergence of this algorithm. However, the curve denoted a in Figure 25.5 corresponds, for each stress component, to the solution from the algorithm (25.68)–(25.69) with 2, 4, and 8 global time steps. The value of N in (25.69) was taken as 2. These solutions are coincident and appear to be correct by comparison with the global product algorithm solutions. The accuracy of the present algorithm with two global time steps exceeds the accuracy of the global product algorithm with 100 time steps. Solutions for this problem with $N \geq 2$ did not result in a marked improvement of the convergence rate of the global algorithm although for problems with greater stress gradients this conclusion may not apply.

ACKNOWLEDGEMENTS

We would like to acknowledge the continued interest and encouragement of Dr G. L. Goudreau of the Lawrence Livermore National Laboratory and Dr T. Khalil of General Motors Research Laboratories. Support for this work from these agencies is also gratefully acknowledged.

REFERENCES

1. B. D. Coleman and M. E. Gurtin, 'Thermodynamics with internal state variables', *J. Chem. Phys.*, **47**, 597–613 (1967).

2. J. Lubliner, 'On the thermodynamic foundations of non-linear solid mechanics', *Int. J. Non-linear Mechanics*, **7**, 237–254 (1972).
3. J. Lubliner, 'On the structure of the rate equations of materials with internal variables', *Acta Mech.*, **17**, 109–119 (1973).
4. J. Mandel, 'Thermodynamics and plasticity', in *Foundations of Continuum Thermodynamics*, (Eds. J. J. Delgado Domingos, M. N. R. Nina, and J. H. Whitelaw), pp. 283–304, John Wiley and Sons, 1973.
5. C. Truesdell, 'The simplest rate theory of pure elasticity', *Comm. Pure Appl. Math.*, **8**, 123–132 (1955).
6. J. G. Oldroyd, 'On the formulation of rheological equations of state', *Proc. Roy. Soc.*, **A200**, 523–541 (1950).
7. B. A. Cotter and R. S. Rivlin 'Tensors associated with time-dependent stress', *Quart. Appl. Math.*, **13**, 177–182 (1955).
8. D. Durban and M. Baruch, 'Incremental behavior of an elasto-plastic continuum', *TAE Report No.* 193, Dept. of Aeronautical Eng., Israel Institute of Technology, Haifa, Israel, Feb. 1975.
9. J. E. Marsden and T. J. R. Hughes, 'Topics in mathematical foundations of elasticity', in *Nonlinear Analysis and Mechanics*, Heriot-Watt Symposium (Ed. R. J. Knops), Vol. II, Pitman Publishing Co., 1978.
10. E. H. Lee and D. T. Liu, 'Finite strain elastic-plastic theory particularly for plane wave analysis', J. Appl. Phys. **38**, 19–27 (1967).
11. S. Nemat-Nasser, 'Decomposition of strain measures and their rates in finite deformation elastoplasticity', *Int. J. Solids Struct.*, **15**, 155–166 (1979).
12. A. E. Green and P. M. Naghdi, 'Some remarks on elastic-plastic deformation at finite strain', *Int. J. Engng. Sci.*, **9**, 1219–1229 (1971).
13. T. J. R. Hughes, I. Levit, and J. Winget, 'Unconditionally stable element-by-element implicit algorithms for heat conduction analysis', *J. Engng. Mechs. Div. of ASCE* (to appear).
14. M. Ortiz, P. M. Pinsky, and R. L. Taylor, 'Unconditionally stable element-by-element algorithms for dynamic problems', *Computer Methods in Applied Mechanics and Engineering*, Vol. 36, pp. 223–239 (1983).
15. M. Ortiz, P. M. Pinsky, and R. L. Taylor, 'Operator split methods for the numerical solution of the elastoplastic dynamic problem', *Computer Methods in Applied Mechanics and Engineering*, Vol. 39, pp. 137–157 (1983).
16. P. M. Pinsky M. Ortiz, and R. L. Taylor, 'Operator split methods in the numerical solution of the finite deformation elastoplastic dynamic problem', *Report No. UCB/SESM*-82/03, Department of Civil Engineering, University of California, Berkeley, California, May, 1982.
17. G. W. Gear, *Numerical Initial Value Problems in Ordinary Differential Equations*, Prentice-Hall, 1971.
18. T. J. R. Hughes and J. Winget, 'Finite rotation effects in numerical integration of rate constitutive equations arising in large-deformation analysis', *Int. J. Num. Meth. Eng.*, **15**, (12) (1980).
19. P. M. Pinsky, M. Ortiz, and K. S. Pister, 'Numerical Integration of Rate Constitutive Equations in Finite Deformation Analysis', Computer Methods in Applied Mechanics and Engineering, Vol. 40, pp. 137–158 (1983).
20. P. M. Pinsky, 'A numerical formulation for the finite deformation problem of solids with rate-independent constitutive equations', *Report No. UCB/SESM*-81/07, Department of Civil Engineering, University of California, Berkeley, California, December, 1981.
21. J. Lubliner, 'A simple theory of plasticity', *Int. J. Solids Struct.*, **10**, 313–319 (1974).
22. R. M. McMeeking and J. R. Rice, 'Finite element formulations for problems of large elastic-plastic deformation', *Int. J. Solids Struct.*, **11**, 601 (1975).

23. B. D. Coleman and W. Noll, 'The thermodynamics of elastic materials with heat conduction and viscosity', *Arch. Ration. Mech. Anal.*, **13**, 167–178 (1963).
24. R. D. Krieg, and S. W. Key, 'Implementation of a time-independent plasticity theory into structural computer programs', in *Constitutive Equations in Viscoplasticity: Computational and Engineering Aspects*, AMD, (Eds. J. A. Stricklin and K. J. Saczalski) Vol. 20, ASME, 125 (1976).
25. J. O. Hallquist, 'NIKE 2D: an implicit, finite-deformation, finite element code for analyzing the static and dynamic response of two-dimensional solids', *Lawrence Livermore National Laboratory Report UCRL*-52678, University of California, Livermore, March 3, 1979.
26. M. Ortiz and P. M. Pinsky, 'Global analysis methods for the solution of the elastoplastic and viscoplastic dynamic problems', *Report No. UCB/SESM*-81/08, Department of Civil Engineering, University of California, Berkeley, California, December, 1981.
27. R. D. Krieg and D. B. Krieg, 'Accuracies of numerical solution methods for the elastic-perfectly plastic model', *ASME J. Pressure Vessel Tech.*, **99**, 510–515 (1977).
28. H. L. Schreyer, R. F. Kulak, and J. M. Kramer, 'Accurate numerical solutions for elastic-plastic models', *ASME J. Pressure Vessel Tech.*, **101**, 226–234 (1979).
29. J. M. Santiago, 'On the accuracy of flow rule approximations used in structural and solid response computer programmes', in *Proceedings of the* 1981 *Army Numerical Analysis And Computers Conference*, ARO Report 81–3 (1981).
30. A. Mendelson, *Plasticity: Theory and Application*, MacMillan, 1968.
31. T. J. R. Hughes and K. S. Pister, 'Consistent linearization in mechanics of solids and structures', *Computers and Structures*, **8**, 391–397 (1978).

Mechanics of Engineering Materials
Edited by C. S. Desai and R. H. Gallagher

Chapter 26

Non-linear Transient Phenomena in Soil Media

J. H. Prevost

26.1 INTRODUCTION

Many practical geotechnical engineering problems involve quasistatic or dynamic loading of saturated soil media. Fluid saturation of an otherwise inviscid porous soil skeleton introduces a time dependence into the response to applied loads. When the rate of loading is much smaller than the pore fluid diffusion rate, the pore fluid pressure remains constant in each material element and the response is said to be 'fully drained'. In that case, the steady state pore fluid pressures only depend on the hydraulic conditions and are independent of the soil skeleton response to external loads, and a single phase description of soil media behaviour is therefore adequate in that case. Similarly, a single phase description of soil media behaviour is also adequate when the loading rate is much higher than the diffusion rate. In that case no flow of the pore fluid can take place, the fluid mass content in each material element remains constant and the fluid follows the motion of the soil skeleton. The response is then said to be 'fully undrained'. However, in intermediate cases in which some flow can take place, there is an interaction between the soil skeleton strains and the pore fluid flow. The solution of these problems requires that saturated soil media be analysed by incorporating the effect of the transient flow of the pore fluid through the soil skeleton voids, and therefore requires that a multiphase continuum formulation and a solution technique for the resulting coupled field equations be available for soil media. The need for such an analytical tool has acquired considerable importance in recent years due to the increased concern with the dynamic behaviour of saturated soil deposits and associated liquefaction of saturated sand deposits under seismic loading conditions. However, despite the extensive literature published in soil dynamics (see e.g., [21, 26, 39] for extensive references) no general analytical technique capable of accounting for all present non-linear effects has been proposed yet, although attempts at presenting a suitable framework have been reported in [8, 9, 20, 36].

It is the purpose of this chapter to review the main features and capabilities of a general analytical model for saturated soil media, developed by the author and reported upon in a number of papers (see e.g., [28, 32]). In that approach, the saturated soil is viewed as a two-phase system consisting of a solid and a fluid phase. Each phase is regarded as a continuum, and each follows its own motion. The model is capable of accounting for both static and dynamic interactions between the two phases.

For completeness, a brief review of the coupled field equations based on mixtures' theories [3, 7, 10] is first presented. The solid and fluid stresses are related to the widely used 'effective stresses' [38] and pore fluid pressures used in soil mechanics, and the constitutive assumptions are stated explicitly. The finite element spatial discretization of the coupled field equations, and the time integration of the resulting non-linear semidiscrete finite element equations are next described. An implicit/explicit predictor/multi-corrector scheme developed by Hughes and co-workers [14, 15] is used for that purpose. The procedure allows for a convenient selection of implicit and explicit elements, and/or for an implicit/explicit splitting of the various operators appearing in the differential equations. It is shown that the procedure is extremely effective in dealing with saturated porous soil media models since (1) the fluid phase is always much stiffer than the solid phase, and therefore would impose too stringent time step restrictions for accuracy if dealt with explicitly; and (2) the solid phase exhibits non-linear behaviour and would impose too many stiffness assemblies and factorizations if dealt with purely implicitly. Numerical results are presented.

As for notation, bold face letters denote vectors, second and fourth-order tensors in three-dimensions [6].

26.2 GOVERNING EQUATIONS

The saturated porous soil medium is viewed herein as a mixture consisting of two phases, namely a solid and a fluid phase. Each phase is regarded as a continuum, and each follows its own motion. The mixture is assumed chemically inert, and to consist of non-polar constituents. It is further assumed that both the solid grains and the fluid are incompressible, and the fluid has no *average* shear viscosity. The balance equations for each constituent then write [28]

Balance of mass

$$\dot{n}^{\mathrm{W}} + (1 - n^{\mathrm{W}})\,\mathrm{div}\,\mathbf{v}^{\mathrm{S}} = 0 \tag{26.1}$$

$$\mathrm{div}\,(n^{\mathrm{W}}(\mathbf{v}^{\mathrm{W}} - \mathbf{v}^{\mathrm{S}})] + \mathrm{div}\,\mathbf{v}^{\mathrm{S}} = 0 \tag{26.2}$$

Balance of momentum

$$\mathrm{div}\boldsymbol{\sigma}'^{\mathrm{S}} - n^{\mathrm{S}}\,\mathrm{grad}\,p_{\mathrm{W}} + \boldsymbol{\xi}\cdot(\mathbf{v}^{\mathrm{S}} - \mathbf{v}^{\mathrm{W}}) + \rho^{\mathrm{S}}\mathbf{b} = \rho^{\mathrm{S}}\mathbf{a}^{\mathrm{S}} \tag{26.3}$$

$$\rho^{\mathrm{W}}(\mathbf{v}^{\mathrm{S}} - \mathbf{v}^{\mathrm{W}})\cdot\mathrm{grad}\,\mathbf{v}^{\mathrm{W}} - n^{\mathrm{W}}\,\mathrm{grad}\,p_{\mathrm{W}} - \boldsymbol{\xi}\cdot(\mathbf{v}^{\mathrm{S}} - \mathbf{v}^{\mathrm{W}}) + \rho^{\mathrm{W}}\mathbf{b} = \rho^{\mathrm{W}}\dot{\mathbf{v}}^{\mathrm{W}} \tag{26.4}$$

when the movement of the solid phase is used as the reference motion. Equation (26.2) is often referred to as the 'storage equation'. In equations (26.1–26.4) a superimposed dot ($\dot{}$) is used to indicate the material derivative following the motion of the solid phase; $\rho^\alpha = n^\alpha \rho_\alpha$ = macroscopic average mass density function, ρ_α = miscroscopic mass density function; n^α = fraction of elemental volume dV occupied by the α-phase (i.e., $n^\alpha = dV^\alpha/dV$). Clearly, n^α is constrained by $\Sigma_\alpha n^\alpha = n^S + n^W = 1$ where the superscript S and W refer to the solid and fluid phase, respectively; n^W = porosity; $\mathbf{v}^\alpha$ and $\mathbf{a}^\alpha$ = (spatial) velocity and acceleration of α-phase; $\mathbf{b}$ = body force per unit mass; σ'^S = (Cauchy) effective stress [38] tensor; p_W = pore fluid pressure; $\boldsymbol{\xi}\cdot(\mathbf{v}^S - \mathbf{v}^W)$ = Stokes' drag (see e.g., [3]). When inertia and convective terms are neglected, equation (26.4) reduces to Darcy's law as

$$n^W(\mathbf{v}^W - \mathbf{v}^S) = (n^W)^2 \boldsymbol{\xi}^{-1}\cdot(\text{grad } p_W - \rho_W \mathbf{b}) \tag{26.5}$$

and thus $\mathbf{k} = (n^W)^2 \rho_W \boldsymbol{\xi}^{-1}$ = permeability tensor (symmetric, positive definite).

26.2.1 Constitutive assumptions

The following constitutive equation is assumed to describe the behaviour of the porous solid skeleton:

$$\mathbf{k}^S : \mathbf{v}^S_{(\ )} = \begin{cases} \dot{\boldsymbol{\sigma}}'^S & \text{small deformations} \\ \overset{\nabla}{\boldsymbol{\sigma}}'^S + \boldsymbol{\sigma}'^S \operatorname{div} \mathbf{v}^S & \text{finite deformations} \end{cases} \tag{26.6}$$

where $\mathbf{v}^S_{(\ )}$ = symmetric part of the solid velocity gradient; $\overset{\nabla}{\boldsymbol{\sigma}}'^S$ = Jaumann [19] rate of effective Cauchy stress, defined by

$$\boldsymbol{\sigma}'^S = \dot{\boldsymbol{\sigma}}'^S - \mathbf{w}^S\cdot\boldsymbol{\sigma}'^S + \boldsymbol{\sigma}'^S\cdot\mathbf{w}^S \tag{26.7}$$

where $\mathbf{w}^S$ = skew-symmetric part of the solid velocity gradient. In equation (26.6) $\mathbf{k}^S$ is an (objective) tensor valued function of, possibly $\boldsymbol{\sigma}'^S$ and the solid deformation gradient. Many non-linear material models of interest can be put in the above form (e.g., all non-linear elastic materials, and many elastoplastic materials). The finite deformation of the constitutive equations above was first proposed by Hill [12] within the context of the plasticity theory. Appropriate expressions for the effective modulus tensor $\mathbf{k}^S$ for soil media are given in [27]. For a linear elastic porous skeleton,

$$k^S_{abcd} = \lambda^S \delta_{ab}\delta_{cd} + \mu^S(\delta_{ac}\delta_{bd} + \delta_{ad}\delta_{bc}) \tag{26.8}$$

where λ^S, μ^S = effective Lamé's moduli, δ_{ab} = Kronecker delta.

Constitutive equations of the above type lead to the following definition for moduli to appear in the *linearized* variational equations

$$\mathbf{k} = \begin{cases} \mathbf{k}^S & \text{small deformations} \\ \mathbf{k}^S + \mathbf{k}^G & \text{finite deformations} \end{cases} \tag{26.9}$$

where $\mathbf{k}^G$ is a tensor arising from geometry, namely,

$$k^G_{abcd} = \tfrac{1}{2}[\sigma'^S_{bd}\delta_{ac} - \sigma'^S_{ac}\delta_{bc} - \sigma'^S_{ac}\delta_{bd} - \sigma'^S_{bc}\delta_{ad}] \tag{26.10}$$

It may be seen that $\mathbf{k}$ possesses the major symmetry which leads to a symmetric tangent stiffness matrix, if and only if $\mathbf{k}^S$ possesses the major symmetry.

26.2.2 Dynamics, wave propagation analysis

The initial boundary value problem consists of the finding the function $\mathbf{u}^S$, $\mathbf{v}^W$, and p_W (function of position and time) satisfying equations (26.1)–(26.4) and (26.6) together with the continuity conditions [equations (26.1) and (26.2)], subject to appropriate initial and boundary conditions. In order to eliminate the pressure unknown and equation (26.2), a penalty-function formulation (17, 22) of the storage equation is used in which equation (26.2) is dropped, and the fluid-pressure is determined from

$$p_W = -\frac{\lambda^W}{n^W}[\operatorname{div}(n^W(\mathbf{v}^W - \mathbf{v}^S) + \operatorname{div}\mathbf{v}^S] \tag{26.11}$$

where $\lambda^W > 0$ is the penalty parameter ($\lambda^W \gg (\lambda^S + 2\mu^S)$).

26.2.3 Remarks

The present formulation follows directly from the general theory of mixtures (see e.g., [3, 7, 10]). It is interesting to remark that equations (26.2) and (26.3) are direct extensions of Biot's equations [2] in the non-linear regime, and upon linearization are equivalent to Biot's equation if the grain compressibility and the apparent mass are neglected. An important difference between the present theory and that of Biot lies in the constitutive assumptions adopted for the fluid phase. Whereas Biot allows for some compressibility of the fluid phase, the present theory treats it as an *incompressible* component. This assumption is valid for all practical situations of interest in soil dynamics [34]. Further, it simplifies the numerical treatment considerably by eliminating the fluid pressure unknown through a penalty type formulation of the storage equation.

26.2.4 Diffusion, consolidation analysis

In that case, inertia and convective terms are neglected and equations (26.1)–(26.4) are combined to yield

$$\operatorname{div}\boldsymbol{\sigma} + \rho\mathbf{b} = 0 \tag{26.12}$$

$$\operatorname{div}\mathbf{v}^S - \operatorname{div}\left[\frac{1}{\rho_W}\mathbf{k}\cdot(\operatorname{grad} p_W - \rho_W\mathbf{b})\right] = 0 \tag{26.13}$$

where $\boldsymbol{\sigma} = \boldsymbol{\sigma}'^{S} - p_W \boldsymbol{\delta}$ = total stress tensor; $\rho = \rho^S + \rho^W = (1 - n^W)\rho_S + n^W \rho_W$ = total mass density function.

The initial boundary value problem consists of finding the functions $\mathbf{u}^S$ and p_W (function of position and time) satisfying equations (26.6). (26.12), and (26.13) subject to appropriate initial and boundary conditions. Numerical treatments of this simplified case have been reported in [30, 33].

26.3 WEAK FORM—SEMIDISCRETE FINITE ELEMENT EQUATIONS

The weak formulation associated with the initial boundary value problems stated in the previous section, is obtained by proceeding along standard lines (see e.g., [1, 40]). The associated matrix problems are obtained by discretizing the domain occupied by the porous medium into non-overlapping finite elements. Associated with this discretization are nodal points at which shape functions are prescribed. The shape functions associated with node A are denoted N^A in the following, and $N^A(\mathbf{x}^B) = \delta_{AB}$, in which $\mathbf{x}^B$ denotes the position vector of node B, δ_{AB} = Kronecker delta. The solution of the Galerkin counterpart of the weak formulation is then expressed in terms of the shape functions and give rise to the following semidiscrete matrix system of equations from (equations (26.2) and (26.3)).

$$\begin{pmatrix} \mathbf{M}^S & \mathbf{0} \\ \mathbf{0} & \mathbf{M}^W \end{pmatrix} \begin{pmatrix} \mathbf{a}^S \\ \mathbf{a}^W \end{pmatrix} + \begin{pmatrix} \boldsymbol{\Xi} + \mathbf{C}^{SS} & -\boldsymbol{\Xi} + \mathbf{C}^{SW} \\ -\boldsymbol{\Xi} + \mathbf{C}^{WS} & \boldsymbol{\Xi} + \mathbf{C}^{WW} \end{pmatrix} \begin{pmatrix} \mathbf{v}^S \\ \mathbf{v}^W \end{pmatrix} = \begin{pmatrix} \mathbf{F}^S \\ \mathbf{F}^W \end{pmatrix} \tag{26.14}$$

where $\mathbf{M}^\alpha$, $\mathbf{a}^\alpha$, $\mathbf{v}^\alpha$, and $\mathbf{F}^\alpha$ represent the (generalized) mass matrix, acceleration, velocity and force vectors, respectively. Several computational simplifications result in using a diagonal mass matrix, and a 'lumped' mass matrix is used throughout. For the two-dimensional four node bilinear isoparametric element,

$$(M_{ab}^{AB})^\alpha = \delta_{ab}\delta^{AB} \int_{\Omega^e} \rho^\alpha N^A \, d\Omega \qquad \text{(no sum on A)} \tag{26.15}$$

where M_{ab}^{AB} is the elemental mass contribution to node A from node B for directions a and $b(=1, 2)$ to the global mass matrix. In equation (26.14) Ξ is a damping matrix arising from the momentum transfer terms in equations (26.2) and (26.3) as

$$\Xi_{ab}^{AB} = \int N^A \xi_{ab} N^B \, d \tag{26.16}$$

$\mathbf{C}^{\alpha\beta}$ $(\alpha, \beta = S, W)$ are damping matrices arising from the fluid contributions to equations (26.2) and (26.3) (from equation (26.11))

$$(\mathbf{C}^{AB})^{\alpha\beta} = \int_{\Omega} \lambda^{W} \frac{n^{\alpha} n^{\beta}}{n^{W}} \begin{bmatrix} N^{A}_{,1} \; N^{B}_{,1} & N^{A}_{,1} \; N^{B}_{,2} \\ N^{A}_{,1} \; N^{B}_{,1} & N^{A}_{,2} \; N^{B}_{,2} \end{bmatrix} \mathrm{d}\Omega \tag{26.17}$$

in two dimensions. The solid force vector $\mathbf{F}^{S}$ is of the following form:

$$\mathbf{F}^{S} = (\mathbf{F}^{ext})^{S} - \mathbf{N}^{S}(\mathbf{d}^{S}) \tag{26.18}$$

where $(\mathbf{F}^{ext})^{S}$ is the vector of external solid forces, $\mathbf{d}^{S}$ = solid nodal displacement vector, and $\mathbf{N}^{S}$ is the vector of effective stress forces, e.g., in two dimensions:

$$(\mathbf{N}^{A})^{S} = \begin{bmatrix} N^{A}_{,1} & \sigma'^{S}_{11} + N^{A}_{,2} & \sigma'^{S}_{12} \\ N^{A}_{,2} & \sigma'^{S}_{22} + N^{A}_{,1} & \sigma'^{S}_{12} \end{bmatrix} \mathrm{d}\Omega \tag{26.19}$$

The fluid force vector $\mathbf{F}^{W}$ is of the following form

$$\mathbf{F}^{W} = (\mathbf{F}^{ext})^{W} - \mathbf{N}^{W}(\mathbf{v}^{S}, \mathbf{v}^{W}) \tag{26.20}$$

where $(\mathbf{F}^{ext})^{W}$ is the vector of external fluid forces, and $\mathbf{N}^{W}(\mathbf{v}^{S}, \mathbf{v}^{W})$ is the vector of convective forces,

$$(N^{A}_{a})^{W} = \int_{\Omega} \rho^{W} N^{A} (v^{W}_{b} - v^{S}_{b}) v^{W}_{a,b} \, \mathrm{d}\Omega \tag{26.21}$$

The initial value problem consists of finding the functions $\mathbf{d}^{S} = \mathbf{d}^{S}(t)$ and $\mathbf{v}^{W} = \mathbf{v}^{W}(t)$, functions of time, satisfying equation (26.14) and the initial conditions

$$\mathbf{d}^{S}(0) = \mathbf{d}^{S}_{0} \qquad \dot{\mathbf{d}}^{S}(0) = \mathbf{v}^{S}_{0}$$
$$\mathbf{v}^{W}(0) = \mathbf{v}^{W}_{0} \tag{26.22}$$

where $\mathbf{d}^{S}_{0}$, $\mathbf{v}^{S}_{0}$ and $\mathbf{v}^{W}_{0}$ are given vectors of initial data.

Due to the presence of the internal forces $\mathbf{N}^{S}$ (equation (26.18)) and $\mathbf{N}^{W}$ (equation (26.20)) in equation (26.14), the system of equations is non-linear. Let

$$\mathbf{K} = \partial \mathbf{N}^{S} / \partial \mathbf{d}^{S} = \mathbf{K}^{S} + \mathbf{K}^{G} \tag{26.23}$$

denote the tangent solid stiffness matrix. It is the 'consistent' linearized operator associated with $\mathbf{N}^{S}$ in the sense of [16], and is in general a non-linear function of $\mathbf{d}^{S}$ and $\mathbf{v}^{S}$, namely from equation (26.9)

$$K^{AB}_{ab} = \int_{\Omega^{e}} B^{A}_{Ia} k^{S}_{IJ} B^{B}_{Jb} \, \mathrm{d}\Omega + \int_{\Omega^{e}} \tilde{B}^{A}_{Ia} k^{G}_{IJ} \tilde{B}^{B}_{Jb} \, \mathrm{d}\Omega \tag{26.24}$$

where the first term is the material tangent part, and the second part the geometric part; k^{S}_{IJ} and k^{G}_{IJ} are the matrix formed from the fourth-order tensors $\mathbf{k}^{S}$ and $\mathbf{k}^{G}$ (equation (26.9)) in the usual manner (see e.g., [1, 37, 40]).

In the two-dimensional case,

$$\mathbf{k}^{\mathrm{G}} = -\begin{pmatrix} \sigma'^{\mathrm{S}}_{11} & 0 & 0 & \sigma'^{\mathrm{S}}_{12} \\ & \sigma'^{\mathrm{S}}_{22} & \sigma'^{\mathrm{S}}_{12} & 0 \\ \text{symmetric} & & \dfrac{\sigma'^{\mathrm{S}}_{11} - \sigma'^{\mathrm{S}}_{22}}{2} & \dfrac{\sigma'^{\mathrm{S}}_{11} + \sigma'^{\mathrm{S}}_{22}}{2} \\ & & & \dfrac{\sigma'^{\mathrm{S}}_{22} - \sigma'^{\mathrm{S}}_{22}}{2} \end{pmatrix} \tag{26.25}$$

and

$$\mathbf{B}^{\mathrm{A}} = \begin{pmatrix} N^{\mathrm{A}}_{,1} & 0 \\ 0 & N^{\mathrm{A}}_{,2} \\ N^{\mathrm{A}}_{,2} & N^{\mathrm{A}}_{,1} \end{pmatrix} \qquad \tilde{\mathbf{B}}^{\mathrm{A}} = \begin{pmatrix} N^{\mathrm{A}}_{,1} & 0 \\ 0 & N^{\mathrm{A}}_{,2} \\ N^{\mathrm{A}}_{,2} & 0 \\ 0 & N^{\mathrm{A}}_{,1} \end{pmatrix} \tag{26.26}$$

and for a linear isotropic elastic solid.

$$\mathbf{k}^{\mathrm{S}} = \lambda^{\mathrm{S}} \begin{pmatrix} 1 & 1 & 0 \\ 1 & 1 & 0 \\ 0 & 0 & 0 \end{pmatrix} + \mu^{\mathrm{S}} \begin{pmatrix} 2 & 0 & 0 \\ 0 & 2 & 0 \\ 0 & 0 & 1 \end{pmatrix} \tag{26.27}$$

It is of importance to note that the matrix $\mathbf{k}^{\mathrm{S}}$ is always split into, roughly speaking, a volumetric and deviatoric part. This is done to facilitate the use of selective integration techniques. The split of $\mathbf{k}^{\mathrm{S}}$ is done to segregate the potentially very stiff part of the matrix which needs to be treated by reduced quadrature in nearly-incompressible applications [17, 22].

26.4 TIME INTEGRATION

Symbolically, the discretized equations of motion (equation (26.14)) can be written as

$$\mathbf{Ma} + \mathbf{Cv} + \mathbf{N}(\mathbf{d}, \mathbf{v}) = \mathbf{F}^{\mathrm{ext}} \tag{26.28}$$

where **M**, **C**, **N**, **a**, **v**, and **d** are defined by equation (26.14) in an obvious manner. Time integration is performed by using the implicit–explicit algorithm of Hughes *et al.* [14, 15, 18], which consists of satisfying the following equations

$$\mathbf{Ma}_{n+1} + \mathbf{Cv}_{n+1} + \mathbf{N}^{\mathrm{I}}(\mathbf{d}_{n+1}, \mathbf{v}_{n+1}) + \mathbf{N}^{\mathrm{E}}(\tilde{\mathbf{d}}_{n+1}, \tilde{\mathbf{v}}_{n+1}) = \mathbf{F}^{\mathrm{ext}}_{n+1} \tag{26.29}$$

$$\begin{aligned} \mathbf{d}_{n+1} &= \tilde{\mathbf{d}}_{n+1} + \alpha \Delta t^2 \mathbf{a}_{n+1} \\ \mathbf{v}_{n+1} &= \tilde{\mathbf{v}}_{n+1} + \beta \Delta t \mathbf{a}_{n+1} \end{aligned} \tag{26.30a}$$

Table 26.1

Integration method	α	β
Central difference	1/2	0
Linear acceleration	1/2	1/6
Trapezoidal	1/2	1/4
Backward difference	3/2	1

where

$$\tilde{\mathbf{d}}_{n+1} = \mathbf{d}_n + \Delta t \mathbf{v}_n + (1 - 2\beta)\frac{\Delta t^2}{2}\mathbf{a}_n$$

$$\tilde{\mathbf{v}}_{n+1} = \mathbf{v}_n + (1 - \alpha)\Delta t \mathbf{a}_n \tag{26.30b}$$

and the superscripts I and E refer to the parts of $\mathbf{N}$ which are treated implicitly and explicitly, respectively.

The notation is as follows: Δt is the time step, $\mathbf{F}_n^{\text{ext}} = \mathbf{F}^{\text{ext}}(t_n)$; $\mathbf{d}_n$, $\mathbf{v}_n$, and $\mathbf{a}_n$ are the approximations to $\mathbf{d}(t_n)$, $\mathbf{v}(t_n)$, and $\mathbf{a}(t_n)$; α and β are parameters which control the accuracy and stability of the method. By appropriate choice of the values of the parameters α and β, various well-known integration methods are obtained as summarized in Table 26.1. The quantities $\tilde{\mathbf{d}}_{n+1}$ and $\tilde{\mathbf{v}}_{n+1}$ are referred to as 'predictor' values, and $\mathbf{d}_{n+1}$ and $\mathbf{v}_{n+1}$ are referred to as 'corrector' values. From equations (26.29) and (26.30), it is apparent that the calculations are rendered partly explicitly by evaluating $\mathbf{N}^{\text{E}}$ in terms of data known from the previous step.

Calculations commence with the given initial data (i.e., $\mathbf{d}_0$ and $\mathbf{v}_0$) and $\mathbf{a}_0$ which is defined by

$$\mathbf{M}\mathbf{a}_0 = \mathbf{F}_0^{\text{ext}} - \mathbf{C}_0\mathbf{v}_0 - \mathbf{N}(\mathbf{d}_0, \mathbf{v}_0) \tag{26.31}$$

since $\mathbf{M}$ is diagonal, the solution of equation (26.31) is rendered trivial. It may be recognized that equations (26.29) and (26.30) correspond to the Newmark formulas [25].

26.4.1 Implementation

At each time step, equations (26.29) and (26.30) constitute a non-linear algebraic problem which is solved by a Newton–Raphson type iterative procedure. The most useful and versatile implementation is to form an 'effective static problem' from equation (26.29) in terms of the unknown $\mathbf{a}_{n+1}$, which is in turn linearized. Let i denote the iteration counter. Within each time step the calculations are performed as summarized in Table 26.2. If additional iterations are to be performed, i is replaced by $i + 1$, and calculations resume at Step 2. When the iterative phase is completed, the solution at step $n + 1$ is defined by the last iterates, n is replaced by $n + 1$, and calculations for the next time step may

Table 26.2 Flow chart

Set iteration counter $i = 0$

1. Predictor phase:

$$\mathbf{d}_{n+1}^{(i)} = \tilde{\mathbf{d}}_{n+1}$$
$$\mathbf{v}_{n+1}^{li} = \tilde{\mathbf{v}}_{n+1}$$
$$\mathbf{a}_{n+1}^{(i)} = \mathbf{0}$$

2. Form residual force:

$$\Delta\mathbf{F} = \mathbf{F}_{n+1}^{\text{ext}} - \mathbf{M}\mathbf{a}_{n+1}^{(i)} - \mathbf{C}\mathbf{v}_{n+1}^{(i)} - \mathbf{N}(\mathbf{d}_{n+1}^{(i)}, \mathbf{v}_{n+1}^{(i)})$$

3. Form effective mass: (reform and factorize only if required)

$$\mathbf{M}^* = \mathbf{M} + \Delta t\alpha\mathbf{C} + \Delta t^2\beta\mathbf{K}^{\text{I}}$$

4. Solution phase:

$$\mathbf{M}^*\Delta\mathbf{a} = \Delta\mathbf{F}$$

5. Corrector phase:

$$\mathbf{a}_{n+1}^{(i+1)} = \mathbf{a}_{n+1}^{(i)} + \Delta a$$
$$\mathbf{v}_{n+1}^{(i+1)} = \tilde{\mathbf{v}}_{n+1} + \Delta t\alpha\mathbf{a}_{n+1}^{(i+1)}$$
$$\mathbf{d}_{n+1}^{(i+1} = \tilde{\mathbf{d}}_{n+1} + \Delta t^2\beta\mathbf{a}_{n+1}^{(i+1)}$$

---

begin. In order to cut down in storage requirements, $\mathbf{d}_n$, $\mathbf{v}_n$, and $\mathbf{a}_n$ are saved during the iterative phase together with $\mathbf{a}_{n+1}^{(i+1)}$, but $\mathbf{v}_{n+1}^{(i+1)}$ and $\mathbf{d}_{n+1}^{(i+1)}$ are not and are computed when needed at the element level. Further, fixed-point iterations are employed.

Note that the **C**-term is always treated 'implicity' (due to the usually large value of λ^{W} compared to $(\lambda^{\text{S}} + 2\mu^{\text{S}})$). In order to cut down on computing costs $\mathbf{M}^*$ is to be reformed and factorized as infrequently as possible. This is achieved by treating the non-linear fluid convective term $\mathbf{N}^{\text{W}}$ and the solid geometric tangent stiffness $\mathbf{K}^{\text{G}}$ 'explicitly'. As for the material tangent stiffness $\mathbf{K}^{\text{S}}$, three options are possible: 'implicit', 'explicit', or 'implicit–explicit' treatment. The choice is to be governed by stability and physical considerations, as follows. For linear analysis (e.g., porous linear elastic solid skeleton), $\mathbf{M}^*$ only needs to be formed and factorized once at step $n = 1$ (if Δt is kept constant), irrespective of whether $\mathbf{K}^{\text{S}}$ is treated implicitly or explicitly. In that case, one might as well achieve unconditional stability, $\mathbf{K}^{\text{S}}$ is treated implicitly, and α and β are selected accordingly. For non-linear analysis (e.g., elastoplastic porous solid skeleton), $\mathbf{K}^{\text{S}}$ depends upon the current state of stress and deformations and thus requires $\mathbf{M}^*$ to be reformed and factorized at each time step. A purely implicit treatment of $\mathbf{K}^{\text{S}}$, thus produces a considerable computational burden. Two options are therefore available: purely 'explicit' treatment or 'implicit–explicit' treatment. In that case, the elastic stiffness is treated 'implicitly' and the non-linear plastic stiffness is treated 'explicitly'. In both cases, Δt is to be chosen according to stability or accuracy considerations. However, if localization-type phenomena due to material instability are to be captured by the numerical solution, $\mathbf{K}^S$ has to be treated *implicitly* (at least close to the localization state) [31].

For 'non-associative' elastoplastic porous skeletons, $\mathbf{K}^S$ is non-symmetric, and thus in a purely 'implicit' treatment $\mathbf{M}^*$ is non-symmetric. However, $\mathbf{M}^*$ still possesses the usual 'band-profile' structure associated with the connectivity of the finite element mesh, and has a symmetric profile. A Crout elimination algorithm [37] is used in Step 4 (Table 26.2) which fully exploits this structure in that zeros outside the profile are neither stored nor operated upon.

26.4.2 Remarks

The algorithm used falls within the category of implicit/explicit predictor/multicorrector algorithms [18]. Evidently, the choice for the implicit–explicit operator splitting can be done differently for various element groups. However, in that case it is important to remember that stability of the algorithm is governed by the explicit group, and thus that the critical time step of the explicit group governs for the system.

26.4.3 Stability conditions

Linearized stability analysis of the algorithm have been reported in [18] for symmetric operators. The important results are summarized as follows: In all cases $\alpha \geq \frac{1}{2}$:

Implicit elements: unconditional stability is achieved if $\beta \geq \alpha/2$, and it is recommended that

$$\beta = (\alpha + \tfrac{1}{2})^{2/4} \tag{26.32}$$

to maximize high-frequency numerical dissipation.

Explicit elements: the time step restriction is

$$\Omega < 2(1 - \xi)/(\alpha + \tfrac{1}{2}) \tag{26.33}$$

to maximize high-frequency numerical dissipation, ξ = modal viscous damping, $\Omega = \omega\Delta t$ = sampling frequency.

Implicit–Explicit elements: the stability of the implicit–explicit scheme is contingent upon the positive-definiteness of the following array:

$$\mathbf{B} = \mathbf{M} + \Delta t(\alpha - \tfrac{1}{2})\mathbf{C} + \Delta t^2\left(\beta - \frac{\alpha}{2}\right)\mathbf{K}^{\mathrm{I}} - \Delta t^2 \frac{\alpha}{2}\mathbf{K}^{\mathrm{E}} \tag{26.34}$$

It should be emphasized that the above stability conditions were derived for symmetric positive definite **M**, **C**, and **K**. Caution should therefore be exercised when applying these results to non-symmetric operators.

It is of interest to note that when $\alpha = \frac{3}{2}$ and $\beta = 1$, the spectral radius of the amplification matrix of the algorithm goes to zero when $\Omega \to \infty$. The high dissipative properties of the algorithm then kill all dynamic transient phenomena and allow 'static' solutions to be obtained 'dynamically'. This is illustrated in [32] and thereafter.

26.5 NUMERICAL EXAMPLES

Equations (26.29) and (26.30) have been incorporated into the computer code DYNA-FLOW [29]. DYNA-FLOW is a finite element analysis program for the static and transient response of linear and non-linear two- and three-dimensional systems. DYNA-FLOW is an expanded version of DIRT II [13]. In particular, DYNA-FLOW offers transient analysis capabilities for both parabolic and hyperbolic initial value problems in both solid and fluid mechanics. There are no restrictions on the number of elements, the number of load cases, and the number or bandwidth of the equations. Despite large system capacity, no loss of efficiency is encountered in solving small problems. In both static and transient analyses, an implicit–explicit predictor-multi-corrector scheme is used. Some features which are available in the program are:

(a) Selective specification of high- and low-speed storage allocations;
(b) Both symmetric and non-symmetric matrix equations solvers;
(c) Eigenvalue/vector solution solver;
(d) Reduced/selective integration procedures, for effective treatment of incompressibility constraints;
(e) Coupled field equation capabilities for treatment of thermoelastic and saturated porous media;
(f) Isoparametric data generation schemes;
(g) Mesh optimization options;
(h) Plotting options;
(i) Interactive options.

The element and material model libraries are modularized and may be easily expanded without alteration of the main code.

The element library contains a two-dimensional element with plane stress/plane strain and axisymmetric options, and full finite deformations may be accounted for. A three-dimensional element is also included. A contact element, a slide-line element, a truss element, and a beam element are available for two- and three-dimensional analysis.

The material library contains a linear elastic model, a linear thermoelastic model, a Newtonian fluid model, and a family of elastoplastic models developed by the author.

In the following a few simple examples are presented which illustrate the versatility and accuracy of the proposed procedure. In these calculations, the four-node bilinear isoparametric element (see e.g., [1] for detailed descriptions) was used with the standard selective integration scheme [17, 22].

26.5.1 One-dimensional wave propagation

A wave propagation analysis [34] shows that the proposed theory predicts one attenuated solid dilatational wave and one attenuated solid rotational wave. Another dilatational (non-dispersive, non-diffusive) wave propagates through the fluid phase with infinite velocity. These waves are the same as those predicted by Biot [2] if the fluid compressibility is neglected [34].

In order to illustrate the essential features of the phenomena associated with the propagation of dilatational waves, numerical results for a simple one-dimensional wave propagation are presented. In the analysis to follow, the initial-value problem consists of a saturated porous medium column, fixed at one end, and subjected to a step loading at the other end. Both the movements of the solid and fluid phases are constrained to take place in the axial direction only. Drainage of the fluid phase can only take place through the free-end of the bar. The total length is 100., and one row of 55 finite elements was used with cross-sectional area 1.0. The mesh is shown in Figure 26.1 and consists of two groups of elements. Group I, $0 \leq x \leq 50.$, consists of 50 equally spaced elements, and Group II, $50. \leq x \leq 100.$ consists of 5 equally spaced elements. The total length of the mesh was selected such that within the space–time window of interest, no reflections would take place in order not to obscure the interpretation of the results. The porous soil skeleton was assumed linear elastic and the material properties were selected such that

$$\rho^S + \rho^W = 10^3; n^W = 0.5; (\lambda^S + 2\mu^S) = 10^7; \lambda^W = 10^{10}; \text{and } k = (n^W)^2 \rho_W / \xi = 10^{-1}$$

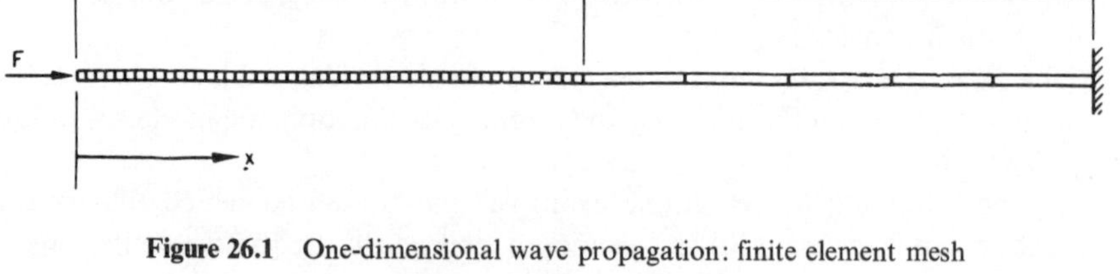

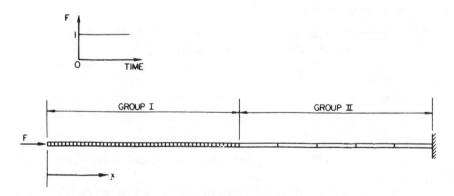

Figure 26.1 One-dimensional wave propagation: finite element mesh

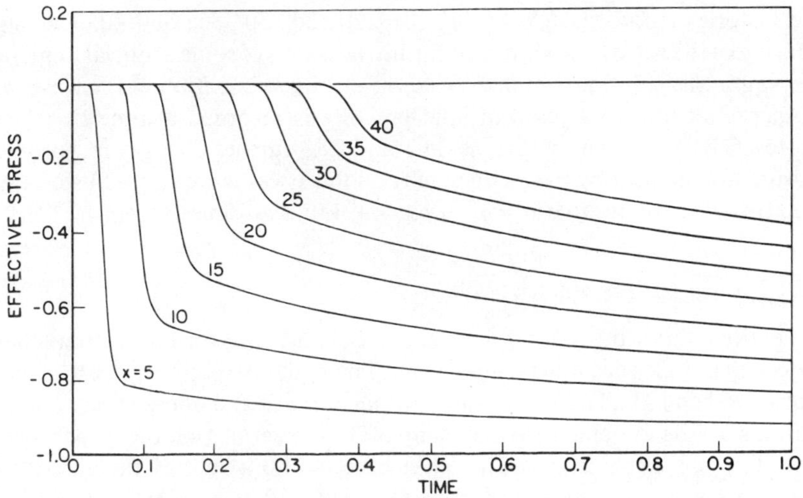

Figure 26.2 Axial effective stress time history—short term solution

26.5.1.1 *Short term solution*: $0 \leq t \leq 1.0$

In this first set of calculations, Group I was treated explicitly whereas Group II was treated implicitly. The following algorithmic parameters were selected: $\alpha = 0.65$ and $\beta = 0.33$ (equation (26.32)). The time integration was performed at the critical time step $\Delta t = 0.0087$ (equation (26.33)).

Figures 26.2 and 26.3 show the axial effective stress and pore fluid pressure

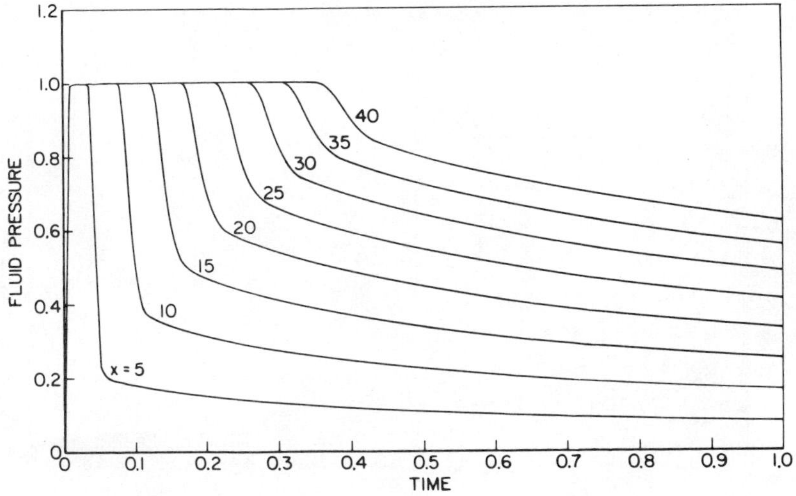

Figure 26.3 Pore-fluid pressure time history—short term solution

time histories recorded at $x = 5.$, $10.$, $15.$, $20.$, $25.$, $30.$, $35.$, and 40. The slight smearing of the wave fronts is due to the introduction of some artificial numerical damping in the solution ($\alpha > 0.5$). Note the strong attenuation of the wave with distance. As expected a jump in fluid-pressure is recorded at time $t = 0^+$ (see Figure 26.3). As shown in Figures 26.2 and 26.3 further changes in pore-fluid pressure are induced by the passage of the solid stress wave and diffusion takes place thereafter. Note that $\sigma = \sigma'^{S} - p_{W} = -1.0$ at all times.

26.5.1.2 *Long term solution*: $t > 1.0$

As mentioned previously, for $\alpha = \frac{3}{2}$ and $\beta = 1$ (Table 26.1) the integration scheme reduces to a backward difference scheme. The high dissipative properties of the algorithm then kill all dynamic transient phenomena and allow purely diffusive solutions to be obtained 'dynamically'. The versatility of the procedure is illustrated in Figures 26.4 and 26.5. At time $t = 1.0$ a shift of the algorithmic parameters was performed, and both Group I and II were treated implicitly for the remaining calculations. The unconditional stability of the algorithm then allowed large time steps and $\Delta t = 1.0$ in the remaining time integrations. Figures 26.4 and 26.5 show the calculated axial effective stress and pore fluid pressure time histories at various locations as diffusion takes place. The procedure should prove most convenient and useful in seismic calculations since it allows both the short term and the long term solutions to be obtained in one single set of calculations at optimal Δt.

Figure 26.6 shows the computed settlement δ of the top surface as a function of time. Numerical agreement with the analytical solution (see e.g., [35]) for that case is shown in Figure 26.6.

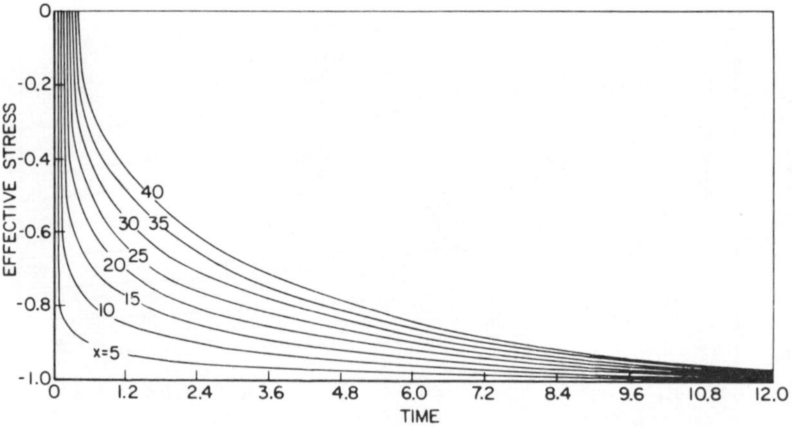

Figure 26.4 Axial effective stress time history—short and long term solutions

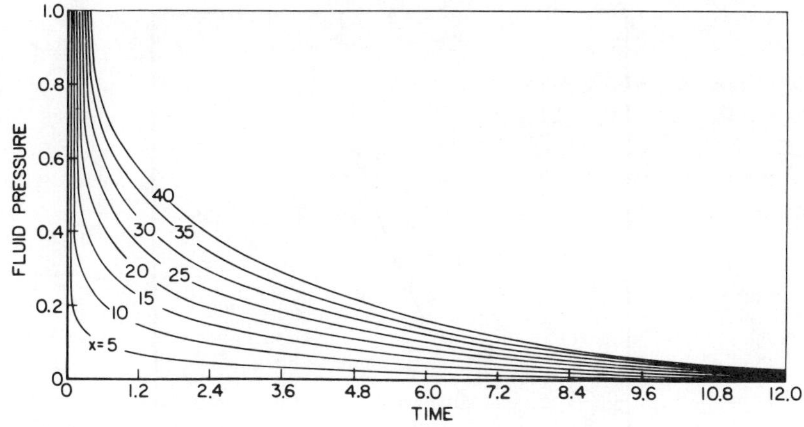

Figure 26.5 Pore-fluid pressure time history—short and long term solutions

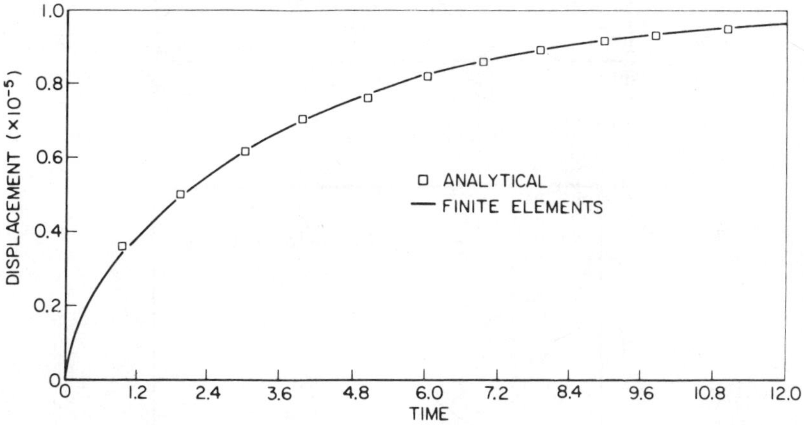

Figure 26.6 Axial displacement time history

26.5.2 One-dimensional non-linear elastic-consolidation

Figure 26.7 illustrates the capability of the proposed method in non-linear finite deformation problems. In those calculations $\alpha = \frac{3}{2}$ and $\beta = 1$. Figure 26.7(a) shows the computed ultimate vertical settlement as a function of the ratio h/E^S where E^S = Young's modulus for the soil skeleton. Figure 26.7(b) shows the computed settlement as a function of the dimensionless time T for $h/E^S = 10^{-4}$ and 1.0. In Figure 26.7(b), $C_V = k/(\rho_W)(\lambda^S + 2\mu^S)$ = one-dimensional consoli-

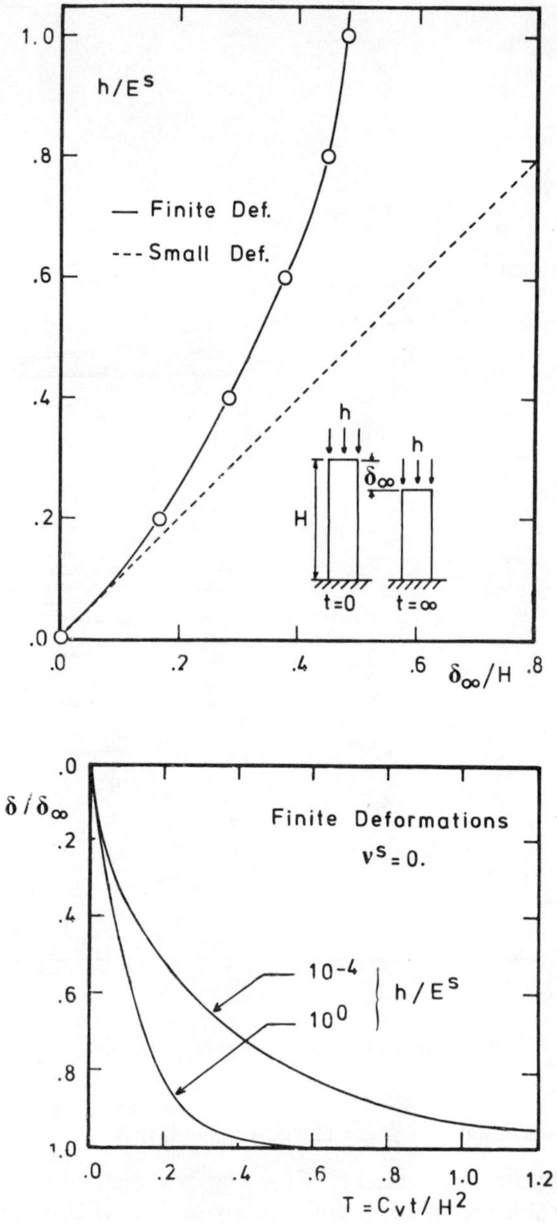

Figure 26.7 One-dimensional non-linear elastic consolidation (finite definitions). (a) Vertical settlement versus h/E^s ratio, (b) vertical settlement versus time

dation coefficient. The curves show that the larger the h/E^S ratio, the more drastic is the difference between finite and small deformation results.

26.6 SUMMARY AND CONCLUSIONS

In this chapter the transient response of saturated soils is analysed. The saturated soil is viewed as a two-phase system whose state is described by the stresses, displacements, velocities, and accelerations within each phase. The coupled field equations based on mixtures' theories are presented, and solved numerically by the use of the finite element method. Time integration is achieved by using an implicit/explicit predictor/multicorrector scheme developed by Hughes and co-workers. The procedure allows for a convenient selection of implicit and explicit elements, and for an implicit/explicit split of the various operators appearing in the differential equations. Numerical results which demonstrate the accuracy and versatility of the proposed procedure are presented.

ACKNOWLEDGEMENTS

This work was supported in part by grants from EXXON Research and Engineering Company, and the National Science Foundation (CEE-8120757). These supports are most gratefully acknowledged. Computer time was provided by Princeton University Computer Center.

REFERENCES

1. K. J. Bathe and E. L. Wilson, *Numerical Methods in Finite Element Analysis*, Prentice-Hall, Englewood Cliffs, NJ, 1976.
2. M. A. Biot, 'Theory of propagation of elastic waves in a fluid-saturated porous solid', *J. Acous. Soc. Am.*, **28**, 168–191 (1956).
3. R. M. Bowen, 'Theory of Mixtures', in *Continuum Physics*, Vol. III, (Ed. A. C. Eringen), pp. 1–127, Academic Press, 1976.
4. A. T. F. Chen, *Plane Strain and Axi-symmetric Primary Consolidation of Saturated Clays* Ph. D. Thesis, Rensselaer Polytechnic Institute, Troy, NY, 1966.
5. C. W. Cryer, 'A comparison of the three-dimensional consolidation theories of Biot and Terzaghi', *Quarterly J. Mech. Appl. Math.*, **16**, 401–412 (1963).
6. A. C. Eringen, *Mechanics of Cotinua*, R. E. Krieger Publ. Co., Huntington, NY, 1980.
7. A. C. Eringen and J. D. Ingram, 'A continuum theory of chemically reacting Media I and II', *Int. J. Eng. Sci.*, **3**, 197–212, 1955, and **5**, 289–322, 1967.
8. J. Ghaboussi and E. L. Wilson, 'Variational formulation of dynamics of fluid-saturated porous elastic solids', *J. Eng. Mech. Div., ASCE*, **98**, No. EM4, 947–963 (1972).
9. J. Ghaboussi and S. U. Dikmen, 'Liquefaction analysis of horizontally layered sands', *J. Geotech. Eng. Div., ASCE*, **104**, No. GT3, 341–356 (1978).
10. A. C. Green and P. M. Naghdi, 'A dynamical theory of interacting continua', *Int. J. Eng. Sci.*, **3**, 231–241 (1965).
11. H. M. Hilber, 'Analysis and design of numerical integration methods in structural

dynamics', *Rep. No. EERC* 76–29, Earthquake Engineering Research Center, University of California, Berkeley, CA, November 1976.

12. R. Hill, 'A general theory of uniqueness and stability in elastic–Plastic solids', *J. Mech. and Phys. of Solids*, **6**, 236–249 (1958).
13. T. J. R. Hughes and J. H. Prevost, 'DIRT II—a nonlinear quasi-static finite element analysis program', California Institute of Technology, Pasadena, CA, August, 1979.
14. T. J. R. Hughes and W. K. Liu, 'Implicit–explicit finite elements in transient analysis: stability theory', *J. Appl. Mech., ASME*, **45**, 371–374 (1978).
15. T. J. R. Hughes and W. K. Liu, 'Implicit–explicit finite elements in transient analysis: implementation and numerical examples', *J. Appl. Mech., ASME*, **45**, 395–398 (1978).
16. T. J. R. Hughes and K. S. Pister, 'Consistent linearization in mechanics of solids', *Computers and Structures*, **8**, 391–397 (1978).
17. T. J. R. Hughes, W. K. Liu, and A. Brooks, 'Finite element analysis of imcompressible viscous flow by the penalty function formulation', *J. Computational Physics*, **30**, (1), 1–60 (1979).
18. T. J. R. Hughes, K. S. Pister, and R. L., Taylor, 'Implicit–explicit finite elements in nonlinear transient analysis', *Comp. Meth. Appl. Mech. Eng.*, **17/18** 159–182 (1979).
19. G. Jaumann, *Grundlagen der Bewegungslehre*, Leipzig, 1903.
20. C. P. Liou, V. L. Specter, and F. E. Richard, 'Numerical Model for Liquefaction', *J. Geot. Eng. Div., ASCE*, **103**, No. GT6, 589–606 (1977).
21. J. Lysmer, 'Analytical procedures in soil dynamics', *Proceedings*, ASCE Geotech. Eng. Div., Specialty Conference on Earthquake Engineering and Soil Dynamics, Pasadena, CA, Vol. 3, pp. 1267–1316, June 1978.
22. D. S. Malkus and T. J. R. Hughes, 'Mixed finite element methods—Reduced and selective integration technique', *Comp. Meth. Appl. Mech. Eng.*, **17/18**, 159–182 (1979).
23. J. Mandel, 'Etude Mathematique de la Consolidation des sols', *Actes du Colloque International de Mechanique*, Poitier, France, **4**, 9–19 (1950).
24. J. Mandel, 'Consolidation des sols (Etude Mathematique)', *Geotechnique*, **3**, 287–299, (1953).
25. N. M. Newmark, 'A method of computation for structural dynamics', *J. Eng. Mech. Div., ASCE*, **85**, No. EM3, 67–94 (1959).
26. V. N. Nikolaeviskii, K. S. Vasniev, A. T. Gorbunov, and G. A. Zotov, *Mechanics of Saturated Porous Media*, 335 pp, Nedra, Moscow, 1970.
27. J. H. Prevost, 'plasticity theory for soil stress-strain Behaviour', *J. Eng. Mech. Div., ASCE*, **104**, No. EM5, 1177–1194 (1978).
28. J. H. Prevost, 'Mechanics of continuous porous media', *Int. J. Eng. Sci.*, **18**, 787–800 (1980).
29. J. H. Prevost, 'DYNA-FLOW—A nonlinear Transient Finite Element Analysis Program', Department of Civil Engineering, Princeton University, 1981.
30. J. H. Prevost, 'Consolidation of anelastic porous media', *J. Eng. Mech. Div., ASCE*, **107**, No. EM1, 169–186 (1981).
31. J. H. Prevost and T. J. R. Hughes, 'Finite-element solution of elastic–plastic boundary value problems', *J. Appl. Mech., ASME*, **48**, No. 1, 1–6 (1981).
32. J. H. Prevost, 'Nonlinear transient phenomena in saturated porous media', *Comp. Meth. Appl. Mech. Eng.*, **30**, 3–18 (1982).
33. J. H. Prevost, 'Implicit–explicit schemes for nonlinear consolidation', *Comp. Meth. Appl. Mech. Eng.*, **39**, 225–239 (1983).
34. J. H. Prevost, 'Wave propagation in fluid-saturated porous media', *J. Eng. Mech. Dif., ASCE*, 1982 (in process).
35. R. F. Scott, 'Principles of Soil Mechanics, Addison-Wesley, 1963.

36. V. L. Streeter, E. B. Whylie, and F. E. Richard, 'Soil motion computations by characteristics method', *J. Geotech. Eng. Div., ASCE*, **100**, No. GT3, 247–263 (1974).
37. R. L. Taylor, 'Computer procedures for finite element analysis', in O. C. Zienkiewicz, *The Finite Element Method*, 3rd. ed., Chapter 24, McGraw-Hill, London, 1977.
38. K. Terzaghi, *Theoretical Soil Mechanics*, Wiley, New York, 1943.
39. H. Van der Kogel, 'Wave Propagation in Saturated Porous Media', Ph. D. Thesis, California Institute of Technology, Pasadena, CA, 1977.
40. O. C. Zienkiewicz, *The Finite Element Method*, 3rd Edition, McGraw-Hill, London, 1977.

Mechanics of Engineering Materials
Edited by C. S. Desai and R. H. Gallagher

Chapter 27

Constitutive Equations for Materials with Memory under Restricted Classes of Deformation

R. S. Rivlin

27.1 INTRODUCTION

In this chapter we take as the point of departure the theory of Green and Rivlin [1] for the formulation of constitutive equations with memory. In this theory the assumption is made that the Cauchy stress at time t depends on the history of the deformation gradients prior to and at time t. Restrictions are first placed on the manner in which the form of this dependence is restricted by the consideration that the superposition on the assumed deformation history of an arbitrary time-dependent rigid rotation causes the stress to be rotated by the amount of this rotation at time t. Further restrictions due to any symmetry the material may possess are then placed on the manner in which the stress can depend on the deformation gradient history. While these restrictions can be rather simply expressed in implicit form, their explicit expression leads, for almost all material symmetries of practical interest, to rather complicated constitutive equations.

It is the object of this chapter to show that the constitutive equations can be greatly simplified by restricting the class of deformations to which they apply. The discussion is generally restricted to isotropic materials.

In Section 27.4 the case when the deformation is one of plane strain or plane stress is considered, both when the material is compressible and when it is incompressible (see also [2]). The results obtained apply not only to isotropic materials, but also to materials which are transversely isotropic with respect to an axis normal to the plane of the deformation. These results are further restricted in Section 27.4 to the situation when the displacement gradients are small compared with unity, but the constitutive equation is nevertheless non-linear.

In Section 27.5 it is shown that if the displacement gradients are small compared with unity the constitutive equation can be considerably simplified if we restrict the class of deformations considered to those for which the time dependence of the displacement field may be written as the product of an arbitrary, but specified, scalar function of time and a time-independent vector field (see also [3]).

A particular case arises in stress relaxation experiments in which the initially undeformed material is deformed at some instant of time, and the deformation is subsequently held constant. This class of deformations is discussed in Sections 27.6 and 27.7 both in the case when the displacement gradients are arbitrarily large and when they are small. In the former case the constitutive equation which is obtained bears a strong similarity to that of finite elasticity, as was shown a long time ago in [4]. Finally, in Section 27.7, the results of some stress relaxation experiments are briefly described, in which tubes are subjected to simultaneous small torsion and extension at some instant of time and the torque and tensile force are measured at subsequent times [5]. The materials of which the tubes are made are PVC and vulcanized GRS compounds containing various amounts of inorganic filler.

27.2 MATERIALS WITH MEMORY

We consider a deformation of a body in which a particle which is initially in vector position $\mathbf{X}$ with respect to the origin of a rectangular cartesian coordinate system x moves to $\mathbf{x}(\tau)$ at time τ. Let X_A and $x_i(\tau)$ $(i, A = 1, 2, 3)$ be the components in the system x of $\mathbf{X}$ and $\mathbf{x}(\tau)$ respectively.

We suppose that the material of the body is such that the Cauchy stress (i.e. true stress) $\boldsymbol{\sigma}(t)$ at time t depends on the history of the deformation gradient matrix $\mathbf{g}(\tau)$ defined by

$$\mathbf{g}(\tau) = \|g_{iA}(\tau)\| = \|x_{i,A}(\tau)\| \tag{27.1}$$

where, A denotes differentiation with respect to X_A. Accordingly $\boldsymbol{\sigma}(t)$ may be expressed as a functional of $\mathbf{g}(\tau)$ thus:

$$\boldsymbol{\sigma}(t) = \mathscr{F}[\mathbf{g}(\tau)], \qquad -\infty < \tau \le t \tag{27.2}$$

Taking (27.2) as their starting point, Green and Rivlin [1] introduced the consideration that the superposition on the assumed deformation of an arbitrary time-dependent rigid rotation causes the stress field $\boldsymbol{\sigma}(t)$ to be rotated by the amount of this rotation at time t. This leads to a restriction on the manner in which $\boldsymbol{\sigma}(t)$ depends on the history of the deformation gradient matrix $\mathbf{g}(t)$. This restriction is expressed by

$$\boldsymbol{\sigma}(t) = \mathbf{g}\mathscr{F}[\mathbf{C}(\tau)]\mathbf{g}^{\dagger}, \qquad -\infty < \tau \le t \tag{27.3}$$

where $\mathbf{C}(\tau)$ is the Cauchy strain matrix at time τ defined by

$$\mathbf{C}(\tau) = \mathbf{g}^{\dagger}(\tau)\mathbf{g}(\tau) \tag{27.4}$$

and

$$\mathbf{g} = \mathbf{g}(t) \tag{27.5}$$

The dagger denotes the transpose and the matrix functional $\mathscr{F}$ in (27.3) is different from that in (27.2).

The main interest in continuum mechanics resides in hereditary materials. We define the lapsed time s by

$$s = t - \tau \tag{27.6}$$

and replace the constitutive assumption (27.2) by

$$\boldsymbol{\sigma}(t) = \mathscr{F}[\mathbf{g}(s)], \qquad 0 \leq s < \infty \tag{27.7}$$

and, correspondingly, (27.3) is replaced by

$$\boldsymbol{\sigma}(t) = \mathbf{g}\mathscr{F}[\mathbf{C}(s)]\mathbf{g}^{\dagger}, \qquad \infty > s \geq 0 \tag{27.8}$$

If the material considered is incompressible the expression (27.8) is replaced by

$$\boldsymbol{\sigma}(t) = \mathbf{g}\mathscr{F}[\mathbf{C}(s)]\mathbf{g}^{\dagger} - p\boldsymbol{\delta} \tag{27.9}$$

where $\boldsymbol{\delta}$ is the 3×3 unit matrix, and the deformation must, of course, satisfy the constraint

$$\det \mathbf{g}(s) = \det \mathbf{C}(s) = 1 \tag{27.10}$$

which expresses the fact that the volume of a material element is not changed by the deformation.

If the material considered has some symmetry, then the manner in which the functionals $\mathscr{F}$ in (27.8) and (27.9) depend on their argument function must satisfy the restriction

$$\mathscr{F}[\mathbf{S}\mathbf{C}(s)\mathbf{S}^{\dagger}] = \mathbf{S}\mathscr{F}[\mathbf{C}(s)]\mathbf{S}^{\dagger} \tag{27.11}$$

for all transformations of $\mathbf{S}$ the group of transformations describing the symmetry of the material. If the material is isotropic then this group is the group of orthogonal transformations.

For any specified material symmetry, the restriction on the form of $\mathscr{F}$ implied by (27.9) can be made explicit. The resulting canonical forms for various material symmetries have been the subject of numerous papers (see, for example, the review articles by Spencer [6] and Rivlin [7]). However, they are generally rather complicated and will not be given here.

It will be useful to introduce the reduced Cauchy strain matrix $\mathbf{E}(s)$ defined by

$$\mathbf{E}(s) = \mathbf{C}(s) - \boldsymbol{\delta} \tag{27.12}$$

It is then evident from (27.8) and (27.9) that $\boldsymbol{\sigma}(t)$ may be expressed in the form

$$\boldsymbol{\sigma}(t) = \mathbf{g}\mathscr{F}[\mathbf{E}(s)]\mathbf{g}^{\dagger}$$

or

$$\boldsymbol{\sigma}(t) = \mathbf{g}\mathscr{F}[\mathbf{E}(s)]\mathbf{g}^{\dagger} - p\boldsymbol{\delta} \tag{27.13}$$

accordingly as the material is, or is not, compressible. In the latter case, the kinematic constraint (27.10) becomes

$$J_1 + 2J_2 + 4J_3 = 0 \tag{27.14}$$

where

$$J_1 = \tfrac{1}{2}\mathrm{tr}\,\mathbf{E}(s), \qquad J_2 = \tfrac{1}{8}[\{\mathrm{tr}\,\mathbf{E}(s)\}^2 - \mathrm{tr}\{\mathbf{E}(s)\}^2], \qquad J_3 = \tfrac{1}{8}\det\mathbf{E}(s) \tag{27.15}$$

27.3 SMALL DEFORMATIONS

We denote the vector displacement at time τ by $\mathbf{u}(\tau)$ and the displacement gradient matrix by $\mathbf{d}(\tau)$. Then,

$$\mathbf{x}(\tau) = \mathbf{X} + \mathbf{u}(\tau)$$

and

$$\mathbf{g}(\tau) = \boldsymbol{\delta} + \mathbf{d}(\tau) \tag{27.16}$$

We now suppose that the displacement gradients are small, i.e.

$$\mathrm{tr}\,[\mathbf{d}^\dagger(\tau)\mathbf{d}(\tau)] \ll 1 \tag{27.17}$$

and neglect terms of higher degree than the first in $\mathbf{d}(\tau)$ in comparison with those of first degree. We then obtain, from (27.4) and (27.12),

$$\mathbf{E}(\tau) = 2\mathbf{e}(\tau) \tag{27.18}$$

where $\mathbf{e}(\tau)$ is the classical infinitesimal strain matrix defined by

$$\mathbf{e}(\tau) = \tfrac{1}{2}\{\mathbf{d}(\tau) + \mathbf{d}^\dagger(\tau)\} \tag{27.19}$$

With (27.18), we can write (27.13) in the form

$$\boldsymbol{\sigma}(t) = \mathscr{F}[\mathbf{e}(s)]$$

or

$$\boldsymbol{\sigma}(t) = \mathscr{F}[\mathbf{e}(s)] - p\boldsymbol{\delta} \tag{27.20}$$

accordingly as the material is compressible or incompressible. The material symmetry restriction (27.11) is now replaced by

$$\mathscr{F}[\mathbf{S}\mathbf{e}(s)\mathbf{S}^\dagger] = \mathbf{S}\mathscr{F}[\mathbf{e}(s)]\mathbf{S}^\dagger \tag{27.21}$$

for all transformations $\mathbf{S}$ of the group of transformations describing the symmetry of the material.

27.4 PLANE STRAIN AND PLANE STRESS

In this section we consider plane strain deformations parallel to the 12-plane of the rectangular cartesian coordinate system x. Then the deformation is described by

$$x_\alpha(\tau) = x_\alpha(X_\beta, \tau), \qquad (\alpha, \beta = 1, 2)$$
$$x_3(\tau) = X_3 \tag{27.22}$$

We define the two-dimensional deformation gradient matrix $\mathbf{g}_2(\tau)$ by

$$\mathbf{g}_2(\tau) = \|g_{\alpha\beta}(\tau)\| = \|x_{\alpha,\beta}(\tau)\| \tag{27.23}$$

where, β denotes the operator $\partial/\partial X_\beta$ $(\beta = 1, 2)$.

With the assumption that the material is of the hereditary type, the corresponding two-dimensional stress matrix at time t, $\boldsymbol{\sigma}_2(t) = \|\sigma_{\alpha\beta}(t)\|$, is given, from (27.13), by expressions of the form

$$\boldsymbol{\sigma}_2(t) = \mathbf{g}_2\mathscr{F}[\mathbf{E}_2(s)]\mathbf{g}_2^\dagger$$

or

$$\boldsymbol{\sigma}_2(t) = \mathbf{g}_2\mathscr{F}[\mathbf{E}_2(s)]\mathbf{g}_2^\dagger - p\boldsymbol{\delta}_2 \tag{27.24}$$

accordingly as the material is compressible or incompressible, where $\mathbf{E}_2(s)$ and $\mathbf{g}_2$ are the 2×2 matrices defined by

$$\mathbf{E}_2(s) = \|E_{\alpha\beta}(s)\|, \qquad \mathbf{g}_2 = \mathbf{g}_2(t) \tag{27.25}$$

In $(27.24)_2$, p is an arbitrary hydrostatic pressure and $\boldsymbol{\delta}_2$ is the 2×2 unit matrix.

The incompressibility condition (27.10) becomes, in this case,

$$\det[\boldsymbol{\delta}_2 + \mathbf{E}_2(s)] = 1 \tag{27.26}$$

This may be rewritten as

$$\operatorname{tr}\mathbf{E}_2(s) + \det\mathbf{E}_2(s) = 0 \tag{27.27}$$

If the material is isotropic, or is transversely isotropic with respect to the 3-axis, then the 2×2 matrix functionals $\mathscr{F}$ in (27.24) must satisfy the relation (cf. (27.11))

$$\mathscr{F}[\mathbf{S}\mathbf{E}_2(s)\mathbf{S}^\dagger] = \mathbf{S}\mathscr{F}[\mathbf{E}_2(s)]\mathbf{S}^\dagger \tag{27.28}$$

for all orthogonal 2×2 matrices $\mathbf{S}$. It follows [2], from considerations similar to those developed by Wineman and Pipkin [8] (see also [2]) in the case of three-dimensional constitutive equations, that $\mathscr{F}$ is an isotropic matrix functional of the scalar functions $\operatorname{tr}\mathbf{E}_2(s)$, $\operatorname{tr}\mathbf{E}_2(s_1)\mathbf{E}_2(s_2)$ and the matrix function $\mathbf{E}_2(s)$ $(s, s_1, s_2 = [0, \infty))$, linear in the matrix function $\mathbf{E}_2(s)$. With the restriction on $\mathbf{g}_2(s)$—and hence on $\mathbf{E}_2(s)$—that it has bounded variation and is continuous, except possibly at a countable number of points at which it has salti, it follows that $\mathscr{F}$ can be expressed in the form [2]

$$\mathscr{F} = \int_0^\infty f(s)\,\mathrm{d}\mathbf{E}_2(s) + h\boldsymbol{\delta}_2 \tag{27.29}$$

where f and h are functionals of $\operatorname{tr}\mathbf{E}_2(s)$ and $\operatorname{tr}\mathbf{E}_2(s_1)\mathbf{E}_2(s_2)$ for $s, s_1, s_2 = [0, \infty)$. In the case when the material is incompressible the term $h\boldsymbol{\delta}_2$ can, of course, be absorbed in the term $p\boldsymbol{\delta}_2$ in $(27.24)_2$. Also, in this case, since

$$\det\mathbf{E}_2(s) = \tfrac{1}{2}\{[\operatorname{tr}\mathbf{E}_2(s)]^2 - \operatorname{tr}[\mathbf{E}_2(s)]^2\} \tag{27.30}$$

so that from (27.27)

$$\operatorname{tr} \mathbf{E}_2(s) = -1 + \{1 + \operatorname{tr}[\mathbf{E}_2(s)]^2\}^{1/2} \tag{27.31}$$

it follows that f depends on $\mathbf{E}_2(s)$ only through the single function $\operatorname{tr} \mathbf{E}_2(s_1)\mathbf{E}_2(s_2)$.

If the displacement gradients are small in the sense described in Section 27.3, then the constitutive equations (27.24) may be written in the form

$$\boldsymbol{\sigma}_2(t) = \mathscr{F}[\mathbf{e}_2(s)]$$

or

$$\boldsymbol{\sigma}_2(t) = \mathscr{F}[\mathbf{e}_2(s)] - p\boldsymbol{\delta}_2 \tag{27.32}$$

accordingly as the material is compressible or incompressible, where $\mathbf{e}_2(s)$ is the 2×2 matrix defined by

$$\mathbf{e}_2(s) = \|e_{\alpha\beta}(s)\| \tag{27.33}$$

These two-dimensional constitutive equations can, of course, be obtained directly from the three-dimensional equations (27.20).

If the material is isotropic—or transversely isotropic—$\mathscr{F}$ in (27.32) may be written (cf. (27.29)) in the form

$$\mathscr{F} = \int_0^\infty f(s)\,\mathrm{d}\mathbf{e}_2(s) + \mathrm{h}\boldsymbol{\delta}_2 \tag{27.34}$$

where f and h are functionals of $\operatorname{tr} \mathbf{e}_2(s)$ and $\operatorname{tr} \mathbf{e}_2(s_1)\mathbf{e}_2(s_2)$. Again, in the incompressible case the term $\mathrm{h}\boldsymbol{\delta}_2$ can be absorbed into the hydrostatic pressure term and the argument $\operatorname{tr} \mathbf{e}_2(s)$ of $f(s)$ can be omitted.

In the case of plane stress in a thin plate, whose major surfaces are initially normal to the 3-axis of the coordinate system x, equation $(27.24)_1$ applies in both the cases when the material is compressible and when it is incompressible. Also, if the material is isotropic, or transversely isotropic with respect to the 3-axis, equation (27.29) applies in both cases. Correspondingly, for small deformations equations $(27.32)_1$ and (27.34) apply in both cases.

27.5 A CLASS OF SMALL DEFORMATIONS IN A COMPRESSIBLE MATERIAL

We now consider (see, also [3]) a class of small deformations described by

$$\mathbf{u}(\tau) = \phi(\tau)\bar{\mathbf{u}} \tag{27.35}$$

where $\bar{\mathbf{u}}$ is independent of the time τ and $\phi(\tau)$ is some specified function of τ. Then,

$$\mathbf{d}(\tau) = \phi(\tau)\bar{\mathbf{d}}, \qquad \mathbf{e}(\tau) = \phi(\tau)\bar{\mathbf{e}} \tag{27.36}$$

where $\bar{\mathbf{d}}$ and $\bar{\mathbf{e}}$ are independent of τ. We note that the class of deformations described by (27.35) will not, in general, be possible in an incompressible material, since if the deformation is isochoric for one value of τ it will not, in general, be isochoric for other values of τ. Consequently, we restrict our discussion for the moment to the case when the material considered is compressible.

For the purposes of this section it is clearer to regard the argument function in (27.20) as a function of the time $\tau(-\infty < \tau \leq t)$ rather than of the lapsed time s. Then for the class of deformations described by (27.35) it follows from (27.20) that

$$\boldsymbol{\sigma}(t) = \mathbf{F}(\bar{\mathbf{e}}, t) \tag{27.37}$$

where the form of the matrix function $\mathbf{F}$ depends on the particular function $\phi(\tau)$ chosen in (27.35).

If the material considered is isotropic, then it follows from (27.21) that the matrix function $\mathbf{F}$ must satisfy the condition

$$\mathbf{F}(\mathbf{S}\bar{\mathbf{e}}\mathbf{S}^{\dagger}, t) = \mathbf{S}\mathbf{F}(\bar{\mathbf{e}}, t)\mathbf{S}^{\dagger} \tag{27.38}$$

for all orthogonal matrices $\mathbf{S}$. This condition is satisfied if and only if $\mathbf{F}$ and hence $\boldsymbol{\sigma}(t)$ has the form

$$\boldsymbol{\sigma}(\mathrm{t}) = \mathbf{F}(\bar{\mathbf{e}}, t) = \alpha_0\boldsymbol{\delta} + \alpha_1\bar{\mathbf{e}} + \alpha_2\bar{\mathbf{e}}^2 \tag{27.39}$$

where the α's are functions of $\operatorname{tr}\bar{\mathbf{e}}$, $\frac{1}{2}\{(\operatorname{tr}\bar{\mathbf{e}})^2 - \operatorname{tr}\bar{\mathbf{e}}^2\}$, $\det\bar{\mathbf{e}}$ and t. The manner in which the α's depend on these quantities is a function of the material considered and of the choice of $\phi(\tau)$ in (27.35).

In the case of plane strain or plane stress in a compressible material, or of plane stress in an incompressible material, the two-dimensional stress $\boldsymbol{\sigma}_2(t)$ is given, for small deformations, by $(27.32)_1$. We now consider the class of such deformations for which the two-dimensional displacement vector $\mathbf{u}_2(\tau)$ is given by

$$\mathbf{u}_2(\tau) = \phi(\tau)\bar{\mathbf{u}}_2 \tag{27.40}$$

where $\bar{\mathbf{u}}_2$ is independent of the time τ and $\phi(\tau)$ is some specified function of τ. Then, paralleling (27.36), we have

$$\mathbf{d}_2(\tau) = \phi(\tau)\bar{\mathbf{d}}_2, \qquad \mathbf{e}_2(\tau) = \phi(\tau)\bar{\mathbf{e}}_2 \tag{27.41}$$

where $\bar{\mathbf{d}}_2$ and $\bar{\mathbf{e}}_2$ are independent of τ. Introducing $(27.41)_2$ into $(27.32)_1$, we obtain

$$\boldsymbol{\sigma}_2(t) = \mathbf{F}_2(\bar{\mathbf{e}}_2, t) \tag{27.42}$$

where $\mathbf{F}_2$ is a 2×2 matrix function, the form of which depends on the particular choice of $\phi(\tau)$ in (27.40).

If the material considered is isotropic, or transversely isotropic with respect to the 3-axis, then $\mathbf{F}_2$, and hence $\boldsymbol{\sigma}_2(t)$, must be expressible in the form

$$\boldsymbol{\sigma}_2(t) = \mathbf{F}_2(\bar{\mathbf{e}}_2, t) = \beta_0\boldsymbol{\delta}_2 + \beta_1\mathbf{e}_2 \tag{27.43}$$

where β_0 and β_1 are functions of $\operatorname{tr}\bar{\mathbf{e}}_2$, $\det\bar{\mathbf{e}}_2$ and t, which depend on the material and on the choice of $\phi(\tau)$ in (27.40).

The result (27.43) can be obtained directly from (27.34). We rewrite (27.34) in the form

$$\mathscr{F} = \int_{-\infty}^{t} f(t-\tau)\mathrm{d}\mathbf{e}_2(\tau) + \mathrm{h}\boldsymbol{\delta}_2 \tag{27.44}$$

where f and h are functionals of $\operatorname{tr}\mathbf{e}_2(\tau)$ and $\operatorname{tr}\mathbf{e}_2(\tau_1)\mathbf{e}_2(\tau_2)$. Then, we introduce (27.41) into (27.44) and obtain

$$\mathscr{F} = \beta_0\boldsymbol{\delta}_2 + \beta_1\bar{\mathbf{e}}_2 \tag{27.45}$$

where β_0 and β_1 are functions of $\operatorname{tr}\bar{\mathbf{e}}_2$, $\operatorname{tr}\bar{\mathbf{e}}_2^2$ and t. Noting that

$$\det\bar{\mathbf{e}}_2 = \tfrac{1}{2}\{(\operatorname{tr}\bar{\mathbf{e}}_2)^2 - \operatorname{tr}\bar{\mathbf{e}}_2^2\} \tag{27.46}$$

we can regard β_0 and β_1 as functions of $\operatorname{tr}\bar{\mathbf{e}}_2$, $\det\bar{\mathbf{e}}_2$ and t.

As a particular case of (27.35) we may consider that the displacement vector increases at a constant rate, so that

$$\mathbf{u}(\tau) = \kappa\tau\bar{\mathbf{u}} \quad (\tau \geq 0), \qquad \mathbf{u}(\tau) = 0 \quad (\tau \leq 0) \tag{27.47}$$

where κ is a constant and $\bar{\mathbf{u}}$ is independent of τ. Then, in the relation (27.39), the α's are functions of $\operatorname{tr}\bar{\mathbf{e}}$, $\frac{1}{2}\{(\operatorname{tr}\bar{\mathbf{e}})^2 - \operatorname{tr}\bar{\mathbf{e}}^2\}$, $\det\bar{\mathbf{e}}$, t, and κ. Similarly, if in the two-dimensional case

$$\mathbf{u}_2(\tau) = \kappa\tau\bar{\mathbf{u}}_2 \quad (\tau \geq 0), \qquad \mathbf{u}_2(\tau) = 0 \quad (\tau \leq 0) \tag{27.48}$$

then β_0 and β_1 in (27.43) are functions of $\operatorname{tr}\bar{\mathbf{e}}_2$, $\det\bar{\mathbf{e}}_2$, t, and κ.

For plane strain of an incompressible material, a two-dimensional deformation of the form (27.40) cannot, in general, satisfy the condition that the deformation must be isochoric for all times. Nevertheless, we can obtain a result analogous to (27.45) by the following argument.

We take as our starting point the expression $(27.24)_2$ for $\boldsymbol{\sigma}_2(t)$ with $\mathscr{F}$ given by (27.29) and absorb the term $h\boldsymbol{\delta}_2$ into the term $p\boldsymbol{\delta}_2$. Also, for clarity we regard $\mathbf{E}_2$ as a function of τ rather than of s. Thus, $\boldsymbol{\sigma}_2(t)$ is given by an expression of the form

$$\boldsymbol{\sigma}_2(t) = \mathbf{g}_2\left\{\int_{-\infty}^{t} f(t-\tau)\mathrm{d}\mathbf{E}_2(\tau)\right\}\mathbf{g}_2^{\dagger} - p\boldsymbol{\delta}_2 \tag{27.49}$$

where f is a scalar functional of $\operatorname{tr}\mathbf{E}_2(\tau_1)\mathbf{E}_2(\tau_2)$.

If the deformations are small then this equation may be replaced by an equation of the form

$$\boldsymbol{\sigma}_2(t) = \int_{-\infty}^{t} f(t-\tau)\mathrm{d}\mathbf{e}_2(\tau) - p\boldsymbol{\delta}_2 \tag{27.50}$$

where f is a functional of $\operatorname{tr}\mathbf{e}_2(\tau_1)\mathbf{e}_2(\tau_2)$.

We now note that the condition that a two-dimensional displacement field $\mathbf{u}_2(\tau)$ satisfies the condition that the deformation be isochoric has the form

$$\operatorname{tr}\mathbf{d}_2(\tau)+0[\mathbf{d}_2(\tau)]^2=0 \tag{27.51}$$

Thus this condition can be strictly satisfied by a displacement field of the form

$$\mathbf{u}_2(\tau)=\phi(\tau)\bar{\mathbf{u}}_2+\mathbf{v}_2(\tau) \tag{27.52}$$

where the magnitude of $\mathbf{v}_2(\tau)$ is small compared with that of $\phi(\tau)\bar{\mathbf{u}}_2$. We accordingly consider this class of deformations. Then, neglecting $\mathbf{v}_2(\tau)$ in comparison with $\phi(\tau)\bar{\mathbf{u}}_2$ and $\mathbf{d}_2(\tau)$ in comparison with $\boldsymbol{\delta}_2$, we obtain from (27.50) an expression for $\boldsymbol{\sigma}_2(t)$ of the form

$$\boldsymbol{\sigma}_2(t)=\beta_1\bar{\mathbf{e}}_2-p\boldsymbol{\delta}_2 \tag{27.53}$$

where β_1 is a function of $\operatorname{tr}\bar{\mathbf{e}}_2(\tau_1)\bar{\mathbf{e}}_2(\tau_2)$ and t.

27.6 STRESS RELAXATION

A particular case of the class of deformations (27.35) is that in which the body is undeformed prior to time $\tau=0$ and in some small time interval at $\tau=0$ the body is deformed, the deformation being subsequently held constant. In this section we will discuss this situation in the case when the material is incompressible and the deformation is not necessarily small.

Let

$$\mathbf{x}=\mathbf{x}(\mathbf{X}),\qquad \mathbf{g}=\mathbf{g}(\mathbf{X}) \tag{27.54}$$

describe the constant deformation in which the body is held. We consider the stress $\boldsymbol{\sigma}(t)$ at time t. Then provided that the material has fading memory and t is sufficiently large so that the details of the deformation process have an insignificant effect on the value of $\boldsymbol{\sigma}(t)$, we may make a constitutive assumption for this class of deformations of the form

$$\boldsymbol{\sigma}(t)=\mathbf{F}(\mathbf{g},t)-p\boldsymbol{\delta} \tag{27.55}$$

The stress at time t depends on the (time-independent) deformation gradient matrix $\mathbf{g}$ and on the time which has elapsed since the deformation was carried out.

Then, from the consideration that the superposition on the assumed deformation of an arbitrary rigid rotation causes the stress field $\boldsymbol{\sigma}(t)$ to be rotated by the amount of this rotation at time t, it follows that $\boldsymbol{\sigma}(t)$ must be expressible in the form

$$\boldsymbol{\sigma}(t)=\mathbf{g}\mathbf{F}(\mathbf{C},t)\mathbf{g}^{\dagger}-p\boldsymbol{\delta} \tag{27.56}$$

where

$$\mathbf{C}=\mathbf{g}^{\dagger}\mathbf{g} \tag{27.57}$$

is the Cauchy strain matrix associated with the time-independent deformation (27.54). The matrix function **F** in (27.56) is, of course, different from that in (27.55).

If the material is isotropic then (27.56) must be expressible in the form [4]

$$\boldsymbol{\sigma}(t) = \alpha_1 \mathbf{c} + \alpha_2 \mathbf{c}^2 - p\boldsymbol{\delta} \tag{27.58}$$

where **c** is the Finger strain defined by

$$\mathbf{c} = \mathbf{g}\mathbf{g}^{\dagger} \tag{27.59}$$

and α_1 and α_2 are functions of I_1, I_2, and t, with

$$I_1 = \operatorname{tr} \mathbf{c}, \qquad I_2 = \tfrac{1}{2}\{(\operatorname{tr} \mathbf{c})^2 - \operatorname{tr} \mathbf{c}^2\} \tag{27.60}$$

We note that the form of this constitutive equation is very similar to that for the Cauchy stress in an isotropic incompressible elastic material. For such a material the strain-energy W, per unit volume, is a function of I_1 and I_2 only and

$$\boldsymbol{\sigma}(t) = 2\left\{\left(\frac{\partial W}{\partial I_1} + I_1 \frac{\partial W}{\partial I_2}\right)\mathbf{c} - \frac{\partial W}{\partial I_2}\mathbf{c}^2\right\} - p\boldsymbol{\delta} \tag{27.61}$$

It follows from the similarity of the constitutive equations (27.58) and (27.61) that many of the results of finite elasticity theory—in particular those for controllable deformations—can be applied to stress relaxation situations in materials with fading memory by making the substitution

$$\frac{\partial W}{\partial I_1} = \tfrac{1}{2}(\alpha_1 + I_1 \alpha_2), \qquad \frac{\partial W}{\partial I_2} = -\tfrac{1}{2}\alpha_2 \tag{27.62}$$

In the case when the material is compressible we can, from considerations similar to those employed in deriving the constitutive equation (27.58) for stress relaxation in an incompressible material, obtain the constitutive equation

$$\boldsymbol{\sigma}(t) = \alpha_0 \boldsymbol{\delta} + \alpha_1 \mathbf{c} + \alpha_2 \mathbf{c}^2 \tag{27.63}$$

Here α_0, α_1, and α_2 are functions of I_1, I_2, I_3, and t, where I_1 and I_2 are defined in (27.60) and $I_3 = \det \mathbf{c}$. However, we note that, except when the deformation considered is homogeneous, it is not, in general, possible to maintain a constant deformation in a compressible material in which stress relaxation is taking place without applying body forces.

An exception arises in the case when the α's in (27.63) have the forms

$$\alpha_0 = \chi(t)\bar{\alpha}_0, \qquad \alpha_1 = \chi(t)\bar{\alpha}_1, \qquad \alpha_2 = \chi(t)\bar{\alpha}_2, \tag{27.64}$$

where the $\bar{\alpha}$'s are independent of t. In this case if no body forces are applied to the body considered and time-independent displacements are specified over the whole or part of the boundary, the remainder of the boundary being foree-free, then it is easy to see that the displacement field throughout the body will be independent of t.

A similar result applies in the incompressible case when the constitutive equation has the form (27.58) and α_1 and α_2 have the forms $(27.64)_{2,3}$.

27.7 STRESS RELAXATION FOR SMALL DEFORMATIONS

The constitutive equation (27.58) may be rewritten in the form

$$\boldsymbol{\sigma}(t) = \alpha_1 \boldsymbol{\varepsilon} + \alpha_2 \boldsymbol{\varepsilon}^2 - p\boldsymbol{\delta} \tag{27.65}$$

where $\boldsymbol{\varepsilon}$ is defined by

$$\boldsymbol{\varepsilon} = \mathbf{c} - \boldsymbol{\delta} \tag{27.66}$$

and α_1 and α_2 are functions of t and of K_1 and K_2 defined by

$$K_1 = \tfrac{1}{8}\{(\operatorname{tr} \boldsymbol{\varepsilon})^2 - \operatorname{tr} \boldsymbol{\varepsilon}^2\}, \qquad K_2 = \tfrac{1}{8}\det \boldsymbol{\varepsilon} \tag{27.67}$$

If the displacement gradients are small compared with unity, we have

$$\boldsymbol{\varepsilon} = 2\mathbf{e} + 0(\mathbf{d}^2) \tag{27.68}$$

Accordingly, we may approximate K_1 and K_2 by

$$K_1 = \tfrac{1}{2}\{(\operatorname{tr} \mathbf{e})^2 - \operatorname{tr} \mathbf{e}^2\}, \qquad K_2 = \det \mathbf{e} \tag{27.69}$$

We note that K_1 and K_2 are of orders 2 and 3 respectively in the displacement gradients. We can now approximate the constitutive equation (27.65) by

$$\boldsymbol{\sigma}(t) = \alpha_1 \mathbf{e} + \alpha_2 \mathbf{e}^2 - p\boldsymbol{\delta} \tag{27.70}$$

where α_1 and α_2 are functions of t and of K_1 and K_2 defined in (27.69).

Provided that α_2 is not very large compared with α_1, we may replace the constitutive equation (27.70) by

$$\boldsymbol{\sigma}(t) = \alpha_1 \mathbf{e} - p\boldsymbol{\delta} \tag{27.71}$$

We now suppose that a tube of incompressible isotropic elastic material with external radius a and internal radius b is subjected simultaneously at time $\tau = 0$ to a small simple extension with fractional extension e and small torsion of amount ψ radians per unit length, and that this deformation is subsequently held constant. Let M and N be the couple and tensile force respectively which are necessary to maintain this deformation. Then, if the constitutive equation (27.71) is valid, M and N are given by

$$M = \pi\psi \int_b^a \alpha_1 r^3 \, dr, \qquad N = 3\pi e \int_b^a \alpha_1 r \, dr, \qquad P = 0 \tag{27.72}$$

Experiments were conducted by Bergen *et al.* [5] on tubes of a GRS vulcanizate containing about 46 per cent by volume of inorganic filler and of four PVC compounds containing from 0 to 54 per cent by volume of inorganic filler. These were subjected simultaneously to simple extensions and torsions of various amounts which were then held constant. The tensile forces and torques were measured for various time t after the deformations were produced. From these measurements, the dependence of β_1 on K_1 and K_2 for various values of t was calculated by using (27.72). For each of the materials, α_1 was found to be substantially independent of K_2 and to decrease with increase of K_1. In the case of each of the PVC compounds the relation between α_1 and $\log K_1$ was found to be substantially linear and the negative slope of the α_1 versus $\log K_1$ line increased as the volume fraction of the filler increased, from approximately zero for the unfilled PVC compound. Furthermore, it appeared from the experiments that β_1 may be expressed with fairly good approximation in the form

$$\alpha_1 = \chi(t)\alpha(J_1) \tag{27.73}$$

Similar conclusions were reached for the filled vulcanized GRS compound, with the exception that the linearity in the relation between α_1 and $\log J_1$ was not obtained.

ACKNOWLEDGEMENT

This paper was written with the support of a grant from the National Science Foundation to Lehigh University.

REFERENCES

1. A. E. Green and R. S. Rivlin, *Arch. Rat'l Mech. Anal.*, **1**, 1 (1957).
2. R. S. Rivlin, *Rend. Sem. Mat. Univ. Padova*, **68**, 279 (1983).
3. R. S. Rivlin, *Arch. Rat'l Mech. Anal.*, **3**, 304 (1959).
4. R. S. Rivlin, *Q, Appl. Math.*, **14**, 83 (1956).
5. J. T. Bergen, D. C. Messersmith, and R. S. Rivlin, *J. Applied Polymer Science*, **3**, 153 (1960).
6. A. J. M. Spencer, '*Theory of invariants*', in *Continuum Mechanics* (Ed. A. C. Erigen), Academic Press, New York, 1971.
7. R. S. Rivlin, '*Notes on the theory of constitutive equations*', in *Materials with Memory* (Eds. D. Graffi), Liguori Editore, Naples, 1979.
8. A. S. Wineman and A. C. Pipkin, *Arch. Rat'l Mech. Anal.*, **17**, 184 (1964).

Mechanics of Engineering Materials
Edited by C. S. Desai and R. H. Gallagher

Chapter 28

An Extension of the Cap Model–Inclusion of Pore Pressure Effects and Kinematic Hardening to Represent an Anisotropic Wet Clay

I. S. Sandler, F. L. DiMaggio, and M. L. Baron

28.1 INTRODUCTION

The cap model falls within the general framework of the classical incremental theory of plasticity for materials which have time- and temperature-independent properties. It is intended to represent the behaviour of geological materials under a wide range of conditions. Only its application to soils will be considered in this chapter.

A constitutive model must (a) be capable of representing important material response characteristics while simple enough to be practical (b) contain material constants which can be determined from routine laboratory or field tests, and (c) lead to sensible mathematical foundations by meeting certain theoretical requirements.

These requirements are that the initial and boundary value problems involving the constitutive model, together with the equations for conservation of mass, momentum, and energy be properly posed. Such formulations have solutions which exist, are unique, and depend continuously on the initial and boundary conditions. These seemingly abstract requirements are of considerable practical importance in numerical solutions of multidimensional continuum problems. One should avoid attempting computer solutions for problems which have no solution as well as for problems with several solutions for which a physically incorrect, though mathematically correct, solution might be obtained. Further, because all numerical solutions are subject to several kinds of error (due to truncation or round-off, the order of accuracy associated with the chosen numerical scheme, and errors in specifying initial and boundary conditions), any solution which is unduly dependent on such errors is highly suspect. Therefore, the continuous dependence of solutions on the input data is directly related to the confidence with which numerical solutions can be obtained for real problems.

Of course, no single constitutive model can represent any material in all situations [1]. Even water, which is probably the most studied and best understood real material known to man, is never described by a single constitutive law to cover all situations. For example, water can sometimes be considered incompressible and sometimes compressible, sometimes inviscid and sometimes newtonian. For some applications surface tension or cavitation, etc., may be modelled. In short, whenever a constitutive model is selected, only those features of material behaviour relevant to the problem at hand should be included.

The cap model has been developed with a number of features which can be included or excluded. These involve anisotropy, kinematic hardening, pore pressure effects, limit surfaces, etc. As will be discussed subsequently, different versions of the model are utilized depending on the soil and application being considered.

All versions of the cap model have been formulated so that the material parameters can be determined knowing only the results of standard and laboratory uniaxial strain and triaxial compression tests.

In order to ensure that continuum problems in which the model is used are properly posed, the model is defined by a convex loading surface and a plastic strain rate obtained from it by a flow rule based on normality [2]. Since these choices are sufficient to satisfy Drucker's postulates [3], [4], the uniqueness, stability, and continuity requirements discussed above are satisfied.

The cap model, based on continuum plasticity theory, is designed to represent global behaviour of soils, as opposed to microscopic or local macroscopic behaviour. In this regard, its use is similar to utilizing a Mises yield surface for metals without invoking the theory of dislocations. Herein lies the difference between a *model* of physical behaviour and a *theory* of physical behaviour.

The test of the adequacy of a constitutive model is its predictive, not its explanatory, capacity. For example, once the material constants of the model have been determined from standard laboratory tests with simple load paths, the model should be capable of predicting stress–strain relations for other laboratory tests in which more complicated load paths are employed. Prediction implies the forecasting of subsequent test results, not fitting them with adjustments to material constants made after the results to be 'predicted' have been obtained.

In this chapter, the development of the family of cap models is first reviewed. Then, a new version, which is appropriate for representation of anisotropic wet clay in triaxial compression tests, is presented. This version incorporates Terzaghi's effective stress theory for pore pressure, and includes a kinematically hardening cap with an initial offset in the direction of vertical compression. In addition to consolidation data on the laboratory-prepared kaolinite clay, several undrained triaxial tests on horizontal and vertical specimens were available for fitting. For simplicity, the model was limited to only nine parameters and was fit and exercised on a hand-held programmable computer to predict the behaviour of various inclined specimens under a series of stress paths. This work was performed as part of a workshop [5] on geotechnical constitutive equations and was completed before the corresponding laboratory data were available for purposes of comparison. As will be exhibited in the sequel, excellent agreement between predictions and experiments was obtained.

28.2 DEVELOPMENT OF CAP MODEL FOR SOILS

28.2.1 Original model

In [6], models of the type illustrated in Figure 28.1 were introduced. Using an associated flow rule, the plastic strain rate vector is perpendicular to the loading

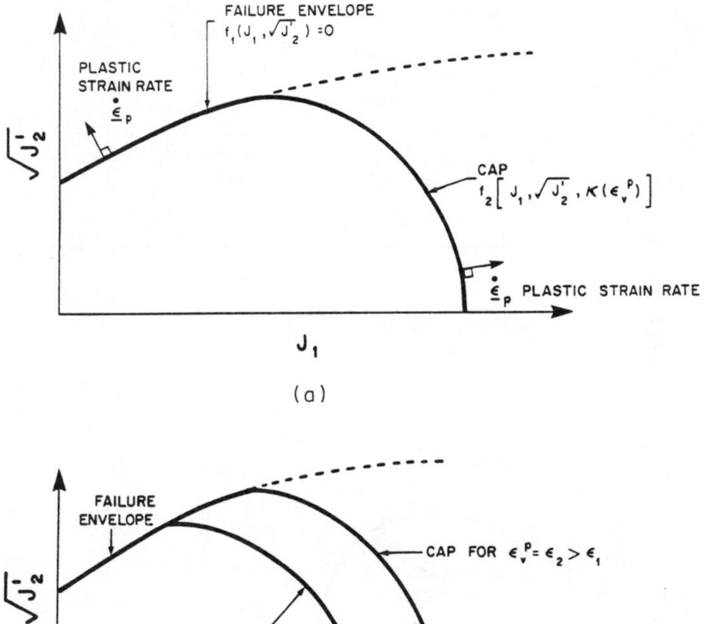

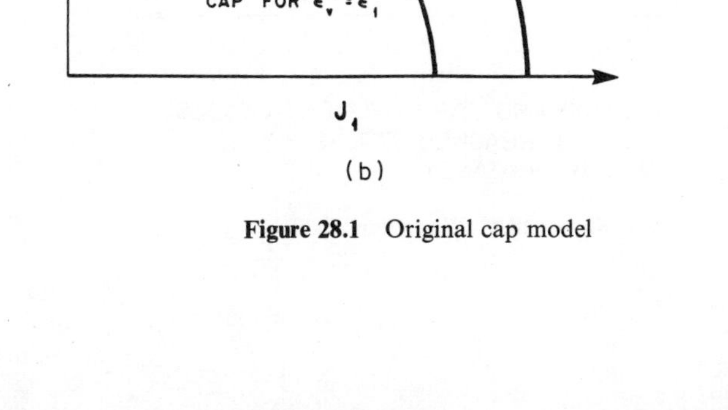

Figure 28.1 Original cap model

surface, which consists of two parts, as shown in Figure 28.1(b). The ideal (non-hardening) part, referred to as the failure envelope, has the form

$$f_1(J_1, \sqrt{J_2'}) = 0 \tag{28.1}$$

in which J_1 and J_2' denote the first and second invariants of stress and stress deviator, respectively. The hardening part, called the cap, may be represented by

$$f_2[J_1, \sqrt{(J_2)}, k(\varepsilon_v^p)] = 0 \tag{28.2}$$

in which k, the hardening parameter, is a function of the plastic volumetric strain ε_v^p. Considering J_1 and ε_p^v positive in compression, the hardening rule is chosen so that the cap moves out or in as ε_v^p increases or decreases, respectively, as in Figure 28.1(b). In addition to permitting inelastic hardening in hydrostatic

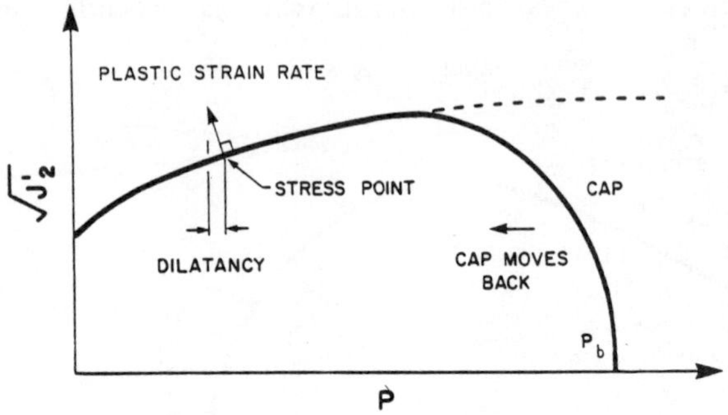

(a) CAP MOVES BACK WHEN FAILURE OCCURS

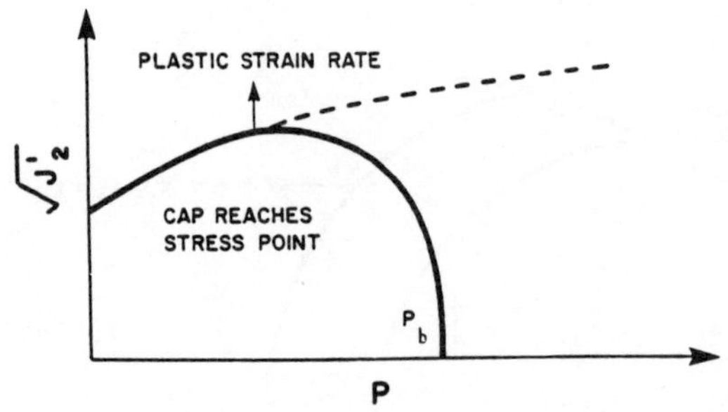

(b) BACKWARD CAP MOVEMENT CEASES WHEN CAP REACHES STRESS POINT, LIMITING DILATANCY

Figure 28.2 Limitation of dilatancy

loading, a cap which moves in this manner causes the amount of dilatancy during shear failure to be limited, as depicted in Figure 28.2. If the stress point reaches the failure envelope, the plastic strain rate vector has a negative volumetric component (dilatancy). This causes the cap to move back until it reaches the stress point, thus limiting further plastic volume increases.

In ref. [6], and many subsequent applications, the failure surface was chosen to exponentially approach a Mises surface for large confining pressures, and the cap was represented by an ellipse. In order to avoid slope discontinuities in the triaxial stress–strain behaviour of the model, the cap was chosen so that its tangent at its intersection with the failure envelope is horizontal (in the J_1, $\sqrt{J_2'}$ plane). Then, referring to Figure 28.3(a), equations (28.1) and (28.2) become

$$\sqrt{J_2'} = A - C\,e^{-BJ_1} \tag{28.3}$$

$$(J - L)^2 + R^2 J_2' = (X - L)^2 \tag{28.4}$$

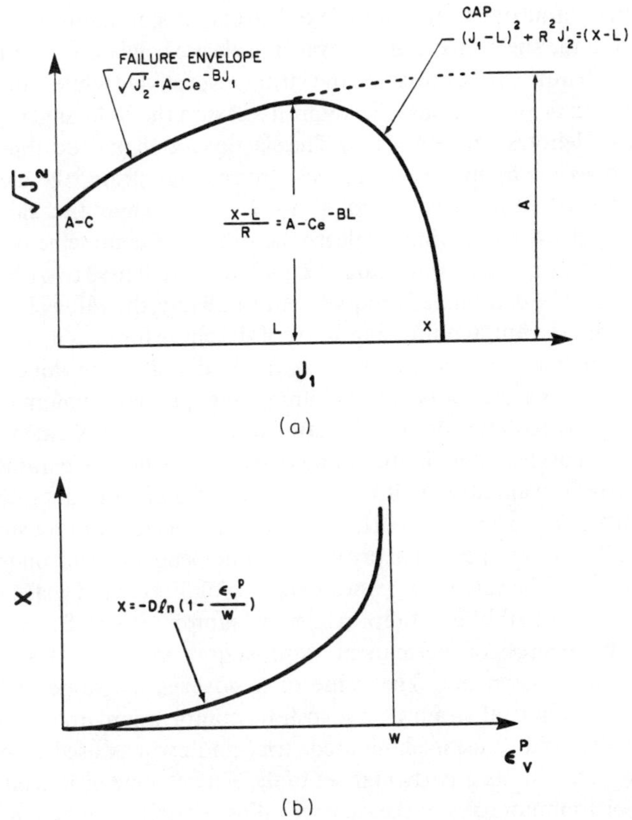

Figure 28.3 Elliptical cap and hardening rule

with X and L related by

$$X - L = R(A - C\,e^{-BL}) \tag{28.5}$$

In conjunction with the cap of equations (28.4) and (28.5), a hardening rule

$$X = -D \ln\left(1 - \frac{\varepsilon_v^p}{W}\right) \tag{28.6}$$

displayed in Figure 28.3(b), was proposed, and has been found appropriate for several soils. If L is thought of as the hardening parameter k, then substitution of equation (28.5) into (28.6) gives an implicit relationship defining $k(\varepsilon_v^p)$.

28.2.1.1 *Selecting parameters for the model*

The plasticity model represented by equations (28.3)–(28.6) contains six parameters A, B, C, D, R, and W. As indicated in Figure 28.4, these may be obtained from laboratory uniaxial strain and triaxial stress tests, in addition to the bulk modulus K and the shear modulus G governing elastic behaviour. In Figure 28.4, σ, and p, and ε denote stress, pressure, and strain respectively while r and z refer to the radial and vertical directions. The quantity next to the right angle symbol is a slope and $f(-)$ denotes function of $(-)$. The elastic moduli may be obtained from the initial slopes of the unloading curves. In uniaxial strain, the slope of the unloading stress–strain curve is equal to $K + \frac{4}{3}G$, while the slope of the corresponding stress path, which reflects the degree of confinement needed to enforce the condition of uniaxial strain, is equal to $2G/K$. These two observations enable K and G to be determined uniquely. Alternatively, the value of $2G$ could be determined from the initial unloading slope of the curve for $\sigma_z - \sigma_r$ versus $\varepsilon_z - \varepsilon_r$ in triaxial compression or from the shear wave speed *in situ*. The value of K could be determined from the slope of the unloading pressure–volumetric strain relationship in hydrostatic compression, or from the value of K calculated from the combination of shear and dilational wave speeds *in situ*. The parameters A, B, and C are usually evaluated by fitting a curve through failure data obtained in triaxial compression. The stress path in loading in uniaxial strain usually tends toward the failure envelope, which provides an independent check on the triaxial compression data. The value of W represents the volume of void space which can be eliminated irreversibly by compressing the sample; it can be evaluated by measuring the degree of permanent compaction in a uniaxial strain or hydrostatic compression test. The value of D governs the slope of the initial loading curve in uniaxial strain or hydrostatic compression, and thus controls how rapidly the void space is eliminated; trial and error is used to obtain the value of D which best fits a particular set of data. The value of R is equal to the ratio of major to minor axes of the quarter-ellipse defining the cap. R is chosen such that the loading slope of $\sigma_z - \sigma_r$ versus p of the model in a uniaxial strain test agrees with measurements. If the slope to be matched is steep, R should be small; as the slope decreases, R increases.

The eight parameter model described above was used in ref. [4] to fit uniaxial and triaxial data for McCormick Ranch sand.

In [7], an algorithm and modular subroutine for the cap model is presented which can easily be incorporated into most dynamic codes. The model has been altered slightly in [7] by replacing L in (28.4) by max $(L, 0)$. This improves the model behaviour somewhat at low stresses [7].

Generalizations of this model have been described which account for strain rate effects [8], anisotropy [9], non-linear (hypoelastic) behaviour [10], and work hardening of the yield envelope in shear (Baladi in [5]). The model is also

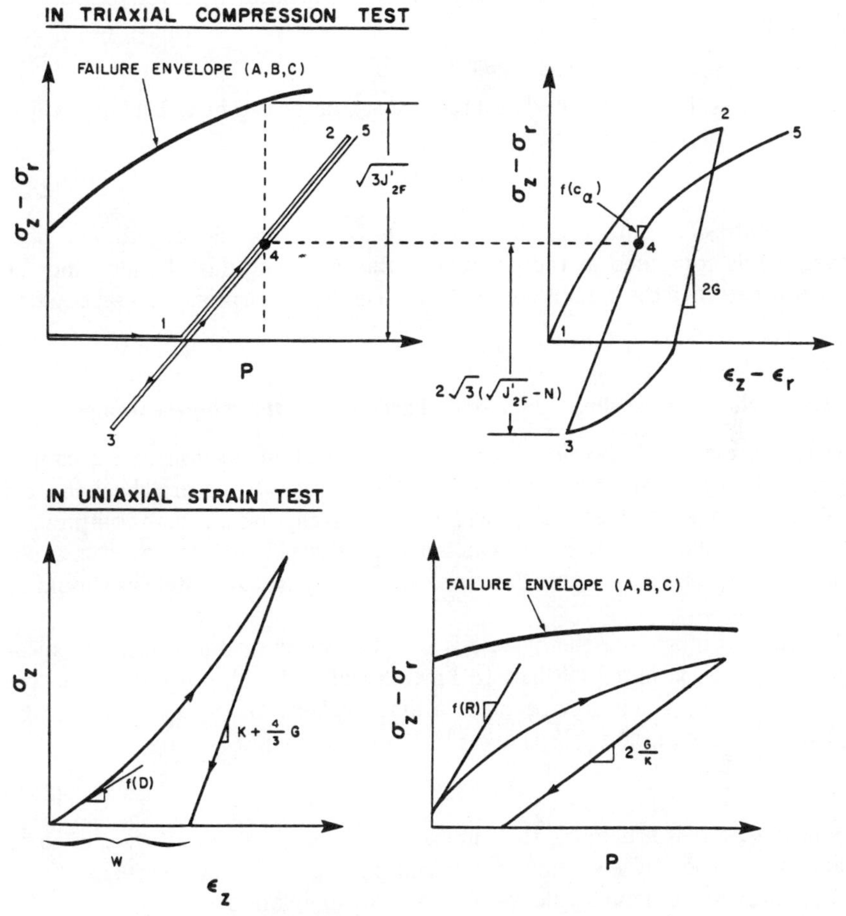

Figure 28.4 Physical meaning of cap parameters

used for rocks by replacing ε_v^p in (28.6) by its maximum previous value (so as to inhibit backward motion of the cap).

28.2.2 Fluid pore pressure—effective stress

The model described in Section 28.2.1 was primarily intended to be applied to dry, granular soils. In order to represent the behaviour of wet soils, the model was amended in ref. [11] using the Terzaghi method of handling pore pressure [12]. According to this approach, the total stress tensor is assumed to be the sum of the fluid pore pressure and the stress supported by the solid matrix. The latter is called the effective stress and is governed by the cap model in the same way as for dry materials.

Letting σ_{ij} be the stress tensor and π the pore pressure, the effective stress σ_{ij}^*, defined as

$$\sigma_{ij}^* = \sigma_{ij} - \pi\delta_{ij} \tag{28.7}$$

(in which δ_{ij} is the Kronecker delta) replaces σ_{ij} in the cap model. Thus the first stress invariant J_1 is replaced by

$$J_1^* = J_1 - 3\pi \tag{28.8}$$

while the stress deviators s_{ij}, and therefore J_2', remain unchanged. The pore pressure π is computed as the product of the bulk modulus of water and the volumetric strain if the latter is compressive; if not the pore pressure is set equal to zero.

28.2.3 Cyclic shear loading—kinematic hardening of the failure envelope

The cap presented is Section 28.2.1 and modified in Section 28.2.2 cannot represent hysteresis in cyclic shear loading. This is not a serious drawback in most applications for explosive loadings which primarily involve one high compressive stress peak of short duration. For earthquake-induced loadings, however, cyclic shear response is a major consideration and the associated hysteresis should be modelled.

In order to reproduce the hysteresis in cyclic shear loading, an initial shear yield surface is permitted to harden kinematically [1, 11], until is reaches the failure envelope, which now becomes a limit surface. This is accomplished by replacing J_2' in equations (28.1) and (28.2), or (28.3) and (28.4) by

$$Q_2' = \tfrac{1}{2}(s_{ij} - \alpha_{ij})^2 \tag{28.9}$$

in which α_{ij}, a tensor defining the translation of the shear yield surface in stress space, is governed by a kinematic hardening rule.

The simplest hardening rule for α_{ij} is the linear relation

$$\dot{\alpha}_{ij} = c_\alpha \dot{e}_{ij}^p \tag{28.10}$$

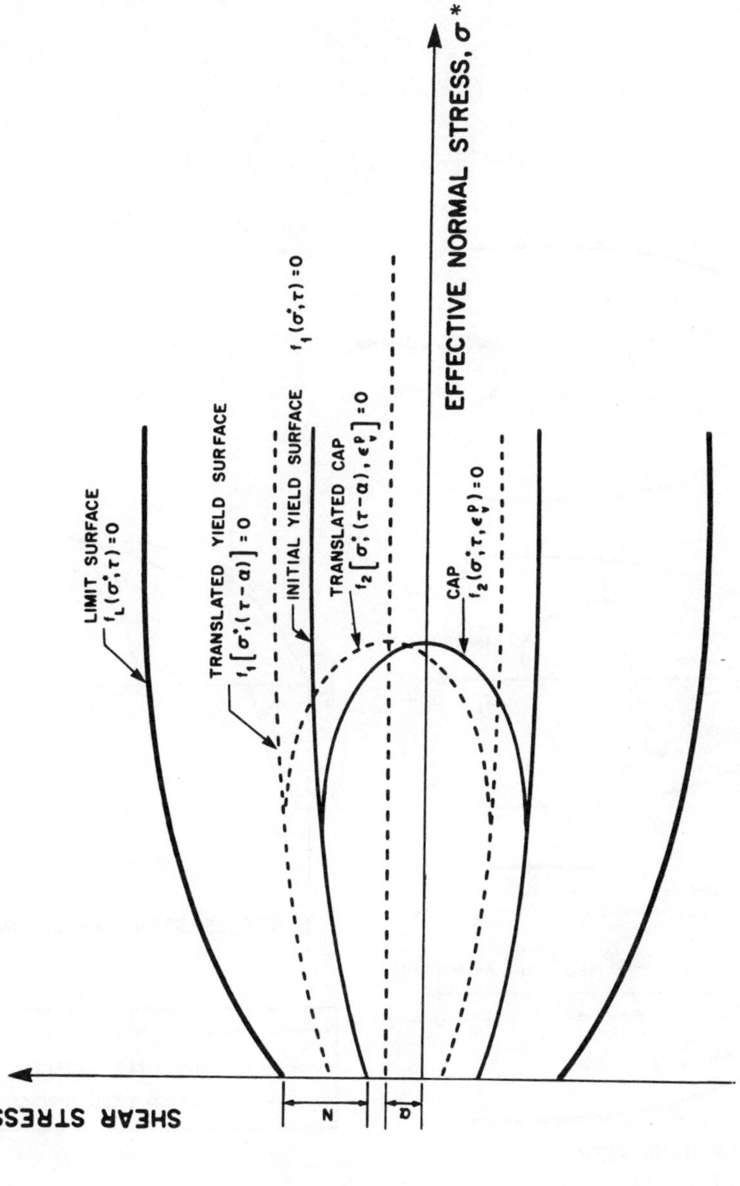

Figure 28.5 Non-linear kinematic cap model subjected to one component of shear and normal stress

in which c_α is a constant and e^{p}_{ij} is the plastic strain deviator, but this does not restrict the loading surface to remain within the failure envelope. This rule is satisfactory, however, for levels of shear stress which do not approach failure.

One of the simplest non-linear rules which can be postulated is

$$\dot{\alpha}_{ij} = c_\alpha F_\alpha \dot{e}^{\mathrm{p}}_{ij} \tag{28.11}$$

in which c_α is a constant and F_α is the scalar function

$$F_\alpha = \max\left(0, 1 - \frac{(s_{ij} - \alpha_{ij})\alpha_{ij}}{2N[\sqrt{(J'_{2F})} - N]}\right) \tag{28.12}$$

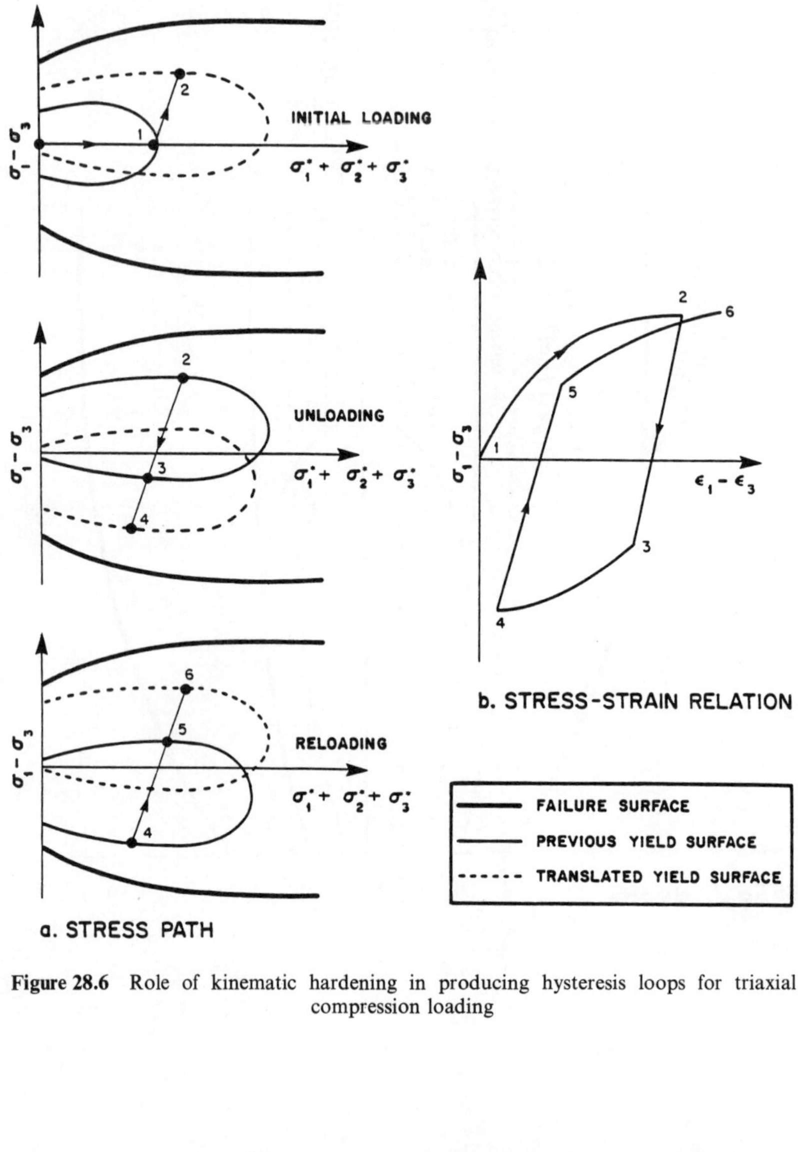

Figure 28.6 Role of kinematic hardening in producing hysteresis loops for triaxial compression loading

Here N is a constant defining the size of the yield surface and

$$f_{\mathrm{L}}(J'_{2F}, J_1) = 0 \tag{28.13}$$

defines the limiting failure surface, which may be in the form of equation 28.3. A translating yield surface is then defined by

$$\sqrt{Q'_2} = J'_{2F} - N \tag{28.14}$$

The non-linear kinematic hardening of the shear yield surface described above is illustrated in Figure 28.5 for the case when only one direct (effective) stress σ and one shear stress τ exist.

The value of F_α is related to the proximity of the yield surface to the failure surface and the location of the stress point on the yield surface. Its value may range from zero to 2. For $\alpha_{ij} = 0$, the right side of equation (28.12) reduces to 1. Therefore c_α is the inelastic slope for initial yielding of the material in shear. F_α decreases for continued yielding and becomes zero as the stress point approaches the limit surface. Unloading from such a stress state toward the hydrostatic axis, $F_\alpha = 2$ upon reyielding of the material.

In Figure 28.6, the manner in which hysteresis loops are produced in this model under cyclic stress in triaxial compression is depicted.

Comparisons of predictions made using the non-linear model described in this section with cyclic triaxial compression test data are exhibited in ref. [11]. Alternate hardening rules have been proposed in [13] and [14] to accomplish similar objectives.

28.2.3.1 *Selecting kinematic hardening parameters*

In order to include kinematic hardening, two additional parameters, c_α and N, must be defined. The slope of the inelastic shear stress–strain relation in a triaxial compression test is an implicit function of c_α; the value of this parameter must be found by trial and error, but a good estimate of c_α is the slope of the shear stress versus inelastic shear strain curve at low amplitude cycles of shear stress. The size of the elastic region in cyclic loading is defined by N, the distance between the failure surface and the initial yield surface along the J'_2 axis. In effect, N specifies the onset of inelastic strain as a fraction of the failure shear stress. The value of N can be estimated by observing the amplitude of shear stress in a cycle which produces only elastic shear strain. If used in conjunction with the 8-parameter model discussed in Section 28.2.1, this model requires 10 material constants to be determined from standard laboratory tests.

28.2.4 Stress anisotropy—kinematic hardening of cap

It is sometimes expedient to permit only the cap to harden kinematically in shear (as well as to expand), while maintaining the shear failure envelope as an ideal yield surface.

The failure envelope will then be of the form of equation (28.1), with equation (28.3) as one example.

Letting

$$Q_1^* = J_1^* - \alpha_0 \tag{28.15}$$

and Q_2' be defined by equation (28.9), the equation of a kinematically hardening cap becomes

$$f(Q_1^*, Q_2', \varepsilon_v^p, e_{ij}^p) = 0 \tag{28.16}$$

if hardening rules are assumed to depend only on the plastic volumetric strain and plastic strain deviator.

In one application (Sandler in [5]), discussed in detail below, a linear failure

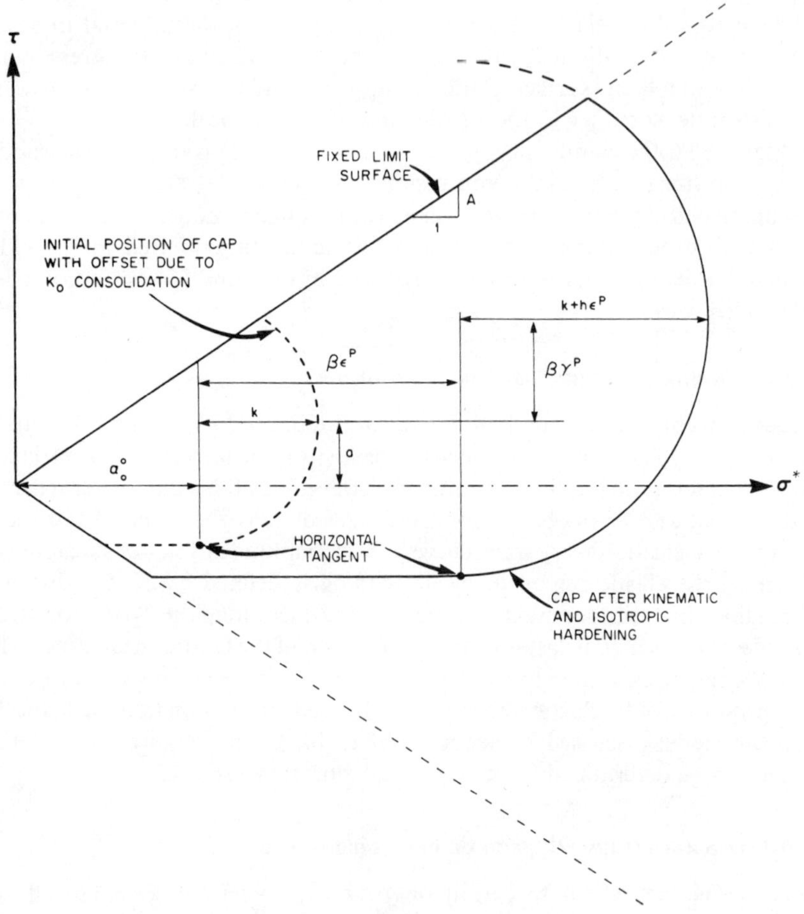

Figure 28.7 Kinematic and isotropic hardening of cap for kaolinite model

surface with no cohesion, was used together with an elliptical kinematically hardening cap with an initial offset and a linear hardening rule. The failure surface was taken as

$$\sqrt{J_2'} = AJ_1 \tag{28.17}$$

while the cap was represented by

$$Q_1^{*2} + R^2 Q_2' = X^2 \tag{28.18}$$

The cap was assumed to isotropically harden according to the rule

$$X^2 = k + h\varepsilon_v^p \tag{28.19}$$

while its kinematic hardening was defined by

$$\alpha_0 = \alpha_0^0 + \beta\varepsilon_v^p \tag{28.20}$$

$$\alpha_{ij} = \alpha_{ij}^0 + \beta e_{ij}^p \tag{28.21}$$

in which k, h, α_0^0, and β are constants and α_{ij}^0 is defined, through the constant a, as the tensor

$$\alpha_{ij} = \begin{bmatrix} a & & \\ & -\dfrac{a}{2} & \\ & & -\dfrac{a}{2} \end{bmatrix} \tag{28.22}$$

The initial offset caused by prestress, defined by α^0 and α_{ij}^0, introduces a form of stress-induced anisotropy in the material.

For the case of one direct stress σ and one shear stress τ (and the corresponding strains ε and γ), the movement of the cap is depicted in Figure 28.7.

When the two elastic constants are added, the model defined by equations (28.17) to (28.22) has nine constants: $A, R, k, h, \alpha_0^0, \beta, a, K, G$. Some like A, R, K, G, and h are determined in a manner similar to that described for the cap without kinematic hardening, discussed in Section 28.2.1 (h plays a role similar to D). The others are obtained by trial and error through comparison of model behaviour and measured results. This process is described in more detail in the following sections.

28.3 PREDICTION OF BEHAVIOUR OF AN ANISOTROPIC WET CLAY

28.3.1 NSF/NERC Workshop

During 28–30 May 1980, a Workshop [5] was held at McGill University. The purpose of this workshop was to discuss and compare predictions, made by eleven teams of predictors, of the stress–strain behaviour of four soils to a series of nine tests. The Predictors, proponents of different constitutive models for soils, submitted predictions *before the workshop* of the performance of each soil to loadings with specified stress paths. At the meeting, the actual test results were revealed for the first time and comparisons could be made with the predicted behaviour.

The team of G. Baladi and I. Sandler utilized cap models. Baladi used a transversely isotropic version of the model described in Section 28.2.1 to fit the stress–strain behaviour of two natural clays and an isotropic version with a strain-hardening failure envelope to model an Ottawa sand. Sandler used the model with a kinematically hardening cap of Section 28.2.4 to study a laboratory-prepared, wet, anisotropic kaolinite clay. In this section, Sandler's predictions are described in detail and compared with the subsequently revealed corresponding laboratory test results.

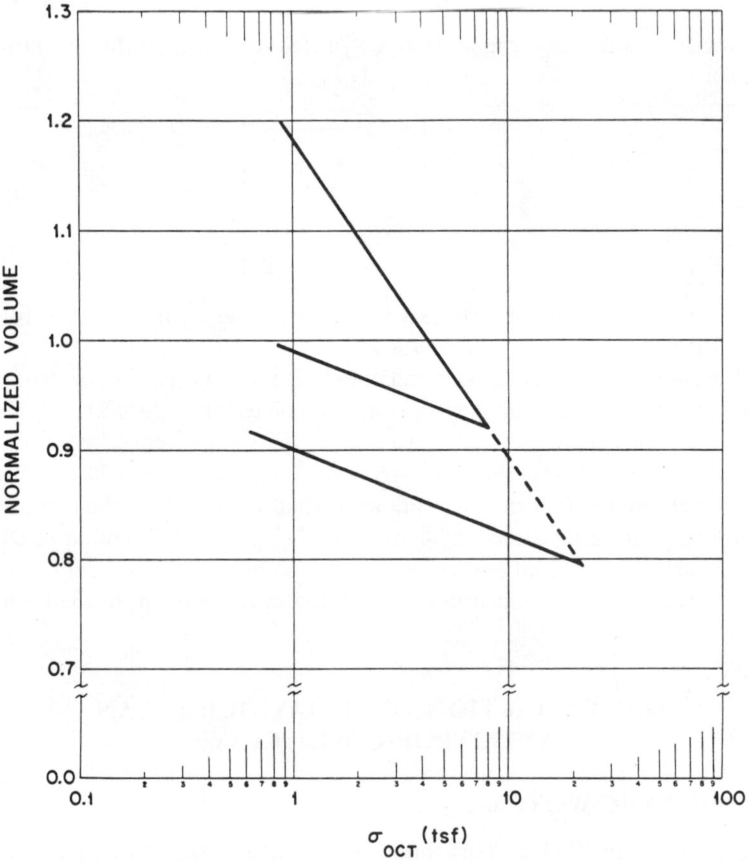

Figure 28.8 K_0-consolidation test

28.3.2 Test data available before predictions

Before each triaxial test, a remoulded kaolinite clay with liquid and plastic limits $W_L = 62.5$ per cent and $W_P = 39$ per cent, specific gravity = 2.61 and ratio of passive to active consolidation stresses $K_0 = 0.48$ was subjected to K_0 consolidation under an effective cell pressure of 40 p.s.i. The average response of the samples is illustrated in Figure 28.8.

The results of the five undrained triaxial tests which were provided in tabular form to each Predictor several months before the workshop are illustrated graphically by the solid curves of Figures 28.9–28.13. In these figures results are shown for undrained conventional triaxial compression and tension tests and compression and extension tests with constant mean stress. The stress paths and

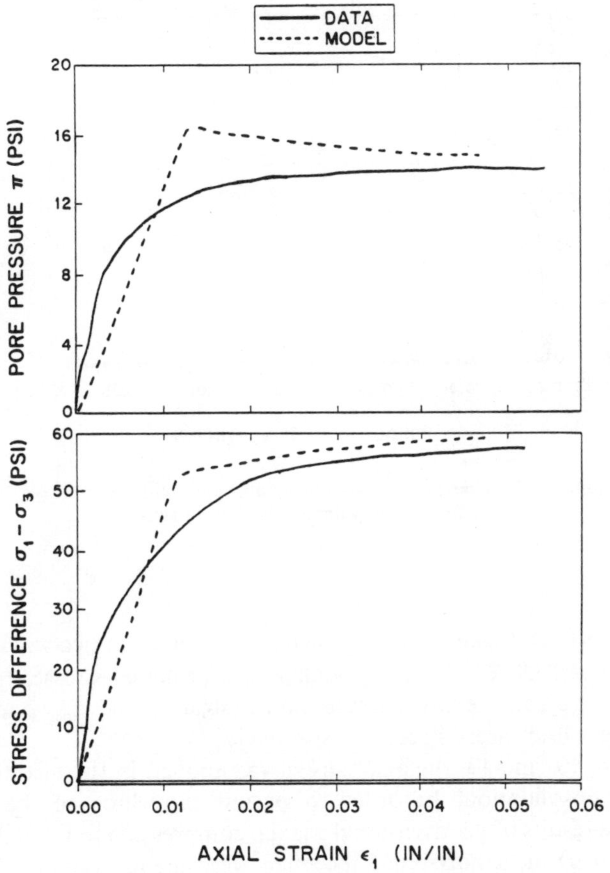

Figure 28.9 Comparison of model behaviour with laboratory data (conventional triaxial compression test)

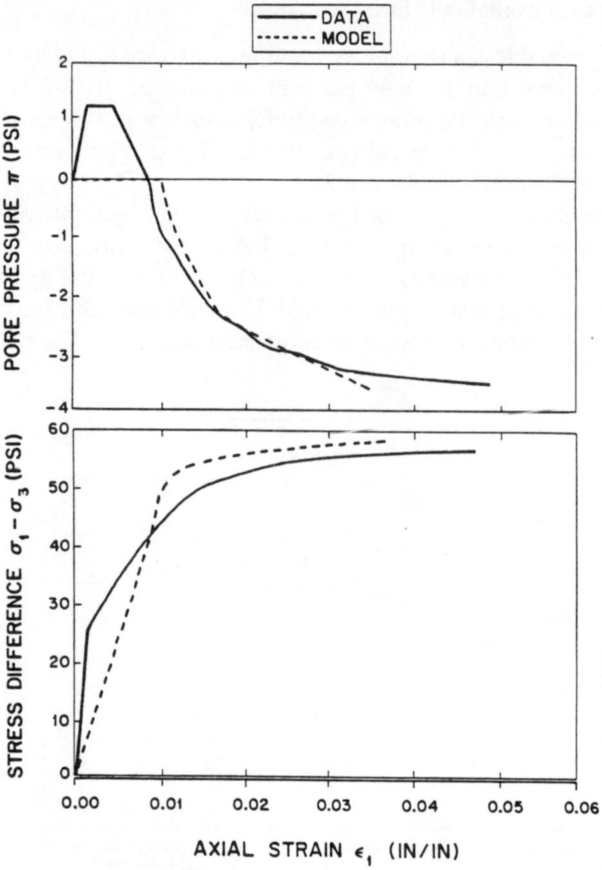

Figure 28.10 Comparison of model behaviour with laboratory data (compression with constant mean stress)

the test numbers (1, 4, 10, and 13) for these tests are shown in two ways in Figure 28.14. In Figures 28.9–28.14, σ_1, σ_2, and ρ_3 are principal stresses (positive if compression) $\varepsilon_1, \varepsilon_2, \varepsilon_3$ are the corresponding strains, and τ_{oct} and σ_{oct} are octahedral shear and direct stresses, respectively.

In Tests 1, 4, 10, and 13, the axial stress was applied in the direction of the original K_0 consolidation. In order to permit modelling of the resulting anisotropy, the results of a conventional triaxial compression test (not designated by a test number) on a horizontal specimen were made available. These are exhibited in Figure 28.13.

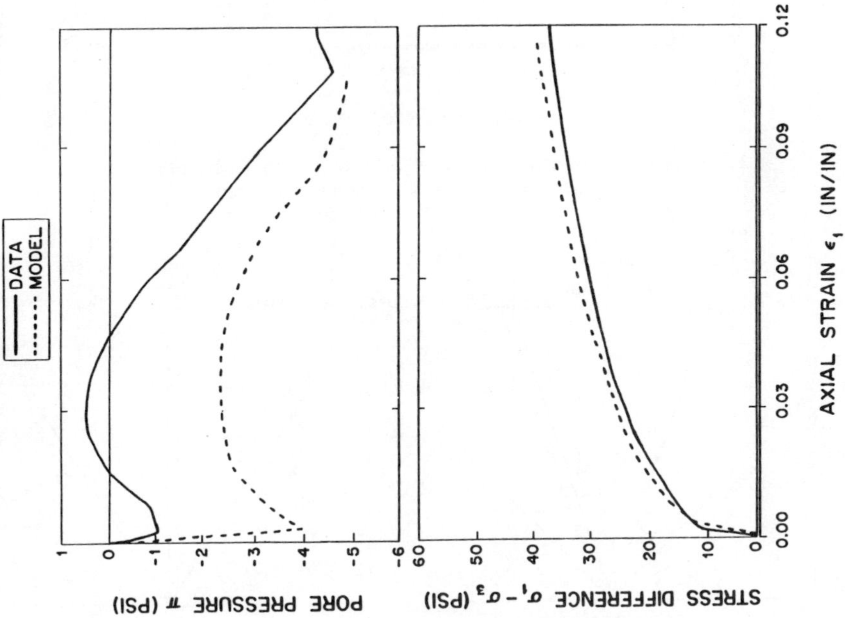

Figure 28.11 Comparison of model behaviour with laboratory data (extension with constant mean stress)

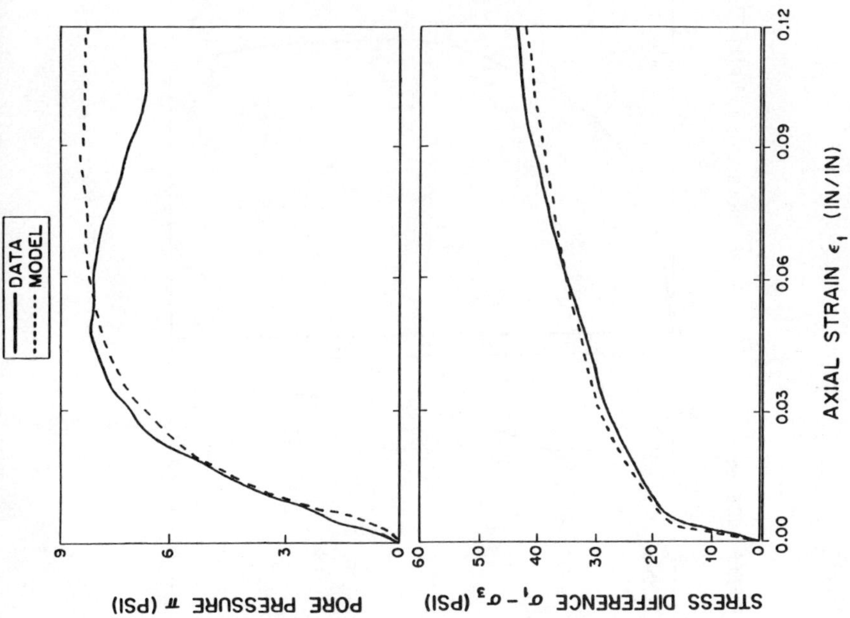

Figure 28.12 Comparison of model behaviour with laboratory data (conventional triaxial extension test)

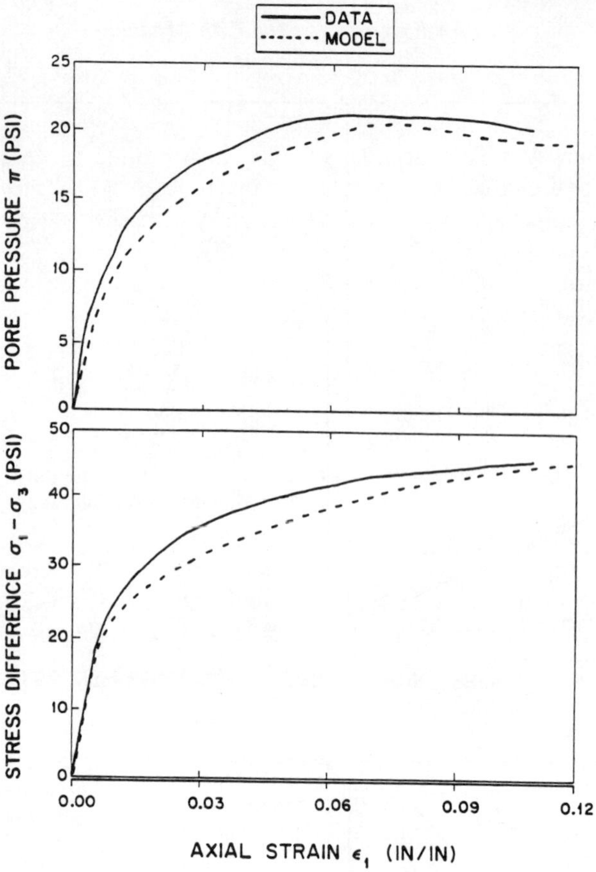

Figure 28.13 Comparison of model behaviour with laboratory data (conventional triaxial on a horizontal specimen)

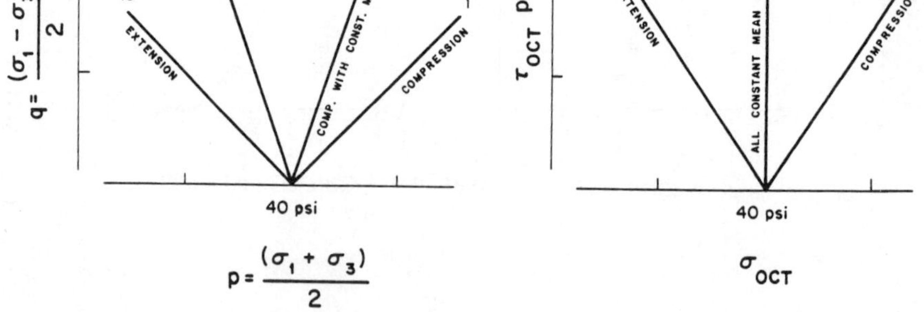

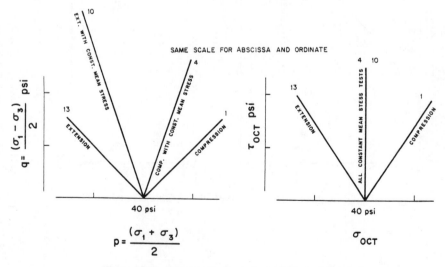

Figure 28.14 Stress paths for tests used for modelling

28.3.3 Choice of model and determination of fitting parameters

Ordinarily, a laboratory remoulded kaolinite clay would be expected to be nearly isotropic. However, because the clay was K_0 consolidated before the triaxial testing, and because the subsequent triaxial tests were known to exhibit substantial anisotropy, the model with a kinematically and isotropically hardening cap (where the cap has an offset in the direction of vertical compression) was assumed. In addition, the material was assumed to be incompressible under the undrained test conditions, i.e., the elastic and plastic volumetric strains were taken equal to each other in magnitude but of opposite sign.

Since all the laboratory tests—those for which data were provided and those to be predicted—were performed in a narrow range of mean pressure in the vicinity of 40 p.s.i, the simple 9-parameter model, with linear failure surface and hardening laws, defined by equations (28.17) to (28.22) was chosen. This made it possible to run the model on a hand-held programmable calculator. All of the model fits of this subsection and the predictions in the next were obtained using only a Texas Instruments TI-59, programmed as shown in the appendix.

The material constant h of equation (28.19) and elastic bulk modulus K were estimated from the average loading and unloading bulk moduli in the consolidation tests. G was determined from the low-stress, nearly linear, shear response in the undrained triaxial tests and A was determined from the failure (large strain) states observed in those tests. The values for these parameters are

$K = 2500$ p.s.i
$G = 1500$ p.s.i.
$h = 2 \times 10^6$ p.s.i.^{2}
$A = 0.25$

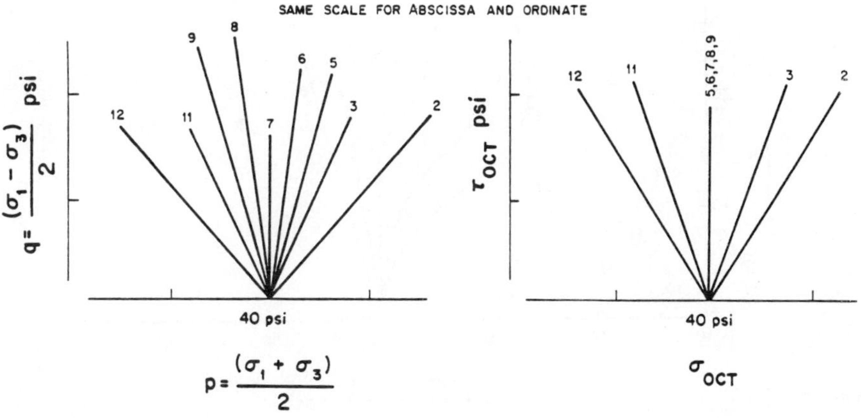

Figure 28.15 Stress paths for tests whose stress–strain curves are to be predicted

Table 28.1 Angle of major principal stress axis to vertical

Test no.	Angle (degrees)
2	15
3	37.5
5	15
6	31.75
7	45
8	58.25
9	75
11	58.25
12	75

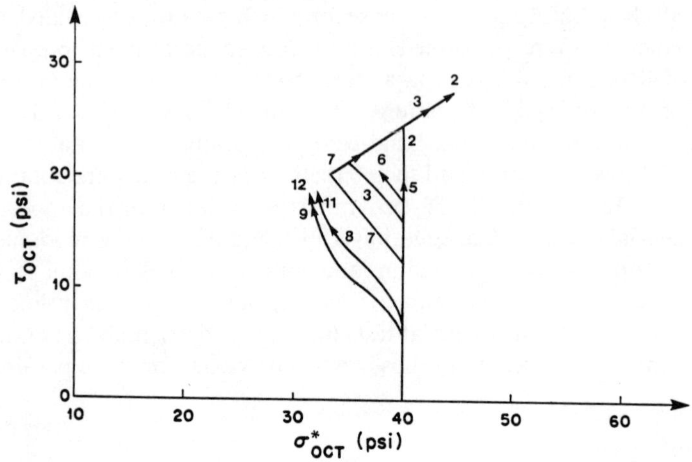

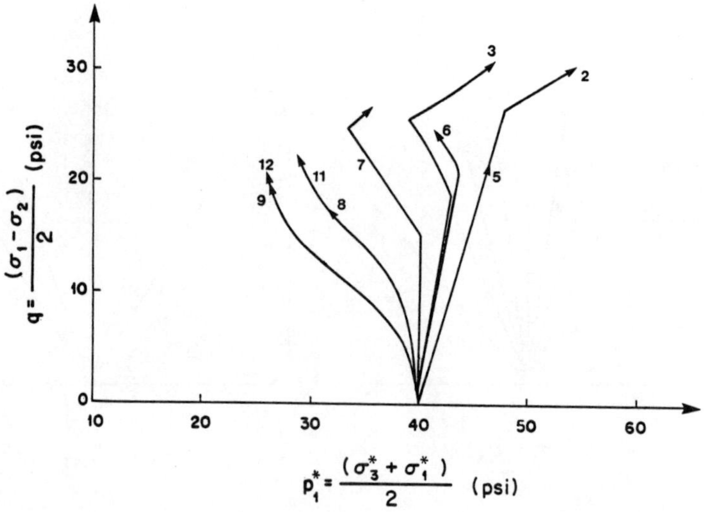

Figure 28.16 Effective stress path in tests whose stress–strain curves are to be predicted

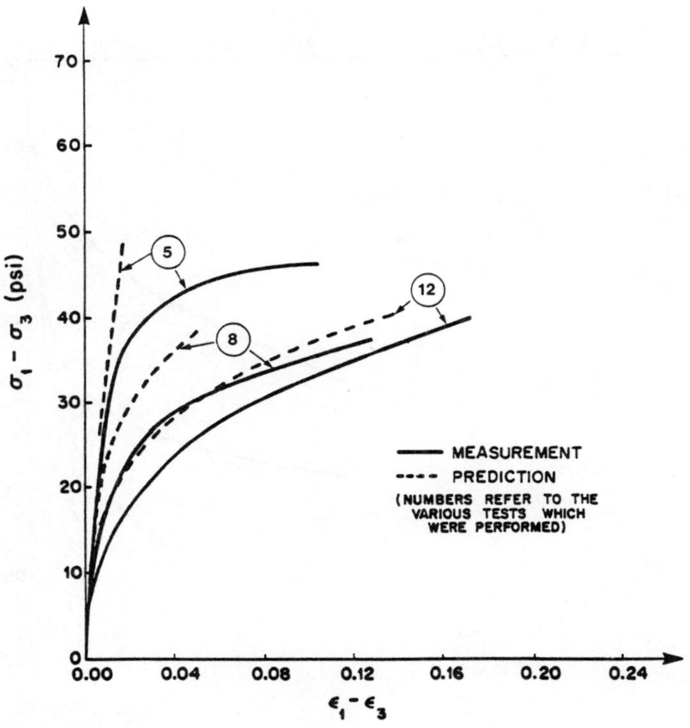

Figure 28.17 Comparison of predicted and observed stress–strain differences for kaolinite

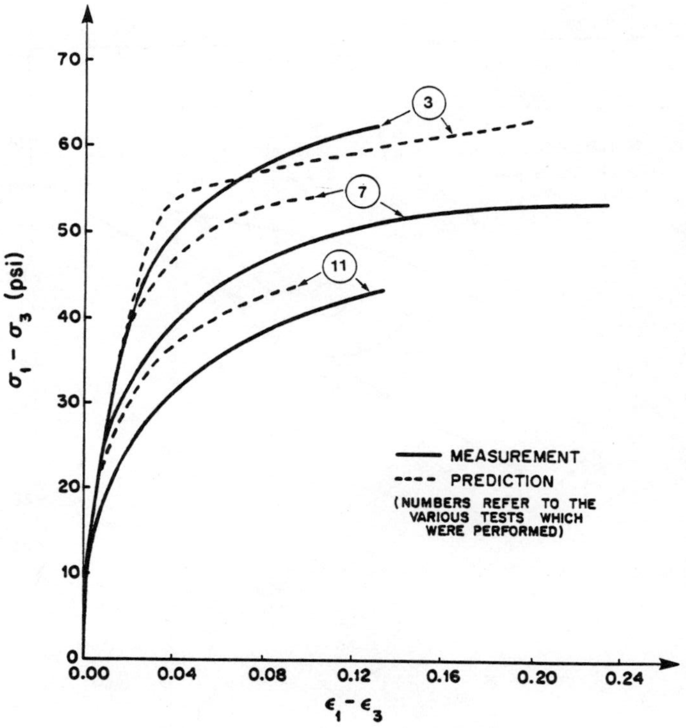

Figure 28.18 Comparison of predicted and observed stress–strain differences for kaolinite

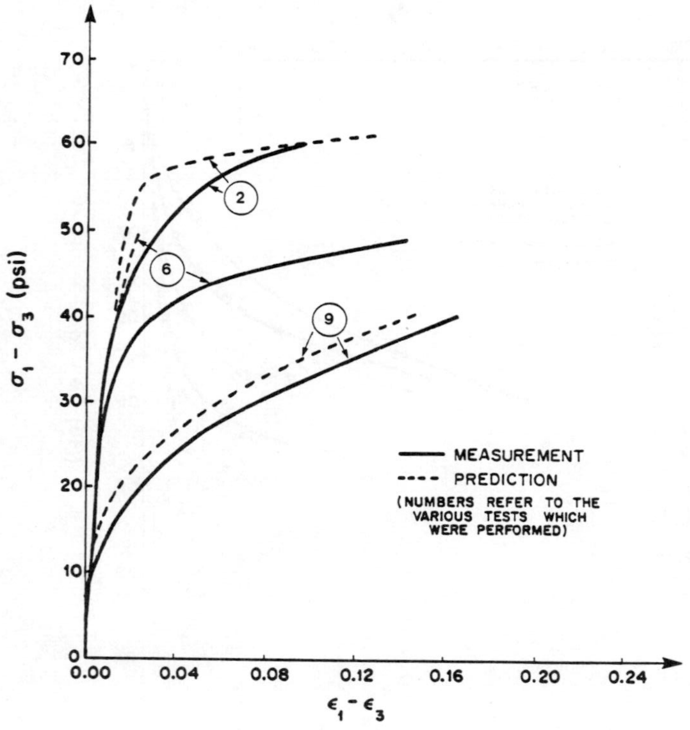

Figure 28.19 Comparison of predicted and observed stress–strain differences for kaolinite

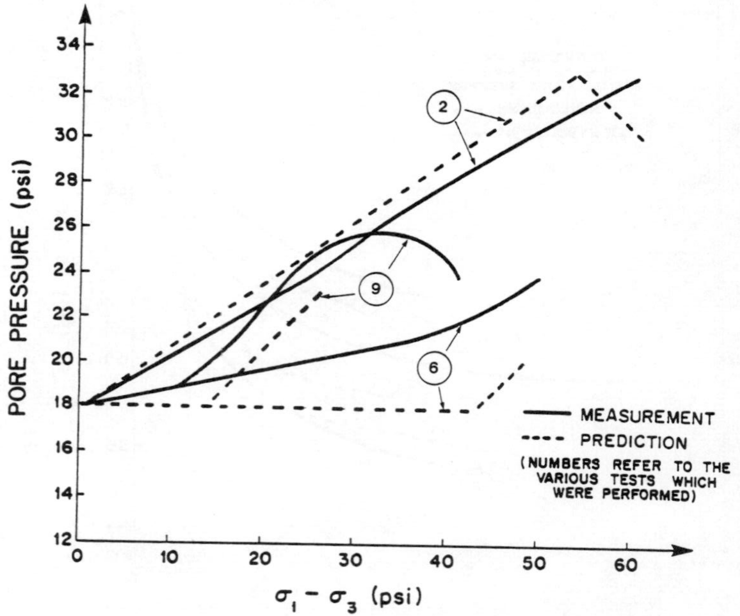

Figure 28.20 Comparison of predicted and observed pore pressure for kaolinite

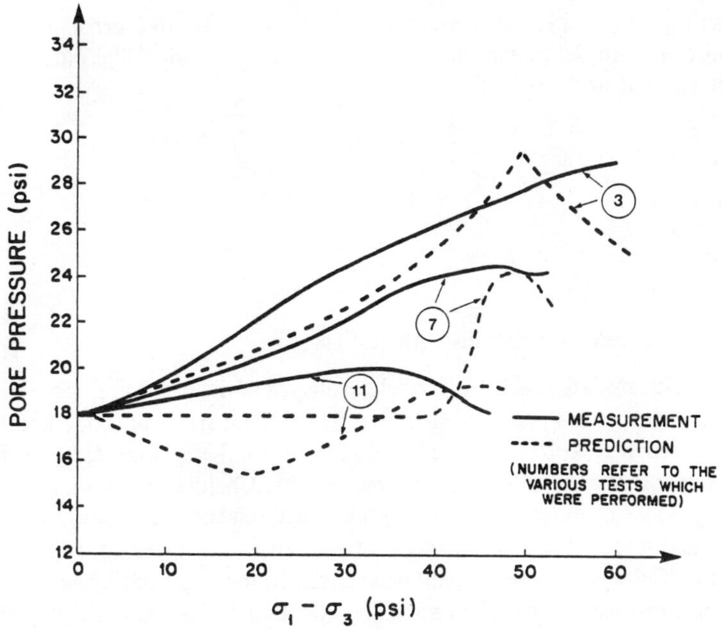

Figure 28.21 Comparison of predicted and observed pore pressure for kaolinite

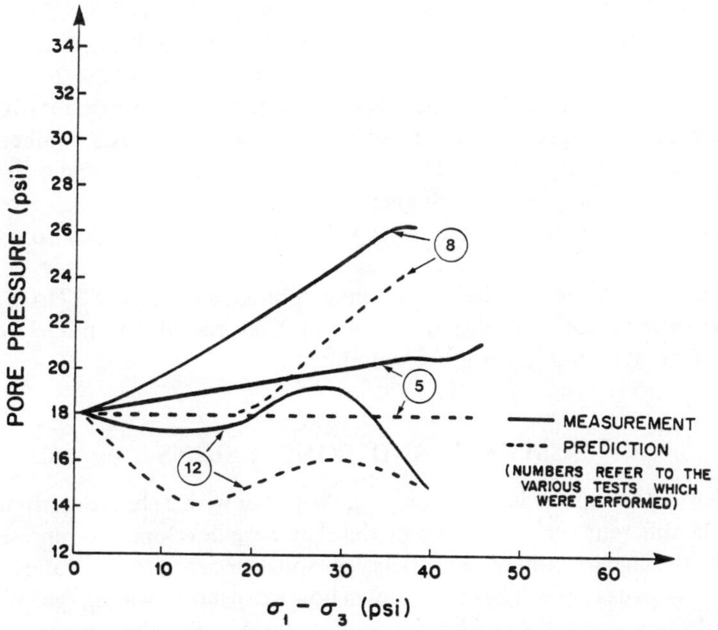

Figure 28.22 Comparison of predicted and observed pore pressure for kaolinite

The remaining five parameters were obtained by trial and error through comparison of the model behaviour with the measured results. The parameters thereby determined are

$$R = 4.5$$
$$k = 8000\ \text{p.s.i.}^2$$
$$\alpha_0^0 = 95\ \text{p.s.i.}$$
$$\beta = 100\ \text{p.s.i.}$$
$$a = 14\ \text{p.s.i.}$$

28.3.4 Predictions and correlations with test results

It was the Predictor's task to use the model developed in the preceding subsection, on the basis of the results provided from the five triaxial stress and one uniaxial strain tests described above, to predict stress–strain behaviour for the nine undrained triaxial tests whose stress paths, provided before the workshop are shown in Figure 28.15 in two ways. The angles between the major principle axis and the vertical, made available with the stress paths, are listed in Table 28.1.

Using the model whose development was described in the preceding subsection and the program listed in the appendix, the effective stress paths displayed in Figure 28.16 (in the format requested by the workshop organizers before the meeting) and the dashed-line pore pressure and stress–strain difference curves of Figures 28.16 to 28.22 were predicted for tests denoted by numbers 2, 3, 5, 6, 7, 8, 9, 11, and 12. The actual tests results, made available at the Workshop after predictions were submitted, are shown in solid lines on Figures 28.17 to 28.22. As can be seen from the figures, the model does a very good job, both qualitatively and quantitatively, to represent kaolinite behaviour, except for tests numbers 5 and 6. Even for these tests, however, the stress–strain curves would bend over (to indicate failure) at only slightly higher stress levels so that the overall representations of these tests are better than would ordinarily be judged from the figures.

It should be emphasized that the predictions exhibited in Figures 28.17 to 22.22 are those submitted prior to the dissemination of the test results. The model from which they were obtained has not been modified.

28.4 SUMMARY AND CONCLUSIONS

A bried review of the cap model has been given, together with a short description of some of the different versions of the model that were developed to represent and predict the behaviour of a variety of soils under a wide range of conditions. As an example, it has been shown how a cap model was merged with Terzaghi's effective stress theory for pore pressure to reflect the behaviour of an anisotropic wet clay in undrained triaxial compression. In order to represent the substantial anisotropy, the model has a kinematically hardening cap with an initial offset in the direction of vertical compression. For simplicity, the model employs only nine parameters and can be fit and exercised on a hand-held programmable calculator. The model fit was based on several undrained triaxial tests on horizontal and vertical specimens together with some consolidation data. It was exercised to predict behaviour of inclined specimens under a series of stress paths. These predictions were made before corresponding laboratory data were available for purposes of comparison. Very good agreement between predictions and experiments was obtained.

In conjunction with previous papers concerning other versions of the model, for different soils and applications, the results presented herein indicate that the cap model is a practical and efficient general framework for predicting soil behaviour in a wide range of situations.

Since the model, when properly formulated, satisfies Drucker's postulates, boundary and initial value problems involving it and the conservation laws are well posed. Thus, in addition to being able to represent and predict static soil behaviour in laboratory tests, it is appropriate as the constitutive component in computer codes for solving large scale problems involving the dynamic response of soils.

REFERENCES

1. D. C. Drucker and L. Palgen, 'On stress–strain relations suitable for cyclic and other loading', *Journal of Applied Mechanics*, **48**, 479–485 (1981).
2. R. Hill, *The Mathematical Theory of Plasticity*, Clarendon Press, Oxford, 1950.
3. D. C. Drucker, 'Some implications of work hardening and ideal plasticity', *Quarterly of Applied Mathematics*, **7**, 411–418 (1950).
4. D. C. Drucker, 'A more fundamental approach to stress–strain relations, *Proc., Ist U.S. Congress of Applied Mechanics, ASME*, 1950, 487–491.
5. *Proc. of the Workshop on Limit Equilibrium, Plasticity and Generalized Stress–Strain in Geotechnical Engineering*, McGill University, May 28–30, 198, ASCE.
6. F. L. DiMaggio and I. S. Sandler, 'Material model for granular soils', *J. Engrg. Mech. Div., ASCE*, **97**, No. EM3 935–950 (1970).
7. I. S. Sandler and D. Rubin, 'An algorithm and a modular subroutine for the cap model', *Int. J. Num. and Analyt. Meth. in Geomech.*, **3** (19) 173–186.
8. F. L. DiMaggio and I. S. Sandler, 'The Effect of Strain Rate on the Constitutive Equations of Rocks', *Tech. Report DNA* 2801*T*, Contract No. DNA001-72-C-000, Weidlinger Associates, New York, NY, Oct., 1971.
9. I. S. Sandler and F. L. DiMaggio, 'Anisotropy in Elastic-Plastic Models of Geological Materials', Progress Report, Contract No. DNA001-73-C-0023, Weidlinger Associates, New York, NY, Nov. 1973.
10. I. S. Sandler, F. L. DiMaggio, and G. Y. Baladi, 'Generalized Cap Model for Geological Materials', *J. Geotech. Engrg. Div., ASCE*, **102**, No. GT7 683–699 (1976).
11. I. S. Sandler and M. L. Baron, 'Recent Developments in the Constitutive Modeling of Geological Materials', *Proc. 3rd Int'l. Conf. on Num. Meth. in Geomechanics*, Aachen, 2–6 April 1979, A. A. Balkema, Rotterdam, 1979.

12. K. Terzaghi, 'The Shearing Resistance of Saturated Soils and the Angle Between the Planes of Shear', *Int. Conf. on Soil Mechanics and Foundations Engineering*, Vol. 1, p. 54, Cambridge, Mass., 1936.
13. Z. Mróz, H. P. Shrivastava, and R. N. Dubey, 'A nonlinear hardening model and its application to cyclic loading', *Acta Mechanica*, **25**, 51–61 (1976).
14. J. G. Ghaboussi and M. Karshenas, 'On the Finite Element Analysis of Certain Nonlinearities in Geomechanics', *Proc. Int'l Conf. on Finite Elements in Nonlinear Solids and Structural Mechanics*, Geilo, Norway, August 1977, Tapir Publisher Co., Trondheim.

APPENDIX CAP MODEL PROGRAM FOR T1-59 CALCULATOR

The next page gives a program description and user instructions for the T1-59 calculator program used to fit and exercise the cap model described in this paper. A listing of the program steps and data are given on the following pages.

PROGRAM DESCRIPTION

'his program calculates undrained triaxial response of a specific kaolinite clay using the cap model. θ is the ngle (to the vertical, in degrees) at which the specimen is loaded, $\sigma_{1,2,3}$ (p.s.i.) are applied stresses, π is the ore fluid pressure, $\varepsilon_{1,2,3}$ are the normal strains, and ε_{13} is the shear strain in the 1–3 plane. (The model is ransversely isotropic with the axis of symmetry assumed in the 1–3 plane).

USER INSTRUCTIONS

TEP	PROCEDURE	ENTER	PRESS	DISPLAY
1	Initialize		RST R/S	0.
2	Enter θ	$\theta°$	R/S	—
3	Enter σ_1, σ_2, σ_3	σ_1, σ_2, σ_3	R/S	—
4	Enter π	π	R/S	σ_1
5	Load to new value of σ_1	σ_1	R/S	σ_2
6	New σ_2	σ_2	R/S	σ_3
7	New σ_3	σ_3	R/S	π
8	Read π Then obtain ε_1	—	R/S	ε_1
9	Read ε_1 Then obtain ε_2	—	R/S	ε_2
10	Read ε_2 Then obtain ε_3	—	R/S	ε_3
11	Read ε_3 Then obtain ε_{13}	—	R/S	ε_{13}
12	Read ε_{13} Then return to step 5 Repeat as desired	—	R/S	σ_1

USER DEFINED KEYS DATA REGISTERS (INV list) LABELS (Op 08)

USER DEFINED KEYS	REG	DATA	REG	DATA
A	1_0	3K	2_0	σ_1
B	1^1	2G	2^1	σ_2
C	1^2	A	2^2	σ_3
D	1^3	R^2	2^3	π
E	1^4	κ	2^4	ε_v^p
A′	1^5	h	2^5	ε_1
B′	1^6	a	2^6	ε_2
C′	1^7	BR^2	2^7	ε_3
D′	1^8	α_0^0	2^8	ε_{13}
E′	9		9	

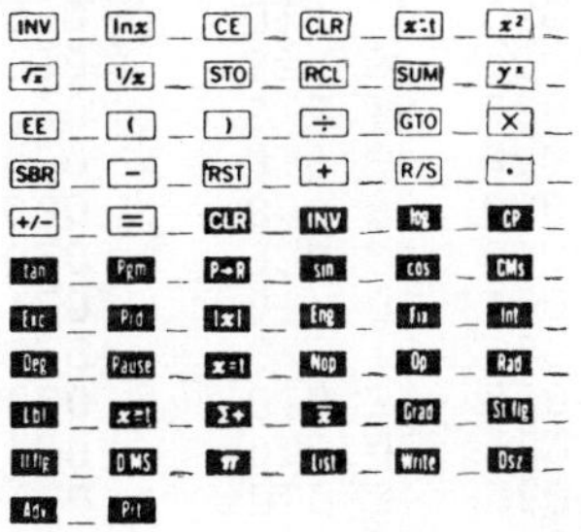

FLAGS 0 1 2 3 4 5 6 7 8 9

 1014966

PROGRAM LISTING

000	01	I
001	52	EE
002	94	+/-
003	02	2
004	55	÷
005	53	(
006	43	RCL
007	10	10
008	85	+
009	43	RCL
010	11	11
011	95	=
012	42	STO
013	09	09
014	22	INV
015	52	EE
016	00	0
017	42	STO
018	24	24
019	42	STO
020	25	25
021	42	STO
022	26	26
023	42	STO
024	27	27
025	42	STO
026	28	28
027	91	R/S
028	42	STO
029	08	08
030	43	RCL
031	20	20
032	91	R/S
033	42	STO
034	20	20
035	42	STO
036	35	35
037	85	+
038	43	RCL
039	21	21
040	91	R/S
041	42	STO
042	21	21
043	42	STO
044	36	36
045	85	+
046	43	RCL
047	22	22
048	91	R/S
049	42	STO
050	22	22
051	42	STO
052	37	37
053	75	-
054	42	STO
055	30	30
056	03	3
057	65	×
058	43	RCL
059	23	23
060	91	R/S
061	42	STO
062	23	23
063	95	=
064	42	STO
065	29	29
066	43	RCL
067	30	30
068	94	+/-
069	55	÷
070	03	3
071	95	=
072	44	SUM
073	35	35
074	44	SUM
075	36	36
076	44	SUM
077	37	37
078	43	RCL
079	35	35
080	42	STO
081	31	31
082	43	RCL
083	36	36
084	42	STO
085	32	32
086	43	RCL
087	37	37
088	42	STO
089	33	33
090	43	RCL
091	16	16
092	55	÷
093	02	2
094	65	×
095	44	SUM
096	36	36
097	94	+/-
098	44	SUM
099	37	37
100	43	RCL
101	08	08
102	38	SIN
103	65	×
104	43	RCL
105	08	08
106	39	COS
107	65	×
108	03	3
109	95	=
110	94	+/-
111	42	STO
112	38	38
113	43	RCL
114	16	16
115	65	×
116	53	(
117	43	RCL
118	08	08
119	39	COS
120	33	X²
121	75	
122	43	RCL
123	08	08
124	38	SIN
125	33	X²
126	55	÷
127	02	2
128	95	=
129	44	SUM
130	37	37
131	22	INV
132	44	SUM
133	35	35
134	76	LBL
135	11	A
136	43	RCL
137	20	20
138	91	R/S
139	42	STO
140	00	00
141	75	-
142	48	EXC
143	20	20
144	85	+
145	42	STO
146	01	01
147	53	(
148	43	RCL
149	21	21
150	91	R/S
151	44	SUM
152	00	00
153	75	-
154	48	EXC
155	21	21
156	54	)
157	42	STO
158	02	02
159	85	+

PROGRAM LISTING (cont'd.)

160	53	(
161	43	RCL
162	22	22
163	91	R/S
164	44	SUM
165	00	00
166	75	-
167	48	EXC
168	22	22
169	54	)
170	42	STO
171	03	03
172	95	=
173	55	÷
174	03	3
175	22	INV
176	49	PRD
177	00	00
178	95	=
179	44	SUM
180	23	23
181	94	+/-
182	44	SUM
183	01	01
184	44	SUM
185	02	02
186	44	SUM
187	03	03
188	43	RCL
189	01	01
190	44	SUM
191	31	31
192	44	SUM
193	35	35
194	43	RCL
195	02	02
196	44	SUM
197	32	32
198	44	SUM
199	36	36
200	43	RCL
201	03	03
202	44	SUM
203	33	33
204	44	SUM
205	37	37
206	43	RCL
207	11	11
208	35	1/X
209	49	PRD
210	01	01
211	49	PRD
212	02	02
213	49	PRD
214	03	03
215	43	RCL
216	01	01
217	44	SUM
218	25	25
219	43	RCL
220	02	02
221	44	SUM
222	26	26
223	43	RCL
224	03	03
225	44	SUM
226	27	27
227	29	CP
228	43	RCL
229	35	35
230	33	X²
231	85	+
232	43	RCL
233	36	36
234	33	X²
235	85	+
236	43	RCL
237	37	37
238	33	X²
239	95	=
240	55	÷
241	02	2
242	85	+
243	43	RCL
244	38	38
245	33	X²
246	95	=
247	65	×
248	43	RCL
249	13	13
250	85	+
251	42	STO
252	30	30
253	53	(
254	43	RCL
255	29	29
256	75	-
257	43	RCL
258	18	18
259	75	-
260	43	RCL
261	17	17
262	55	÷
263	43	RCL
264	13	13
265	65	×
266	43	RCL
267	24	24
268	85	+
269	42	STO
270	19	19
271	50	I×I
272	54	)
273	33	X²
274	55	÷
275	04	4
276	75	-
277	43	RCL
278	14	14
279	75	-
280	43	RCL
281	15	15
282	65	×
283	43	RCL
284	24	24
285	95	=
286	32	X:T
287	77	GE
288	34	√X
289	32	X:T
290	42	STO
291	39	39
292	43	RCL
293	19	19
294	32	X:T
295	22	INV
296	77	GE
297	35	1/X
298	01	1
299	75	-
300	43	RCL
301	39	39
302	55	÷
303	43	RCL
304	30	30
305	95	=
306	34	√X
307	35	1/X
308	75	-
309	01	1
310	95	=
311	55	÷
312	43	RCL
313	17	17
314	95	=
315	42	STO
316	39	39
317	61	GTO
318	32	X:T
319	76	LBL

PROGRAM LISTING (cont'd.)

320	35	1/X	360	36	PGM	400	95	=	440	65	×
321	00	0	361	08	08	401	44	SUM	441	43	RCL
322	36	PGM	362	15	E	402	23	23	442	13	13
323	08	08	363	42	STO	403	76	LBL	443	95	=
324	11	A	364	39	39	404	32	X⇌T	444	49	PRD
325	43	RCL	365	65	×	405	43	RCL	445	04	04
326	39	39	366	06	6	406	39	39	446	49	PRD
327	55	÷	367	65	×	407	65	×	447	05	05
328	53	(	368	42	STO	408	43	RCL	448	49	PRD
329	43	RCL	369	08	08	409	17	17	449	06	06
330	19	19	370	53	(	410	85	+	450	49	PRD
331	65	×	371	43	RCL	411	01	1	451	07	07
332	02	2	372	10	10	412	95	=	452	43	RCL
333	65	×	373	85	+	413	35	1/X	453	04	04
334	53	(	374	43	RCL	414	49	PRD	454	44	SUM
335	24	CE	375	17	17	415	35	35	455	25	25
336	65	×	376	55	÷	416	49	PRD	456	43	RCL
337	43	RCL	377	43	RCL	417	36	36	457	05	05
338	10	10	378	13	13	418	49	PRD	458	44	SUM
339	85	+	379	54	)	419	37	37	459	26	26
340	43	RCL	380	85	+	420	49	PRD	460	43	RCL
341	15	15	381	01	1	421	38	38	461	06	06
342	54	)	382	95	=	422	43	RCL	462	44	SUM
343	85	+	383	35	1/X	423	35	35	463	27	27
344	43	RCL	384	65	×	424	42	STO	464	43	RCL
345	17	17	385	43	RCL	425	04	04	465	07	07
346	65	×	386	19	19	426	43	RCL	466	44	SUM
347	43	RCL	387	65	×	427	36	36	467	28	28
348	30	30	388	43	RCL	428	42	STO	468	76	LBL
349	55	÷	389	08	08	429	05	05	469	34	√X
350	03	3	390	65	×	430	43	RCL	470	43	RCL
351	95	=	391	44	SUM	431	37	37	471	31	31
352	36	PGM	392	24	24	432	42	STO	472	42	STO
353	08	08	393	43	RCL	433	06	06	473	01	01
354	12	B	394	10	10	434	43	RCL	474	33	X²
355	43	RCL	395	55	÷	435	38	38	475	85	+
356	09	09	396	22	INV	436	42	STO	476	43	RCL
357	36	PGM	397	44	SUM	437	07	07	477	32	32
358	08	08	398	29	29	438	43	RCL	478	42	STO
359	14	14	399	03	3	439	39	39	479	02	02

PROGRAM LISTING (cont'd.)

480	33	X²	520	55	÷	560	91	R/S	600	65	×
481	85	+	521	43	RCL	561	43	RCL	601	43	RCL
482	43	RCL	522	12	12	562	28	28	602	15	15
483	33	33	523	33	X²	563	91	R/S	603	94	+/-
484	42	STO	524	55	÷	564	61	GTO	604	75	-
485	03	03	525	43	RCL	565	11	A	605	43	RCL
486	33	X²	526	29	29	566	76	LBL	606	14	14
487	95	=	527	94	+/-	567	16	A'	607	85	+
488	55	÷	528	95	=	568	53	(	608	53	(
489	02	2	529	49	PRD	569	53	(	609	43	RCL
490	95	=	530	01	01	570	53	(	610	19	19
491	34	√X	531	49	PRD	571	42	STO	611	65	×
492	55	÷	532	02	02	572	39	39	612	32	X⇌T
493	43	RCL	533	49	PRD	573	65	×	613	54	)
494	12	12	534	03	03	574	06	6	614	33	X²
495	95	=	535	43	RCL	575	65	×	615	85	+
496	32	X⇌T	536	01	01	576	32	X⇌T	616	43	RCL
497	43	RCL	537	44	SUM	577	53	(	617	30	30
498	29	29	538	25	25	578	43	RCL	618	55	÷
499	77	GE	539	43	RCL	579	10	10	619	53	(
500	33	X²	540	02	02	580	85	+	620	01	1
501	32	X⇌T	541	44	SUM	581	43	RCL	621	85	+
502	75	-	542	26	26	582	17	17	622	43	RCL
503	48	EXC	543	43	RCL	583	55	÷	623	39	39
504	29	29	544	03	03	584	43	RCL	624	65	×
505	95	=	545	44	SUM	585	13	13	625	43	RCL
506	55	÷	546	27	27	586	54	)	626	17	17
507	03	3	547	76	LBL	587	85	+	627	54	)
508	94	+/-	548	33	X²	588	01	1	628	33	X²
509	65	×	549	43	RCL	589	54	)	629	54	)
510	44	SUM	550	23	23	590	35	1/X	630	92	RTN
511	23	23	551	91	R/S	591	65	×	631	00	0
512	03	3	552	43	RCL	592	32	X⇌T	632	00	0
513	55	÷	553	25	25	593	65	×	633	00	0
514	43	RCL	554	91	R/S	594	43	RCL	634	00	0
515	10	10	555	43	RCL	595	19	19	635	00	0
516	55	÷	556	26	26	596	85	+	636	00	0
517	44	SUM	557	91	R/S	597	43	RCL	637	00	0
518	24	24	558	43	RCL	598	24	24	638	00	0
519	06	6	559	27	27	599	54	)	639	00	0

DATA REGISTERS

57.93666667	00
.0006988732	01
-.0003087876	02
-.0003900855	03
.0022479071	04
-.0008523458	05
-.0013955612	06
-.0005149964	07
.0000295253	08
.0000009524	09
7500.	10
3000.	11
0.25	12
20.25	13
8000.	14
2000000.	15
14.	16
2025.	17
95.	18
26.19273684	19
93.44	20
42.25	21
38.12	22
16.8485936	23
-.0004352292	24
.0164751035	25
-.0070570796	26
-.0094180239	27
-.0008183664	28
123.2642192	29
8588.675389	30
35.50333333	31
-15.68666667	32
-19.81666667	33
0.	34
22.558511	35
-8.553579806	36
-14.0049312	37
-5.168163362	38
.0000049209	39

Mechanics of Engineering Materials
Edited by C. S. Desai and R. H. Gallagher

Chapter 29

Strain Softening Stress–Strain Relations for Concrete

S. P. Shah

29.1 INTRODUCTION

Stress–strain relationship of concrete subjected to short-term, monotonically increasing uniaxial compressive loading is characterized by an increasing non-linearity with an increase in stress, having a peak stress (often termed failure or compressive strength of concrete) and a strain softening or a descending part. A frequently used relationship between uniaxial stress and axial strain is given by:

$$Y = \frac{AX + (D-1)X^2}{1 + (A-2)X + DX^2} \tag{29.1}$$

where Y and X refer to stress and strain non-dimensionalized with respect to the corresponding values at the peak stress, and A and D are constants which can be determined by curve fitting. Note that A depends primarily on the ascending part whereas the constant D depends on the descending part of the stress–strain curve.

For the uniaxial case, there is no difference if the constitutive relation is expressed as above, or in terms of equivalent secant stiffness relation:

$$\sigma = E_s \varepsilon \tag{29.2}$$

where E_s is a function of strain for a given strain. It can also be expressed in terms of strain energy density function:

$$W(\varepsilon) = \int \sigma \, d\varepsilon \quad \text{and} \quad \sigma = \frac{\partial W}{\partial \varepsilon} \tag{29.3}$$

or in terms of incremental stress–strain relationships:

$$\dot{\sigma} = E_t \dot{\varepsilon} \quad \text{and} \quad E_t = \frac{\partial Y}{\partial X} \frac{\sigma_0}{\varepsilon_0} \tag{29.4}$$

Relationships (29.2), (29.3), and (29.4) can be easily derived from equation (29.1) for which one needs to know the constants A, and D and the stress and the strain at the peak point.

Even for uniaxial case, there is little information available on the strain softening part or, equivalently, on the parameter D, and how that parameter varies with the composition of concrete. In addition, lateral strains are rarely measured beyond the peak stress. Thus, almost no information is available for the descending portion of the axial stress versus lateral strain relationship.

One of the reasons for the scarcity of data is the difficulty in experimentally obtaining the post-peak response of concrete. In this paper, some of the experimental problems associated with determining the descending part of the stress–strain curves are described. Two novel methods of overcoming these problems are also presented.

Design of many concrete structures is based on the uniaxial response of concrete. Since sufficient information has not been readily available to define the entire stress–strain response, the current design practice does not explicitly include the descending part. However, the author and others have shown that for many design applications, such as reinforced concrete columns subjected to seismic loading, explicit use of accurate uniaxial response is critical for rational computations and predictions [1]. This is especially true for high strength concrete where the straightforward extension of the empirical equations developed for lower strength concrete may not be valid [2].

For concrete subjected to multiaxial loading,† the constitutive relationships expressed in terms of the total stress–strain relationship (equation (29.1)), the secant formulations (equation (29.2)), the strain energy formulation (equation (29.3)), or the incremental formulation (equation (29.4)) are not generally equivalent unless certain integrability conditions are satisfied and the existence of total differential is guaranteed [3, 4]. It can be shown that if the constitutive relations are derived from the energy potential then the path-independent reversible behaviour (hyperelasticity) is assured. Otherwise the constitutive models may be less restrictive and various formulations may not be fully equivalent. None the less, depending upon the end-use, and the desired simplicity, it might be acceptable to use less general forms of constitutive models for multiaxial loading.

To predict the response of concrete subjected to monotonically increasing loading, several constitutive models have recently been proposed [3, 4]. These models have been classified into elastic models, hyperelastic, hypoelastic, elastic ideally plastic, and hardening-softening models. Few, if any, of these models include strain softening and associated volume dilation in a realistic but simple manner. Many of the constitutive models represent non-linear behaviour of concrete by describing the elastic constants in terms of non-linear scalar functions of stress or strain tensors. For example, Palaniswamy and Shah [5] expressed the bulk modulus (K) and Poisson's ratio (ν) as functions of the first two principal

†The discussion here is restricted to compressive, monotonically increasing, short-term, proportional loading.

stress invariants (I_1, I_2). Comparison with the experimental data for triaxial loading showed that their model predicted stress–strain curves quite well only up to about 80 per cent of the peak maximum compressive stress. A total stress–strain model based on secant Young's modulus (E_s) and Poisson's ratio has been presented by Ottosen [6]. The secant modulus was expressed as a function of a non-linearity index (which depends on the failure criteria) and the second invariant of the deviatoric stresses (J_2). A satisfactory expression of the Poisson's ratio in the strain softening region was not given because of the lack of data.

If it is assumed that the hydrostatic and deviatoric responses are decoupled (that is, the strain energy density function does not depend on the third stress or strain invariant) then the bulk modulus can be expressed as a function of only the first stress or strain invariants and the shear modulus (G) as a function of only the second deviatoric stress or strain invariant (τ_{oct} or γ_{oct}). Relationships based on decoupled secant bulk and shear moduli have been suggested in references [7 and 8]. Although such models may not be too erroneous for biaxial loading up to the peak load, sufficient, data are not available to derive expressions for bulk modulus and shear modulus in the post-peak region. Note that it is possible to include interaction between hydrostatic and deviatoric response by assuming that the bulk and shear moduli are functions of both I_1 and J_2 [9]. However, the resulting stress–strain relations will not be as general as when it is assumed that the strain energy density is a function of all three invariants [4].

In this paper, an orthotropic constitutive model is described. The model describes in a realistic and simple way the non-linear response including the strain softening. The parameters for the model were derived from the test results. These tests were designed to determine the descending part of the stress–strain curves of concrete subjected to multiaxial loading.

29.2 EXPERIMENTAL TECHNIQUES FOR OBTAINING STRAIN SOFTENING RESPONSE

Difficulties of obtaining the post-peak behaviour of concrete under both uniaxial compression and tension have been recognized for a long time [10, 11]. When a concrete specimen is loaded in a testing machine, both the specimen and the machine deform. When the stress in the specimen reaches its peak value, the specimen requires a reduction in load which results in a release of stored strain energy by the machine. This released strain energy can cause uncontrolled sudden failure of the specimen. As a result, a stable descending part will not be observed. Everything else being equal, the stiffer the testing machine the smaller is the amount of released energy and the lesser will be the extent of the machine–specimen interaction. By consideration of static stability, it can be shown that if the stiffness of the machine is greater than the absolute value of the stiffness of the specimen in the descending part, then a stable descending part can be obtained. The absolute value of the stiffness of the specimen after the peak (the average

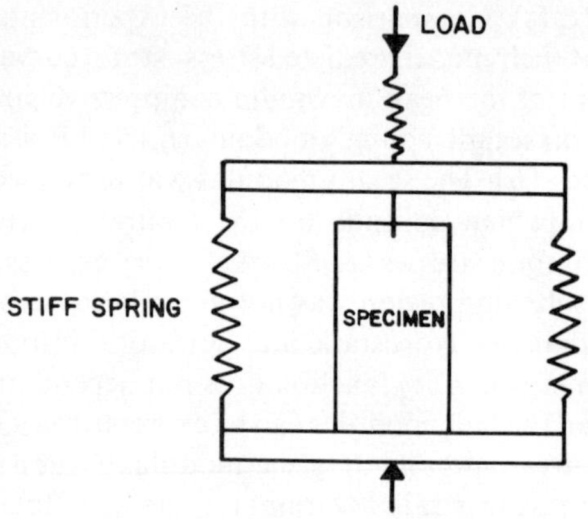

Figure 29.1 Stiff testing machine analogue

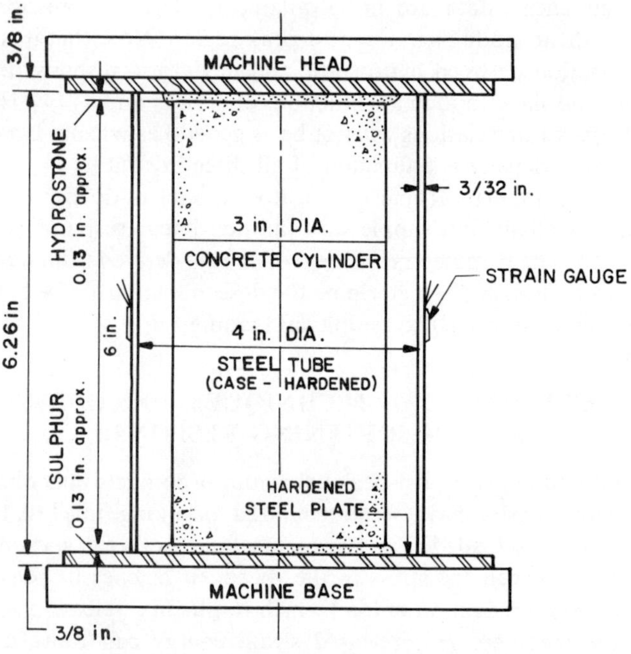

Figure 29.2 Test set-up for uniaxial compression

slope of the load–displacement curve of the specimen in the descending part) can be decreased by using a smaller diameter specimen, or by having a specimen with a less steep descending part (lower strength, lower strain rate, or smaller length–diameter ratio). Thus, for a given testing machine a stable failure can be achieved, for example, by using a sufficiently smaller diameter specimen. In addition to the static-stability considerations, the effects of the strain energy release rate, the velocity of crack propagation and other time related effects must also be considered.

Several attempts have been made in the past to obtain the complete stress–strain curves of concrete and rock by augmenting the stiffness of the testing machine (Figure 29.1). More recently a simple method has been devised to eliminate the strain energy release of the testing machine during unloading of the specimen [12]. Concrete cylinders were loaded in parallel with a steel tube whose diameter was larger than that of the specimen (Figure 29.2). The thickness of the tube was such that the total load (sum of the load carried by tube and by the concrete cylinder) was always increasing. The tube was made from a special alloy, case-hardened steel so that if displayed a linear stress–strain curve up to the strain of 0.006. Using this method, complete stress–strain curves of normal weight concrete of compressive strength of up to 12 000 p.s.i. and of lightweight concrete of up to 8000 p.s.i. compressive strength were observed (Figure 29.3). From the data, values of constants *A* and *D* needed to characterize the stress–strain relations according to equation (29.1) were statistically determined in terms of the compress strength of concrete [12]. Although the method is relatively simple and gives a very reproducible descending portion of the stress–

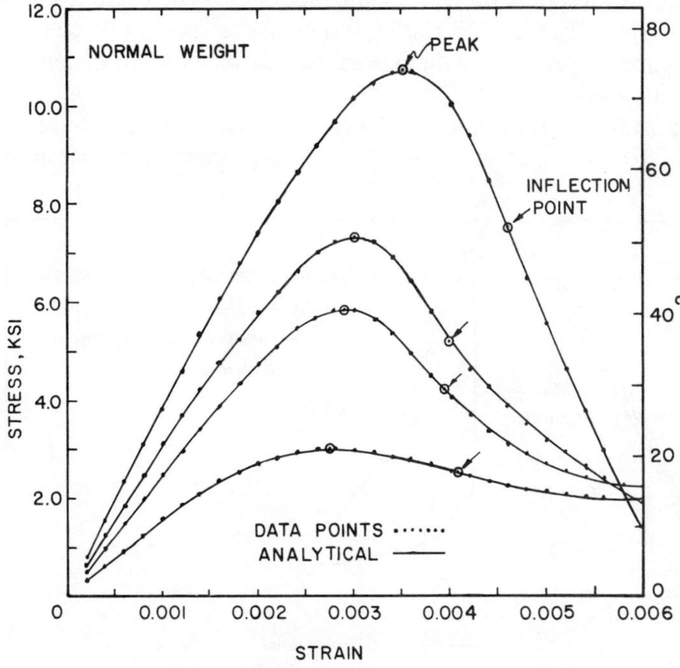

Figure 29.3 Typical stress–strain curves for normal weight concrete

strain curve it has certain limitations. The maximum strain that can be obtained is limited to the elastic range of the tubing material. The size of the concrete cylinder that can be tested is limited by the reduced capacity of the testing machine.

29.3 CLOSED-LOOP TESTING SYSTEM

An alternate approach is to use a closed-loop servo-controlled testing machine. The feedback control of such a machine can be programmed so that any experimental output can be chosen and continuously controlled as an independent variable. During an uncontrolled failure of a concrete specimen, the axial compressive strain sharply increases. However, if a closed-loop testing machine is programmed to maintain a constant rate of increase of axial strain, then the axial load applied by the machine could be reduced appropriately to maintain this constant rate even beyond the peak load.

Even with the closed-loop testing machines, the relative stiffness of the testing system may be critical in determining the mode of failure of the specimen. In order to avoid uncontrolled failure, the load on the specimen must be reduced at a sufficiently fast rate to control the rate of the strain energy release of the testing system. The speed at which a testing system can respond to a given signal depends on the frequency response or the response time of the system (electronic as well as hydraulic system). For a given testing system, if the energy release rate is faster than the frequency response of the system then the machine-specimen interaction cannot be avoided.

The importance of the relative stiffness of the system for a closed-loop testing machine is demonstrated in Figure 29.4. In this figure, stress–strain curves of

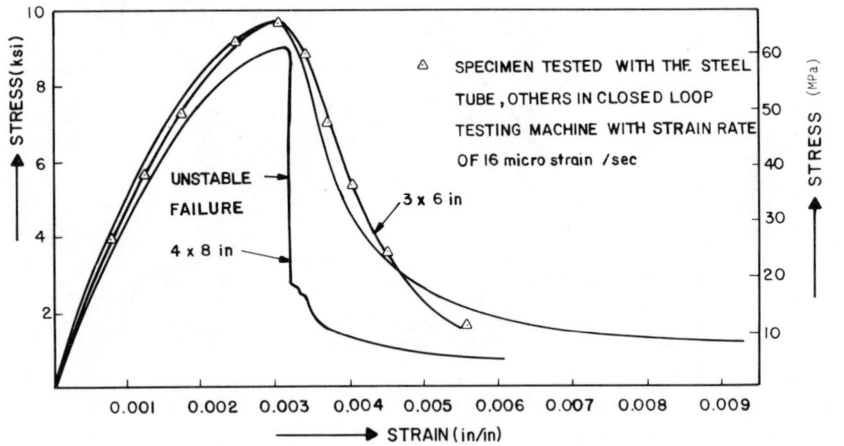

Figure 29.4 Specimen size effect and comparison of testing method on stress–strain curves

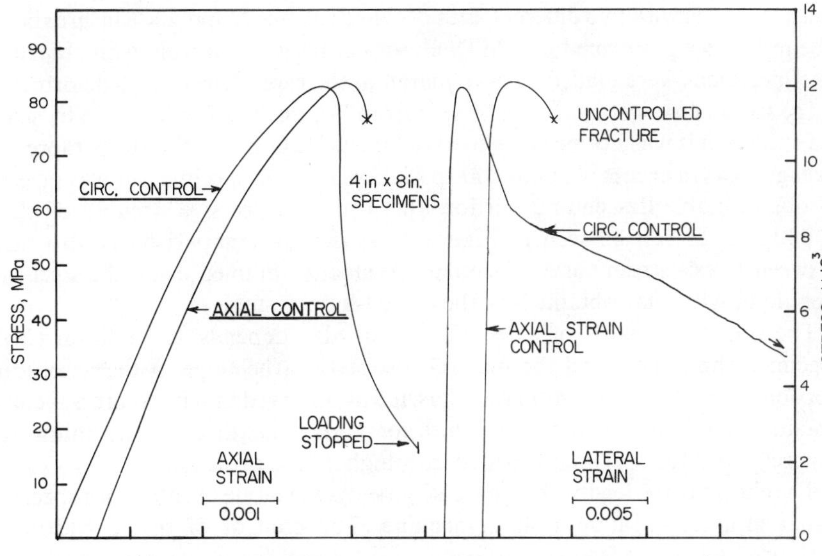

Figure 29.5 Stress–strain curves obtained with axial or circumferential strain control

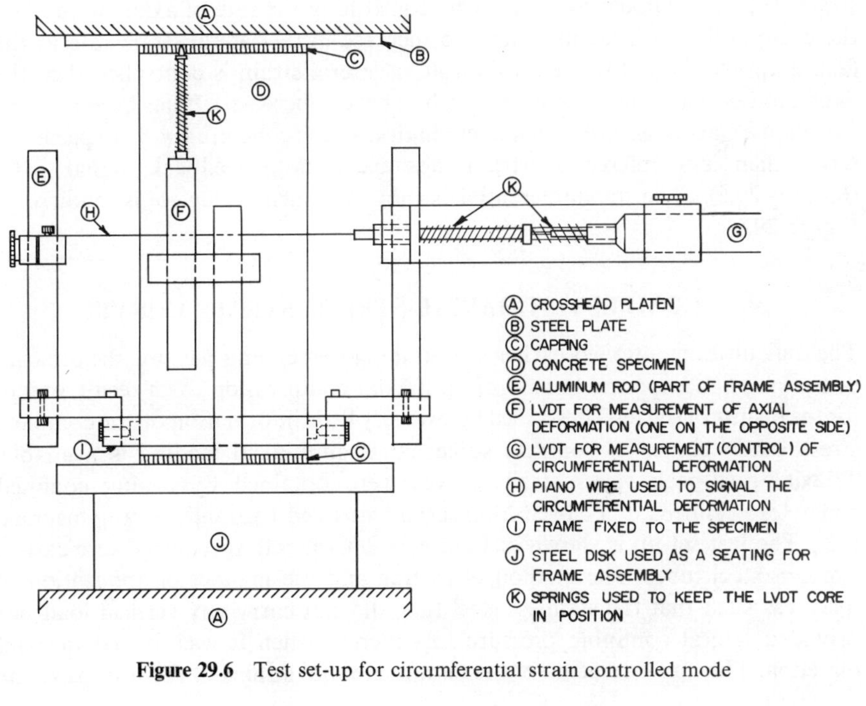

Figure 29.6 Test set-up for circumferential strain controlled mode

concrete made with two different sizes of cylinders: 3 × 6, and 4 × 8 in. are shown. The tests were performed in a MTS closed-loop servo-control testing machine. The specimens were loaded so as to maintain the rate of the overall deformation of the specimen constant. It can be seen from Figure 29.4 that for 4 × 8 in. size of the specimen it was not possible to avoid unstable failure when the compressive strength was in excess of about 8000 p.s.i. For 3 × 6 in. specimens it was possible to obtain stable descending portion up to the compressive strength of about 10 000 p.s.i. as can be seen in Figure 29.4 where a comparison is also made between stress–strain curve of similar specimens obtained using the steel-tube technique with that obtained by the closed-loop system.

The relative stiffness of a testing system also depends, in addition to the specimen dimensions, on the material properties (the slope of the descending portion of the stress–strain curve). Thus, it was observed that it is more difficult to obtain a stable and controlled failure for high strength concrete, lightweight concrete, and for specimens loaded at a higher rate of strain.

The choice of the feedback signal also governs the mode of failure of a specimen tested in a closed-loop testing machine. The control of brittle fracture is optimized when the feedback transducer is located for maximum sensitivity in detecting failure. Failure of concrete systems subjected to uniaxial compression is generally observed to be by vertical splitting cracks. Thus, the rate of increase of circumferential strain (lateral strain) is likely to be greater than that of axial strain. When specimens are loaded by controlling the rate of axial strain, then there can still be a large uncontrolled increase in the lateral strain rate as the failure approaches. However, if the rate of lateral strain is controlled then the likelihood of obtaining a stable descending part is increased. It has been possible to obtain a stable fracture even for very high strength concrete 4 × 8 in. specimens when their circumferential strain was used as a feedback signal [11], (Figure 29.5). The circumferential strain measuring device is shown in Figure 29.6.

29.4 COMPLETE TRIAXIAL STRESS–STRAIN CURVES

The difficulties mentioned earlier in obtaining descending part for the uniaxial compression, compound in case of multiaxial compression. As a result, except for the results of the tests described below, very little information on the complete stress–strain curves of concrete subjected to multiaxial loading is available. Triaxial post-peak stress–strain curves were obtained by testing confined concrete specimens in a closed-loop servocontrolled hydraulic testing machine [13]. The test set-up is shown in Figure 29.7. Concrete specimens were cast in stainless steel tubes. The method of casting and the manner of application of load was such that the stainless steel tube did not carry any vertical load but provided lateral confining pressure to concrete when it was loaded in axial direction. The measurements of the circumferential strains of the tube gave the

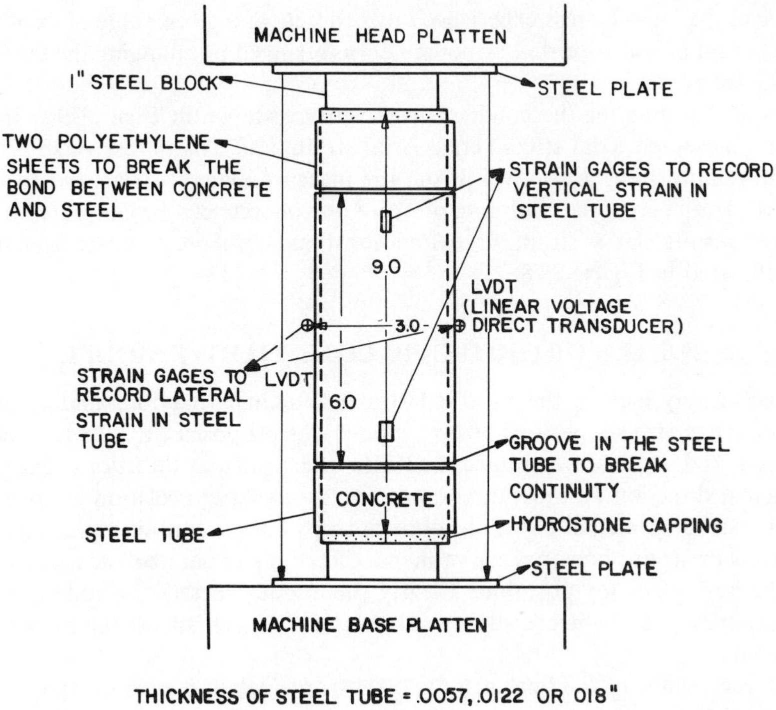

Figure 29.7 Test set-up for triaxial compression

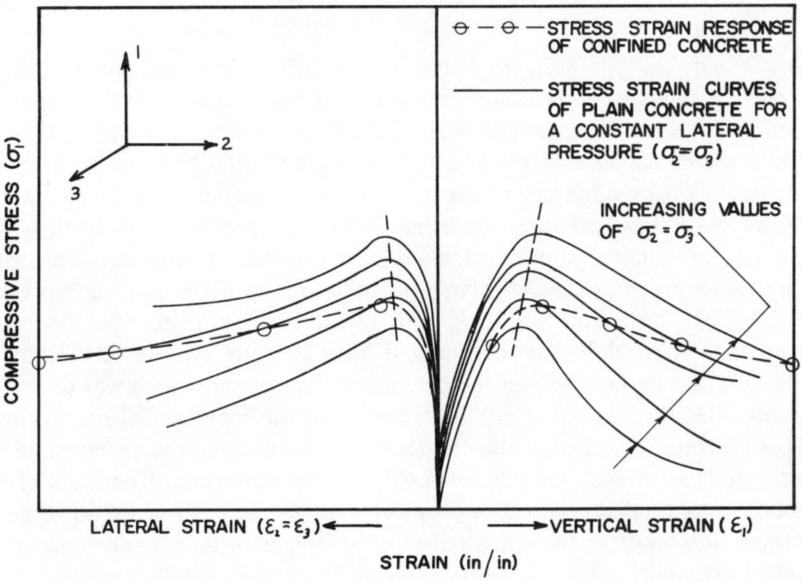

Figure 29.8 Schematic representation for behaviour of confined concrete

value of the lateral stress experienced by concrete at a given value of axial load. The lateral pressure exerted on concrete was changed by changing the thickness of the tube.

Typical results for the confined specimen are shown in Figure 29.8. In this figure, measured axial stress versus axial strain, and axial stress versus lateral strain relationships are shown. From the measured lateral strain and the steel stress–strain curve, the confining pressure on concrete can be calculated. From the test results, stress–strain curves for a constant lateral pressure were calculated as indicated in Figure 29.8.

29.5 AN ORTHOTROPIC CONSTITUTIVE MODEL

To accurately include the observed strain softening behaviour under triaxial compressive stresses, an orthotropic model was proposed [13]. In this model, stresses and strains are transformed into the principal directions. For each principal direction an individual non-linear stress–strain relation is proposed. Note that the principal directions of orthotropy vary according to current state of stress or strain; however, the principal directions remain collinear according to the basic isotropy postulate. Clearly this model does not include the path dependency and the irreversibility, which are characteristic of the response of concrete.

At each state of loading, a total orthotropic relation relating the current principal stresses and strains can be given in the following form:

$$Y = \frac{A_i X_i + (D_i - 1)X_i^2}{1 - (A_i - 2)X_i + D_i X_i^2} \tag{29.5}$$

where $Y = \sigma_1/\sigma_p$, $X_i = \varepsilon_i/\varepsilon_{ip}$, σ_1 is the most principal compressive stress, σ_p is the most principal compressive strength, ε_i is the strain in the ith principal direction, ε_{ip} is the strain at the peak in the ith direction; $A_i = E_i/E_{ip}$, E_i is the initial slope of the σ_1–ε_i curve and $E_{ip} = \sigma_p/\varepsilon_{ip}$ and D_i is the parameter which controls the descending part of the σ_i–ε_i curve. Note that the $\sigma_1 \geq \sigma_2 \geq \sigma_3$ are the principal stresses, compression being positive. To predict stress–strain curves using equation (29.5) four parameters are needed: (1) the most principal compressive strength, (2) the three principal strains at the peak compressive strength, (3), the initial slope, E_i and (29.4) the parameter D_i. The peak compressive strength can be determined from strength criteria. A three-parameter failure criteria based on the octahedral theory and which was expressed in terms of the three invariants of the stress tensor was found sufficiently accurate [13]. The three principal strains at the peak were expressed in terms of the three principal stresses at failure based on the regressional analysis of the available experimental data. Constants A_i and D_i were expressed in terms of the three invariants of the stress tensor. The details of formulations are given in refs. [13], and [14].

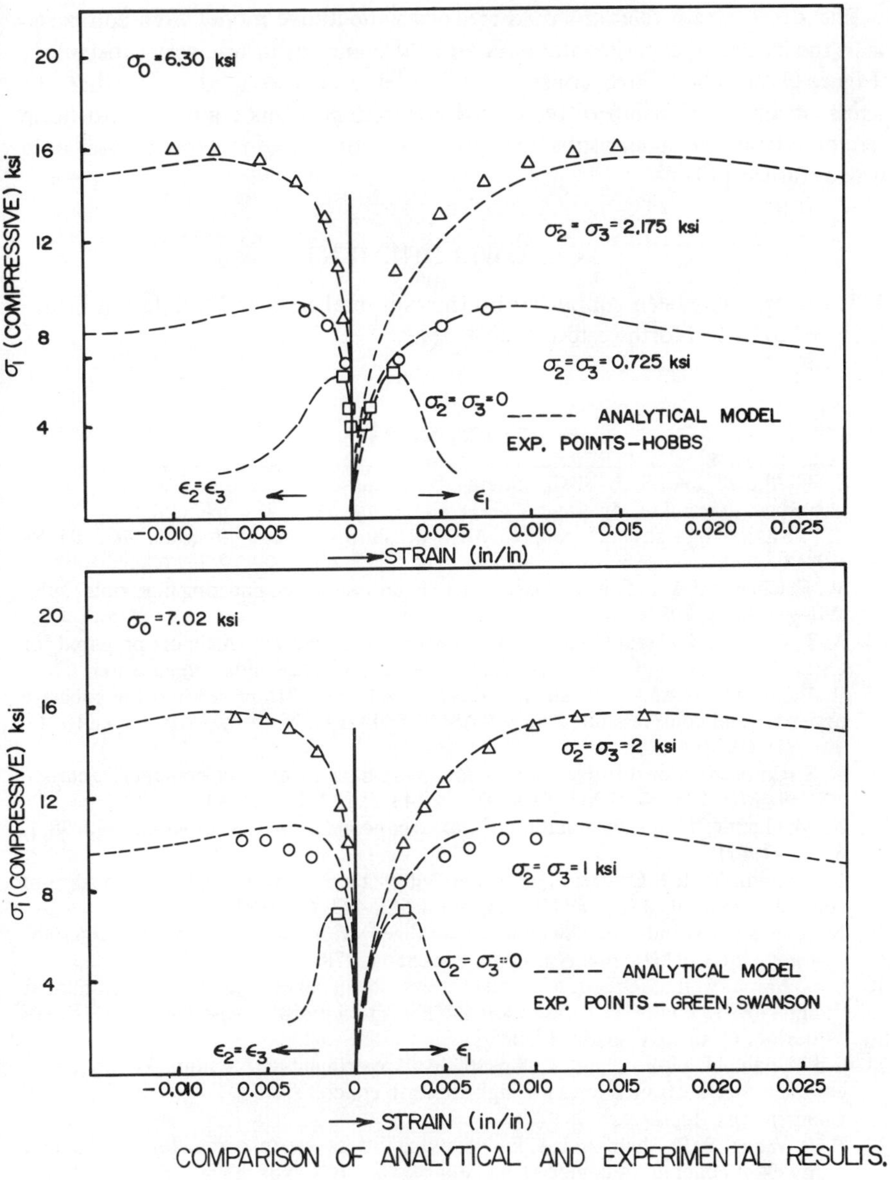

Figure 29.9 Comparison of analytical and experimental results

The stress–strain relations predicted with the above model were compared with the available experimental data and the comparison was very satisfactory (Figure 29.9). The above constitutive model was also used to predict the stress–strain relationship of reinforced concrete columns confined with hoop reinforcement and again, satisfactory comparison with the experimental data was obtained [1].

ACKNOWLEDGEMENT

This research has been supported by the National Science Foundation Grant CEE-8120774 to Northwestern University.

REFERENCES

1. S. H. Ahmad and S. P. Shah, 'Stress–strain curves of confined concrete', to be published, *Journal of American Concrete Institute*, **79**, 6, 484–490 (1982).
2. S. P. Shah, 'High strength concrete—a workshop summary', *Journal, ACI*, 94–98 (1981).
3. W. F. Chen and A. F. Saleeb, 'Constitutive equations for engineering materials', John Wiley & Sons, 1982.
4. K. J. William, 'General Concepts of Constitutive Equations', A report prepared for CEB Committee on Constitutive Equations of Concrete, under preparation.
5. R. Palaniswamy and S. P. Shah, 'Fracture and stress–strain relation for concrete under triaxial compression', Journal, ASCE-STD, **100**, No. ST5, Proc. Paper 10547, 901–916 (1974).
6. N. S. Ottosen, 'Constitutive models for short-time loading of concrete', *Journal, ASCE-EMD*, **105**, No. EM1, Proc. Paper 14375, 127–141 (1979).
7. K. M. Gerstle, 'Simple formulation of biaxial concrete behaviour', *Journal, ACI*, **78**, 1, 62–68 (1981).
8. L. Cedolin, Y. R. J. Crutzen, and S. dei Poli, 'Triaxial stress–strain relationship of concrete', *Journal, ASCE-EMD*, **103**, No. EM3, 423–429 (1977).
9. M. D. Kostovos and J. B. Newman, 'Generalized stress–strain relations for concrete', *Journal, ASCE-EMD*, **104**, No. EM4, 845–856 (1978).
10. S. H. Ahmad and S. P. Shah, 'Complete stress–strain curve of concrete and nonlinear design', *Proceedings, CSCE-ASCE-ACI-CEB International Symposium*, University of Waterloo, Ontario, Canada, 1980.
11. S. P. Shah, U. Gokoz, and F. Ansari, 'An experimental technique for obtaining complete stress–strain curves for high strength concrete', *ASTM Journal of Cement, Concrete and Aggregates*, **3**, 1 (1981).
12. P. T. Wang, S. P. Shah, and A. E. Naaman, 'Stress–strain curves for normal and lightweight concrete in compression', *Journal of ACI*, Nov. 1978.
13. S. H. Ahmad and S. P. Shah, 'Complete triaxial stress–strain curves for concrete', *Journal, ASCE-STD*, **108**, No. ST4 (1982).
14. S. H. Ahmad and S. P. Shah, 'Orthotropic Model for Complete Stress–Strain Curves of Concrete Under Multiaxial Stresses', Proceedings, RILEM-CEB Symposium on Concrete Under Multiaxial Conditions, May 22–24, 1984, Toulouse, France.

Mechanics of Engineering Materials
Edited by C. S. Desai and R. H. Gallagher

Chapter 30

Endochronic Plasticity: Physical Basis and Applications

K. C. Valanis and C. F. Lee

30.1 INTRODUCTION

In 1971 Valanis proposed endochronic plasticity as an alternative theory to classical plasticity for the description and hopefully, the prediction of rate independent yet history dependent response of materials. The two theories differ in certain important respects in that (a) the endochronic theory does not require the concept of yield surface for its development, (b) the physical assumptions that underlie the theory have their origin in irreversible thermodynamics of internal variables, and (c) the material memory is defined in terms of an intrinsic time scale which is a material property.

Since that time endochronic plasticity has undergone a significant evolution especially in the 'proper' definition of the intrinsic time scale ζ. See Section 30.2 for discussion.

In Section 30.3 we use the current theory to predict the cyclic behaviour of steels under constant and variable strain amplitudes. In Section 30.4 we address ourselves to the question of the physical basis of the endochronic theory by using a single slip mechanism as the means of plastic deformation in a cubic crystal. We show that the increment in intrinsic time is equal to the algebraic value of the increment of the slip displacement and that the evolution equation for slip is of the same type as that of the evolution equation of internal variables that underlie the development of the endochronic theory.

30.2 BRIEF REVIEW OF THE ENDOCHRONIC THEORY

The point of departure of endochronic plasticity from classical plasticity is that the former does not require the concept of yield surface for its development. In its early stages, the theory rested on the notion that the stress response of dissipative materials is a function of the strain path. When the material behaviour considered is rate-independent, the path in question must also be rate independent. The early version of the endochronic theory [1], was constructed in terms of a path in the strain space ε_{ij}. The distance along the path between the two strain states P and P' is denoted by $d\zeta$. If **P**, a fourth order positive definite tensor, is the metric of the space, then,

$$d\zeta^2 = P_{ijkl} d\varepsilon_{ij} d\varepsilon_{kl} \tag{30.1}$$

The tensor **P** is a material property in the sense that in general it will vary from material to material. Since successive strain states on a strain path are distinct and $d\zeta$ is always positive, the latter can serve as a time measure which is a property of the material at hand, since **P** is such. The length of the path ζ is then an intrinsic time scale where 'time' is used here in a very general sense. The stress at point P is not determined simply by the strain at P, but by the history of the strain along the path. Materials for which the stress is a function of the history of strain with respect to an intrinsic time scale, have been called 'endochronic' by the first author and the theory of the mechanical response of such materials is called 'endochronic theory' [1].

In the subsequent applications, it was found necessary to define an intrinsic time scale Z which is related to the intrinsic time measure ζ, by the relation:

$$dZ = \frac{d\zeta}{f} \tag{30.2}$$

where f is a function of the *history* of strain. The function f, generally considered to be a function of ζ, is related proportionally to the degree of internal friction in a material. If a material hardens, $f(\zeta)$ increases with ζ; if it softens, f decreases with ζ; and is constant otherwise. For convenience in the following discussion, we will call a material stable if f is a constant.

The formulation of the theory does not depend on an explicit definition of Z. Thus one can envision a constitutive equation, applicable to a large class of materials, from which an explicit response, pertaining to a sub-class, can be obtained by simply choosing the appropriate form of Z [1, 2].

In the case of linear isotropic theory the constitutive equations so derived can be decomposed into deviatoric and hydrostatic parts. The deviatoric stress **s** is related to the history of the deviatoric strain **e** by the linear functional relation:

$$\mathbf{s} = 2\int_0^Z \mu(Z - Z') \frac{\partial \mathbf{e}}{\partial Z'} dZ' \tag{30.3}$$

where in the reference configuration, **s** is zero, $Z = 0$, and the shear modulus, $\mu(Z)$ is given by a Dirichlet series, i.e.,

$$\mu(Z) = \lambda_0 + \sum_{r=1}^{n} \lambda_r e^{-\beta_r Z} \tag{30.4}$$

where λ_0, λ_r, and β_r are positive constants and n is finite or infinite depending on the number of internal variables. The hydrostatic stress σ_H, is related to the history of volumetric strain, θ, in a similar fashion by the linear functional relation:

$$\sigma_H = \int_0^Z K(Z - Z') \frac{\partial \theta}{\partial Z'} dZ' \tag{30.5}$$

where $\sigma_H = \sigma_{kk}/3$ and $\theta = \varepsilon_{kk}$, in the usua notation where the summation convention is employed. The bulk modulus, $K(Z)$ is given by a Dirichlet series of the form of equation (30.4). Note again that $\sigma_H = 0$ in the reference configuration.

For further details of the derivation of equations above see [1], where both $\mu(Z)$ and $K(Z)$ are composed of finite sums of positive exponentially decaying terms. In particular, $\mu(0)$ and $K(0)$ are the shear and bulk elastic moduli, respectively.

The early version of the endochronic theory has been applied with success to a number of problems of practical interest (e.g. refs [1, 3, 4]). Despite this fact, it failed to predict closed hysteresis loops for 'small' unloading–reloading processes in one-dimensional conditions as pointed out by Valanis [5]. For such deformation histories, the theory predicts a slope at the reloading point that is smaller than the unloading slope at the same point. This feature of the theory is at odds with the observed behaviour of most metals. Various criticisms were levelled at the theory on these grounds by Sandler [6] and Rivlin [7, 8]. For an extensive discussion of the substance of these criticisms see Hsieh [9] and Valanis [10]. In this regard a recent report by Edelstein [11] discusses the development of the endochronic theory prior to our introduction of the singular kernel, which is treated in the following parts of this section.

It was shown that the openness of the hysteresis loops is due to the fact that the intrinsic time rate of dissipation at the onset of unloading is equal to the intrinsic time rate of dissipation upon continuation of loading. However, experience shows that most rate-insensitive materials initially unload in an elastic manner and, therefore, with essentially zero rate of dissipation. In view of this, the discrepancy between prediction and observation was bound to arise [12].

It was subsequently demonstrated [12] that if the measure of intrinsic time is defined in terms of the increment of *plastic* strain, the rate of dissipation at the onset of unloading and reloading is, in fact, zero. Therefore it was appropriate to adopt the plastic strain increment as the measure of intrinsic time. Moreover, the constitutive equations (30.3) and (30.5) are recast in a form whereby the stress is related to the history of *plastic* strain. This was done by the first author recently [12]. This model was used to prove mathematically the existence of yield surface under special conditions and that kinematic hardening rules are a consequence, of the theory.

Of greater theoretical and practical consequence, however, is the fact that new measure of intrinsic time makes feasible the complete elimination of the yield surface by shrinking its size to zero and thereby reducing the surface to a point. This is done by introducing weakly singular kernel functions in the linear functional representation of stress in terms of history of plastic strain by allowing the kernel functions to possess an integrable singularity at the origin (i.e. $Z=0$). On the basis of the above considerations, endochronic constitutive equations of isotropic materials, which exhibit yielding immediately upon application of loading, are as follows

$$\mathbf{s}=2\int_0^{Z_\mathrm{D}}\rho(z_\mathrm{D}-z'_\mathrm{D})\frac{\partial \mathbf{e}^\mathrm{p}}{\partial Z'_\mathrm{D}}\,\mathrm{d}Z'_\mathrm{D},\qquad \rho(0)=\infty \tag{30.6}$$

$$\sigma_{kk}=3\int_0^{Z_\mathrm{H}}\kappa(Z_\mathrm{H}-Z'_\mathrm{H})\frac{\partial \varepsilon^\mathrm{p}_{kk}}{\partial Z'_\mathrm{H}}\,\mathrm{d}Z'_{\mathrm{H}'}\,\kappa(0)=\infty \tag{30.7}$$

and

$$\int_0^{Z_\mathrm{H}}\kappa(Z'_\mathrm{H})\,\mathrm{d}Z'_\mathrm{H}<\infty;\qquad \int_0^{Z_\mathrm{D}}\rho(Z'_\mathrm{D})\mathrm{d}Z'_\mathrm{D}<\infty, \tag{30.7a, b}$$

for all finite Z_H and Z_D where D and H denote the deviatoric and hydrostatic state, respectively. Also

$$\mathrm{d}\mathbf{e}^\mathrm{p}=\mathrm{d}\mathbf{e}-\frac{\mathrm{d}\mathbf{s}}{2\mu_1} \tag{30.8}$$

$$\mathrm{d}\varepsilon^\mathrm{p}_{kk}=\mathrm{d}\varepsilon_{kk}-\frac{\mathrm{d}\sigma_{kk}}{3K_1} \tag{30.9}$$

where μ_1 and K_1 are the appropriate elastic moduli. The intrinsic time scale increments $\mathrm{d}Z_\mathrm{H}$ and $\mathrm{d}Z_\mathrm{D}$ are related to the intrinsic time measured by the equations:

$$\mathrm{d}Z_\mathrm{D}=\mathrm{d}Z_\mathrm{D}/f_\mathrm{D}(\zeta_\mathrm{D}) \tag{30.10}$$

$$\mathrm{d}Z_\mathrm{H}=\mathrm{d}\zeta_\mathrm{H}/f_\mathrm{H}(\zeta_\mathrm{H}) \tag{30.11}$$

where

$$\mathrm{d}\zeta_\mathrm{D}=|\mathrm{d}e^\mathrm{p}_{ij}\mathrm{d}e^\mathrm{p}_{ij}|^{1/2} \tag{30.12}$$

$$\mathrm{d}\zeta_\mathrm{H}=|\mathrm{d}\varepsilon^\mathrm{p}_{kk}| \tag{30.13}$$

Here $|\cdot|$ denotes the absolute value. Other more general definitions are possible see ref. [5]. More generally

$$\rho(Z_\mathrm{D})=\frac{\rho_0}{Z^\alpha}\sum_{n=1}^{N}R_n\exp[-k_nZ_\mathrm{D}],$$

$$\kappa(Z_\mathrm{H})=\frac{k_0}{Z^\omega}\sum_{n=1}^{M}P_n\exp[-\kappa_nZ_\mathrm{H}] \tag{30.14, 15}$$

where all the constants are positive and finite, M, N may be finite or infinite, and α and ω are bounded by the inequalities

$$0 < (\alpha, \omega) < 1 \tag{30.16}$$

See ref. [5] for detailed discussion. The above equations summarize the new model of the isotropic endochronic theory. It is of interest that this type of kernel has been proposed in viscoelasticity by the Russian school. See ref. [13].

In conclusion, two significant results are accomplised: (1) The slope of the deviatoric (or hydrostatic) stress–strain curve at points of unloading and reloading or strain rate reversal is always elastic, i.e., equal to the slope at the origin of the appropriate stress–strain curve. (2) The hysteresis loops in the first quadrant of the stress–strain space are always closed. For details see refs. [5, 14].

30.2.1 Constitutive relations in tension–torsion

The constitutive equations that apply in this specific case are found from equations (30.6) and (30.7) and are given below:

$$\tau = 2\int_0^{Z_D} \rho(Z_D - Z'_D)\frac{\partial \eta^p}{\partial Z'_D}\,dZ'_D \tag{30.17}$$

$$\sigma_1 = 2\int_0^{Z_D} \rho(Z_D - Z'_D)\frac{\partial}{\partial z'_D}(\varepsilon_1^p - \varepsilon_2^p)\,dZ'_D \tag{30.18}$$

$$\sigma_1 = 3\int_0^{Z_D} \kappa(Z_H - Z'_H)\frac{\partial}{\partial Z'_H}(\varepsilon_1^p + 2\varepsilon_2^p)\,dZ'_H \tag{30.19}$$

where ε_i^p and σ_i are the axial plastic strains and stresses, respectively, along the axes x_i and $\varepsilon_2^p = \varepsilon_3^p$ to satisfy the condition of isotropy. Also τ and η^p stand for s_{12} and e_{12}^p, respectively, in the notation of equation (30.6).

Because in the experiments to be investigated the hydrostatic strain was not measured we shall proceed to make the usual (approximate) assumption of elastic hydrostatic response, in which case equations (30.7) and (30.16) do not apply, but instead the plastic incompressibility condition

$$\varepsilon_1^p + 2\varepsilon_2^p = 0 \tag{30.20}$$

is used. In the following, we will omit the subscripts D and H.

In light of the above hypotheses and in view of equations (30.17) and (30.20) the appropriate constitutive equations in tension–torsion are the following:

$$\tau = 2\int_0^{Z} \rho(Z - Z')\frac{\partial \eta^p}{\partial Z'}\,dZ' \tag{30.21a}$$

$$\sigma_1 = \int_0^{Z} E(Z - Z')\frac{\partial \varepsilon_1^p}{\partial Z'}\,dZ' \tag{30.21b}$$

$$\varepsilon_1 + 2\varepsilon_2 = \frac{\sigma_1}{3K_1} \tag{30.21c}$$

where

$$E(Z) = 3\rho(Z) \tag{30.22}$$

$$dZ = dZ_D = \frac{d\zeta}{f(\zeta)} \tag{30.23}$$

$$d\zeta = d\zeta_D = |[\tfrac{2}{3}(d\varepsilon_1^p - d\varepsilon_2^p)^2 + 2(d\eta^p)^2]^{1/2}| \tag{30.24a}$$

Alternatively, $d\zeta$ can be expressed in terms of the engineering shear strain $\gamma^p = 2\eta^p$, in which case, upon using equation (30.20).

$$d\zeta = |[\tfrac{3}{2}(d\varepsilon^p)^2 + \tfrac{1}{2}(d\gamma^p)^2]^{1/2}| \tag{30.24b}$$

Here $\varepsilon^p = \varepsilon_1^p$.

In the applications that follow we will use the above equations in the study of cyclic response to a variety of test conditions.

30.3 APPLICATION TO THE CYCLIC PLASTICITY

Uniaxial response

In this section we shall treat the class of metals, specifically, mild and Grade 60 steels, where

$$\rho(Z) = \frac{\rho_0}{Z^\alpha} \tag{30.25a}$$

$$f(\zeta) = 1 \tag{30.25b}$$

It follows from equations (30.23) and (30.24b) that

$$dZ = \sqrt{(3/2)}|d\varepsilon^p| \tag{30.26}$$

In view of equation (30.22)

$$E(Z) = E_0 Z^{-\alpha}, \qquad E_0 = 3\rho_0 \tag{30.27}$$

We remark that the stress response of steels of this type to any type of plastic strain history is completely defined in terms of *two* material constants ρ_0 and α. We also point out that equation (30.27) gives rise to the Ramberg–Osgood equation for the monotonic tensile response of a number of metals. For *a method of determining more general types of functions $\rho(Z)$ and $f(\zeta)$, see ref.* [15].

Upon substitution of equation (30.27) in equation (30.21b) and at the completion of n reversals the following relation applies where Z_i denotes the value of Z at the point where the ith reversal has been completed and $Z_0 \overset{\text{def}}{=} 0$.

$$\sigma = \sqrt{(2/3)}\frac{E_0}{1-\alpha}\left[Z^{1-\alpha} + 2\sum_{i=1}^{n}(-1)^i(Z - Z_i)^{1-\alpha}\right] \tag{30.28}$$

Cyclic shear response

In this case, we use equations (30.23) and (30.24b) to obtain the relation

$$dZ = d\zeta = |d\gamma^{p}|/\surd(2) \tag{30.29}$$

which is the shear counterpart of equation (3.20). In the fashion outlined above, the cyclic shear response is given by the equation

$$\tau = \frac{\surd(2)\rho_0}{1-\alpha}\left[Z^{1-\alpha} + 2\sum_{i=1}^{n}(-1)^{i}(Z - Z_i)^{1-\alpha}\right] \tag{30.30}$$

We remark that equations (30.28) and (30.30) obey the linear homogeneous transformation between indicated stresses and strains given below:

$$\tau = \sigma/\surd(3), \qquad \gamma^{p} = \surd(3)\varepsilon^{p} \qquad (30.31 \text{ and } 30.32)$$

We also note that the relations (30.31) and (30.32) hold for *all* forms of kernel function $\rho(Z)$ and material function $f(\zeta)$.

To apply the theory, we use the experimental results on normalized mild steel obtained by Jhansal and Topper [16] and Grade 60 steel obtained by Dafalias and Popov [17].

30.31 Applications to normalized mild steel

30.3.1.1 *Constant uniaxial strain amplitude*

We consider the class of metals whose asymptotic stress response to sustained cyclic strain excitation at constant strain amplitude is a periodic stress history with constant amplitude. Specifically in a uniaxial test of this type, the axial stress amplitude $\Delta\varepsilon^{p}$ is also constant, following equation (30.18). Thus

$$\Delta\varepsilon^{p} = \Delta\varepsilon - \frac{\Delta\sigma}{E_1} \tag{30.33}$$

where $\Delta\varepsilon$ is the axial strain amplitude and E_1 is Young's modulus. As a result, the value of Z during cyclic tension and compression can be found from equation (30.26).

Specifically, the value of Z such that $Z_n \leq Z < Z_{n+1}$, where Z_n is Z at the nth reversal is given by equation (30.34a, b), where

$$Z = \surd(3/2)(2n\Delta\varepsilon^{p} \pm \varepsilon^{p}) \tag{30.34a}$$

in particular

$$Z = \surd(3/2)(2n - 1)\Delta\varepsilon^{p} \tag{30.34b}$$

where in equation (30.34a) minus is used for n odd and plus for n even. Thus using equations (30.34a, b) and substituting for Z in equation (30.28) we obtain the closed form solution for stress response given below

$$\sigma(\varepsilon^p) = (2/3)^{\alpha/2} \frac{E_0}{1-\alpha} (\Delta\varepsilon^p)^{1-\alpha} F_n(\alpha, x) \tag{30.35}$$

$$x = \varepsilon^p / \Delta\varepsilon^p \tag{30.36a}$$

$$F_n(\alpha, x) = (2n \pm x)^{1-\alpha} + 2 \sum_{i=1}^{n} (-1)^i (2n - 2i + 1 \pm x)^{1-\alpha} \tag{30.36b}$$

where the plus or minus signs depend on whether n is even or odd. The algebraic value of the peak stress (i.e., stress amplitude for any n) is found from equation (30.36b) by setting $x = 1$ for n odd or $x = -1$ for n even in equation (30.36b), i.e.,

$$F_n(\alpha) = (2n - 1)^{1-\alpha} + 2 \sum_{i=1}^{n} (-1)^i (2n - 2i)^{1-\alpha} \tag{30.36c}$$

It can be shown that, in the limit of $n \to \infty$, F_n converges to a constant $F_\infty(\alpha)$, where F_∞ varies with α but is essentially close to unity. For instance, for $\alpha = 0.864$, F_∞ is equal to 1.03 [2]. Thus the asymptotic value of $\Delta\sigma$ as n tends to infinity is given by the equation

$$\Delta\sigma = (2/3)^{\alpha/2} \frac{E_0}{1-\alpha} (\Delta\varepsilon^p)^{1-\alpha} F_\infty(\alpha) \tag{30.37}$$

This is the equation of the cyclic stress–(plastic) strain curve.

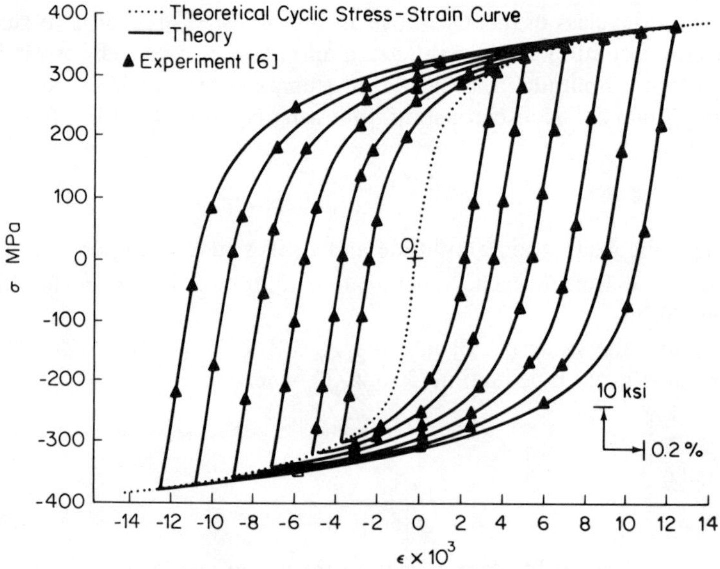

Figure 30.1

Cyclic response in shear can be found in a similar fashion or by using equations (31.31) and (30.32).

At this point we test the theory against experimental data on normalized mild steel [16]. In ref. [16], a set of stable uniaxial hysteresis loops corresponding to various constant strain amplitudes was presented in the stress–strain space. *A propos* of the ensuing theoretical predictions we note that the *geometric shape* of the loops is given by equation (30.35), whereas the peak stresses are given by equation (30.37). We also note that there are only *two* undetermined parameters in these equations, α and E_0. The *form* of equation (30.37) was corroborated in ref. [2] where a semi-logarithmic plot of the experimental values of $\Delta\sigma$ versus $\Delta\varepsilon^p$ gave rise to a linear relation. The plot also determines α and E_0 which were found to be 0.864 (a pure number) and 107.6 MPa (15.61 k.s.i.), respectively. These values are than used in equation (30.35), and the shape of the stable hysteresis loops is thereby obtained for large n (say > 25). Agreement between theory and experiment is shown in Figure 30.1.

We make, in passing, an observation of historical interest. Equation (30.37) agrees with the empirical relationship proposed by Landgraff *et al.* [18] for steels, i.e.,

$$\Delta\sigma \sim (\Delta\varepsilon^p)^{1-\alpha}$$

where $1-\alpha$ ranges from 0.12 to 0.17. In our case, $1-\alpha = 0.136$.

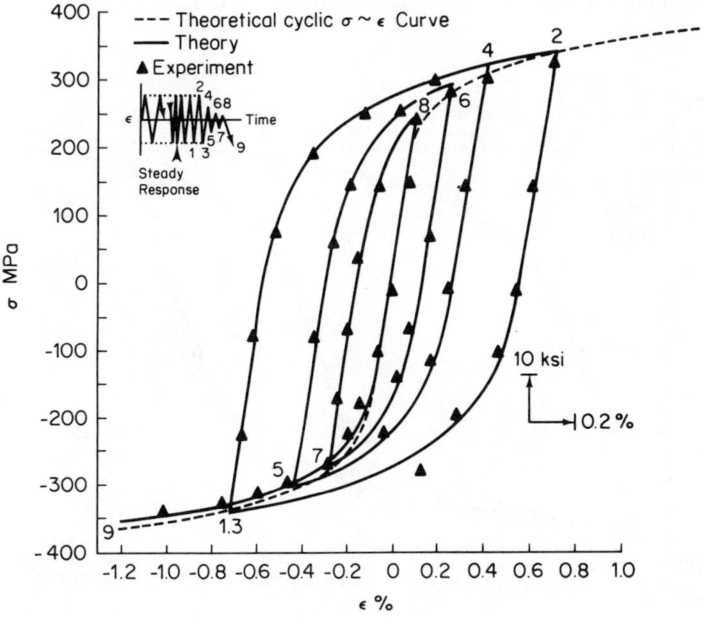

Figure 30.2

30.3.1.2 *Variable uniaxial strain amplitudes*

To extend the experimentally verified domain of validity of the theory, we test it under conditions of variable uniaxial strain amplitude histories. The stress response to such histories is found by using equations (30.35) and (30.36a, b). The analytical results are compared with the experimental data on normalized mild steel [16]. The experiment consists of a *constant* uniaxial strain amplitude cyclic test (until stable hysteresis loops are reached) followed by a *variable* uniaxial strain amplitude test. The experimental data are shown in Figure 30.2. Despite the complexity of the history, agreement between theory and experiment is obtained and shown in Figure 30.2.

30.3.1.3 *Other complex histories*

A strain history of practical importance is shown in Figure 30.3 where a cyclic strain history at a fixed strain amplitude is followed by another at a lower strain amplitude. The experimental results are shown in Figure 30.3. The theoretical results obtained are also shown in Figure 30.3. Again agreement between theory and experiment is demonstrated.

It is important to observe that the decreasing effect of the previous history on the stress response to a periodic strain history (cyclic test at constant strain amplitude) is the natural consequence of the monotonically decaying kernel function used in the present theory, i.e., in equations (30.25a). This type of kernel

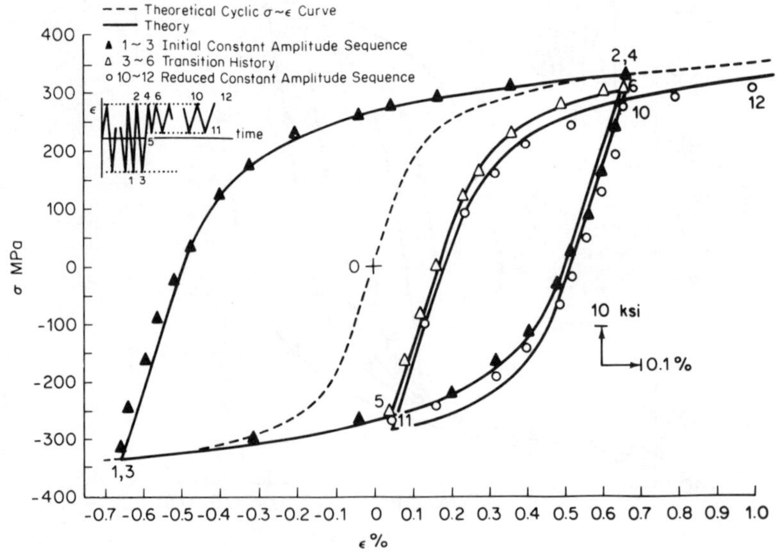

Figure 30.3

does indeed impart to the material a fading memory with respect to the endochronic time scale.

30.3.2 Application to Grade 60 steel

In Figure 30.4 a specimen of Grade 60 steel was tested by using random uniaxial cyclic strain history. Due to lack of information regarding the method of preparation of specimen (i.e., its prehistory) and its hysteresis loop geometry at constant strain amplitude, it is difficult to obtain the form of kernel function $\rho(Z)$ and of the material function $f(\zeta)$ using only the data in Figure 30.4. However, ignoring the stress response during the first quarter cycle, we can still predict the experimental data of the subsequent history by using the present theory with only two material parameters E_0 and α.

In the following, we used equation (30.28). The material parameters E_0 and α are determined by using the procedure in ref. [15]. We find $E_0 = 264.2$ MPa and $\alpha = 0.82$. Ensuing agreement between theory and experiment is shown in Figure 30.4. It is worthy of note that predicted values of onset of unloading and reloading are always equal to Young's modulus value (2.06×10^5 MPa). However, in ref. [17] those slopes were assigned a smaller value (1.68×10^5 MPa) than the Young's modulus to account for 'softening'. This has not been necessary in the present case.

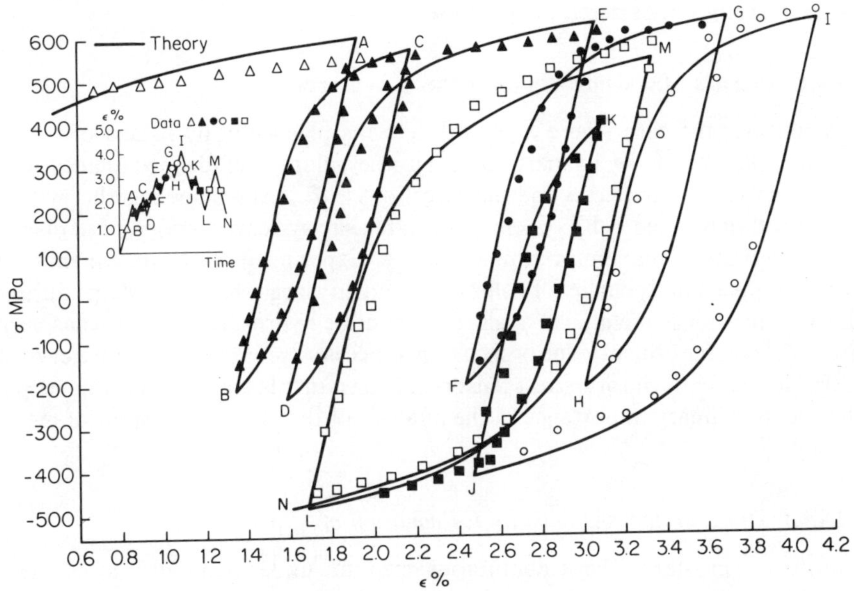

Figure 30.4

30.3.3 Conclusions

On the basis of the results presented in Section 30.3, we conclude that the integral constitutive equations of the endochronic type are suitable for the analytical prediction of cyclic response of steels under a variety of conditions. These equations are derived from the endochronic theory of plasticity of isotropic materials with an intrinsic time scale defined in the *plastic* strain space.

Also noteworthy is the fact that a constitutive equation with *two* material constants, which are easily determined, can predict accurately and with computational ease the stress (strain) response of a material to a variety of general strain (stress) histories, without a need for special memory rules often (discussed in the literature. The cyclic ratcheting phenomenon is also predicted by the present theory in routine fashion [19].

30.4 SLIP AS A PHYSICAL BASIS OF ENDOCHRONIC PLASTICITY

In this section we gain insight into a physical nature of the endochronic theory and more specifically, the definition of intrinsic time by addressing the case where material deformation takes place by the mechanism of slip, which is a discontinuous tangential displacement along a plane (the slip plane). Specifically we limit the analysis to cubic symmetric crystals and isothermal small strain fields, where rotations due to slip are negligible. Also interactions between slip systems are not considered in the present paper.

30.4.1 Analysis of a single slip mechanism in a crystal

We consider the case where a crystal deforms plastically by means of translational slip of one of its parts over another along well-defined planes and directions, called slip planes and slip directions [20]. A slip plane together with a slip direction is called a slip system. Most often slip systems correspond to planes and directions of maximum atomic density. Experiments have shown that (1) generally very small changes of volume accompany large changes of shape during the deformation of metallic crystals; (2) moderate hydrostatic pressure has very little effect on the critical value of shear stress necessary for plastic flow in a crystal [20]. Hence, the deviatoric stress tensor and constant plastic volume are concepts that are of primary importance in the analysis of the plastic behaviour of metal crystals.

30.4.1.1 *Slip criterion and constitutive equation of slip*

Consider a crystal to be a continuous medium under a uniform stress field denoted by the tensor $\boldsymbol{\sigma}$. Let there exist a predetermined slip plane with unit normal $\bar{n}$. (See Figure 30.5.) Let $\bar{T}^{(n)}$ be the stress vector on the slip plane; then

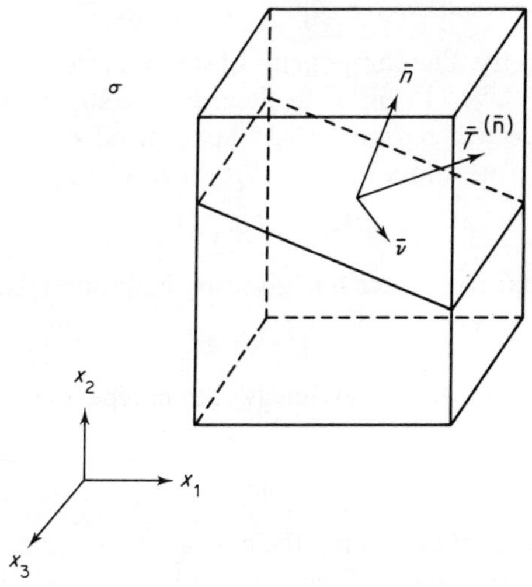

Figure 30.5

$T^{(n)} = \sigma_{ji} n_j$. The shear stress vector $\bar{T}^s$ on the slip plane is

$$T_i^s = T_{kj} n_k (\delta_{ij} - n_i n_j) \tag{30.38}$$

where T_{ij} is the deviatoric stress tensor. If there exists a preferred slip direction with unit vector $\bar{\nu}$ where $\bar{\nu} \perp \bar{n}$ and $\bar{\nu}$ is a fixed unit vector along the slip line, then the resolved shear stress vector in the direction $\bar{\nu}$ is

$$T_i^s \nu_i = T_{ki} n_k \nu_i \tag{30.39}$$

According to Schmid's law, slip will be activated when the magnitude of the resolved shear stress reached a critical value b such that

$$|T_i^s \nu_i| = b \tag{30.40a}$$

where b is a constant and independent of the normal stress on the slip plane. See ref. [20]. We note that equation (30.40a) is only a necessary condition.

The necessary and sufficient conditions are that equation (30.40a) is satisfied and equation (30.40b) is also true.

$$\frac{\mathrm{d}}{\mathrm{d}t} |T_i^s \nu_i| = 0 \tag{30.40b}$$

Hence in the event that either $|T_i^s \nu_i| < b$ or $|T_i^s \nu_i| = b$ but $\mathrm{d}/\mathrm{d}t |T_i^s \nu_i| < 0$, slip will not be activated. Treating b at least initially as a constant (i.e. no hardening) and $\bar{\nu}$ as a fixed vector is reasonable, since no 'geometric softening' is expected in a small strain field [20].

30.4.1.2 *Constitutive equation for slip*

Let $\bar{v}$ be the slip velocity. The components of slip velocity in the direction of slip $\bar{\nu}$ can be written as $(v_k \nu^k)\nu_i$. In order to describe the slip motion, we propose a constitutive equation which asserts that 'the resolved shear stress vector in the direction of slip is proportional to the slip velocity', i.e.,

$$(T_k^s \nu_k)\nu_i = \eta(v_k \nu_k)\nu_i \tag{30.41}$$

here the proportional parameter η is given by equation (30.42).

$$\eta = b/v \tag{30.42}$$

The resulting shear response is obviously rate independent.

30.4.1.3 *A representation of plastic strain*

Since slip has the effect of deforming the crystal by means of relative rigid body slippage between two parts, the rate of work done $\dot{W}^P$ due to slip can be expressed by equation (30.43):

$$\dot{W}^P = T_i^s v_i a(n) \tag{30.43}$$

where $a(n)$ is the area of the slip plane. Substituting equation (30.39) in the above equation, and letting the rate of plastic work done per unit crystal volume V be $\dot{w}^P$, one obtains

$$\dot{w}^P = T_{ij} n_j \nu_i (v_k \nu_k) a(n)/V \tag{30.44}$$

We note that $\bar{n}, \bar{v}$, and $a(n)$ are *internal* quantities which are not measurable on the surface on the deformed crystal.

We define, therefore, a tesnor $\mathbf{q}$, such that its time derivative $\dot{\mathbf{q}}$ is given by equation (30.45)

$$\dot{q}_{ij} = (n_i \nu_j + n_j \nu_i)(v_k \nu_k) a(n)/2V \tag{30.45}$$

The tensor $\mathbf{q}$ is obviously an internal variable. Defining the norm of $\dot{\mathbf{q}}$ by the relation $\|\dot{\mathbf{q}}\| = (\dot{q}_{ij}\dot{q}_{ij})^{1/2}$, then

$$\|\dot{\mathbf{q}}\| = \tfrac{1}{2} v \frac{a(n)}{V} \tag{30.46}$$

Hence

$$\|\mathrm{d}\mathbf{q}\| = \frac{1}{\sqrt{2}} |\mathrm{d}\bar{u}^P(n)| \frac{a(n)}{V} \tag{30.47}$$

where $\mathrm{d}\bar{u}^P(n) = \bar{v}\,\mathrm{d}t$ is the incremental slip displacement vector on the slip plane $\bar{n}$. Substitution of equation (30.45) into equation (30.44) gives

$$\dot{w}^P = T_{ij}\dot{q}_{ij} \tag{30.48}$$

Consider a homogeneous plastic strain rate $\dot{\boldsymbol{\varepsilon}}^{\mathrm{p}}$, determined by measurement on the surfaces of the crystal. If $\dot{\boldsymbol{\varepsilon}}^{\mathrm{p}}$ is due to the slip motion of $(\bar{n}, \bar{v})$, activated by $\boldsymbol{\sigma}$, then $\mathrm{tr}\,\dot{\boldsymbol{\varepsilon}}^{\mathrm{p}}$ is equal to zero and the rate of plastic work $\dot{w}^{\mathrm{p}}$, per unit volume, is

$$\dot{w}^{\mathrm{p}} = T_{ij}\dot{\varepsilon}^{\mathrm{p}}_{ij} \tag{30.49}$$

Comparison of equations (30.48) and (30.49) shows that

$$\dot{\varepsilon}^{\mathrm{p}}_{ij} = \dot{q}_{ij} \tag{30.50}$$

equation (30.50) is an important mathematical statement to the effect that the tensor $\dot{\varepsilon}^{\mathrm{p}}_{ij}$ plays the role of an internal variable. Since the slip motion on the plane $\bar{n}$ is not an affine deformation, we did not calculate the plastic strain rate field by spatial differentiation of the velocity field. However, we note that if we call the value of $(v_k v_k)a(n)/V$ an 'engineering shear strain rate' of the system $(\bar{n}, \bar{v})$, then the result of equation (30.50) is the same as that obtained by Bishop [21] and Taylor [22] who assumed that the slip motion on the slip plane is an affine deformation and that a strain rate tensor can be calculated by differentiation of a continuous displacement field in the coordinate system $(\bar{v}, \bar{n}, \bar{v}xn)$. After projecting this particular strain rate tensor on to the reference coordinate system (x_1, x_2, x_3), one obtained equation (30.50). However the story here is quite different.

The analysis in this section is obviously a very important step toward the understanding of the nature of the internal variables used in the endochronic theory of plasticity. The fact that the plastic strain $\mathbf{q}$ is traceless (i.e., $\mathrm{tr}\,\mathbf{q} = 0$), allows for the development of a plastic theory with elastic hydrostatic response.

30.4.2 Internal variable theory of plasticity due to a single slip mechanism in a cubic-symmetric crystal

In this section we study the plastic behaviour of a cubic-symmetric crystal due to single slip. Consider a crystal to be a continuous medium undergoing plastic deformation due to an applied uniform stress field $\boldsymbol{\sigma}$. The corresponding measurable homogeneous strain field can only represent the *average strain* field over the whole volume. In the presence of plastic deformation, the current stress $\boldsymbol{\sigma}$ cannot be a single valued function of the current strain state $\boldsymbol{\sigma}$. This can be observed to be the case in a stress history of cyclic straining [23]. Hence, from the phenomenological point of view, it is necessary to introduce a set of internal variables $\mathbf{q}^{\alpha}$ which represent the state of the current material microstructure [1]. The stress $\boldsymbol{\sigma}$ is then determined by the current state variables $(\boldsymbol{\varepsilon}, \mathbf{q}^{\alpha})$. The strain field can be obtained by superposition of a homogeneous elastic strain $\boldsymbol{\varepsilon}^{\mathrm{e}}$ and a homogeneous plastic strain $\boldsymbol{\varepsilon}^{\mathrm{p}}$. If $\boldsymbol{\varepsilon}^{\mathrm{p}}$ is totally due to a single slip in the crystal, then as a result of equation (30.50), $\mathbf{q}$ represents $\boldsymbol{\varepsilon}^{\mathrm{p}}$.

30.4.2.1 *Constitutive formulation in terms of the free energy*

Let the Helmholtz free energy ψ be a quadratic function of the elastic strain tensor $(\boldsymbol{\varepsilon}-\mathbf{q})$. For a cubic-symmetric crystal, ψ is of the form

$$\psi=\tfrac{1}{2}D_{ijkl}(\varepsilon_{ij}-q_{ij})(\varepsilon_{kl}-q_{kl}) \tag{30.51}$$

where $\mathbf{D}$ is a cubic-symmetric, fourth order material tensor in the coordinate system (x_1, x_2, x_3) corresponding to the symmetric axes of the material. It was shown that ψ can be decomposed into a hydrostatic part ψ_H and a deviatoric part ψ_D [24]. That is

$$\psi_H=\tfrac{1}{2}A\varepsilon_H^2 \tag{30.52a}$$

$$\psi_D=\tfrac{1}{2}D_{ijkl}(e_{ij}-q_{ij})(e_{kl}-q_{kl}) \tag{30.52b}$$

where $A=D_{iikk}$. A is a bulk modulus. Equation (30.52a) indicates that the hydrostatic response is elastic. No plastic dilatation occurs because of the fact that $\mathbf{q}$ is a deviatoric tensor. Two central results of the internal variable theory under isothermal conditions [1] are shown in equations (30.53) and (30.54):

$$\mathbf{T}=\frac{\partial\psi_D}{\partial\mathbf{e}} \tag{30.53}$$

$$\theta\dot{\gamma}=-\frac{\partial\psi_D}{\partial\mathbf{q}}\cdot\dot{\mathbf{q}}\geq 0 \tag{30.54}$$

where θ is the absolute temperature, $\dot{\gamma}$ is the rate of change of irreversible entrophy and $\theta\dot{\gamma}$ is the rate of energy dissipation per unit volume. From equations (30.52b) and (30.53) we obtain the relation

$$\mathbf{T}=-\frac{\partial\psi_D}{\partial\mathbf{q}} \tag{30.55}$$

Substitution of equation (30.55) into (30.54) gives the relation

$$\theta\dot{\gamma}=\mathbf{T}\cdot\dot{\mathbf{q}}\geq 0 \tag{30.56}$$

Comparison of equations (30.48) and (30.56), yields

$$\theta\dot{\gamma}=\dot{w}^{\mathrm{p}}\geq 0 \tag{30.57}$$

Equation (30.57) *is a mathematical statement of the very important result that the rate of energy dissipation* $\theta\dot{\gamma}$ *is exactly the rate of plastic work* $\dot{w}^{\mathrm{p}}$ *due to the slip mechanism in a crystal.*

30.4.2.2 *Evolution equation*

The determination of form of the evolution equation amounts to finding a relationship between $\mathbf{T}$ and $\dot{\mathbf{q}}$. Eliminating $v_k v_k$ from equations (30.45) and

(30.41), we obtain the relation

$$\eta \dot{q}_{ij} = (n_i v_j + n_j v_i) T^{s}_{k} v_k a(n)/2V \tag{30.58a}$$

Since $T^{s}_{k} v_k$ can be found from equation (30.39), equation (30.58a) becomes

$$\lambda \dot{q}_{ij} = N_{ijkl} T_{kl} \tag{30.58b}$$

where

$$N_{ijkl} = \tfrac{1}{2}(n_i v_j + n_j v_i)(n_k v_1 + n_1 v_k) \tag{30.58c}$$

and

$$\lambda = 2V\eta/a(n) \tag{30.58d}$$

Evidently **N** is a projection that projects the stress space **T** into a plastic strain rate space $\dot{\mathbf{q}}$. This projector **N** shows that the plastic strain rate is strongly dependent on the directionality of $(\bar{n}, \bar{v})$ in the crystal. One can readily prove that

$$T_{ij} N_{ijkl} T_{kl} = 2(T^{s}_{i} v_i)^2 \geq 0 \tag{30.59}$$

for arbitrary **T**. The equal sign applies when $\bar{T}^{s} \perp \bar{v}$. Thus equation (30.59) demonstrates that **N** is a positive semi-definite tensor. At this point we recall equation (30.58b) to give the evolution equation of $\dot{\mathbf{q}}$ which is obtained by substituting equation (30.55) into equation (30.58b):

$$\lambda \dot{\mathbf{q}} + \mathbf{N} \cdot \frac{\partial \psi_{\mathrm{D}}}{\partial \mathbf{q}} = 0 \tag{30.60}$$

30.4.2.3 *Endochronic time*

As indicated by equation (30.42) and (30.58d) λ is inversely proportional to the magnitude of slip velocity $\bar{v}$. Let the norm of $\mathrm{d}\mathbf{q}$, i.e. $\|\mathrm{d}\mathbf{q}\|$, be denoted by $\mathrm{d}\zeta$, as shown in equation (30.61).

$$\mathrm{d}\zeta = \|\mathrm{d}\mathbf{q}\| = \|\mathrm{d}\boldsymbol{\varepsilon}^{\mathrm{p}}\| \tag{30.61}$$

Then by using equation (30.47) we have

$$\mathrm{d}\zeta = \frac{1}{\sqrt{2}} \frac{a(n)}{V} |\mathrm{d}\bar{u}^{\mathrm{p}}| \tag{30.62}$$

Substitution of equation (30.62) into equation (30.60), gives the result

$$\sqrt{2} b \frac{\mathrm{d}\mathbf{q}}{\mathrm{d}\zeta} + \mathbf{N} \cdot \frac{\partial \Psi_{\mathrm{D}}}{\partial \mathbf{q}} = 0 \tag{30.63}$$

It is shown in equation (30.63) that the evolution equation for **q** is, in terms of the 'intrinsic time' $\mathrm{d}\zeta$, where $\mathrm{d}\zeta$ is given by the norm of the increment of plastic strain, as proposed on phenomenological grounds in ref. [12].

Equations (30.53) and (30.63) are the basis of the constitutive response of the crystal and are sufficient to determine the deviatoric stress tensor given the history of total strain. This consists of an elastic strain and a plastic counterpart which is generated by a mechanism of single slip in the crystal. Note that the hydrostatic response is elastic.

The above analysis can be generalized into multiple slip systems. For details see ref. [24].

30.4.2.4 *Conclusion*

In this section we apply the internal variable theory to the deformation of single crystals. In the case of single slip only one internal variable is present. We show that the internal variable is a tensor representation of the part of the plastic strain arising from translational slip along a slip plane and in a specified direction, thus giving a physical significance to this variable in the case of plastic deformation of single crystals.

A three-dimensional equation, governing the evolution of the internal variable is then derived on the basis of a scalar constitutive equation which governs the onset and continuation of the slip motion. Even though the equation is of an anisotropic character, it is essentially the same type as the endochronic evolution equation proposed on a phenomenological basis [5, 12].

When slip is the physical basis of plastic deformation, then the measure of intrinsic time is the norm of the increment of the plastic strain tensor as proposed by Valanis on thermodynamic considerations pertaining to zero dissipation rate at the onset of unloading [12].

REFERENCES

1. K. C. Valanis, 'A theory of viscoplasticity without a yield surface, Part I: General theory; Part II: Application to mechanical behaviour of metals', *Archives of Mechanics*, **23**, 517–551 (1971).
2. K. C. Valanis and C. F. Lee, 'Some recent developments of the endochronic theory with applications', *Nuclear Engr. and Design*, **69**, 327–344 (1982).
3. K. C. Valanis, 'Effect of prior deformation on cyclic response of metals', *J. of Appl Mech. Trans. ASME*, **41**, 441–447 (1974).
4. K. C. Valanis and H. C. Wu, 'Endochronic representation of cyclic creep and relaxation of metals', *J. Appl. Mech., Trans. ASME*, **42**, 67–73 (1975).
5. K. C. Valanis, 'Endochronic theory with proper hysteresis loop closure properties', *Topical Report, SSS-R*-80-4182, System, Science and Software, San Diego, California, USA, August 1979.
6. I. S. Sandler, 'On the uniqueness and stability of endochronic theories of material behaviour', *J. Appl. Mech., Trans. ASME*, **45**, 263–266 (1978).
7. R. S. Rivlin, '"Some comments" on the endochronic theory of plasticity', *Int. J. Solids Structures*, **17**, 231–248 (1981).
8. R. S. Rivlin, '"Comments on" On the Substance of Rivlin's Remarks on the Endochronic Theory, By K. C. Valanis', *Int. J. Solids Structures*, **17**, 267–268 (1981).

9. B. J. Hsieh, 'On the uniqueness and stability of endochronic theory', *J. Appl. Mech., Trans. ASME*, **47**, 748–754 (1980).
10. K. C. Valanis, 'On the substance of Rivlin's Remarks on the Endochronic Theory', *Int. J. Solids Structures*, **17**, 249–265 (1981).
11. W. S. Edelstein, 'A review of some endochronic theories and their applications', Argonne Natl. Lab. *Technical Mem. ANL-CT*-82-22, 1982.
12. K. C. Valanis, 'Fundamental consequences of a new intrinsic time measure: plasticity as a limit of the endochronic theory', *Arch. of Mech.*, **32**, 171–191 (1980).
13. Yu. N. Rabotnov, *Creep Problems in Structural Members*, pp. 127–130, North-Holland Pub. Co., 1969.
14. K. C. Valanis and H. E. Read, 'A new endochronic plasticity model for soils', in *Soil Mechanics—Transient and Cyclic Loads*, Chapter 14 (Eds. G. N. Pande and O. C. Zienkiewicz), John Wiley & Sons Ltd., New York, 1982.
15. K. C. Valanis and C. F. Lee, 'Endochronic Theory of Cyclic Plasticity with Applications', Univ. of Cincinnati, College of Eng., *Research Annals*, **83**, No. ADM-111 (1983).
16. H. R. Jhansale and T. H. Topper, 'Engineering Analysis of the inelastic stress response of a structure metal under variable cyclic strain', in *ASTM STP* 519, *Cyclic Stress–* Strain Behaviour-Analysis, Experimentation, and Failure Prediction, pp. 246–270, 1973.
17. Y. F. Dafalias and E. P. Popov, 'Plastic internal variables formalism of cyclic plasticity', *J. of Appl. Mech., Trans. ASME*, **98**, 645–651 (1976).
18. R. W. Landgraf, J. Morrow, and T. Endo, 'Determination of the cyclic stress–strain curve', *J. of Materials*, **4**, 176–188 (1969).
19. K. C. Valanis and C. F. Lee, 'Some recent developments in the endochronic theory with application to cyclic histories', in *NASA Symposium on Nonlinear Constitutive Relations for High Temperature Applications*, at University of Akron, Akron, Ohio, May 1982.
20. E. Schmid and W. Boas, *Plasticity of Crystals*, F. A. Hughes & Co. Limited, London, 1950.
21. J. F. W. Bishop, '*A theoretical examination of the plastic deformation of crystal by glids*', *Phil. Mag.*, Ser. 7, 44, 51–64 (1953).
22. G. I. Taylor, 'Analysis of plastic strain in a cubic crystal', in *The Scientific Papers of Sir Geoffrey Ingram Taylor* (Ed. G. K. Batchelor), Cambridge Univ. Press, Vol. I, Paper 28, 1958.
23. C. F. Feltne and C. Laird, 'Cyclic stress–strain response of F.C.C. Metals and alloys—I and II, '*Acta Metallurgica*, **15**, 1621–1653 (1967).
24. K. C. Valanis and C. F. Lee, 'The concept of endochronic time applied to the mechanism of slip', Univ. of Cincinnati, College of Eng., *Research Annals*, **83**, No. ADM-110 (1983).

Mechanics of Engineering Materials
Edited by C. S. Desai and R. H. Gallagher

Chapter 31

Constitutive Relationships for Composite Materials Through Micromechanics

S. Valliappan and J. I. Curiskis

31.1 INTRODUCTION

A composite material may be defined as a heterogeneous medium consisting of two or more materials combined together to produce a new material which exhibits the best strength and deformation characteristics of its constituents and probably some of its own individual characteristics resulting from the process of mixing. Composites themselves may be broadly classified as dispersion-strengthened composites, particulate composites and fibre-reinforced composites. In dispersion-strengthened composite materials, the major load-bearing constituent is the matrix and the purpose of dispersion is mainly to maintain the strength of the matrix under varying conditions. In particulate composites or particle-reinforced composite materials, both the matrix and the dispersed particles share the load. In fibre-reinforced composite materials, the fibre is the primary load-bearing constituent and the function of the matrix is only to act as a binder and to transmit load to the reinforcing fibres. Because of their load-bearing capacity, particles and fibers are attractive constituents for reinforcements in composites and hence they are widely used in various industries. Hence, the efficient use of composite materials necessitates a proper understanding of deformation and fracture behaviour of these materials under applied loads. In the case of particle and fibre reinforced composite materials, the stress (deformation) analysis becomes complicated due to the interaction between the matrix and the reinforcement as well as the interaction between the reinforcing constituents. This difficult situation is further compounded in the case of fibre-reinforced composites due to the nature of the three-dimensional problem. To a certain extent, continuous fibres can be analysed using 'law of mixtures'. However, discontinuous fibres pose problems related to stress concentrations at fibre ends because of the discontinuities.

Due to the heterogeneous nature of composite materials it is necessary to adopt a micromechanics analysis to study its mechanical behaviour in terms of those of constituent materials. The primary object of a micromechanics analysis is to determine the constitutive relationships for the composite (which is a heterogeneous medium) such that it can be subsequently treated as a macroscopically homogeneous material. In addition to this, it is also necessary to examine the local behaviour which determines the areas of stress concentration, the onset of failure and initiation of cracks. These phenomena of load transfer and stress concentration are controlled by a number of parameters depending on the properties of constituent materials and the nature of the interface and its bonding. In order to identify the important individual effects, it is best to adopt numerical techniques to determine the effect of various parameters such as aspect ratio, moduli ratio, and volumetric ratio. From the micromechanics point of view, short discontinuous fibers can be effectively studied numerically isolating a single representative volume element and treating one feature of the problem in each study. The representative volume element (RVE) is the smallest region or sample of the composite material which is structurally typical of the lamina and over which the distribution of stresses and strains is macroscopically uniform and typical of the lamina as a whole.

In this chapter, the linear and non-linear constitutive relationships for both continuous and discontinuous fibre composites have been obtained using three-dimensional finite element analysis. The analysis is based on representative volume element and the composite properties are defined from a number of uniaxial load cases and the associated (constraint) boundary conditions.

31.2 REVIEW OF PREVIOUS MICROMECHANICS ANALYSES

The growth of fibre-reinforced composite technology has been quite rapid in recent years and hence the amount of published literature is extensive. Due to the extensive nature of the published work available in books, conference proceedings, and journals, it is not the intention of the authors to review all the previous analyses in this chapter. Only the studies relevant to the authors' approach to micromechanics analysis will be reviewed in this section. For general discussion on composite materials, readers can refer to the references [1 to 3].

In almost all studies of the mechanical behaviour of composites, certain simplifying assumptions are generally made. Such assumptions include that the constituent materials are isotropic or transversely isotropic and are either linear elastic or elastoplastic. The fibres are either continuous or discontinuous, parallel to each other within the matrix, regularly spaced, have identical geometry with regard to cross-section and length, and perfect bonding exists between the constituents. As a result of the fabrication process, real composites may violate some of these assumptions; for example, the arrangement and alignment of the fibres may be imperfect. Despite these limitations, significant information concerning the local stress states within the matrix and at the interface has been obtained from various studies.

Most micromechanics analyses in the past have been concerned with the prediction of elastic constants only. Among the various investigations, the ones which were based on energy theorems of classical elasticity are by Hashin and Rosen [4], Hill [5], Whitney and Riley [6]. Adams and Doner [7, 8] utilized the finite difference technique for the analysis of unidirectional composite subjected to longitudinal shear loading and transverse normal loading. Foye [9] was the first to use the finite element method for the micromechanics analysis of composites. He investigated square, diamond, and hexagonal arrays with continuous filaments of various cross-sectional shapes. The longitudinal and transverse elastic constants were obtained by the superposition of macroscopic triaxial stress states. The shear constant was obtained from a longitudinal-shear loading. Later he extended the method [10] to include rectangular array geometry. Adams and Tsai [11] used plane strain finite element analysis to study the influence of random filament packing. Goanker [12] has used the finite element method to study the stress state in a plane composite with a rectangular array of discontinuities.

Micromechanics analyses of composites to study their elastoplastic behaviour are not as extensive as the elastic analyses. A full review has been given by Adams [13]. Owen *et al.* [14] applied the finite element technique to the elastoplastic analysis of single and multiple fibre systems. The systems were modelled using axisymmetric and plane strain elements. The assumption of plane strain implies modelling of discontinuous plates embedded in a matrix. Adams [15] considered the problem of a rectangular array of continuous filaments embedded in an elastoplastic matrix and subjected to transverse loads. A plane strain condition was assumed. He obtained stress–strain curves and octahedral shear stress contours. Later, Adams [16] extended this analysis to allow for crack initiation and propagation in the matrix. Foye [17 to 19] used the initial strain method of elastoplastic analysis and a generalized plane strain condition for the study of continuous filament composites subjected to single longitudinal, transverse and longitudinal-shear loadings as well as combined loadings. The boundary conditions on the borders of the representative volume element were enforced by an 'integral force' method in which the constraint equations were established between the relevant nodal forces. He predicted the composite macroscopic response for the single loadings and the development of plastic zones. Agarwal *et al.* [20] employed an axisymmetric finite element analysis for short fibre composites under longitudinal loading. The matrix behaviour was assumed to be bilinear and von Mises yield criterion was adopted. A row and column or banded symmetric array of fibres and a symmetric staggered array of overlapping fibres surrounded by matrix material were considered for the analysis. The boundary conditions on the surface of the cylindrical RVE (representative volume element) were simulated by a modified superposition method employing trial and error estimates of the prescribed normal displacements. For the staggered array, the tangential displacements along the cylindrical surface were also prescribed.

The general approach to the study of load-transfer role of the interface has been the prediction of the state of stress at the interface. The consideration of an imperfect interface is a difficult problem. Most of the analyses available so far are of simpler forms for a single fibre model. Amirbayat and Hearle [21 to 23] considered a single fibre in a cylindrical matrix under longitudinal loading and derived expressions for the shear stress distributions for three interfacial conditions—no bond, no friction; perfect bond; no bond, limited friction. They also combined the slippage and perfect bond systems utilizing a shear failure criterion for the interface to simulate crack initiation, propagation, and arrest in a composite system. Owen and Lyness [24] used the initial stress approach of the elastoplastic analysis to investigate the bond failure in a single fibre model under longitudinal loading. For the two-dimensional analysis, they developed a special bond element to model imperfect interface allowing delamination and slippage failure modes. Conway *et al.* [25] used a point-matching technique to study the bond stresses for a matrix sheet reinforced by a repeating pattern of overlapping rigid, one-dimensional fibres subjected to longitudinal tensile loading Later Chang *et al.* [26] extended the method to a three-dimensional fibre array geometry. Agarwal [27] studied the role of the interface or interphase in short fibre composites under longitudinal tensile loading. A perfect bonding between the interphase and the fibre and matrix was assumed in the axisymmetric analysis of a row column fibre array.

With all the analyses discussed above, certain deficiencies exist, primarily due to the two-dimensional modelling whether it be plane strain or axisymmetric and in certain cases due to the prescription of the boundary conditions. In this paper, a fully three-dimensional formulation is presented for the micromechanics analysis of fibre-reinforced composites. The appropriate boundary conditions for the solution of a RVE have been used on the basis of linear constraint equations formulated in terms of nodal displacements.

31.3 MICROMECHANICS ANALYSIS OF COMPOSITES

The main aims of the analysis are (1) to predict the macroscopic properties (both linear elastic and elastoplastic) of the composite material and (2) to obtain the internal stress distribution. In order to achieve these aims, finite element method has been used as the numerical technique.

It is needless to say that the mechanical behaviour of a composite depends upon the physical properties and the geometrical arrangement of the constituent materials. It is relatively easy to define the physical properties of the constituents. However, it is not simple to define the geometry of the packing and hence some sort of idealization of composite geometry is required in order to obtain a RVE for detailed analysis. Similarly the various loading cases have to be identified so that the relevant macroscopic properties can be deduced in a straightforward manner from the finite element analysis.

31.3.1 Composite geometry

The standard geometric idealizations of continuous fibre composites are doubly infinite arrays of uniform, parallel, cylindrical inclusions in rectangular or

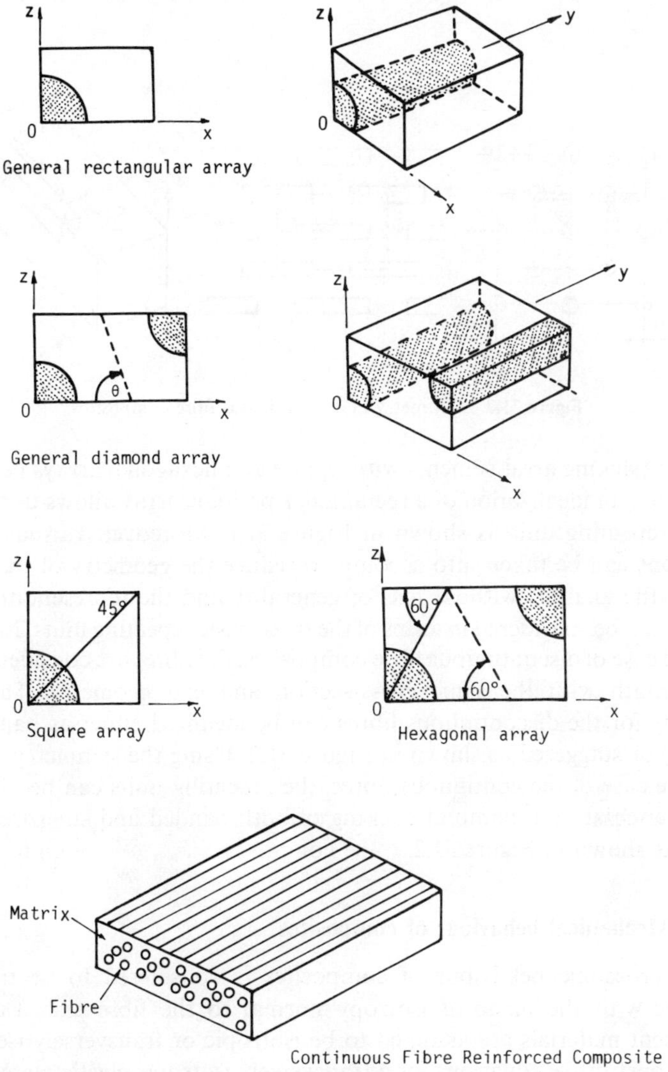

Figure 31.1 Typical repeating units of RVE

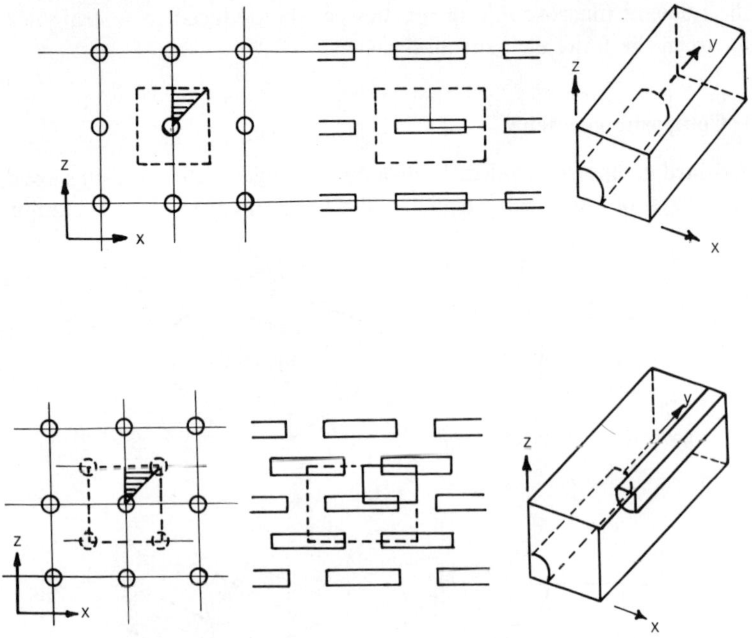

Figure 31.2 Geometry of discontinuous fibre composites

diamond packing arrangements with square and hexagonal arrays being special cases. Such an idealization of a rectangular periodic array allows us to isolate a typical repeating unit as shown in Figure 31.1. Moreover, various symmetry conditions can be taken into account to reduce the geometry of the repeating unit for the analysis without loss of generality and the representative volume element can be considered in terms of the three basic repeating units (Figure 31.1).

In the case of discontinuous fibre composites, the fibres are considered to be of equal length with the same cross-section and end geometry. The packing geometry for the discontinuous fibres can be idealized either as banded (row–column) or staggered as shown in Figure 31.2. Using the symmetry conditions, as in the case of the continuous fibres, the repeating units can be obtained for the rectangular and diamond packing of both banded and staggered arrangements as shown in Figure 31.2.

31.3.2 Mechanical behaviour of composites

The macroscopic behaviour of composites is considered to be transversely isotropic with the plane of isotropy normal to the fibre axis. Further, the constituent materials are assumed to be isotropic or transversely isotropic.

The constitutive equations for a transversely isotropic elastic material can be written as

$$\begin{bmatrix} \varepsilon_x \\ \varepsilon_y \\ \varepsilon_z \\ \gamma_{xy} \\ \gamma_{yz} \\ \gamma_{zx} \end{bmatrix} = \begin{bmatrix} \frac{1}{E_T} & -\frac{\nu_{LT}}{E_L} & -\frac{\nu_{TT}}{E_T} & & & \\ -\frac{\nu_{LT}}{E_L} & \frac{1}{E_L} & -\frac{\nu_{LT}}{E_L} & & & \\ -\frac{\nu_{TT}}{E_T} & -\frac{\nu_{LT}}{E_L} & \frac{1}{E_T} & & & \\ & & & \frac{1}{G_{LT}} & 0 & 0 \\ & & & 0 & \frac{1}{G_{LT}} & 0 \\ & & & 0 & 0 & \frac{2(1+\nu_{TT})}{E_T} \end{bmatrix} \begin{bmatrix} \sigma_x \\ \sigma_y \\ \sigma_z \\ \tau_{xy} \\ \tau_{yz} \\ \tau_{zx} \end{bmatrix}$$

As can be seen from the above equation, there are only five independent elastic constants—E_T, E_L, ν_{LT}, ν_{TT}, and G_{LT}. These five constants can be determined from three basic independent uniaxial loadings—longitudinal, transverse, and longitudinal shear. It should be noted that a square array fibre packing geometry leads to a composite with six independent elastic constants.

31.3.3 Symmetry conditions for repeating units

If the geometric idealizations of the repeating units as illustrated in Figure 31.1 and 31.2 are adopted for analysis with uniaxial loadings, then it is necessary to impose certain conditions on the symmetric boundaries in order to prohibit the type of deformations which may result in gaps between the units (Figure 31.3).

For the case of longitudinal loading, the six faces of a rectangular prismatic repeat unit, after deformation should retain their rectangular shape. Thus the normal displacement of these boundary faces is constrained since this displacement must be same for all points on the respective faces. A further constraint condition due to the implication that the essential array geometry is maintained even after loading, is that the displacements in x and z directions (Figure 31.3) of any point within the repeat unit must be such that the point remains on the radial plane passing through the axis of the closest fibres. This is illustrated in Figure 31.3 and it can be noted that the minimum RVE is reduced to transverse cross-section ABD for the square and hexagonal arrays. The normal displacements of points on AB and BD are constant.

In the case of transverse loading, similar to the case of longitudinal loading, the normal displacement is constrained on all the six boundary faces of the quadrant repeating unit. However, it should be realized that reduced geometry adopted

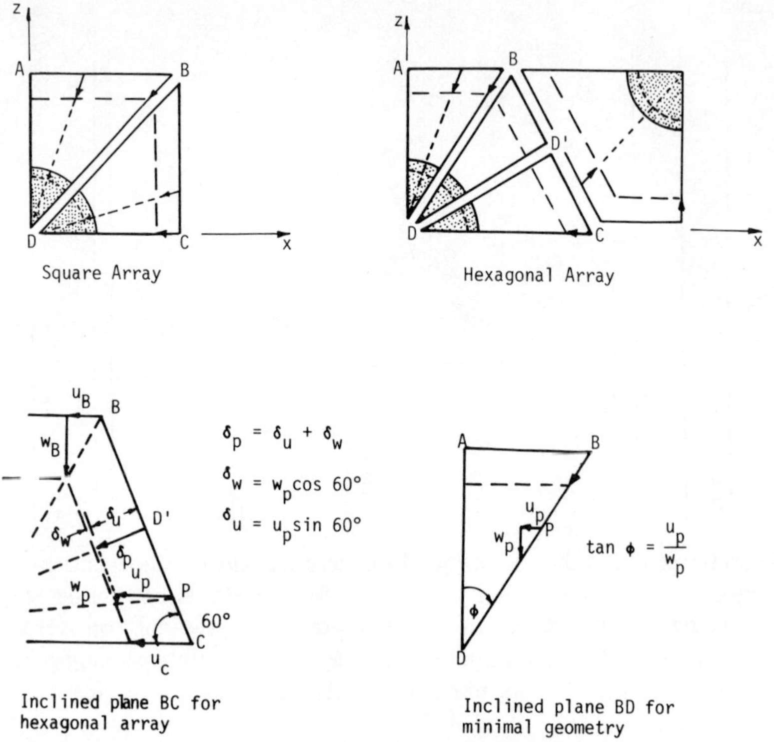

Figure 31.3 Constraint boundary conditions

before is not possible for this loading and hence a complete quadrant unit has to be used.

Regarding the longitudinal-shear loading, the displacements in y direction must be same for any point on the loaded boundary planes. On the intermediate planes parallel to the boundary, a constraint condition is defined such that equivalent transverse sections normal to y-axis deform identically. This condition implies that the RVE is the quadrant unit for continuous fibre composites and twice this unit in y direction for the case of discontinuous fibre composites, with the constraint condition applied at the opposite boundary faces normal to y-axis.

As discussed above, the macroscopic behaviour of the composite is defined in terms of three independent loadings on RVE with the associated boundary conditions. For the rectangular prismatic RVE shown in Figure 31.4(a), the three loading cases are illustrated in Figure 31.4(b), (c), (d). In order to illustrate some of the boundary constraints, let us refer to Figure 31.4(a). The points I, J, K, P, Q, R correspond to any node on the respective boundary faces and the points L, M, N are arbitrarily chosen nodes on the free faces. The normal displacements of the

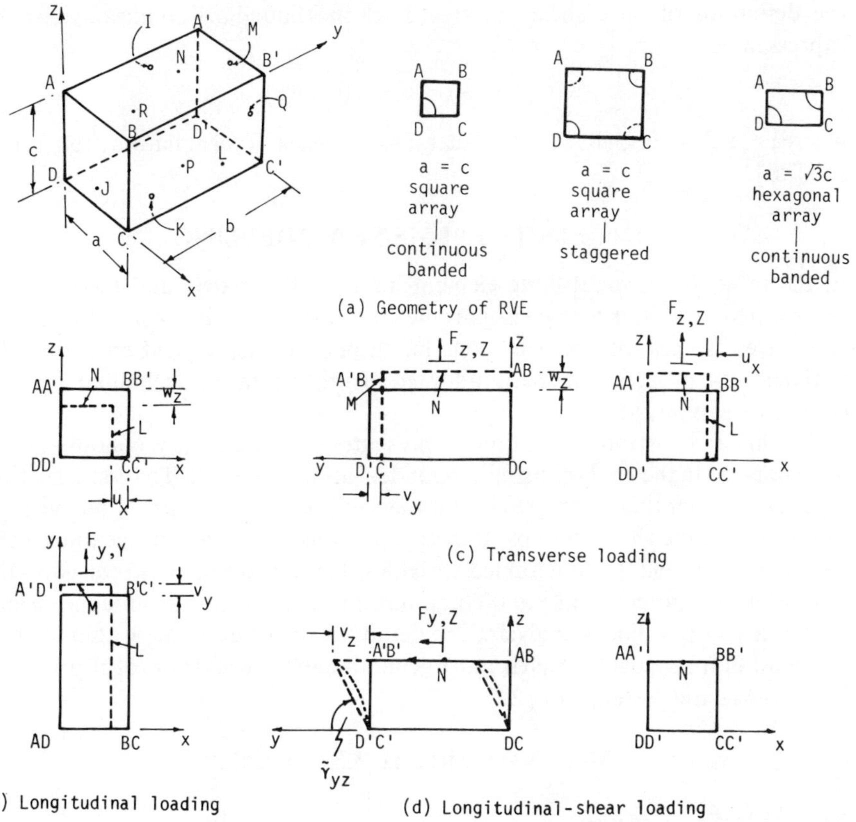

Figure 31.4 Rectangular prismatic RVE and associated uniaxial loading cases

three faces defined by the cartesian axes are fixed for the cases of longitudinal and transverse loading. On the free faces, due to constraint conditions the normal displacement of any point such as P, Q, R are equal to that of the chosen node L, M, N respectively. Thus a set of linear constraint equations is established. For example, the constraint conditions in terms of the nodal displacements can be defined as follows:

For longitudinal and transverse loading,

$$u_I = 0; \quad v_J = 0; \quad w_K = 0$$

$$u_P = u_L; \; v_Q = v_M; \; w_R = w_N$$

For the case of longitudinal shear loading, the four boundary faces parallel to y-axis have zero x and z displacements. Also, plane DCC′D′ has a prescribed y displacement of zero and all nodes on ABB′A′ are constrained such that their y displacements are set equal to that of the independent node N, thus satisfying the definition of pure shear. In this case, the boundary conditions can be expressed as

$$u_{\bar{i}}=0; \quad v_K=0; \quad w_{\bar{i}}=0; \quad v_R=v_N$$

where i ∈{I, J, K, P, Q, R, L, M} full details of constraint formulation are given in ref. [28].

31.4 FINITE ELEMENT MODELLING

In the three-dimensional finite element analysis, the matrix and the fibre are represented by 15 and 20 node isoparametric elements. The non-linear behaviour considered is elastoplastic behaviour based on von Mises' yield criterion. The initial stress process has been adopted as the iterative technique for the elastoplastic analysis.

The linear constraint equations representing the boundary conditions are incorporated in the analysis using the condensation technique. The condensation procedure rather than the more elegant Lagrange multipliers was chosen because in this approach, the system of stiffness equations is reduced in size due to the redundancy among the constrained variables. This feature can yield considerable savings in computer storage and computation time and this is quite important in the case of non-linear analysis. The details of the solution algorithm for this condensation approach which is more general than the usual approach are given by Curiskis and Valliappan [29].

31.5 NUMERICAL EXAMPLES

31.5.1 Elastic properties

The elastic constants of glass-epoxy composites with continuous fibre reinforcements have been determined using the micromechanics approach discussed in the previous section. The results for volume fractions 0.1 and 0.7 with relevant comparisons from Foye [30] are summarized in Table 31.1 for square array and in Table 31.2 for hexagonal array.

Table 31.1

	Square array					
	Volume fraction = 0.1			Volume fraction = 0.7		
Materials constants	Present	Foye (DEM)	Foye (Best mixtures)	Present	Foye (DEM)	Foye (Best mixtures)
$E_L \times 10^6$ p.s.i.	1.51	1.51	1.51	7.58	7.58	7.58
ν_{LT}	0.32	0.32	0.325	0.23	0.23	0.24
$E_T \times 10^6$ p.s.i.	0.64	0.64	0.64	3.23	3.23	3.2
ν_{TT}	0.43	0.43	0.43	0.21	0.20	0.30
$G_{LT} \times 10^6$ p.s.i.	0.23	0.23	0.23	1.04	1.04	1.04

Table 31.2

Materials Constants	Hexagonal array Volume fraction = 0.1 Present	Foye (DEM)	Foye (Best mixtures)	Volume fraction = 0.7 Present	Foye (DEM)	Foye (Best Mixtures)
$E_L \times 10^6$ p.s.i.	1.51	1.51	1.51	7.58	7.58	7.58
ν_{LT}	0.32	0.32	0.325	0.24	0.24	0.245
$E_T \times 10^6$ p.s.i.	0.63	0.63	0.63	2.36	2.36	2.50
ν_{TT}	0.44	0.44	0.43	0.35	0.35	0.30
$G_{LT} \times 10^6$ p.s.i.	0.23	0.23	0.23	0.88	0.88	0.90

The properties of the constituents assumed for the analysis are:

Epoxy: $E = 0.5 \times 10^6$ p.s.i. $\quad \nu = 0.34$

Glass: $E = 10.6 \times 10^6$ p.s.i. $\quad \nu = 0.20$

The results indicate that for the same constituent materials and volume fraction, the square array yields a higher modulus (E_T) but lower associated Poisson's ratio (ν_{TT}) than the hexagonal array. Comparison, with Foye's two-dimensional finite element approach (DEM) and 'best' analytical estimates show that the present constraint formulation gives similar values for the elastic constants. Foye's 'best' closed form expressions for E_T and ν_{TT} in the case of hexagonal packing are inadequate because the derivation is based on mixtures estimate with an effective Poisson's ratio. It should be noted that the value of ν_{TT} for square packing has dropped off substantially for increasing volume fraction whereas the drop-off is not very much in the case of hexagonal packing.

31.5.2 Non-linear properties

31.5.2.1 *Boron–epoxy composite—continuous reinforcement*

The boron filaments are considered to be continuous and packed in a square array. The volume fraction was assumed to be 0.4. The properties of the boron were assumed to be elastic whereas the stress–strain curve for epoxy matrix was considered to be linear elastic–plastic with two slopes in the plastic range (Figure 31.5). The constituent properties are as follows:

Boron: $E = 60.0 \times 10^6$ p.s.i. $\quad \nu = 0.20$

Epoxy: $E = 0.5 \times 10^6$ p.s.i. $\quad \nu = 0.34$

$\bar{\sigma}_0 = 5$ k.s.i. $\quad E_{p,1} = 0.75 \times 10^6$ p.s.i.

$\bar{\sigma}_1 = 20$ k.s.i. $\quad E_{p,2} = 4.54 \times 10^4$ p.s.i.

where $\bar{\sigma}_0$ and $\bar{\sigma}_1$, are initial and post-yield stresses and $E_{p,1}$ and $E_{p,2}$ are initial and post-yield slopes of the effective stress-plastic strain curves.

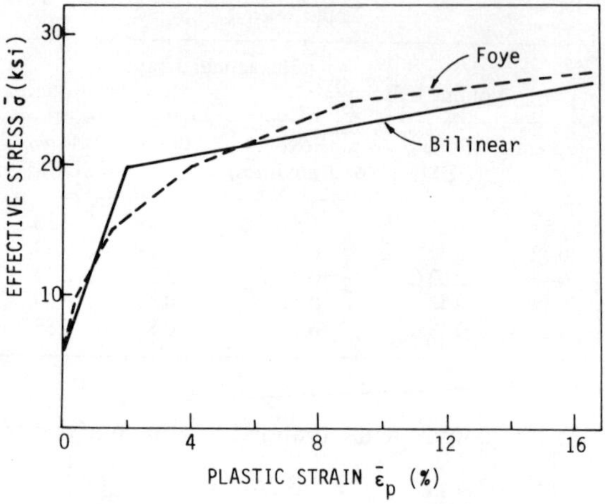

Figure 31.5 Stress–strain curve for matrix

The predicted elastic constants and initial yield stresses for the composite are given below:

$$E_L = 24.3 \times 10^6 \text{ p.s.i.} \qquad \nu_{LT} = 0.27$$
$$E_T = 1.35 \times 10^6 \text{ p.s.i.} \qquad \nu_{TT} = 0.37$$
$$G_{LT} = 0.43 \times 10^6 \text{ p.s.i.}$$
$$\bar{\sigma}_y = 245 \text{ k.s.i.} \qquad \bar{\sigma}_x = -4.36 \text{ k.s.i.} \qquad \bar{\tau} = 1.95 \text{ k.s.i.}$$

These values compare favourably with the results obtained by Foye [17, 18].

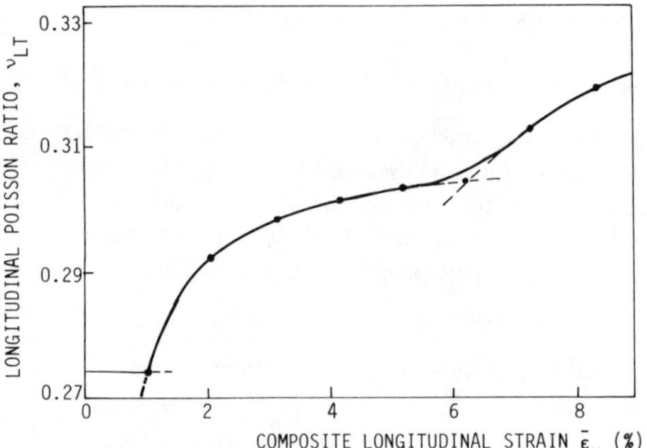

Figure 31.6 Variation of ν_{LT}

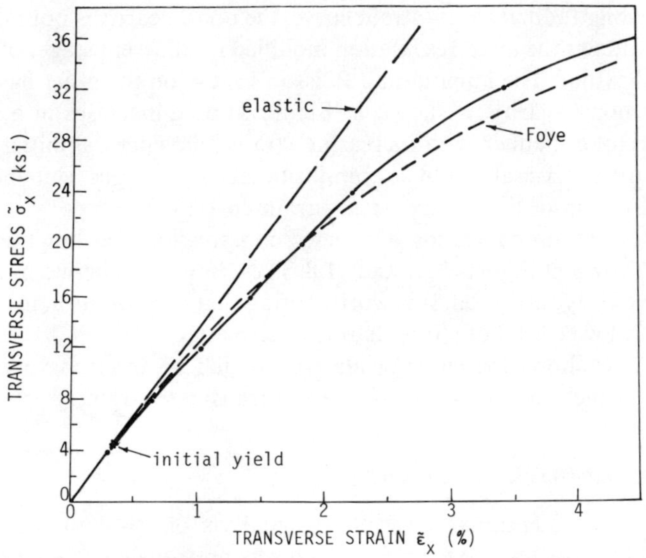

Figure 31.7 Transverse stress–strain curve for composite

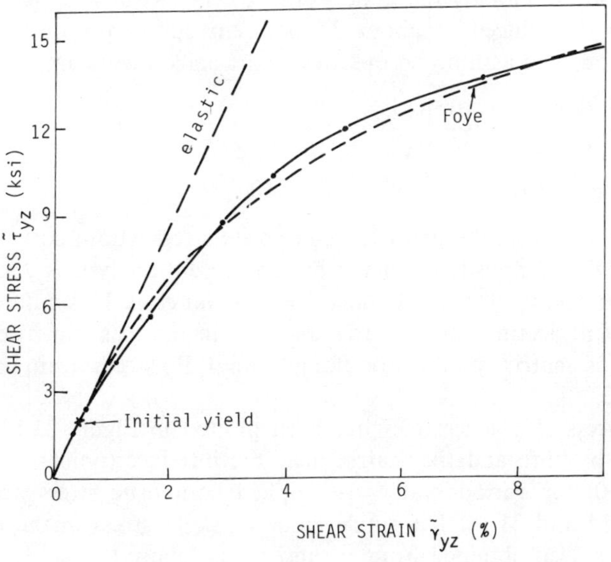

Figure 31.8 Longitudinal shear stress–strain curve for composite

For the longitudinal stress–strain curve, the non-linearity is not marked since the curve follows the mixtures relation modified for different stages of non-linear matrix behaviour. The longitudinal Poisson's ratio, on the other hand, exhibits a marked non-linearity as shown in Figure 31.6. It increases in a monotonic and asymptotic manner with increasing strain following the initial composite yield. A further rise at about a composite strain of 6 per cent indicates the post-yield region of the matrix stress–strain curve.

The stress–strain curves for transverse compression and longitudinal shear stress are shown in Figures 31.7 and 31.8. The comparison between these results and those of Foye are good. It is worth noting that in both cases, initial yielding occurs at a lower level of stress than for the matrix.

Figure 31.9 shows the effect of matrix yielding on the transverse Poisson's ratio (ν_{TT}), which increases with increasing transverse strain.

31.5.2.2 *Discontinuous reinforcement*

For the three-dimensional elastoplastic analysis of discontinuous fibre-reinforcement composites, the problem which was previously analysed by Agarwal *et al.* [20] on the basis of axisymmetric idealization is considered. A row–column arrangement of fibres in longitudinal direction with square packing in transverse plane has been adopted. For the purpose of illustration of the process, only longitudinal loading has been taken into account in presenting the results.

The assumed geometry of the problem is shown in Figure 31.10 and the dimensions are essentially the same as those of Agarwal *et al.* [20] except the fibre volume is reduced by about 7.5 per cent due to the three-dimensional representation. The assumed properties of the constituents are:

$$\text{Fibre: } E = 11.8 \times 10^6 \text{ p.s.i.} \qquad \nu = 0.2$$

$$\text{Matrix: } E = 0.4 \times 10^6 \text{ p.s.i.} \qquad \nu = 0.35$$

$$\bar{\sigma}_0 = 8 \text{ k.s.i.} \qquad E_p = 100 \text{ p.s.i.}$$

Figure 31.11 shows the predicted composite stress–strain curve. The elastic modulus of the composite predicted by the present analysis is 2.85×10^6 p.s.i. compared to 3.58×10^6 p.s.i. obtained by Agarwal *et al.* This difference reflects the effect of neglecting some matrix material in the axisymmetric modelling. The effect of matrix yielding on longitudinal Poisson's ratio is shown in Figure 31.12.

The progress of plastic zones has been plotted in Figure 31.13. The shear stresses in the fibre and the matrix near the interface (points A_f and A_m in Figure 31.10) for various values of applied composite stress are shown in Figures 31.14 and 31.15. The single value of shear stress in the longitudinal direction has been obtained from τ_{xy} and τ_{yz} as follows:

$$\tau_{ry} = \tau_{xy} \cos\theta + \tau_{yz} \sin\theta$$

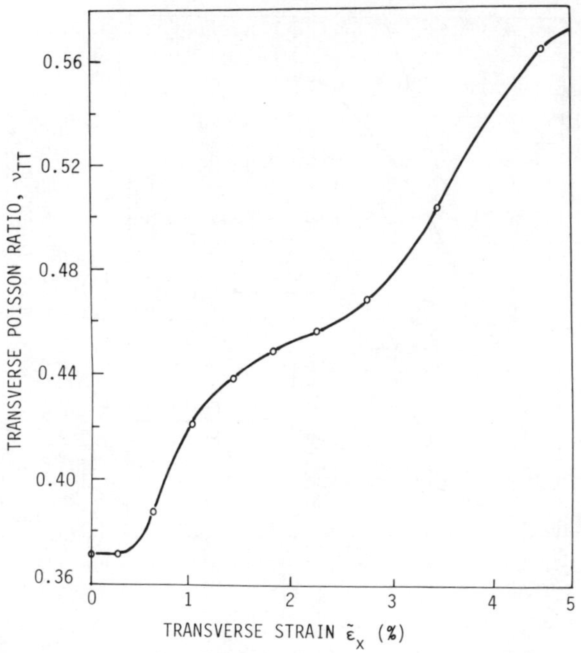

Figure 31.9 Variation of ν_{TT}

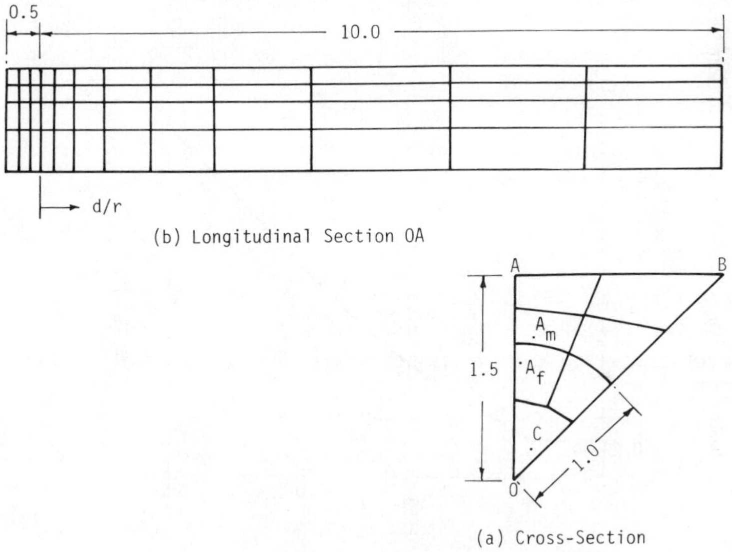

Figure 31.10 Discretization of short-fibre composite

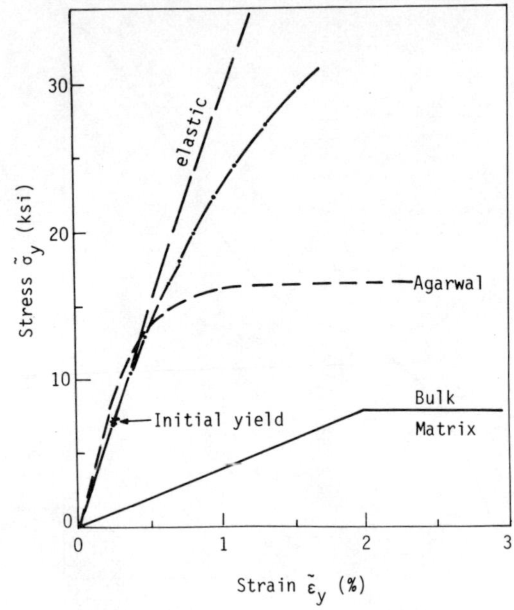

Figure 31.11 Stress–strain curve for composite

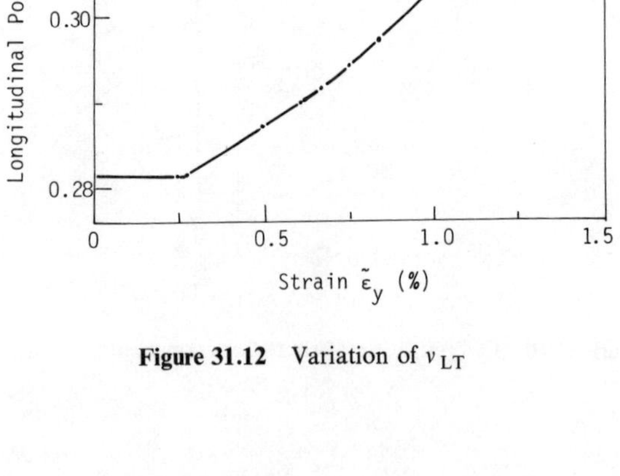

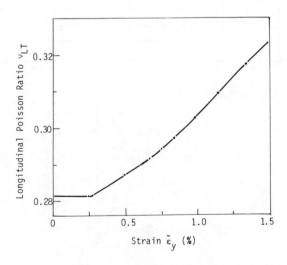

Figure 31.12 Variation of ν_{LT}

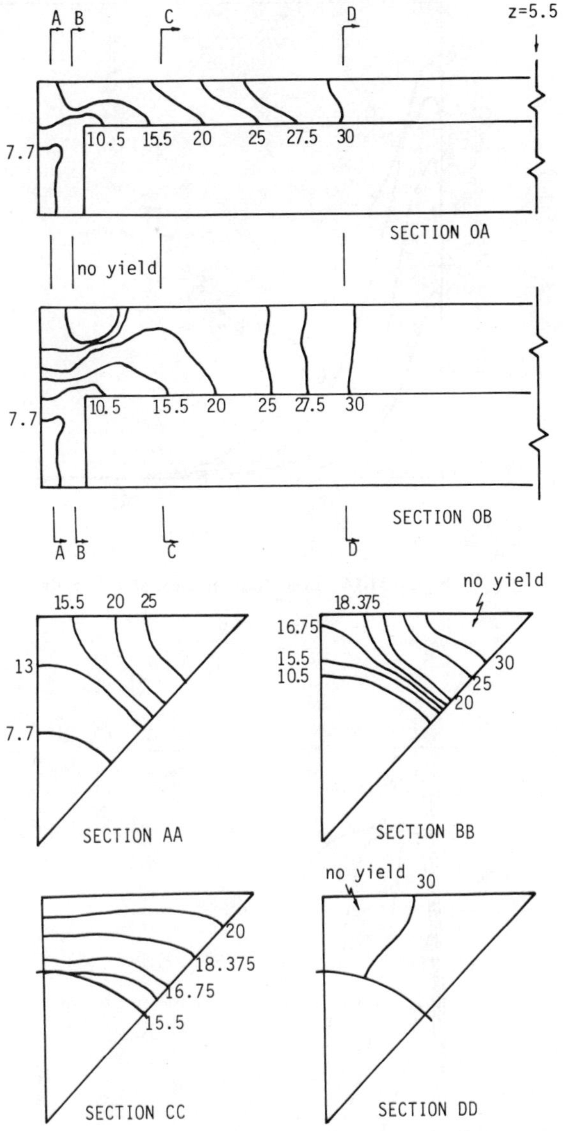

Figure 31.13 Spread of plastic zones (note: yield stress in k.s.i.)

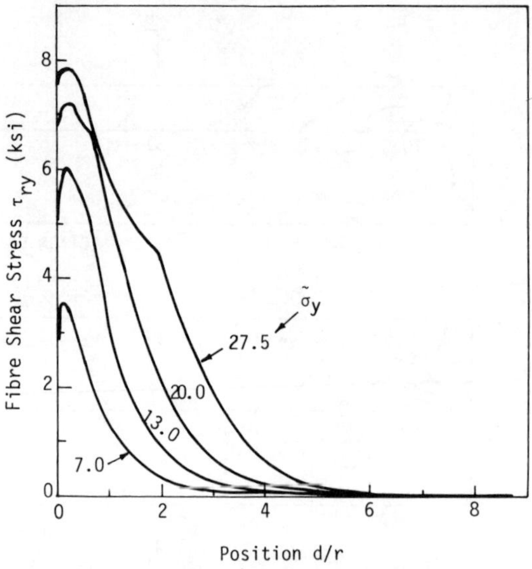

Figure 31.14 Longitudinal shear stress in fibre

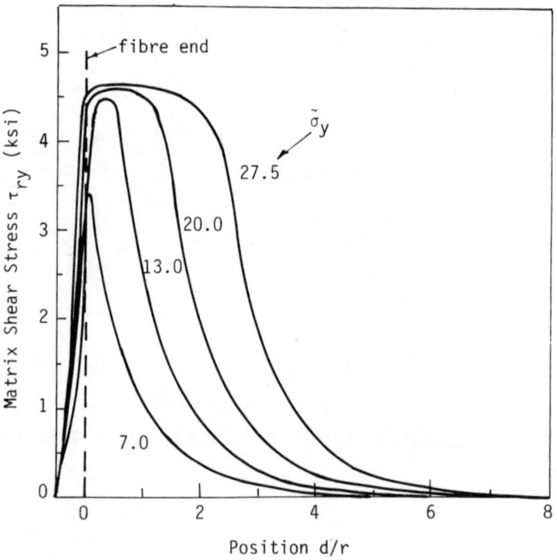

Figure 31.15 Longitudinal shear stress in matrix

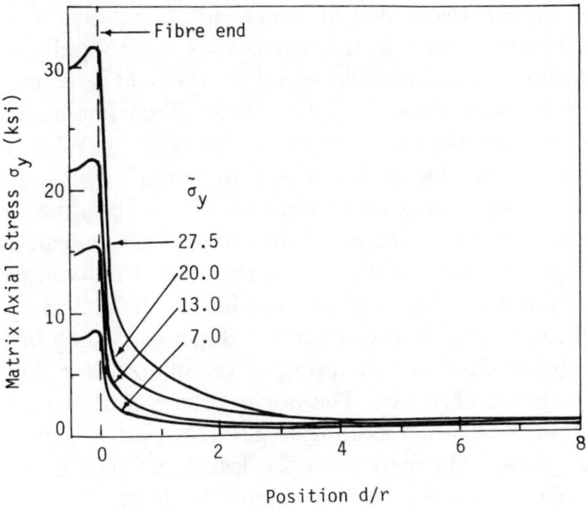

Figure 31.16 Axial stress in matrix

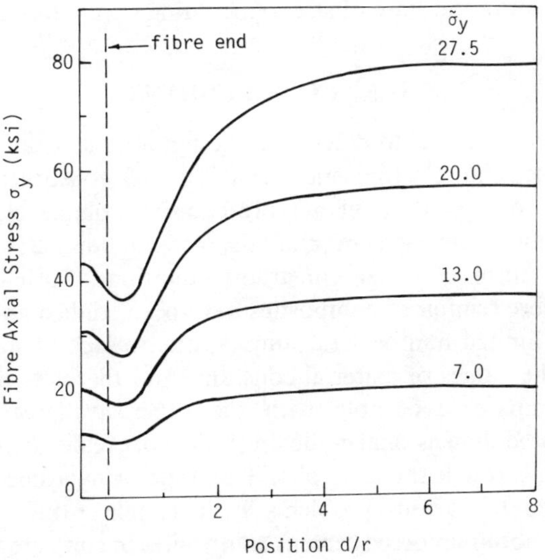

Figure 31.17 Axial stress in fibre

where θ is the angle between line of integrating points and x axis. It can be observed that the shear stress in the matrix rises more rapidly to a maximum value near the fibre end and then drops off. As the load is increased and as the yielding zone progresses, there is only a spread of maximum shear stress over a greater distance rather than any increase in the maximum value. The maximum value of the shear stress obtained in this study compares favourably with the value obtained by Agarwal *et al.* In the case of the fibre, the values of shear stress increase as the load is increased and there is no plateau as observed in the case of matrix. It is interesting to note the discontinuity in the shear stress curves near the interface, due to plastic yielding in the matrix.

The distribution of axial stress at various stages of loading has been plotted in Figures 31.16 and 31.17 for the point in the matrix near the interface and for the point near the fibre axis. The normal stress near the fibre axis drops slightly in the matrix at earlier loading stages. However, it is pronounced when the yielding progresses. The matrix stress is less than would be expected for the corresponding elastic case. Further, for higher loads the maximum axial stress at the centre of fibre is slightly higher than would be observed for the corresponding elastic case. This indicates that the matrix yielding transfers a greater load-carrying role to the fibre. It can be observed from Figure 31.16 that a high axial stress can be supported in the end gap region due to the triaxial stress state.

At this stage, it should be pointed out that these curves differ significantly from those presented by Agarwal *et al.* This type of difference reflects the importance of a three-dimensional analysis of the problem wherein a more realistic prediction of the state of stress (including shear stresses) can be made.

31.6 CONCLUSIONS

A micromechanics analysis to determine the material constants of composites based on elastic and elastoplastic behaviour of constituent materials, is presented. Due to separation of a representative volume element from the composite for the finite element analysis, special boundary conditions are imposed in the form of linear constraint equations. Both continuous and discontinuous fibre reinforced composites have been studied and the results are presented for a limited number of examples, due to lack of space. The results which include the values of material constants and the plots of internal stress distribution compare favourably with the other available solutions. The importance of three-dimensional modelling for a more realistic stress distribution has been shown through the examples. This type of micromechanics analysis has been applied to various problems such as pile groups in geotechnical engineering and fibre-reinforced cement composites in concrete technology. The results from those investigations indicate the validity and applicability of this approach to a variety of engineering problems.

REFERENCES

1. J. E. Ashton, J. C. Halpin, and P. H. Petit, *Primer on Composite Materials: Analysis*, Technomic, Stamford, Conn., 1969.
2. L. J. Broutman and R. H. Krock (eds.), *Composite Materials*, Academic Press, New York, 1975.
3. R. M. Jones, *Mechanics of Composite Materials*, Scripta Book Co., Washington, DC, 1972.
4. Z. Hashin and B. W. Rosen, 'The elastic moduli of fibre-reinforced materials', *J. App. Mech.* **31E**, 223–232 (1964).
5. R. Hill, 'Theory of mechanical properties of fibre-strengthened materials, Part I: Elastic behaviour', *J. Mech. Phys. Solids*, **12**, 199–212 (1964).
6. J. M. Whitney and M. B. Riley, 'Elastic properties of fibre-reinforced composite materials', *A.I.A.A.*, **4**, 1537–1542 (1966).
7. D. F. Adams and D. R. Doner, 'Longitudinal shear loading of a unidirectional composite', *J. Comp. Mat.*, **1**, 4–17 (1967).
8. D. F. Adams and D. R. Doner, 'Transverse normal loading of a unidirectional composite', *J. Comp. Mat.*, **1**, 152–164 (1967).
9. R. L. Foye, 'An evaluation of various engineering estimates of the transverse properties of unidirectional composites', *Advanced Fibrous Reinforced Composites*, 10th S.A.M.P.E., San Diego, Nov. 1966.
10. R. L. Foye, 'Stress Concentrations and Stiffness Estimates for Rectangular Reinforcing Arrays', *J. Comp. Mat.*, **4**, 562–566 (1970).
11. D. F. Adams and S. W. Tsai, 'The influence of random filament packing on the transverse stiffness of unidirectional composites', *J. Comp. Mat.*, **3**, 368–381 (1969).
12. G. H. Goanker, 'Uniaxial loading in an elastic continuum with doubly periodic array of material discontinuities', *J. App. Mech.*, **36** (1), 134–139 (1969).
13. D. F. Adams, 'Elastoplastic behaviour of composites', *Composite Materials*, Vol. 2 (Eds. L. J. Broutman and R. H. Krock), pp. 170–208, Academic Press, New York, 1975.
14. D. R. J. Owen, J. Halbeche, and O. C. Zienkiewicz, 'Elastic–plastic analysis of fibre-reinforced materials', *Fibre Sci. Tech.*, **1**, 185–207 (1969).
15. D. F. Adams, 'Inelastic analysis of unidirectional composite subjected to transverse normal loading', *J. Comp. Mat.*, **4**, 310–328 (1970).
16. D. F. Adams, 'A mircomechanical analysis of crack propagation in an elastic composite material', *Fibre Sci. Tech.*, **7**, 237–256 (1974).
17. R. L. Foye and D. J. Baker, 'Design/Analysis Methods for Advanced Composite Structures', *Technical Report AFML-TR*-70-299, Wright–Patterson Air Force Base, Ohio, 1971.
18. R.L. Foye, 'Theoretical post-yield behaviour of composite laminates, Part I—Inelastic micromechanics', *J. Comp. Mat.*, **7**, 178–193 (1973).
19. R. L. Foye, 'Theoretical post-yield behaviour of composite laminates, Part II—Inelastic macromechanics', *J. Comp Mat.*, **7**, 310–319 (1973).
20. B. D. Agarwal, J. M. Lifshitz, and L. J. Broutman, 'Elastic-plastic finite element analysis of short fibre composites', *Fibre Sci. Tech.*, **7**, 45–62 (1974).
21. J. Amirbayat and J. W. S. Hearle, 'Properties of unit composites as determined by the properties of the interface, Part I: Mechanism of matrix-fibre load transfer', *Fibre Sci. Tech.*, **2**, 123–141 (1969).
22. J. Amirbayat and J. W. S. Hearle, 'Properties of unit composites as determined by the properties of interface, Part II: Effect of fibre length and slippage on the modulus of unit composites', *Fibre Sci. Tech.*, **2**, 143–153 (1969).
23. J. Amirbayat and J. W. S. Hearle, 'Properties of unit composites as determined by the

properties of the interface, Part III: Experimental study of unit composites without a perfect bond between the phases', *Fibre Sci. Tech.*, **2**, 223–239 (1970).

24. D. R. J. Owen and J. F. Lyness, 'Investigation of bond failure in fibre-reinforced materials by the finite element method', *Fibre Sci. Tech.*, **5**, 129–141 (1972).
25. H. D. Conway, W. W. Chu, and C. I. Chang, 'Effect on bond stresses of partial bond failure of overlapping fibres in a composite material', *Fibre Sci. Tech.*, **2**, 289–297 (1970).
26. C. I. Chang, H. D. Conway, and T. C. Weaver, 'The elastic constants and bond stresses for a three-dimensional composite reinforcement by discontinuous fibres', *Fibre Sci. Tech.*, **5**, 143–162 (1972).
27. B. D. Agarwal, 'Micromechanics analysis of composite materials using finite element methods', *Report* C00-1794-16, Illinois Institute of Technology, Chicago, 1972.
28. J. I. Curiskis, *A Study of the Micromechanics of Fibrous-Reinforced Composite Materials using Finite Element Techniques*, Ph. D. thesis, University of New South Wales, 1978.
29. J. I. Curiskis and S. Valliappan, 'A solution algorithm for linear constraint equations in finite element analysis', *Computers and Structures*, **8**, 117–124 (1978).
30. R. L. Foye, 'Lectures Notes—Composite Materials Workshop', University of Toronto, Ontario, 1969.

Mechanics of Engineering Materials
Edited by C. S. Desai and R. H. Gallagher

Chapter 32

Choice of Models for Geotechnical Predictions

D. M. Wood

32.1 INTRODUCTION

The designer of a geotechnical structure has to ensure that the structure will not fail under any conceivable loading. Analysis of the failure of geotechnical structures has for long been based on study of appropriate stress fields and collapse mechanisms backed by the theory of plasticity, with suitable parameters being introduced for the strength of the soil.

Having made sure that the structure will not reach such a collapse or ultimate limit state, the designer needs next to ensure that it will not reach a serviceability limit state either—in other words that the deformations under working loads will be so excessive as either to prevent the structure from serving its design purpose or to present an alarming appearance to the non-technical public. Prediction of deformations under working loads requires knowledge of the way the soil will behave before failure. The search for this knowledge, and the attempt to build appropriate constitutive models for soils has now been in progress for several decades.

Because there exists a large theoretical framework that has been constructed for the prediction of the behaviour of elastic materials, predictions for geotechnical structures have often been made using simple elastic models. It is, however, well known that the behaviour of most soils can hardly be described as elastic and as a result procedures have been developed to lead towards rational choice of elastic parameters. The stress path approaches to prediction of settlements of foundations [1, 2], require triaxial tests to be carried out on representative samples of soil following the stress changes that have been estimated to be relevant for selected elements of soil beneath the foundation. From the response observed in these tests elastic parameters can be deduced which may then be applied to the prediction of settlements.

This stress path method can in principle be extended to a wider range of geotechnical problems. The necessary steps are:

(1) to identify critical soil elements around a structure;
(2) to estimate the stress paths followed by these elements as the structure is constructed or loaded;
(3) to perform laboratory tests on soil samples following these paths; and
(4) to estimate the response of the geotechnical structure from the results of these tests.

The problems of attempting to use this method are, however, that the actual stress path followed will be influenced by the soil behaviour that is being investigated; that it may not be possible to follow the relevant stress paths in laboratory tests; and the method of working from the results of the tests to the estimated performance of the structure may not be obvious.

Nevertheless, the stress path method does impose a certain discipline on engineers which should help them to identify the shortcomings of the tests that they are able to perform, and to assess the value of the constitutive models of soil behaviour which they may wish to use in order to make their predictions.

32.2 STRESS PATHS

Stress paths for a number of typical geotechnical problems will be discussed in order that their great variety may be appreciated. A standard soil stress history will be assumed in these examples: one-dimensional compression followed by a small amount of one-dimensional unloading, so that the initial state of the soil is lightly over-consolidated [3]. In general total and effective stress paths will be shown so that the effect of drainage on the effective stress paths followed by clay elements can be illustrated.

32.2.1 Element on centreline beneath circular load (Figure 32.1)

Elements on the centreline beneath a circular load—such as an oil storage tank—will be subjected to an axially symmetric system of stresses, with principal stresses vertical and horizontal. The stress path is therefore one that can be followed precisely in triaxial apparatus and the path can be sketched in a triaxial stress space such as Figure 32.1(b). With principal stresses $\sigma_1,\sigma_2,\sigma_3$ the stress parameters in Figure 32.1(b) are the deviator stress $q=\sigma_1-\sigma_3$ (in this case $\sigma_2=\sigma_3$) and the mean normal stress $p=\frac{1}{3}(\sigma_1+2\sigma_3)$. The total and effective mean normal stresses p and p' are separated by a horizontal distance equal to the pore pressure in the soil.

The effective stresses at the end of one-dimensional compression (a') and after unloading (b') are shown as is the initial total stress (b). The total stress path (bc) is likely to involve an increase in both q and p. For a lightly over-consolidated clay the effective stress path ($b'c'$) is likely to move up and then to the left as positive pore pressures are generated as the clay yields [3, 4]. When drainage occurs the total stresses may perhaps not change significantly (cd) but the effective mean

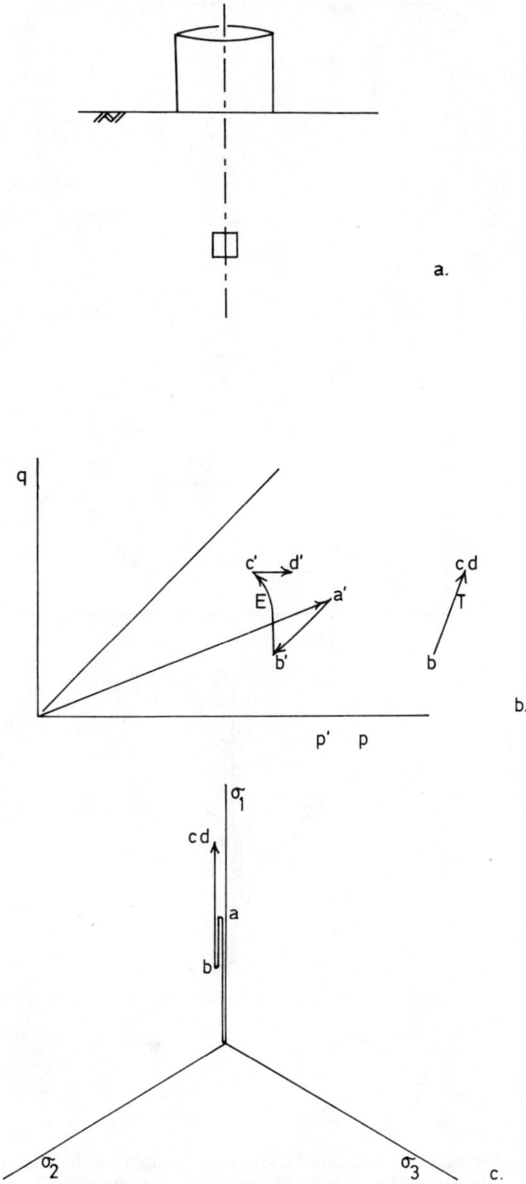

Figure 32.1 Element on centreline beneath circular load. (a) Key sketch; (b) total (*T*) and effective (*E*) stress paths in $q:p$ and $q:p'$ space for loading (*bc*) and subsequent consolidation (*cd*); (c) stress path in deviatoric view of principal stress space

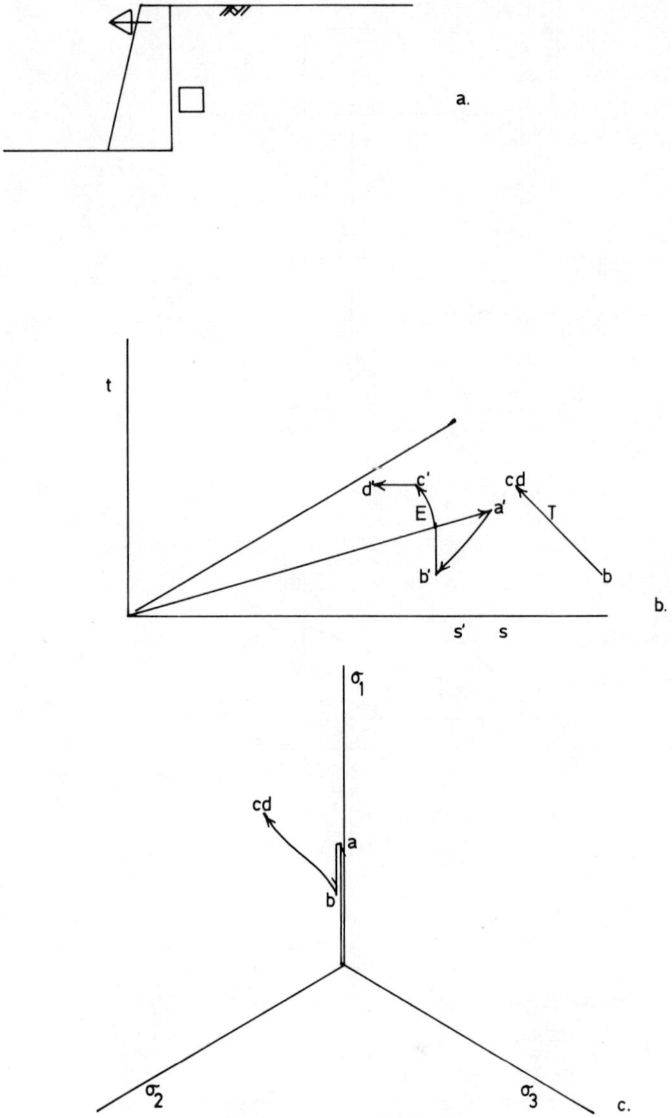

Figure 32.2 Element behind long retaining wall deforming actively (a) key sketch; (b) total (T) and effective (E) stress paths in $t:s$ and $t:s'$ space for shearing (bc) and subsequent consolidation (cd); (c) stress path in deviatoric view of principal stress space

normal stress will increase ($c'd'$) as the pore pressures return to their initial values and the soil hardens.

The stress path is shown in a deviatoric view of principal stress space (seen down the line $\sigma_1 = \sigma_2 = \sigma_3$) in Figure 32.1(c). Because of the axial symmetry of the stress changes the stress path lies entirely on the projection of the σ_1 axis. Since the existence of pore pressures merely displaces a stress state in a direction parallel to the line $\sigma_1 = \sigma_2 = \sigma_3$ there is no distinction between total and effective stress paths in this diagram.

32.2.2 Element behind retaining wall (Figure 32.2)

Axially symmetric stress paths are of rare field occurrence. Conditions of plane strain are found more frequently. In plane strain the stress in the direction normal to the plane of straining (σ_2) is not an independent stress variable—it takes up whatever value is necessary to guarantee zero strain in that direction. It is convenient, then, to sketch stress paths in $t:s$ or $t:s'$ space, where $t = \frac{1}{2}(\sigma_1 - \sigma_3)$, $s = \frac{1}{2}(\sigma_1 + \sigma_3)$, $s' = \frac{1}{2}(\sigma_1' + \sigma_3')$, and σ_1 and σ_3 are the principal stresses in the plane of shearing—frequently the vertical and horizontal stresses.

Total and effective stress paths for an element of soil behind a retaining wall that is deforming actively are shown in Figure 32.2(b). In this case the total stress path involves a decrease in horizontal total stress with no change of vertical total stress and consequently $\delta s = -\delta t(bc)$. The undrained effective stress path ($b'c'$) is forced as in Figure 32.1(b) to move up and to the left but the effect of drainage is to move the effective stress state *further* to the left ($c'd'$) and hence nearer to failure. This drainage is likely to be associated with sucking in of water and hence softening of the soil.

The stress path can be sketched in the deviatoric view of principal stress space, Figure 32.2(c). Because a condition of plane strain is being imposed, the stress

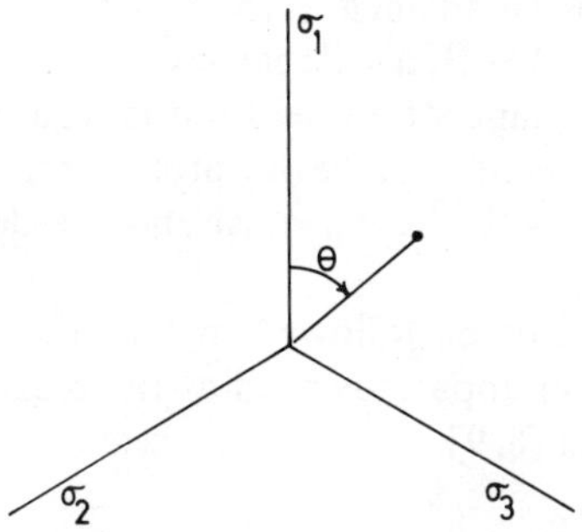

Figure 32.3 Key sketch for definition of angle θ defining orientation of stress vector in deviatoric view of principal stress space

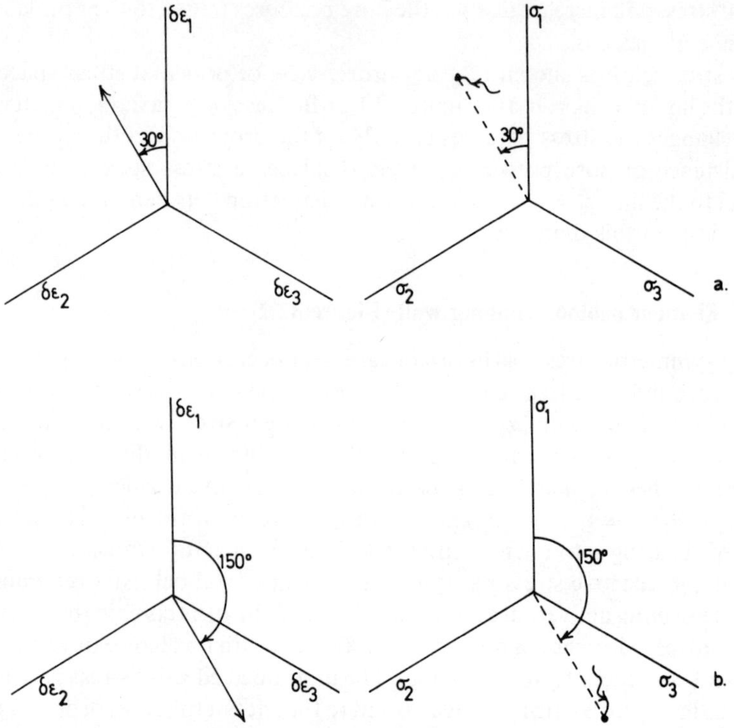

Figure 32.4 Expected ultimate alignment of deviatoric stress vector with deviatoric strain increment vector for (a) active constant volume plane strain shearing $\theta \rightarrow -30°$; (b) passive constant volume plane strain shearing $\theta \rightarrow +150°$

path leaves the projection of the σ_1 axis and may perhaps move towards the radius $\theta = -30°$—where θ is an angle in the deviatoric plane measured clockwise from the σ_1 axis (Figure 32.3). (If a soil reaches the critical state [5] and shears at constant volume and in plane strain; and if it is regarded as a plastic material, then association of the directions of the deviatoric strain increment vector (forced to make an angle $-30°$ with the $\delta\varepsilon_1$ axis) and the deviatoric stress vector is expected (Figure 32.4(a)).)

Clearly this path cannot be followed in the triaxial apparatus. It can be followed in a plane strain apparatus such as the biaxial apparatus [6, 7] or in a true triaxial apparatus [8, 9].

32.2.3 Elements adjacent to excavation (Figure 32.5)

A long excavation is also a situation of plane strain. The stress path for element A (Figure 32.5(b)) adjacent to the wall of the trench is very similar to that for the

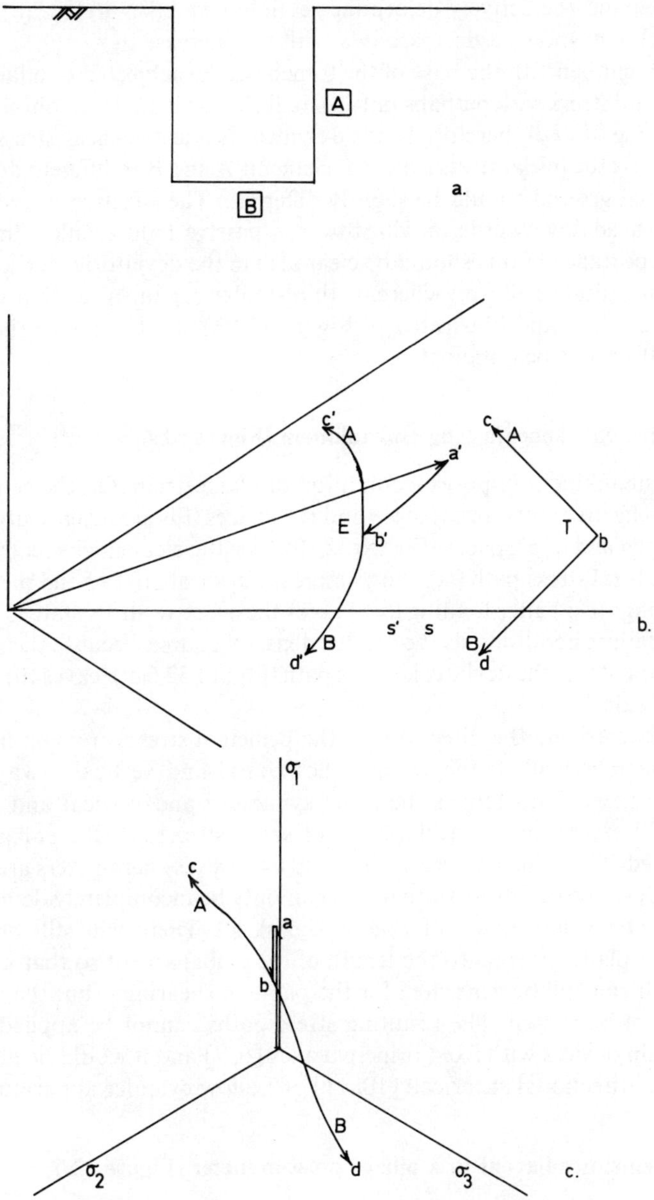

Figure 32.5 Elements beside (A) and beneath (B) long excavation. (a) Key sketch; (b) total (T) and effective (E) stress paths in $t:s$ and $t:s'$ space; (c) stress paths in deviatoric view of principal stress space

element behind the actively deforming retaining wall (Figure 32.2(b)): the total stress path (*bc*) shows a decrease in *s* with the increase in *t*.

An element beneath the base of the trench will be subjected to a large drop in vertical total stress with perhaps only a small change in the horizontal stress. The total stress path (*bd*) therefore heads downwards and the shear stress *t* changes sign. (Strictly the initial stress states for elements A and B at different depths in the undisturbed ground should be slightly different.) The effective stress path ($b'd'$) must also head downwards, moving towards 'passive' failure, failure in extension.

The importance of this should be clear also in the deviatoric view of principal stress space (Figure 32.5(c)) where, with a similar argument to that used in the previous section, and illustrated in Figure 32.4(b), a tendency to the direction $\theta = +150°$ might be anticipated.

32.2.4 Elements beneath long embankment (Figure 32.6)

A long embankment imposes a condition of plane strain. On the centreline the total and effective stress paths in $t:s$ and $t:s'$ spaces (Figure 32.6(b)) are similar to those in $q:p$ and $q:p'$ spaces (Figure 32.1(b)) for the elements beneath a circular load. The total stress path (*bc*) shows increase in shear stress *t* and mean stress *s*, and drainage ($c'd'$) after loading ($b'c'$) takes the effective stress state further away from the failure condition as the soil hardens. Of course, because the situation is one of plane strain the deviatoric stress path (Figure 32.6(c)) leaves the projection of the σ_1 axis.

On the centreline the directions of the principal stresses remain fixed as the embankment is loaded: they remain horizontal and vertical. Away from the centreline there is no longer the same symmerty and vertical and horizontal planes will experience a development of shear stresses as the embankment is constructed. The principal axes rotate, and extra stress parameters are needed in order to describe the stress state which can only be incompletely depicted in the principal stress sub-spaces of Figure 32.6(b), (c). There will still be no shear stresses on planes normal to the length of the embankment so that a $t:s$ or $t:s'$ stress path can still be generated for the plane of shearing—but the rotation of axes cannot be shown. The resulting stress paths cannot be applied in simple plane strain devices with fixed principal axes [6, 7] and it would be necessary to resort to a directional shear cell [10, 11] or hollow cylinder apparatus [12, 13].

32.2.5 Elements adjacent to a pile or pressuremeter (Figure 32.7)

The assessment of the stress paths relevant for elements of soil affected by the installation of a pile is a much more complex problem than those so far considered. However, some success has been achieved [14] in comparing the driving of a cylindrical pile to the expansion of a long cylindrical cavity which is the process involved in the pressuremeter test.

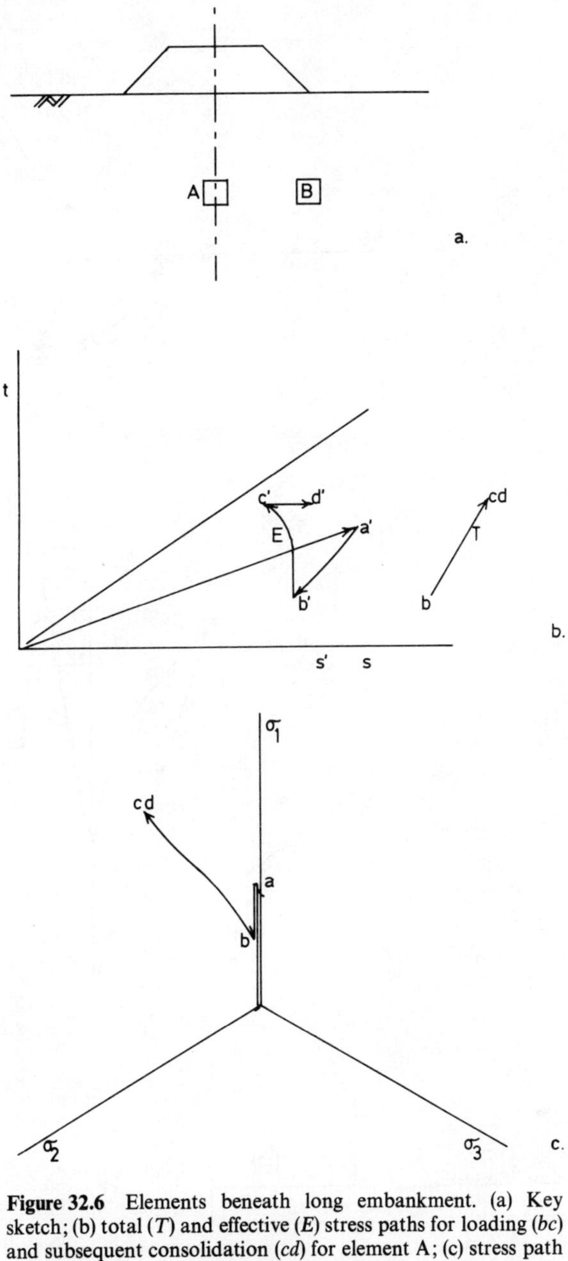

Figure 32.6 Elements beneath long embankment. (a) Key sketch; (b) total (*T*) and effective (*E*) stress paths for loading (*bc*) and subsequent consolidation (*cd*) for element A; (c) stress path in deviatoric view of principal stress space for element A

Note that at element **B** rotation of principal axes occurs and this aspect of a stress path cannot be indicated in views of principal stress space such as (b) and (c)

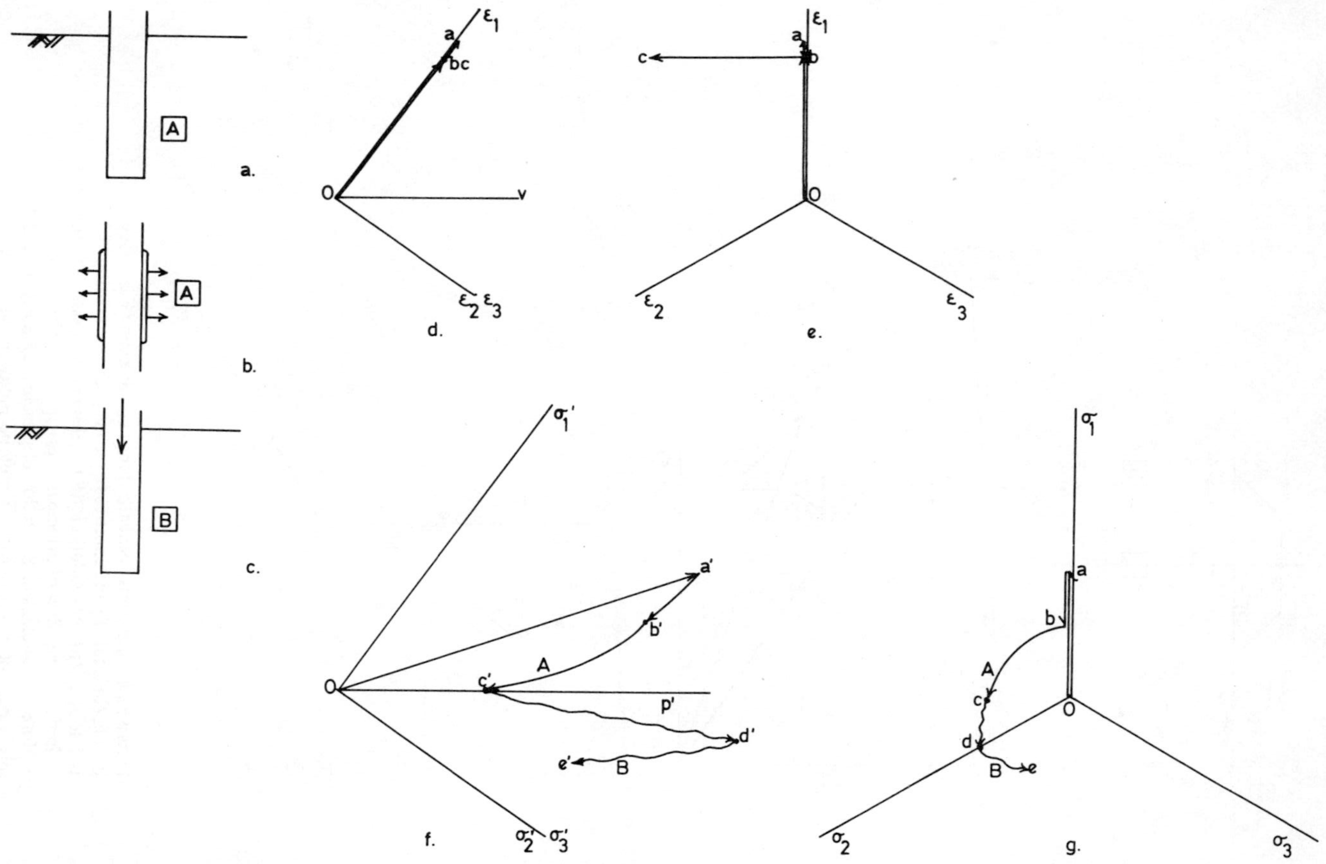

Figure 32.7 (a) Installation of a cylindrical pile; (b) expansion of cylindrical pressuremeter cavity; (c) loading of pile; (d) and (e) strain path in two orthogonal views of principal strain space for element A in (a) and (b): one-dimensional compression (O*a*); one-dimensional unloading (*ab*); constant volume plane strain shearing in plane normal to direction of one-dimensional compression (*bc*); (f) and (g) stress path in two orthogonal views of principal stress space for element A: one-dimensional compression (O*a*); one-dimensional unloading (*ab*); constant volume plane strain shearing (*bc*); for subsequent consolidation (*cd*); and for element B: rapid loading of pile (*de*).

Note that loading of the pile (element B) involves rotation of principal axes: this aspect of the stress path cannot be indicated in views of principal stress space such as (f) and (g).

This process is one in which the strain path is clearly defined [15]. Initial one-dimensional compression and unloading of the soil (O*ab* in Figures 32.7(d), (e)) is followed by constant volume plane strain shearing with the plane of strain normal to the original direction of one-dimensional compression (*bc*). In Figures 32.7(d), (e) two orthogonal views of principal strain space are shown in third angle projection. Figure 32.7(e) is a deviatoric view down the line $\varepsilon_1 = \varepsilon_2 = \varepsilon_3$. Figure 32.7(d) is an orthogonal view which contains this line and also the ε_1 axis in true view, whereas the ε_2 and ε_3 axes are both foreshortened. A similar pair of views of stress space, Figures 32.7(f), (g) is used to indicate the expected resulting stress path. The left-hand view, Figure 32.7(f), is a view *on to* the familiar 'triaxial plane' ($\sigma_2 = \sigma_3$)—though, of course, the stress paths that are being depicted do not necessarily lie in this plane.

Constant volume shearing tends to lead to a build-up in pore pressure and hence a drop in mean effective stress. The path $b'c'$ in Figure 32.7(f) moves to the left. Using a similar argument to that used in sections 32.2.2 and 32.2.3, the deviatoric stress state in Figure 32.7(g) can be expected to move round to the position $\theta = -90°$ as shown in response to the fixed direction of strain increment that is being imposed.

Once the pile has been installed dissipation of pore pressures may occur with the vertical effective stress regaining something approaching its initial value ($c'd'$) but with the radial stress being left as the major principal stress (Figure 32.7(g)) [14]. Evidently the further the discussion of this particular problem proceeds the more speculatory will be the estimated paths.

When the pile is eventually loaded rapidly the elements of soil adjacent to the pile will be subjected to conditions approaching those of simple shear [16] and the associated rotation of principal axes that is bound to be important if the response of the pile to this loading is to be assessed cannot be shown in the principal stress space of Figures 32.7(f), (g) although the principal stress path may be something like that shown ($d'e'$, de) with, once again, a build up of pore pressure, and a drop in mean normal effective stress.

Conditions have now become so complex that it would be pointless to contemplate imposing such a complete stress path on a single element of soil in the laboratory and it would make a great deal more sense to carry out some careful model tests either at one gravity or, using a centrifuge [17], at many gravities.

32.3 EXPERIMENTAL RESULTS AND SIMPLE SOIL MODEL

It has already been mentioned that the stress paths that can be applied in the triaxial test are confined to the diameter of the deviatoric plane $\sigma_2 = \sigma_3 (\theta = 0°$ or $180°)$ (Figure 32.8). Triaxial compression corresponds to $\theta = 0°$ and is the more

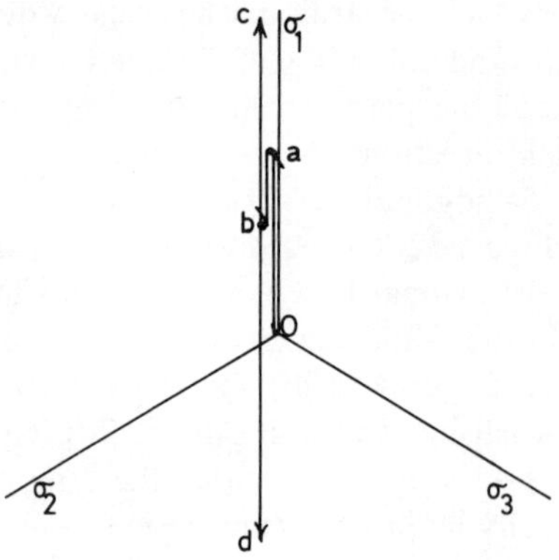

Figure 32.8 Stress paths available in triaxial apparatus: compression (*bc*) and extension (*bd*) tests on one-dimensionally lightly over-consolidated (O*ab*) soil shown in deviatoric view of principal stress space

common mode of testing. Triaxial extension corresponds to $\theta = 180°$ and, being slightly more complicated to achieve, is less frequently used.

Most constitutive models for soil behaviour have been generated from study of the results of triaxial compression tests. One particular series of simple elastic–plastic models is the set of so-called Cam clay models [18, 19]. The essential features of these models will be described briefly here.

32.3.1 Cam clay models

The soil is assumed to behave as an elastic–plastic material with a yield locus dividing elastic from plastic behaviour. In the modified Cam clay model [19] an elliptical yield locus in $q{:}p'$ space is assumed (Figure 32.9(a)).

The soil is assumed to obey the principle of normality so that appropriate plastic strain increment vectors are directed normally to the yield locus which serves also as a plastic potential. (With stress parameters q and p' defined for conditions of axial symmetry $q=\sigma_1=\sigma_3$, $p'=\frac{1}{3}(\sigma_1'+2\sigma_3')$ the appropriate associated strain parameters are a shear strain $\varepsilon=\frac{2}{3}(\varepsilon_1-\varepsilon_3)$ and the volumetric strain $v=\varepsilon_1+2\varepsilon_3$.)

The soil exhibits a series of critical states at which continued shearing can continue without further change in volume or effective stresses when the stress state reaches the top of the yield locus and normality indicates that plastic volumetric strains can no longer occur.

Hardening of the soil is linked with the isotropic normal compression of the soil, so that the stress space of Figure 32.9(a) has to be associated with the compression space of Figure 32.9(b) where V is the specific volume of the soil—

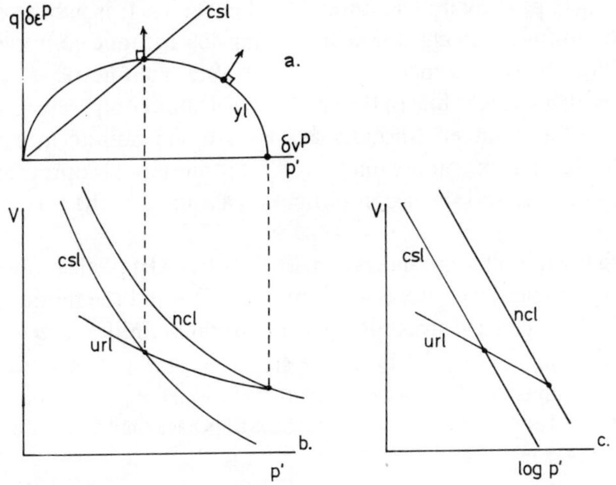

Figure 32.9 Summary of features of Cam clay model. (a) Elliptical yield locus (*yl*) in $q:p'$ space with normality of associated plastic strain increments $\delta\varepsilon^p$ and δv^p; csl = critical state line linking stress states at which shearing can continue without further change of effective stresses or volume; (b) yield locus is associated with an elastic unloading–reloading line (url) in $V:p'$ space and hardening of the soil is associated with movement of the tip of the unloading–reloading line down the normal compression line (ncl); csl = critical state line. (c) normal compression line (ncl) and critical state line (csl) are straight and parallel in semi-logarithmic $V:\log p'$ space: unloading–reloading line (url) also straight in this space

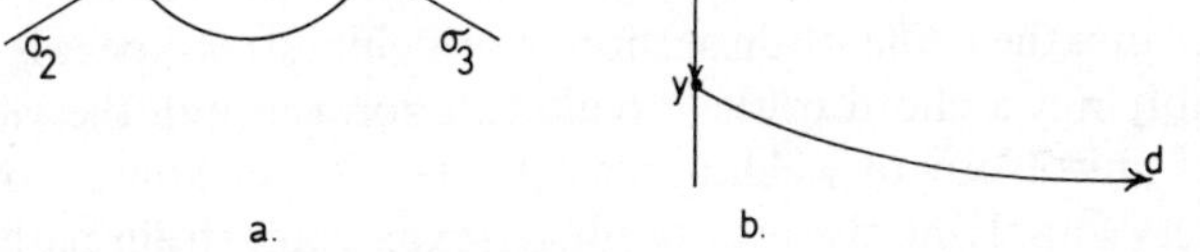

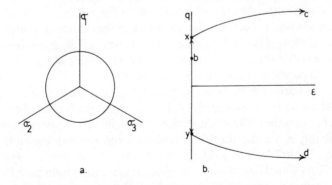

Figure 32.10 (a) Isotropic hardening of Cam clay yield surface implies circular constant p' section in deviatoric view of principal effective stress space; (b) isotropic hardening of Cam clay yield surface implies absence of Bauschinger effect in undrained triaxial tests: yield occurs at same value of q in compression (x) and extension (y)—for triaxial extension ($\theta = 180°$) q has been plotted as though it were negative

the volume occupied by unit volume of soil particles. It is assumed that normal compression follows a straight line in the semi-logarithmic compression space of Figure 32.9(c). It is assumed that recoverable volumetric strains are also associated with a straight line in the semi-logarithmic compression space. A shear modulus may be assumed (though this is assumed infinite in early Cam clay models) in order that the model may be used for numerical computations. No link between volumetric and shear components is assumed in the elastic region inside the yield locus.

Although the model was conceived with data from triaxial compression tests—mostly on isotropically compressed samples—in mind the model can be readily extended to more general stress using a consistent definition of $q = \{\frac{1}{2}[(\sigma_2 - \sigma_3)^2 + (\sigma_3 - \sigma_1)^2 + (\sigma_1 - \sigma_2)^2]\}^{1/2}$ and $p' = \frac{1}{3}(\sigma_1' + \sigma_2' + \sigma_3')$. The yield locus in $q{:}p'$ space then becomes an ellipsoidal yield surface in $\sigma_1'{:}\sigma_2'{:}\sigma_3'$ space and deviatoric sections, at constant p', become circles: the soil is assumed to harden isotropically (Figure 32.10(a)).

Consequently, if a one-dimensionally lightly over-consolidated sample of Cam clay is subjected to undrained triaxial compression and extension tests the response, neglecting elastic shear strains, is as shown in Figure 32.10(b). The soil shows no Bauschinger effect and yield occurs at the same value of q in both compression and extension.

32.3.2 Experimental results

Undrained triaxial compression and extension tests were performed on spestone kaolin by Nadarajah [20]. Typical results are shown in Figure 32.11: effective stress paths in Figure 32.11(a) and stress–strain curves in Figure 32.11(b). The effective stress paths should be vertical where the soil is behaving isotropically and elastically (inside the Cam clay yield surface) but there is significant deviation from the vertical particularly in the extension tests. It is clear also from the stress–strain curves that the assumption of isotropic hardening also does not provide a good correspondence with the observed behaviour.

Similar remarks can be made with regard to undrained plane strain (biaxial) tests performed by Sketchley [21] also on spestone kaolin (Figure 32.12)—though, because the plane strain section of principal stress space is not a diameter and probably not a chord of the circular section through the yield surface the evidence of anisotropy of yielding may not be so immediately obvious.

The effect of applying the pressuremeter expansion strain path (*bc*) of Figure 32.7 to normally compressed Boston blue clay in a true triaxial apparatus is shown in Figure 32.13 together with the path predicted by the Cam clay model [22, 23]. The stress paths are shown in the two orthogonal views of principal stress space used previously (e.g. in Figure 32.7) but also in a general $q{:}p'$ space where q and p' are here defined as above in terms of all three principal stresses.

It is clear that the Cam clay model is not particularly successful at matching the

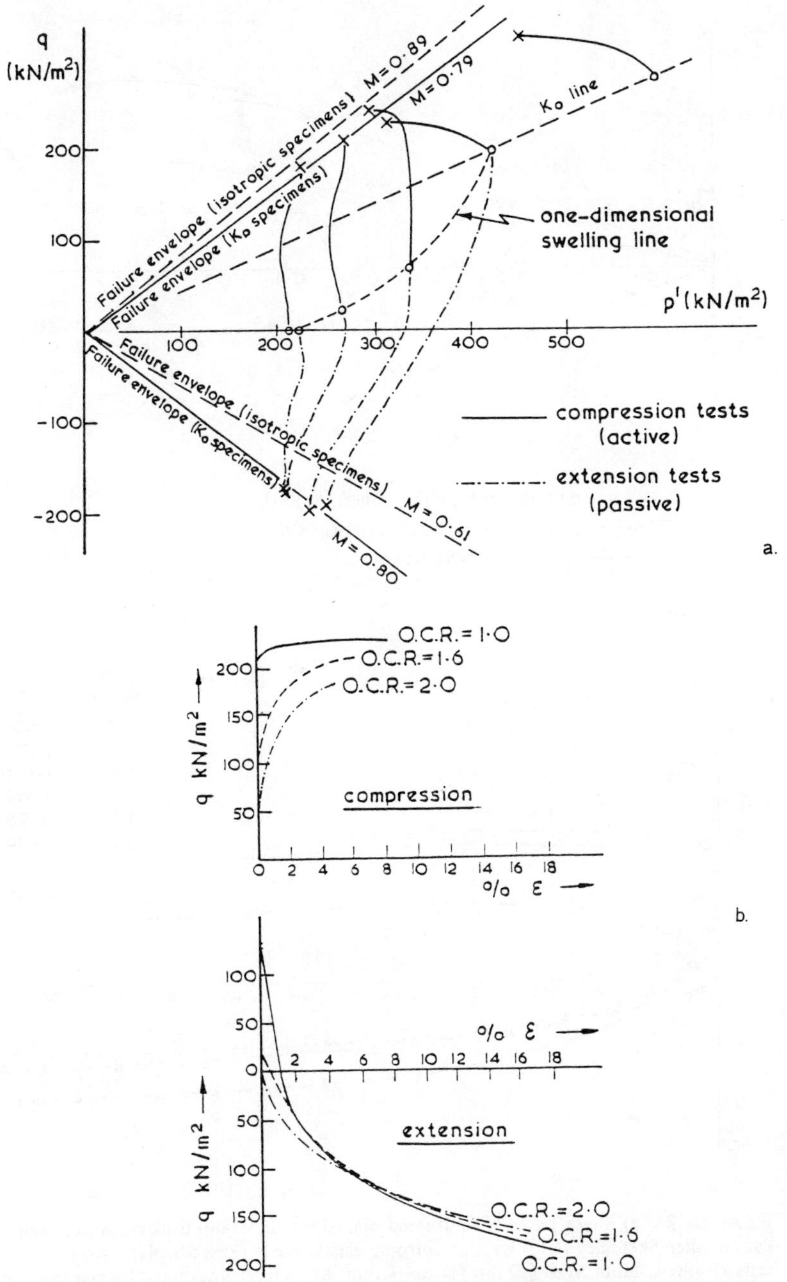

Figure 32.11 (a) Stress paths of undrained triaxial tests on one-dimensionally (K_0) consolidated kaolin (after Nadarajah [20]) (b) Stress–strain curves from undrained triaxial tests on K_0 consolidated kaolin (after Nadarajah [20])

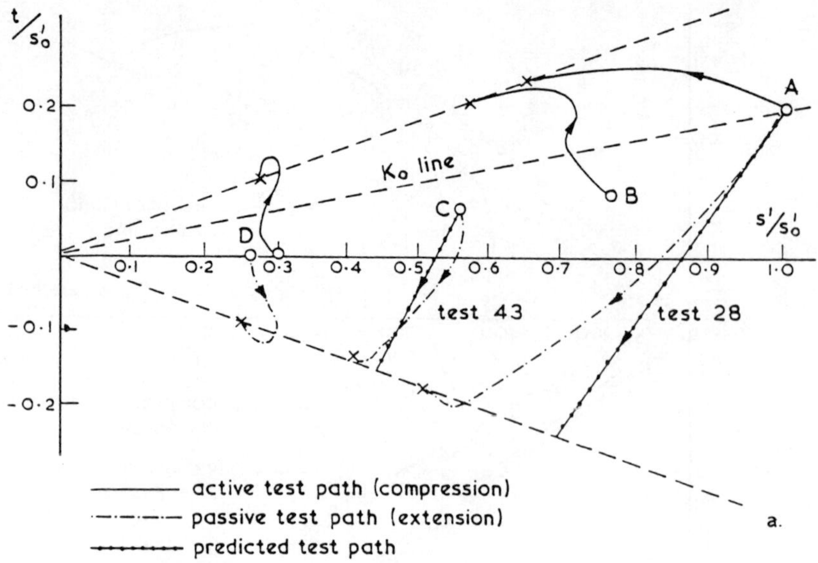

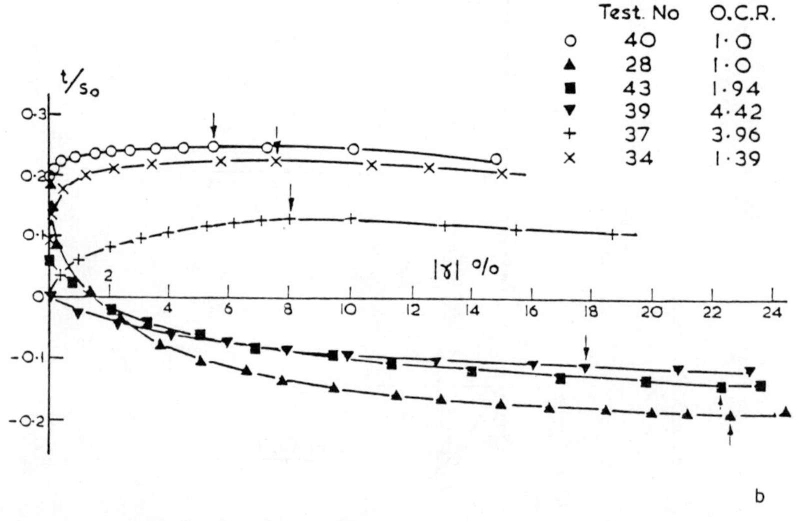

Figure 32.12 (a) Stress paths of undrained biaxial (plane strain) tests on K_0 consolidated kaolin (after Sketchley [21]) with anisotropic elastic predictions for plane strain extension tests 28 and 43 (after Mair [27]) (b) stress–strain curves from undrained biaxial tests on K_0 consolidated kaolin (after Sketchley [21])

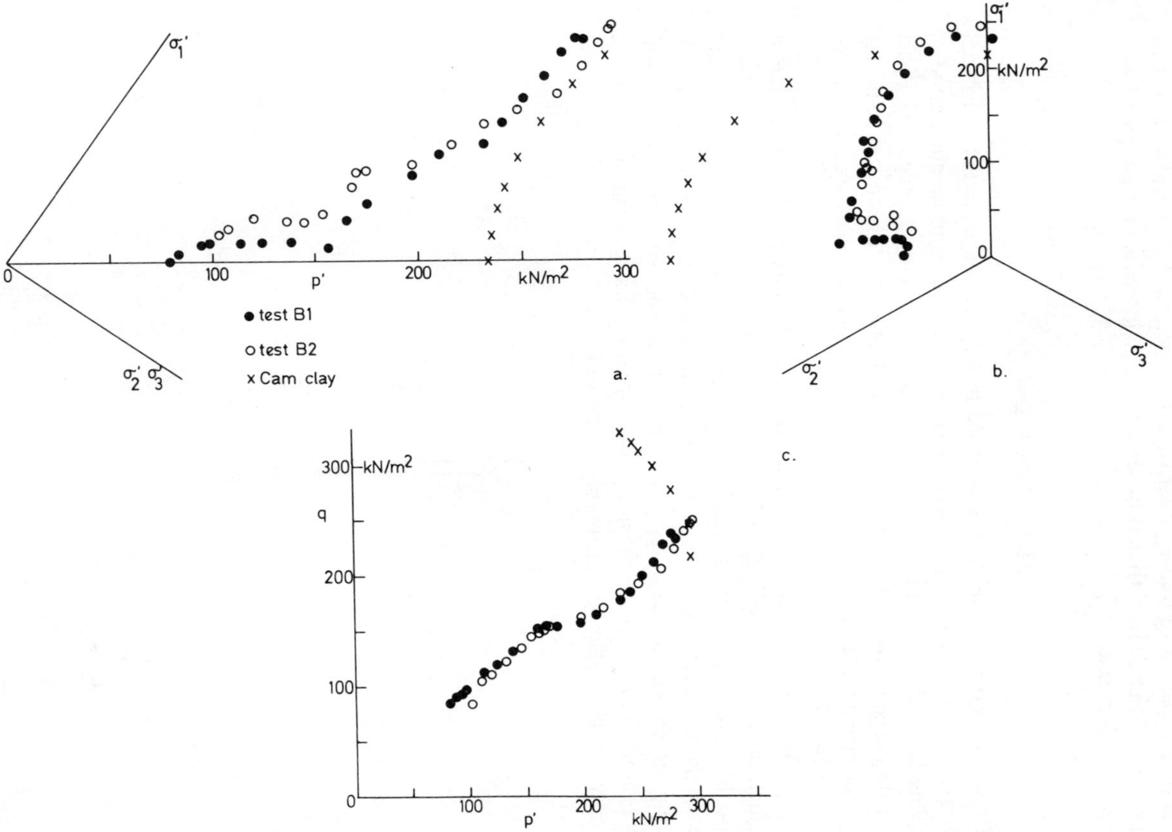

Figure 32.13 Comparison of results of true triaxial tests on Boston blue clay with predictions of Cam clay model for strain path of Figure 32.7(d) and (e): (a) and (b) orthogonal views of principal effective stress space; (c) $q:p'$ space

observed experimental path: the isotropically hardened yield surface is too fat and the path is forced to move rather rapidly round to the critical state with a much smaller decrease in p' than is observed and an increase in q compared with the observed decrease.

32.4 DISCUSSION

The Cam clay models have been generated largely from the results of triaxial compression tests on soft clay samples and can be expected to be quite good in predicting field behaviour if the most important soil elements are engaging a part of the putative isotropic yield surface close to triaxial compression, but to be less good if the stress paths are predominantly exploring other parts of stress space (Figure 32.14).

Thus predictions for situations such as the circular loading of Figure 32.1, and the embankment loading of Figure 32.6 may be quite good because the behaviour is probably predominantly dependent on the response of elements of soil immediately beneath the loading [24, 25, 4] even though the stress path beneath the long embankment deviates somewhat from the condition of triaxial compression. Predictions for the active deformation of a retaining wall (Figure 32.2) may similarly also be reasonably successful.

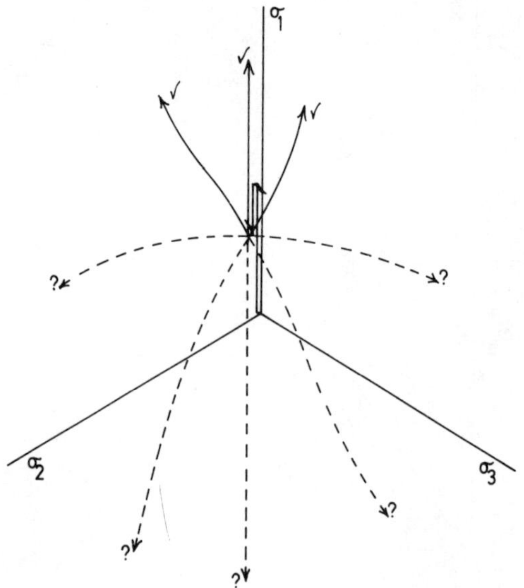

Figure 32.14 Quality of predictions of simple Cam clay model depends on area of stress space that is being explored: quality is likely to be better for paths that do not deviate dramatically from triaxial compression than for those that do

However, passive plane strain loading—behind a strutted wall, for example, or beneath a long trench (Figure 32.5) is in a very different direction from triaxial compression and less success would be expected. The complicated case of pile installation (Figure 32.7) has already been discussed and the actual lack of success in theoretical matching of the experimental observations has been noted.

Wood [9] used the Cam clay model to make predictions for a number of true triaxial tests on spestone kaolin exploring a deviatoric plane with $p' = \text{constant}$. He concludes that the model is not satisfactory for stress paths where the stress ratio q/p' is smaller than its previous maximum value. For these tests at constant p' this is equivalent to saying that the model is reasonably satisfactory provided the hypothesized isotropic Cam clay yield locus is being expanded.

That conclusion is similar to a suggestion made by Drucker [26] that for metals 'a crude isotropic hardening would be obtained if plastic strain increments of the order of total elastic strains were ignored.... The yield surface for additional strains of, say, 1 per cent will resemble an expanded version of the initial yield surface far more than the translating and shrinking yield surfaces defined by small deformations of the order of 2×10^{-3} per cent.' Evidently the figures would be an order of magnitude higher for soils—but the message may be that the predictions of a simple isotropic model may be reasonable if the strains are large.

In practice, if a designer is faced with stress paths that clearly explore a region of stress space remote from that in which the simple model was constructed he has a number of choices.

(1) He can accept that a simple model will give a crude estimate of the behaviour to be expected. The Cam clay model will be too stiff if the stress path spends most of its time inside the yield surface but may perhaps not be too bad if the path moves well into the region of plastic yielding.

(2) He can adopt a modification of the simple model to match his particular application. Thus, for example, Mair [27] who was interested in using a Cam clay model in finite element predictions of deformations observed in model tests of tunnels in soft clay injected anisotropic elastic properties into the description of the response of the soil inside the yield surface in order to be able more closely to match the experimental data of Sketchley [21]. The predictions of the stress path shown in Figure 32.12 are made using such an anisotropic elastic model.

If this approach is adopted then it is important to be aware that the modification made may be particular to a certain application. Thus although anisotropic elasticity provides a reasonable match for the extension (passive) stress paths in Figure 32.12 the assumption of elasticity implies that a reversal of the direction of undrained shearing would cause the same stress path to be retraced back to the starting point. In fact, it is quite likely that the response to such a reversal of shearing might be close to isotropic elastic: undrained cyclic loading tests show a relatively small increase in pore pressure per cycle after the soil has been taken once into compression and once into extension [28]. Consequently Mair's modification would be helpful only for monotonic passive shearing.

(3) He can adopt a more complex model. If the behaviour inside the yield surface is really going to be of great importance then a much more elaborate model is going to be required—perhaps one involving kinematic hardening [29] or an eroding memory of previous history [23, 30]. However, the adoption of a complex model must indicate that the engineer has a greater confidence in the knowledge of the soil conditions at his site and of the properties of the soils that are to be used in making the predictions of response. If the density of boreholes, or the quality of sampling, or the sophistication of testing is low then there can be little justification for attempting to make use of an elaborate constitutive model for the soil.

32.5 CONCLUSION

Consideration of stress paths imposes an important discipline on the geotechnical designer even if it is not intended to use a stress path method to make predictions of the performance of geotechnical structures.

In selecting a soil model for use in making geotechnical predictions it is important to be aware of the laboratory tests that may have been used in building the model so that caution can be exercised if it is clear that the prediction involves extrapolation into a region which is unknown or unfamiliar to the premisses of the model.

Equally, of course, there is little point in making use of a complex model for engineering predictions—which will inevitably make the numerical analysis more lengthy and costly and make the requirements for site investigation and selection of soil parameters more stringent—if consideration of the stress paths shows that a simple model based on the results of a rather limited range of laboratory tests should be quite adequate.

SYMBOLS

p	Mean normal stress $\frac{1}{3}(\sigma_1 + \sigma_2 + \sigma_3)$ or $\frac{1}{3}(\sigma_1 + 2\sigma_3)$ for $\sigma_2 = \sigma_3$.
q	Deviator stress $\{\frac{1}{2}[(\sigma_2 - \sigma_3)^2 + (\sigma_3 - \sigma_1)^2 + (\sigma_1 - \sigma_2)^2]\}^{1/2}$ or $(\sigma_1 - \sigma_3)$ for $\sigma_2 = \sigma_3$.
s	Mean stress in plane strain $\frac{1}{2}(\sigma_1 + \sigma_3)$.
t	Shear stress in plane strain $\frac{1}{2}(\sigma_1 - \sigma_3)$.
v	Volumetric strain $\varepsilon_1 + \varepsilon_2 + \varepsilon_3$ or $\varepsilon_1 + 2\varepsilon_3$ for $\varepsilon_2 = \varepsilon_3$.
V	Specific volume (volume occupied by unit volume of soil particles).
γ	Shear strain in plane strain $(\varepsilon_1 - \varepsilon_3)$.
$\varepsilon_1, \varepsilon_2, \varepsilon_3$	Principal strains.
ε	Deviatoric strain $\frac{2}{3}(\varepsilon_1 - \varepsilon_3)$ for $\varepsilon_2 = \varepsilon_3$.

θ	Angle in deviatoric stress plane $\tan^{-1}[\sqrt{(3)}(\sigma_3 - \sigma_2)/(2\sigma_1 - \sigma_2 - \sigma_3)]$.
$\sigma_1, \sigma_2, \sigma_3$	Principal stresses.
δ	Indicates small increment, $'$ indicates effective stress parameter, ^{p} indicates plastic quantity.

REFERENCES

1. T. W. Lambe, 'Methods of estimating settlement', *Proc. ASCE*, **90**, SM5, 43–67 (1964).
2. E. H. Davis and H. G. Poulos, 'The use of elastic theory for settlement prediction under three-dimensional conditions', *Geotechnique*, **18**, 1, 67–91 (1968).
3. R. H. G. Parry and C. P. Wroth, 'Shear stress–strain properties of soft clay', chap. 4 in *Soft Clay Engineering* (eds. E. W. Brand and R. P. Brenner), pp. 309–364, Elsevier Scientific Publishing Co., 1981.
4. D. M. Wood, 'Yielding in soft clay at Bäckebol, Sweden', *Geotechnique*, **30**, 1, 49–65 (1980).
5. A. N. Schofield and C. P. Wroth, *Critical State Soil Mechanics*, pp. 310. McGraw-Hill, London, 1968.
6. E. C. Hambly, 'Plane strain behaviour of remoulded normally consolidated kaolin', *Geotechnique*, **22**, 2, 301–317 (1972).
7. C. J. Sketchley and P. L. Bransby, 'The behaviour of an overconsolidated clay in plane strain', *Proc. 8th Int. Conf. Soil Mech.*, Moscow, **1**, 2, 377–384 (1973).
8. J. A. Pearce, 'A new true triaxial apparatus', in *Stress strain behaviour of soils, Proc. Roscoe Memorial Symp.*, pp. 330–339, Foulis & Co., 1971.
9. D. M. Wood, 'Explorations of principal stress space with kaolin in a true triaxial apparatus', *Geotechnique*, **25**, 4, 783–797 (1975).
10. J. R. F. Arthur, K. S. Chua, and T. Dunstan, 'Induced anisotropy in a sand', *Geotechnique*, **27**, 1, 13–30 (1977).
11. J. R. F. Arthur, K. S. Chua, T. Dunstan, and J. I. Rodriguez del C., 'Principal stress rotation: a missing parameter', *Proc. ASCE*, **106**, GT4, 419–433 (1980).
12. B. B. Broms and A. O. Casbarian, 'Effects of rotation of the principal stress axes and of the intermediate principal stress on the shear strength', *Proc. 6th Int. Conf. Soil Mech.*, Montreal, **1**, 179–183 (1965).
13. A. S. Saada, 'Testing of anisotropic clay soils', *Proc. ASCE*, **96**, SM5, 1847–1852 (1970).
14. M. F. Randolph, J. P. Carter, and C. P. Wroth, 'Driven piles in clay—the effects of installation and subsequent consolidation', *Geotechnique*, **29**, 4, 361–393 (1979).
15. D. M. Wood and C. P. Wroth, 'Some laboratory experiments related to the results of pressuremeter tests', *Geotechnique*, **27**, 2, 181–201 (1977).
16. M. F. Randolph and C. P. Wroth, 'Application of the failure state in undrained simple shear to the shaft capacity of driven piles', *Geotechnique*, **31**, 1, 143–157 (1981).
17. A. N. Schofield 'Cambridge Geotechnical Centrifuge operations', *Geotechnique*, **30**, 3, 227–268 (1980).
18. K. H. Roscoe and A. N. Schofield, 'Mechanical behaviour of an idealised 'wet' clay, *Proc. 2nd Eur. Conf. Soil Mech.*, Wiesbaden, **1**, 47–54 (1963).
19. K. H. Roscoe and J. B. Burland, 'On the generalised stress–strain behaviour of 'wet' clay', in *Engineering Plasticity* (Eds. J. Heyman and F. A. Leckie), pp. 535– 609, Cambridge University Press, 1968.
20. V. Nadarajah, *Stress–strain Properties of Lightly Over-consolidated Clays*. Ph.D. thesis, Cambridge University, 1973.

21. C. J. Sketchley, *The Behaviour of Kaolin in Plane Strain.* Ph.D. thesis, Cambridge University, 1973.
22. D. M. Wood, 'Discussion: Driven piles in clay—the effects of installation and subsequent consolidation', *Geotechnique*, **31**, 2, 291–293 (1981).
23. D. M. Wood, 'True triaxial tests on Boston blue clay', *Proc. 10th Int. Conf. on Soil Mechs. and Foundations Eng.*, Stockholm, **1**, 825–830 (1981).
24. D. M. Wood, 'Author's reply: Yielding in soft clay at Bäckebol, Sweden', *Geotechnique*, **31**, 4, 572–573 (1981).
25. C. P. Wroth, 'The predicted performance of a soft clay under a trial embankment loading based on the Cam clay model', chap. 6 in *Finite Elements in Geomechanics* (Ed. G. Gudehus), pp. 191–208, J. Wiley, 1977.
26. D. C. Drucker, 'Concepts of path independence and material stability for soils', *Proc. IUTAM Symposium on Rheology and Soil Mechanics*, Grenoble, pp. 23–46, Springer-Verlag, 1964.
27. R. J. Mair, *Centrifugal Modelling of Tunnel Construction in Soft Clay.* Ph.D. Thesis, Cambridge University, 1979.
28. D. M. Wood, 'Laboratory investigations of the behaviour of soils under cyclic loading: A review', chap. 20 in *Soil Mechanics—Transient and Cyclic Loads* (Eds. G. N. Pande and O. C. Zienkiewicz), pp. 513–582 John Wiley and Sons Ltd., 1982.
29. Z. Mróz, V. A Norris, and O. C. Zienkiewicz, 'An anisotropic hardening model for soils and its application to cyclic loading', *International Journal for Numerical and Analytical Methods in Geomechanics*, **2**, 3, 203–221 (1978).
30. K. C. Valanis and H. E. Read, 'A new endochronic plasticity model for soils', chap. 14 in *Soil Mechanics—Transient and Cyclic Loads* (Eds. G. N. Pande and O. C. Zienkiewicz), pp. 375–417, John Wiley and Sons, 1982.

Mechanics of Engineering Materials
Edited by C. S. Desai and R. H. Gallagher

Chapter 33

Generalized Plasticity Formulation and Applications to Geomechanics

O. C. Zienkiewicz and Z. Mróz

33.1 INTRODUCTION

The subject of modelling the non-linear behaviour of real materials is shrouded in mystique and complexity and a fruitful field for theoretical investigations. Whilst almost any constitutive law can now be incorporated into numerical analysis procedures to deal with realistic boundary value situations the greatest difficulties arise in the identification of *realistic constitutive models.* Here the complexity of various theoretical forms has to be understood by the practitioner who, often from meagre experimental information has to obtain engineering predictions. Nowhere is the situation so serious as in the field of geomechanics where the highly variable material behaves at times in a most complex manner.

What is therefore necessary is first the simplication of general models and then the development of model types capable of predicting the most important aspects of behaviour. In this paper we shall thus be concerned with the development of a very general class of elastoplastic models to which Section 33.2 of the chapter is devoted. In Section 33.3 some applications to geomechanical modelling will be introduced in which a simple method of predicting cyclic soil behaviour is discussed.

The modelling considered will in the main concentrate on time independent behaviour—characteristic of a large class of geomechanical problems. As numerical formulation is now capable of dealing with large deformations of the material we shall focus on the current configuration and Cauchy stresses (true stresses) σ_{ij}.

The constitutive laws will be presented in terms of co-rotational stress increments of the Jaumann–Zaremba type related to the Cauchy stress by

$$\mathrm{d}\sigma_{ij} = \mathrm{d}\overset{\nabla}{\sigma}_{ij} + \sigma_{ik}\mathrm{d}\omega_{kj} + \sigma_{jk}\mathrm{d}\omega_{ki} \tag{33.1}$$

where

$$\mathrm{d}\omega_{ij} = \tfrac{1}{2}(V_{i,j} - V_{j,i})\,\mathrm{d}t \tag{33.2}$$

is the rotation increment with V_i being the velocity of a particle,

The deformation (strain) increments are similarly defined as

$$d\varepsilon_{ij} = \tfrac{1}{2}(V_{i,j} + V_{j,i})\,dt \tag{33.3}$$

In what follows we shall drop the $^{\nabla}$ superscript as all the stresses will be assumed co-rotational.

33.2 TIME INDEPENDENT NON-LINEARITY AND PLASTICITY DEFINITIONS

Quite generally all linear and non-linear behaviour of solids can be described by some incremental relationship of the form

$$d\sigma_{ij} = D_{ijkl}(d\varepsilon_{kl} - d\varepsilon_{k}^{0}) \tag{33.4}$$

where $d\varepsilon_{kl}^{0}$ is an increment of a strain either not associated with stress changes or time-dependent (creep) strain. At the present stage we shall not take it into consideration.

The constitutive tensor D_{ijkl} can in general be dependent on

(a) State of stress and strain ($\boldsymbol{\sigma}, \boldsymbol{\varepsilon}$),
(b) History dependent (degradation) parameters α defined usually by an auxiliary relation

$$d\alpha = Q_{ij}\,d\varepsilon_{ij} \tag{33.5}$$

(c) The direction of loading and unloading.

At this stage it is convenient to define two general possibilities of material behaviour. These are

(1) *Elastic behaviour* in which all processes are reversible and conducted without loss (or gain) of energy;
(2) *Plastic behaviour* in which reversibility does not occur and where energy is expended during a process of a deformation cycle.

In common usage the two processes are often distinguished simply by the existence or non-existence of permanent deformation after a process of stress application and removal.

In the so-called 'rational' mechanics literature the two classes are often associated with the word elastic by adding a hyper or hypo prefix.† This may be due to the dislike of the word 'plastic' this being associated with the classic 'theory of plasticity'. The broader definition given here for plasticity should not raise this objection.

†'Hypo' is associated with under, below, deficient, while 'hyper' is the precise opposite according to the *Oxford Dictionary*.

33.2.1 Elasticity

The distinction between plastic and elastic deformation outlined above cannot be made as a result of a simple test with monotonically increasing loading. Indeed, for such loads predictions based on elastic and plastic models may well be identical, which accounts for an extensive use of non-linear elasticity for the modelling of certain phases of soil behaviour.

In a uniaxial situation elastic behaviour can be characterized by a scalar form of equation (33.4), i.e.

$$\mathrm{d}\sigma = D\,\mathrm{d}\varepsilon \tag{33.6}$$

where any single valued $D = D(\sigma)$ or $D = D(\varepsilon)$ is permissible. This is illustrated by a single curve of Figure 33.1.

For general stress/strain relations of a multiaxial kind considerable restrictions must be placed on $D(\sigma)$ or $D(\varepsilon)$ to ensure path independence of deformation and energy conservation. As such restrictions do not seem to be widely known [1] it is useful to recapitulate them here.

Consider first the situation given by equation (33.4), the inverse of this being written as

$$\mathrm{d}\varepsilon_{ij} = C_{ijkl}\,\mathrm{d}\sigma_{kl} \tag{33.7}$$

and to ensure *path independence* of the deformation ε_{ij} it is necessary for (33.7) to be capable of being written as a simple differential as

$$\mathrm{d}\varepsilon_{ij} = \frac{\partial \varepsilon_{ij}}{\partial \sigma_{kl}}\,\mathrm{d}\sigma_{kl} \tag{33.8}$$

Thus

$$C_{ijkl} = \frac{\partial \varepsilon_{ij}}{\partial \sigma_{kl}} \tag{33.9}$$

must be preserved. This condition can be verified if all coefficients obey the

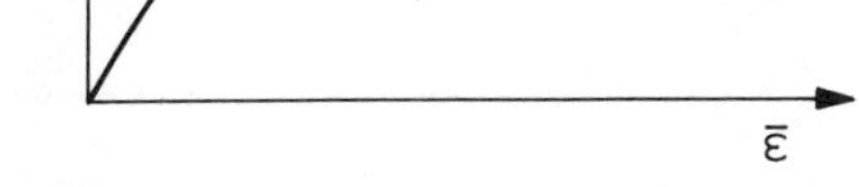

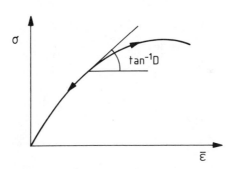

Figure 33.1 Non-linear elastic response

requirement that

$$\frac{\partial C_{ijkl}}{\partial \sigma_{mn}} = \frac{\partial C_{ijmn}}{\partial \sigma_{kl}} \tag{33.10}$$

It will be observed that many so-called non-linear elastic laws used in soil mechanics do not satisfy this reversibility condition [1].

The second requirement is that of energy conservation in small cycles of stress. Without proof we state the well known symmetry requirement

$$C_{ijkl} = C_{klij} \tag{33.11}$$

needed for this condition.

Both requirements are satisfied simply if a strain energy potential can be defined.

33.2.2 Plasticity generalized

The type of one-dimensional response shown in Figure 33.2 is characteristic of real materials with irreversible plastic behaviour. Now clearly we have a dependence of $\mathbf{D}$ on $\boldsymbol{\sigma}$ (or $\boldsymbol{\varepsilon}$) and on some history parameters α. Further, even in the uniaxial example the value of D does not depend uniquely on the stress and history (state) parameters but may take different values for loading and unloading. We can thus write for an uniaxial problem the *most general description* as $d\sigma = D_L\, d\varepsilon$ for loading and $d\sigma = D_U\, d\varepsilon$ for unloading with

$$\begin{aligned} D_{L/U} &= D_{L/U}(\sigma, \alpha) \\ d\alpha &= Q(\sigma, \alpha)\, d\varepsilon \end{aligned} \tag{33.12}$$

Here the direction of 'loading' and 'unloading' could be defined for instance by the sign of $d\varepsilon$

$$\begin{aligned} d\varepsilon > 0 \quad &\text{loading} \\ d\varepsilon < 0 \quad &\text{unloading} \end{aligned} \tag{33.13}$$

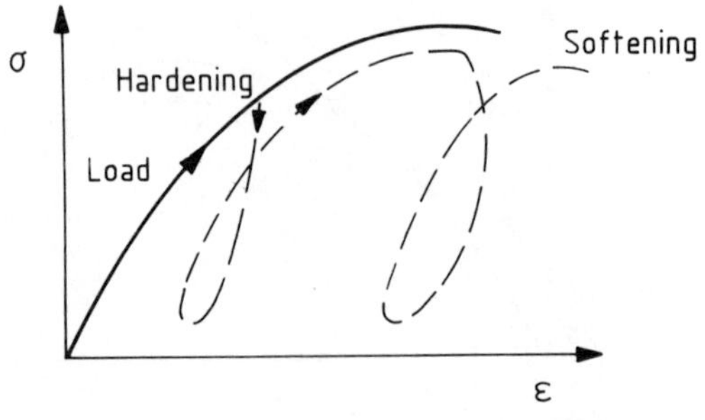

Figure 33.2 Elastoplastic response $D_L \neq D_U$

or alternatively by the sign of the work increment

$$\begin{aligned} \sigma \mathrm{d}\varepsilon > 0 \quad & \text{loading} \\ \sigma \mathrm{d}\varepsilon < 0 \quad & \text{unloading} \end{aligned} \tag{33.14}$$

In the above description no separation into 'elastic' and 'plastic' strains was attempted and in general this is unnecessary—though at times convenient.

The sign of the moduli can be positive or negative depending whether hardening or softening behaviour is encountered.

If, for instance, there are indications of elastic behaviour we can always write

$$\begin{aligned} \mathrm{d}\varepsilon^{\mathrm{e}} &= D_{\mathrm{e}}^{-1}\,\mathrm{d}\sigma \\ \mathrm{d}\varepsilon_{\mathrm{L}}^{\mathrm{p}} &= D_{\mathrm{Lp}}^{-1}\,\mathrm{d}\sigma; \quad \mathrm{d}\varepsilon_{\mathrm{U}}^{\mathrm{p}} = D_{\mathrm{Up}}^{-1}\mathrm{d}\sigma \end{aligned} \tag{33.15}$$

with

$$\mathrm{d}\varepsilon = \mathrm{d}\varepsilon^{\mathrm{e}} - \mathrm{d}\varepsilon^{\mathrm{p}}$$

In the above tangent moduli differ only in the plastic component and frequently we assume that $D_{\mathrm{Up}}^{-1} = 0$ giving no plastic strains when unloading.

The simple uniaxial descriptions need to be expanded in a multiaxial stress space Figure 33.3. Here if we consider a stress point P a direction has to be specified in the stress space which distinguishes between loading and unloading. This is conveniently given by a unit vector $\mathbf{n}$ normal to a plane t.

We shall now specify that loading/unloading directions are determined by the sign of the projection of the stress increment, i.e.

$$\begin{aligned} \mathbf{n}^T \mathrm{d}\boldsymbol{\sigma} > 0 \quad & \text{loading} \\ \mathbf{n}^T \mathrm{d}\boldsymbol{\sigma} < 0 \quad & \text{unloading} \end{aligned} \tag{33.16}$$

This foils if strain softening occurs. A more useful definition replaces in above $\mathrm{d}\boldsymbol{\sigma}$ by $\mathrm{d}\boldsymbol{\sigma}^{\mathrm{e}}$ defined as

$$\mathrm{d}\boldsymbol{\sigma}^{\mathrm{e}} = \mathbf{D}_{\mathrm{e}}\mathrm{d}\varepsilon$$

For stress increments along the plane t, i.e. when $\mathbf{n}^{\mathrm{T}}\,\mathrm{d}\boldsymbol{\sigma} = 0$ purely elastic

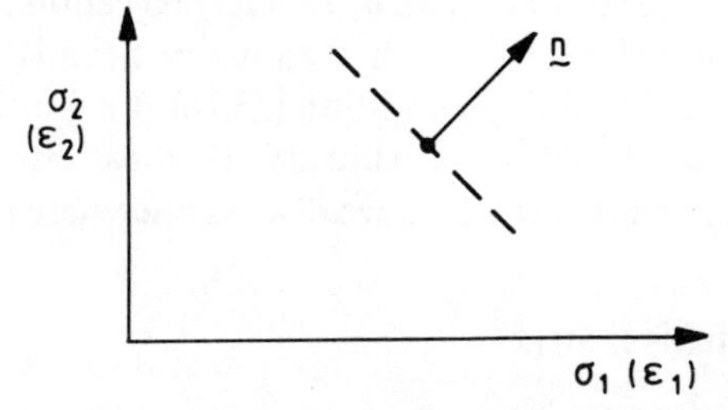

Figure 33.3 Behaviour in multidimensional stress space

deformations occur and then

$$d\boldsymbol{\varepsilon} = \mathbf{D}_e^{-1}\, d\boldsymbol{\sigma} \tag{33.17}$$

Now, in general, if a loading occurs for a stress change $d\boldsymbol{\sigma}$ we have

$$d\boldsymbol{\varepsilon} = d\boldsymbol{\varepsilon}^e + d\boldsymbol{\varepsilon}^p = \mathbf{D}_e^{-1}\, d\boldsymbol{\sigma} + \mathbf{D}_{Lp}^{-1}\, d\boldsymbol{\sigma} \tag{33.18}$$

and for unloading

$$d\boldsymbol{\varepsilon} = d\boldsymbol{\varepsilon}^e + d\boldsymbol{\varepsilon}^p = \mathbf{D}_e^{-1}\, d\boldsymbol{\sigma} + \mathbf{D}_{Up}^{-1}\, d\boldsymbol{\sigma}$$

In the above all the moduli can, as before depend on values of $\boldsymbol{\sigma}$ and state parameters α, but before continuing it is necessary to observe that an arbitrary choice of moduli might lead to inconsistencies. In particular we note that when $\mathbf{n}^T d\boldsymbol{\sigma} = 0$ both (33.17) and (33.18) have to apply to ensure no jumps in the behaviour pattern. This is readily ensured by writing

$$D_{Lp}^{-1} = \mathbf{n}_L \mathbf{n}^T / K_L$$

and

$$D_{Up}^{-1} = \mathbf{n}_U \mathbf{n}^T / K_U \tag{33.19}$$

where K_L and K_U are loading or unloading scalars and $\mathbf{n}_L$ and $\mathbf{n}_U$ are arbitrary directions (vectors) of unit magnitude.

It will be recognized immediately that $K_{L/U}$ represent the well known plastic moduli (now specified in both loading and unloading directions) and $\mathbf{n}_L$ and $\mathbf{n}_U$ are directions of the plastic strain vector.

Complete elastoplastic behaviour can be thus specified by giving at each point of the stress space $\boldsymbol{\sigma}$ (and for given α)

(1) A set of three directions $\mathbf{n}, \mathbf{n}_L, \mathbf{n}_U$,
(2) Specifying K_L and K_U as appropriate functions of $\boldsymbol{\sigma}$ and α, and
(3) Specifying the elastic matrix $\mathbf{D}^e$

This specification is a generalization of the more restrictive hypo-elastic [2] or classical plasticity assumptions [3] and allows a much greater flexibility.

The range of values of K is $\pm\infty$ with $K = 0$ for ideal plasticity, $K = \infty$ for zero plastic deformation, and $K < 0$ for strain softening.

We shall also place the restriction on $\mathbf{n}_L$ that its projection on $\mathbf{n}$ (i.e., $\mathbf{n}^T\mathbf{n}_L$) must always be positive (and similarly that $\mathbf{n}^T\mathbf{n}_U$ must be negative). This ensures the validity of the loading/unloading condition (33.16) previously specified.

It should also be noted that some arbitrary specifications of $K_{K/L}$ can lead to unrealistic ratcheting behaviour which we discuss elsewhere [see Chapter 22].

33.2.3 Tangent modulus matrix

For computation purposes it is convenient to obtain the combined elastoplastic matrices for loading (and unloading) in the form

$$d\boldsymbol{\sigma} = D_{Lep}\, d\varepsilon \quad \text{or} \quad d\boldsymbol{\sigma} = \mathbf{D}_{Uep}\, d\varepsilon \tag{33.20}$$

This can be done by suitable manipulation (relegated to Appendix 1) and we can write

$$\mathbf{D}_{\mathrm{Lep}} = \mathbf{D}_{\mathrm{e}} - \frac{\mathbf{D}_{\mathrm{e}}\mathbf{n}_{\mathrm{L}}\mathbf{n}^{\mathrm{T}}\mathbf{D}_{\mathrm{e}}}{K_{\mathrm{L}} + \mathbf{n}^{\mathrm{T}}\mathbf{D}_{\mathrm{e}}\mathbf{n}_{\mathrm{L}}} \tag{33.21}$$

with similar expression for elastoplastic unloading.

We note that $n_{\mathrm{L}} \neq n$ results in non-symmetric tangent moduli in a similar manner as non-associative plasticity.

33.2.4 Classical plasticity—Bounding models

In classical plasticity the behaviour is specified by surfaces in the $\boldsymbol{\sigma}$ space which enclose elastic regions.

The yield surface

$$F(\boldsymbol{\sigma}, \alpha) = 0 \tag{33.22}$$

defines the loading and unloading directions giving the unit normal vector as

$$\mathbf{n} = \frac{\partial F}{\partial \boldsymbol{\sigma}} \Bigg/ \left[\frac{\partial F^{\mathrm{T}}}{\partial \boldsymbol{\sigma}}\frac{\partial F}{\partial \boldsymbol{\sigma}}\right]^{1/2} \tag{33.23}$$

Similarly the flow direction n_{L} is defined by the plastic potential $Q(\boldsymbol{\sigma}, \alpha)$ with

$$\mathbf{n}_{\mathrm{L}} = \frac{\partial Q}{\partial \boldsymbol{\sigma}} \Bigg/ \left[\frac{\partial Q^{\mathrm{T}}}{\partial \boldsymbol{\sigma}}\frac{\partial Q}{\partial \boldsymbol{\sigma}}\right]^{1/2} \tag{33.24}$$

The hardening modulus K_{L} follows from above definitions if we postulate that the stress point P remain on the yield surface during 'loading', i.e. that

$$\mathrm{d}F = \frac{\partial F}{\partial \boldsymbol{\sigma}}\mathrm{d}\boldsymbol{\sigma} + \frac{\partial F}{\partial \alpha}\frac{\partial \alpha}{\partial \varepsilon^{\mathrm{p}}}\mathrm{d}\varepsilon^{\mathrm{p}} = 0 \tag{33.25}$$

Putting

$$\mathrm{d}\varepsilon^{\mathrm{p}} = \mathbf{n}_{\mathrm{L}}\mathbf{n}\,\mathrm{d}\boldsymbol{\sigma}/K_{\mathrm{L}} \tag{33.26}$$

gives after some manipulation

$$K = \frac{\partial F}{\partial \alpha}\frac{\partial \alpha}{\partial \varepsilon^{\mathrm{p}}} \quad \mathbf{n}_{\mathrm{L}} \Bigg/ \left[\frac{\partial F^{\mathrm{T}}}{\partial \boldsymbol{\sigma}}\frac{\partial F}{\partial \boldsymbol{\sigma}}\right]^{1/2} \tag{33.27}$$

The two descriptions thus can be identified. For problems of ideal plasticity the yield surface presents a realistic and physically important boundary in the stress space and the classic description is very advantageous. However, if different hardening/softening behaviour patterns are to be described the complexity of description in classical terms can be quite formidable and a more direct description of moduli is convenient.

A very useful artifice for model description has been introduced by Dafalias

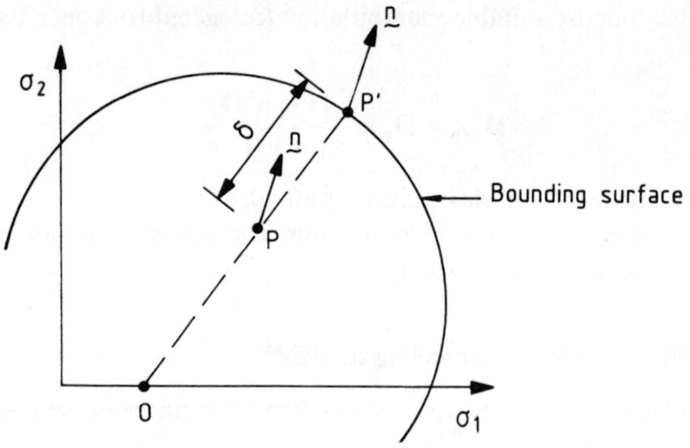

Figure 33.4 Bounding plasticity concept. Rule of conjugate points P – P′. **n** at P and P′ made identical, $K^{P'}$ derived by some interpolation rule from K^{P}. $K^{P'} \to K^{P}$ as $\delta \to 0$

and Popov [4] and Krieg [5]. Here classic plasticity is used to describe a 'bounding surface' Figure 33.4 and for all other points P the loading and unloading direction as well as moduli are defined directly by some interpolation rule from values on the bounding surface.

For instance for every internal point P we can by some rule (such as shown in Figure 33.4 by connecting a specified pole O to P) associate a point P′ on the bounding surface and then take the directions $\mathbf{n}_L$ and $\mathbf{n}$ at P as being those of P′ with a suitable interpolation defining K_L and K_U at P in terms of the values of P′.

Many soil models have been so derived [6, 7] with considerable success and application to metals are widely used.

33.2.5 Viscoplasticity and endochronic models

In the general form of constitutive relations given by equation 33.4 we assumed the existence of initial strains $\boldsymbol{\varepsilon}^0$ caused by creep effects, etc. Indeed, we could in the general case assume that

$$d\varepsilon_{kl}^0 = g_{kl}\,dt + q_{kl}\,d\alpha \tag{33.28}$$

where $\mathbf{g} = \mathbf{g}(\boldsymbol{\sigma}, \boldsymbol{\alpha})$ and $\mathbf{q} = \mathbf{q}(\boldsymbol{\sigma}, \boldsymbol{\alpha})$ give the strain rates due to changes of time and some state parameter (or parameters).

For time dependent problems the inclusion of creep effects in the manner described above is natural and forms the basis of viscoplastic formulations now widely used in geomechanics [8, 9]. On occasion such formulation can be very effectively used for the solution of plasticity problems—this usage introduced by the present writer is, however, not always most efficient and is not recommended for dynamic phenomena.

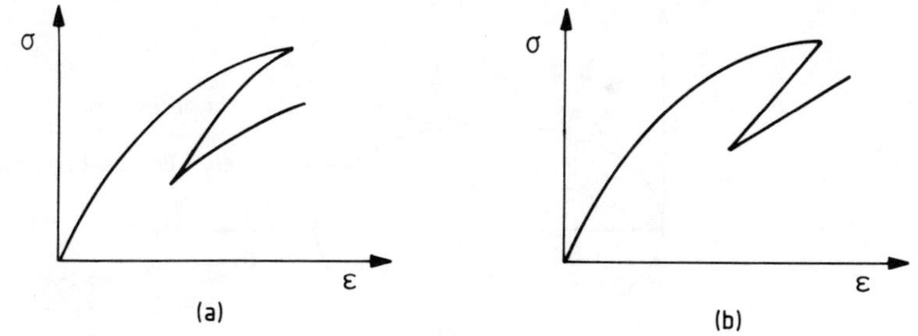

Figure 33.5 History dependent and endochronic models (a) and (b)

The introduction of the rate with respect to the parameter α is an alternative to the generalized plasticity description given earlier. It is the basis of the so-called endochronic model introduced by Valanis [10] and applied by some to geomechanics [11, 12]. With

$$\mathrm{d}\alpha = Q_{ij}\,\mathrm{d}\varepsilon_{ij} \tag{33.29}$$

permanent deformations can be obtained and thus plastic behaviour introduced.

For continuous Q_{ij} a different path can be followed in loading and unloading—unfortunately showing a continuous tangent as in Figure 33.5(a). Alternatively a sign change in Q is often introduced in its definition using for instance the absolute value of strain path increment

$$\mathrm{d}\alpha = \left| \sqrt{\mathrm{d}\varepsilon_{ij}\alpha\varepsilon_{ij}} \right| \tag{33.30}$$

This does allow a different unloading direction but in a limited sense only compared with the generalized plasticity formulation, Figure 33.5(b).

A very successful use of such an 'endochronic' form has been introduced by the author [13, 14] in conjunction with standard plasticity to account for densification strains leading to liquefaction under cyclic loading. Developments of this procedure have been made recently by Finn [15].

33.3 APPLICATIONS TO GEOMECHANICS

33.3.1 General—Critical state models

For nearly monotonic loading the most successful types of model used in geomechanics are based on classic hardening plasticity in the manner introduced by Drucker, Gibson, and Henkel in their now classic paper of 1957 [16]. Many specializations of the model have been made since its original introduction mainly with regard to the shape of the yield surface in space [8, 9, 17, 18] and to generalize the behaviour to both sands and clays [19, 20] but all basically assume a closed elliptic (or nearly elliptic) yield surface in space which hardens or

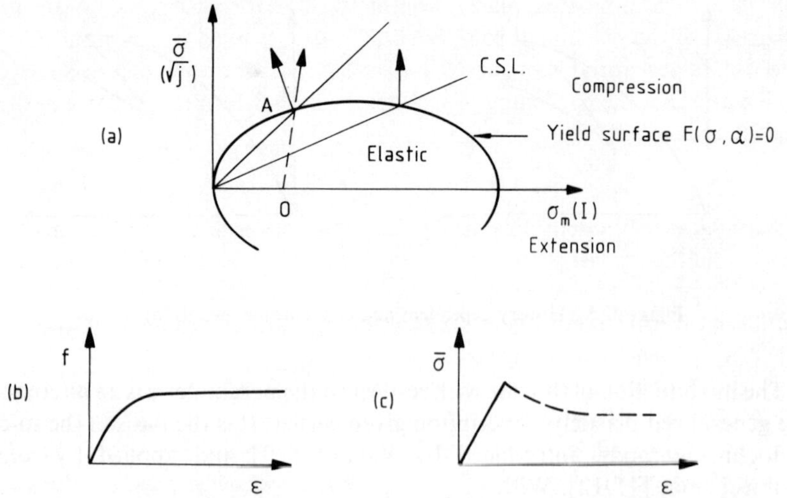

Figure 33.6 Classical plasticity, critical surface model with volumetric and deviatoric hardening [20] $\alpha = \varepsilon p_{vol} + f(\bar{\varepsilon}^p)$. (a) Yield surface in stress space, (b) deviatoric hardening function, (c) stress–strain behaviour in drained loading along OA

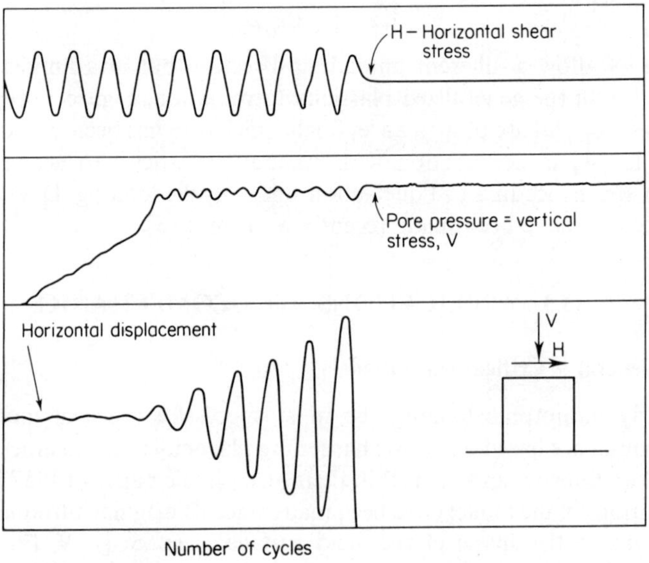

Figure 33.7 Typical behaviour of soil under cyclic loading

softens with accumulated, plastic, volumetric, and deviatoric [20] strain with the first playing the prominent role. We illustrate this surface in Figure 33.6 for the two triaxial test variables σ_m and $\bar{\sigma}$ and assume that the flow rule is associative, i.e. $\mathbf{n}_L = \mathbf{n}$ and that the hardening parameter α which determines the size of the yield surface is

$$\alpha = \varepsilon_p^v + f(|\bar{\varepsilon}_p|) \tag{33.31}$$

where ε^v is the volumetric plastic strain, $|\bar{\varepsilon}_p|$ the second strain invariant of plastic strain, and that f is a function which leads to a constant value with accumulating $|\bar{\varepsilon}_p|$ as shown in Figure 33.6(b). The model described shows now a softening behaviour illustrated in Figure 33.6(c) in a monotonic drained test approaching the 'critical state' line asymptotically.

With a suitable choice of parameters describing the surface and hardening rules behaviour of soils under monotonic loads can be adequately described for drained, undrained, and over-consolidated conditions. Up to this point there appears little motivation to introduce the more general plasticity formulation described earlier (although there is some inadequacy in the over-consolidated domain above the critical state line). The need for model improvement arises immediately when cyclic or transient load conditions are encountered. Here the observed fact is that

(a) Under drained conditions permanent and continuing 'compaction' (i.e. volumetric change) occurs when the shear stress is cycled with magnitude well in the elastic zone of the critical state model and

(b) Under undrained conditions this compaction or densification results in a permanent pore pressure increase which in extreme conditions can lead to a complete loss of strength (liquefaction). (Figure 33.7.)

Many modifications of various degrees of complexity have been introduced to the basic model to account for such phenomena. One of these modifications based on a semi-endochronic formulation has already been mentioned [13, 14]. While

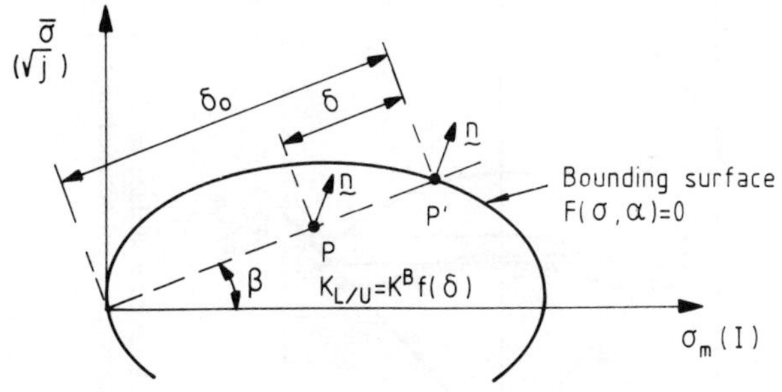

Figure 33.8 A bounding surface critical state model

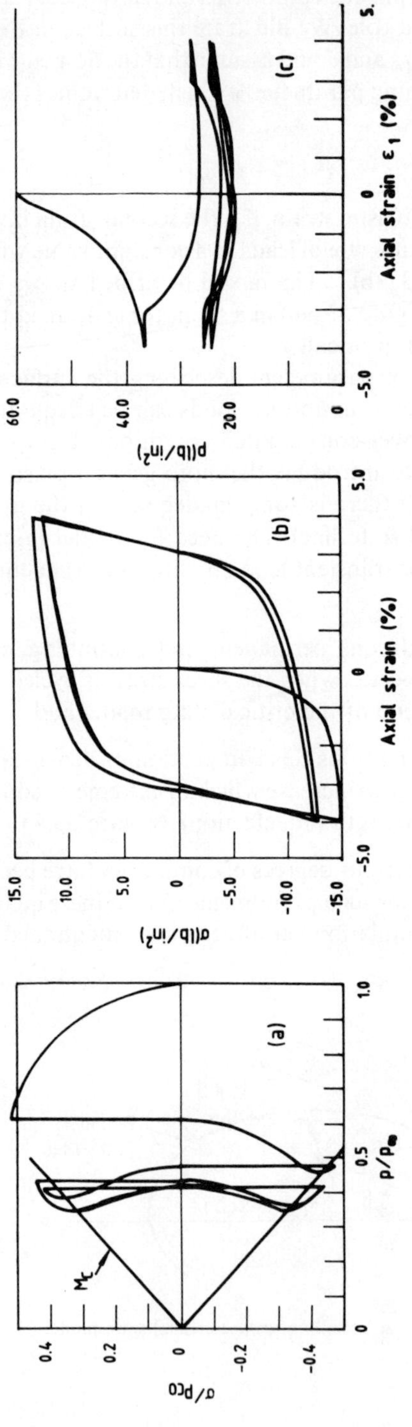

Figure 33.9 Bounding surface model with interpolation of equation (33.32) [6]. Controlled strain cyclic triaxial loading (± 4 per cent), undrained. (a) Effective stress path, (b) octahedral stress–axial strain, (c) effective mean stress–axial strain. Material data: $M = 0.95$, $e_0 = 1.0$, $\kappa = 0.05$, $\lambda = 0.26$, $m = 0.2$

the essential phenomena of densification are reasonably well modelled by this artifice it is not totally satisfactory in prediction of increasing shear strains and failure. For this reason more complex kinematically hardening models have been introduced with limited success [21 to 23].

The simplest path of achieving the desired effect is, however, through the use of a generalized plasticity-bounding surface model. Here the first attempt was made by Dafalias and Herrmann [6] using the homologous points as defined in Figure 33.8 and an interpolation rule

$$K^{(\mathrm{p})} = K^{(\mathrm{p}')} + \hat{K}\left(\frac{\delta}{\delta_0 - \delta}\right)^m f(\beta) \tag{33.32}$$

where $K^{\mathrm{p}'}$ is the modulus of the homologous point and $\hat{K}$ and arbitrary large positive modulus thus ensuring that generally hardening occurs (rather than softening characterized by a negative K) at interior points to the bounding surface.

This model gives an improvement on the standard critical surface model but still fails to predict complete liquefaction. In Figure 33.9 results of a triaxial undrained strain-controlled cyclic behaviour of a soil are presented showing respectively

(a) The stress path,
(b) the decreasing stiffness of the sample,
(c) The decrease of effective mean stress (or increase of pore pressure) with the cyclic amplitude of strain.

33.3.2 An alternative bounding surface model

A simple modification of the bounding surface model given above allows a considerable extension of the modelling power. Now in place of the interpolation given in equation (33.32) a simpler one is introduced preserving at all points the sign of K but simply increasing in magnitude. In this as in the Dafalias model of ref. [16] plasticity is allowed only on loading and the interpolation suggested is simply

$$K = K^{\mathrm{p}'}\left(\frac{\delta_0}{\delta_0 - \delta}\right)^{\gamma} \tag{33.33}$$

In Figure 33.10 we show the results of the previous test recomputed with the new model showing a more realistic behaviour.

In Figure 33.11 we give a quantitative comparison of performance of the above model with tests presented at a recent workshop for model evaluation [24]. The agreement of results is quite remarkable.

Figure 33.12 presents the experimental as well as model results showing the variation of effective stress versus the number of cycles. Clearly the new model is

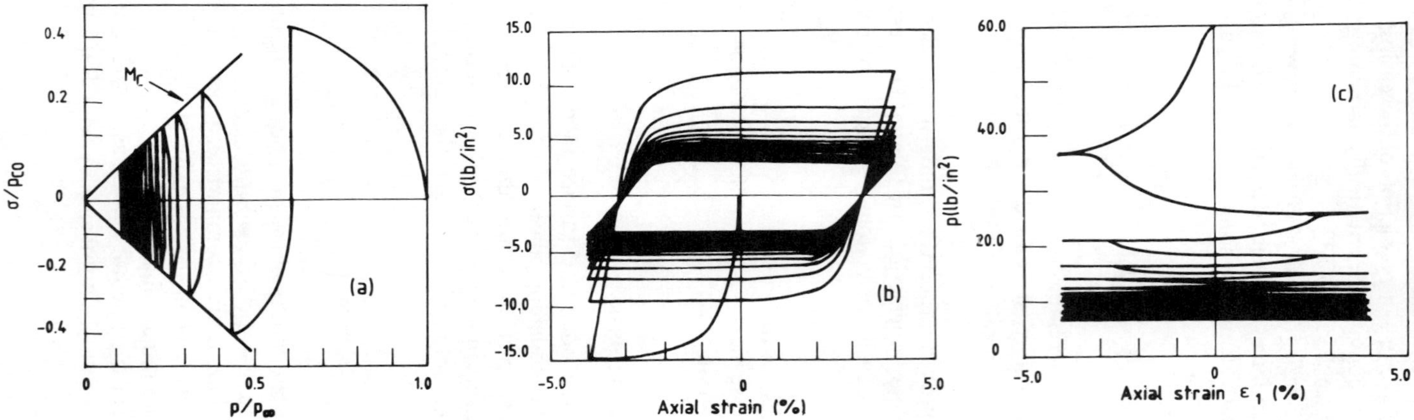

Figure 33.10 Bounding surface model with interpolation of equation (33.33) (Present) controlled strain cyclic triaxial loading (± 4 per cent) undrained. (a) Effective stress path, (b) octahedral stress–axial strain, (c) effective mean stress–axial strain. ($\gamma = 2.5$)

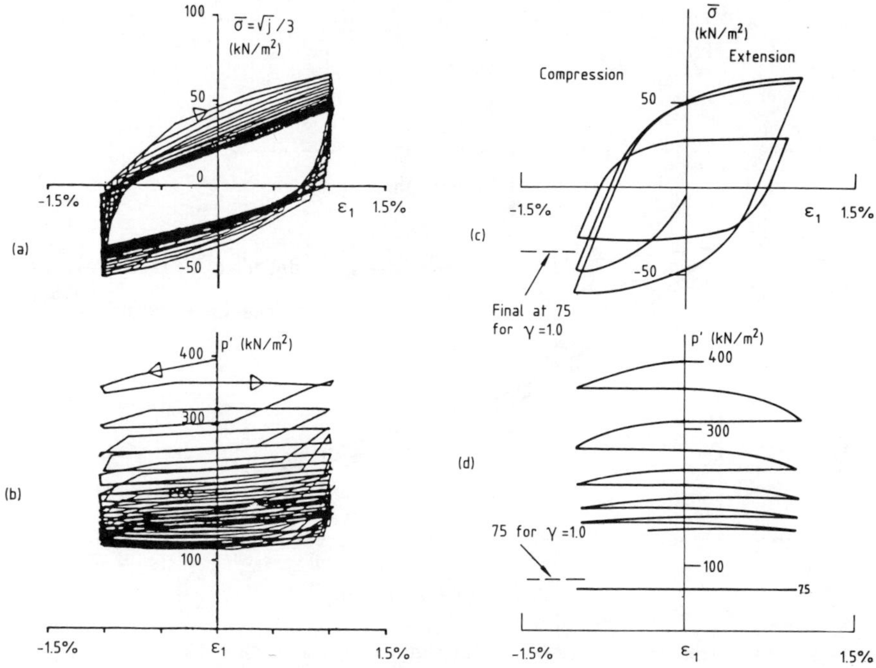

Figure 33.11 Triaxial, undrained, strain-controlled test on kaolin, Grenoble, 1982 (± 1.0 per cent strain). (a) Experimental results (octahedral stress–axial strain), (b) experimental results (mean effective stress–axial strain), (c) present bounding surface model (octahedral stress–axial strain), (d) present bounding surface model (mean effective stress–axial strain). (Material data: $\phi' = 19°$, $\lambda = 0.2$, $\kappa = 0.03$, $G = 10.000$ kN/m^2)

superior to alternatives and yet requires only the introduction of a single new parameter in addition to standard critical state behaviour.

The very simple behaviour of the model allows its extension to include both loading and unloading plastic moduli. Even in monotonic tests it appears to behave in a superior manner to the 'standard' critical state model by avoiding the entry of the stress point into the softening 'super critical' region. Here further refinement is of course possible such as the inclusion of shear hardening in the manner of equation (33.31) or, indeed, by specifying an alternative way of obtaining 'homologous' points. A recent suggestion by Naylor [25] of simply shifting the origin of interpolation (0 of Figure 33.7) could be promising and simply introduces one further parameter.

A model very similar to the one just described but formulated in a more complex manner has recently been introduced by Pande and Pietruszczak [26] and again shows similar behaviour.

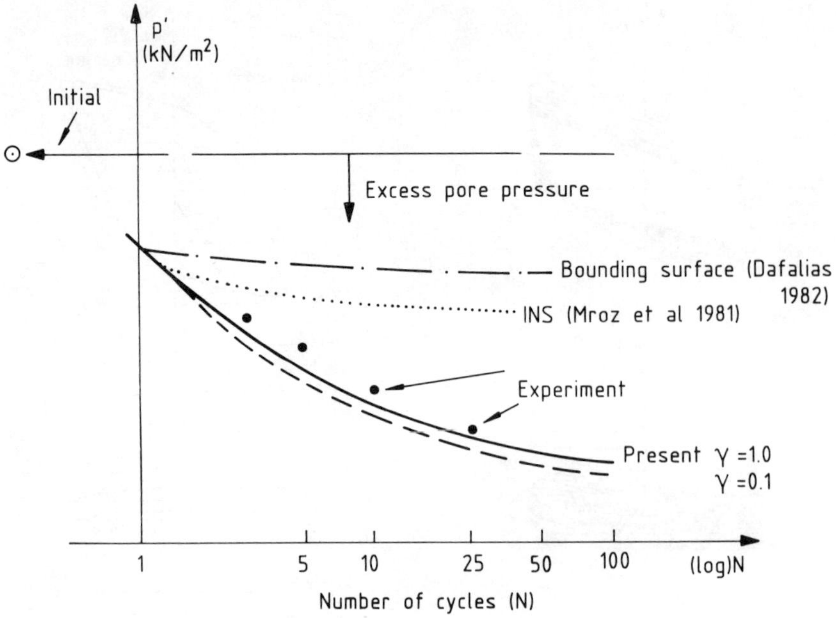

Figure 33.12 Triaxial undrained strain-controlled test on kaolin, Grenoble, 1982 ($\pm$ 1.0 per cent strain). Mean effective stress–cycles. Experiment and predictions

33.3.3 Performance of models in transient dynamic analysis

In a series of publications [13, 14, 27, 28], the author has demonstrated how the Biot theory of porous media can be generalized to non-linear situations and how with an efficient algorithm numerical results can be obtained for realistic problems. The program DIANA-G has been developed in this context to deal with fully coupled soil–pore fluid phenomena and is capable of dealing with short as well as long duration problems by an automatic time step adjustment. Figure 33.13 show typical results of such a calculation for a somewhat artificial problem in which a transient load occurring in 0.3 s is followed by an earthquake of 8 s duration and finally a consolidation phase of 60×10^3 s.

In Figure 33.14 a much simpler situation is analysed in which the behaviour of a sand layer which liquefies under the action of El Centro earthquake [27, 29] is to be followed. Figure 33.15(a), (b), (c) show the development of pore pressures, its time history, and shear stresses for three models

(a) The endochronic densification of ref. [13, 14].
(b) The 'reflecting surface' model of ref. [26] (similar to the present new model), and
(c) A degrading CSM of Carter *et al.* [30].

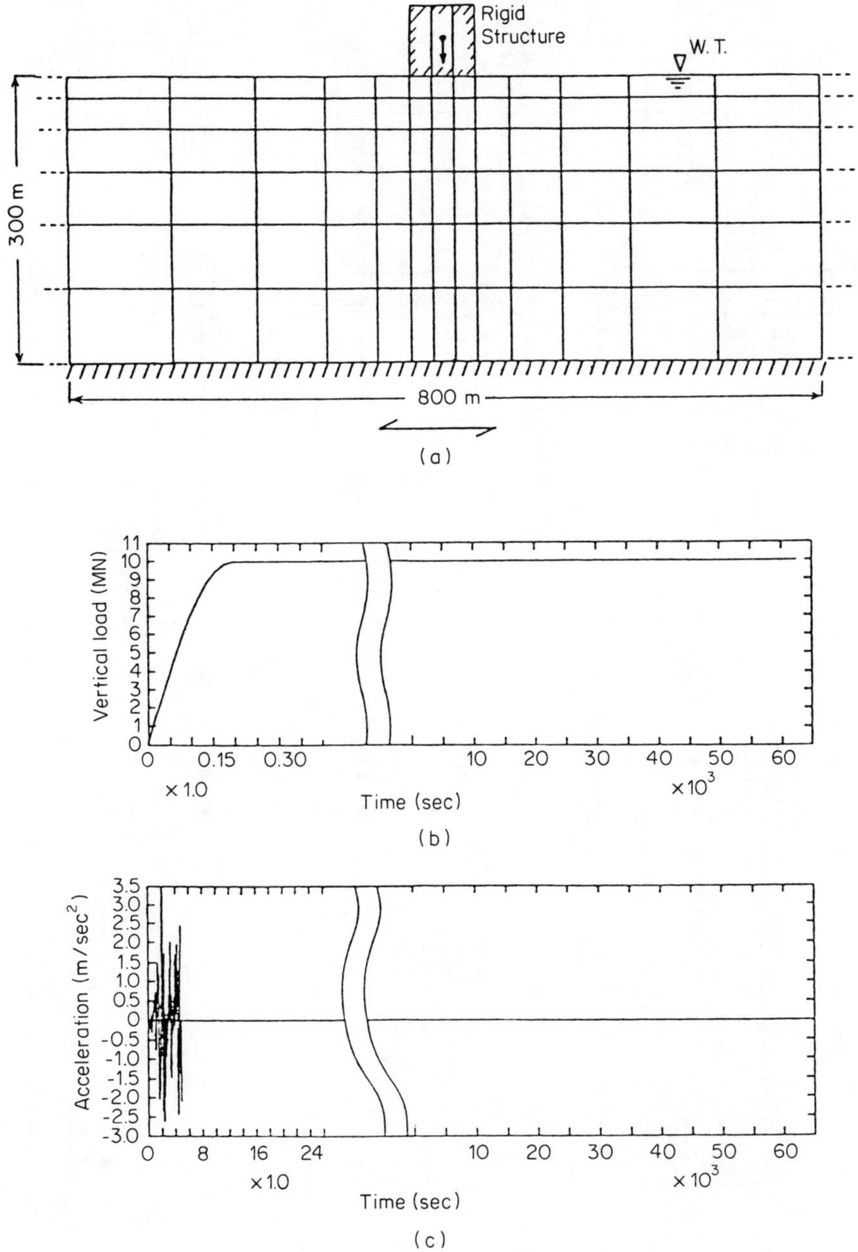
Rigid
Structure
W. T.
300 m
800 m
(a)
Vertical load (MN)
11
10
9
8
7
6
5
4
3
2
1
0
0 0.15 0.30
10 20 30 40 50 60
×1.0
×10^3
Time (sec)
(b)
Acceleration (m/sec^2)
3.5
3.0
2.5
2.0
1.5
1.0
0.5
0
-0.5
-1.0
-1.5
-2.0
-2.5
-3.0
0 8 16 24
10 20 30 40 50 60
×1.0
×10^3
Time (sec)
(c)

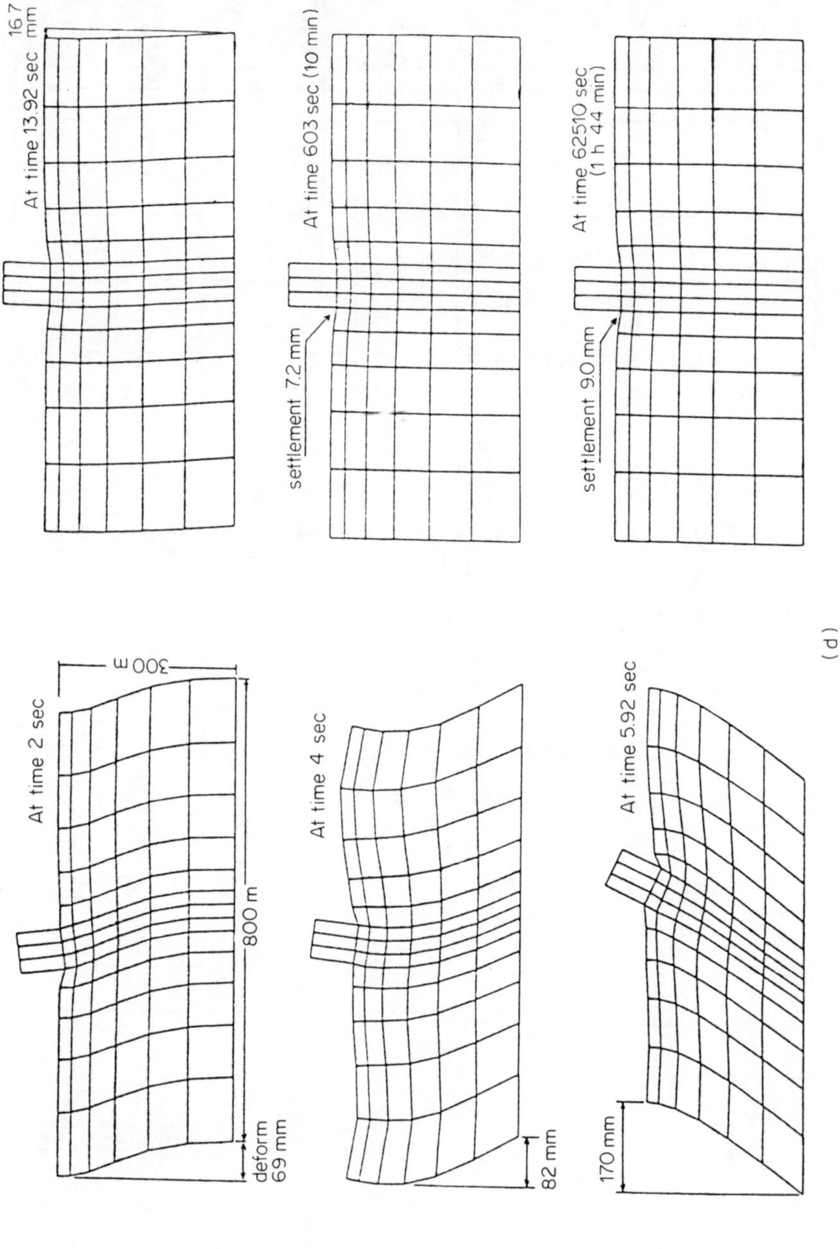
At time 13.92 sec
16.7 mm
At time 603 sec (10 min)
settlement 7.2 mm
At time 62510 sec (1 h 44 min)
settlement 9.0 mm
At time 2 sec
300 m
800 m
deform 69 mm
At time 4 sec
82 mm
At time 5.92 sec
170 mm
(d)

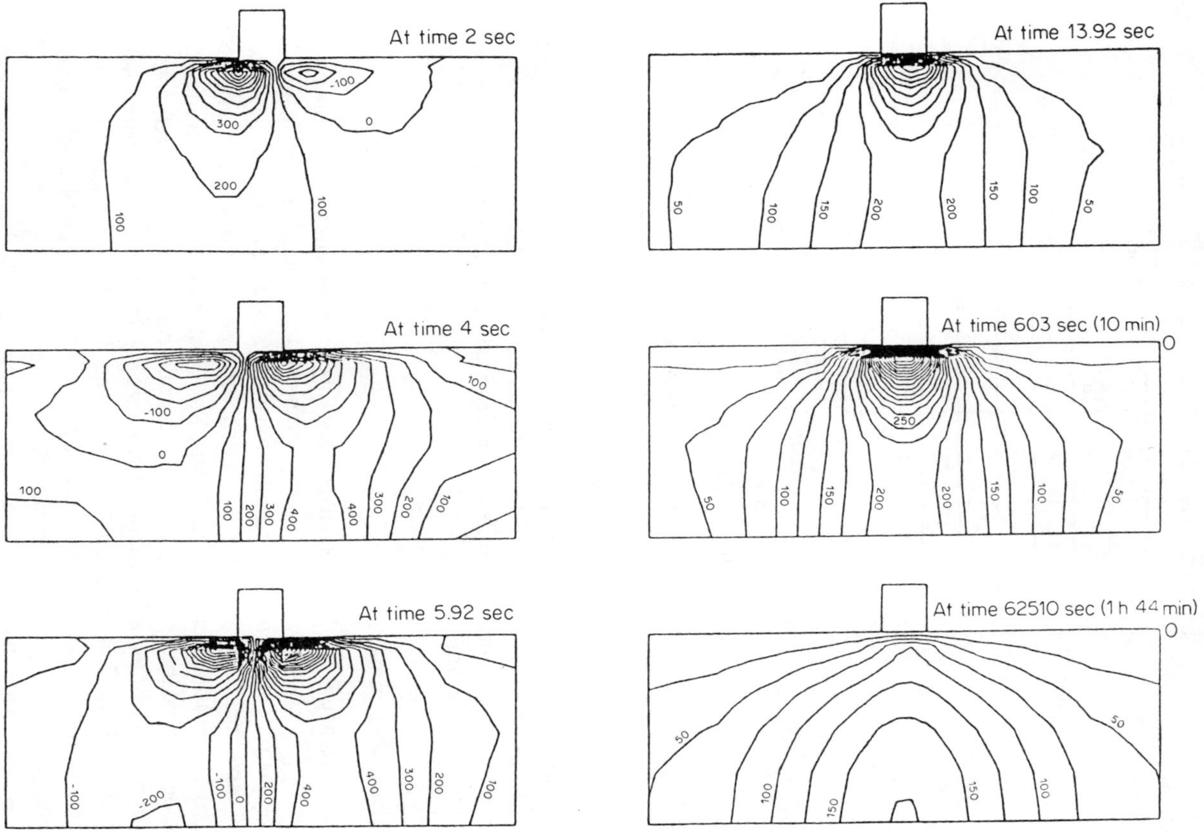

(e)

Figure 33.13 A coupled response of soil with drainage to rapid and slow excitation. (a) Configuration of foundation and structure, (b) vertical load history, (c) simultaneous earthquake history, (d) deformations at various times, (e) pore pressure distribution in various times

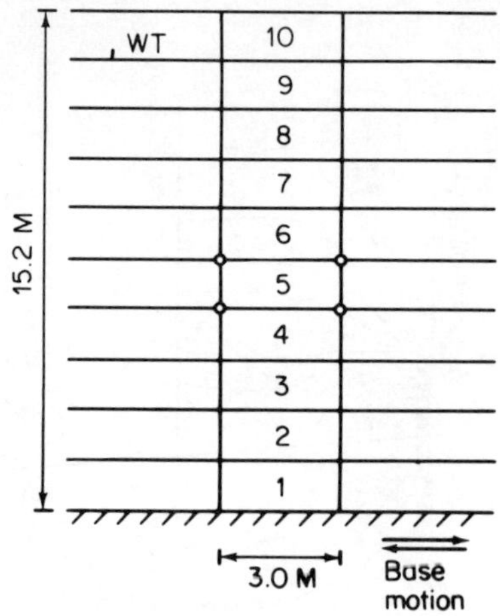

Figure 33.14 A horizontal sand layer subject to the E1 Centro motion at base (NS component × 0.16) geometry and discretization

In this example the first two models perform very similarly while the third is not satisfactory. Obviously the 'endochronic' densification model requires a more complex identification process while the generalized plasticity one needs only a single additional parameter.

33.3.4 Concluding remarks

The theory presented for the generalized plasticity forms opens many doors and possibilities. We show here how a very simple interpolation results in good modelling of cyclic soil behaviour but obviously further refinements can be made to follow experiments more closely. A philosophical point, however, arises in the general problem of material (soil) modelling. Some believe that the model should be based on rational interpretation of materials microstructure. Others refine the formulation on phenomenological observations only. It occurs to the writer that perhaps the middle course should be followed in which reasonably close phenomenological descriptions are used together with an attempt at obtaining more complex responses by a microstructure technique.

Such an approach has been used successfully in describing anisotropic hardening phenomena in metals by a simple overlaying of ideal plasticity models [31].

In an alternative context Pande and Sharma [32] obtain the response of soils to simple rotation of principal stress axes by superposing effects of a random distribution of planes.

Such techniques may well be the most efficient way of refining models for complex soil behaviour by superposition. It is possible to visualize an overlay of several isotropic generalized plasticity models responding satisfactorily to anisotropic hardening behaviour [33].

ACKNOWLEDGEMENTS

The authors are grateful to Dr S. Pietruszczak and Mr K. H. Leung for carrying out model assessment computations. Thanks are also due to Mr T. Shiomi and Mr Leung for carrying out the solution of some boundary value problems quoted.

APPENDIX 1 DERIVATION OF ELASTOPLASTIC MATRIX

From equations (33.18) and (33.19) we have for elastoplastic loading

$$\mathrm{d}\boldsymbol{\varepsilon} = \mathbf{D}_{\mathrm{e}}^{-1}\,\mathrm{d}\boldsymbol{\sigma} + \mathbf{n}_{\mathrm{L}}\mathbf{n}^{\mathrm{T}}\mathrm{d}\boldsymbol{\sigma}/K_{\mathrm{L}} \tag{33.34}$$

Substitute in above

$$\mathbf{n}^{\mathrm{T}}\,\mathrm{d}\boldsymbol{\sigma}/K_{\mathrm{L}} \equiv \lambda \tag{33.35}$$

and premultiply by $\mathbf{n}^{\mathrm{T}}\mathbf{D}^{\mathrm{e}}$ giving

$$^{\dagger}\ \mathbf{n}^{\mathrm{T}}\mathbf{D}_{\mathrm{e}}\,\mathrm{d}\boldsymbol{\varepsilon} = \mathbf{n}^{\mathrm{T}}\,\mathrm{d}\boldsymbol{\sigma} + \mathbf{n}^{\mathrm{T}}\mathbf{D}_{\mathrm{e}}\mathbf{n}_{\mathrm{L}}\lambda \tag{33.36}$$

or

$$\mathbf{n}^{\mathrm{T}}\mathbf{D}_{\mathrm{e}}\,\mathrm{d}\boldsymbol{\varepsilon} = \lambda(K_{\mathrm{L}} + \mathbf{n}^{\mathrm{T}}\mathbf{D}_{\mathrm{e}}\mathbf{n}_{\mathrm{L}}) \tag{33.37}$$

Thus

$$\lambda = \frac{\mathbf{n}^{\mathrm{T}}D_{\mathrm{e}}\,\mathrm{d}\boldsymbol{\varepsilon}}{K_{\mathrm{L}} + \mathbf{n}^{\mathrm{T}}\mathbf{D}_{\mathrm{e}}\mathbf{n}_{\mathrm{L}}} \tag{33.38}$$

Substitution into (33.34) and (33.35) gives

$$\mathrm{d}\boldsymbol{\varepsilon} = \mathbf{D}_{\mathrm{e}}^{-1}\,\mathrm{d}\boldsymbol{\sigma} + \frac{\mathbf{n}_{\mathrm{L}}\mathbf{n}^{\mathrm{T}}\mathbf{D}_{\mathrm{e}}\,\mathrm{d}\boldsymbol{\varepsilon}}{K_{\mathrm{L}} + \mathbf{n}^{\mathrm{T}}\mathbf{D}_{\mathrm{e}}\mathbf{n}_{\mathrm{L}}}$$

or on solving

$$\mathrm{d}\boldsymbol{\sigma} = \left(\mathbf{D}_{\mathrm{e}} - \frac{\mathbf{D}_{\mathrm{e}}\mathbf{n}_{\mathrm{L}}\mathbf{n}^{\mathrm{T}}\mathbf{D}_{\mathrm{e}}}{K_{\mathrm{L}} + \mathbf{n}^{\mathrm{T}}\mathbf{D}\mathbf{n}_{\mathrm{L}}}\right)\mathrm{d}\boldsymbol{\varepsilon} \tag{33.39}$$

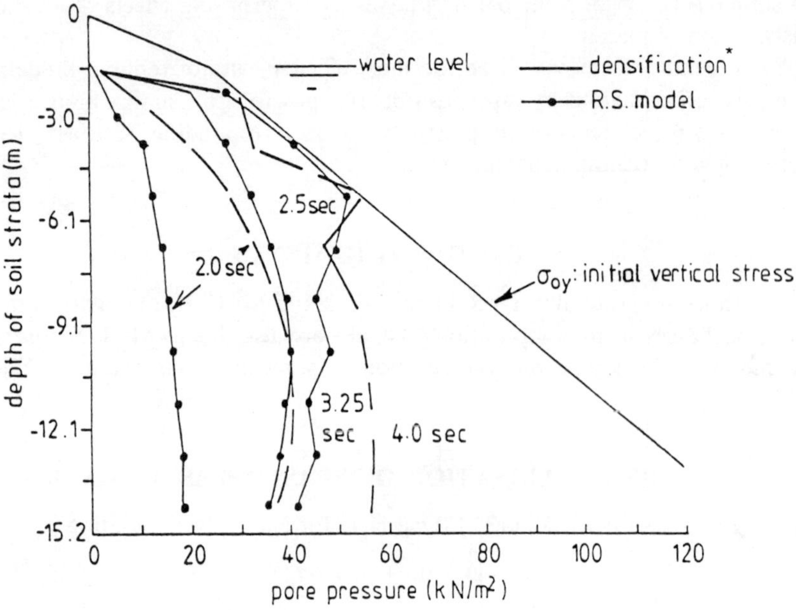

*After Zienkiewicz *et al.* (1982)

(a)

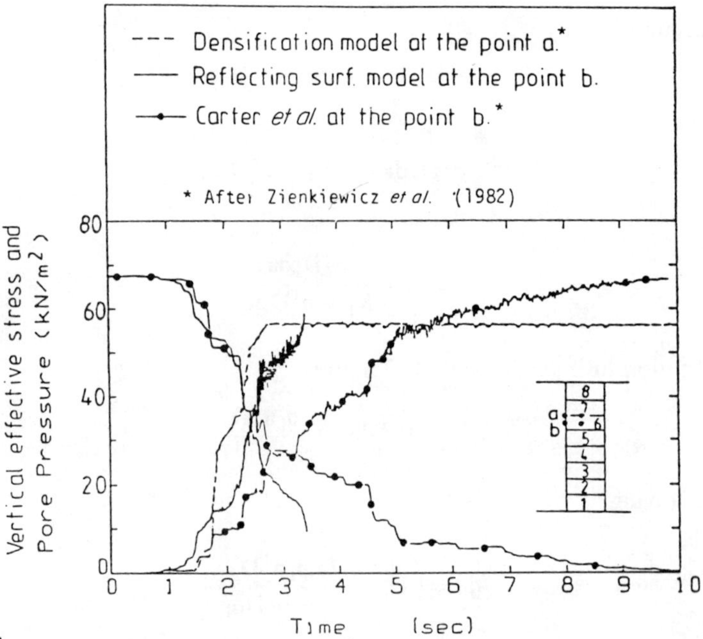

(b)

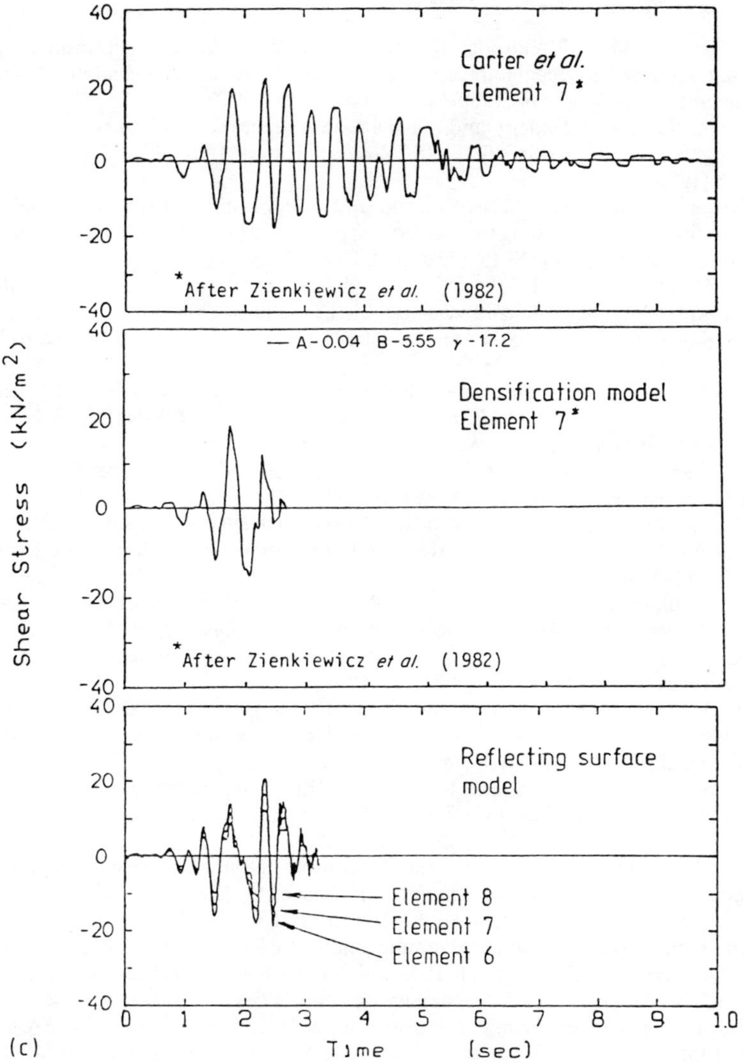

(c)

Figure 33.15 A horizontal sand layer of Figure 33.14. (a) Pore pressure distribution at various times, (b) history of pore pressure build-up with different soil models, (c) shear stress at level of element 7 for different soil models

REFERENCES

1. M. Zytyzski, M. F. Randolph, R. Nova, and C. P. Wroth, 'On modelling the unloading-reloading behaviour of Soils, *Int. J. for Numerical and Analytical Methods in Geomechanics*, **2**, 87–94 (1980).
2. Z. Mróz, 'On hypo-elasticity and plasticity approaches to constitutive modelling of inelastic behaviour of soils'. *Int. J. for Num. and Anal. Methods in Geomechanics*, **4**, 45–55 (1980).
3. Z. Mróz and V. Norris, 'Elasto-plastic and viscoplastic constitutive models with application to cyclic loading for soils', ch. 8, pp. 173–217, *Soil Mechanics—Transient and Cyclic Loads* (Ed. G. N. Pande and O. C. Zienkiewicz), John Wiley & Sons, 1982.
4. Y. F. Dafalias and E. P. Popov, 'A model of nonlinearly hardening materials for complex loadings', *Acta Mechanica*, **21**, 173–192 (1975).
5. R. D. Krieg, 'A practical two-surface plasticity theory', *J. Appl. Mech.*, **E42**, 641–646 (1975).
6. Y. F. Defalias and L. R. Herrmann, 'Bounding surface formulation in soil plasticity', ch. 10, pp. 253–282, *Soil Mechanics—Transient and Cyclic Loads* (Ed. G. N. Pande and O. C. Zienkiwicz), John Wiley & Sons, 1982.
7. O. C. Zienkiewicz, 'Generalized plasticity and some models for geomechanics', *Proc. NATO Symposium on Soil Mechanics*, Reidel, 1982.
8. O. C. Zienkiewicz and I. C. Cormeau, 'Visco-Plasticity—plasticity and creep in elastic solids—a unified numerical solution approach. *Int. J. Num. Meth. Engng.*, **8**, 821–845 (1974).
9. O. C. Zienkiewicz, C. Humpheson, and R. W. Lewis, 'Associated and non-associated viscoplasticity and plasticity in soil mechanics', *Geotechnique*, **25**, 671–89 (1975).
10. K. C. Valanis, 'A theory of viscoplasticity without a yield surface', *Arch. of Mech.* (Polish Academy) **23**, 517–55 (1971).
11. K. C. Valanis and H. E. Read, 'A new endochronic plasticity model for soils, Ch. 14, pp. 375–468 *Soil Mechanics—Transient and Cyclic Loads* (Eds. G. N. Pande and O. C. Zienkiewicz), John Wiley & Sons, 1982.
12. Z. P. Bazant, A. M. Ansel, and R. J. Krizek, 'Endochronic models for soils', Ch. 15, pp. 419–438, *Soil Mechanics—Transient and Cyclic Loads* (Ed. G. N. Pande and O. C. Zienkiewicz), John Wiley & Sons, 1982.
13. O. C. Zienkiewicz, C. T. Chang, and E. Hinton, 'Nonlinear seismic response and liquefaction', *Int. J. Num. and Anal. Meth. Geomech.*, **2**, 381–404 (1978).
14. O. C. Zienkiewicz, K. H. Leung, E. Hinton, and C. T. Chang, 'Liquefaction and permanent deformation under dynamic conditions—Numerical solutions and constitutive relations. Ch. 5, pp. 71–103, *Soil Mechanics—Transient and Cyclic Loads* (Ed. G. N. Pande and O. C. Zienkiewicz), Jonh Wiley & Sons, 1982.
15. W. D. L. Finn, 'Endochronic description of seismic response of soils. *Proc. Conf. Constitutive Laws for Engineering Materials*, Tucson, Arizona, Jan. 1983.
16. D. C. Drucker, R. E. Gibson, and D. J. Henkel, 'Soil mechanics and work hardening theories of plasticity', *Proc. Am. Soc. Civ. Eng.*, **122**, 388 and 346, 1957.
17. K. H. Roscoe and J. B. Burland, 'On the generalized behaviour of wet clay', in *Engineering Plasticity* (Eds. J. Heyman and F. A. Leckie), pp. 535–609, Cambridge Univ. Press, 1968.
18. S. Pietruszczak and Z. Mróz, 'Description of mechanical behaviour of anisotropically consolidated clays', *Proc. Euromech Coll., Anisotropy in Mechanics*', *Grenoble*, Noordhoff Int. Publ. 1979.
19. P. V. Lade and J. M. Duncan, 'Elastoplastic stress–strain theory for cohesionless soil, *J. Geotech Eng. Div., Proc. ASCE*, **101**, GT10, 1037–1053, (1975).

20. P. Wilde, 'Two invariant depending models of granula media', *Arch. Mech. Stos.*, **29**, 799–809 (1977).
21. Z. Mróz, V. A. Norris, and O. C. Zienkiewicz, 'An anisotropic hardening model for soils and its application to cyclic loading', *Int. J. Num. Anal. Meth. Geom*, **2**, 203–221 (1978).
22. Z. Mróz, V. A. Norris, and O. C. Zienkiewicz, 'Application of an anisotropic hardening model in the analysis of elastoplastic deformation of soils', *Geotechnique*, **29**, 1–34 (1979).
23. Z. Mróz, V. A. Norris, and O. C. Zienkiewicz, 'An anisotropic critical state model for soils subjected to cyclic loading', *Geotechnique*, **31**, 451–469 (1981).
24. Proceedings of Int. Workshop on Constitutive Behaviour of Soils. (Ed. G. Gudehus), Grenoble, 1982.
25. D. J. Naylor, 'A continuous plasticity version of the C. S. M'., *CR*/422/82, Inst of Num. Meth. Engng. Swansea, Sept. 1982.
26. G. N. Pande and S. Pietruszczak, 'Reflecting surface model for soils', pp. 56–64, *Proc. Int. Symp. on Numerical Models in Geomechanics*, A. A. Balkems, Rotterdam, 1982.
27. O. C. Zienkiewicz, O. C. Leung, and E. Hinton, 'Earthquake response behaviour of soils with drainage', *Proc. 3rd Conf. on Numerical Methods in Geomechanics*, Edmonton, 1982.
28. O. C. Zienkiewicz and T. Shiomi, 'Dynamic behaviour of saturated porous media. The generalized Biot formulation and its numerical solution', *CR*/431/82. Inst. Nu. Meth. Engng. Swansea, October 1982. *Int. J. Num. and Anal. Meth. Geomech.*, **8**, 71–96, 1984.
29. T. Shiomi, S. Pietruszczak, and G. N. Pande, 'A liquefaction study of sand layers using the reflecting surface model', pp. 411–418, *Proc. Int. Symp. on Numerical Models in Geomechanics*, A. A. Balkema, Rotterdam, 1982.
30. J. P. Carter, J. R. Booker, and C. P. Wroth, 'A critical state model for cyclic loading, Ch. 9, pp. 219–252, *Soil Mechanics—Transient and Cyclic Loads* (Ed. G. N. Pande and O. C. Zienkiewicz), John. Wiley & Sons, 1982.
31. O. C. Zienkiewicz, G. C. Nayak, and D. R. J. Owen, 'Composite and "overlay" models in numerical analysis of elastoplastic continua', *Foundations of Plasticity* (Ed. A. Sawczuk), 107–122, Noordhoff Press, 1972.
32. G. N. Pande and K. G. Sharma, 'Time dependent multi-laminate model for clays', Conf. on Implementation of computer procedure and stress–strain laws in geotechnical engineering. Acorn Press, Durham, USA, 1980.
33. O. C. Zienkiewicz, V. A. Norris, and D. J. Naylor, 'Plasticity and viscoplasticity in soil mechanics with special reference to cyclic loading problems', *Proc. Int. Con. on Finite Elements in Nonlinear Solid and Structural Mechanics*, Geilo, Norway, August 1977, pp. 455–485, Vol. 2, Tapir Press, Norwegian Institute of Technology, Trondheim.

Subject Index